DEPARTMENT OF THE ARMY
U.S. Army Corps of Engineers
Washington, DC 20314-1000

EM 1110-2-1100
(Change 2)

CECW-CE

Manual
No. 1110-2-1100

1 April 2008

Engineering and Design
COASTAL ENGINEERING MANUAL

1. **Purpose.** The purpose of the *Coastal Engineering Manual* (CEM) is to provide a comprehensive technical coastal engineering document. It includes the basic principles of coastal processes, methods for computing coastal planning and design parameters, and guidance on how to formulate and conduct studies in support of coastal flooding, shore protection, and navigation projects. This Change 2 to EM 1110-2-1100, 1 April 2008, includes the following changes and updates:

   a. Part I-1. References were checked and some were deleted (Engineer Manuals that are no longer in the USACE inventory).

   b. Part I-4. Minor changes were made in the text to better reflect the contents of subsequent parts of the CEM.

   c. Part II-1. Figure II-1-9 has been revised; Equations II-1-128, II-1-160, and II-1-161 have been corrected.

   d. Part II-2. Equations II-2-4, II-2-5, and II-2-32 have been corrected along with other errors reported by various users.

   e. Part II-5. References were checked and some were deleted (Engineer Manuals that are no longer in the USACE inventory).

   f. Part II-6. The value of "e" used in Eq. II-6-28 has been corrected.

   g. Part II-7. The table of contents was corrected. A new section, II-7-11, Note to Users, Vessel Buoyancy, was added at the end of the chapter.

   h. Part III-3. Corrections have been made to format and spelling. Different plots were added to Figures III-3-24 and III-3-26.

   i. Part IV-1. Corrections have been made to references.

   j. Part V-1. Citation of an Engineer Regulation has been corrected.

   k. Part V-2. Citation of references has been changed, web pages with sources of wind and wave data have been added. Some minor text changes have also been made.

   l. Part V-3. Citations of unpublished reports or personal communications have been deleted, and links to other figures or parts of the CEM have been checked and corrected.

   m. Part V-4. Minor text changes, corrections to references and Figure V-4-1.

   n. Part V-5. Links to other parts of the CEM that were planned but never written have been deleted.

2. **Applicability**. This manual applies to all HQUSACE elements and all USACE commands having Civil Works and military design responsibilities.

3. Discussion. The CEM is divided into five parts in two major subdivisions: science-based and engineering-based. The first four parts of the CEM and Appendix A compose the science-based subdivision:

Part I, "Introduction"
Part II, "Coastal Hydrodynamics"
Part III, "Coastal Sediment Processes"
Part IV, "Coastal Geology"
Appendix A, "Glossary"

The engineering-based subdivision is oriented toward a project-type approach, Part V, "Coastal Project Planning and Design."

4. Distribution Statement. Approved for public release, distribution unlimited.

5. Note to Users. Revised chapters are dated 1 April 2008. Readers need to download the entire new chapters and discard earlier versions in their possession.

FOR THE COMMANDER:

STEPHEN L. HILL
Colonel, Corps of Engineers
Chief of Staff

Table of Contents

Page

II-1-1. Introduction II-1-1

II-1-2. Regular Waves II-1-3
  a. Introduction II-1-3
  b. Definition of wave parameters II-1-4
  c. Linear wave theory II-1-5
    (1) Introduction II-1-5
    (2) Wave celerity, length, and period II-1-6
    (3) The sinusoidal wave profile II-1-9
    (4) Some useful functions II-1-9
    (5) Local fluid velocities and accelerations II-1-12
    (6) Water particle displacements II-1-13
    (7) Subsurface pressure II-1-21
    (8) Group velocity II-1-22
    (9) Wave energy and power II-1-26
    (10) Summary of linear wave theory II-1-29
  d. Nonlinear wave theories II-1-30
    (1) Introduction II-1-30
    (2) Stokes finite-amplitude wave theory II-1-32
    (3) Subsurface pressure II-1-34
    (4) Maximum wave steepness II-1-35
  e. Other wave theories II-1-36
    (1) Introduction II-1-36
    (2) Nonlinear shallow-water wave theories II-1-36
    (3) Korteweg and de Vries and Boussinesq wave theories II-1-36
    (4) Cnoidal wave theory II-1-37
    (5) Solitary wave theory II-1-41
    (6) Stream-function wave theory II-1-50
    (7) Fourier approximation -- Fenton's theory II-1-50
  f. Wave breaking II-1-56
  g. Validity of wave theories II-1-56

II-1-3. Irregular Waves II-1-59
  a. Introduction II-1-59
  b. Wave train (wave-by-wave) analysis II-1-65
    (1) Introduction II-1-65
    (2) Zero-crossing method II-1-65
    (3) Definition of wave parameters II-1-66
    (4) Significant wave height II-1-68
    (5) Short-term random sea state parameters II-1-68
    (6) Probability distributions for a sea state II-1-70
    (7) Wave height distribution II-1-74

(8) Wave period distribution II-1-75

(9) Joint distribution of wave heights and periods II-1-76

 c. Spectral analysis II-1-77

(1) Introduction II-1-77

(2) Description of wave spectral analysis II-1-80

(3) Examples of frequency spectra II-1-85

(4) Wave spectrum and its parameters II-1-85

(5) Relationship between $H_{1/3}$, H_s and H_{m0} in shallow water II-1-88

(6) Parametric spectrum models II-1-88

(7) Directional spectra II-1-93

(8) Wave groups and groupiness factors II-1-95

(9) Random wave simulation II-1-97

(10) Kinematics and dynamics of irregular waves II-1-98

II-1-4. References and Bibliography II-1-99

II-1-5. Definitions of Symbols II-1-116

II-1-6. Acknowledgments II-1-121

List of Tables

Page

Table II-1-1.   Classification of Water Waves II-1-8

Table II-1-2.   Boundary Value Problem of Water Wave Theories (Dean 1968) II-1-50

List of Figures

Page

Figure II-1-1.　Definition of terms - elementary, sinusoidal, progressive wave II-1-4

Figure II-1-2.　Local fluid velocities and accelerations II-1-13

Figure II-1-3.　Profiles of particle velocity and acceleration by Airy theory in relation
to the surface elevation II-1-14

Figure II-1-4.　Water particle displacements from mean position for shallow-water
and deepwater waves II-1-17

Figure II-1-5.　Variation of wave parameters with d/L_0 (Dean and Dalrymple 1991) II-1-23

Figure II-1-6.　Characteristics of a wave group formed by the addition of sinusoids
with different periods II-1-25

Figure II-1-7.　Variation of the ratios of group and phase velocities to deepwater
phase speed using linear theory (Sarpkaya and Isaacson 1981) II-1-27

Figure II-1-8.　Variation of shoaling coefficient with wave steepness (Sakai and
Battjes 1980) II-1-30

Figure II-1-9.　Summary of linear (Airy) wave theory - wave characteristics II-1-31

Figure II-1-10.　Wave profile shape of different progressive gravity waves II-1-34

Figure II-1-11.　Normalized surface profile of the cnoidal wave (Wiegel 1960).
For definition of variables see Part II-1-2.e.(3) II-1-40

Figure II-1-12.　Normalized surface profile of the cnoidal wave for higher values of
k^2 and X/L (Wiegel 1960) II-1-40

Figure II-1-13.　k^2 versus L^2H/d^3, and k^2 versus $T\sqrt{g/d}$ and H/d (Wiegel 1960) II-1-41

Figure II-1-14.　Relationship among L^2H/d^3 and the square of the elliptic modulus
(k^2), y_c/H, y_t/H, and K(k) (Wiegel 1960) II-1-42

Figure II-1-15.　Relationships among $T\sqrt{g/d}$, L^2H/d^3, and H/d (Wiegel 1960) II-1-43

Figure II-1-16.　Relationship between cnoidal wave velocity and L^2H/d^3 (Wiegel 1960) II-1-44

Figure II-1-17.　Functions M and N in solitary wave theory (Munk 1949) II-1-49

Figure II-1-18.　Surface elevation, horizontal velocity and pressure in 10-m
depth (using Fenton's theory in ACES) II-1-55

Figure II-1-19. Influence of a uniform current on the maximum wave height (Dalrymple and Dean 1975) II-1-57

Figure II-1-20. Ranges of suitability of various wave theories (Le Méhauté 1976) II-1-58

Figure II-1-21. Grouping of wind waves based on universal parameter and limiting height for steep waves II-1-59

Figure II-1-22. Radar image of the sea surface in the entrance to San Francisco Bay II-1-61

Figure II-1-23. Measured sea surface velocity in the entrance to San Francisco Bay II-1-62

Figure II-1-24. Representations of an ocean wave II-1-63

Figure II-1-25. Wave profile of irregular sea state from site measurements II-1-63

Figure II-1-26. Definition of wave parameters for a random sea state II-1-64

Figure II-1-27. Definition sketch of a random wave process (Ochi 1973) II-1-64

Figure II-1-28. Gaussian probability density and cumulative probability distribution II-1-71

Figure II-1-29. Rayleigh probability density and cumulative probability distribution ($x = \alpha$ corresponds to the mode) II-1-71

Figure II-1-30. Histograms of the normalized (a) wave heights and (b) wave periods with theoretical distributions (Chakrabarti 1987) II-1-73

Figure II-1-31. Surface elevation time series of a regular wave and its spectrum (Briggs et al. 1993) II-1-79

Figure II-1-32. Surface elevation time series of an irregular wave and its spectrum (Briggs et al. 1993) II-1-80

Figure II-1-33. Schematic for a two-dimensional wave spectrum $E(f,\theta)$ II-1-81

Figure II-1-34. Directional spectrum and its frequency and direction spectrum (Briggs et al. 1993) II-1-82

Figure II-1-35. Sketches of wave spectral energy and energy density (Chakrabarti 1987) II-1-84

Figure II-1-36. Definitions of one- and two-sided wave spectra (Chakrabarti 1987) II-1-85

Figure II-1-37. Energy density and frequency relationship (Chakrabarti 1987) II-1-86

Figure II-1-38. Comparison of the PM and JONSWAP spectra (Chakrabarti 1987) II-1-87

Figure II-1-39. Definition sketch for Ochi-Hubble spectrum (Ochi and Hubble 1976) II-1-90

Figure II-1-40.   Variation of H_s/H_{mo} as a function of relative depth $\bar{d}$ and significant
steepness (Thompson and Vincent 1985) II-1-91

Figure II-1-41.   Identification and description of wave groups through ordered
statistics (Goda 1976) II-1-96

Chapter II-1
Water Wave Mechanics

II-1-1. Introduction

a. Waves on the surface of the ocean with periods of 3 to 25 sec are primarily generated by winds and are a fundamental feature of coastal regions of the world. Other wave motions exist on the ocean including internal waves, tides, and edge waves. For the remainder of this chapter, unless otherwise indicated, the term waves will apply only to surface gravity waves in the wind wave range of 3 to 25 sec.

b. Knowledge of these waves and the forces they generate is essential for the design of coastal projects since they are the major factor that determines the geometry of beaches, the planning and design of marinas, waterways, shore protection measures, hydraulic structures, and other civil and military coastal works. Estimates of wave conditions are needed in almost all coastal engineering studies. The purpose of this chapter is to give engineers theories and mathematical formulae for describing ocean surface waves and the forces, accelerations, and velocities due to them. This chapter is organized into two sections: *Regular Waves* and *Irregular Waves.*

c. In the *Regular Waves* section, the objective is to provide a detailed understanding of the mechanics of a wave field through examination of waves of constant height and period. In the *Irregular Waves* section, the objective is to describe statistical methods for analyzing irregular waves (wave systems where successive waves may have differing periods and heights) which are more descriptive of the waves seen in nature.

d. In looking at the sea surface, it is typically irregular and three-dimensional (3-D). The sea surface changes in time, and thus, it is unsteady. At this time, this complex, time-varying 3-D surface cannot be adequately described in its full complexity; neither can the velocities, pressures, and accelerations of the underlying water required for engineering calculations. In order to arrive at estimates of the required parameters, a number of simplifying assumptions must be made to make the problems tractable, reliable and helpful through comparison to experiments and observations. Some of the assumptions and approximations that are made to describe the 3-D, time-dependent complex sea surface in a simpler fashion for engineering works may be unrealistic, but necessary for mathematical reasons.

e. The *Regular Waves* section of this chapter begins with the simplest mathematical representation assuming ocean waves are two-dimensional (2-D), small in amplitude, sinusoidal, and progressively definable by their wave height and period in a given water depth. In this simplest representation of ocean waves, wave motions and displacements, kinematics (that is, wave velocities and accelerations), and dynamics (that is, wave pressures and resulting forces and moments) will be determined for engineering design estimates. When wave height becomes larger, the simple treatment may not be adequate. The next part of the *Regular Waves* section considers 2-D approximation of the ocean surface to deviate from a pure sinusoid. This representation requires using more mathematically complicated theories. These theories become nonlinear and allow formulation of waves that are not of purely sinusoidal in shape; for example, waves having the flatter troughs and peaked crests typically seen in shallow coastal waters when waves are relatively high.

f. The *Irregular Waves* section of this chapter is devoted to an alternative description of ocean waves. Statistical methods for describing the natural time-dependent three-dimensional characteristics of real wave systems are presented. A complete 3-D representation of ocean waves requires considering the sea surface as an irregular wave train with random characteristics. To quantify this randomness of ocean waves, the *Irregular Waves* section employs statistical and probabilistic theories. Even with this approach, simplifications are required. One approach is to transform the sea surface using Fourier theory into summation of simple sine waves and then to define a wave's characteristics in terms of its spectrum. This allows treatment of the variability of waves with respect to period and direction of travel. The second approach is to describe a wave record at a point as a sequence of individual waves with different heights and periods and then to consider the variability of the wave field in terms of the probability of individual waves.

g. At the present time, practicing coastal engineers must use a combination of these approaches to obtain information for design. For example, information from the *Irregular Waves* section will be used to determine the expected range of wave conditions and directional distributions of wave energy in order to select an individual wave height and period for the problem under study. Then procedures from the *Regular Waves* section will be used to characterize the kinematics and dynamics that might be expected. However, it should be noted that the procedures for selecting and using irregular wave conditions remain an area of some uncertainty.

h. The major generating force for waves is the wind acting on the air-sea interface. A significant amount of wave energy is dissipated in the nearshore region and on beaches. Wave energy forms beaches; sorts bottom sediments on the shore face; transports bottom materials onshore, offshore, and alongshore; and exerts forces upon coastal structures. A basic understanding of the fundamental physical processes in the generation and propagation of surface waves must precede any attempt to understand complex water motion in seas, lakes and waterways. The *Regular Waves* section of this chapter outlines the fundamental principles governing the mechanics of wave motion essential in the planning and design of coastal works. The *Irregular Waves* section of this chapter discusses the applicable statistical and probabilistic theories.

i. Detailed descriptions of the basic equations for water mechanics are available in several textbooks (see for example, Kinsman 1965; Stoker 1957; Ippen 1966; Le Méhauté 1976; Phillips 1977; Crapper 1984; Mei 1991; Dean and Dalrymple 1991). The Regular Waves section of this chapter provides only an introduction to wave mechanics, and it focuses on simple water wave theories for coastal engineers. Methods are discussed for estimating wave surface profiles, water particle motion, wave energy, and wave transformations due to interaction with the bottom and with structures.

j. The simplest wave theory is the *first-order, small-amplitude*, or *Airy* wave theory which will hereafter be called *linear theory*. Many engineering problems can be handled with ease and reasonable accuracy by this theory. For convenience, prediction methods in coastal engineering generally have been based on simple waves. For some situations, simple theories provide acceptable estimates of wave conditions.

k. When waves become large or travel toward shore into shallow water, higher-order wave theories are often required to describe wave phenomena. These theories represent *nonlinear waves*. The linear theory that is valid when waves are infinitesimally small and their motion is small also provides some insight for finite-amplitude periodic waves (nonlinear). However, the linear theory cannot account for the fact that wave crests are higher above the mean water line than the troughs are below the mean water line. Results obtained from the various theories should be carefully interpreted for use in the design of coastal projects or for the description of coastal environment.

l. Any basic physical description of a water wave involves both its surface form and the water motion beneath the surface. A wave that can be described in simple mathematical terms is called a *simple wave*. Waves comprised of several components and difficult to describe in form or motion are termed *wave trains* or *complex waves*. Sinusoidal or monochromatic waves are examples of simple waves, since their surface profile can be described by a single sine or cosine function. A wave is *periodic* if its motion and surface profile recur in equal intervals of time termed the *wave period*. A wave form that moves horizontally relative to a fixed point is called a *progressive wave* and the direction in which it moves is termed the *direction of wave propagation*. A progressive wave is called *wave of permanent form* if it propagates without experiencing any change in shape.

m. Water waves are considered *oscillatory* or *nearly oscillatory* if the motion described by the water particles is circular orbits that are closed or nearly closed for each wave period. The linear theory represents pure oscillatory waves. Waves defined by finite-amplitude wave theories are not pure oscillatory waves but still periodic since the fluid is moved in the direction of wave advance by each successive wave. This motion is termed *mass transport* of the waves. When water particles advance with the wave and do not return to their original position, the wave is called a *wave of translation*. A solitary wave is an example of a wave of translation.

n. It is important in coastal practice to differentiate between two types of surface waves. These are *seas* and *swells*. Seas refer to short-period waves still being created by winds. Swells refer to waves that have moved out of the generating area. In general, swells are more regular waves with well-defined long crests and relatively long periods.

o. The growth of wind-generated oceanic waves is not indefinite. The point when waves stop growing is termed a *fully developed sea* condition. Wind energy is imparted to the water leading to the growth of waves; however, after a point, the energy imparted to the waters is dissipated by wave breaking. Seas are short-crested and irregular and their periods are within the 3- to 25- sec range. Seas usually have shorter periods and lengths, and their surface appears much more disturbed than for swells. Waves assume a more orderly state with the appearance of definite crests and troughs when they are no longer under the influence of winds (swell).

p. To an observer at a large distance from a storm, swells originating in a storm area will appear to be almost unidirectional (i.e., they propagate in a predominant direction) and long-crested (i.e., they have well-defined and distinctly separated crests). Although waves of different periods existed originally together in the generation area (seas), in time the various wave components in the sea separate from one another. Longer period waves move faster and reach distant sites first. Shorter period components may reach the site several days later. In the wave generation area, energy is transferred from shorter period waves to the longer waves. Waves can travel hundreds or thousands of kilometers without much loss of energy. However, some wave energy is dissipated internally within the fluid, by interaction with the air above, by turbulence upon breaking, and by percolation and friction with the seabed. Short-period components lose their energy more readily than long-period components. As a consequence of these processes, the periods of swell waves tend to be somewhat longer than seas. Swells typically have periods greater than 10 sec.

II-1-2. Regular Waves

a. Introduction. Wave theories are approximations to reality. They may describe some phenomena well under certain conditions that satisfy the assumptions made in their derivation. They may fail to describe other phenomena that violate those assumptions. In adopting a theory, care must be taken to ensure that the wave phenomenon of interest is described reasonably well by the theory adopted, since shore protection design depends on the ability to predict wave surface profiles and water motion, and on the accuracy of such predictions.

b. Definition of wave parameters.

(1) A progressive wave may be represented by the variables x (spatial) and t (temporal) or by their combination (phase), defined as $\theta = kx - \omega t$, where k and ω are described in the following paragraphs. The values of θ vary between 0 and 2π. Since the θ-representation is a simple and compact notation, it will be used in this chapter. Figure II-1-1 depicts parameters that define a simple, progressive wave as it passes a fixed point in the ocean. A simple, periodic wave of permanent form propagating over a horizontal bottom may be completely characterized by the wave height H wavelength L and water depth d.

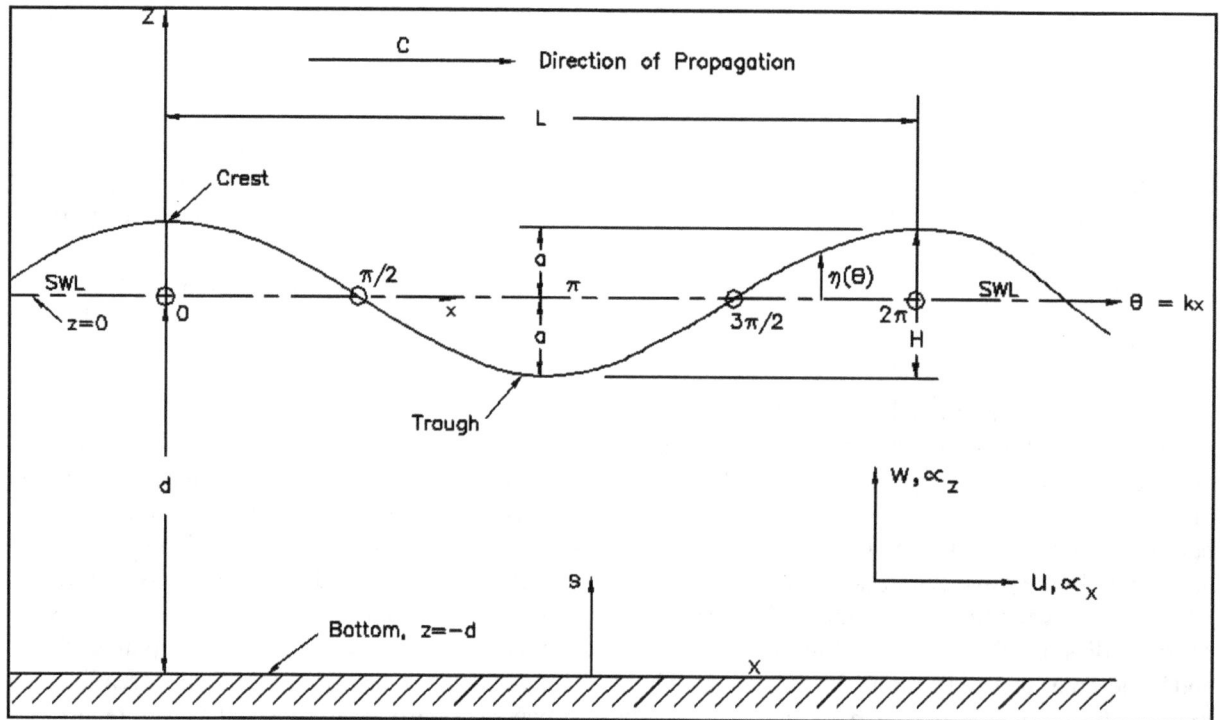

Figure II-1-1. **Definition of terms - elementary, sinusoidal, progressive wave**

(2) As shown in Figure II-1-1, the highest point of the wave is the *crest* and the lowest point is the *trough*. For linear or small-amplitude waves, the height of the crest above the still-water level (SWL) and the distance of the trough below the SWL are each equal to the wave amplitude a. Therefore $a = H/2$, where H = *the wave height*. The time interval between the passage of two successive wave crests or troughs at a given point is the *wave period* T. The *wavelength* L is the horizontal distance between two identical points on two successive wave crests or two successive wave troughs.

(3) Other wave parameters include $\omega = 2\pi/T$ the *angular* or *radian frequency*, the *wave number* $k = 2\pi/L$, the *phase velocity* or *wave celerity* $C = L/T = \omega/k$, the *wave steepness* $\epsilon = H/L$, the *relative depth* d/L, and the *relative wave height* H/d. These are the most common parameters encountered in coastal practice. Wave motion can be defined in terms of dimensionless parameters H/L, H/d, and d/L; these are often used in practice. The dimensionless parameters ka and kd, preferred in research works, can be substituted for H/L and d/L, respectively, since these differ only by a constant factor 2π from those preferred by engineers.

c.   Linear wave theory.

(1) Introduction.

(a) The most elementary wave theory is the *small-amplitude* or *linear wave theory*. This theory, developed by Airy (1845), is easy to apply, and gives a reasonable approximation of wave characteristics for a wide range of wave parameters. A more complete theoretical description of waves may be obtained as the sum of many successive approximations, where each additional term in the series is a correction to preceding terms. For some situations, waves are better described by these higher-order theories, which are usually referred to as *finite-amplitude wave theories* (Mei 1991, Dean and Dalrymple 1991). Although there are limitations to its applicability, linear theory can still be useful provided the assumptions made in developing this simple theory are not grossly violated.

(b) The assumptions made in developing the linear wave theory are:

- The fluid is homogeneous and incompressible; therefore, the density ρ is a constant.

- Surface tension can be neglected.

- Coriolis effect due to the earth's rotation can be neglected.

- Pressure at the free surface is uniform and constant.

- The fluid is ideal or inviscid (lacks viscosity).

- The particular wave being considered does not interact with any other water motions. The flow is irrotational so that water particles do not rotate (only normal forces are important and shearing forces are negligible).

- The bed is a horizontal, fixed, impermeable boundary, which implies that the vertical velocity at the bed is zero.

- The wave amplitude is small and the waveform is invariant in time and space.

- Waves are plane or long-crested (two-dimensional).

(c) The first three assumptions are valid for virtually all coastal engineering problems. It is necessary to relax the fourth, fifth, and sixth assumptions for some specialized problems not considered in this manual. Relaxing the three final assumptions is essential in many problems, and is considered later in this chapter.

(d) The assumption of irrotationality stated as the sixth assumption above allows the use of a mathematical function termed the *velocity potential* Φ. The velocity potential is a scaler function whose gradient (i.e., the rate of change of Φ relative to the x-and z-coordinates in two dimensions where x = horizontal, z = vertical) at any point in fluid is the velocity vector. Thus,

$$u = \frac{\partial \Phi}{\partial x} \tag{II-1-1}$$

is the fluid velocity in the x-direction, and

$$w = \frac{\partial \Phi}{\partial z} \tag{II-1-2}$$

is the fluid velocity in the z-direction. Φ has the units of length squared divided by time. Consequently, if $\Phi(x, z, t)$ is known over the flow field, then fluid particle velocity components u and w can be found.

(e) The incompressible assumption (a) above implies that there is another mathematical function termed the *stream function* Ψ. Some wave theories are formulated in terms of the stream function Ψ, which is orthogonal to the potential function Φ. Lines of constant values of the potential function (equipotential lines) and lines of constant values of the stream function are mutually perpendicular or orthogonal. Consequently, if Φ is known, Ψ can be found, or vice versa, using the equations

$$\frac{\partial \Phi}{\partial x} = \frac{\partial \Psi}{\partial z} \qquad (II-1-3)$$

$$\frac{\partial \Phi}{\partial z} = -\frac{\partial \Psi}{\partial x} \qquad (II-1-4)$$

termed the *Cauchy-Riemann conditions* (Whitham 1974; Milne-Thompson 1976). Both Φ and Ψ satisfy the *Laplace equation* which governs the flow of an *ideal fluid* (inviscid and incompressible fluid). Thus, under the assumptions outlined above, the Laplace equation governs the flow beneath waves. The Laplace equation in two dimensions with x = horizontal, and z = vertical axes in terms of velocity potential Φ is given by

$$\frac{\partial^2 \Phi}{\partial x^2} + \frac{\partial^2 \Phi}{\partial z^2} = 0 \qquad (II-1-5)$$

(f) In terms of the stream function, Ψ, Laplace's equation becomes

$$\frac{\partial^2 \Psi}{\partial x^2} + \frac{\partial^2 \Psi}{\partial z^2} = 0 \qquad (II-1-6)$$

(g) The linear theory formulation is usually developed in terms of the potential function, Φ.

In applying the seventh assumption to waves in water of varying depth (encountered when waves approach a beach), the local depth is usually used. This can be justified, but not without difficulty, for most practical cases in which the bottom slope is flatter than about 1 on 10. A progressive wave moving into shallow water will change its shape significantly. Effects due to the wave transformations are addressed in Parts II-3 and II-4.

(h) The most fundamental description of a simple sinusoidal oscillatory wave is by its length L (the horizontal distance between corresponding points on two successive waves), height H (the vertical distance to its crest from the preceding trough), period T (the time for two successive crests to pass a given point), and depth d (the distance from the bed to SWL).

(i) Figure II-1-1 shows a two-dimensional, simple progressive wave propagating in the positive x-direction, using the symbols presented above. The symbol η denotes the displacement of the water surface relative to the SWL and is a function of x and time t. At the wave crest, η is equal to the amplitude of the wave a, or one-half the wave height $H/2$.

(2) Wave celerity, length, and period.

(a) The speed at which a wave form propagates is termed the *phase velocity* or *wave celerity C*. Since the distance traveled by a wave during one wave period is equal to one wavelength, wave celerity can be related to the wave period and length by

$$C = \frac{L}{T} \tag{II-1-7}$$

(b) An expression relating wave celerity to wavelength and water depth is given by

$$C = \sqrt{\frac{gL}{2\pi} \tanh\left(\frac{2\pi d}{L}\right)} \tag{II-1-8}$$

(c) Equation II-1-8 is termed the *dispersion relation* since it indicates that waves with different periods travel at different speeds. For a situation where more than one wave is present, the longer period wave will travel faster. From Equation II-1-7, it is seen that Equation II-1-8 can be written as

$$C = \frac{gT}{2\pi} \tanh\left(\frac{2\pi d}{L}\right) \tag{II-1-9}$$

(d) The values $2\pi/L$ and $2\pi/T$ are called the *wave number k* and the *wave angular frequency ω*, respectively. From Equation II-1-7 and II-1-9, an expression for wavelength as a function of depth and wave period may be obtained as

$$L = \frac{gT^2}{2\pi} \tanh\left(\frac{2\pi d}{L}\right) = \frac{gT}{\omega} \tanh\left(kd\right) \tag{II-1-10}$$

(e) Use of Equation II-1-10 involves some difficulty since the unknown L appears on both sides of the equation. Tabulated values of d/L and d/L_0 (SPM 1984) where L_0 is the deepwater wavelength may be used to simplify the solution of Equation II-1-10. Eckart (1952) gives an approximate expression for Equation II-1-10, which is correct to within about 10 percent. This expression is given by

$$L \approx \frac{gT^2}{2\pi} \sqrt{\tanh\left(\frac{4\pi^2 d}{T^2 g}\right)} \tag{II-1-11}$$

(f) Equation II-1-11 explicitly gives L in terms of wave period T and is sufficiently accurate for many engineering calculations. The maximum error 10 percent occurs when d/L ≈ 1/2. There are several other approximations for solving Equation II-1-10 (Hunt 1979; Venezian and Demirbilek 1979; Wu and Thornton 1986; Fenton and McKee 1990).

(g) Gravity waves may also be classified by the water depth in which they travel. The following classifications are made according to the magnitude of d/L and the resulting limiting values taken by the function *tanh (2πd/L)*. Note that as the argument of the hyperbolic tangent kd = 2πd/L gets large, the tanh (kd) approaches 1, and for small values of kd, tanh (kd) ≈ kd.

(h) Water waves are classified in Table II-1-1 based on the relative depth criterion *d/L*.

Table II-1-1
Classification of Water Waves

Classification	d/L	kd	tanh (kd)
Deep water	1/2 to ∞	π to ∞	≈1
Transitional	1/20 to 1/2	π/10 to π	tanh (kd)
Shallow water	0 to 1/20	0 to π/10	≈ kd

(i) In deep water, tanh (kd) approaches unity, Equations II-1-7 and II-1-8 reduce to

$$C_0 = \sqrt{\frac{gL_0}{2\pi}} = \frac{L_0}{T} \qquad\qquad\text{(II-1-12)}$$

and Equation II-1-9 becomes

$$C_0 = \frac{gT}{2\pi} \qquad\qquad\text{(II-1-13)}$$

(j) Although *deep water* actually occurs at an infinite depth, *tanh (kd)*, for most practical purposes, approaches unity at a much smaller d/L. For a relative depth of one-half (i.e., when the depth is one-half the wavelength), tanh $(2\pi d/L) = 0.9964$.

(k) When the relative depth d/L is greater than one-half, the wave characteristics are virtually independent of depth. Deepwater conditions are indicated by the subscript 0 as in L_o and C_o except that the period T remains constant and independent of depth for oscillatory waves, and therefore, the subscript for wave period is omitted (Ippen 1966). In the SI system (System International or metric system of units) where units of meters and seconds are used, the constant $g/2\pi$ is equal to 1.56 m/s^2, and

$$C_0 = \frac{gT}{2\pi} = \frac{9.8}{2\pi} T = 1.56T \ m/s \qquad\qquad\text{(II-1-14)}$$

and

$$L_0 = \frac{gT^2}{2\pi} = \frac{9.8}{2\pi} T^2 = 1.56T^2 \ m \qquad\qquad\text{(II-1-15)}$$

(l) If units of feet and seconds are specified, the constant $g/2\pi$ is equal to 5.12 ft/s^2, and

$$C_0 = \frac{gT}{2\pi} = 5.12T \ ft/s \qquad\qquad\text{(II-1-16)}$$

and

$$L_0 = \frac{gT^2}{2\pi} = 5.12T^2 \ ft \qquad\qquad\text{(II-1-17)}$$

(m) If Equations II-1-14 and II-1-15 are used to compute wave celerity when the relative depth is $d/L = 0.25$, the resulting error will be about 9 percent. It is evident that a relative depth of 0.5 is a satisfactory boundary separating deepwater waves from waves in water of *transitional depth*. If a wave is traveling in *transitional* depths, Equations II-1-8 and II-1-9 must be used without simplification. As a rule of thumb, Equation II-1-8 and II-1-9 must be used when the relative depth is between 0.5 and 0.04.

(n) When the relative water depth becomes shallow, i.e., $2\pi d/L < 1/4$ or $d/L < 1/25$, Equation II-1-8 can be simplified to

$$C = \sqrt{gd} \tag{II-1-18}$$

(o) Waves sufficiently long such that Equation II-1-18 may be applied are termed long waves. This relation is attributed to Lagrange. Thus, when a wave travels in shallow water, wave celerity depends only on water depth.

(p) In summary, as a wind wave passes from deep water to the beach its speed and length are first only a function of its period (or frequency); then as the depth becomes shallower relative to its length, the length and speed are dependent upon both depth and period; and finally the wave reaches a point where its length and speed are dependent only on depth (and not frequency).

(3) The sinusoidal wave profile. The equation describing the free surface as a function of time t and horizontal distance x for a simple sinusoidal wave can be shown to be

$$\eta = a \cos (kx - \omega t) = \frac{H}{2} \cos \left(\frac{2\pi x}{L} - \frac{2\pi t}{T} \right) = a \cos \theta \tag{II-1-19}$$

where η is the elevation of the water surface relative to the SWL, and $H/2$ is one-half the wave height equal to the wave amplitude a. This expression represents a periodic, sinusoidal, progressive wave traveling in the positive x-direction. For a wave moving in the negative x-direction, the minus sign before $2\pi t/T$ is replaced with a plus sign. When $\theta = (2\pi x/L - 2\pi t/T)$ equals 0, $\pi/2$, π, $3\pi/2$, the corresponding values of η are $H/2$, 0, $-H/2$, and 0, respectively (Figure II-1-1).

(4) Some useful functions.

(a) Dividing Equation II-1-9 by Equation II-1-13, and Equation II-1-10 by Equation II-1-15 yields,

$$\frac{C}{C_0} = \frac{L}{L_0} = \tanh \left(\frac{2\pi d}{L} \right) = \tanh kd \tag{II-1-20}$$

(b) If both sides of Equation II-1-20 are multiplied by d/L, it becomes

$$\frac{d}{L_0} = \frac{d}{L} \tanh \left(\frac{2\pi d}{L} \right) = \frac{d}{L} \tanh kd \tag{II-1-21}$$

(c) The terms d/L_o and d/L and other useful functions such as $kd = 2\pi d/L$ and $tanh\ (kd)$ have been tabulated by Wiegel (1954) as a function of d/L_o (see also SPM 1984, Appendix C, Tables C-1 and C-2). These functions simplify the solution of wave problems described by the linear theory and are summarized in Figure II-1-5. An example problem illustrating the use of linear wave theory equations and the figures and tables mentioned follows.

EXAMPLE PROBLEM II-1-1

FIND:

The wave celerities C and lengths L corresponding to depths d = 200 meters (656 ft) and d = 3 m (9.8 ft).

GIVEN:

A wave with a period T = 10 seconds is propagated shoreward over a uniformly sloping shelf from a depth d = 200 m (656 ft) to a depth d = 3 m (9.8 ft).

SOLUTION:

Using Equation II-1-15,

$$L_0 = \frac{gT^2}{2\pi} = \frac{9.8\ T^2}{2\pi} = 1.56\ T^2\ m\ (5.12\ T^2\ ft)$$

$$L_0 = 1.56T^2 = 1.56(10)^2 = 156\ m\ (512\ ft)$$

For d = 200 m

$$\frac{d}{L_0} = \frac{200}{156} = 1.2821$$

Note that for values of

$$\frac{d}{L_0} > 1.0$$

$$\frac{d}{L_0} = \frac{d}{L}$$

therefore,

$$L = L_0 = 156\ m\ (512\ ft)\ (deepwater\ wave,\ since \frac{d}{L} > \frac{1}{2})$$

which is in agreement with Figure II-1-5.

By Equation II-1-7

$$C = \frac{L}{T} = \frac{156}{T}$$

$$C = \frac{156}{10} = 15.6\ m/s\ (51.2\ ft/s)$$

For d = 3 m

$$\frac{d}{L_0} = \frac{3}{156} = 0.0192$$

Example Problem II-1-1 (Continued)

Example Problem II-1-1 (Concluded)

By trial-and-error solution (Equation II-1-21) with d/L_o it is found that

$$\frac{d}{L} = 0.05641$$

hence

$$L = \frac{3}{0.05641} = 53.2 \ m \ (174 \ ft) \left(\text{transitional depth, since} \frac{1}{25} < \frac{d}{L} < \frac{1}{2} \right)$$

$$C = \frac{L}{T} = \frac{53.2}{10} = 5.32 \ m/s \ (17.4 \ ft/s)$$

An approximate value of L can also be found by using Equation II-1-11

$$L \approx \frac{gT^2}{2\pi} \sqrt{\tanh\left(\frac{4\pi^2}{T^2}\frac{d}{g}\right)}$$

which can be written in terms of L_o as

$$L \approx L_0 \sqrt{\tanh\left(\frac{2\pi d}{L_0}\right)}$$

therefore

$$L \approx 156 \sqrt{\tanh\left(\frac{2\pi(3)}{156}\right)}$$

$$L \approx 156 \sqrt{\tanh(0.1208)}$$

$$L \approx 156 \sqrt{0.1202} = 54.1 \ m \ (177.5 \ ft)$$

which compares with $L = 53.3$ m obtained using Equations II-1-8, II-1-9, or II-1-21. The error in this case is 1.5 percent. Note that Figure II-1-5 or Plate C-1 (SPM 1984) could also have been used to determine d/L.

(5) Local fluid velocities and accelerations.

(a)   In wave force studies, the local fluid velocities and accelerations for various values of z and t during the passage of a wave must often be found. The horizontal component u and the vertical component w of the local fluid velocity are given by the following equations (with θ, x, and t as defined in Figure II-1-1):

$$u = \frac{H}{2} \frac{gT}{L} \frac{\cosh[2\pi(z+d)/L]}{\cosh(2\pi d/L)} \cos\theta \qquad (\text{II-1-22})$$

$$w = \frac{H}{2} \frac{gT}{L} \frac{\sinh[2\pi(z+d)/L]}{\cosh(2\pi d/L)} \sin\theta \qquad (\text{II-1-23})$$

(b)   These equations express the local fluid velocity components any distance $(z+d)$ above the bottom. The velocities are periodic in both x and t. For a given value of the phase angle $\theta = (2\pi x/L - 2\pi t/T)$, the hyperbolic functions *cosh* and *sinh*, as functions of z result in an approximate exponential decay of the magnitude of velocity components with increasing distance below the free surface. The maximum positive horizontal velocity occurs when $\theta = 0$, 2π, etc., while the maximum horizontal velocity in the negative direction occurs when $\theta = \pi$, 3π, etc. On the other hand, the maximum positive vertical velocity occurs when $\theta = \pi/2$, $5\pi/2$, etc., and the maximum vertical velocity in the negative direction occurs when $\theta = 3\pi/2$, $7\pi/2$, etc. Fluid particle velocities under a wave train are shown in Figure II-1-2.

(c)   The local fluid particle accelerations are obtained from Equations II-1-22 and II-1-23 by differentiating each equation with respect to t. Thus,

$$\alpha_x = \frac{g\pi H}{L} \frac{\cosh[2\pi(z+d)/L]}{\cosh(2\pi d/L)} \sin\theta = \frac{\partial u}{\partial t} \qquad (\text{II-1-24})$$

$$\alpha_z = -\frac{g\pi H}{L} \frac{\sinh[2\pi(z+d)/L]}{\cosh(2\pi d/L)} \cos\theta = \frac{\partial w}{\partial t} \qquad (\text{II-1-25})$$

(d)   Positive and negative values of the horizontal and vertical fluid accelerations for various values of θ are shown in Figure II-1-2.

(e)   Figure II-1-2, a sketch of the local fluid motion, indicates that the fluid under the crest moves in the direction of wave propagation and returns during passage of the trough. Linear theory does not predict any net mass transport; hence, the sketch shows only an oscillatory fluid motion. Figure II-1-3 depicts profiles of the surface elevation, particle velocities, and accelerations by the linear wave theory. The following problem illustrates the computations required to determine local fluid velocities and accelerations resulting from wave motions.

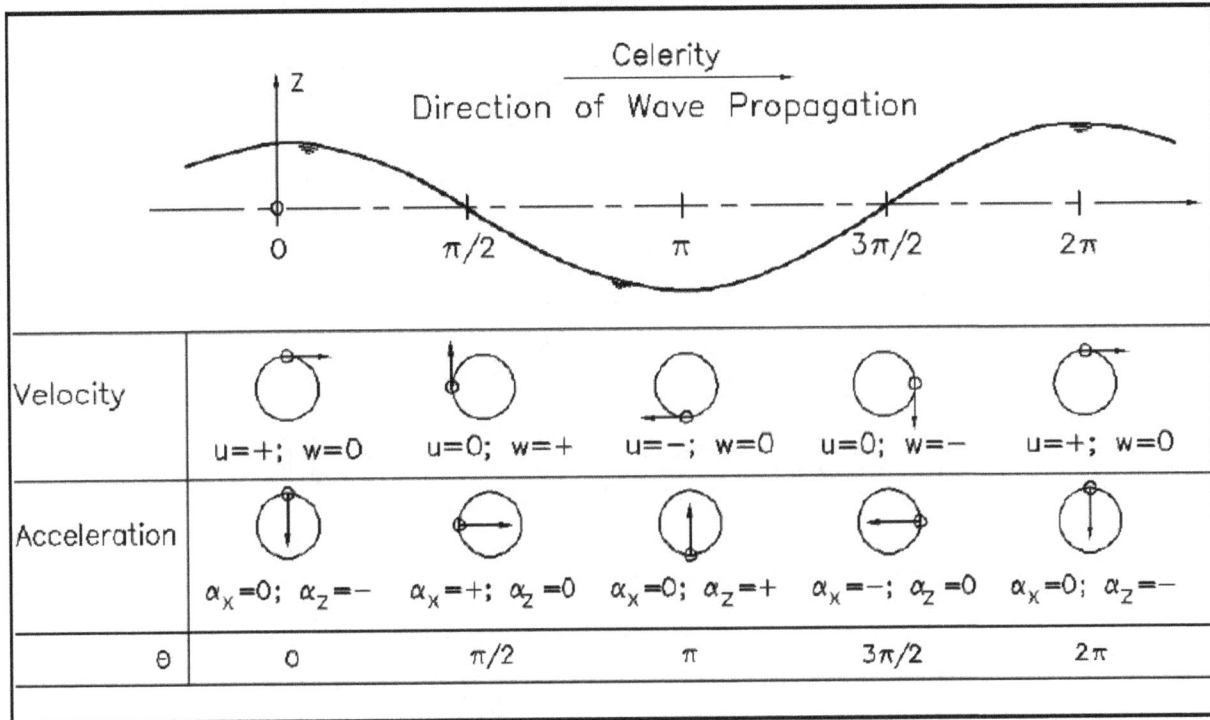

Figure II-1-2. Local fluid velocities and accelerations

(6) Water particle displacements.

(a) Another important aspect of linear wave theory deals with the displacement of individual water particles within the wave. Water particles generally move in elliptical paths in shallow or transitional depth water and in circular paths in deep water (Figure II-1-4). If the mean particle position is considered to be at the center of the ellipse or circle, then vertical particle displacement with respect to the mean position cannot exceed one-half the wave height. Thus, since the wave height is assumed to be small, the displacement of any fluid particle from its mean position must be small. Integration of Equations II-1-22 and II-1-23 gives the horizontal and vertical particle displacements from the mean position, respectively (Figure II-1-4).

(b) Fluid particle displacements are

$$\xi = -\frac{HgT^2}{4\pi L} \frac{\cosh\left(\frac{2\pi(z+d)}{L}\right)}{\cosh\left(\frac{2\pi d}{L}\right)} \sin \theta \qquad \text{(II-1-26)}$$

$$\zeta = +\frac{HgT^2}{4\pi L} \frac{\sinh\left(\frac{2\pi(z+d)}{L}\right)}{\cosh\left(\frac{2\pi d}{L}\right)} \cos \theta \qquad \text{(II-1-27)}$$

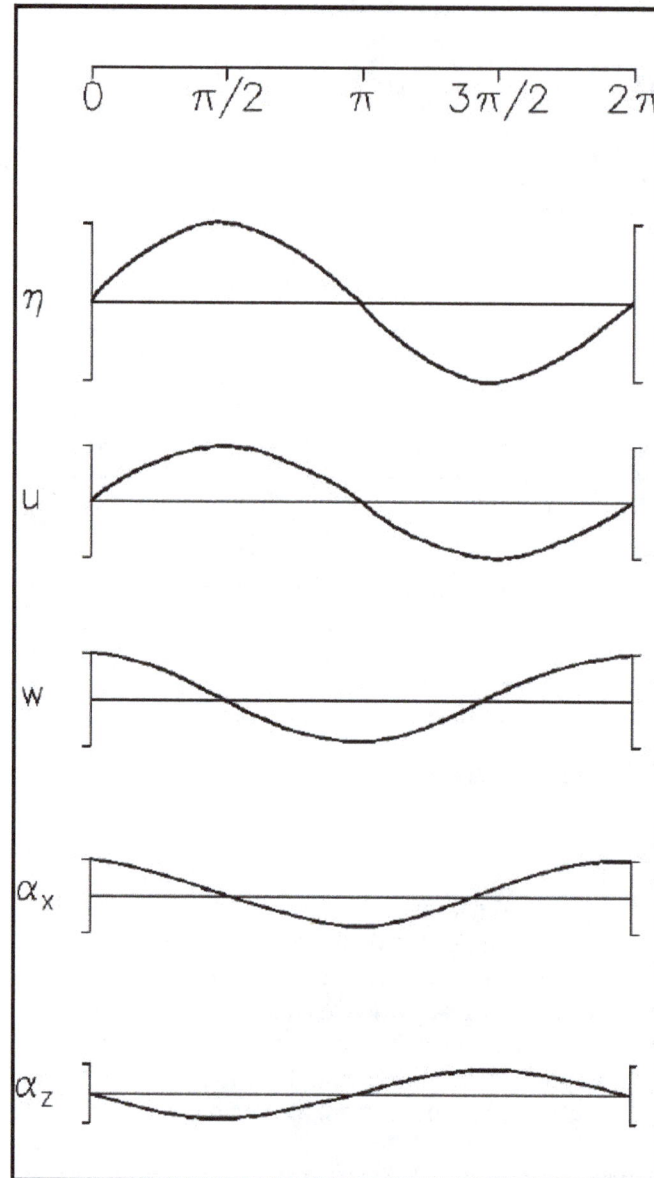

Figure II-1-3. Profiles of particle velocity and acceleration by Airy theory in relation to the surface elevation

where ξ is the horizontal displacement of the water particle from its mean position and ζ is the vertical displacement from its mean position (Figure II-1-4). The above equations can be simplified by using the relationship

$$\left(\frac{2\pi}{T}\right)^2 = \frac{2\pi g}{L} \tanh \frac{2\pi d}{L} \qquad \text{(II-1-28)}$$

EXAMPLE PROBLEM II-1-2

FIND:

The local horizontal and vertical velocities u and w, and accelerations α_x and α_z at an elevation $z = -5$ m (or $z = -16.4$ ft) below the SWL when $\theta = 2\pi x/L - 2\pi t/T = \pi/3$ (or 60^0).

GIVEN:

A wave with a period $T = 8$ sec, in a water depth $d = 15$ m (49 ft), and a height $H = 5.5$ m (18.0 ft).

SOLUTION:

Calculate

$$L_0 = 1.56T^2 = 1.56(8)^2 = 99.8 \ m \ (327 \ ft)$$

$$\frac{d}{L_0} = \frac{15}{99.8} = 0.1503$$

By trial-and-error solution or using Figure II-1-5 for $d/L_0 = 0.1503$, we find

$$\frac{d}{L} = 0.1835$$

and

$$\cosh\frac{2\pi d}{L} = 1.742$$

hence

$$L = \frac{15}{0.1835} = 81.7 \ m \ (268 \ ft)$$

Evaluation of the constant terms in Equations II-1-22 to II-1-25 gives

$$\frac{HgT}{2L} \frac{1}{\cosh(2\pi d/L)} = \frac{5.5 \ (9.8)(8)}{2 \ (81.7)} \frac{1}{1.742} = 1.515$$

$$\frac{Hg\pi}{L} \frac{1}{\cosh(2\pi d/L)} = \frac{5.5 \ (9.8)(3.1416)}{81.7} \frac{1}{1.742} = 1.190$$

Substitution into Equation II-1-22 gives

$$u = 1.515 \cosh\left[\frac{2\pi(15 - 5)}{81.7}\right] [\cos 60^0]$$

$$= 1.515 \ [\cosh(0.7691)] \ (0.500)$$

Example Problem II-1-2 (Continued)

Example Problem II-1-2 (Concluded)

From the above known information, we find

$$\frac{2\pi d}{L} = 0.7691$$

and values of hyperbolic functions become

$$\cosh(0.7691) = 1.3106$$

and

$$\sinh(0.7691) = 0.8472$$

Therefore, fluid particle velocities are

$$u = 1.515(1.1306)(0.500) = 0.99 \ m/s \ (3.26 \ ft/s)$$

$$w = 1.515(0.8472)(0.866) = 1.11 \ m/s \ (3.65 \ ft/s)$$

and fluid particle accelerations are

$$\alpha_x = 1.190(1.3106)(0.866) = 1.35 \ m/s^2 \ (4.43 \ ft/s^2)$$

$$\alpha_z = -1.190(0.8472)(0.500) = -0.50 \ m/s^2 \ (1.65 \ ft/s^2)$$

(c) Thus,

$$\xi = -\frac{H}{2} \frac{\cosh\left(\frac{2\pi(z+d)}{L}\right)}{\sinh\left(\frac{2\pi d}{L}\right)} \sin \theta \qquad \text{(II-1-29)}$$

$$\zeta = +\frac{H}{2} \frac{\sinh\left(\frac{2\pi(z+d)}{L}\right)}{\sinh\left(\frac{2\pi d}{L}\right)} \cos \theta \qquad \text{(II-1-30)}$$

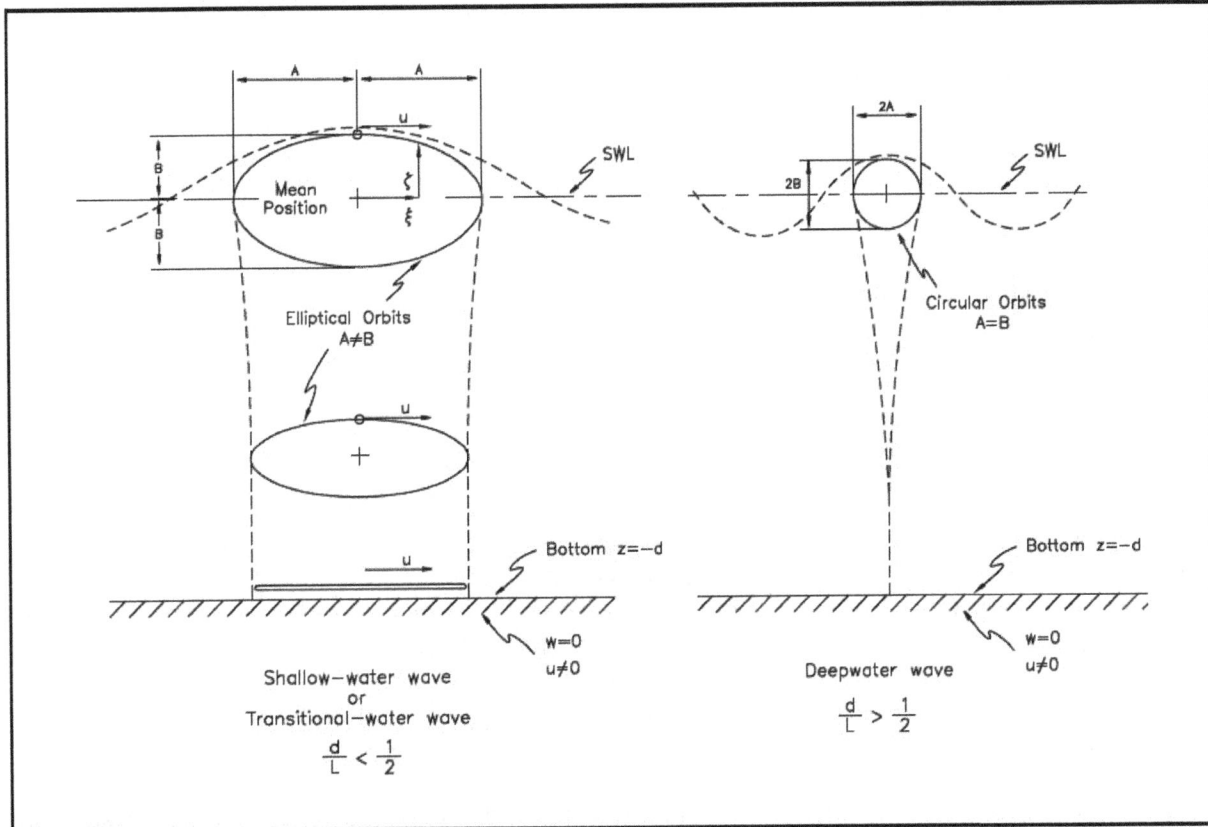

Figure II-1-4. Water particle displacements from mean position for shallow-water and deepwater waves

(d) Writing Equations II-1-29 and II-1-30 in the forms,

$$\sin^2 \theta = \left[\frac{\xi}{a} \frac{\sinh\left(\frac{2\pi d}{L}\right)}{\cosh\left(\frac{2\pi(z+d)}{L}\right)} \right]^2 \qquad (\text{II-1-31})$$

$$\cos^2 \theta = \left[\frac{\zeta}{a} \frac{\sinh\left(\frac{2\pi d}{L}\right)}{\sinh\left(\frac{2\pi(z+d)}{L}\right)} \right]^2 \qquad (\text{II-1-32})$$

and adding gives

$$\frac{\xi^2}{A^2} + \frac{\zeta^2}{B^2} = 1 \qquad (\text{II-1-33})$$

in which A and B are

$$A = \frac{H}{2} \frac{\cosh\left(\frac{2\pi(z+d)}{L}\right)}{\sinh\left(\frac{2\pi d}{L}\right)} \tag{II-1-34}$$

$$B = \frac{H}{2} \frac{\sinh\left(\frac{2\pi(z+d)}{L}\right)}{\sinh\left(\frac{2\pi d}{L}\right)} \tag{II-1-35}$$

(e) Equation II-1-33 is the equation of an ellipse with a major- (horizontal) semi-axis equal to A and a minor (vertical) semi-axis equal to B. The lengths of A and B are measures of the horizontal and vertical displacements of the water particles (see Figure II-1-4). Thus, the water particles are predicted to move in closed orbits by linear wave theory; i.e., a fluid particle returns to its initial position after each wave cycle. Comparing laboratory measurements of particle orbits with this theory shows that particle orbits are not completely closed. This difference between linear theory and observations is due to the mass transport phenomenon, which is discussed later in this chapter. It shows that linear theory is inadequate to explain wave motion completely.

(f) Examination of Equations II-1-34 and II-1-35 shows that for deepwater conditions, A and B are equal and particle paths are circular (Figure II-1-4). These equations become

$$A = B = \frac{H}{2} e^{\left(\frac{2\pi z}{L}\right)} \quad \text{for} \quad \frac{d}{L} > \frac{1}{2} \text{ (i.e., deepwater limit)} \tag{II-1-36}$$

(g) For shallow-water conditions (d/L < 1/25), the equations become

$$A = \frac{H}{2} \frac{L}{2\pi d} \tag{II-1-37}$$

and

$$B = \frac{H}{2}\left(1 + \frac{z}{d}\right) \tag{II-1-38}$$

EXAMPLE PROBLEM II-1-3

FIND:

(a) The maximum horizontal and vertical displacement of a water particle from its mean position when $z = 0$ and $z = -d$.

(b) The maximum water particle displacement at an elevation $z = -7.5$ m (-24.6 ft) when the wave is in infinitely deep water.

(c) For the deepwater conditions of (b) above, show that the particle displacements are small relative to the wave height when $z = -L_0/2$.

GIVEN:

A wave in a depth $d = 12$ m (39.4 ft), height $H = 3$ m (9.8 ft), and a period $T = 10$ sec. The corresponding deepwater wave height is $H_0 = 3.13$ m (10.27 ft).

SOLUTION:

(a)

$$L_0 = 1.56T^2 = 1.56(10)^2 = 156 \ m \ (512 \ ft)$$

$$\frac{d}{L_0} = \frac{12}{156} = 0.0769$$

From hand calculators, we find

$$\sinh\left(\frac{2\pi d}{L}\right) = 0.8306$$

$$\tanh\left(\frac{2\pi d}{L}\right) = 0.6389$$

When $z = 0$, Equation II-1-34 reduces to

$$A = \frac{H}{2}\frac{1}{\tanh\left(\frac{2\pi d}{L}\right)}$$

and Equation II-1-35 reduces to

$$B = \frac{H}{2}$$

Thus

$$A = \frac{3}{2}\frac{1}{(0.6389)} = 2.35 \ m \ (7.70 \ ft)$$

$$B = \frac{H}{2} = \frac{3}{2} = 1.5 \ m \ (4.92 \ ft)$$

Example Problem II-1-3 (Continued)

Example Problem II-1-3 (Concluded)

When $z = -d$,

$$A = \frac{H}{2 \sinh\left(\frac{2\pi d}{L}\right)} = \frac{3}{2(0.8306)} = 1.81 \ m \ (5.92 \ ft)$$

and $B = 0$.

(b) With $H_0 = 3.13$ m and $z = -7.5$ m (-24.6 ft), evaluate the exponent of e for use in Equation II-1-36, noting that $L = L_0$.

$$\frac{2\pi z}{L} = \frac{2\pi(-7.5)}{156} = -0.302$$

thus,

$$e^{-0.302} = 0.739$$

Therefore,

$$A = B = \frac{H_0}{2} e^{\left(\frac{2\pi z}{L}\right)} = \frac{3.13}{2}(0.739) = 1.16 \ m \ (3.79 \ ft)$$

The maximum displacement or diameter of the orbit circle would be $2(1.16) = 2.32$ m (7.61 ft) when $z = -7.5$ m.

(c) At a depth corresponding to the half wavelength from the MWL, we have

$$z = -\frac{L_0}{2} = \frac{-156}{2} = -78.0 \ m \ (255.9 \ ft)$$

$$\frac{2\pi z}{L} = \frac{2\pi(-78)}{156} = -3.142$$

Therefore

$$e^{-3.142} = 0.043$$

and

$$A = B = \frac{H_0}{2} e^{\left(\frac{2\pi z}{L}\right)} = \frac{3.13}{2}(0.043) = 0.067 \ m \ (0.221 \ ft)$$

Thus, the maximum displacement of the particle is 0.067 m, which is small when compared with the deepwater height, $H_0 = 3.13$ m (10.45 ft).

(h) Thus, in deep water, the water particle orbits are circular as indicated by Equation II-1-36 (see Figure II-1-4). Equations II-1-37 and II-1-38 show that in transitional and shallow water, the orbits are elliptical. The more shallow the water, the flatter the ellipse. The amplitude of the water particle displacement decreases exponentially with depth and in deepwater regions becomes small relative to the wave height at a depth equal to one-half the wavelength below the free surface; i.e., when $z = L_0/2$.

(i) Water particle displacements and orbits based on linear theory are illustrated in Figure II-1-4. For shallow regions, horizontal particle displacement near the bottom can be large. In fact, this is apparent in offshore regions seaward of the breaker zone where wave action and turbulence lift bottom sediments into suspension. The vertical displacement of water particles varies from a minimum of zero at the bottom to a maximum equal to one-half the wave height at the surface.

(7) Subsurface pressure.

(a) Subsurface pressure under a wave is the sum of two contributing components, dynamic and static pressures, and is given by

$$p' = \frac{\rho g H \cosh\left[\frac{2\pi(z+d)}{L}\right]}{2\cosh\left(\frac{2\pi d}{L}\right)} \cos\theta - \rho g z + p_a \tag{II-1-39}$$

where p' is the total or absolute pressure, p_a is the atmospheric pressure, and ρ is the mass density of water (for salt water, $\rho = 1,025$ kg/m³ or 2.0 slugs/ft³, for fresh water, $\rho = 1,000$ kg/m³ or 1.94 slugs/ft³). The first term of Equation II-1-39 represents a dynamic component due to acceleration, while the second term is the static component of pressure. For convenience, the pressure is usually taken as the gauge pressure defined as

$$p = p' - p_a = \frac{\rho g H \cosh\left[\frac{2\pi(z+d)}{L}\right]}{2\cosh\left(\frac{2\pi d}{L}\right)} \cos\theta - \rho g z \tag{II-1-40}$$

(b) Equation II-1-40 can be written as

$$p = \rho g \eta \frac{\cosh\left[\frac{2\pi(z+d)}{L}\right]}{\cosh\left(\frac{2\pi d}{L}\right)} - \rho g z \tag{II-1-41}$$

since

$$\eta = \frac{H}{2}\cos\left(\frac{2\pi x}{L} - \frac{2\pi t}{T}\right) = \frac{H}{2}\cos\theta \tag{II-1-42}$$

(c) The ratio

$$K_z = \frac{\cosh\left[\dfrac{2\pi(z + d)}{L}\right]}{\cosh\left(\dfrac{2\pi d}{L}\right)}$$

(II-1-43)

is termed *the pressure response factor*. Hence, Equation II-1-41 can be written as

$$p = \rho g(\eta K_z - z)$$

(II-1-44)

(d) The pressure response factor K for the pressure at the bottom when $z = -d$,

$$K_z = K = \frac{1}{\cosh\left(\dfrac{2\pi d}{L}\right)}$$

(II-1-45)

is presented as function of d/L_0 in the tables (SPM 1984); see also Figure II-1-5. This figure is a convenient graphic means to determine intermediate and shallow-water values of the bottom pressure response factor K, the ratio C/C_0 ($=L/L_0 = k_0/k$), and a number of other variables commonly occurring in water wave calculations.

(e) It is often necessary to determine the height of surface waves based on subsurface measurements of pressure. For this purpose, it is convenient to rewrite Equation II-1-44 as

$$\eta = \frac{N(p + \rho g z)}{\rho g K_z}$$

(II-1-46)

where z is the depth below the SWL of the pressure gauge, and N a correction factor equal to unity if the linear theory applies.

(f) Chakrabarti (1987) presents measurements that correlate measured dynamic pressure in the water column (s in his notation is the elevation above the seabed) with linear wave theory. These laboratory measurements include a number of water depths, wave periods, and wave heights. The best agreement between the theory and these measurements occurs in deep water. Shallow-water pressure measurements for steep water waves deviate significantly from the linear wave theory predictions. The example problem hereafter illustrates the use of pertinent equations for finding wave heights from pressure measurements based on linear theory.

(8) Group velocity.

(a) It is desirable to know how fast wave energy is moving. One way to determine this is to look at the speed of wave groups that represents propagation of wave energy in space and time. The speed a group of

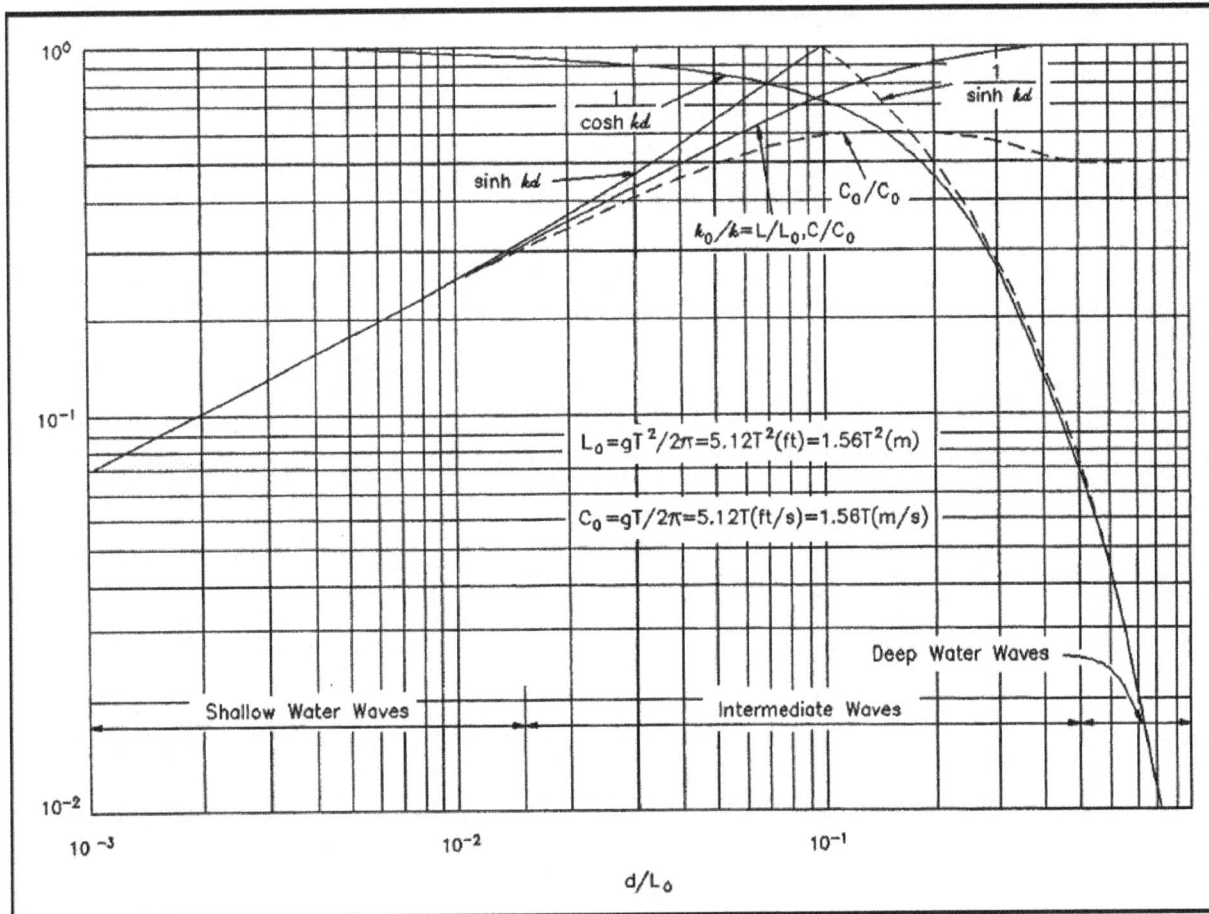

Figure II-1-5. Variation of wave parameters with d/L_0 (Dean and Dalrymple 1991)

waves or a wave train travels is generally not identical to the speed with which individual waves within the group travel. The group speed is termed the *group velocity* C_g; the individual wave speed is the *phase velocity* or *wave celerity* given by Equations II-1-8 or II-1-9. For waves propagating in deep or transitional water with gravity as the primary restoring force, the group velocity will be less than the phase velocity. For those waves, propagated primarily under the influence of surface tension (i.e., capillary waves), the group velocity may exceed the velocity of an individual wave.

(b) The concept of group velocity can be described by considering the interaction of two sinusoidal wave trains moving in the same direction with slightly different wavelengths and periods. The equation of the water surface is given by

$$\eta = \eta_1 + \eta_2 = \frac{H}{2}\cos\left(\frac{2\pi x}{L_1} - \frac{2\pi t}{T_1}\right) + \frac{H}{2}\cos\left(\frac{2\pi x}{L_2} - \frac{2\pi t}{T_2}\right) \qquad \text{(II-1-47)}$$

where η_1 and η_2 are the two components. They may be summed since superposition of solutions is permissible when the linear wave theory is used. For simplicity, the heights of both wave components have been assumed equal. Since the wavelengths of the two component waves, L_1 and L_2, have been assumed slightly different for some values of x at a given time, the two components will be in phase and the wave height observed will be 2H; for some other values of x, the two waves will be completely out of phase and

EXAMPLE PROBLEM II-1-4

FIND:

The height of the wave H assuming that linear theory applies and the average frequency corresponds to the average wave amplitude.

GIVEN:

An average maximum pressure $p = 124$ kilonewtons per square meter is measured by a subsurface pressure gauge located in salt water 0.6 meter (1.97 ft) above the bed in depth $d = 12$ m (39 ft). The average frequency $f = 0.06666$ cycles per second (Hertz).

SOLUTION:

$$T = \frac{1}{f} = \frac{1}{(0.0666)} \approx 15 \ s$$

$$L_0 = 1.56T^2 = 1.56(15)^2 = 351 \ m \ (1152 \ ft)$$

$$\frac{d}{L_0} = \frac{12}{351} \approx 0.0342$$

From Figure II-1-5, entering with d/L_0,

$$\frac{d}{L} = 0.07651$$

hence,

$$L = \frac{12}{(0.07651)} = 156.8 \ m \ (515 \ ft)$$

and

$$\cosh\left(\frac{2\pi d}{L}\right) = 1.1178$$

Therefore, from Equation II-1-43

$$K_z = \frac{\cosh\left[\dfrac{2\pi(z+d)}{L}\right]}{\cosh\left(\dfrac{2\pi d}{L}\right)} = \frac{\cosh\left[\dfrac{2\pi(-11.4+12)}{156.8}\right]}{1.1178} = 0.8949$$

Since $\eta = a = H/2$ when the pressure is maximum (under the wave crest), and $N = 1.0$ since linear theory is assumed valid,

$$\frac{H}{2} = \frac{N(p + \rho g z)}{\rho g K_z} = \frac{1.0 \ [124 + (10.06)(-11.4)]}{(10.06)(0.8949)} = 1.04 \ m \ (3.44 \ ft)$$

Therefore,

$$H = 2(1.04) = 2.08 \ m \ (6.3 \ ft)$$

Note that the value of K in Figure II-1-5 or SPM (1984) could not be used since the pressure was not measured at the bottom.

the resultant wave height will be zero. The surface profile made up of the sum of the two sinusoidal waves is given by Equation II-1-47 and is shown in Figure II-1-6. The waves shown in Figure II-1-6 appear to be traveling in groups described by the equation of the envelope curves

$$\eta_{envelope} = \pm H \cos\left[\pi\left(\frac{L_2 - L_1}{L_1 L_2}\right)x - \pi\left(\frac{T_2 - T_1}{T_1 T_2}\right)t\right]$$
(II-1-48)

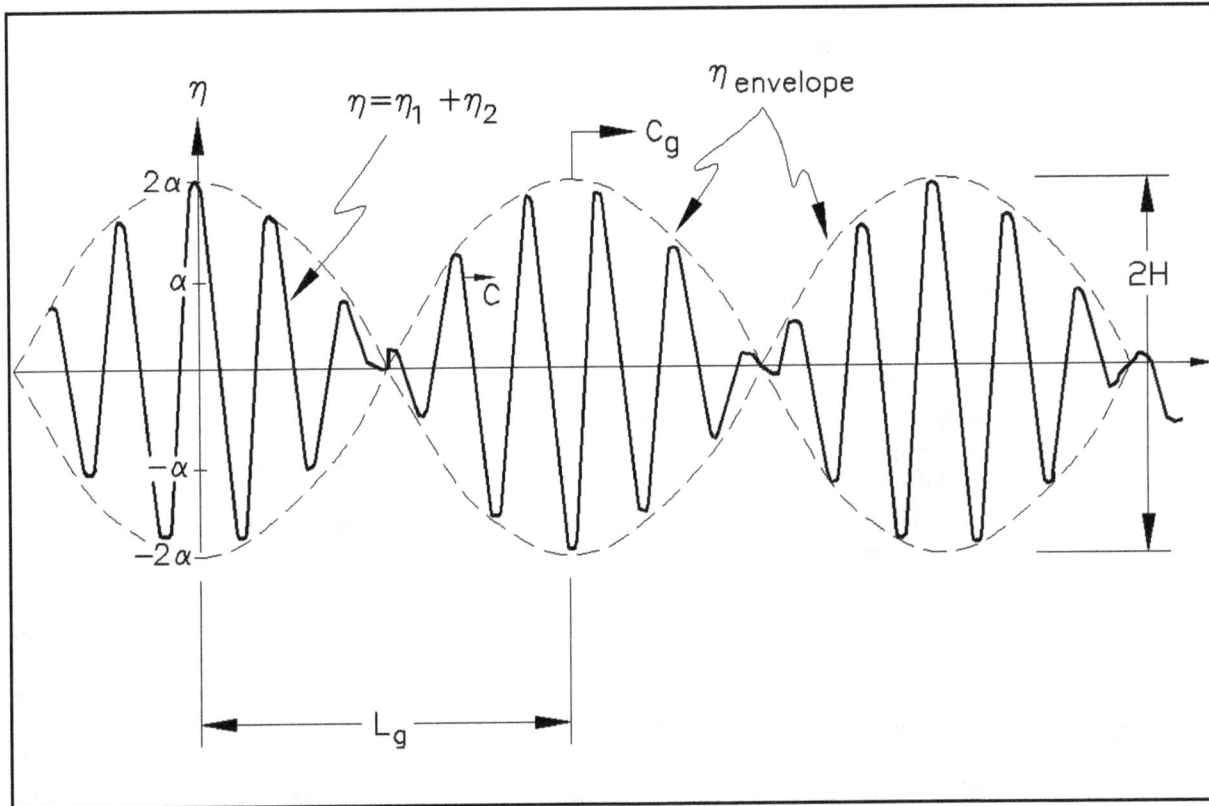

Figure II-1-6. Characteristics of a wave group formed by the addition of sinusoids with different periods

(c) It is the speed of these groups (i.e., the velocity of propagation of the envelope curves) defined in Equation II-1-48 that represents the group velocity. The limiting speed of the wave groups as they become large (i.e., as the wavelength L_1 approaches L_2 and consequently the wave period T_1 approaches T_2) is the group velocity and can be shown to be equal to

$$C_g = \frac{1}{2}\frac{L}{T}\left[1 + \frac{\dfrac{4\pi d}{L}}{\sinh\left(\dfrac{4\pi d}{L}\right)}\right] = nC$$
(II-1-49)

where

$$n = \frac{1}{2}\left[1 + \frac{\dfrac{4\pi d}{L}}{\sinh\left(\dfrac{4\pi d}{L}\right)}\right]$$

(II-1-50)

(d) In deep water, the term $(4\pi d/L)/\sinh(4\pi d/L)$ is approximately zero and $n = 1/2$, giving

$$C_{g_0} = \frac{1}{2}\frac{L_0}{T} = \frac{1}{2}C_0 \text{ (deep water)}$$

(II-1-51)

or the group velocity is one-half the phase velocity.

(e) In shallow water, $\sinh(4\pi d/L \approx 4\pi d/L)$ and

$$C_{g_s} = \frac{L}{T} = C \approx \sqrt{gd} \text{ (shallow water)}$$

(II-1-52)

hence, the group and phase velocities are equal. Thus, in shallow water, because wave celerity is determined by the depth, all component waves in a wave train will travel at the same speed precluding the alternate reinforcing and canceling of components. In deep and transitional water, wave celerity depends on wavelength; hence, slightly longer waves travel slightly faster and produce the small phase differences resulting in wave groups. These waves are said to be *dispersive* or propagating in a *dispersive medium*; i.e., in a medium where their celerity is dependent on wavelength.

(f) The variation of the ratios of group and phase velocities to the deepwater phase velocity C_g/C_0 and C/C_0, respectively are given as a function of the depth relative to the deep water wavelength d/L_0 in Figure II-1-7. The two curves merge together for small values of depth and C_g reaches a maximum before tending asymptotically toward $C/2$.

(g) Outside of shallow water, the phase velocity of gravity waves is greater than the group velocity. An observer that follows a group of waves at group velocity will see waves that originate at the rear of the group move forward through the group traveling at the phase velocity and disappear at the front of the wave group.

(h) Group velocity is important because it is with this velocity that wave energy is propagated. Although mathematically the group velocity can be shown rigorously from the interference of two or more waves (Lamb 1945), the physical significance is not as obvious as it is in the method based on the consideration of wave energy. Therefore an additional explanation of group velocity is provided on wave energy and energy transmission.

(9) Wave energy and power.

(a) The total energy of a wave system is the sum of its kinetic energy and its potential energy. The kinetic energy is that part of the total energy due to water particle velocities associated with wave motion. The kinetic energy per unit length of wave crest for a wave defined with the linear theory can be found from

$$\bar{E}_k = \int_x^{x+L}\int_{-d}^{\eta}\rho\frac{u^2 + w^2}{2}\,dz\,dx$$

(II-1-53)

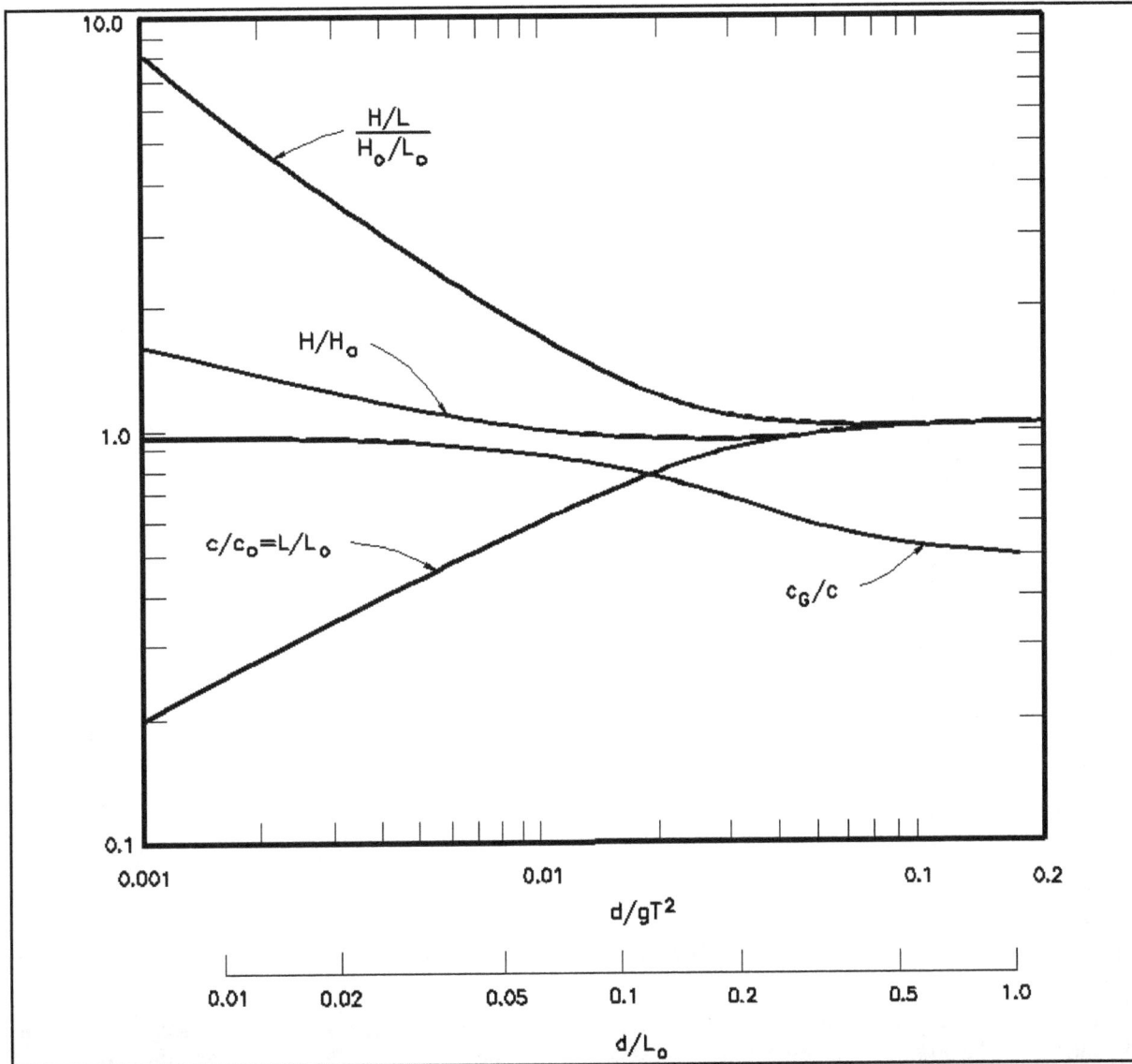

Figure II-1-7. Variation of the ratios of group and phase velocities to deepwater phase speed using linear theory (Sarpkaya and Isaacson 1981)

which, upon integration, gives

$$\bar{E}_k = \frac{1}{16} \rho \, g \, H^2 \, L \tag{II-1-54}$$

(b) Potential energy is that part of the energy resulting from part of the fluid mass being above the trough: the wave crest. The potential energy per unit length of wave crest for a linear wave is given by

$$\bar{E}_p = \int_x^{x+L} \rho \, g \left[\frac{(\eta + d)^2}{2} - \frac{d^2}{2} \right] dx \tag{II-1-55}$$

which, upon integration, gives

$$\bar{E}_p = \frac{1}{16} \rho \, g \, H^2 \, L \tag{II-1-56}$$

(c) According to the Airy theory, if the potential energy is determined relative to SWL, and all waves are propagated in the same direction, potential and kinetic energy components are equal, and the total wave energy in one wavelength per unit crest width is given by

$$E = E_k + E_p = \frac{\rho g H^2 L}{16} + \frac{\rho g H^2 L}{16} = \frac{\rho g H^2 L}{8} \tag{II-1-57}$$

where subscripts k and p refer to kinetic and potential energies. Total average wave energy per unit surface area, termed the *specific energy* or *energy density*, is given by

$$\bar{E} = \frac{E}{L} = \frac{\rho g H^2}{8} \tag{II-1-58}$$

(d) *Wave energy flux* is the rate at which energy is transmitted in the direction of wave propagation across a vertical plan perpendicular to the direction of wave advance and extending down the entire depth. Assuming linear theory holds, the average energy flux per unit wave crest width transmitted across a vertical plane perpendicular to the direction of wave advance is

$$\bar{P} = \frac{1}{T}\int_{t}^{t+r} \int_{-d}^{\eta} p \, u \, dz \, dt \tag{II-1-59}$$

which, upon integration, gives

$$\bar{P} = \bar{E}nC = \bar{E}C_g \tag{II-1-60}$$

where $\bar{P}$ is frequently called *wave power*, and the variable n has been defined earlier in Equation II-1-50.

(e) If a vertical plane is taken other than perpendicular to the direction of wave advance, $\bar{P} = E \, C_g \sin \theta$, where θ is the angle between the plane across which the energy is being transmitted and the direction of wave advance.

(f) For deep and shallow water, Equation II-1-60 becomes

$$\bar{P}_0 = \frac{1}{2} \bar{E}_0 \, C_o \quad \text{(deep water)} \tag{II-1-61}$$

$$\bar{P} = \bar{E}C_g = \bar{E}C \quad \text{(shallow water)} \tag{II-1-62}$$

(g) An energy balance for a region through which waves are passing will reveal that, for steady state, the amount of energy entering the region will equal the amount leaving the region provided no energy is added or removed. Therefore, when the waves are moving so that their crests are parallel to the bottom contours

$$\bar{E}_0 \, n_0 \, C_0 = \bar{E}nC \tag{II-1-63}$$

or since

$$n_0 = \frac{1}{2} \tag{II-1-64}$$

$$\frac{1}{2} \bar{E}_0 \, C_0 = \bar{E} n C \tag{II-1-65}$$

(h) When the wave crests are not parallel to the bottom contours, some parts of the wave will be traveling at different speeds and the wave will be refracted; in this case Equation II-1-65 does not apply (see Parts II-3 and II-4). The rate of energy transmission is important for coastal design, and it requires knowledge of C_g to determine how fast waves move toward shore. The mean rate of energy transmission associated with waves propagating into an area of calm water provides a different physical description of the concept of group velocity.

(i) Equation II-1-65 establishes a relationship between the ratio of the wave height at some arbitrary depth and the deepwater wave height. This ratio, known as the *shoaling coefficient* (see Part II-3 for detail derivation), is dependent on the wave steepness. The variation of shoaling coefficient with wave steepness as a function of relative water depth d/L_0 is shown in Figure II-1-8. Wave shoaling and other related nearshore processes are described in detail in Parts II-3 and II-4.

(10) Summary of linear wave theory.

(a) Equations describing water surface profile particle velocities, particle accelerations, and particle displacements for linear (Airy) theory are summarized in Figure II-1-9. The Corps of Engineers' microcomputer package of computer programs (ACES; Leenknecht et al. 1992) include several software applications for calculating the linear wave theory and associated parameters. Detailed descriptions of the ACES and CMS software to the linear wave theory may be found in the ACES and CMS documentation.

(b) Other wave phenomena can be explained using linear wave theory. For example, observed decreases and increases in the mean water level, termed wave setdown and wave setup, are in essence nonlinear quantities since they are proportional to wave height squared. These nonlinear quantities may be explained using the concept of radiation stresses obtained from linear theory. Maximum wave setdown occurs just seaward of the breaker line. Wave setup occurs between the breaker line and the shoreline and can increase the mean water level significantly. Wave setdown and setup and their estimation are discussed in Part II-4.

(c) *Radiation stresses* are the forces per unit area that arise because of the excess momentum flux due to the presence of waves. In simple terms, there is more momentum flow in the direction of wave advance because the velocity U is in the direction of wave propagation under the wave crest when the instantaneous water surface is high (wave crest) and in the opposite direction when the water surface is low (wave trough). Also, the pressure stress acting under the wave crest is greater than the pressure stress under the wave trough leading to a net stress over a wave period. Radiation stresses arise because of the finite amplitude (height) of the waves. Interestingly, small-amplitude (linear) wave theory can be used to reasonably approximate radiation stresses and explain effects such as wave set down, wave setup, and the generation of longshore currents.

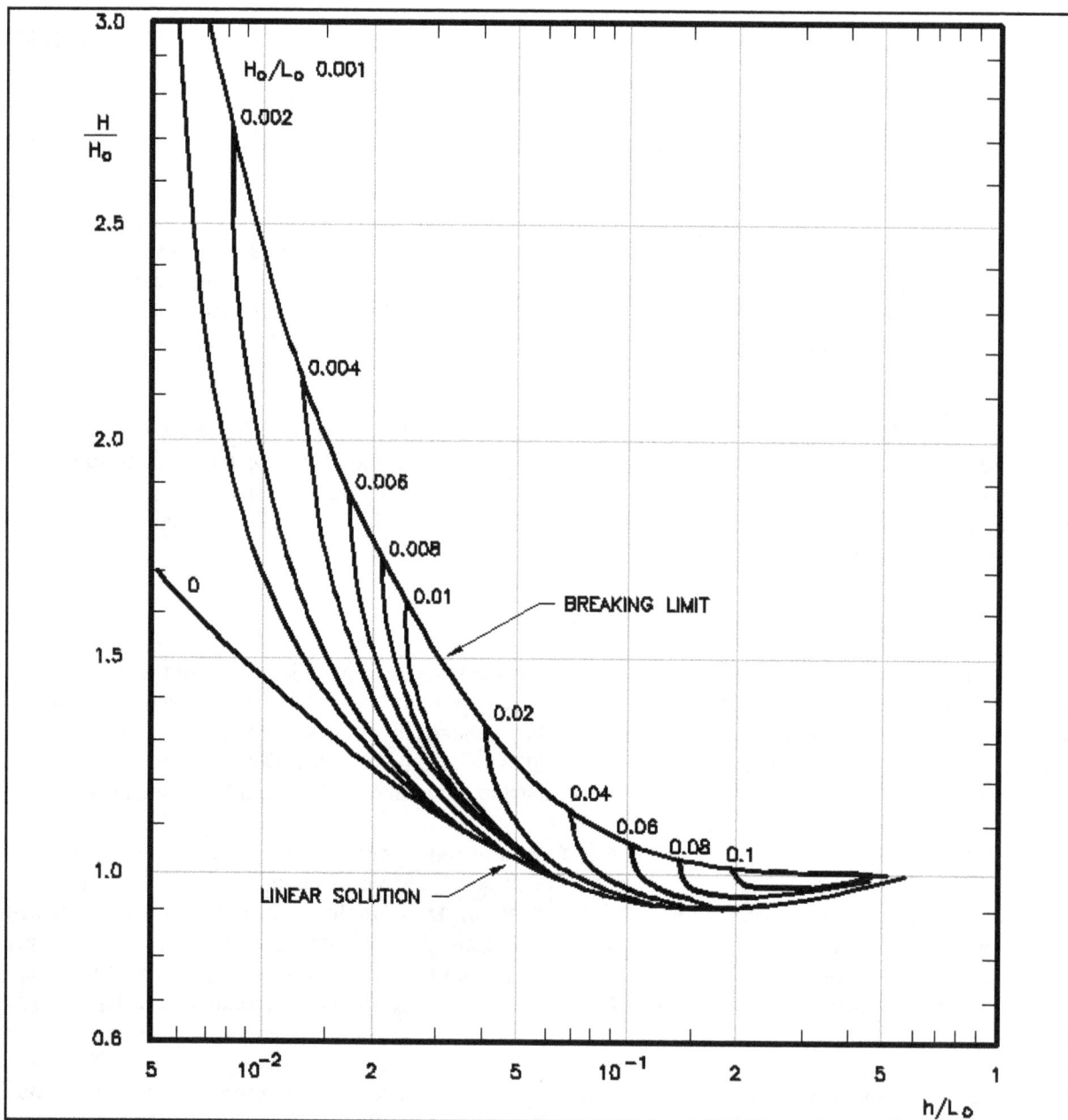

Figure II-1-8. Variation of shoaling coefficient with wave steepness (Sakai and Battjes 1980)

d. Nonlinear wave theories.

(1) Introduction.

(a) Linear waves as well as finite-amplitude waves may be described by specifying two dimensionless parameters, the wave steepness H/L and the relative water depth d/L. The relative water depth has been discussed extensively earlier in this chapter with regard to linear waves. The *Relative depth* determines whether waves are dispersive or nondispersive and whether the celerity, length, and height are influenced by water depth. *Wave steepness* is a measure of how large a wave is relative to its height and whether the linear wave assumption is valid. Large values of the wave steepness suggest that the small-amplitude

Relative Depth	Shallow Water $\dfrac{d}{L} < \dfrac{1}{20}$ $kd < \dfrac{\pi}{10}$	Transitional Water $\dfrac{1}{20} < \dfrac{d}{L} < \dfrac{1}{2}$ $\dfrac{\pi}{10} < kd < \dfrac{\pi}{2}$	Deep Water $\dfrac{d}{L} > \dfrac{1}{2}$ $kd > \dfrac{\pi}{2}$
1. Wave profile	Same As >	$\eta = \dfrac{H}{2} \cos\left[\dfrac{2\pi x}{L} - \dfrac{2\pi t}{T}\right] = \dfrac{H}{2}\cos\theta$	< Same As
2. Wave celerity	$C = \dfrac{L}{T} = \sqrt{gd}$	$C = \dfrac{L}{T} = \dfrac{gT}{2\pi}\tanh\left(\dfrac{2\pi d}{L}\right)$	$C = C_0 = \dfrac{L}{T} = \dfrac{gT}{2\pi}$
3. Wavelength	$L = T\sqrt{gd} = CT$	$L = \dfrac{gT^2}{2\pi}\tanh\left(\dfrac{2\pi d}{L}\right)$	$L = L_0 = \dfrac{gT^2}{2\pi} = C_0 T$
4. Group velocity	$C_g = C = \sqrt{gd}$	$C_g = nC = \dfrac{1}{2}\left[1 + \dfrac{4\pi d/L}{\sinh(4\pi d/L)}\right]C$	$C_g = \dfrac{1}{2}C = \dfrac{gT}{4\pi}$
5. Water particle velocity (a) Horizontal (b) Vertical	$u = \dfrac{H}{2}\sqrt{\dfrac{g}{d}}\cos\theta$ $w = \dfrac{H\pi}{T}\left(1 + \dfrac{z}{d}\right)\sin\theta$	$u = \dfrac{H}{2}\dfrac{gT}{L}\dfrac{\cosh[2\pi(z+d)/L]}{\cosh(2\pi d/L)}\cos\theta$ $w = \dfrac{H}{2}\dfrac{gT}{L}\dfrac{\sinh[2\pi(z+d)/L]}{\cosh(2\pi d/L)}\sin\theta$	$u = \dfrac{\pi H}{T}e^{\left(\frac{2\pi z}{L}\right)}\cos\theta$ $w = \dfrac{\pi H}{T}e^{\left(\frac{2\pi z}{L}\right)}\sin\theta$
6. Water particle accelerations (a) Horizontal (b) Vertical	$a_x = \dfrac{H\pi}{T}\sqrt{\dfrac{g}{d}}\sin\theta$ $a_z = -2H\left(\dfrac{\pi}{T}\right)^2\left(1+\dfrac{z}{d}\right)\cos\theta$	$a_x = \dfrac{g\pi H}{L}\dfrac{\cosh[2\pi(z+d)/L]}{\cosh(2\pi d/L)}\sin\theta$ $a_z = -\dfrac{g\pi H}{L}\dfrac{\sinh[2\pi(z+d)/L]}{\cosh(2\pi d/L)}\cos\theta$	$a_x = 2H\left(\dfrac{\pi}{T}\right)^2 e^{\left(\frac{2\pi z}{L}\right)}\sin\theta$ $a_z = -2H\left(\dfrac{\pi}{T}\right)^2 e^{\left(\frac{2\pi z}{L}\right)}\cos\theta$
7. Water particle displacements (a) Horizontal (b) Vertical	$\xi = -\dfrac{HT}{4\pi}\sqrt{\dfrac{g}{d}}\sin\theta$ $\zeta = \dfrac{H}{2}\left(1+\dfrac{z}{d}\right)\cos\theta$	$\xi = -\dfrac{H}{2}\dfrac{\cosh[2\pi(z+d)/L]}{\sinh(2\pi d/L)}\sin\theta$ $\zeta = \dfrac{H}{2}\dfrac{\sinh[2\pi(z+d)/L]}{\sinh(2\pi d/L)}\cos\theta$	$\xi = -\dfrac{H}{2}e^{\left(\frac{2\pi z}{L}\right)}\sin\theta$ $\zeta = \dfrac{H}{2}e^{\left(\frac{2\pi z}{L}\right)}\cos\theta$
8. Subsurface pressure	$p = \rho g(\eta - z)$	$p = \rho g\eta\dfrac{\cosh[2\pi(z+d)/L]}{\cosh(2\pi d/L)} - \rho gz$	$p = \rho g\eta\, e^{\left(\frac{2\pi z}{L}\right)} - \rho gz$

Figure II-1-9. Summary of linear (Airy) wave theory - wave characteristics

assumption may be questionable. A third dimensionless parameter, which may be used to replace either the wave steepness or relative water depth, may be defined as the ratio of wave steepness to relative water depth. Thus,

$$\frac{H/L}{d/L} = \frac{H}{d} \qquad\qquad\qquad (\text{II-1-66})$$

which is termed the *relative wave height*. Like the wave steepness, large values of the relative wave height indicate that the small-amplitude assumption may not be valid. A fourth dimensionless parameter often used to assess the relevance of various wave theories is termed the *Ursell number*. The Ursell number is given by

$$U_R = \left(\frac{L}{d}\right)^2 \frac{H}{d} = \frac{L^2 H}{d^3} \tag{II-1-67}$$

(b) The value of the Ursell number is often used to select a wave theory to describe a wave with given L and H (or T and H) in a given water depth d. High values of U_R indicate large, finite-amplitude, long waves in shallow water that may necessitate the use of nonlinear wave theory, to be discussed next.

(c) The linear or small-amplitude wave theory described in the preceding sections provides a useful first approximation to the wave motion. Ocean waves are generally not small in amplitude. In fact, from an engineering point of view it is usually the large waves that are of interest since they result in the largest forces and greatest sediment movement. In order to approach the complete solution of ocean waves more closely, a perturbation solution using successive approximations may be developed to improve the linear theory solution of the hydrodynamic equations for gravity waves. Each order wave theory in the perturbation expansion serves as a correction and the net result is often a better agreement between theoretical and observed waves. The extended theories can also describe phenomena such as *mass transport* where there is a small net forward movement of the water during the passage of a wave. These higher-order or extended solutions for gravity waves are often called *nonlinear wave theories*.

(d) Development of the nonlinear wave theories has evolved for a better description of surface gravity waves. These include *cnoidal*, *solitary*, and *Stokes* theories. However, the development of a Fourier-series approximation by Fenton in recent years has superseded the previous historical developments. Since earlier theories are still frequently referenced, these will first be summarized in this section, but Fenton's theory is recommended for regular waves in all coastal applications.

(2) Stokes finite-amplitude wave theory.

(a) Since the pioneering work of Stokes (1847, 1880) most extension studies (De 1955; Bretschneider 1960; Skjelbreia and Hendrickson 1961; Laitone 1960, 1962, 1965; Chappelear 1962; Fenton 1985) in wave perturbation theory have assumed the wave slope ka is small where k is the wave number and a the amplitude of the wave. The perturbation solution, developed as a power series in terms of $\epsilon = ka$, is expected to converge as more and more terms are considered in the expansion. Convergence does not occur for steep waves unless a different perturbation parameter from that of Stokes is chosen (Schwartz 1974; Cokelet 1977; Williams 1981, 1985).

(b) The fifth-order Stokes finite-amplitude wave theory is widely used in practical applications both in deep- and shallow-water wave studies. A formulation of Stokes fifth-order theory with good convergence properties has recently been provided (Fenton 1985). Fenton's fifth-order Stokes theory is computationally efficient, and includes closed-form asymptotic expressions for both deep- and shallow-water limits. Kinematics and pressure predictions obtained from this theory compare with laboratory and field measurements better than other nonlinear theories.

(c) In general, the perturbation expansion for velocity potential Φ may be written as

$$\Phi = \epsilon \Phi_1 + \epsilon^2 \Phi_2 + \dots \tag{II-1-68}$$

in which $\epsilon = ka$ is the *perturbation expansion parameter*. Each term in the series is smaller than the preceding term by a factor of order ka. In this expansion, Φ_1 is the first-order theory (linear theory), Φ_2 is the second-order theory, and so on.

(d) Substituting Equation II-1-68 and similar expressions for other wave variables (i.e., surface elevation η, velocities u and w, pressure p, etc.) into the appropriate governing equations and boundary conditions describing the wave motion yields a series of higher-order solutions for ocean waves. Equating the coefficients of equal powers of ka gives recurrence relations for each order solution. A characteristic of the perturbation expansion is that each order theory is expressed in terms of the preceding lower order theories (Phillips 1977; Dean and Dalrymple 1991; Mei 1991). The first-order Stokes theory is the linear (Airy) theory.

(e) The Stokes expansion method is formally valid under the conditions that $H/d \ll (kd)^2$ for $kd < 1$ and $H/L \ll 1$ (Peregrine 1972). In terms of the Ursell number U_R these requirements can be met only for $U_R < 79$. This condition restricts the wave heights in shallow water and the Stokes theory is not generally applicable to shallow water. For example, the maximum wave height in shallow water allowed by the second-order Stokes theory is about one-half of the water depth (Fenton 1985). The mathematics of higher-order Stokes theories is cumbersome and is not presented here. See Ippen (1966) for a detailed derivation of the Stokes second-order theory.

(f) In the higher-order Stokes solutions, superharmonic components (i.e., higher frequency components at two, three, four, etc. times the fundamental frequency) arise. These are superposed on the fundamental component predicted by linear theory. Hence, wave crests are steeper and troughs are flatter than the sinusoidal profile (Figure II-1-10). The fifth-order Stokes expansion shows a secondary crest in the wave trough for high-amplitude waves (Peregrine 1972; Fenton 1985). In addition, particle paths for Stokes waves are no longer closed orbits and there is a *drift* or *mass transport* in the direction of wave propagation.

(g) The linear dispersion relation is still valid to second order, and both wavelength and celerity are independent of wave height to this order. At third and higher orders, wave celerity and wavelength depend on wave height, and therefore, for a given wave period, celerity and length are greater for higher waves. Some limitations are imposed on the finite-amplitude Stokes theory in shallow water both by the water depth and amplitude nonlinearities. For steeper waves in shallow water, higher-order terms in Stokes expansion may increase in magnitude to become comparable or larger than the fundamental frequency component (Fenton 1985; Chakrabarti 1987). When this occurs, the Stokes perturbation becomes invalid.

(h) Higher-order Stokes theories include aperiodic (i.e., not periodic) terms in the expressions for water particle displacements. These terms arise from the product of time and a constant depending on the wave period and depth, and give rise to a continuously increasing net particle displacement in the direction of wave propagation. The distance a particle is displaced during one wave period when divided by the wave period gives a mean drift velocity $\bar{U}(z)$, called the *mass transport velocity*. To second-order, the mass transport velocity is

$$\bar{U}(z) = \left(\frac{\pi H}{L} \right)^2 \frac{C}{2} \frac{\cosh\,[4\pi(z + d)/L]}{\sinh^2\,(2\pi d/L)} \qquad \text{(II-1-69)}$$

indicating that there is a net transport of fluid by waves in the direction of wave propagation. If the mass transport leads to an accumulation of mass in any region, the free surface must rise, thus generating a pressure gradient. A current, formed in response to this pressure gradient, will reestablish the distribution of mass.

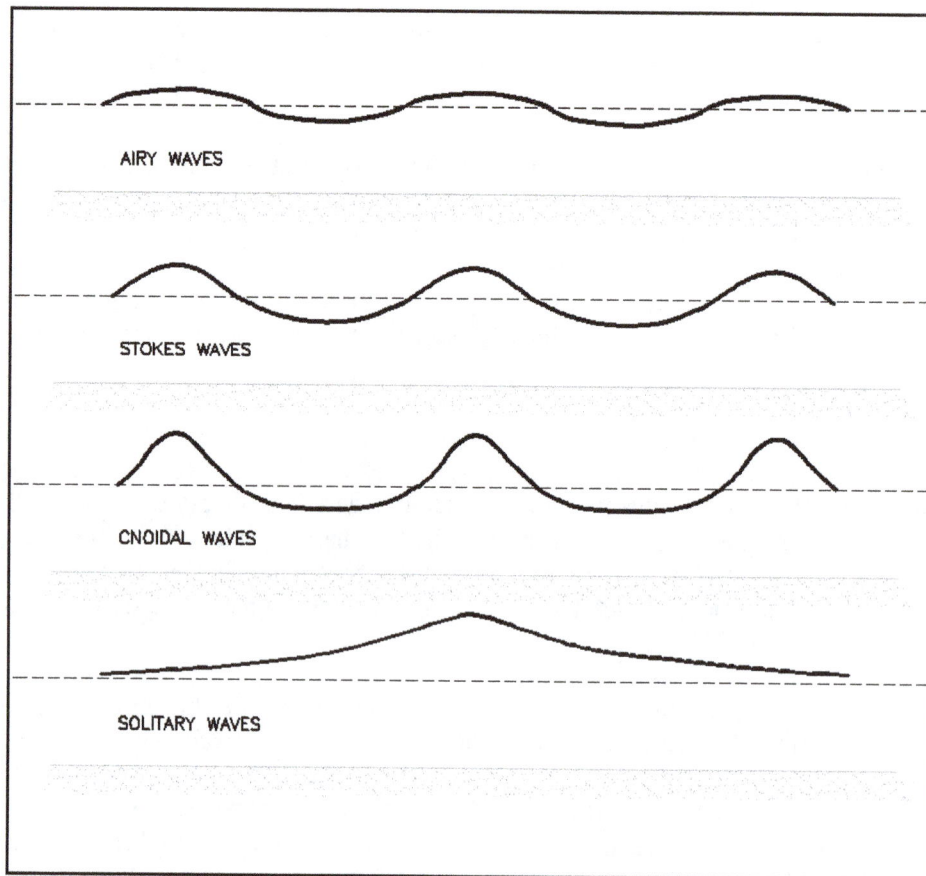

Figure II-1-10. Wave profile shape of different progressive gravity waves

(i) Following Stokes, using higher-order wave theories, both theoretical and experimental studies of mass transport have been conducted (Miche 1944; Ursell 1953; Longuet-Higgins 1953; Russell and Osorio 1958; Isaacson 1978). Results of two-dimensional wave tank experiments where a return flow existed in these studies show that the vertical distribution of the mass transport velocity is modified so that the net transport of water across a vertical plane is zero. For additional information on mass transport, see Dean and Dalrymple (1991).

(3) Subsurface pressure.

(a) Higher-order Stokes theories introduce corrections to the linear wave theory, and often provide more accurate estimates of the wave kinematics and dynamics. For example, the second-order Stokes theory gives the pressure at any distance below the fluid surface as

$$p = \rho g \frac{H}{2} \frac{\cosh[2\pi(z+d)/L]}{\cosh(2\pi d/L)} \cos\theta - \rho g z$$

$$+ \frac{3}{8}\rho g \frac{\pi H^2}{L} \frac{\tanh(2\pi d/L)}{\sinh^2(2\pi d/L)} \left(\frac{\cosh[4\pi(z+d)/L]}{\sinh^2(2\pi d/L)} - \frac{1}{3} \right) \cos 2\theta \qquad \text{(II-1-70)}$$

$$- \frac{1}{8}\rho g \frac{\pi H^2}{L} \frac{\tanh(2\pi d/L)}{\sinh^2(2\pi d/L)} \left(\cosh\frac{4\pi(z+d)}{L} - 1 \right)$$

(b) The terms proportional to the wave height squared in the above equation represent corrections by the second-order theory to the pressure from the linear wave theory. The third term is the steady component of pressure that corresponds to time-independent terms mentioned earlier.

(c) A direct byproduct of the high-order Stokes expansion is that it provides means for comparing different orders of resulting theories, all of which are approximations. Such comparison is useful to obtain insight about the choice of a theory for a particular problem. Nonetheless, it should be kept in mind that linear (or first-order) theory applies to a wave that is symmetrical about the SWL and has water particles that move in closed orbits. On the other hand, Stokes' higher-order theories predict a wave form that is asymmetrical about the SWL but still symmetrical about a vertical line through the crest and has water particle orbits that are open (Figure II-1-10).

(4) Maximum wave steepness.

(a) A progressive gravity wave is physically limited in height by depth and wavelength. The upper limit or breaking wave height in deep water is a function of the wavelength and, in shallow and transitional water, is a function of both depth and wavelength.

(b) Stokes (1880) predicted theoretically that a wave would remain stable only if the water particle velocity at the crest was less than the wave celerity or phase velocity. If the wave height were to become so large that the water particle velocity at the crest exceeded the wave celerity, the wave would become unstable and break. Stokes found that a wave having a crest angle less than 120 deg would break (angle between two lines tangent to the surface profile at the wave crest). The possibility of the existence of a wave having a crest angle equal to 120 deg is known (Lamb 1945). Michell (1893) found that in deep water the theoretical limit for wave steepness is

$$\left(\frac{H_0}{L_0} \right)_{max} = 0.142 \approx \frac{1}{7} \qquad \text{(II-1-71)}$$

Havelock (1918) confirmed Michell's finding.

(c) Miche (1944) gives the limiting steepness for waves traveling in depths less than $L_0/2$ without a change in form as

$$\left(\frac{H}{L} \right)_{max} = \left(\frac{H_0}{L_0} \right)_{max} \tanh\left(\frac{2\pi d}{L} \right) = 0.142 \tanh\left(\frac{2\pi d}{L} \right) \qquad \text{(II-1-72)}$$

Laboratory measurements indicate that Equation II-1-72 is in agreement with an envelope curve to laboratory observations (Dean and Dalrymple 1991).

e. Other wave theories.

(1) Introduction.

(a) Extension of the Stokes theory to higher orders has become common with computers, but the mathematics involved is still tedious. Variations of the Stokes theory have been developed in the last three decades oriented toward computer implementation. For example, Dean (1965) used the stream function in place of the velocity potential to develop the stream function theory. Dean (1974) did a limited comparison of measured horizontal particle velocity in a wave tank with the tenth-order stream function theory and several other theories. Forty cases were tabulated in dimensionless form to facilitate application of this theory.

(b) Others (Dalrymple 1974a; Chaplin 1980; Reinecker and Fenton 1981) developed variations of the stream function theory using different numerical methods. Their studies included currents. For near-breaking waves, Cokelet (1977) extended the method of Schwartz (1974) for steep waves for the full range of water depth and wave heights. Using a 110th-order theory for waves up to breaking, Cokelet successfully computed the wave profile, wave celerity, and various integral properties of waves, including the mean momentum, momentum flux, kinetic and potential energy, and radiation stress.

(2) Nonlinear shallow-water wave theories.

(a) Stokes' finite amplitude wave theory is applicable when the depth to wavelength ratio d/L is greater than about 1/8 or $kd > 0.78$ or $U_r < 79$. For longer waves a different theory must be used (Peregrine 1976). As waves move into shallow water, portions of the wave travel faster because of amplitude dispersion or waves travel faster because they are in deeper water. Waves also feel the effects of frequency dispersion less in shallow water, e.g., their speed is less and less influenced by water depth.

(b) For the mathematical representation of waves in shallow water, a different perturbation parameter should be used to account for the combined influence of amplitude and frequency dispersion (Whitham 1974; Miles 1981; Mei 1991). This can be achieved by constructing two perturbation parameters whose ratio is equivalent to the Ursell parameters (Peregrine 1972). The set of equations obtained in this manner are termed the *nonlinear shallow-water wave equations.* Some common wave theories based on these equations are briefly described in the following sections.

(3) Korteweg and de Vries and Boussinesq wave theories.

(a) Various shallow-water equations can be derived by assuming the pressure to be hydrostatic so that vertical water particle accelerations are small and imposing a horizontal velocity on the flow to make it steady with respect to the moving reference frame. The horizontal velocity might be the velocity at the SWL, at the bottom, or the velocity averaged over the depth. If equations are written in terms of depth-averaged velocity $\bar{u}$ they become:

$$\frac{\partial \eta}{\partial t} + \frac{\partial}{\partial x}(d + \eta)\bar{u} = 0$$

$$\frac{\partial \bar{u}}{\partial t} + \bar{u}\frac{\partial \bar{u}}{\partial x} + g\frac{\partial \eta}{\partial x} = \frac{1}{3}d^2\frac{\partial^3 \bar{u}}{\partial x^2 \partial t} \tag{II-1-73}$$

which are termed the *Boussinesq equations* (Whitham 1967; Peregrine 1972; Mei 1991). Originally, Boussinesq used the horizontal velocity at the bottom. Eliminating $\bar{u}$ yields (Miles 1979, 1980, 1981)

$$\frac{\partial^2 \eta}{\partial t^2} - gd\frac{\partial^2 \eta}{\partial x^2} = gd\frac{\partial^2}{\partial x^2}\left(\frac{3}{2}\frac{\eta^2}{d} + \frac{1}{3}d^2\frac{\partial^2 \eta}{\partial x^2} \right) \tag{II-1-74}$$

A periodic solution to Equation II-1-74 is of the form

$$\begin{aligned} \eta &= a\,e^{\,i(kx-\omega t)} = a\,\cos\theta \\ \bar{u} &= U_0\,e^{\,i(kx-\omega t)} = U_0\,\cos\theta \end{aligned} \tag{II-1-75}$$

which has a dispersion relation and an approximation to it given by

$$C = \frac{C_s}{\left[1 + \frac{1}{3}(kd)^2\right]^{1/2}} \approx C_s\left[1 - \frac{1}{3}(kd)^2 + \ldots\right] \tag{II-1-76}$$

The term $1/3\,(kd)^2$ in Equation II-1-76 represents the dispersion of wave motion.

(b) The most elementary solution of the Boussinesq equation is the *solitary wave* (Russell 1844; Fenton 1972; Miles 1980). A solitary wave is a wave with only crest and a surface profile lying entirely above the SWL. Fenton's solution gives the maximum solitary wave height, $H_{max} = 0.85$ d and maximum propagation speed $C^2_{max} = 1.7$ gd. Earlier research studies using the solitary waves obtained $H_{max} = 0.78$ d and $C^2_{max} = 1.56$ gd. The maximum solitary-amplitude wave is frequently used to calculate the height of breaking waves in shallow water. However, subsequent research has shown that the highest solitary wave is not necessarily the most energetic (Longuet-Higgins and Fenton 1974).

(4) Cnoidal wave theory.

(a) Korteweg and de Vries (1895) developed a wave theory termed the *cnoidal theory*. The cnoidal theory is applicable to finite-amplitude shallow-water waves and includes both nonlinearity and dispersion effects. Cnoidal theory is based on the Boussinesq, but is restricted to waves progressing in only one direction. The theory is defined in terms of the *Jacobian elliptic function, cn*, hence the name cnoidal. Cnoidal waves are periodic with sharp crests separated by wide flat troughs (Figure II-1-10).

(b) The approximate range of validity of the cnoidal theory is $d/L < 1/8$ when the Ursell number $U_R > 20$. As wavelength becomes long and approaches infinity, cnoidal wave theory reduces to the solitary wave theory, which is described in the next section. Also, as the ratio of wave height to water depth becomes small (infinitesimal wave height), the wave profile approaches the sinusoidal profile predicted by the linear theory.

(c) Cnoidal waves have been studied extensively by many investigators (Keulegan and Patterson 1940; Keller 1948; Laitone 1962) who developed first- through third-order approximations to the cnoidal wave theory. Wiegel (1960) summarized the principal results in a more usable form by presenting such wave characteristics as length, celerity, and period in tabular and graphical form to facilitate application of cnoidal theory.

(d) Wiegel (1964) further simplified the earlier works for engineering applications. Recent additional improvements to the theory have been made (Miles 1981; Fenton 1972, 1979). Using a Rayleigh-Boussinesq series, Fenton (1979) developed a generalized recursion relationship for the KdV solution of any order. Fenton's fifth- and ninth-order approximations are frequently used in practice. A summary of formulas of the cnoidal wave theory are provided below. See Fenton (1979), Fenton and McKee 1990), and Miles (1981) for a more comprehensive theoretical presentation.

(e) Long, finite-amplitude waves of permanent form propagating in shallow water may be described by cnoidal wave theory. The existence in shallow water of such long waves of permanent form may have first been recognized by Boussinesq (1871). However, the theory was originally developed by Korteweg and de Vries (1895).

(f) Because local particle velocities, local particle accelerations, wave energy, and wave power for cnoidal waves are difficult to describe such descriptions are not included here, but can be obtained in graphical form from Wiegel (1960, 1964). Wave characteristics are described in parametric form in terms of the modules k of the *elliptic integrals*. While k itself has no physical significance, it is used to express the relationships between various wave parameters. Tabular presentations of the elliptic integrals and other important functions can be obtained from the above references. The ordinate of the water surface y_s measured above the bottom is given by

$$y_s = y_t + H \ cn^2 \left[2K(k) \left(\frac{x}{L} - \frac{t}{T} \right) \ , \ k \right]$$

(II-1-77)

where

y_t = distance from the bottom to the wave trough

H = trough to crest wave height

cn = elliptic cosine function

$K(k)$ = complete elliptic integral of the first kind

k = modulus of the elliptic integrals

(g) The argument of cn^2 is frequently denoted simply by ( ); thus, Equation II-1-77 above can be written as

$$y_s = y_t + H \ cn^2(\ )$$

(II-1-78)

(h) The elliptic cosine is a periodic function where $cn^2 [2K(k) ((x/L) - (t/T))]$ has a maximum amplitude equal to unity. The modulus k is defined over the range 0 and 1. When $k = 0$, the wave profile becomes a sinusoid, as in the linear theory; when $k = 1$, the wave profile becomes that of a solitary wave.

(i) The distance from the bottom to the wave trough y_t, as used in Equations II-1-77 and II-1-78, is given by

$$\frac{y_t}{d} = \frac{y_c}{d} - \frac{H}{d} = \frac{16d^2}{3L^2} K(k) \ [K(k) - E(k)] + 1 - \frac{H}{d}$$

(II-1-79)

where y_c is the distance from the bottom to the crest, and $E(k)$ the complete elliptic integral of the second kind. Wavelength is given by

$$L = \sqrt{\frac{16d^3}{3H}} \, k \, K(k) \tag{II-1-80}$$

and wave period by

$$T\sqrt{\frac{g}{d}} = \sqrt{\frac{16y_t}{3H}} \, \frac{d}{y_t} \left[\frac{k \, K(k)}{1 + \dfrac{H}{y_t \, k^2} \left(\dfrac{1}{2} - \dfrac{E(k)}{K(k)} \right)} \right] \tag{II-1-81}$$

Note that cnoidal waves are periodic and of permanent form; thus $L = CT$ (see Figure II-1-10).

(j) Pressure under a cnoidal wave at any elevation y above the bottom depends on the local fluid velocity, and is therefore complex. However, it may be approximated in a hydrostatic form as

$$p = \rho g \, (y_s - y) \tag{II-1-82}$$

i.e., the pressure distribution may be assumed to vary linearly from $\rho g y_s$ at the bed to zero at the surface.

(k) Wave profiles obtained from different wave theories are sketched in Figure II-1-10 for comparison. The linear profile is symmetric about the SWL. The Stokes wave has higher more peaked crests and shorter, flatter troughs. The cnoidal wave crests are higher above the SWL than the troughs are below the SWL. Cnoidal troughs are longer and flatter and crests are sharper and steeper than Stokes waves. The solitary wave, a form of the cnoidal wave described in the next section, has all of its profile above the SWL.

(l) Figures II-1-11 and II-1-12 show the dimensionless cnoidal wave surface profiles for various values of the square of the modulus of the elliptic integrals k^2, while Figures II-1-13 to II-1-16 present dimensionless plots of the parameters which characterize cnoidal waves. The ordinates of Figures II-1-13 and II-1-14 should be read with care, since values of k^2 are extremely close to 1.0 ($k^2 = 1 - 10^{-1} = 1 - 0.1 = 0.90$). It is the exponent α of $k^2 = 1 - 10^{-\alpha}$ that varies along the vertical axis of Figures II-1-13 and II-1-14.

(m) Ideally, shoaling computations might be performed using a higher-order cnoidal wave theory since this theory is able to describe wave motion in relatively shallow water. Simple, completely satisfactory procedures for applying cnoidal wave theory are not available. Although linear wave theory is often used, cnoidal theory may be applied for practical situations using Figures such as II-1-11 to II-1-16. The following problem illustrates the use of these figures.

(n) There are two limits to the cnoidal wave theory. The first occurs when the period of the function cn is infinite when $k = 1$. This corresponds to a solitary wave. As the wavelength becomes infinite, the cnoidal theory approaches the solitary wave theory. The second limit occurs for $k = 0$ where the cnoidal wave approaches the sinusoidal wave. This happens when the wave height is small compared to water depth and the cnoidal theory reduces to the linear theory.

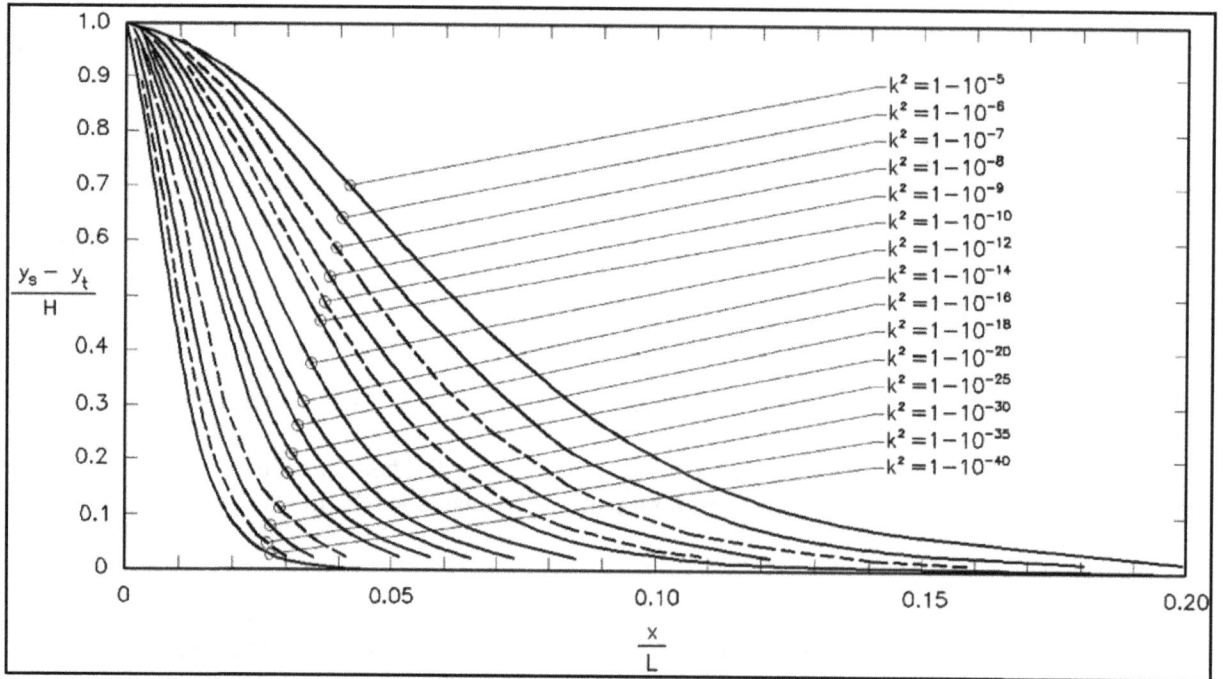

Figure II-1-11. Normalized surface profile of the cnoidal wave (Wiegel 1960). For definition of variables see Part II-1-2.e.(3)

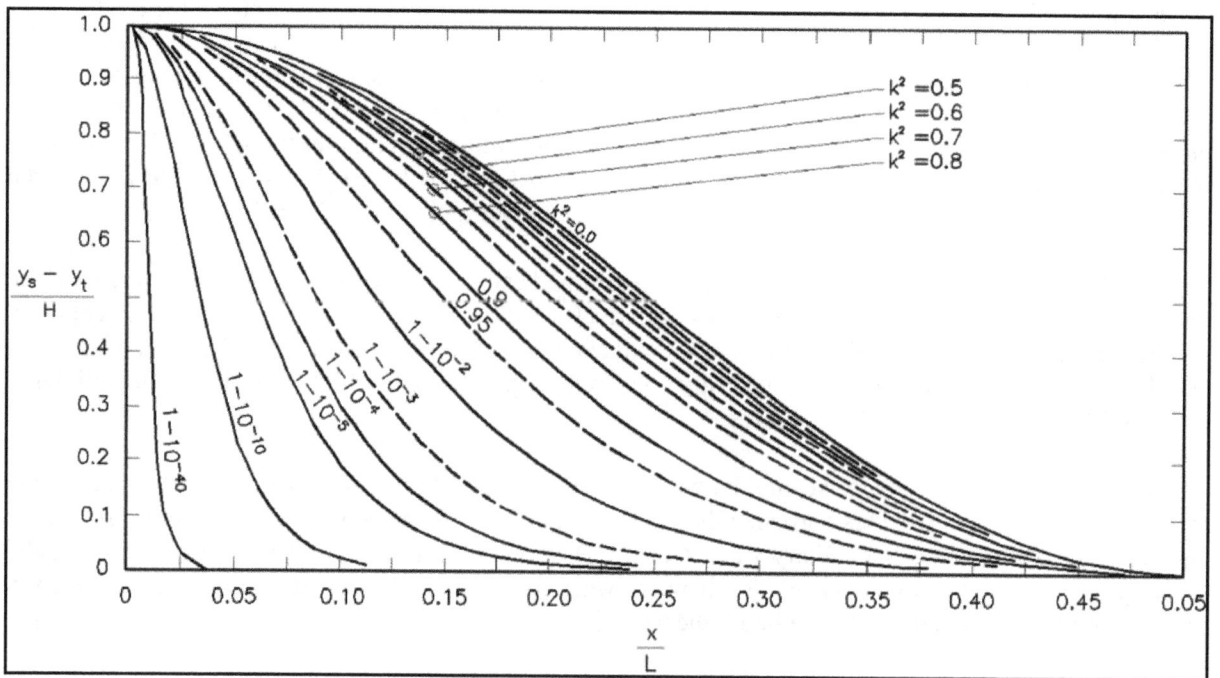

Figure II-1-12. Normalized surface profile of the cnoidal wave for higher values of k^2 and X/L (Wiegel 1960)

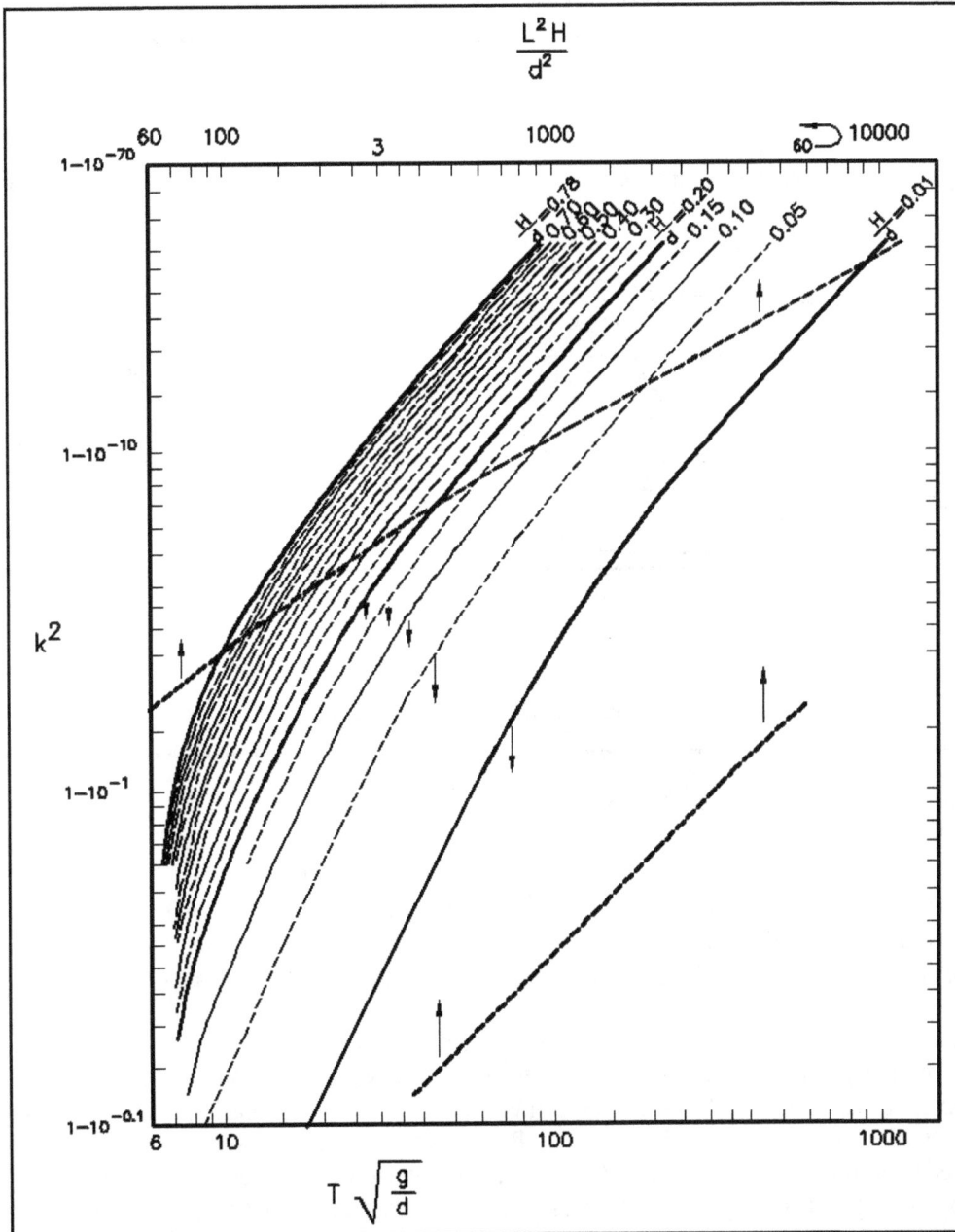

Figure II-1-13. k^2 versus L^2H/d^3, and k^2 versus $T\sqrt{g/d}$ and H/d (Wiegel 1960)

(5) Solitary wave theory.

(a) Waves considered in the previous sections were oscillatory or nearly oscillatory waves. The water particles move backward and forward with the passage of each wave, and a distinct wave crest and wave trough are evident. A solitary wave is neither oscillatory nor does it exhibit a trough. In the pure sense, the solitary wave form lies entirely above the still-water level. The solitary wave is a wave of translation because the water particles are displaced a distance in the direction of wave propagation as the wave passes.

(b) The *solitary wave* was discovered by Russell (1844). Boussinesq (1871), Rayleigh (1876), Keller (1948), and Munk (1949) performed pioneering theoretical studies of solitary waves. More recent analyses

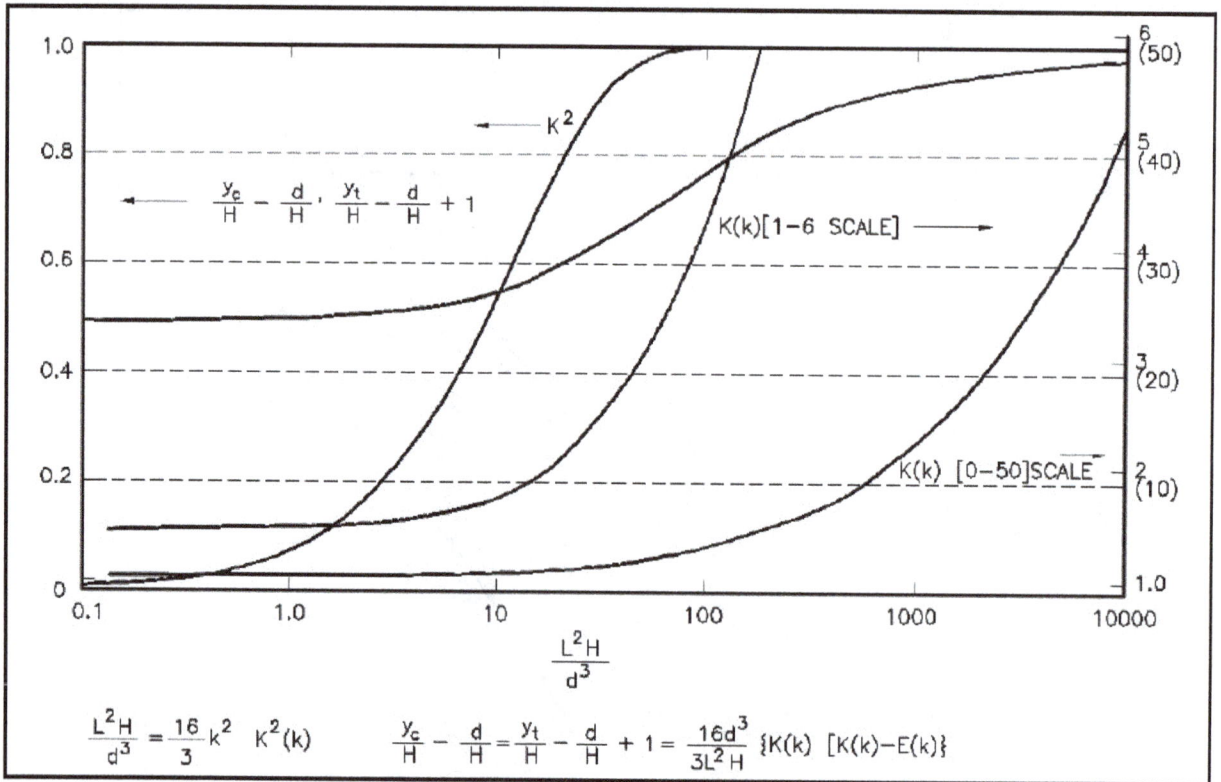

Figure II-1-14. Relationship among L^2H/d^3 and the square of the elliptic modulus (k^2), y_c/H, y_t/H, and K(k) (Wiegel 1960)

of solitary waves were performed by Fenton (1972), Longuet-Higgins and Fenton (1974), and Byatt-Smith and Longuet-Higgins (1976). The first systematic observations and experiments on solitary waves can probably be attributed to Russell (1838, 1844), who first recognized the existence of a solitary wave.

(c)   In nature it is difficult to form a truly solitary wave, because at the trailing edge of the wave there are usually small dispersive waves. However, long waves such as tsunamis and waves resulting from large displacements of water caused by such phenomena as landslides and earthquakes sometimes behave approximately like solitary waves. When an oscillatory wave moves into shallow water, it may often be approximated by a solitary wave (Munk 1949). As an oscillatory wave moves into shoaling water, the wave amplitude becomes progressively higher, the crests become shorter and more pointed, and the trough becomes longer and flatter.

(d)   Because both wavelength and period of solitary waves are infinite, only one parameter H/d is needed to specify a wave. To lowest order, the solitary wave profile varies as $sech^2q$ (Wiegel 1964), where $q = (3H/d)^{1/2} (x-Ct)/2d$ and the free-surface elevation, particle velocities, and pressure may be expressed as

$$\frac{\eta}{H} = \frac{u}{\sqrt{gd}\,\frac{H}{d}} \qquad\qquad\qquad\qquad\qquad\text{(II-1-83)}$$

$$\frac{u}{\sqrt{gd}}\frac{H}{d} = \frac{\Delta p}{\rho g H} \qquad\qquad\qquad\qquad\qquad\text{(II-1-84)}$$

$$T\sqrt{\frac{g}{d}} = \sqrt{\frac{16d}{3H}} \; \frac{k \; K\langle k\rangle}{1+\frac{H}{d\,k^2}\left(\frac{1}{2} - \frac{E\langle k\rangle}{K\langle k\rangle}\right)}$$

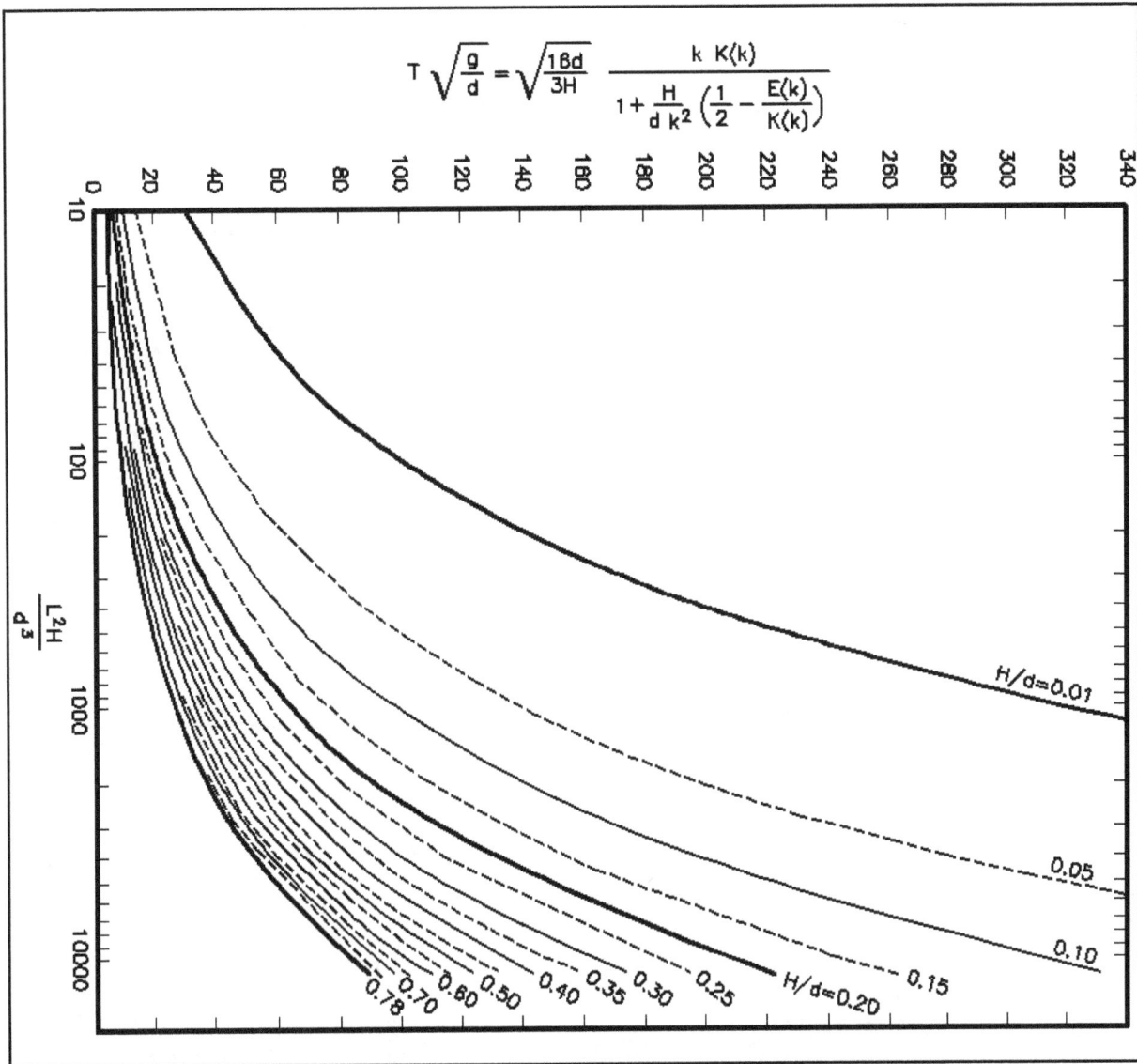

Figure II-1-15. Relationships among $T\sqrt{g/d}$, L^2H/d^3, and H/d (Wiegel 1960)

$$\frac{\Delta p}{\rho g H} = sech^2 \, q \qquad\qquad\qquad\qquad\qquad\qquad\qquad\qquad \text{(II-1-85)}$$

where Δp is the difference in pressure at a point due to the presence of the solitary wave.

(e)   To second approximation (Fenton 1972), this difference is given by

$$\frac{\Delta p}{\rho g H} = 1 - \frac{3}{4}\frac{H}{d}\left[1 - \left(\frac{Y_s}{d}\right)^2\right] \qquad\qquad\qquad\qquad \text{(II-1-86)}$$

where y_s = the height of the surface profile above the bottom. The wave height H required to produce Δp on the seabed can be estimated from

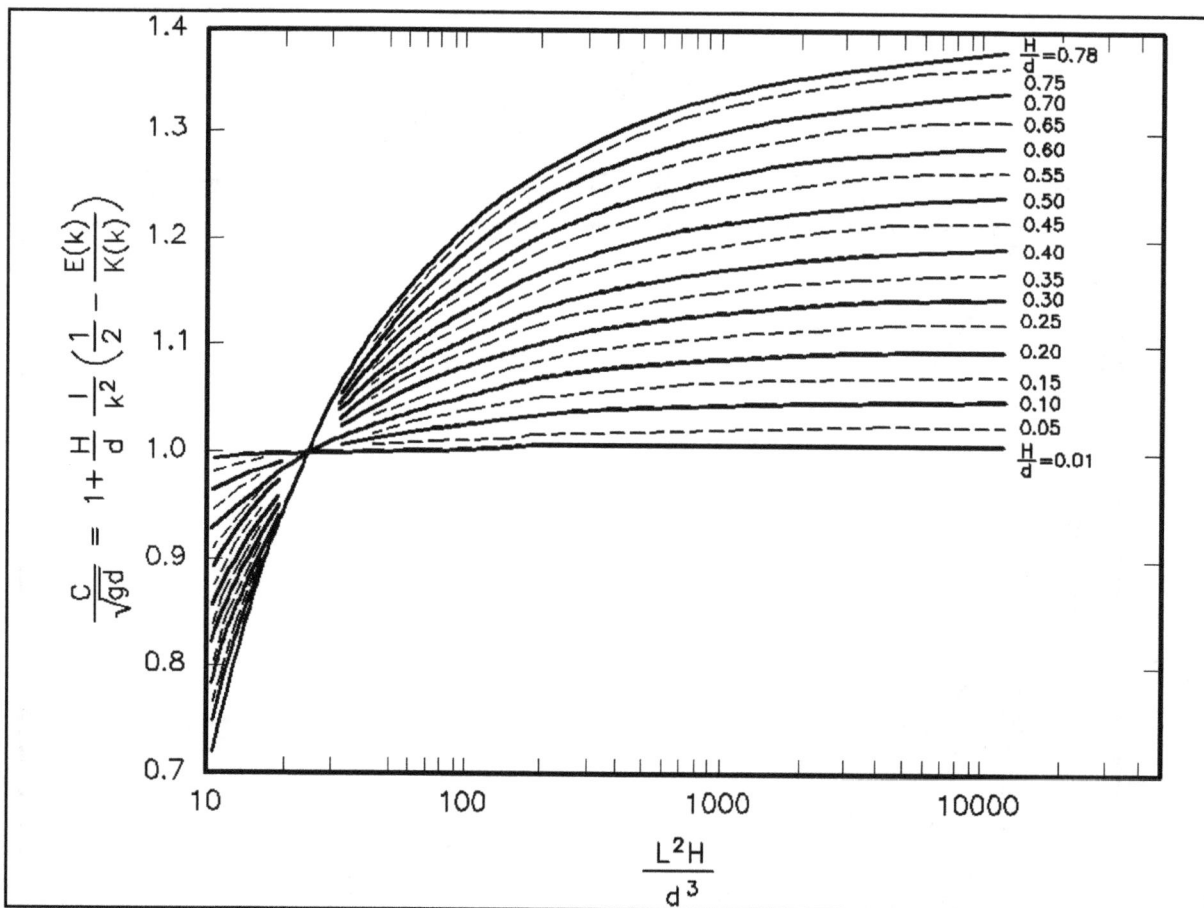

Figure II-1-16. Relationship between cnoidal wave velocity and L^2H/d^3 (Wiegel 1960)

$$\frac{\Delta p}{\rho g H} = \frac{1}{2} + \frac{1}{2} \sqrt{1 - \frac{3\Delta p}{\rho g d}} \qquad \text{(II-1-87)}$$

(f) Since the solitary wave has horizontal particle velocities only in the direction of wave advance, there is a net displacement of fluid in the direction of wave propagation.

(g) The solitary wave is a limiting case of the cnoidal wave. When $k^2 = 1$, $K(k) = K(1) = \infty$, and the elliptic cosine reduces to the hyperbolic secant function and the water surface y_s measured above the bottom reduces to

$$y_s = d + H \, sech^2 \left[\sqrt{\frac{3}{4} \frac{H}{d^3}} \, (x - Ct) \right] \qquad \text{(II-1-88)}$$

(h) The free surface is given by

$$\eta = H \, sech^2 \left[\sqrt{\frac{3}{4} \frac{H}{d^3}} \, (x - Ct) \right] \qquad \text{(II-1-89)}$$

EXAMPLE PROBLEM II-1-5

FIND:

(a) Using cnoidal wave theory, find the wavelength L and compare this length with the length determined using Airy theory.

(b) Determine the celerity C. Compare this celerity with the celerity determined using Airy theory.

(c) Determine the distance above the bottom of the wave crest y_c and wave trough y_t.

(d) Determine the wave profile.

GIVEN:

A wave traveling in water depth $d = 3$ m (9.84 ft), with a period $T = 15$ sec, and a height $H = 1.0$ m (3.3 ft).

SOLUTION:

(a) Calculate

$$\frac{H}{d} = \frac{1}{3} = 0.33$$

and

$$T \sqrt{\frac{g}{d}} = 15 \sqrt{\frac{9.8}{3}} = 27.11$$

From Figure II-1-13, enter H/d and T to determine the square of the modulus of the complete elliptical integrals, k^2:

$$k^2 = 1 - 10^{-5}$$

Entering both Figures II-1-13 and II-1-14 with the value of k^2 gives

$$\frac{L^2 H}{d^3} = 290$$

or

$$L = \sqrt{\frac{290 \, d^3}{H}} = \sqrt{\frac{290 \, (3)^3}{1}}$$

Example Problem II-1-5 (Continued)

Example Problem II-1-5 (Continued)

which gives L = 88.5 m (290.3 ft). The wavelength from the linear (Airy) theory is

$$L = \frac{gT^2}{2\pi} \tanh\left(\frac{2\pi d}{L}\right) = 80.6 \ m \ (264.5 \ ft)$$

To check whether the wave conditions are in the range for which cnoidal wave theory is valid, calculate d/L and the *Ursell number* $= L^2H/d^3$:

$$\frac{d}{L} = \frac{3}{88.5} = 0.0339 < \frac{1}{8} \quad \text{O.K.}$$

$$\frac{L^2H}{d^3} = \frac{1}{\left(\frac{d}{L}\right)^2}\left(\frac{H}{d}\right) = 290 > 26 \quad \text{O.K.}$$

Therefore, cnoidal theory is applicable.

   (b) Wave celerity is given by

$$C = \frac{L}{T} = \frac{88.5}{15} = 5.90 \ m/s \ (19.36 \ ft/s)$$

while the linear theory predicts

$$C = \frac{L}{T} = \frac{80.6}{15} = 5.37 \ m/s \ (17.63 \ ft/s)$$

Thus, if it is assumed that the wave period is the same for cnoidal and Airy theories, then

$$\frac{C_{cnoidal}}{C_{Airy}} = \frac{L_{cnoidal}}{L_{Airy}} \approx 1$$

   (c) The percentage of the wave height above the SWL may be determined from Figure II-1-11 or II-1-12. Entering these figures with $L^2H/D^3 = 290$, the value of $(y_c - d)/H$ is found to be 0.865, or 86.5 percent. Therefore,

$$y_c = 0.865 \ H + d$$

$$y_c = 0.865(1) + 3 = 0.865 + 3 = 3.865 \ m \ (12.68 \ ft)$$

Example Problem II-1-5 (Continued)

Example Problem II-1-5 (Concluded)

Also from Figure II-1-11 or II-1-12,

$$\frac{(y_t - d)}{H} + 1 = 0.865$$

thus,

$$y_t = (0.865 - 1)(1) + 3 = 2.865 \ m \ (9.40 \ ft)$$

(d) The dimensionless wave profile is given in Figures II-1-11 and II-1-12 for $k^2 = 1 - 10^{-5}$. The results obtained in (c) above can also be checked by using Figures II-1-11 and II-1-12. For the wave profile obtained with $k^2 = 1 - 10^{-5}$, the SWL is approximately 0.14H above the wave trough or 0.86H below the wave crest.

The results for the wave celerity determined under (b) above can now be checked with the aid of Figure II-1-16. Calculate

$$\frac{H}{y_t} = \frac{(1)}{2.865} = 0.349$$

Entering Figure II-1-16 with

$$\frac{L^2 H}{d^3} = \frac{(1)}{2.865} = 0.349$$

and

$$\frac{H}{y_t} = 0.349$$

it is found that

$$\frac{C}{\sqrt{g \ y_t}} = 1.126$$

Therefore,

$$C = 1.126 \ \sqrt{(9.8)(2.865)} = 5.97 \ m/s \ (19.57 \ ft/s)$$

The differences between this number and the 5.90 m /sec (18.38 ft/s) calculated under (b) above is the result of small errors in reading the curves.

where the origin of x is at the wave crest. The volume of water within the wave above the still-water level per unit crest width is

$$V = \left[\frac{16}{3} d^3 H \right]^{\frac{1}{2}}$$

(II-1-90)

(i)   An equal amount of water per unit crest length is transported forward past a vertical plane that is perpendicular to the direction of wave advance. Several relations have been presented to determine the celerity of a solitary wave; these equations differ depending on the degree of approximation. Laboratory measurements suggest that the simple expression

$$C = \sqrt{g(H + d)}$$

(II-1-91)

gives a reasonably accurate approximation to the celerity of solitary wave.

(j)   The water particle velocities for a solitary wave (Munk 1949), are

$$u = CN \frac{1 + \cos(My/d) \cosh(Mx/d)}{[\cos(My/d) + \cosh(Mx/D)]^2}$$

(II-1-92)

$$w = CN \frac{\sin(My/d) \sinh(Mx/d)}{[\cos(My/d) + \cosh(Mx/D)]^2}$$

(II-1-93)

where M and N are the functions of H/d shown in Figure II-1-17, and y is measured from the bottom. The expression for horizontal velocity u is often used to predict wave forces on marine structures situated in shallow water. The maximum velocity u_{max} occurs when x and t are both equal to zero; hence,

$$u_{max} = \frac{CN}{1 + \cos(My/d)}$$

(II-1-94)

(k)   Total energy in a solitary wave is about evenly divided between kinetic and potential energy. Total wave energy per unit crest width is

$$E = \frac{8}{3\sqrt{3}} \rho g H^{\frac{3}{2}} d^{\frac{3}{2}}$$

(II-1-95)

and the pressure beneath a solitary wave depends on the local fluid velocity, as does the pressure under a cnoidal wave; however, it may be approximated by

$$p = \rho g (y_s - y)$$

(II-1-96)

(l)   Equation II-1-96 is identical to that used to approximate the pressure beneath a cnoidal wave.

(m)   As a solitary wave moves into shoaling water it eventually becomes unstable and breaks. A solitary wave breaks when the water particle velocity at the wave crest becomes equal to the wave celerity. This occurs when (Miles 1980, 1981)

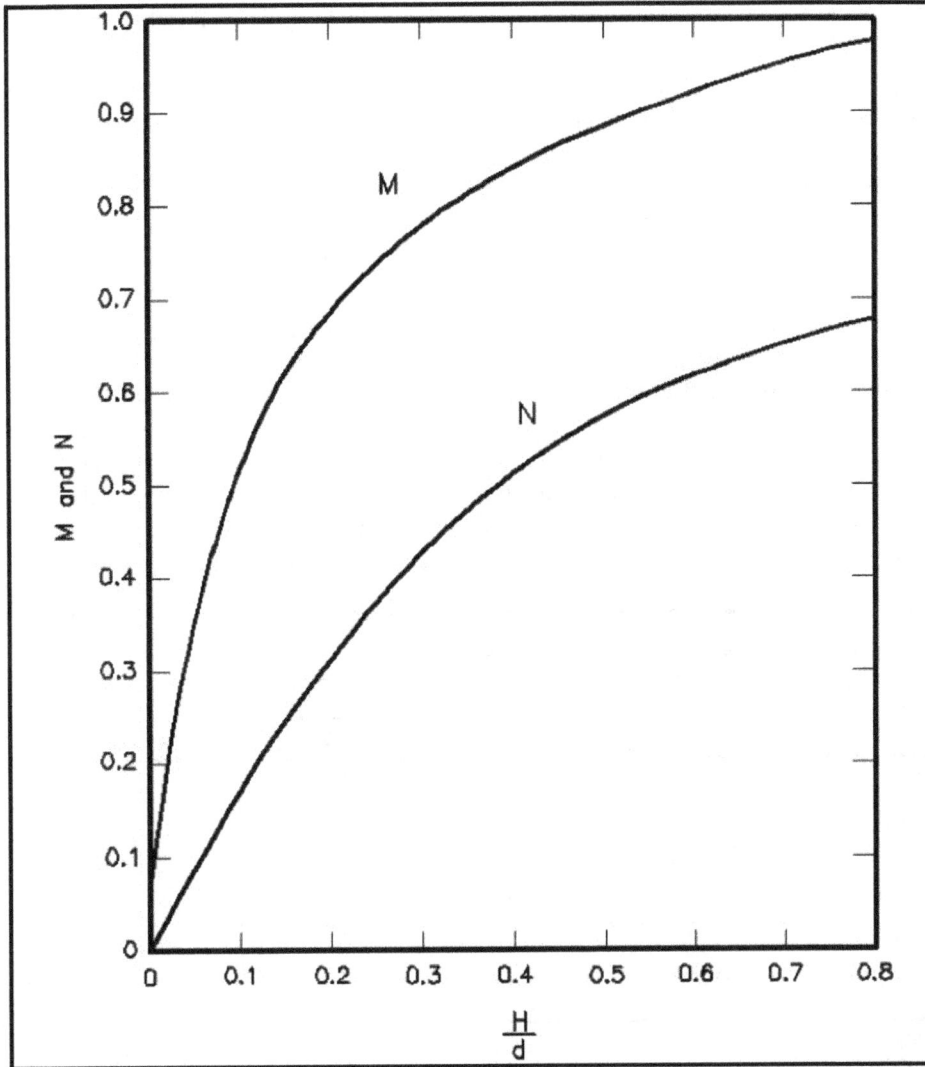

Figure II-1-17. Functions M and N in solitary wave theory (Munk 1949)

$$\left(\frac{H}{d}\right)_{max} = 0.78 \tag{II-1-97}$$

(n) Laboratory studies have shown that the value of $(H/d)_{max} = 0.78$ agrees better with observations for oscillatory waves than for solitary waves and that the nearshore slope has a substantial effect on this ratio. Other factors such as bottom roughness may also be involved. Tests of periodic waves with periods from 1 to 6 sec on slopes of $m = 0.0, 0.05, 0.10,$ and 0.20 have shown (SPM 1984) that H_b/d_b ratios are approximately equal to 0.83, 1.05, 1.19, and 1.32, respectively. Tests of single solitary waves on slopes from $m = 0.01$ to $m = 0.20$ (SPM 1984) indicate an empirical relationship between the slope and the breaker height-to-water depth ratio given by

$$\frac{H_b}{d_b} = 0.75 + 25m - 112m^2 + 3870m^3 \tag{II-1-98}$$

in which waves did not break when the slope m was greater than about 0.18 and that as the slope increased the breaking position moved closer to the shoreline. This accounts for the large values of H_b/d_b for large slopes; i.e., as $d_b \rightarrow 0$. For some conditions, Equations II-1-97 and II-1-98 are unsatisfactory for predicting breaking depth. Further discussion of the breaking of waves with experimental results is provided in Part II-4.

(6) Stream-function wave theory. Numerical approximations to solutions of hydrodynamic equations describing wave motion have been proposed and developed. Some common theories and associated equations are listed in Table II-1-2. The approach by Dean (1965, 1974), termed a *symmetric, stream-function theory*, is a nonlinear wave theory that is similar to higher order Stokes' theories. Both are constructed of sums of *sine* or *cosine* functions that satisfy the original differential equation (Laplace equation). The theory, however, determines the coefficient of each higher order term so that a best fit, in the least squares sense, is obtained to the theoretically posed, dynamic, free-surface boundary condition. Assumptions made in the theory are identical to those made in the development of the higher order Stokes' solutions. Consequently, some of the same limitations are inherent in the stream-function theory, and it represents an alternative solution to the equations used to approximate the wave phenomena. However, the stream-function representation had successfully predicted the wave phenomena observed in some laboratory wave studies (Dean and Dalrymple 1991), and thus it may possibly describe naturally occurring wave phenomena.

Table II-1-2
Boundary Value Problem of Water Wave Theories (Dean 1968)

Theory		Exactly Satisfies			
		DE	BBC	KFSBC	DFSBC
Linear wave theory		X	X	-	-
Third-order Stokes		X	X	-	-
Fifth-order Stokes		X	X	-	-
First-order cnoidal		-	X	-	-
Second-order cnoidal		-	X	-	-
Stream function theory	numerical wave	X	X	X	-

DE = Differential equation.
BBC = Bottom boundary condition.
KFSBC = Kinematic free surface boundary condition.
DFSBC = Dynamic free surface boundary condition.
X = Exactly satisfies.

(7) Fourier approximation -- Fenton's theory.

(a) Fenton's Fourier series theory, another theory developed in recent years (Fenton 1988), is somewhat similar to Dean's stream function theory, but it appears to describe oceanic waves at all water depths better than all previous similar theories.

(b) The long, tedious computations involved in evaluating the terms of the series expansions that make up the higher order stream-function theory of Dean had in the past limited its use to either tabular or graphical presentations of the solutions. These tables, their use, and their range of validity may be found elsewhere (Dean 1974).

(c) Stokes and cnoidal wave theories yield good approximations for waves over a wide range of depths if high-order expansions are employed. Engineering practice has relied on the Stokes fifth-order theory (Skjelbreia and Hendrickson 1961), and the stream function theory (Dean 1974). These theories are applicable to deepwater applications. An accurate steady wave theory may be developed by numerically solving the full nonlinear equations with results that are applicable for short waves (deep water) and for long waves (shallow water). This is the Fourier approximation method. The method is termed *Fenton's theory* here. Any periodic function can be approximated by Fourier series, provided the coefficients of the series can be found. In principal, the coefficients are found numerically. Using this approach, Chappelear (1961) developed a Fourier series solution by adopting the velocity potential as the primary field variable. Dean (1965, 1974) developed the stream function theory. The solutions by both Chappelear and Dean successively correct an initial estimate to minimize errors in the nonlinear free-surface boundary conditions.

(d)   A simple Fourier approximation wave theory was introduced by Rienecker and Fenton (1981) and was subsequently improved by Fenton (1985, 1988; Fenton and McKee 1990). It is an improved numerical theory that has a range of applicability broader than the Stokes and cnoidal theories. Details of the theory are given by Reinecker and Fenton (1981) and Fenton (1985, 1988; Fenton and McKee 1990). Sobey et al. (1987) recast Fenton's work into a standardized format including currents in the formulation up to fifth order. The theory has been implemented to calculate wave kinematics and the loading of offshore structures (Demirbilek 1985). For coastal applications, a PC-based computer code of Fenton's theory is available in the Automated Coastal Engineering System (ACES) (Leenknecht, Szuwalski, and Sherlock 1992). A brief description of Fenton's theory is given here; details are provided in ACES.

(e)   Fenton's Fourier approximation wave theory satisfies field equations and boundary conditions to a specified level of accuracy. The hydrodynamic equations governing the problem are identical to those used in Stokes' theory (Table II-1-2). Various approximations introduced in earlier developments are indicated in the table. Like other theories, Fenton's theory adopts the same field equation and boundary conditions. There are three major differences between Fenton's theory and the others. First, Fenton's theory is valid for deep- and shallow-water depths, and any of the two quantities' wave height, period or energy flux can be specified to obtain a solution. Second, the Fourier coefficients are computed numerically with efficient algorithms. Third, the expansion parameter for the Fourier coefficients is $\epsilon = kH/2$ rather than $\epsilon = ka$, which is used in Stokes theories. The coefficients are found numerically from simultaneous algebraic equations by satisfying two nonlinear free-surface boundary conditions and the dispersion relationship. Finding the coefficients requires that wave height, wave period, water depth, and either the Eulerian current or the depth-averaged mass transport velocity be specified.

(f)   In Fenton's theory, the governing field equation describing wave motion is the two-dimensional (x,z in the Cartesian frame) Laplace's equation, which in essence is an expression of the conservation of mass:

$$\nabla^2 \Psi = \frac{\partial^2 \Psi}{\partial x^2} + \frac{\partial^2 \Psi}{\partial z^2} = 0 \qquad\qquad \text{(II-1-99)}$$

where Ψ is the stream function. Ψ is a periodic function that describes wave motion in space and time, which also relates to the flow rate.

(g)   Wave motion is a boundary-value problem, and its solution requires specifying realistic boundary conditions. These boundary conditions are usually imposed at the free surface and sea bottom. Since the seabed is often impermeable, flow rate through the sea bottom must be zero. Therefore, the bottom boundary condition may be stated in terms of Ψ as

$$\Psi(x,-d) = 0 \quad \text{at } z = -d \qquad\qquad \text{(II-1-100)}$$

(h)   Two boundary conditions, *kinematic* and *dynamic*, are needed at the free surface. The kinematic condition states that water particles on the free surface remain there, and consequently, flow rate through the surface boundary must be zero. The net flow Q between the sea surface and seabed may be specified as

$$\Psi(x,\eta) = -Q \qquad \text{at } z = \eta \tag{II-1-101}$$

where η is the sea surface elevation. The dynamic free-surface boundary condition is an expression of specifying the pressure at the free surface that is constant and equal to the atmospheric pressure. In terms of the stream function Ψ this condition may be stated as

$$\frac{1}{2}\left\{\left(\frac{\partial\Psi}{\partial x}\right)^2 + \left(\frac{\partial\Psi}{\partial z}\right)^2\right\} + g\eta = R \qquad \text{at } z = \eta \tag{II-1-102}$$

in which R is the Bernoulli constant.

(i)   The boundary-value problem for wave motion as formulated above is complete. The time-dependency may be removed from the problem formulation by simply adapting a coordinate system that moves with the same velocity as the wave phase speed (Fenton 1988; Fenton and McKee 1990; Sobey et al. 1987). This is equivalent to introducing an underlying current relative to which the wave motion is measured. The current (also called *Stokes' drift velocity* or *Eulerian current*) causes a Doppler shift of the apparent wave period measured relative to a stationary observer or gauge. The underlying current velocity must therefore also be known in order to solve the wave problem in the steady (moving) reference frame.

(j)   Fenton's solution method uses the Fourier cosine series in kx to the governing equations. It is clearly an approximation, but very accurate, since results of this theory appear not to be restricted to any water depths. $\epsilon = kH/2$ is the expansion parameter replacing ka in the Stokes wave theory. The dependent variable is the stream function Ψ represented by a Fourier cosine series in kx, expressed up to the Nth order as

$$\Psi(x,z) = -\bar{u}(z+d) + \left(\frac{g}{k^3}\right)^{\frac{1}{2}} \sum_{j=1}^{N} B_j \frac{\sinh jk(z+d)}{\cosh jkd} \cos jkx \tag{II-1-103}$$

where the B_j are dimensionless Fourier coefficients. The truncation limit of the series N determines the order of the theory. The nonlinear free-surface boundary conditions are satisfied at each of $M+1$ equi-spaced points on the surface. Wave height, wave period, water depth, and either the mean Eulerian velocity or the Stokes drift velocity must be specified to obtain a solution.

(k)   The solution is obtained by numerically computing the N Fourier coefficients that satisfy a system of simultaneous equations. The numerical solution solves a set of $2M+6$ algebraic equations to find unknown Fourier coefficients. The problem is uniquely specified when $M = N$ and overspecified when $M > N$. Earlier wave theories based on stream function consider the overspecified case and used a least-squares method to find the coefficients. Fenton was the first to consider the uniquely specified case and used the collocation method to produce the most accurate and computationally efficient solution valid for any water depth.

(l)   An initial estimate is required to determine the $M+N+6$ variables. The linear theory provides this initial estimate for deep water. In relatively shallow water, additional Fourier components are introduced. An alternative method is used in the shallow-water case by increasing the wave height in a number of steps. Smaller heights are used as starting solutions for subsequent higher wave heights. This approach eliminates the triple-crested waves reported by others (Huang and Hudspeth 1984; Dalrymple and Solana 1986).

(m)   Sobey et al. (1987) compared several numerical methods for steady water wave problems, including Fenton's. Their comparison indicated that accurate results may be obtained with Fourier series of 10 to 20 terms, even for waves close to breaking. Comparisons with other numerical methods and experimental data (Fenton and McKee 1990; Sobey 1990) showed that results from Fenton's theory and experiments agree consistently and better than results from other theories for a wide range of wave height, wave period, and water depth. Based on these comparisons, Fenton and McKee (1990) define the regions of validity of Stokes and cnoidal wave theory as

$$\frac{L}{d} = 21.5 \, e^{\left(-1.87\frac{H}{d}\right)}$$ (II-1-104)

(n) The cnoidal theory should be used for wavelengths longer than those defined in this equation. For shorter waves, Stokes' theory is applicable. Fenton's theory can be used over the entire range, including obtaining realistic solutions for waves near breaking.

(o) In water of finite depth, the greatest (unbroken) wave that could prevail as a function of both wavelength and depth is determined by Fenton and McKee (1990) as

$$\frac{H}{d} = \frac{0.141063\frac{L}{d} + 0.0095721\left(\frac{L}{d}\right)^2 + 0.0077829\left(\frac{L}{d}\right)^3}{1.0 + 0.078834\frac{L}{d} + 0.0317567\left(\frac{L}{d}\right)^2 + 0.0093407\left(\frac{L}{d}\right)^3} a$$ (II-1-105)

(p) The leading term in the numerator of this equation is the familiar steepness limit for short waves in deep water. For large values of L/d (i.e., shallow-water waves), the ratio of cubic terms in the above equation approaches the familiar 0.8 value, a limit for depth-induced breaking of the solitary waves. Therefore, the above equation may also be used as a guide to delineate unrealistic waves in a given water depth.

(q) The formulas for wave kinematics, dynamics, and wave integral properties for Fenton's theory have been derived and summarized (Sobey et al.1987; Klopman 1990). Only the engineering quantities of interest including water particle velocities, accelerations, pressure, and water surface elevation defined relative to a Eulerian reference frame are provided here.

(r) The horizontal and vertical components of the fluid particle velocity are

$$u(x,z) = \frac{\partial \Psi}{\partial z} = -\bar{u} + \left(\frac{g}{k}\right)^{\frac{1}{2}} \sum_{j=1}^{N} jB_j \, \frac{\cosh jk(z+d)}{\cosh jkd} \, \cos jkx$$ (II-1-106)

$$w(x,z) = -\frac{\partial \Psi}{\partial x} = \left(\frac{g}{k}\right)^{\frac{1}{2}} \sum_{j=1}^{N} jB_j \, \frac{\sinh jk(z+d)}{\cosh jkd} \, \sin jkx$$ (II-1-107)

(s) Fluid particle accelerations in the horizontal and vertical directions are found by differentiating the velocities and using the continuity equation. These component accelerations are

$$a_x(x,z) = \frac{Du}{Dt} = u\frac{\partial u}{\partial x} + w\frac{\partial u}{\partial z}$$ (II-1-108)

$$a_z(x,z) = \frac{Dw}{Dt} = u\frac{\partial w}{\partial x} + w\frac{\partial w}{\partial z} = u\frac{\partial u}{\partial z} - w\frac{\partial u}{\partial x}$$

where

$$\frac{\partial u}{\partial x} = -\left(\frac{g}{k}\right)^{\frac{1}{2}} \sum_{j=1}^{N} j^2 B_j \frac{\cosh jk(z+d)}{\cosh jkd} \sin jkx \tag{II-1-109}$$

$$\frac{\partial u}{\partial z} = \left(\frac{g}{k}\right)^{\frac{1}{2}} \sum_{j=1}^{N} j^2 B_j \frac{\sinh jk(z+d)}{\cosh jkd} \cos jkx \tag{II-1-110}$$

(t)   The instantaneous water surface elevation $\eta(x)$ and water particle pressure are given by

$$\eta(x) = \frac{1}{2} a_N \cos Nkx + \sum_{j=1}^{N-1} a_j \cos jkx$$

$$p(x,z) = \rho(R-gd-gz) - \frac{1}{2}\rho(u^2 + w^2) \tag{II-1-111}$$

(u)   Integral properties of periodic gravity waves, including wave potential and kinetic energy, wave momentum and impulse, wave energy flux and wave power, and wave radiation stresses obtained by Klopman (1990) and Sobey et al. (1987) are listed in the Leenknecht, Szuwalski, and Sherlock (1992) documentation.

(v)   A computer program developed by Fenton (1988) has recently been implemented in the ACES package. The ACES implementation facilitates use of Fenton's theory to applications in deep water and finite-depth water. It uses Fourier series of up to 25 terms to describe a wave train and provides information about various wave quantities. The output includes wave estimates for common engineering parameters including water surface elevation, wave particle kinematics, and wave integral properties as functions of wave height, period, water depth, and position in the wave form.

(w)   The wave is assumed to co-exist on a uniform co-flowing current, taken either as the mean Eulerian current or mean mass transport velocity. At a given point in the water column, wave kinematics are tabulated over two wavelengths, and vertical distribution of the selected kinematics under the wave crest are graphically displayed. ACES implementation of Fenton's theory and its input/output requirements, computations, and examples are described in detail in the ACES documentation manual (Leenknecht, Szuwalski, and Sherlock 1992).

(x)   Figure II-1-18 illustrates the application of Fenton's theory. This case represents shallow-water (10-m) conditions and wave height and period of 5 m and 10 sec, respectively. Surface elevation, horizontal velocity, and pressure over two wavelengths is shown graphically in Figure II-1-18. The ACES documentation includes guidance on proper use of Fenton's theory.

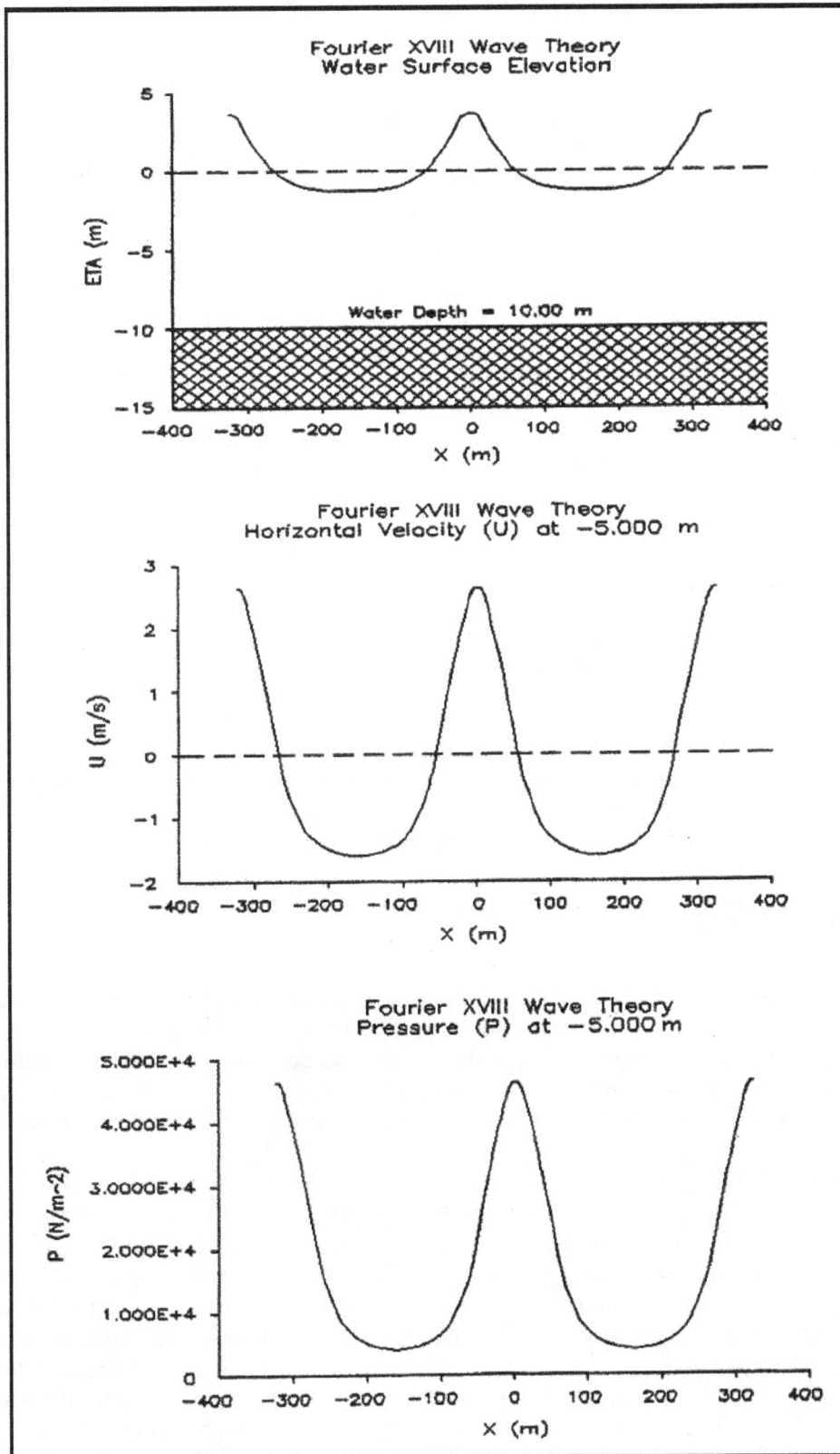

Figure II-1-18. Surface elevation, horizontal velocity, and pressure in 10-m depth (using Fenton's theory in ACES)

f.   Wave breaking.

(1)   Wave height is limited by both depth and wavelength. For a given water depth and wave period, there is a maximum height limit above which the wave becomes unstable and breaks. This upper limit of wave height, called *breaking wave height*, is in deep water a function of the wavelength. In shallow and transitional water it is a function of both depth and wavelength. Wave breaking is a complex phenomenon and it is one of the areas in wave mechanics that has been investigated extensively both experimentally and numerically.

(2)   Researchers have made some progress over the last three decades in the numerical modeling of waves close to breaking (Longuet-Higgins and Fenton 1974; Longuet-Higgins 1974; 1976; Schwartz 1974; Dalrymple and Dean 1975; Byatt-Smith and Longuet-Higgins 1976; Peregrine 1976; Cokelet 1977; Longuet-Higgins and Fox 1977; Longuet-Higgins 1985; Williams 1981; 1985). These studies suggest the limiting wave steepness to be $H/L = 0.141$ in deep water and $H/d = 0.83$ for solitary waves in shallow water with a corresponding solitary wave celerity of $c/(gd)^{1/2} = 1.29$.

(3)   Dalrymple and Dean (1975) investigated the maximum wave height in the presence of a steady uniform current using the stream function theory. Figure II-1-19 shows the influence of a uniform current on the maximum wave height where T_c is the wave period in a fixed reference frame and U is the current speed.

(4)   The treatment of wave breaking in the propagation of waves is discussed in Part II-3. Information about wave breaking in deep and shoaling water and its relation to nearshore processes is provided in Part II-4.

g.   Validity of wave theories.

(1)   To ensure their proper use, the range of validity for various wave theories described in this chapter must be established. Very high-order Stokes theories provide a reference against which the accuracy of various theories may be tested. Nonlinear wave theories better describe mass transport, wave breaking, shoaling, reflection, transmission, and other nonlinear characteristics. Therefore, the usage of the linear theory has to be carefully evaluated for final design estimates in coastal practice. It is often imperative in coastal projects to use nonlinear wave theories.

(2)   Wave amplitude and period may sometimes be estimated from empirical data. When data are lacking or inadequate, uncertainty in wave height and period estimates can give rise to a greater uncertainty in the ultimate answer than does neglecting the effect of nonlinear processes. The additional effort necessary for using nonlinear theories may not be justified when large uncertainties exist in the wave data used for design. Otherwise, nonlinear wave theories usually provide safer and more accurate estimates.

(3)   Dean (1968, 1974) presented an analysis by defining the regions of validity of wave theories in terms of parameters H/T^2 and d/T^2 since T^2 is proportional to the wavelength. Le Méhauté (1976) presented a slightly different analysis (Figure II-1-20) to illustrate the approximate limits of validity for several wave theories, including the third- and fourth-order theories of Stokes. In Figure II-1-20, the fourth-order Stokes theory may be replaced with more popular fifth-order theory, since the latter is often used in applications. Both Le Méhauté and Dean recommend cnoidal theory for shallow-water waves of low steepness, and Stokes' higher order theories for steep waves in deep water. Linear theory is recommended for small steepness H/T^2 and small U_R values. For low steepness waves in transitional and deep water, linear theory is adequate but other wave theories may also be used in this region. Fenton's theory is appropriate for most of the wave parameter domain. For given values of H, d, and T, Figure II-1-20 should be used as a guide to select an appropriate wave theory.

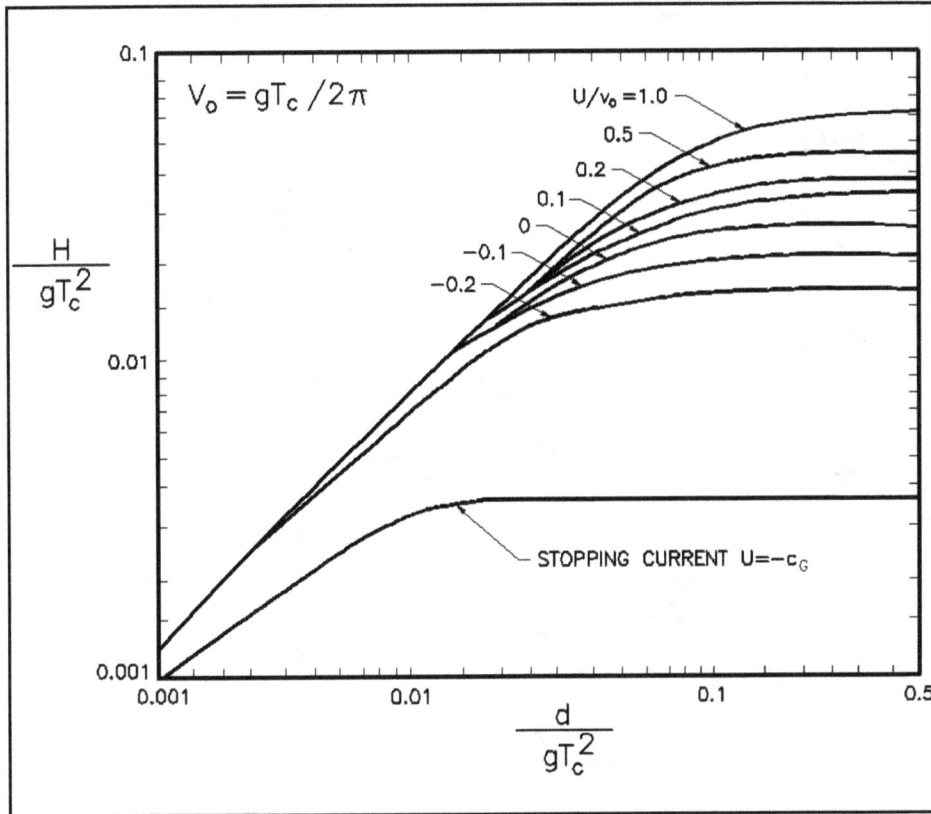

Figure II-1-19. Influence of a uniform current on the maximum wave height (Dalrymple and Dean 1975)

(4) It is necessary to know the limiting value of wave heights and wave steepness at different water depths to establish range of validity of any wave theory that uses a Stokes-type expansion. This is customarily done by comparing the magnitude of each successive term in the expansion. Each should be smaller than the term preceding it. For example, if the second term is to be less than 1 percent of the first term in the Stokes second-order theory, the limiting wave steepness is

$$\frac{H}{L} \leq \frac{1}{80} \frac{\sinh^3 kd}{\cosh kd \ (3 + 2 \sinh^2 kd)} \tag{II-1-112}$$

(5) If the third-order term is to be less than 1 percent of the second-order term, the limiting wave steepness is

$$\frac{H}{L} \leq \frac{1}{7} \frac{\sinh^3 kd}{\sqrt{1 + 8 \cosh^3 kd}} \tag{II-1-113}$$

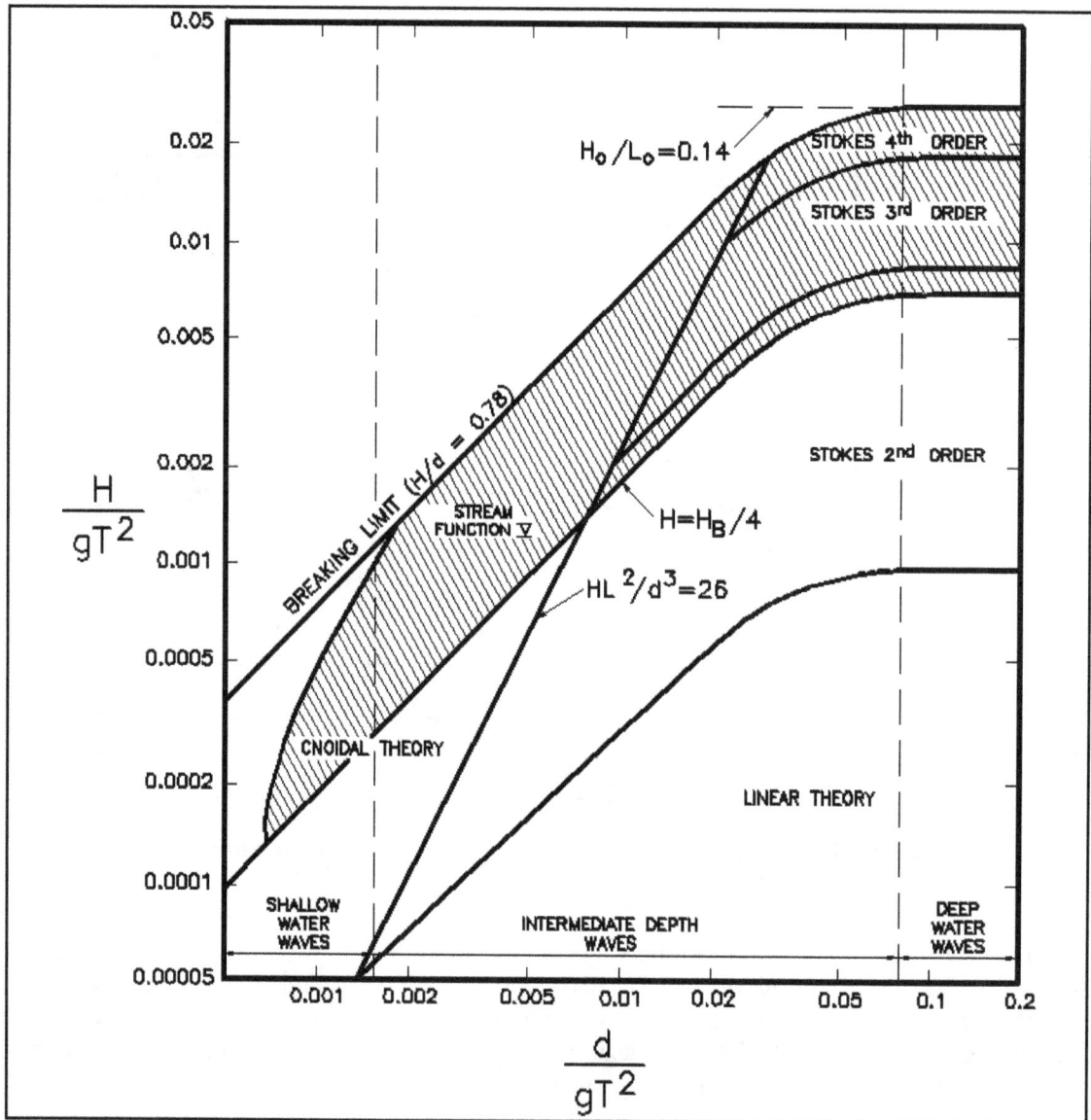

Figure II-1-20. Ranges of suitability of various wave theories (Le Méhauté 1976)

(6)   Similarly, using the fifth-order expansion, the asymptotes to Stokes third-order theory are $H/L_0 <$ 0.1 and $H/d < 3/4(kd)^2$ for deep water and shallow water, respectively. This allows the range of Stokes' theory to be expanded by adding successively smaller areas to the domain of linear theory in Figure II-1-20 until the breaking limit is reached. The fifth-order Stokes theory gets close enough to the breaking limit, and higher order solutions may not be warranted. Laitone (1962) suggests a shallow-water limit on Stokes' theory by setting the Ursell number U_R equal to 20. For an Ursell number of approximately 20, Stokes' theory approaches the cnoidal theory.

(7)   The magnitude of the Ursell number U_R (sometimes also called the *Stokes number*) shown in Figure II-1-20 may be used to establish the boundaries of regions where a particular wave theory should be used. Stokes (1847) noted that this parameter should be small for long waves. An alternative, named the *Universal parameter* (U_p), has recently been suggested (Goda 1983) for classification of wave theories.

(8)   Limits of validity of the nonlinear (higher-order) wave theories established by Cokelet (1977) and Williams (1981), are shown in Figure II-1-21. Regions where Stokes waves (short waves) and cnoidal and solitary waves (long waves) are valid are also shown in this figure. The breaking limit for solitary waves $H_b^W = 0.833$ established by Williams (1981) and the limiting height designated as H_b^F determined by Cokelet (1977) are also shown on Figure II-1-21. The line between short and long waves corresponds to a value of the Ursell number $U_R \approx 79$. This theoretical partition agrees with data from Van Dorn (1966).

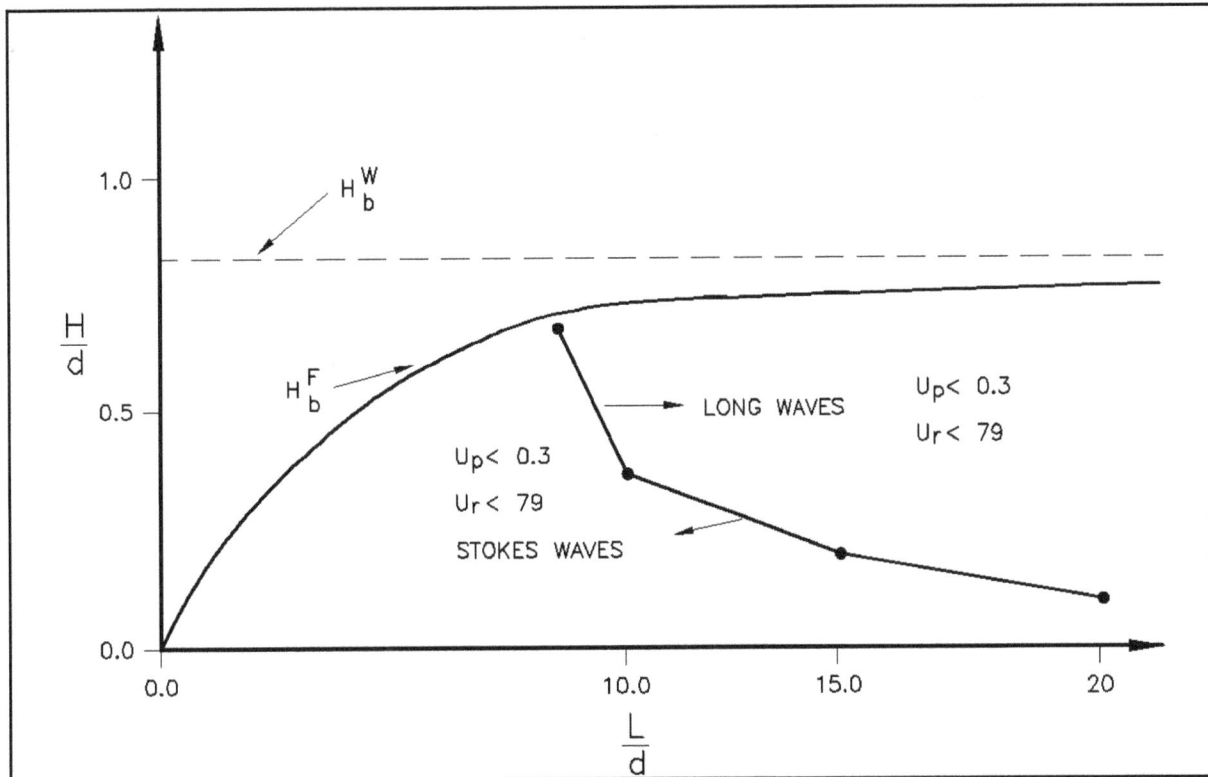

Figure II-1-21. Grouping of wind waves based on universal parameter and limiting height for steep waves

II-1-3. Irregular Waves

   a.   Introduction.

   (1)   In the first part of this chapter, waves on the sea surface were assumed to be nearly sinusoidal with constant height, period and direction (i.e., monochromatic waves). Visual observation of the sea surface (as in the radar image of the entrance to San Francisco Bay in Figure II-1-22) and measurements (such as in Figure II-1-23) indicate that the sea surface is composed of waves of varying heights and periods moving in differing directions. In the first part of this chapter, wave height, period, and direction could be treated as deterministic quantities. Once we recognize the fundamental variability of the sea surface, it becomes necessary to treat the characteristics of the sea surface in statistical terms. This complicates the analysis but more realistically describes the sea surface. The term *irregular waves* will be used to denote natural sea states in which the wave characteristics are expected to have a statistical variability in contrast to *monochromatic waves*, where the properties may be assumed constant. Monochromatic waves may be generated in the laboratory but are rare in nature. "Swell" describes the natural waves that appear most like monochromatic waves in deep water, but swell, too, is fundamentally irregular in nature. We note that the sea state in nature during a storm is always short-crested and irregular. Waves that have travelled far from

EXAMPLE PROBLEM II-1-6

FIND:

Applicable wave theory for waves in (a) and (b). Which of these waves is a long wave?

GIVEN:

(a). d = 15 m, H = 12.2 m, T = 12 sec; (b). d = 150 m, H = 30 m, T = 16 sec.

SOLUTION:

(a) Calculate dimensionless parameters necessary for using Figure II-1-20. These are

$$\frac{d}{gT^2} \approx 0.01$$

$$\frac{H}{gT^2} \approx 0.009$$

$$\frac{H}{d} = 0.8$$

$$\sqrt{gd} \approx 12 \ \frac{m}{sec}$$

$$U_R \approx 55$$

From Figure II-1-20, the applicable theory is cnoidal.

(b) In a similar fashion, compute

$$\frac{d}{gT^2} \approx 0.06$$

$$\frac{H}{gT^2} \approx 0.01$$

$$\frac{H}{d} = 0.2$$

$$\sqrt{gd} \approx 40 \ \frac{m}{sec}$$

$$U_R \approx 1.5$$

With these values, Figure II-1-20 indicates the applicable theory is Stokes third- or fifth-order. It is noted that the linear theory is also applicable.

Based on the values of Ursell parameter, neither wave (a) or (b) is a true long wave. Wave (a) may be considered a long wave in comparison to wave (b).

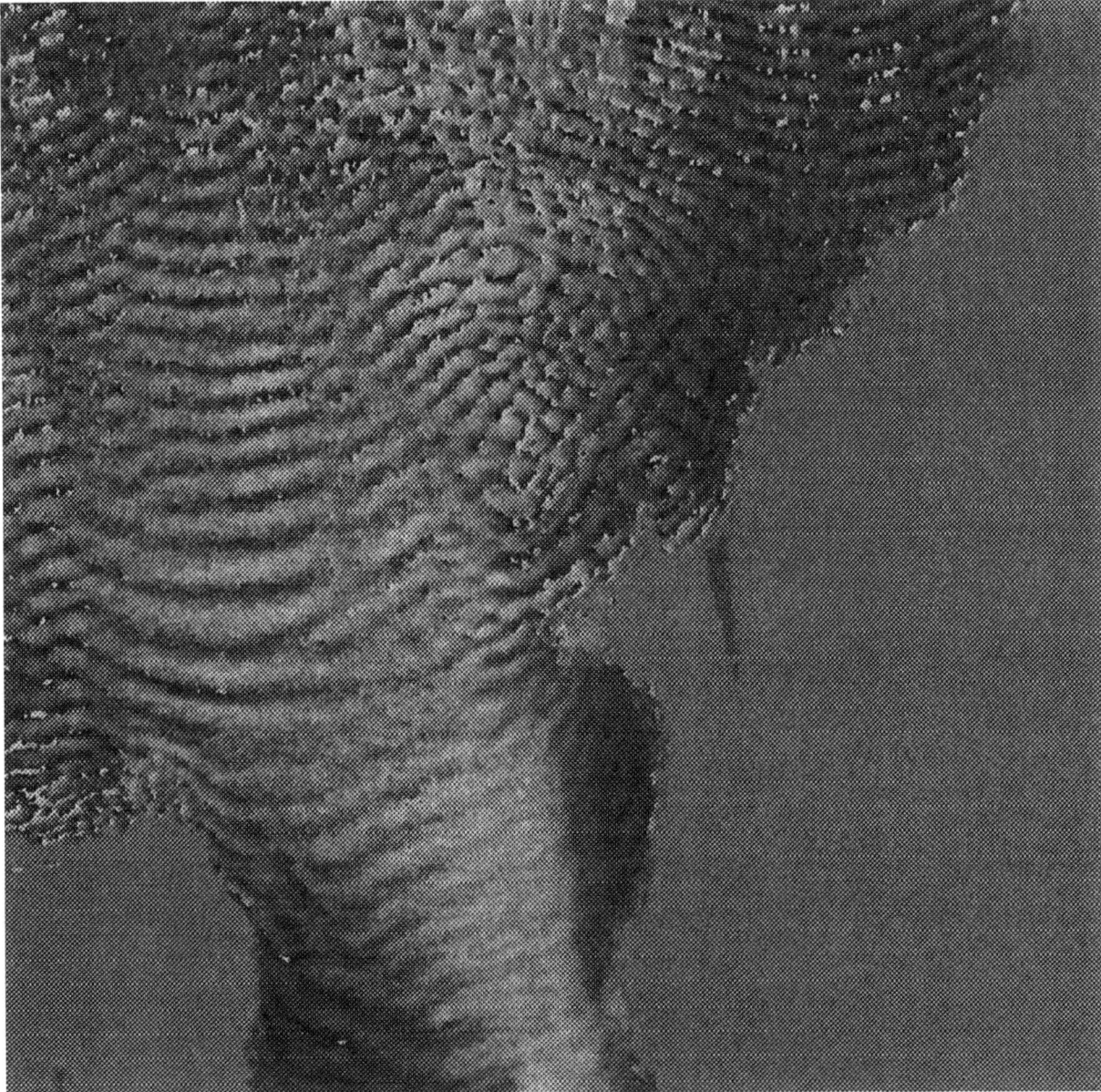

Figure II-1-22. Radar image of the sea surface in the entrance to San Francisco Bay

the region of generation are called *swells*. These waves have a much more limited range of variability, sometimes appearing almost monochromatic and long-crested.

(2) When the wind is blowing and the waves are growing in response, the sea surface tends to be confused: a wide range of heights and periods is observed and the length of individual wave crests may only be a wave length or two in extent (short-crested). Such waves are called wind seas, or often, just sea. Long-period waves that have traveled far from their region of origin tend to be more uniform in height, period, and direction and have long individual crests, often many wave lengths in extent (i.e., long-crested). These are termed swell. A sea state may consist of just sea or just swell or may be a combination of both.

(3) The ocean surface is often a combination of many wave components. These individual components were generated by the wind in different regions of the ocean and have propagated to the point of observation.

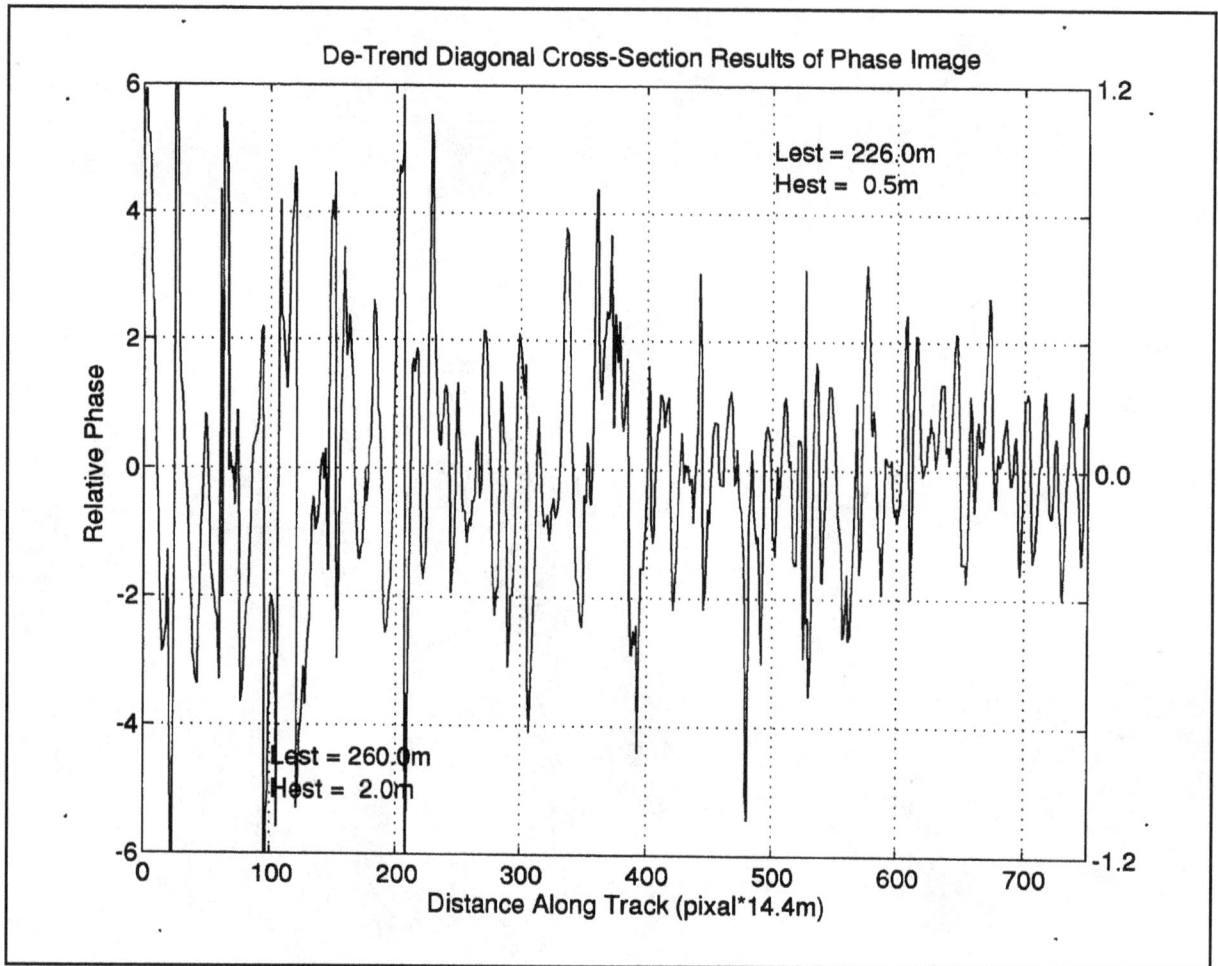

Figure II-1-23. Measured sea surface velocity in the entrance to San Francisco Bay

If a recorder were to measure waves at a fixed location on the ocean, a non-repeating wave profile would be seen and the wave surface record would be rather irregular and random (Figure II-1-23). Although individual waves can be identified, there is significant variability in height and period from wave to wave. Consequently, definitions of wave height, period, and duration must be statistical and simply indicate the severity of wave conditions.

(4)   Wave profiles are depicted in Figure II-1-24 for different sea conditions. Figure II-1-25 shows a typical wave surface elevation time series measured for an irregular sea state. Important features of the field-recorded waves and wave parameters to be used in describing irregular waves later in this section are defined in Figures II-1-26 and II-1-27. We note that the sea state in nature during a storm is always short-crested and irregular. Waves that have traveled far from the region of generation are called *swells*. These waves have much more limited range of variability sometimes appearing monochromatic and long-crested.

(5)   This part of Part II-1 will develop methods for describing and analyzing natural sea states. The concept of *significant wave height*, which has been found to be a very useful index to characterize the heights of the waves on the sea surface, will be introduced. *Peak period* and *mean wave direction* which characterize the dominant periodicity and direction of the waves, will be defined. However, these parameterizations of the sea surface in some sense only index how big some of the waves are. When using irregular wave heights in engineering, the engineer must always recognize that larger and smaller (also longer and shorter) waves

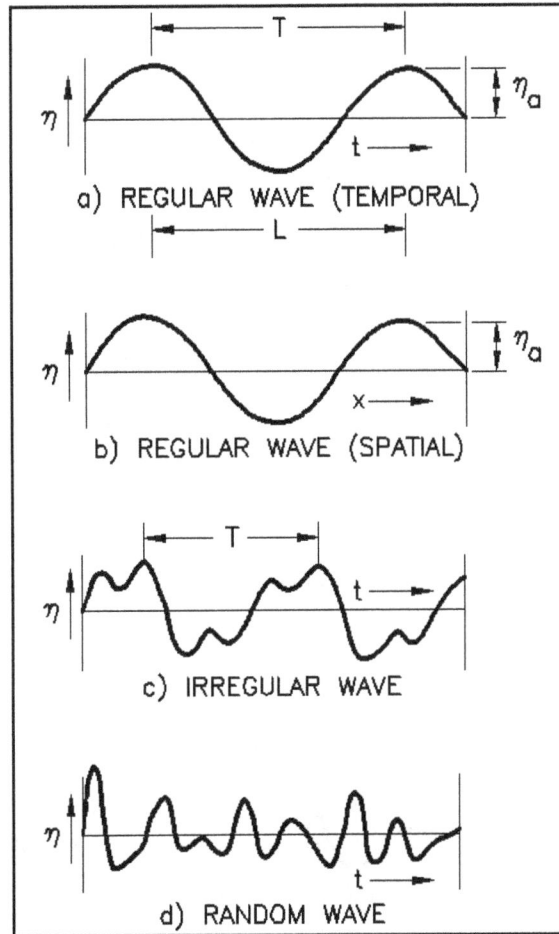

Figure II-1-24. Representations of an ocean wave

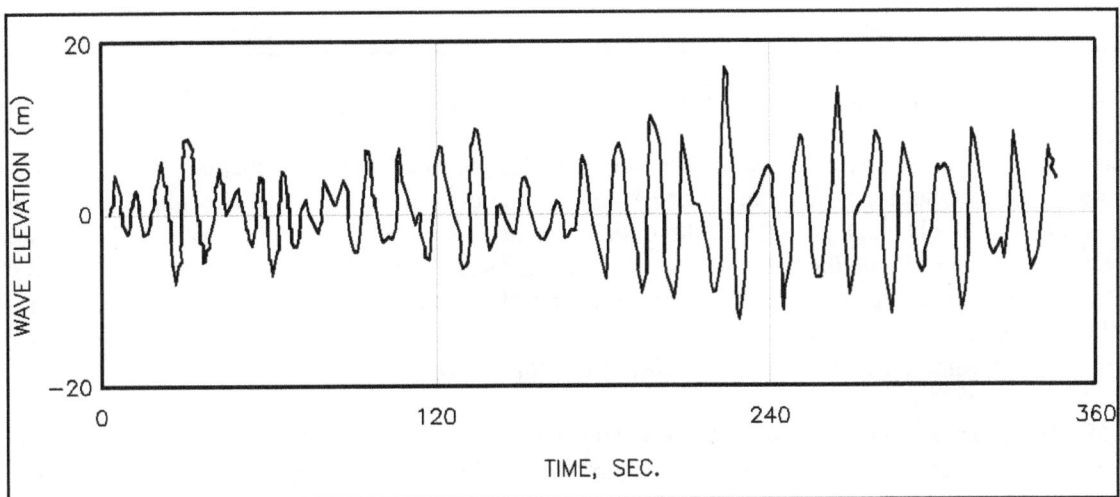

Figure II-1-25. Wave profile of irregular sea state from site measurements

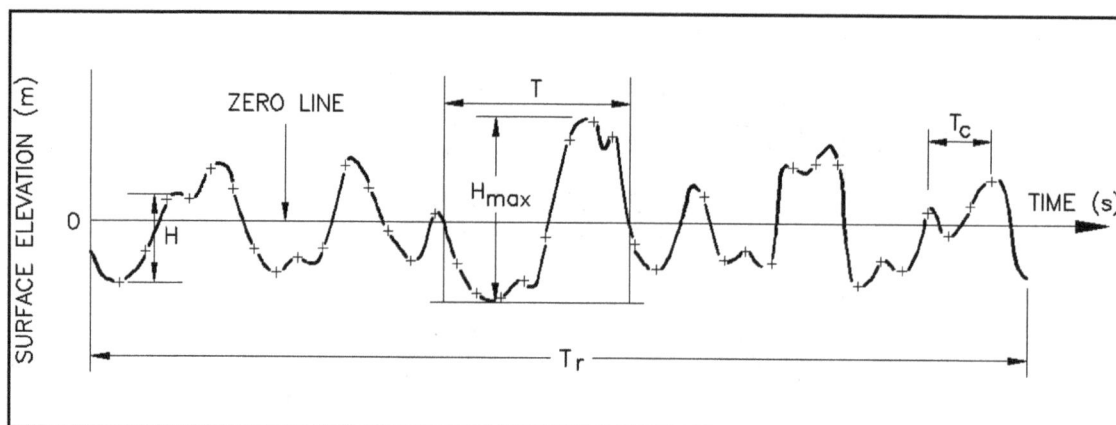

Figure II-1-26. Definition of wave parameters for a random sea state

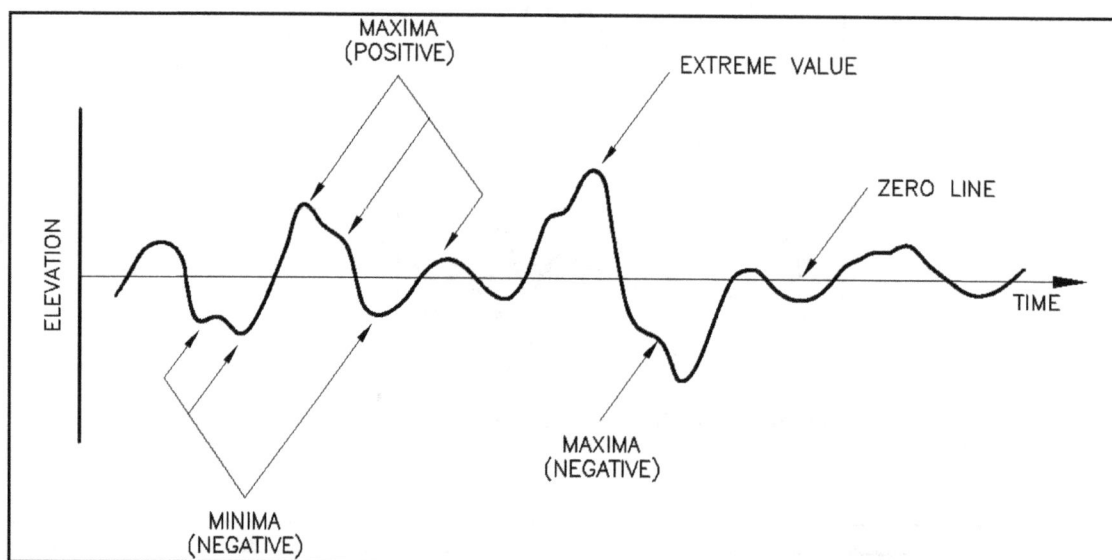

Figure II-1-27. Definition sketch of a random wave process (Ochi 1973)

are present. The monochromatic wave theories described in the first part of this chapter will be seen to have two major uses. One use is to estimate the kinematics and dynamics associated with a wave with the significant wave height, peak period, and direction. The other is when an individual wave has been isolated in a wave record to estimate the velocities, accelerations, forces, etc., associated with that individual wave event. The engineer must recognize that the implication of the statistical nature of irregular waves implies that the kinematics and dynamics likewise require statistical treatment. IAHR (1986) provides a detailed description of parameters and terminology used with irregular waves.

(6)   Two approaches exist for treating irregular waves: *spectral methods* and *wave-by-wave (wave train) analysis*. Spectral approaches are based on the Fourier Transform of the sea surface. Indeed this is currently the most mathematically appropriate approach for analyzing a time-dependent, three-dimensional sea surface record. Unfortunately, it is exceedingly complex and at present few measurements are available that could fully tap the potential of this method. However, simplified forms of this approach have been proven to be very useful. The other approach used is wave-by-wave analysis. In this analysis method, a time-history of the sea surface at a point is used, the undulations are identified as waves, and statistics of the record are

developed. This is a very natural introduction to irregular waves and will be presented first before the more complicated spectral approach is presented. The primary drawback to the wave-by-wave analysis is that it cannot tell anything about the direction of the waves. Indeed, what appears to be a single wave at a point may actually be the local superposition of two smaller waves from different directions that happen to be intersecting at that time. Disadvantages of the spectral approach are the fact that it is linear and can distort the representation of nonlinear waves.

b. Wave train (wave-by-wave) analysis.

(1) Introduction.

(a) Wave train analysis requires direct measurements of irregular seas. A typical irregular wave record obtained from a wave-measuring device is shown in Figure II-1-25. The recorded wave traces have to be of finite length with the sea surface sampled at a set interval (typically every second). The time-history of sea surface elevation at a point is a random-appearing signal exhibiting many maxima and minima (Figures II-1-26 and II-1-27). It is necessary to develop a criterion for identifying individual waves in the record.

(b) In a wave-by-wave analysis, undulation in the time-history of the surface must be divided into a series of segments, which will then be considered as individual waves. The height and period of each wave will be measured. Once this is done for every segment of the record, statistical characteristics of the record can be estimated, and the statistics of the record are compiled.

(c) Knowing the statistics of one record can be useful in itself, particularly if the record is important (such as the observation of waves at a site when a structure failed). However, it would be helpful to know whether the statistical characteristics of individual wave records followed any consistent pattern. Statistics of the sea state could be predicted knowing only a little about the wave conditions. It would be very useful if the distribution of wave characteristics in a wave record followed a known statistical distribution. After defining characteristics of individual records, the larger statistical question will be addressed.

(d) In the time-domain analysis of irregular or random seas, wave height and period, wavelength, wave crest, and trough have to be carefully defined for the analysis to be performed. The definitions provided earlier in the regular wave section of this chapter assumed that the crest of a wave is any maximum in the wave record, while the trough can be any minimum. However, these definitions may fail when two crests occur within an intervening trough lying below the mean water line. Also, there is not a unique definition for wave period, since it can be taken as the time interval between either two neighboring wave troughs or two crests. Other more common definitions of wave period are the time interval between successive crossings of the mean water level by the water surface in a downward direction called *zero down-crossing period* or *zero up-crossing period* for the period deduced from successive up-crossings.

(2) Zero-crossing method.

(a) The adopted engineering procedure is the zero-crossing technique, where a wave is defined when the surface elevation crosses the zero-line or the mean water level (MWL) upward and continues until the next crossing point. This is the *zero-upcrossing* method. When a wave is defined by the downward crossing of the zero-line by the surface elevation, the method is the *zero-downcrossing*.

(b) The *zero-crossing wave height* is the difference in water surface elevation of the highest crest and lowest trough between successive zero-crossings. The definition of wave height depends on the choice of trough occurring before or after the crest. Here, a wave will be identified as an event between two successive zero-upcrossings and wave periods and heights are defined accordingly. Note that there can be differences between the definitions of wave parameters obtained by the zero up- and down-crossing methods for description of irregular sea states.

(c) Both methods usually yield statistically similar mean values of wave parameters. There seems to be some preference for the zero-downcrossing method (IAHR 1986). The downcrossing method may be preferred due to the definition of wave height used in this method (the vertical distance from a wave trough to the following crest). It has been suggested that this definition of wave height may be better suited for extreme waves (IAHR 1986).

(d) Using these definitions of wave parameters for an irregular sea state, it is seen in Figures II-1-26 and II-1-27 that, unlike the regular (monochromatic) sinusoidal waves, the periods and heights of irregular waves are not constant with time, changing from wave to wave. Wave-by-wave analysis determines wave properties by finding average statistical quantities (i.e., heights and periods) of the individual wave components present in the wave record. Wave records must be of sufficient length to contain several hundred waves for the calculated statistics to be reliable.

(e) Wave train analysis is essentially a manual process of identifying the heights and periods of the individual wave components followed by a simple counting of zero-crossings and wave crests in the wave record. The process begins by dissecting the entire record into a series of subsets for which individual wave heights and periods are then noted for every zero down-crossing or up-crossing, depending on the method selected. In the interest of reducing manual effort, it is customary to define wave height as the vertical distance between the highest and lowest points, while wave period is defined as the horizontal distance between two successive zero-crossing points (Figures II-1-26 and II-1-27). In this analysis, all local maxima and minima not crossing the zero-line have to be discarded. From this information, several wave statistical parameters are subsequently calculated. Computer programs are available to do this (IAHR 1986).

(3) Definition of wave parameters.

(a) Determination of wave statistics involves the actual processing of wave information using the principles of statistical theory. A highly desirable goal is to produce some statistical estimates from the analyzed time-series data to describe an irregular sea state in a simple parametric form. For engineering, it is necessary to have a few simple parameters that in some sense tell us how severe the sea state is and a way to estimate or predict what the statistical characteristics of a wave record might be had it been measured and saved. Fortunately, millions of wave records have been observed and a theoretical/empirical basis has evolved to describe the behavior of the statistics of individual records.

(b) For parameterization, there are many short-term candidate parameters which may be used to define statistics of irregular sea states. Two of the most important parameters necessary for adequately quantifying a given sea state are characteristic height H and characteristic period T. Other parameters related to the combined characteristics of H and T, may also be used in the parametric representation of irregular seas.

(c) Characteristic wave height for an irregular sea state may be defined in several ways. These include the *mean height*, the *root-mean-square height*, and the mean height of the highest one-third of all waves known as the *significant height*. Among these, the most commonly used is the significant height, denoted as H_s or $H_{1/3}$. Significant wave height has been found to be very similar to the estimated visual height by an experienced observer (Kinsman 1965). The characteristic period could be the *mean period*, or *average zero-crossing period*, etc.

(d) Other statistical quantities are commonly ascribed to sea states in the related literature and practice. For example, the mean of all the measured wave heights in the entire record analyzed is called the *mean wave height $\overline{H}$*. The largest wave height in the record is the *maximum wave height H_{max}*. The root-mean-square of all the measured wave heights is the *rms wave height* H_{rms}. The average height of the largest $1/n$ of all waves in the record is the $H_{1/n}$ where n = 10, 11, 12, 13,..., 99, 100 are common values. For instance, $H_{1/10}$ is the mean height of the highest one-tenth waves. In coastal projects, engineers are faced with designing for the maximum expected, the highest possible waves, or some other equivalent wave height. From one wave record measured at a point, these heights may be estimated by ordering waves from the largest to the smallest and assigning to them a number from *1* to *N*. The significant wave height $H_{1/3}$ or H_s will be the average of the first (highest) *N/3* waves.

(e) The probability that a wave height is greater (less) than or equal to a design wave height H_d may be found from

$$P(H > H_d) = \frac{m}{N}$$

$$P(H \le H_d) = 1 - \frac{m}{N}$$

(II-1-114)

where *m* is the number of waves higher than H_d. For an individual observed wave record the probability distribution $P(H > H_d)$ can be formulated in tabular form and possibly fitted by some well-known distribution. The root-mean-square wave height H_{rms} may be computed as

$$H_{rms} = \sqrt{\frac{1}{N} \sum_{j=1}^{N} H_j^2}$$

(II-1-115)

in which H_j denote the ordered individual wave heights in the record.

(f) Probability distributions discussed in the irregular wave section of the CEM refer to <u>short term wave statistics</u>. This subject concerns the probability that a wave of a given height will occur given that we know the statistics of the sea surface over a 16- to 60-min period. A short-term wave statistics question might be, for example, "If we have measured the waves for 15 min and found that H_s is 2m, what is the chance that a wave of 4 m may occur?" This must be contrasted to <u>long-term wave statistics</u>. To obtain long-term wave statistics, a 15-min record may have been recorded (and statistics of each record computed) every 3 hr for 10 years (about 29,000 records) and the statistics of the set of 29,000 significant wave heights compiled. A long-term wave statistics question might be, "If the mean <u>significant</u> wave height may be 2m with a standard deviation of 0.75m, what is the chance that once in 10 years the significant wave height will exceed 4 m?" These are two entirely different statistical questions and must be treated differently.

(g) A similar approach can be used for the wave period. The mean zero-crossing period is called the *zero-crossing period* T_z. The average wave period between two neighboring wave crests is the wave crest period T_c. Therefore, in the time domain wave record analysis, the average wave period may also be obtained from the total length of *record length* T_r, either using T_z or T_c (Tucker 1963). These periods are related to T_r by

$$T_z = \frac{T_r}{N_z}$$

(II-1-116)

$$T_c = \frac{T_r}{N_c}$$

where N_z and N_c are the number of zero-upcrossings and crests in the wave record, respectively. We emphasize that in Tucker's method of wave train analysis, crests are defined by zero-crossing. Note also by definition of these periods that $T_c \leq T_z$.

(h)   The list of definitions stated above is not all-inclusive, and several other statistical quantities may be obtained from a wave train analysis (Ochi 1973; IAHR 1986). For example, the rms surface elevation η_{rms} (described later in the short-term sea states section) (σ in IAHR list) defines the standard deviation of the surface elevation, and the significant wave height H_s is related to η_{rms} by

$$H_s = 3.8 \; \eta_{rms} \approx 4 \; \eta_{rms} \qquad\qquad\qquad \text{(II-1-117)}$$

(4)   Significant wave height.

(a)   The *significant wave height H_s* (or $H_{1/3}$) is the most important quantity used describing a sea state and thus, is discussed further here for completeness. The concept of significant wave height was first introduced by Sverdrup and Munk (1947). It may be determined directly from a wave record in a number of ways. The most frequently used approach in wave-by-wave analysis is to rank waves in a wave record and then choose the highest one-third waves. The average of the chosen waves defines the significant wave height as

$$H_s = \frac{1}{\frac{N}{3}} \sum_{i=1}^{N/3} H_i \qquad\qquad\qquad \text{(II-1-118)}$$

where N is the number of individual wave heights H_i in a record ranked highest to lowest.

(b)   Sverdrup and Munk (1947) defined significant wave height in this fashion because they were attempting to correlate what sailors reported to what was measured. Hence, this is an empirically driven definition. Today, when wave measuring is generally automated, some other parameter might be appropriate, but significant wave height remains in recognition of its historical precedence and because it has a fairly tangible connection to what observers report when they try to reduce the complexity of the sea surface to one number. It is important to recognize that it is a statistical construct based only on the height distribution. Knowing the significant height from a record tells us nothing about period or direction.

(5)   Short-term random sea state parameters.

(a)   It is well-known that any periodic signal $\eta(t)$ with a zero mean value can be separated into its frequency components using the standard *Fourier analysis*. Periodic wave records may generally be treated as random processes governed by laws of the probability theory. If the wave record is a random signal, the term used is *random waves*. For a great many purposes, ocean wave records may be considered random (Rice 1944-1945, Kinsman 1965, Phillips 1977, Price and Bishop 1974).

(b)   The statistical properties of a random signal like the wave surface profile may be obtained from a set of many simultaneous observations called an *ensemble* or set of signals $\{\eta_1(t), \; \eta_2(t), \; \eta_3(t),...\}$, but not from a single observation. A single observation even infinitely long may not be sufficient for determining the spatial variability of wave statistics. An ensemble consists of different realizations or measurements of the process $\eta(t)$ at some known locations. To determine wave properties from the process $\eta(t)$, certain assumptions related to its time and spatial variation must be made.

(c) First, it would be necessary to assume that the process described by the wave record (i.e., a sea state), say $\eta(t)$, is *stationary*, which means that the statistical properties of $\eta(t)$ are independent of the origin of time measurement. Since the statistics of stationary processes are time-invariant, there is no drift with time in the statistical behavior of $\eta(t)$. The stationarity requirement is necessary as we shall see later for developing a *probability distribution* for waves, which is the fraction or percentage of time an *event* or *process* (say, the sea state depicted in time series of the wave surface profile) is not exceeded. The probability distribution may be obtained by taking $\eta_1(t_1)$, $\eta_2(t_1)$, $\eta_3(t_1)$,..., as variables, independent of the instant t_1. If in addition, $\eta(t)$ can be measured at different locations and the properties of $\eta(t)$ are invariant or do not depend on location of measurements, the process may then be assumed *homogenous*. In reality, $\eta(t)$ may be assumed stationary and homogenous only for a limited duration at the location data are gathered. Wind waves may be considered approximately stationary for only a few hours (3 hr or less), beyond which their properties are expected to change.

(d) Second, the process $\eta(t)$ is assumed to be *ergodic*, which means that any measured record of the process say $\eta_1(t)$ is typical of all other possible realizations, and therefore, the average of a single record in an ensemble is the same as the average across the ensemble. For an ergodic process, the sample mean from the ensemble approaches the real mean μ, and the sample variance approaches the variance σ of the process (sea state). The ergodicity of $\eta(t)$ implies that the measured realization of $\eta(t)$, say $\eta_1(t_1)$ is typical of all other possible realizations $\eta_2(t_1)$, $\eta_3(t_1)$, ..., all measured at one instant t_1. The concept of ergodicity permits derivation of various useful statistical information from a single record, eliminating the need for multiple recordings at different sites. The assumptions of stationarity and ergodicity are the backbones of developing wave statistics from wave measurements. It is implicitly assumed that such hypotheses exist in reality, and are valid, particularly for the sea state.

(e) To apply these concepts to ocean waves, consider an ensemble of records representing the sea state by $\eta(t)$ over a finite time T. The *mean* or *expected value* of the sea state, denoted by $\bar{\eta}$, or μ_η, or $E[\eta]$, is defined as

$$\mu_\eta = E[\eta(t)] = \frac{1}{\tau} \int_{-\frac{\tau}{2}}^{\frac{\tau}{2}} \eta(t)\ dt \qquad\qquad\text{(II-1-119)}$$

where the symbol E denotes the expected value of $\eta(t)$. Similarly, the mean-square of η corresponds to the second moment of η, denoted by $E[\eta^2]$. The standard deviation σ_η or the root-mean-square value of the process is the square root of this. The *variance* of η, represented by σ_η^2 may be expressed in terms of the variance of the process V as

$$\sigma_\eta^2 = V[\eta(t)] = E[\eta^2] - \mu_\eta^2 \qquad\qquad\text{(II-1-120)}$$

(f) The *standard deviation* σ_η is the square root of the variance, also called the second central moment of $\eta(t)$. The standard deviation characterizes the spread in the values of $\eta(t)$ about its mean.

(g) The *autocorrelation* or *autocovariance function* of the sea state is denoted by R_η, relating the value of η at time t to its value at a later time $t+\tau$. This is defined as

$$R_\eta(t,\ t+\tau) = E[\eta(t)\ \eta(t+\tau)] \qquad\qquad\text{(II-1-121)}$$

(h) The value of R_η gives an indication of the correlation of the signal with itself for various time lags τ, and so it is a measure of the temporal variation of $\eta(t)$ with time. If the signal is perfectly correlated with itself for zero lag τ, its autocorrelation coefficient, defined as

$$\rho_\eta = \frac{E[\eta(t)\ \eta(t+\tau)]}{E[\eta^2]} = \frac{R_\eta}{E[\eta^2]} \qquad \text{(II-1-122)}$$

will be equal to 1.

(i)   For two different random signals η_1 and η_2, the *cross-correlation coefficient R* may be defined as

$$R = E[\eta_1(t)\ \eta_2(t+\delta t)] = \frac{1}{\tau}\int_{-\frac{\tau}{2}}^{\frac{\tau}{2}} \eta_1(t)\ \eta_2(t+\delta t)\ dt \qquad \text{(II-1-123)}$$

which measures the degree of correlation between two signals. This concept is useful for example in relating wave velocities and pressures obtained at two separate locations during wave gauge measurements in coastal projects. Note that the process $\eta(t)$ is stationary if μ_η and σ_η are constant for all values of t, and that R is a function only of $\tau = t_2 - t_1$.

(j)   Assuming that the water surface elevation $\eta(t)$ is a stationary random process, representing a sea state over the duration of several hours, we will next focus our attention on defining the probabilistic properties of ocean waves. The probabilistic representation of sea state is useful in practice for two reasons. First, it allows the designer to choose wave parameters within a limit that will yield an acceptable level of risk. Second, a probabilistic-based design criterion may result in substantial cost savings by considering uncertainties in the wave estimates. Therefore, an overview of the probability laws and distributions for ocean waves follows.

(6)   Probability distributions for a sea state.

(a)   As noted earlier, irregular sea states are random signals. For engineers to effectively use irregular waves in design, properties of the individual wave records must follow some probability laws so that wave statistics can readily be obtained analytically. Rice (1944-1945) developed the statistical theory of random signals for electrical noise analysis. Longuet-Higgins (1952) applied this theory to the random water surface elevation of ocean waves to describe their statistics using certain simplified assumptions. Longuet-Higgins found that the parameters of a random wave signal follow known probability laws.

(b)   The *probability distribution P(x)* is the fraction of events that a particular event is not exceeded. It can be obtained directly from a plot of the proportion of values less than a particular value versus the particular value of the variable x_0, and is given by

$$P(x) = prob\{x \le x_0\} \qquad \text{(II-1-124)}$$

(c)   The *probability density p(x)* is the fraction of events that a particular event is expected to occur and thus, it represents the rate of change of a distribution and may be obtained by simply differentiating $P(x)$ with respect to its argument x.

(d)   The two most commonly used probability distributions in the study of random ocean waves are the *Gaussian* (Figure II-1-28) and *Rayleigh* distributions (Figure II-1-29). The Gaussian distribution is particularly suited for describing the short-term probabilities of the sea surface elevation η. Its probability density is given by

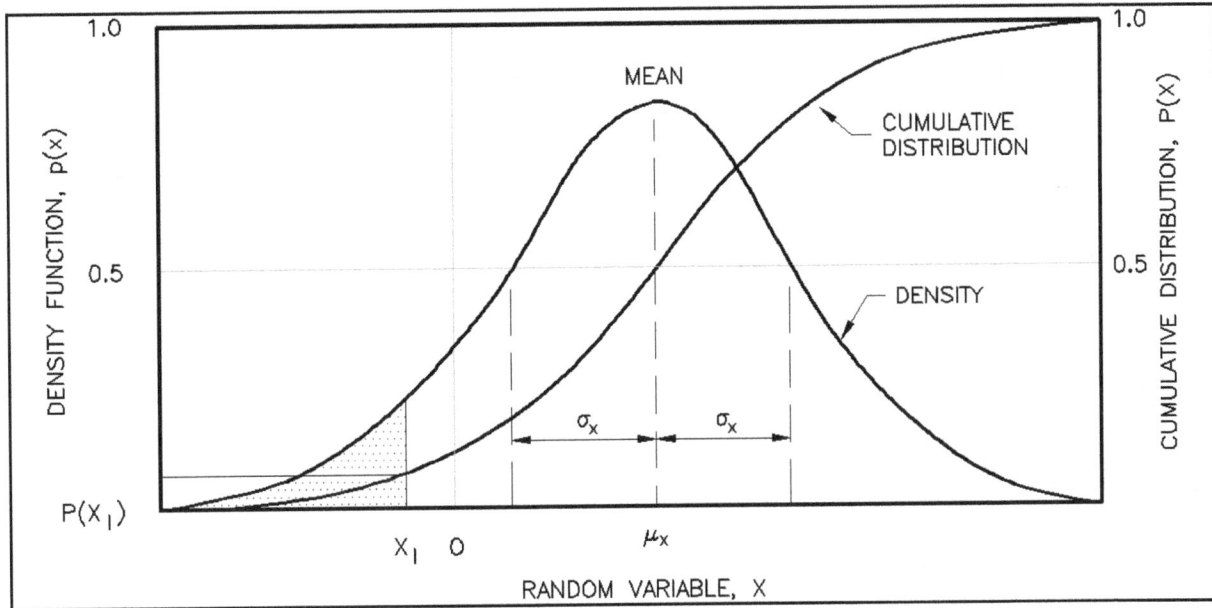

Figure II-1-28.   The Gaussian probability density and cumulative probability distribution

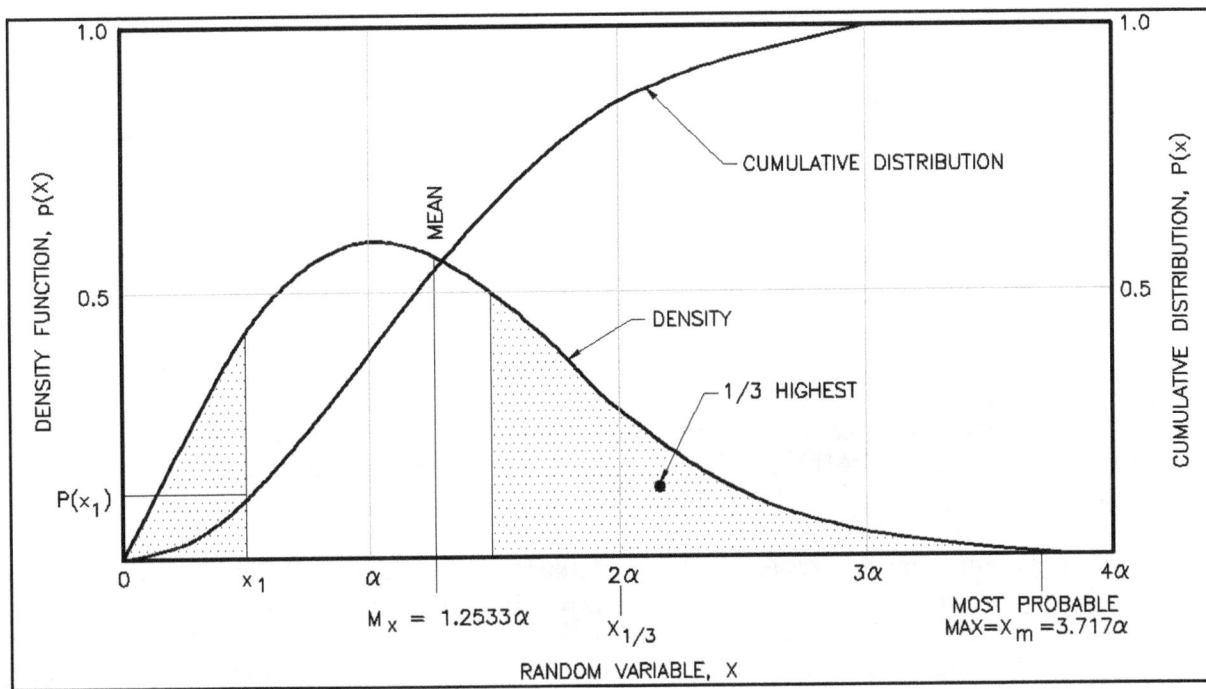

Figure II-1-29.   The Rayleigh probability density and cumulative probability distribution ($x = \alpha$ corresponds to the mode)

$$p(x) = \frac{1}{\sigma_x \sqrt{2\pi}} e^{-\left(\frac{(x-\mu_x)^2}{2\sigma_x^2}\right)}$$

(II-1-125)

where μ_x is the mean of x and σ_x is the standard deviation. The Gaussian cumulative probability or probability distribution denoted by $P(x)$ in Figure II-1-28, is the integral of p(x). A closed form of this integral is not possible. Therefore, Gaussian distribution is often tabulated as the normal distribution with the mean μ_x and standard deviation σ_x in handbooks (e.g., Abramowitz and Stegun (1965)), and is written as

$$p(x) = N(\mu_x, \sigma_x) \qquad P(x) = \Phi\left[\frac{x-\mu_x}{\sigma_x}\right]$$

(II-1-126)

For zero mean ($\mu_x = 0$) and unit standard deviation ($\sigma_x = 1$), the Gaussian probability density and distributions reduce to

$$p(x) = \frac{1}{\sqrt{2\pi}} e^{\left(-\frac{x^2}{2}\right)}$$

$$\Phi(x) = \int_0^x p(y) \, dy$$

(II-1-127)

where the last integral is the *error function*.

(e)   The probability of exceedence $Q(x)$ may be expressed in terms of the probability of non-exceedence P(x) as

$$Q[x(t) > x_1] = 1 - P[x(t) < x_1] = 1 - \Phi\left[\frac{x-\mu_{x_1}}{\sigma_x}\right]$$

(II-1-128)

(f)   This is the probability that x will exceed x_1 over the time period t, and is shown as the shaded area in the bottom lower end of Figure II-1-28. The probability of exceedence is an important design parameter in risk-based design.

(g)   In engineering practice, we are normally concerned with wave height rather than surface elevation. However, to define wave height distribution, we only need to examine the statistics of the slowly varying envelope of the surface elevation $\eta(t)$. With this approach, Longuet-Higgins (1952) found from statistical theory that both wave amplitudes and heights follow the Rayleigh distribution shown in Figure II-1-29. Note that this distribution can never be negative, it decays asymptotically to zero for large x, but never reaches zero. The probability density function of the Rayleigh distribution and its cumulative probability are given by

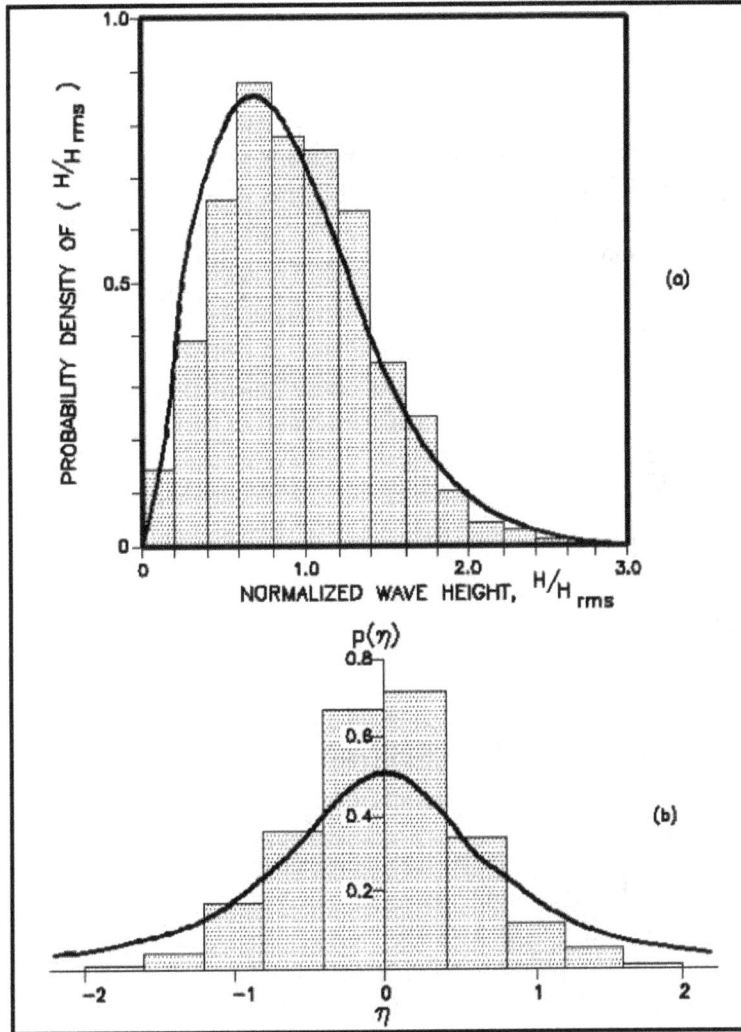

Figure II-1-30. Histograms of the normalized (a) wave heights, and (b) wave periods with theoretical distributions (Chakrabarti 1987)

$$p(x) = \left\{ \frac{\pi x}{2\mu_x^2} e^{-\frac{\pi}{4}\left(\frac{x}{\mu_x}\right)^2} \quad \textit{for } x \geq 0 \right\}$$

$$P(x) = \left\{ 1 - e^{-\frac{\pi}{4}\left(\frac{x}{\mu_x}\right)^2} \quad \textit{for } x \geq 0 \right\}$$

$$(\text{II-1-129})$$

where μ_x is the mean. These are displayed in Figure II-1-29 in which the density function is offset to the right and has only positive values. The distributions used for wave heights, wave periods, and their joint relations are described next.

(7)   Wave height distribution.

(a)   The heights of individual waves may be regarded as a stochastic variable represented by a probability distribution function. From an observed wave record, such a function can be obtained from a histogram of wave heights normalized with the mean heights in several wave records measured at a point (Figure II-1-30). Thompson (1977) indicated how well coastal wave records follow the Rayleigh distribution. If wave energy is concentrated in a very narrow range of wave period, the maxima of the wave profile will coincide with the wave crests and the minima with the troughs. This is termed a *narrow-band condition*. Under the narrow-band condition, wave heights are represented by the following Rayleigh distribution (Longuet-Higgins 1952, 1975b, 1983)

$$p(H) = \frac{2H}{H_{rms}^2} \exp\left[-\frac{H^2}{H_{rms}^2}\right]$$

$$P(H) = 1 - \exp\left[-\frac{H^2}{H_{rms}^2}\right]$$

(II-1-130)

(b)   The significant wave height $H_{1/3}$ is the centroid of the area for $H \geq H_*$ under the density function where $H > H_*$ corresponds to waves in the highest one-third range as shown in Figure II-1-29, that is

$$P(H_*) = 1 - \frac{1}{3} = 1 - e^{\left(-\frac{H_*^2}{H_{rms}^2}\right)}$$

(II-1-131)

from which we find $H_* = 1.05 H_{rms}$. Various estimates of wave heights may then be obtained upon integration of the above equation using certain mathematical properties of the Error function (Abramowitz and Stegun 1965). We find

$$H_{1/3} \approx 4.00 \sqrt{m_0} = 1.416 \, H_{rms}$$

$$H_{1/10} = 1.27 \, H_{1/3} = 1.80 \, H_{rms} = 5.091 \sqrt{m_0}$$

$$H_{1/100} = 1.67 \, H_{1/3} = 2.36 \, H_{rms} = 6.672 \sqrt{m_0}$$

$$H_{max} = 1.86 \, H_{1/3} \quad \textit{(for 1000 wave cycles in the record)}$$

(II-1-132)

(c)   The *most probable maximum wave height* in a record containing N waves is related to the rms wave height (Longuet-Higgins 1952) by

$$H_{max} = \left[\sqrt{\log N} + \frac{0.2886}{\sqrt{\log N}} - \frac{0.247}{(\log N)^{3/2}}\right] H_{rms}$$

(II-1-133)

(d)   The value of H_{max} obtained in this manner can be projected to a longer period of time by adjusting the value of N based on the mean zero-upcrossing period (Tucker 1963).

(e)   The fact that the statistics of wave height for wave records in general follows a Rayleigh distribution is of great significance in coastal engineering. For instance, an engineer may have information from a hindcast (see Part II-2) that the significant height for a storm is 10 m. Assuming that the Rayleigh distribution describes the wave record, the engineer can estimate that the 10-percent wave will be 12.7 m and that the H_{max} (assuming 1,000 waves in the record) will be 18.6 m. Often measured ocean wave records are analyzed spectrally (see "Spectral Analysis" section later in this chapter) by the instrument package and only condensed information is reported via satellite to a data bank, with no other information retained. The inherent assumption made is that the Rayleigh distribution is adequate.

(f)   Theoretical relationships derived from the Rayleigh distribution generally agree well with the values determined directly from the records. The Rayleigh probability distribution density function is compared with a histogram of the measured deepwater wave heights in Figure II-1-30 (Chakrabarti 1987). Clearly the Rayleigh distribution fits this data well, even though the frequency spectra of ocean waves may not always be narrow-banded as assumed in the Rayleigh distribution. Field measurements sometimes deviate from the Rayleigh distribution, and the deviation appears to increase with increasing wave heights, and decrease as the wave spectrum becomes sharply peaked. The effect of bandwidth on wave height distribution has been accounted for theoretically (Tayfun 1983).

(g)   Deepwater wave height measurements from different oceans have been found to closely obey a Rayleigh distribution (Tayfun 1983a,b; Forristall 1984; Myrhaug and Kjeldsen 1986). This is not true for shallow-water waves, which are strongly modulated by the bathymetric effects combined with the amplitude nonlinearities. The wave energy spectrum of the shallow-water waves is not narrow-banded and may substantially deviate from the Rayleigh distribution especially for high frequencies. In general, the Rayleigh distribution tends to overpredict the larger wave heights in all depths.

(h)   In summary, the Rayleigh distribution is generally adequate, except for near-coastal wave records in which it may overestimate the number of large waves. Investigations of shallow-water wave records from numerous studies indicate that the distribution deviates from the Rayleigh, and other distributions have been shown to fit individual observations better (SPM 1984). The primary cause for the deviation is that the large waves suggested in the Rayleigh distribution break in shallow water. Unfortunately, there is no universally accepted distribution for waves in shallow water. As a result, the Rayleigh is frequently used with the knowledge that the large waves are not likely.

(8)   Wave period distribution.

(a)   Longuet-Higgins (1962) and Bretschneider (1969) derived the wave period distribution function assuming the wave period squared follows a Rayleigh distribution. This distribution is very similar to the normal distribution with a mean period given by

$$T_{0,1} = \frac{m_0}{m_1} \qquad \text{(II-1-134)}$$

where the moments are defined in terms of cyclic frequency (i.e., Hertz). The probability density of wave period T is given by (Bretschneider 1969)

$$p(T) = 2.7 \frac{T^3}{\bar{T}} \exp\left[-0.675\tau^4\right]$$

(II-1-135)

$$\tau = \frac{T}{\bar{T}}$$

(b) A different probability density distribution of the wave period has been derived by Longuet-Higgins (1962). This is given by

$$p(\tau) = \frac{1}{2(1 + \tau^2)^{3/2}}$$

(II-1-136)

$$\tau = \frac{T - T_{0,1}}{\upsilon T_{0,1}} \quad ; \quad \nu = \frac{m_0 m_2 - m_1^2}{m_1^2}$$

where ν is the *spectral width parameter* and m_0, m_1, and m_2 are *moments* of the wave spectrum, which will be defined later. This probability density function is symmetric about $\tau = 0$ where it is maximum, and is similar to the normal distribution with a mean equal to $T_{0,1}$. This distribution fits field measurements reasonably well, and is often used in offshore design. In general, probability density for the wave period is narrower than that of wave height, and the spread lies mainly in the range 0.5 to 2.0 times the mean wave period.

(c) Various characteristic wave periods are related. This relationship may be stated in a general way as

$$T_{max} \approx T_{1/3} \approx C\bar{T}$$

(II-1-137)

where the coefficient C varies between 1.1 and 1.3.

(9) Joint distribution of wave heights and periods.

(a) If there were no relation between wave height and wave period, then the joint distribution between wave height and wave period can simply be obtained from the individual probability distributions of the height and period by

$$p(H,T) = p(H)\, p(T)$$

(II-1-138)

(b) The distribution p(H,T) so obtained is inappropriate for ocean waves, since their heights and periods are correlated. For the joint distribution of wave height-period pairs, Longuet-Higgins (1975b) considered wave heights and periods also representable by a narrow-band spectrum. He derived the joint distribution assuming wave heights and periods are correlated, a more suitable assumption for real sea states.

(c) The probability density function of wave period may be obtained directly from the joint distribution, provided that a measure of the spectrum width is included in the latter. Under this condition, the distribution of wave period is simply the marginal probability density function of the joint distribution of H and T. This is done by integrating $p(H,T)$ for the full range of H from 0 to ∞. Likewise, the distribution for wave heights may be obtained by integrating $p(H,T)$ for the full range of periods. The joint distribution derived by Longuet-Higgins (1975b) was later modified (Longuet-Higgins 1983), and is given by

$$p(H,T) = \frac{\pi}{4} \frac{f(\nu)}{\left(\frac{H_*}{T_*}\right)^2} \exp\left\{-\frac{\pi H_*^2}{4}\left[1 + \frac{1 - \frac{\sqrt{1 + \nu^2}}{T_*}}{\nu^2}\right]\right\}$$

(II-1-139)

$$H_* = \frac{H}{\overline{H}} \quad ; \quad T_* = \frac{T}{\overline{T}_z} \quad ; \quad f(\nu) = \frac{2(1 + \nu^2)}{\nu + \frac{\nu}{\sqrt{1 + \nu^2}}}$$

with ν as the spectral width parameter. The period $\overline{T}_z$ is the mean zero-upcrossing period and its relation to the mean wave period $\overline{T}$ and mean crest period $\overline{T}_c$ defined in terms of moments of spectrum is as follows:

$$\overline{T}_z = 2\pi \sqrt{\frac{m_0}{m_2}} \quad ;$$

(II-1-140)

$$\overline{T} = 2\pi \frac{m_0}{m_1} \quad ; \quad \overline{T}_c = 2\pi \sqrt{\frac{m_2}{m_4}}$$

(d)   The *most probable maximum period* associated with any given H_* is

$$T_*^{max} = \frac{2\sqrt{1 + \nu^2}}{1 + \sqrt{1 + \frac{16\nu^2}{\pi H_*^2}}}$$

(II-1-141)

(e)   Chakrabarti and Cooley (1977) investigated the applicability of the joint distribution and determined that it fits field data provided the spectrum is narrow-banded and has a single peak. A different theoretical model has been suggested by Cavanie et al. (1978), and it also compares well with the field data.

c.   *Spectral analysis.*

(1)   Introduction.

(a)   In the period 1950-1960, Rice's (1944-1945) work on signal processing was extended to ocean waves (Kinsman 1965; Phillips 1977). In pinciple, the time-history of surface elevation (such as in Figures II-1-31 and II-1-32) was recognized to be similar to a noise record. By assuming that it is a discrete sample of a continuous process, the principles of Fourier analysis could be extended to describe the record. The power of Fourier representation is such that given a series of time snapshots of measurements of a three-dimensional surface, a full mathematical representation of the surface and its history may be obtained. Unfortunately, this is a lot of information. As an example, the image in Figure II-1-22 of the entrance to San Francisco Bay is one snapshot of the surface current field and represents nearly 1 million sample points. To understand the time variation of the field it would be reasonable to do this every 2 sec or so for an hour. The result is about 1.8 billion sample points that would need to be Fourier transformed. Although, this is computationally feasible such a measurement cannot be made on a routine basis and it is not clear how the information could be condensed into a form for practical engineering. However, the utility of the spectral analysis approach is that it uses a reduced dimensional approach that is powerful and useful. This section will discuss the underlying approach to using spectral representations in engineering, discuss the basic approach for the simplified spectral approaches, and describe how the spectral information can be used. However, the underlying statistical theory and assumptions will only be touched upon and details of the derivations will only be referenced.

(b) The easiest place to begin is with a nonrigorous discussion of what a spectral analysis of a single-point measurement of the surface can produce and then generalize it to the case of a sea surface. The following sections would then describe of the procedure.

(c) Considering a single-point time-history of surface elevation such as in Figures II-1-25, II-1-31, and II-1-32, spectral analysis proceeds from viewing the record as the variation of the surface from the mean and recognizes that this variation consists of several periodicities. In contrast to the wave-by-wave approach, which seeks to define individual waves, the spectral analysis seeks to describe the distribution of the variance with respect to the frequency of the signal. By convention, the distribution of the variance with frequency is written as $E(f)$ or $S(f)$ with the underlying assumption that the function is continuous in frequency space. The reason for this assumption is that all observations are discretely sampled in time, and thus, the analysis should produce estimates as discrete frequencies which are then statistically smoothed to estimate a continuum. Although $E(f)$ is actually a measurement of variance, it is often called the *one-dimensional* or *frequency energy spectrum* because (assuming linear wave theory) the energy of the wave field may be estimated by multiplying $E(f)$ by ρg.

(d) Figures II-1-31 (a regular wave) and II-1-32 (an irregular wave) provide two wave records and their spectrum. One immediate value of the spectral approach is that it tells the engineer what frequencies have significant energy content and thus acts somewhat analogous to the height-period diagram. The primary disadvantage of spectral analysis is that information on individual waves is lost. If a specific record is analyzed, it is possible to retain information about the phases of the record (derived by the analysis), which allows reconstruction of waves. But this is not routinely done.

(e) The surface can be envisioned not as individual waves but as a three-dimensional surface, which represents a displacement from the mean and the variance to be periodic in time and space. The simplest spectral representation is to consider $E(f, \theta)$, which represents how the variance is distributed in frequency f and direction θ (Figure II-1-33). $E(f, \theta)$ is called the 2-D or directional energy spectrum because it can be multiplied by ρg to obtain wave energy. The advantage of this representation is that it tells the engineer about the direction in which the wave energy is moving. A directional spectrum is displayed in Figure II-1-34 with its frequency and direction spectrums.

(f) The power of spectral analysis of waves comes from three major factors. First, the approach is easily implemented on a microchip and packaged with the gauging instrument. Second, the principal successful theories for describing wave generation by the wind and for modelling the evolution of natural sea states in coastal regions are based on spectral theory. Third, it is currently the only widely used approach for measuring wave direction. A final factor is that Fourier or spectral analysis of wave-like phenomena has an enormous technical literature and statistical basis that can be readily drawn upon.

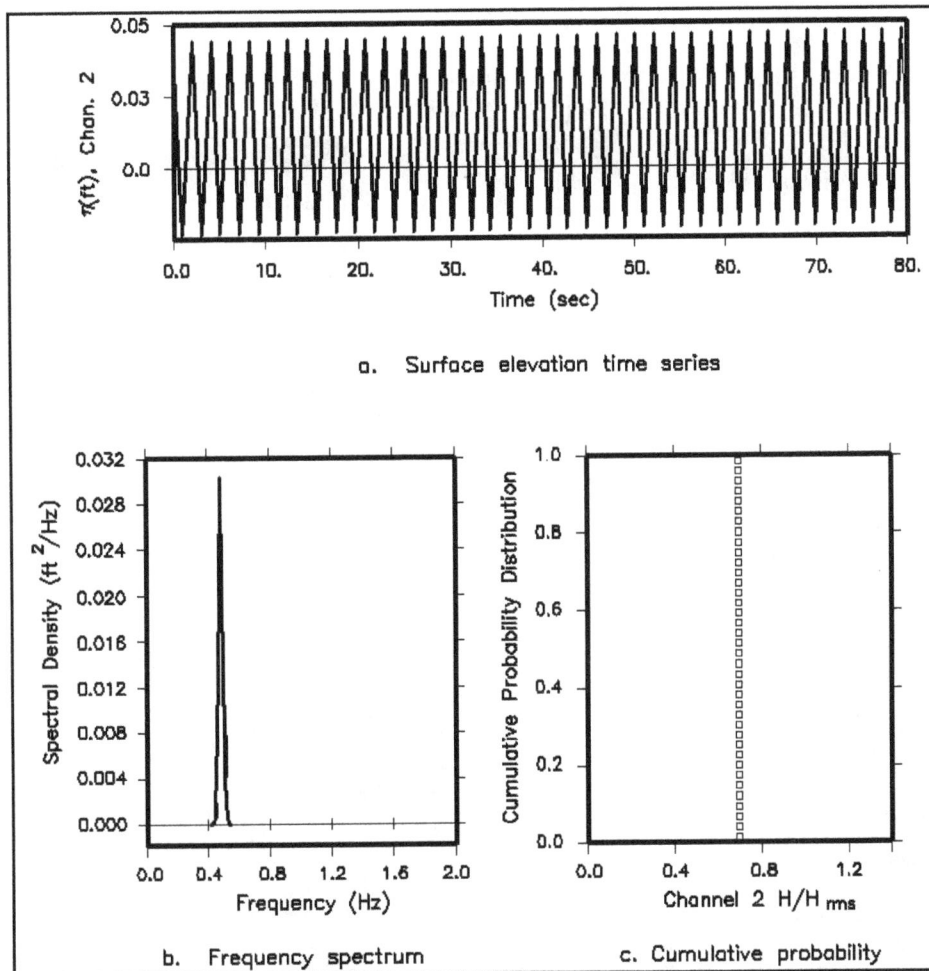

Figure II-1-31. Surface elevation time series of a regular wave and its spectrum (Briggs et al. 1993)

(g)   Before proceeding to the details of how a wave spectrum is derived from a record, it is important to touch upon some statistical assumptions that are important in analyzing a wave record spectrally. Many of these assumptions also hold for making a wave-by-wave analysis useful as well. First of all, wave records are finite in length (typically 17-68 min long) and are made up of samples of surface elevation at a discrete sampling interval (typically 0.5-2.0 sec). For the wave records to be of general use, the general characteristics of the record should not be expected to change much if the record was a little shorter or longer, if the sampling was started some fraction of time earlier or later, or if the records were collected a short distance away. In addition, it is desirable that there not be any underlying trend in the data.

(h)   If the above assumptions are not reasonably valid, it implies that the underlying process is unstable and may not be characterized by a simple statistical approach. Fortunately, most of the time in ocean and coastal areas, the underlying processes are not changing too fast and these assumptions reasonably hold. In principal the statistical goal is to assume that there is some underlying statistical process for which we have obtained an observation. The observation is processed in such a way that the statistics of the underlying process are obtained.

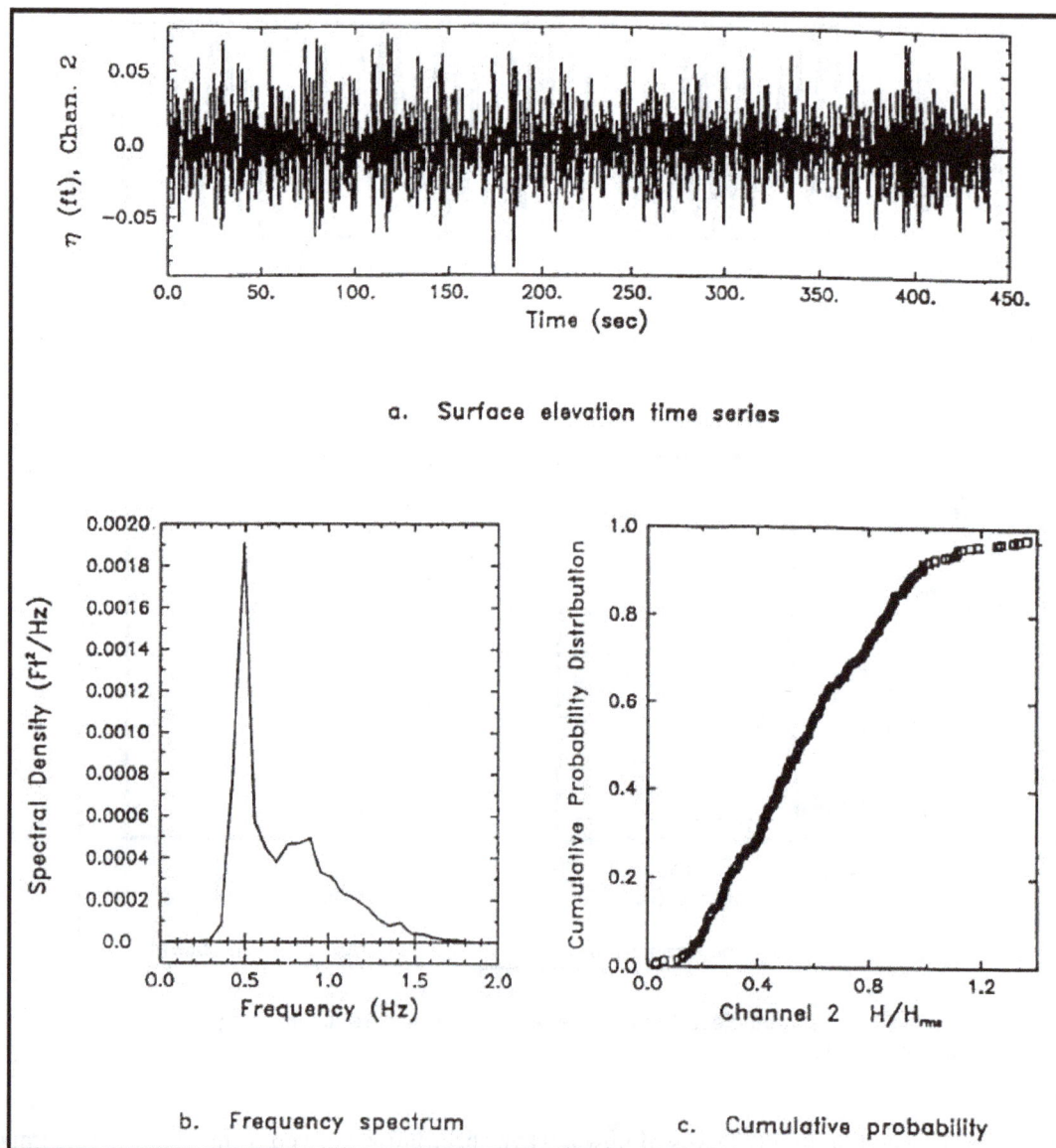

a. Surface elevation time series

b. Frequency spectrum

c. Cumulative probability

Figure II-1-32. Surface elevation time series of an irregular wave and its spectrum (Briggs et al. 1993)

(2) Description of wave spectral analysis.

(a) Unlike the wave train or wave-by-wave analysis, the spectral analysis method determines the distribution of wave energy and average statistics for each wave frequency by converting time series of the wave record into a wave spectrum. This is essentially a transformation from time-domain to the frequency-domain, and is accomplished most conveniently using a mathematical tool known as the Fast Fourier Transform (FFT) technique (Cooley and Tukey 1965). Here we will treat analysis of the time recording of the surface at a point, in order to obtain a frequency spectrum of the record. In a later section, we will describe how to obtain a frequency-directional spectrum.

(b) The *wave energy spectral density E(f)* or simply the *wave spectrum* may be obtained directly from a continuous time series of the surface $\eta(t)$ with the aid of the Fourier analysis. Using a Fourier analysis, the

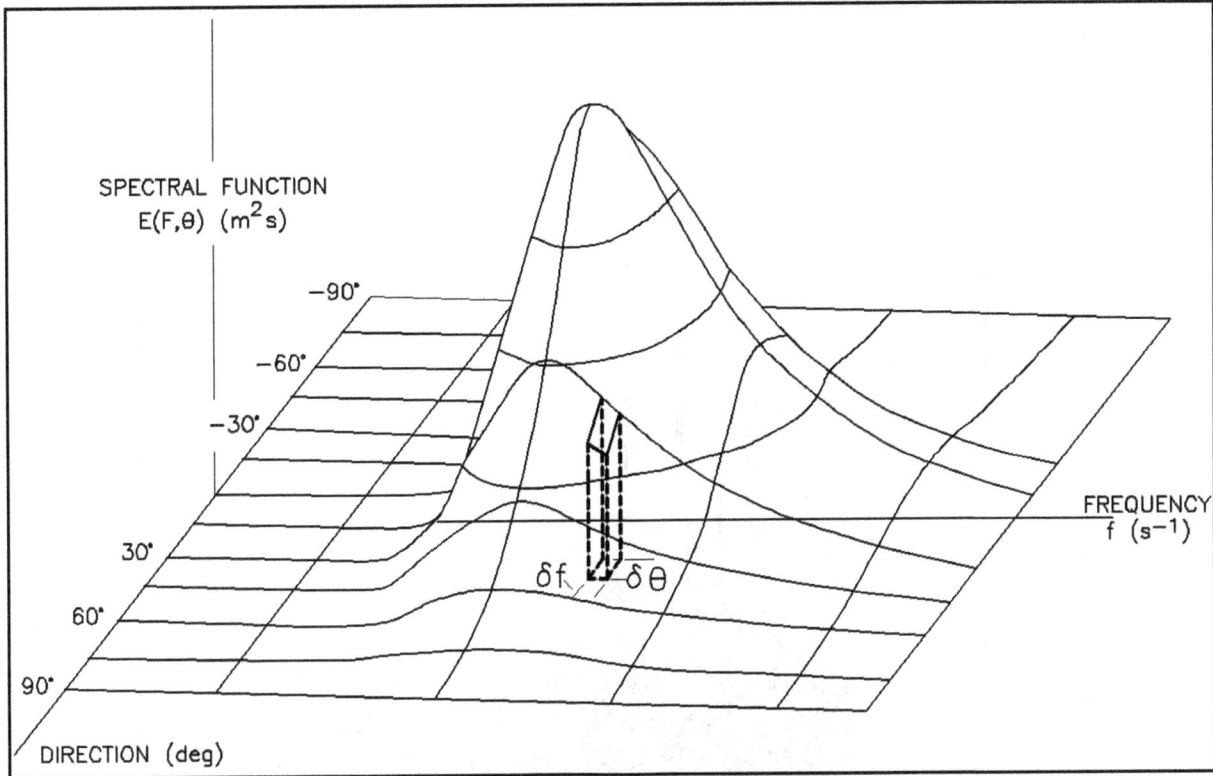

Figure II-1-33. A schematic for a two-dimensional wave spectrum E(f,θ)

wave profile time trace can be written as an infinite sum of sinusoids of amplitude A_n, frequency ω_n, and relative phase ϵ_n, that is

$$\eta(t) = \sum_{n=0}^{\infty} A_n \cos (\omega_n t - \epsilon_n)$$

$$= \sum_{n=0}^{\infty} a_n \cos n\omega t + b_n \sin n\omega t$$

(II-1-142)

(c)   The coefficients a_n and b_n in the above equation may be determined explicitly from the orthogonality properties of circular functions. Note that a_0 is the mean of the record. Because real observations are of finite length, the finite Fourier transform is used and the number of terms in the sum n is a finite value.

(d)   The *covariance* of $\eta(t)$ is related to the wave energy spectrum. This is defined in terms of the squares of component amplitudes as

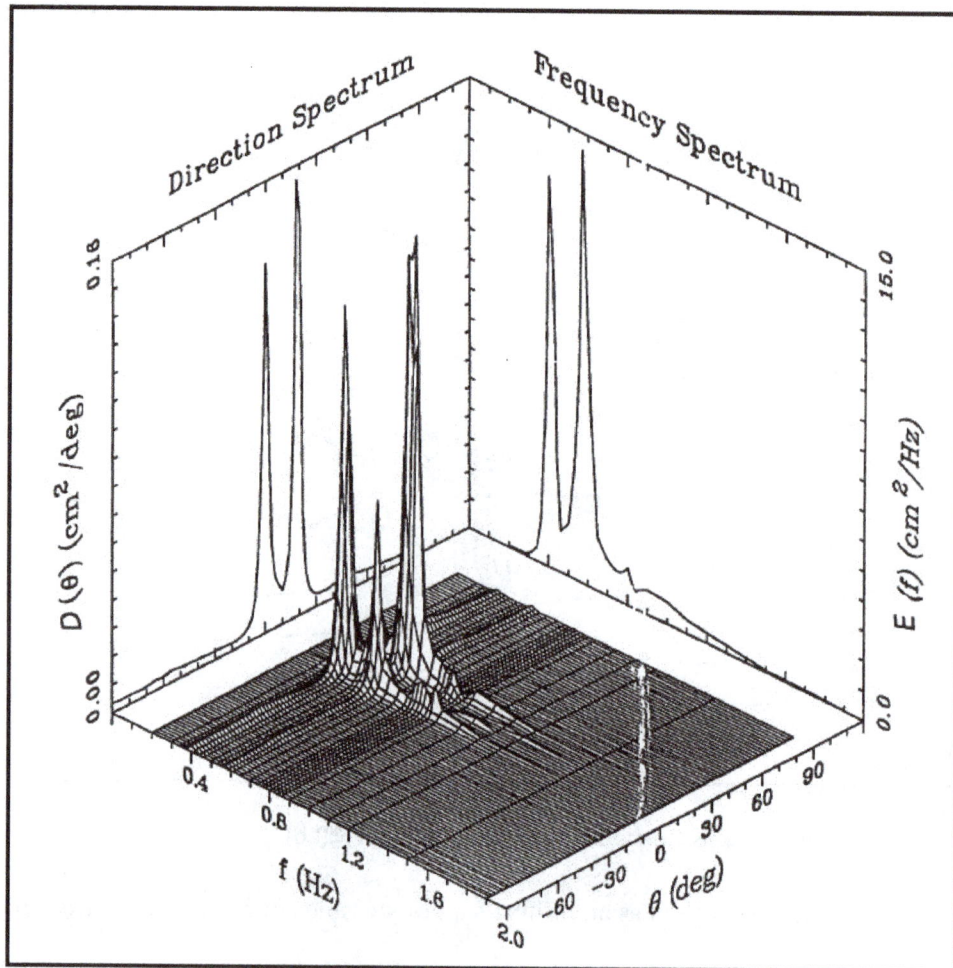

Figure II-1-34. A directional spectrum and its frequency and direction spectrum (Briggs et al 1993)

$$\overline{\eta}^2(t) = \sum_{0}^{\infty} A_n^2 \, \Delta f$$

$$A_n^2 = \frac{1}{2}\sqrt{a_n^2 + b_n^2}$$

(II-1-143)

$$\epsilon_n = \tan^{-1} \frac{b_n}{a_n}$$

(e) By induction, an estimate of the continued energy spectrum of $\eta(t)$ may be obtained by

$$E(f) = \frac{1}{T_r}\left[\sum_{n=0}^{N} \eta(n\Delta t) \, e^{2\pi i f(n\Delta t)} \, \Delta t\right]^2$$

(II-1-144)

where T_r is the *record length* and Δt is the *sampling interval*.

(f)   There are numerous intricacies involved in the application of these discrete formulas, ranging from the length of time series necessary to digitizing frequency and many others. For unfamiliar users, most computer library systems now have *FFT (Finite Fourier Transform) algorithms* available to perform the above computations. Part VII-3 of the CEM provides a discussion of the methods. Some general guidelines are provided next.

(g)   In actual practice, the total data length is divided into M smaller segments with equal number of data points N. By letting N be a power of 2 for computational efficiency, the result then is averaged over the M sections. In an FFT, the variables M, N, and Δt have to be independently selected, though T_r and Δt are fixed for a given record so that the total number of data points can be obtained from these values. Therefore, the only choice that has to be made is the number of sections M. Traditionally, the most common values of N used range from *512* to *2,048*, while the value of M is usually *8* or greater. Since T_r is dependent on N, M, and Δt as $T_r = M N \Delta t$, then higher N and M values in general yield better resolution and high confidence in the estimate of spectra. The larger the N, the more spiky or irregular the spectrum, and the smaller the N, the smoother the spectrum (Cooley and Tukey 1965; Chakrabarti 1987).

(h)   To better understand the wave spectrum by the FFT method, consider first the wave surface profile of a single-amplitude and frequency wave given by a sinusoidal function as

$$\eta(t) = a \sin \omega t \tag{II-1-145}$$

where a and ω are the amplitude and frequency of the sine wave. The variance of this wave over the wave period of 2π is

$$\sigma^2 = \overline{[\eta(t)]^2} = \frac{1}{2\pi} \int_0^{2\pi} a^2 \sin^2 2\pi f t \; d(2\pi f t)$$

$$= \frac{a^2}{2} = 2 \int_0^\infty E^1(f) \; df = \int_{-\infty}^\infty E^2(f) \; df \tag{II-1-146}$$

(i)   Thus the quantity $a^2/2$ represents the contribution to the variance σ^2 associated with the component frequency $\omega = 2\pi f$ (Figure II-1-35). The connection between the variance, wave energy, and the wave energy spectrum is now more obvious since these all are proportional to the wave amplitude (or height) squared. For consistency of units, an equality between these quantities requires that the wave spectrum not include the ρg term.

(j)   The difference between a *two-sided spectrum* E^2 and a *one-sided spectrum* E^1 as illustrated in Figure II-1-36 is quite important. Note that the two-sided spectrum is symmetric about the origin, covering both negative and positive frequencies to account for all wave energy from $-\infty$ to $+\infty$. But, it is customary in ocean engineering to present the spectrum as a one-sided spectrum. This requires that the spectral density ordinates of E^2 be doubled in value if only the positive frequencies are considered. This is the reason for introducing a factor of two in Equation II-1-146. This definition will be used subsequently throughout Part II-1; thus, it is henceforth understood that $E(f)$ refers to E^1 (Figures II-1-35 and II-1-36).

(k)   By an intuitive extension of this simple wave, the variance of a random signal with zero mean may be considered to be made up of contributions with all possible frequencies. For a random signal using the above equations, we find

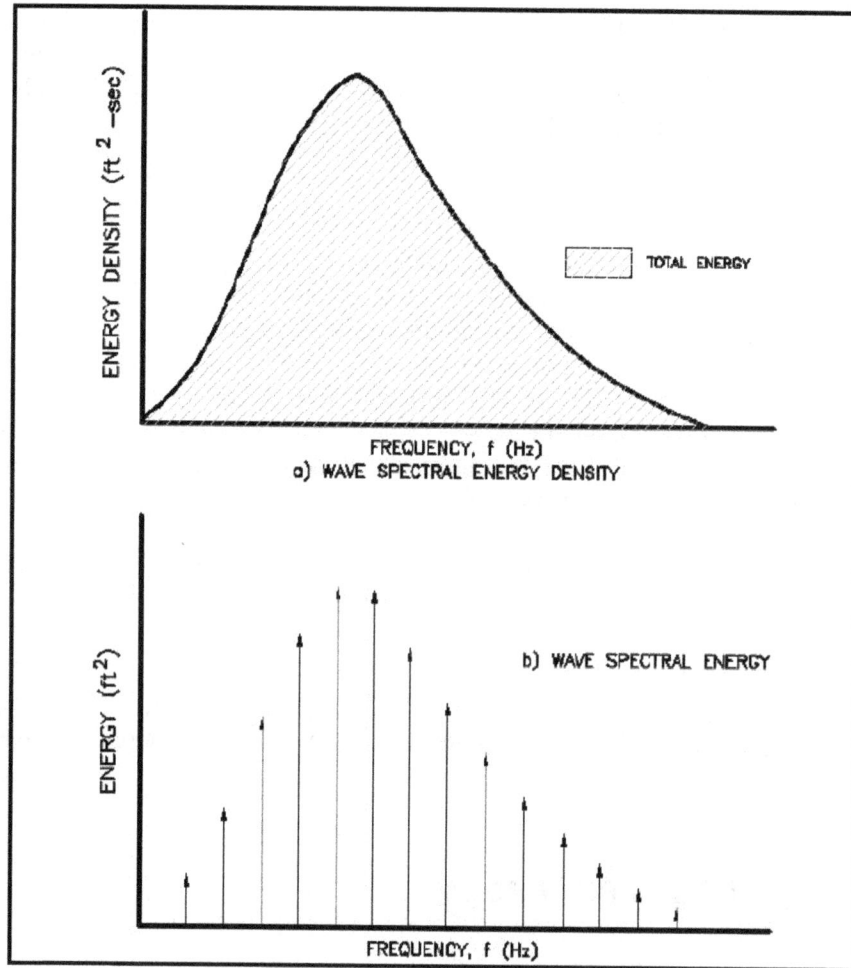

Figure II-1-35. Sketches of wave spectral energy and energy density (Chakrabarti 1987)

$$\sigma_\eta^2 = \sum_{n=1}^{\infty} \frac{a_n^2}{2} = \int_0^\infty E(f)\, df = m_0 \qquad\qquad \text{(II-1-147)}$$

where m_0 is the zero-th moment of the spectrum. Physically, m_0 represents the area under the curve of $E(f)$. The area under the spectral density represents the variance of a random signal whether the one-sided or two-sided spectrum is used.

(l) The moments of a spectrum can be obtained by

$$m_i = \int_0^\infty f^i\, E(f)\, df \qquad i = 0,1,2,\dots \qquad\qquad \text{(II-1-148)}$$

(m) We now use the above definition of the variance of a random signal to provide a third definition of the significant wave height. As stated earlier, this gives an estimate of the significant wave height by the wave spectrum. For Rayleigh distributed wave heights, H_s may be approximated (Longuet-Higgins 1952) by

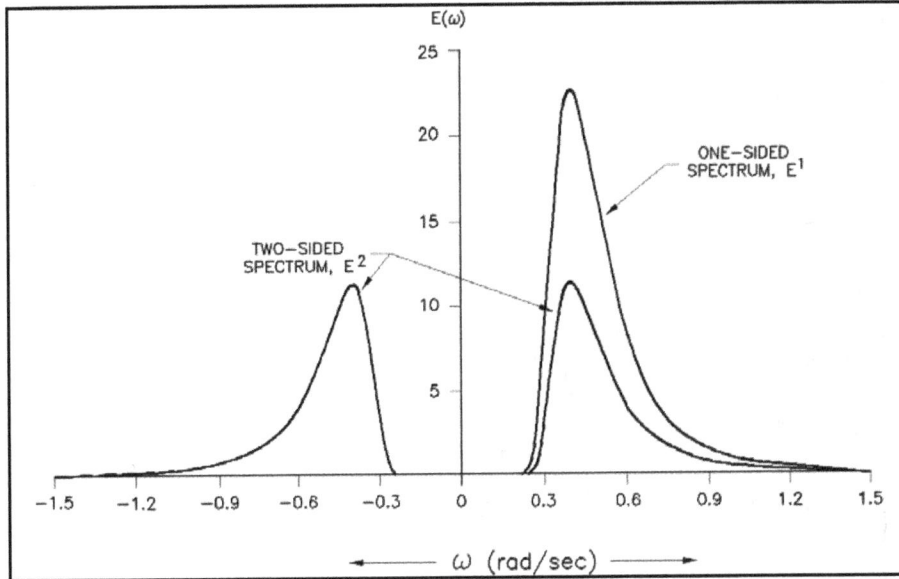

Figure II-1-36. Definition of one- and two-side wave spectrum (Chakrabarti 1987)

$$H_s = 3.8 \sqrt{m_0} \approx 4 \sqrt{m_0} \qquad\qquad\qquad\qquad \text{(II-1-149)}$$

(n) Therefore, the zero-th moment m_0, which is the total area under the wave energy density spectrum, defines the significant wave height for a given $E(f)$ (Figure II-1-37).

(3) Examples of frequency spectra. The frequency spectrum is normally plotted as energy density on the ordinate versus frequency on the abscissa (Figures II-1-31 through II-1-37). In principal, the form of $E(f)$ can be quite variable. However, some generalizations are possible. First of all, during strong wind events, the spectrum tends to have a strong central peak and a fairly predictable shape. For swell that has propagated a long distance from the source of generation, waves tend to have a single sharp peak. Waves in shallow water near breaking tend to have a sharp peak at the peak frequency f_p and have a series of smaller peaks at frequencies $2 f_p$, $3 f_p$, etc., which are harmonics of the main wave. The presence of harmonics indicates that the wave has the sharp crest and flat trough of highly nonsinusoidal waves often found near breaking. To complicate matters, Thompson (1977) has shown that about two-thirds of U.S. coastal wave records have more than one peak, indicating the presence of multiple wave trains. These wave trains most likely originated from different areas and have different directions of propagation. Moreover, it is possible to have a single-peak spectrum, which consists of two trains of waves of about the same frequency but different directions of propagation. In order to sort these issues out, observations of the directional spectrum are required. Figures II-1-31, II-1-32, and II-1-35 include examples of different frequency spectra providing some indication of their range of variability.

(4) Wave spectrum and its parameters.

(a) Two parameters are frequently used in the probability distribution for waves. These are the *spectral width* ν and the *spectral bandwidth* ϵ, and are used to determine the narrowness of a wave spectra. These parameters range from 0 to 1, and may be approximated in terms of spectral moments by

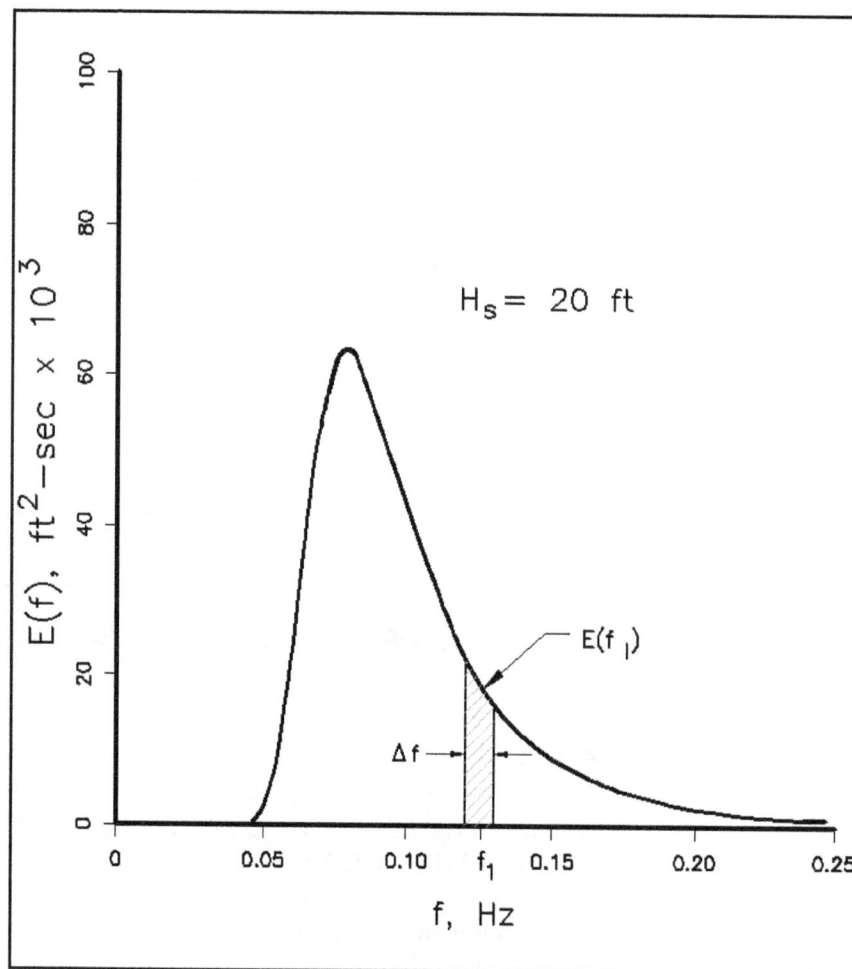

Figure II-1-37. Energy density and frequency relationship (Chakrabarti 1987)

$$\nu = \sqrt{\frac{m_0 \, m_2}{m_1^2} - 1}$$

$$\epsilon = \sqrt{1 - \frac{m_2^2}{m_0 \, m_4}}$$

(II-1-150)

(b)   For a narrow-band spectrum, both ν and ϵ must be close to 0 (Figure II-1-38). For example, for the two most common empirical spectra, the *Pierson-Moskowitz (PM)* spectrum (Pierson and Moskowitz 1964) and the *JONSWAP* spectrum (Hasselmann et al. 1973), which are discussed in the next section, $\nu = 0.425$ and 0.389, respectively, with $\epsilon = 1$ for both. Natural ocean waves, therefore, have a broad-banded spectrum.

(c)   The values of ϵ obtained from a wave energy spectrum are generally not considered as the sole indication of how broad the spectra are. This is due to the amplification of the noise present in the wave energy spectral density at higher frequencies that enters into the calculation of the higher moments m_2 and m_4 in the above equation for ϵ. Goda (1974) proposed a *spectral peakedness parameter* called Q_p defined as

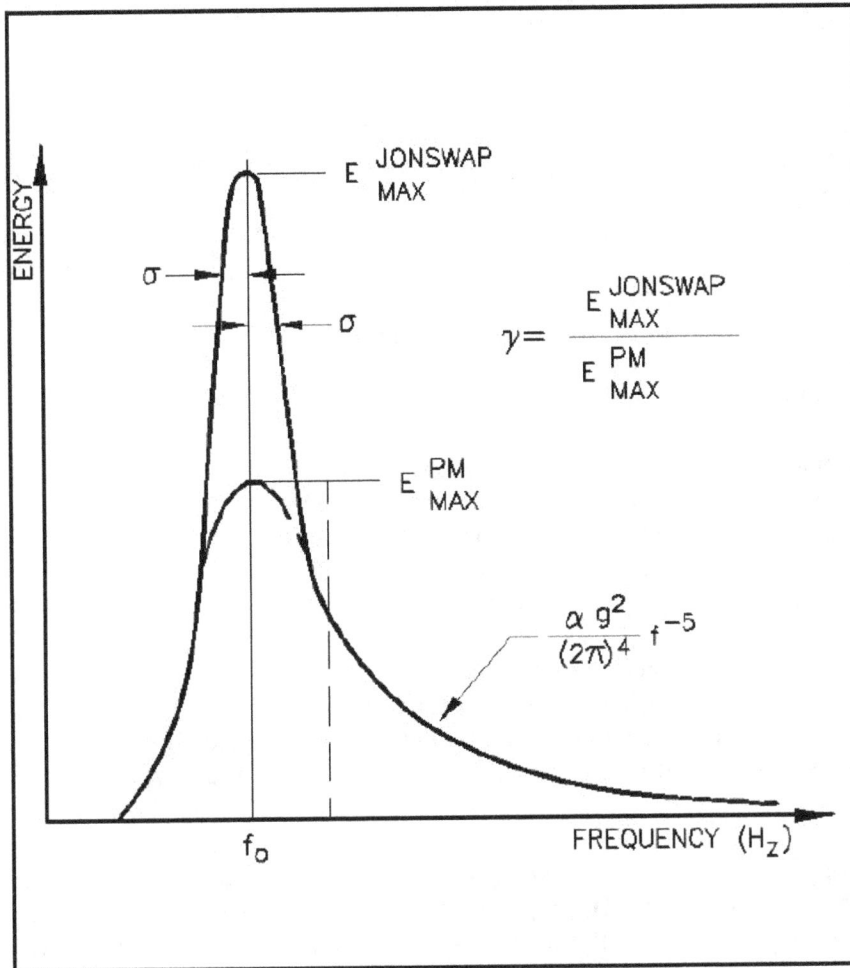

Figure II-1-38. Comparison of the PM and JONSWAP spectra (Chakrabarti 1987)

$$Q_p = \frac{2}{m_0^2} \int_0^\infty f\, E^2(f)\ df \qquad \text{(II-1-151)}$$

which depends only on the first moment of the energy density spectrum, and is not directly related to ϵ. In general, a small ϵ implies that Q_p is large, and a large ϵ means Q_p is small.

(d) Approximate relations for most common wave parameters by the statistical analysis are

$$H_S = 4.0 \sqrt{m_0} \qquad ; \qquad H_{1/10} = 5.1 \sqrt{m_0}$$

$$T_z = \sqrt{\frac{m_0}{m_2}} \qquad ; \qquad T_c = \sqrt{\frac{m_2}{m_4}} \qquad \text{(II-1-152)}$$

$$\bar{\eta} = \sqrt{m_0} \qquad ; \qquad \epsilon = \sqrt{1 - \frac{m_2^2}{m_0 m_4}}$$

(e) In deep and intermediate water depths, the significant wave height obtained by the spectral analysis using the above equation is usually greater than that from the wave train analysis. The zero-crossing period from the spectral method is only an approximation, while the period associated with the largest wave energy known as the *peak period T_p*, can only be obtained via the spectral analysis. In the spectral representation of swell waves, there is a single value of the peak period and wave energy decays at frequencies to either side. The spectra for storm waves is sometimes multi-peaked. One peak (not always the highest) corresponds to the swell occurring at lower frequencies. One and sometimes more peaks are associated with storm waves occurring at comparatively higher frequencies. In a double-peaked spectra for storm waves, the zero-crossing period generally occurs at higher frequencies than the peak period. In a multi-peaked spectrum, the zero-crossing period is not a measure of the frequency where peak energy occurs.

(5) Relationships among $H_{1/3}$, H_s, and H_{m0} in shallow water.

(a) By conception, significant height is the average height of the third-highest waves in a record of time period. By tradition, wave height is defined as the distance from crest to trough. Significant wave height H_s can be estimated from a wave-by-wave analysis in which case it is denoted $H_{1/3}$, but more often is estimated from the variance of the record or the integral of the variance in the spectrum in which case it is denoted H_{m0}. Therefore, H_s in Equation II-1-152 should be replaced with H_{m0} when the latter definition of H_s is implied. While $H_{1/3}$ is a direct measure of H_s, H_{m0} is only an estimate of the significant wave height which under many circumstances is accurate. In general in deep water $H_{1/3}$ and H_{m0} are very close in value and are both considered good estimates of H_s. All modern wave forecast models predict H_{m0} and the standard output of most wave gauge records is H_{m0}. Few routine field gauging programs actually compute and report $H_{1/3}$ and report as H_s with no indication of how it was derived. Where $H_{1/3}$ and H_s are equivalent, this is of little concern.

(b) Thompson and Vincent (1985) investigated how $H_{1/3}$ and H_{m0} vary in very shallow water near breaking. They found that the ratio $H_{1/3}/H_{m0}$ varied systematically across the surf zone, approaching a maximum near breaking. Thompson and Vincent displayed the results in terms of a nomogram (Figure II-1-40). For steep waves, $H_{1/3}/H_{m0}$ increased from 1 to about 1.1, then decreased to less than 1 after breaking. For low steepness waves, the ratio increased from 1 before breaking to as much as 1.3-1.4 at breaking, then decreased afterwards. Thompson and Vincent explained this systematic variation in the following way. As low steepness waves shoal prior to breaking, the wave shape systematically changes from being near sinusoidal to a wave shape that has a very flat trough with a very pronounced crest. Although the shape of the wave is significantly different from the sine wave in shallow water, the variance of the surface elevation is about the same, it is just arranged over the wave length differently from a sine wave. After breaking, the wave is more like a bore, and as a result the $H_{1/3}$ can be smaller (by about 10 percent) than H_{m0}.

(c) The critical importance of this research is in interpreting wave data near the surf zone. It is of fundamental importance for the engineer to understand what estimate of significant height he is using and what estimate is needed. As an example, if the data from a gauge is actually H_{m0} and the waves are near breaking, the proper estimate of H_s is given by $H_{1/3}$. Given the steepness and relative depth, $H_{1/3}$ may be estimated from H_{m0} by Figure II-1-40. Numerically modelled waves near the surf zone are frequently equivalent to H_{m0}. In this case, H_s will be closer to $H_{1/3}$ and the nomogram should be used to estimate H_s.

(6) Parametric spectrum models.

(a) In general, the spectrum of the sea surface does not follow any specific mathematical form. However, under certain wind conditions the spectrum does have a specific shape. A series of empirical expressions have been found which can be fit to the spectrum of the sea surface elevation. These are called *parametric spectrum models*, and are useful for routine engineering applications. A brief description of these follows.

(b) There are many forms of wave energy spectra used in practice, which are based on one or more parameters such as wind speed, significant wave height, wave period, shape factors, etc. Phillips (1958) developed an equation for the equilibrium range of the spectrum for a fully-developed sea in deep water, which became the basis of most subsequent developments. Phillips' equilibrium range is often written in terms of the *angular frequency* ω and is of the form

$$E(\omega) = \alpha g^2 \omega^{-5} \tag{II-1-153}$$

where α is the *Phillips' constant* (= .0081) and g the gravitational acceleration.

(c) One commonly used spectrum in wave hindcasting and forecasting projects is the single-parameter spectrum of *Pierson-Moskowitz PM* (Pierson and Moskowitz 1964). An extension of the PM spectrum is the *JONSWAP* spectrum (Hasselmann et al. 1973, 1976); this is a five-parameter spectrum, although three of these parameters are usually held constant. The relationship between PM and JONSWAP spectra is shown in Figure II-1-38. Other commonly used two-parameter wave spectra forms, including those proposed by Bretschneider (1959), ISSC (1964), Scott (1965), ITTC (1966), Liu (1971), Mitsuyasu (1972), Goda (1985a), and Bouws et al. (1985) are essentially derivatives of the *PM* and *JONSWAP* spectra. A six-parameter wave spectrum has been developed by Ochi and Hubble (1976). The utility of this spectrum is that it is capable of describing multi-peaks in the energy spectrum in a sea state mixed with swell (Figure II-1-39). Only the parametric wave spectra forms most often used in coastal engineering will be briefly discussed here.

(d) The equilibrium form of the *PM* spectrum for fully-developed seas may be expressed in terms of wave frequency f and wind speed U_w as

$$E(f) = \frac{0.0081 g^2}{(2\pi)^4 f^5} \exp\left(-0.74 \left[\frac{2\pi U_w f}{g}\right]^{-4}\right) \tag{II-1-154}$$

where U_w is the wind speed at 19.5 m above mean sea level. The *PM* spectrum describes a *fully-developed sea* with one parameter, the wind speed, and assumes that both the fetch and duration are infinite. This idealization is justified when wind blows over a large area at a constant speed without substantial change in its direction for tens of hours.

(e) The *JONSWAP* spectrum for *fetch-limited seas* was obtained from the Joint North Sea Wave Project - JONSWAP (Hasselmann et al. 1973) and may be expressed as

$$E(f) = \frac{\alpha g^2}{(2\pi)^4 f^5} \exp\left[-1.25 \left(\frac{f}{f_p}\right)^{-4}\right] \gamma^{\exp\left[-\frac{\left(\frac{f}{f_p}-1\right)^2}{2\sigma^2}\right]} \tag{II-1-155}$$

$$f_p = 3.5 \left[\frac{g^2 F}{U_{10}^3}\right]^{-0.33} \quad ; \quad \alpha = 0.076 \left[\frac{gF}{U_{10}^2}\right]^{-0.22} \quad ; \quad 1 \le \gamma \le 7$$

$$\sigma = 0.07 \quad for \ f \le f_p \quad and \quad \sigma = 0.09 \quad for \ f > f_p$$

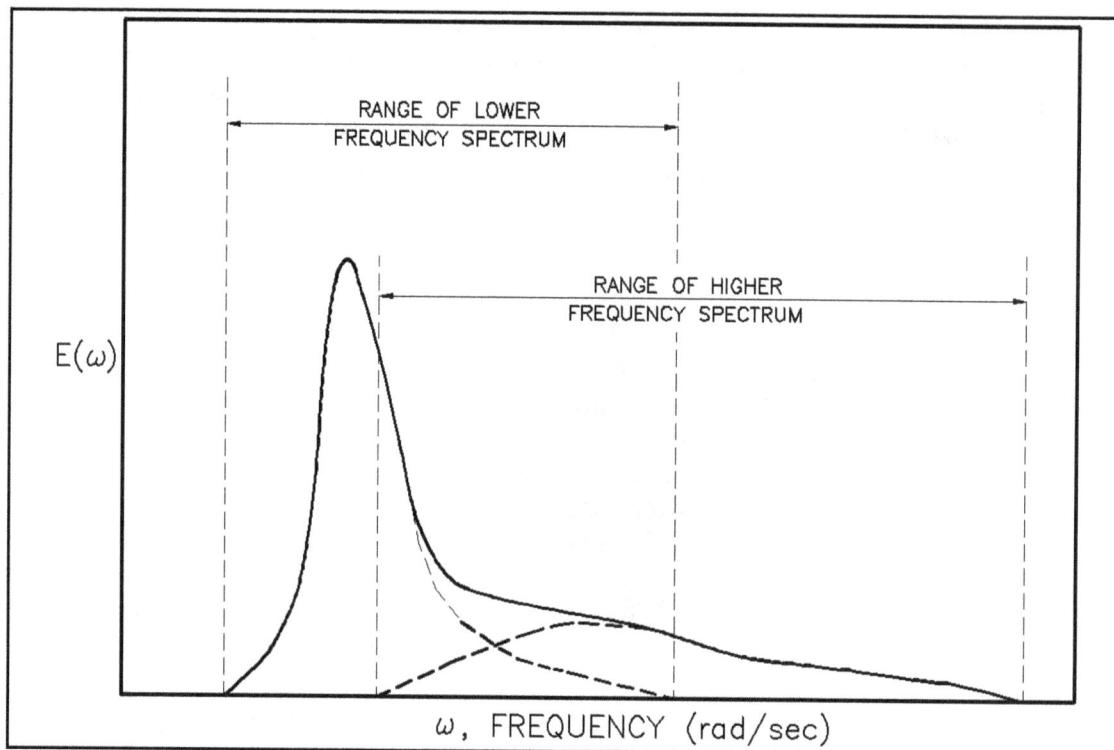

Figure II-1-39. Definition sketch for Ochi-Hubble spectrum (Ochi and Hubble 1976)

(f)   In this equation, α is the scaling parameter, γ the peak enhancement factor, f_p the frequency at the spectral peak, U_{10} the wind speed at the elevation 10 m above the sea surface, F the fetch length. Figure II-1-38 qualitatively illustrates the relationship between JONSWAP and PM spectra. The JONSWAP spectrum can also be fitted mathematically to observed spectra by iteratively solving for d, γ, f_m, and σ.

(g)   A six-parameter spectrum developed by Ochi and Hubble (1976) is the only wave spectrum which exhibits two peaks (Figure II-1-39), one associated with underlying swell (lower frequency components) and the other with locally generated waves (higher frequency components). It is defined as

$$E(\omega) = \frac{1}{4} \sum_{j=1}^{2} \frac{\left(\dfrac{4\lambda_j + 1}{4} \omega_{0j}^4 \right)^{\lambda_j}}{\Gamma(\lambda_j)} \frac{H_{sj}^2}{\omega^{4\lambda_j+1}} \exp\left[-\frac{4\lambda_j+1}{4} \left(\frac{\omega_{0j}}{\omega} \right)^4 \right] \tag{II-1-156}$$

where H_{s1}, ω_{01}, and λ_1 are the significant wave height, modal frequency, and shape factor for the lower-frequency components while H_{s2}, ω_{02}, and λ_2 correspond to the higher frequency components (Figure II-1-39). The value of λ_1 is usually much higher than λ_2. For the most probable value of ω_{01}, it can be shown that $\lambda_1 = 2.72$, while λ_2 is related to H_s in feet as

$$\lambda_2 = 1.82 \ e^{\left(-0.027 H_s \right)} \tag{II-1-157}$$

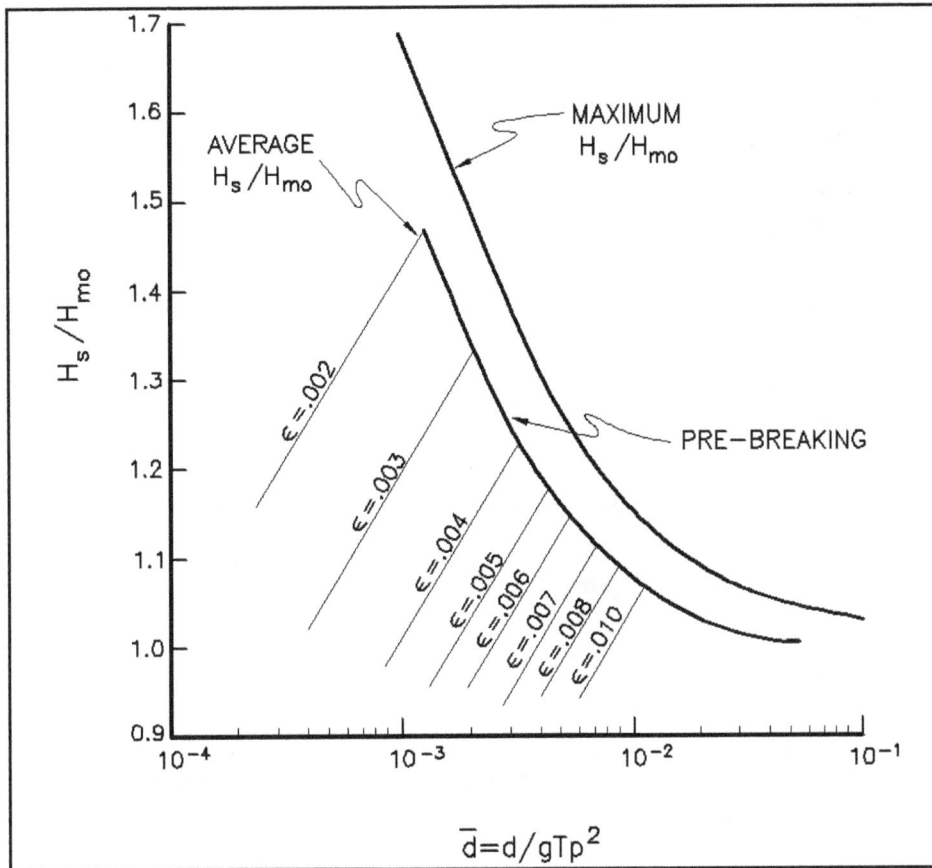

Figure II-1-40. Variation of H_s/H_{mo} as a function of relative depth $\bar{d}$ and significant steepness (Thompson and Vincent 1985)

(h)   The parameters λ_j control the shape and the sharpness of the spectral peak of the Ochi-Hubble spectral model if in either spectral component (i.e., sea or swell) the values of H_{sj} and ω_{0j} are held constant. Therefore, λ_1 and λ_2 are called the *spectral shape parameters*. On the assumption of a narrow-bandedness of the entire Ochi-Hubble spectrum, an equivalent significant wave height may be calculated by

$$H_s = \sqrt{H_{s1}^2 + H_{s2}^2}$$
(II-1-158)

Note that for $\lambda_1 = 1$ and $\lambda_2 = 0$, the *PM* spectra may be recovered from this equation.

(i)   In shallow water, the wave spectrum deviates from the standard spectra forms presented so far, and at frequencies above the peak, the spectrum no longer decays as f^5. Kitaigorodoskii et al. (1975) showed that the equilibrium range is proportional to -3 power of the wave number, and thus, the form of the spectrum is of f^3 in the high-frequency range. This change is attributed to the effect of water depth on wave spectrum and to the interaction between spectral components. Bouws et al. (1984) proposed a variation to the *JONSWAP* energy spectrum for representing wave spectra in finite-depth water. The spectrum so obtained, the product of *JONSWAP* and the *Kitaigorodoskii depth function* accounting for the influence of the water depth, is called the *TMA spectrum* after the names of three sources of data used in its development (Texel, Marsen, and Arsloe).

(j)   Kitaigorodoskii et al. (1975) obtained the form of depth dependence as

$$\Phi(\omega,d) = \frac{\left[k^{-3}\dfrac{\partial k}{\partial \omega}\right]_{d=finite}}{\left[k^{-3}\dfrac{\partial k}{\partial \omega}\right]_{d=\infty}} s \qquad\qquad (\text{II-1-159})$$

(k)   Thus, Φ is a *weighing factor* of the quantity in the bracket, which is determined from the ratio of the quantity evaluated for finite and infinite water depth cases. Using the linear wave theory, the above equation has been approximated by Kitaigorodoskii et. al. (1975) as

$$\Phi(\omega^*,d) \approx \begin{cases} \dfrac{1}{2}\omega^{*2} & \text{for } \omega^* \le 1 \\[3mm] 1 - \dfrac{1}{2}(2 - \omega^*)^2 & \text{for } \omega^* > 1 \end{cases} \qquad\qquad (\text{II-1-160})$$

(l)   The *TMA* spectrum was intended for wave hindcasting and forecasting in water of finite depth. This spectrum is a modification of the *JONSWAP* spectrum simply by substituting Kitaigorodoskii's expression for effects of the finite depth equilibrium function. By using the linear wave theory, we find the following complete form of the *TMA* spectrum:

$$S_{TMA}(\omega,d) = S_{JONSWAP}(\omega)\ \Phi(\omega^*,d)$$

$$\Phi(\omega^*,d) = \frac{1}{[f(\omega^*)]^{\frac{1}{2}}}\left[1 + \frac{K}{\sinh K}\right]^{-1} \quad ; \quad \omega^* = \omega\sqrt{\frac{d}{g}} \qquad\qquad (\text{II-1-161})$$

$$f(\omega^*) = \tanh^{-1}[k(\omega^*)d] \quad ; \quad K = 2\omega^{*2}\, f(\omega^*)$$

(m)   In effect, this substitution transforms the decay or slope of the spectral density function of the *JONSWAP* spectrum in the high-frequency side from ω^5 to ω^3 type dependence during the shoaling process approximated by linear wave theory. Bouws et. al (1984) present equations for α, γ, and σ. As with the JONSWAP, the equation may be iteratively fit to an observed spectrum and α, γ, f_m , and σ may be estimated.

(n)   The *PM*, *JONSWAP*, and *TMA* spectra can be estimated if something about the wind, depth and fetch are known. Furthermore, these spectral equations can be used as target spectra whose parameters can be varied to fit observed spectra which may have been measured. In the first situation, the value of the parameterization is in making an educated guess at what the spectrum may have looked like. The value in the second case is for ease of analytical representation. However, very often today engineering analyses are made on the basis of numerical simulations of a specific event by use of a numerical model (see Part II-2). In this case, the model estimates the spectrum and a parametric form is not required.

(7)   Directional spectra.

(a)   The wave spectra described so far have been one-dimensional frequency spectra. Wave direction does not appear in these representations, and thus variation of wave energy with wave direction was not considered. However, the sea surface is often composed of many waves coming from different directions. In addition to wave frequency, the mathematical form of the sea state spectrum corresponding to this situation should therefore include the wave direction θ. Each wave frequency may then consist of waves from different directions θ. The wave spectra so obtained are *two-dimensional*, and are denoted by $E(f, \theta)$. Figures II-1-33 and II-1-34 display directional spectra.

(b)   Measurement of a directional spectrum typically involves measurement of either the same hydrodynamic parameter (such as surface elevation or pressure) at a series of nearby locations (within one to tens of meters) or different parameters (such as pressure and two components of horizontal velocity) at the same point. These records are then cross-correlated through a cross-spectral analysis and a directional spectrum is estimated. In general, the more parameters or more locations involved, the higher the quality of the directional spectrum obtained. The procedures for converting measurements into estimates of the directional spectrum are outside the scope of this chapter. Part VII-3 of the CEM and Dean and Dalrymple (1991) provide some additional details on this subject.

(c)   The major systems routinely employed at the present time for measuring directional spectra include directional buoys, arrays of pressure or velocity gauges, and the p-U-V technique. With directional buoys, pitch-roll-and-heave or heave-and-tilt methods are used. Most directional buoys are emplaced in deeper water. Arrays of pressure gauges or velocity gauges arranged in a variety of shapes (linear, cross, star, pentagon, triangle, rectangle, etc.) are also used, but these are usually restricted to shallower water. The p-U-V technique uses a pressure gauge and a horizontal component current meter almost co-located to measure the wave field. This can be used in shallow or in deeper water if there is something to attach it to near the surface. Other techniques include arrays of surface-piercing wires, triaxial current meters, acoustic doppler current meters, and radars.

(d)   A mathematical description of the directional sea `state is feasible by assuming that the sea state can be considered as a superposition of a large number of regular sinusoidal wave components with different frequencies and directions. With this assumption, the representation of a spectrum in frequency and direction becomes a direct extension of the frequency spectrum alone, allowing the use of *FFT* method. It is often convenient to express the wave spectrum $E(F, \theta)$ describing the angular distribution of wave energy at respective frequencies by

$$E(f,\theta) = E(f) \; G(f,\theta) \tag{II-1-162}$$

where the function $G(f, \theta)$ is a dimensionless quantity, and is known as the *directional spreading function*. Other acronyms for $G(f, \theta)$ are the *spreading function, angular distribution function*, and the *directional distribution*.

(e)   The one-dimensional spectra may be obtained by integrating the associated directional spectra over θ as

$$E(f) = \int_{-\pi}^{\pi} E(f,\theta) \; d\theta \tag{II-1-163}$$

(f)   It therefore follows from the above last two equations that $G(f, \theta)$ must satisfy

$$\int_{-\pi}^{\pi} G(f,\theta) \; d\theta \; = \; 1 \tag{II-1-164}$$

(g)   The functional form of $G(f, \theta)$ has no universal shape and several proposed formulas are available. In the most convenient simplification of $G(f, \theta)$, it is customary to consider G to be independent of frequency f such that we have

$$G(\theta) \; = \; \frac{2}{\pi} \; \cos^2\theta \quad for \quad |\theta| \; < \; 90° \tag{II-1-165}$$

(h)   This cosine-squared distribution is due to St. Denis and Pierson (1953), and testing with field data shows that it reproduces the directional distribution of wave energy. Longuet-Higgins (1962) found the cosine-power form

$$G(\theta) \; = \; C(s) \; \cos^{2s} \frac{\theta \, - \, \bar{\theta}}{2}$$

$$C(s) \; = \; \frac{\sqrt{\pi}}{2\pi} \; \frac{\Gamma(s \, + \, 1)}{\Gamma\left(s \, + \, \dfrac{1}{2} \right)} \tag{II-1-166}$$

where θ is the principal (central) direction for the spectrum, s is a controlling parameter for the angular distribution that determines the peakedness of the directional spreading, $C(s)$ is a constant satisfying the normalization condition, θ is a counterclockwise measured angle from the principal wave direction, and Γ is the Gamma function.

(i)   Mitsuyasu et al. (1975), Goda and Suzuki (1976), and Holthuijsen (1983) have shown that for wind waves, the parameter s varies with wave frequency and is related to the stage of wave development (i.e., wind speed and fetch) by

$$s \; = \; \begin{cases} s_{max} \left(\dfrac{f}{f_p} \right)^{5} & for \; f \leq f_p \\[4mm] s_{max} \left(\dfrac{f}{f_p} \right)^{-2.5} & for \; f > f_p \end{cases} \tag{II-1-167}$$

where s_{max} and f_p are defined as

$$s_{max} \; = \; 11.5 \left(\frac{2\pi f_p U}{g} \right)^{-2.5}$$

$$\frac{2\pi f_p U}{g} \; = \; 18.8 \left(\frac{gF}{U^2} \right)^{-0.33} \tag{II-1-168}$$

(j)   In the above equations, U is the wind speed at the 10-m elevation above the sea surface and F is the fetch length. These equations remain to be validated with field data for wind waves. The parameter s for shallow-water waves may also vary spatially during wave transformation. This is due to refraction. A large value greater that 50, may be necessary if dependence of s_{max} on refraction is of concern. For deepwater applications where wind waves are jointly present with swells in deep water, Goda and Suzuki (1976) proposed the following values for s_{max}: 10 for wind waves, and 25 for swells present with wind waves of relatively large steepness, and 75 for swells with wind waves of small steepness. Under simple wind wave conditions, the spreading function may be approximated by the equations provided. They are typical of deepwater wind seas for which the wind has been constant. If the wind has shifted in direction, if there is significant swell, or if the waves are in shallow water, the directional distribution may be different than the shape functions presented.

(8) Wave groups and groupiness factors.

(a) Measurements of waves usually show a tendency of grouping between waves that is; high waves; often seem to be grouped together. Examination of the sea surface profile records indicates that wave heights are not uniform and they occur in successive groups of higher or lower waves. The interest in wave groups is stimulated by the fact that wave grouping and associated nonlinear effects play an important role in the long-period oscillation of moored vessels (Demirbilek 1988, 1989; Faltinsen and Demirbilek 1989), surf beats, irregular wave runup, resonant interaction between structures (Demirbilek and Halvorsen 1985; Demirbilek, Moe, and Yttervoil 1987;), and other irregular fluctuations of the mean water level nearshore (Goda 1985b; 1987). Unfortunately there is no way to predict grouping.

(b) Wave grouping is an important research topic and there are several ways to quantify wave grouping. These include the smoothed instantaneous wave energy history analysis (Funke and Mansard 1980), the concept of the run of wave heights (Goda 1976), and the Hilbert transform. A short exposition of the wave grouping analysis is provided here.

(c) The length of wave grouping can be described by counting the number of waves exceeding a specified value of the wave height which could be the significant, mean, or other wave height. The succession of high wave heights is called *a run* or *a run length* with an associated wave number j_1. The definition sketch for two wave groups is shown in Figure II-1-41 with the threshold wave height limit set at $H = H_c$. The recurrence interval or repetition length above the threshold value of wave height is called the *total run* denoted by j_2.

(d) The group occurrence for N waves with k number of lags between waves in a sequence in a record may be defined in terms of a correlation coefficient. The correlation coefficient R_H so defined will describe the correlation between wave heights as a function of the mean μ and standard deviation σ and is given by

$$R_H = \frac{1}{\sigma_0} \frac{1}{N-k} \sum_{i=1}^{N-k} (H_i - \mu)(H_{i+k} - \mu)$$

$$\sigma_0 = \frac{1}{N} \sum_{i=1}^{N} (H_i - \mu)^2$$

(II-1-169)

(e) Thus, R_H varies with the number of lags k between waves. If the succeeding waves are uncorrelated, then $R_H \rightarrow 0$ as $N \rightarrow \infty$. Real wave data indicate that $R_H(1) \approx 0.20$ to 0.40 while $R_H(k) \approx 0$ for $k > 1$. Furthermore, a positive value of R_H suggests that large waves tend to be succeeded by large waves, and small waves by other small waves.

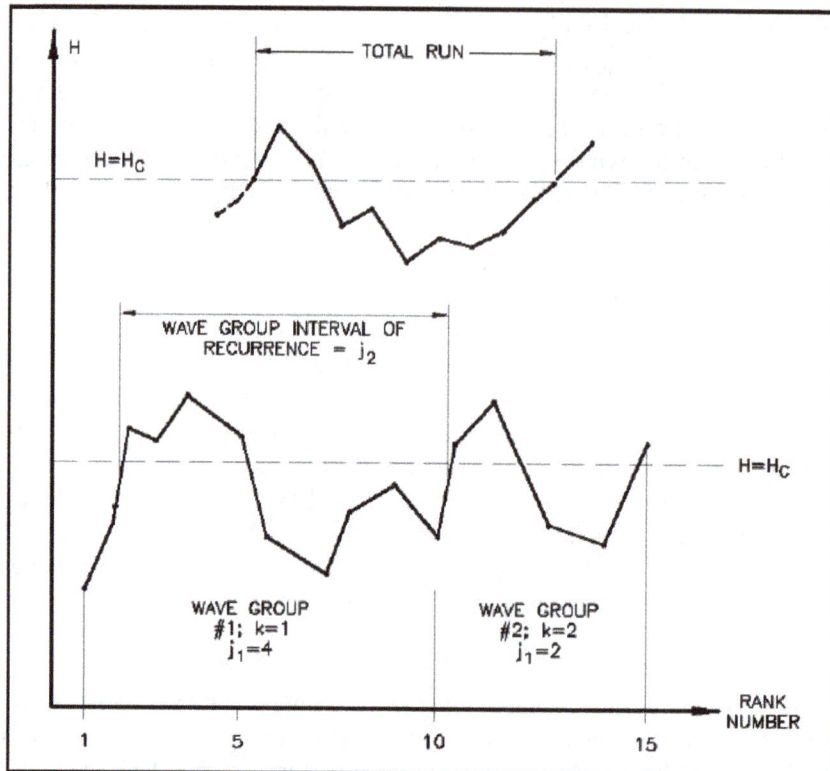

Figure II-1-41. Identification and description of wave groups through ordered statistics (Goda 1976)

(f)   Assuming that successive wave heights are uncorrelated, the probability of a run length j_1 is (Goda 1976)

$$P(j_1) = p^{(j_1-1)}(1-p) \qquad \text{(II-1-170)}$$

in which p is the occurrence probability for $H > H_c$. The mean and standard deviation of j_1 are

$$\mu_{j_1} = \frac{1}{q} \quad ; \quad q = 1 - p \quad ; \quad \sigma_{j_1} = \frac{\sqrt{1-q}}{q}$$

$$p = p(H > H_c) = \exp\left[-\frac{1}{8}\eta_c^2\right] \quad ; \quad \eta_c = \frac{H_c}{\sigma_\eta} \qquad \text{(II-1-171)}$$

(g)   The probability of a *total run* with the length j_2 can be derived by mathematical induction as

$$\mu_{j_2} = \frac{1}{p} + \frac{1}{q} \quad ; \quad \sigma_{j_2} = \sqrt{\frac{p}{q^2} + \frac{q}{p^2}} \qquad \text{(II-1-172)}$$

where it has been assumed that successive wave heights are uncorrelated. Successive wave heights of the real ocean waves are mutually correlated, and the degree of correlation depends on the sharpness of the spectral peak. The effect of spectral bandwidth on wave height distribution has been considered by Kimura (1980), Tayfun (1983a), and Longuet-Higgins (1984). Tayfun has shown that the parameter that best describes the spectral peakedness is the correlation coefficient of the wave envelope, relating wave height variation between successive wave heights. This coefficient R_{HH} may be calculated as (Tayfun 1983)

$$R_{HH} = \frac{E(\lambda) - (1-\lambda^2)\frac{K(\lambda)}{2} - \frac{\pi}{4}}{1 - \frac{\pi}{4}}$$

$$\lambda(\bar{T}) = \frac{1}{m_0}\sqrt{A^2 + B^2}$$

(II-1-173)

$$A = \int_0^\infty E(f)\cos 2\pi f\bar{T}\,df \quad ; \quad B = \int_0^\infty E(f)\sin 2\pi f\bar{T}\,df$$

(h)   By further assuming that Rayleigh distribution is suitable for the consecutive wave heights, the joint probability density function $p(H_1, H_2)$ for two successive wave heights H_1 and H_2 in the wave group may then be established. See Tayfun (1983) for details.

(i)   The correlation coefficient R_{HH} takes a value of about 0.2 for wind waves and 0.6 or greater for swells (Goda 1976), a clear indication that wind waves rarely develop significant grouping of high waves. Su (1984) has shown that the wave group containing the highest wave in a record is often longer than the ordinary groups of high waves, and that the extreme wave usually consists of three high waves with the highest greater than the significant wave height. Wave groups and their characteristics have been investigated by analyzing the successive wave groups (Goda 1976 and Kimura 1980).

(j)   Wave grouping and its consequences are of significant concern, but there is little guidance and few practical formulae for use in practical engineering. The engineer needs to be aware of its existence and, for designs that would be sensitive to grouping-related phenomena, attempt to evaluate its importance to the problem of concern. This may involve performing numerical simulations or physical model simulations in which a wide variety of wave conditions are tested and are designed to include those with high levels of groupiness. The procedures for this lie beyond the scope of the CEM.

(9)   *Random wave simulation.*

(a)   Given a one-dimensional parametric spectrum model or an actual wave energy density spectrum, it is sometimes necessary to use these quantities to calculate the height, period, and phase angle of a wave at a particular frequency. Such an approach for simulating random waves from a known wave spectra is sometimes termed the *deterministic spectral amplitude method*, since individual wave components in this superposition method are deterministic (Borgman 1967). The method is also called the *random phase method* because the phases of individual components are randomly chosen (Borgman 1969). Random waves simulated by this approach may not satisfy the condition of a Gaussian sea unless $N \rightarrow \infty$ in the limit. In practice, for $200 \le N \le 1200$ components, the spectrum can be duplicated accurately.

(b)   The wave profile generated by simulation methods is used in a number of engineering applications in spite of requiring a large number of components and considerable computer time. For example, random wave simulation is frequently used during modeling studies in a wave tank for duplicating a required target wave energy density spectrum. Random wave profiles are also extensively used in numerical models for calculating structural loads and responses due to a random sea. The simulation method permits direct prediction of the wave particle kinematics at any location in a specified water depth for given wave height-period pair and random phase angle. The ARMA algorithms (Spanos 1983) and digital simulation methods

(Hudspeth and Chen 1979) are two alternatives for simulating random waves from a given one-dimensional spectrum.

(c)   There are two ways for simulating wave surface profiles from known wave spectra; deterministic and non-deterministic spectral amplitude methods. In the deterministic spectral simulation method, the wave height, period, and phase angle associated with a frequency f_1 whose corresponding energy density is $E(f_1)$ may be obtained from

$$H(f_1) = H\big|_{f_1} = 2\sqrt{2E(f_1)\,\Delta f}$$

$$T(f_1) = T\big|_{f_1} = \frac{1}{f_1} \tag{II-1-174}$$

$$\epsilon(f_1) = \epsilon\big|_{f_1} = 2\pi r_N$$

where the phase angle ϵ is arbitrary since r_N is a random number between zero and one. The time series of the wave profile at a point x and time t may be computed by (Tucker et al. 1984)

$$\eta(x,t) = \sum_{n=1}^{N} H(n)\cos\big[k(n)x - 2\pi f(n)t + \epsilon(n)\big] \tag{II-1-175}$$

where $k(n) = 2\pi/L(n)$, and $L(n)$ is the wavelength corresponding to the nth frequency $f(n)$; N the total number of frequency bands of width Δf. It is not required to divide the spectrum curve equally, except that doing so greatly facilitates computations. The value of wave height is sensitive to the choice of Δf, but as long as Δf is small, this method produces a satisfactory random wave profile. The use of the equal increments, Δf, requires N to be greater than 50 to assure randomness and duplicating the spectrum accurately.

(d)   In the non-deterministic spectral amplitude method, the wave surface profile is represented in terms of two independent Fourier coefficients. These Gaussian distributed random variables a_n and b_n with zero mean and variance of $E(f)\,\Delta f$ are then obtained from

$$\eta(x,t) = \sum_{n=1}^{N} a_n\cos\big[k(n)x - 2\pi f(n)t\big]$$
$$+ \sum_{n=1}^{N} b_n\sin\big[k(n)x - 2\pi f(n)t\big] \tag{II-1-176}$$

(e)   In essence, an amplitude and a phase for individual components are replaced by two amplitudes, the coefficients of cosine and sine terms in the wave profile. This random coefficient scheme may yield a realistic representation of a Gaussian sea, provided that N is large for a true random sea. This method differs from the deterministic spectral amplitude approach by ensuring that sea state is Gaussian. Elgar et al. (1985) have considered simultaneous simulation of both narrow and broad-banded spectra using more than 1000 Fourier components, and concluded that both simulation methods yield similar statistics. These approaches may be extended to the two-dimensional case. This is beyond the scope of the CEM.

(10) Kinematics and dynamics of irregular waves. In the above sections of the CEM we have considered definition of irregular wave parameters and development of methods to measure them and use them analytically. Velocities, pressures, accelerations, and forces under irregular waves are estimated analytically in three ways. In the first, an individual wave is measured by either a wave-by-wave analysis or constructed synthetically (such as choosing, H_s, T_r, and a direction) and monochromatic theory is used to estimate the

desired quantities at a given wave phase (Faltinsen and Demirbilek 1989). In the second, pressure, velocity, and acceleration spectra are estimated by applying linear theory to translate the surface elevation spectra to the desired parameter (Dean and Dalrymple 1991). Finally, the random wave simulation technique may be used to synthetically generate a surface time history and corresponding kinematic and dynamic properties (Borgman 1990). Of the three methods, the last may provide the most realistic results, but it is also the most complex approach. These methods lie beyond the CEM and generally require the assistance of a knowledgeable oceanographic engineer.

II-1-4. References and Bibliography

Abramowitz and Stegun 1965
Abramowitz, M. and Stegun, I. A. 1965. *Handbook of Mathematical Functions*, Dover Pub., New York.

Airy 1845
Airy, G. B. 1845. *Tides and Waves*, Encyc. Metrop., Article 192, pp 241-396.

Barthel et al. 1983
Barthel, V. et al. 1983. "Group Bounded Long Waves in Physical Models," *Intl. Jour. Ocean Engr.*, Vol 10, pp 261-294.

Battjes 1970
Battjes, J. A. 1970. "Long-Term Wave Height Distribution at Seven Stations Around the British Isles," Rep. No.A44, Natl. Inst. Ocean., Godalming, U.K.

Battjes 1972
Battjes, J. A. 1972. "Set-Up due to Irregular Waves," *Proceedings of the 13th Coastal Engineering Conference*, American Society of Civil Engineers, pp 1993-2004.

Bendat and Piersol 1971
Bendat, J. S. and Piersol, A. G. 1971. *Random Data: Analysis and Measurement Procedures*, Wiley-Interscience.

Borgman 1963
Borgman, L. E. 1963. "Risk Criteria," *ASCE Jour. Waterw., Port, Coastal and Ocean Engr.*, Vol 89, pp 1-35.

Borgman 1967
Borgman, L. E. 1967. "Spectral Analysis of Ocean Wave Forces on Piling," *ASCE Jour. Waterw., Port, Coastal and Ocean Engr.*, Vol 93, pp 129-156.

Borgman 1969
Borgman, L. E. 1969. "Ocean wave Simulation for Engineering Design," *ASCE Jour. Waterw., Port, Coastal and Ocean Engr.*, Vol 95, pp 557-583.

Borgman 1975
Borgman, L. E. 1975. "Extremal Statistics in Ocean Engineering," *Proc. Civil Engr. in the Oceans*, Vol 1, pp 117-133.

Borgman 1990
Borgman, L. E. 1990. "Irregular Ocean waves: Kinematics and Forces," *The Sea*, B. Le Méhauté and D. M. Hanes, ed., Part A, Vol 9, pp 121-168.

Boussinesq 1871
Boussinesq, J. 1871. "Theorie de L'intumescence Liquide Appelee Onde Solitaire ou de Translation se Propageant dans un Canal Rectangulaire," *Comptes Rendus Acad. Sci. Paris,* Vol 72, pp 755-759.

Bouws, Gunther, Rorenthal, and Vincent 1984
Bouws, E., Gunther, H., Rorenthal, W., and Vincent, C. L. 1984. "Similarity of the Wind Wave Spectrum in Finite Depth Water, Part I-Spectral Form," *Jour. Geophys. Res.,* Vol 89, No. C1, pp 975-986.

Bouws, Gunther, and Vincent 1985
Bouws, E., Gunther, H., and Vincent, C. L. 1985. "Similarity of the wind wave spectrum in finite-depth water, Part I-Spectral form," *J. Geophys. Res.,* Vol 85, No. C3, pp 1524-1530.

Bowen 1969
Bowen, A. J. 1969. "The Generation of Longshore Currents on a Plain Beach," *Jour. Marine Res.,* Vol 27, pp 206-215.

Bretschneider 1959
Bretschneider, C. L. 1959. "Wave Variability and Wave Spectra for Wind-Generated Gravity Waves," U.S. Army Corps of Engrs., Beach Erosion Board, Tech. Memo. No. 118.

Bretschneider 1961
Bretschneider, C. L. 1961. "One-Dimensional Gravity Wave Spectrum," *Ocean Wave Spectra*, Prentice-Hall, Englewood Cliffs, NJ.

Bretschneider 1960
Bretschneider, C. L. 1960. "A Theory for Waves of Finite Height," *Proc. 7th Coastal Engr. Conf.,* Vol 1, pp 146-183.

Bretschneider 1969
Bretschneider, C. L. 1969. "Wave Forecasting," Chapter 11 in *Handbook of Ocean and Underwater Engineering*, J. J. Myers et al., eds., McGraw-Hill Book Co., New York.

Briggs et al. 1993
Briggs, M. J., Thompson, E. F., Green, D. R., and Lillycrop, L. S. 1993. "Laboratory Description of Harbor Idealized Tests," Vol 1, U.S. Army Engineer Waterways Experiment Station, CERC-93-1, Vicksburg, MS.

Byatt-Smith and Longuet-Higgins 1976
Byatt-Smith, J. G. B., and Longuet-Higgins, M. S. 1976. "On the Speed and Profile of Steep Solitary Waves," *Proc. Roy. Soc. London, Series* A, Vol 350, pp 175-189.

Cartwright and Longuet-Higgins 1956
Cartwright, D. E., and Longuet-Higgins, M. S. 1956. "The Statistical Distribution of the Maxima of a Random Function," *Proc. Roy. Soc. London, Series A,* Vol 237, pp 212-232.

Cavanie et al. 1978
Cavanie, A. et al. 1978. "A Statistical Relationship Between Individual Heights and Periods of Storm Waves," *Proc. BOSS'76 Conf.,* Vol 2, pp 354-360.

Chakrabarti 1987
Chakrabarti, S. K. 1987. "Hydrodynamics of Offshore Structures," WIT Press, Southampton, UK.

Chakrabarti and Cooley 1977
Chakrabarti, S. K. and Cooley, R. P. 1977. "Statistical Distribution of Periods and Heights of Ocean Waves," *Jour. Gephys. Res.,* Vol 82, pp 1363-1368.

Challenor and Srokosz 1984
Challenor, P. G., and Srokosz, M. A. 1984. "Extraction of Wave Period from Altimeter Data," *Proc. Workshop on Radar Alti. Data,* Vol 1, pp 121-124.

Chaplin 1980
Chaplin, J. R. 1980. "Developments of Stream-Function Wave Theory," *Coastal Engr.,* Vol 3, pp 179-205.

Chappelear 1961
Chappelear, J. E. 1961. "Direct Numerical Calculation of Wave Properties," *Journal of Geophysical Research,* Vol 66, pp 501-508.

Chappelear 1962
Chappelear, J. E. 1962. "Shallow Water Waves," *Journal of Geophysical Research,* Vol 67, pp 4693-4704.

Cialone et al. 1991
Cialone, M. A., Mark, D. J., Chou, L. W., Leenknecht, D. A., Davis, J. A., Lillycrop. L. S., Jensen, R.E., Thompson, Gravens, M. B., Rosati, J. D., Wise, R. A., Kraus, N. C., and Larson, P. M. 1991. "Coastal Modeling System (CMS) User's Manual." Instruction Report CERC-91-1, U.S. Army Engineer Waterways Experiment Station, Coastal Engineering Research Center, Vicksburg, MS.

Cokelet 1977
Cokelet, E. D. 1977. "Steep Gravity Waves in Water of Arbitrary Uniform Depth," *Phil. Trans. Roy. Soc. London,* Series A, Vol 286, pp 183-230.

Collins 1970
Collins, J. I. 1970. "Probabilities of Breaking Wave Characteristics," *Proceedings of the 12th Coastal Engineering Conference,* American Society of Civil Engineers, pp 399-412.

Collins 1972
Collins, J. I. 1972. "Prediction of Shallow-Water Wave Spectra," *Journal of Geophysical Research,* Vol 77, pp 2693-2707.

Cooley and Tukey 1965
Cooley, J. W. and Tukey, J. W. 1965. "An Algorithm for the Machine Computation of Complex Fourier Series," *Math. Comp.,* Vol 19, pp 297-301.

Crapper 1984
Crapper, G. D. 1984. *Introduction to Water Waves,* John Wiley & Sons, New York.

Dalrymple 1974a
Dalrymple, R. A. 1974. "A Finite Amplitude Wave on a Linear Shear Current," *Journal of Geophysical Research*, Vol 79, pp 4498-4504.

Dalrymple 1974b
Dalrymple, R. A. 1974. "Models for Nonlinear Water Waves in Shear Currents," *Proc. 6th Offshore Tech. Conf.*, Paper No. 2114.

Dalrymple and Dean 1975
Dalrymple, R. A. and Dean R. G. 1975. "Waves of Maximum Height on Uniform Currents," *ASCE Jour. Waterw., Port, coastal and Ocean Engr.*, Vol 101, pp 259-268.

Dalrymple and Solana 1986
Dalrymple, R. A., and Solana, P. 1986. "Nonuniqueness in stream function wave theory, *J. Waterway, Port, Coastal and Ocean Engr., ASCE*, Vol 112, pp 333-337.

De 1955
De, S. C. 1955. "Contribution to the Theory of Stokes Waves," *Proc. Phil. Soc.*, Vol 51, pp 713-736.

Dean 1965
Dean, R. G. 1965. "Stream Function Representation of Nonlinear Ocean Waves," *Journal of Geophysical Research*, Vol 70, pp 4561-4572.

Dean 1968
Dean, R. G. 1968. "Relative Validity of Water Wave Theories," *Proc. Civil Engr. in Ocean, ASCE*, Vol 1, pp1-30.

Dean 1974
Dean, R. G. 1974. *Evaluation and Development of Water Wave Theories for Engineering Applications*, Coastal Engineering Research Center, U.S. Army Engineer Waterways Experiment Station, Special Report No.1.

Dean and Dalrymple 1991
Dean, R. G. and Dalrymple, R. A. 1991. *Water Wave Mechanics for Engineers and Scientists*, World Scientific Pub. Co., Teaneck, NJ.

Demirbilek 1977
Demirbilek, Z. 1977. "Transverse Oscillations of a Circular Cylinder in Uniform Flow," Thesis Rept., Naval Postgraduate School, Monterey, Calif.

Demirbilek 1985
Demirbilek, Z. 1985. "A Computer Program for Wave Kinematics by Fenton's Theory," Proprietary of Conoco Inc., prepared for the Jolliet Project. Also in Conoco's Riser and Mooring Systems Analysis Programs, Houston, TX.

Demirbilek 1988
Demirbilek, Z. 1988. "Forces on Marine Risers in a Coexisting Environment," *J. Waterway, Port, Coastal, and Ocean Engineering*, Vol 114, No. 3, pp 346-362.

Demirbilek 1989
Demirbilek, Z. 1989. *Tension Leg Platform: A State-Of-The-Art Review*, New York, pp 335.

Demirbilek and Halvorsen 1985
Demirbilek, Z. and Halvorsen, T. 1985. "Hydrodynamic Forces on Multitube Production Risers Exposed to Currents and Waves," *Trans. ASME, J. Energy Resources Technology,* Vol 107, No. 2, pp 226-234.

Demirbilek, Moe, and Yttervoll 1987
Demirbilek, Z., Moe, G., and Yttervoll, P. 1987. "Morison's Formula: Relative Velocity versus Independent Flow Fields Formulations for a Case Representing Fluid Damping," *Proc. Intl. Symposium on Offshore Mech. and Arctic Engineering,* Vol II, pp 25-32, Houston, TX.

Eckart 1952
Eckart, C. 1952. "The Propagation of Gravity Waves From Deep to Shallow Water," Natl. Bur. Standards, Circular 521, Washington, DC, pp 165-173.

Elgar et al. 1985
Elgar, S. et al. 1985. "Wave Group Statistics from Numerical Simulation of a Random Sea," *Jour. Applied Ocean Res.,* Vol 7, pp 93-96.

Faltinsen 1990
Faltinsen, O. M. 1990. *Sea Loads on Ships and Offshore Structures*, Cambridge Press, New York, p 328.

Faltinsen and Demirbilek 1989
Faltinsen, O. M., and Demirbilek, Z. 1989. "Hydrodynamics of Tension Leg Platforms - TLP's," *Tension Leg Platform: A State-of-the-art Review*, Z. Demirbilek, ed., ASCE Publ., New York.

Fenton 1972
Fenton, J. D. 1972. "A Ninth-Order Solution for Solitary Waves," *Jour. Fluid Mech.,* Vol 53, pp 257-271.

Fenton 1979
Fenton, J. D. 1979. "A Higher-Order Cnoidal Wave Theory," *Jour. Fluid Mech.,* Vol 94, pp 129-161.

Fenton 1985
Fenton, J. D. 1985. "A Fifth-Order Stokes Theory for Steady Waves," *ASCE Jour. Waterw., Port, Coastal and Ocean Engr.,* Vol 111, pp 216-234.

Fenton 1988
Fenton, J. D. 1988. "The Numerical solution of Steady Water Wave Problem," *Jour. Comp. and Geo.,* Vol 14, pp 357-368.

Fenton and McKee 1990
Fenton, J. D. and McKee, W. D. 1990. "On Calculating the Lengths of Water Waves," *Coastal Engr.,* Vol 14, pp 499-513.

Forristall 1984
Forristall, G. Z. 1984. "The Distribution of Measured and Simulated Wave Heights as a Function of Spectral Shape," *Journal of Geophysical Research*, Vol 89, No. C6, pp 10547-10552.

Funke and Mansard 1980
Funke, E. R. and Mansard, E. P. D. 1980. "On the Synthesis of Realistic Sea State," *Proc. 17th Coastal Engr. Conf.,* Vol 2, pp 2974-2991.

Goda 1974
Goda, Y. 1974. "Estimation of wave Statistics from Spectra Information," *Proc. Intl. Symp. on Ocean Wave Meas. and Anal., ASCE,* Vol 1, pp 320-337.

Goda 1975
Goda, Y. 1975. "Irregular Wave Deformation in the Surf Zone," *Coastal Engineering in Japan,* Vol 18, pp 13-26.

Goda 1976
Goda, Y. 1976. "On Wave Groups," *Proc. BOSS'76,* Vol 1, pp 115-128.

Goda 1978
Goda, Y. 1978. "The Observed Joint Distribution of Periods and Heights of Sea Waves," *Proc. 16th Coastal Engr. Conf.,* Vol 2, pp 227-246.

Goda 1983
Goda, Y. 1983. "A Unified Nonlinearity Parameter of Water Waves," *Rept. Port and Harbor Res. Inst. of Japan,* Vol 22, No. 3, pp 3-30.

Goda 1985a
Goda, Y. 1985a. "Numerical Examination of Several Statistical Parameters of Sea Waves," *Rept. Port and Harbor Res. Inst.,* Vol 24, No. 4, pp 65-102.

Goda 1985b
Goda, Y. 1985b. *Random Seas and Design of Maritime Structures,* Univ. of Tokyo Press, Tokyo, Japan.

Goda 1987
Goda, Y. 1987. "Statistical Variability of Sea State Parameters as a Function of Wave Spectrum," *Proc. IAHR Seminar on Wave Anal. and Gen. in Lab. Basins,* Lausanne, Switzerland.

Goda and Suzuki 1976
Goda, Y. and Suzuki, Y. 1976. "Estimation of Incident and Reflected Waves in Random Wave Experiments," *Proc. 15th Coastal Engr. Conf.,* Vol 1, pp 828-845.

Gumbel 1958
Gumbel, E. J. 1958. *Statistics of Extremes,* Columbia Univ. Press, New York.

Hasselmann et al. 1973
Hasselmann et al. 1973. "Measurements of Wind-Wave Growth and Swell Decay During the Joing North Sea Wave Project (JONSWAP)," *Deutsche Hydrograph.* Zeit., Erganzungsheft Reihe A (8⁰), No. 12.

Hasselmann et al. 1976
Hasselmann, K. et al. 1976. "A Parametric Wave Prediction Model," *Jour. Phys. Ocean.,* Vol 6, pp 200-228.

Havelock 1918
Havelock, T. H. 1918. "Periodic Irrotational Waves of Finite Height," *Trans. Royal Soc. of London,* Vol 95 series A, pp 37-51.

Hedges 1981
Hedges, T. S. 1981. "Some Effects of Currents on Wave Spectra," *Proc., 1st Indian Conf. on Ocean Engr.,* Vol 1, pp 30-35.

Hedges 1987
Hedges, T. S. 1987. "Combinations of Waves and Currents: An Introduction," *Proc. Inst. Civil Engrs.,* Vol 82, pp 567-585.

Hoffman and Walden 1977
Hoffman, D., and Walden, D. A. 1977. "Environmental Wave Data for Determining Hull Structural Loadings," SSC-268, Final Tech. rep., Ship Struc. Comm., U.S. Coast Guard, Washington, DC.

Holthuijsen 1983
Holthuijsen, L. H. 1983. "Observation of the Directional Distribution of Ocean-Wave Energy in Fetch-Limited Conditions," *Jour. Phys. Ocean.,* Vol 13, No. 2, pp 191-207.

Horikawa and Kuo 1966
Horikawa, K., and Kuo, C. T. 1966. "A Study of Wave Transformation Inside Surf Zone," *Proc. 10th Coastal Engr. Conf.,* Vol 1, pp 217-233.

Huang et al. 1981
Huang, N. E., et al. 1981. "A Unified Two-Parameter Wave Spectral Model for a General Sea State," *Jour. Fluid Mech.,* Vol 112, pp 203-224.

Huang and Hudspeth 1984
Huang, M-C., and Hudspeth, R. T. 1984. "Stream Function Solutions for Steady Water Waves," *Continental Shelf Res.,* Vol 3, pp 175-190.

Huang and Long 1980
Huang, N. E., and Long, S. R. 1980. "An Experimental Study of the Surface Elevation Probability Distribution and Statistics of Wind-Generated Waves," *Jour. Fluid Mech.,* Vol 101, pp 179-200.

Hudspeth and Chen 1979
Hudspeth, R. T., and Chen, M. C. 1979. "Digital Simulation of Nonlinear Random Waves," *ASCE Jour. Waterw., Port, Coastal and Ocean Engr.,* Vol 105, pp 67-85.

Hughes and Borgman 1987
Hughes, S. A., and Borgman, L. E. 1987. "Beta-Rayleigh Probability Distribution for Shallow Water Wave Heights," *Coastal Hydro.,* R. A. Dalrymple, ed., Vol 1, pp 17-31.

Hunt 1979
Hunt, J. N. 1979. "Direct Solution of Wave Dispersion Equation," *ASCE Jour. Waterw., Port, Coastal and Ocean Engr.,* Vol 105, pp 457-459.

IAHR 1986
IAHR 1986. *List of Sea State Parameters*, Intl. Assoc. Hydr. Res., Suppl. to Bull. No. 52, Brussels.

Ippen 1966
Ippen, A.T., 1966. *Estuary and Coastline Hydrodynamics,* McGraw-Hill Book Company, Inc, New York.

Isaacson 1978
Isaacson, M. de St. Q. 1978. "Mass Transport in Shallow Water Waves," *ASCE Jour. Waterw., Port, Coastal and Ocean Engr.,* Vol 104, No. 2, pp 215-225.

ISSC 1964
ISSC 1964. *Proceedings of the Second International Ship Structures Congress*, Delft, The Netherlands.

ITTC 1966
ITTC 1966. "Recommendations of the 11th International Towing Tank Conference," *Proc. 11th ITTC,* Tokyo, Japan.

Jasper 1956
Jasper, N. H. 1956. "Statistical Distribution Patterns of Ocean Waves and Wave-Induced Ship Stress and Motions with Engineering Applications," *Trans. SNAME,* Vol 64.

Jonsson 1990
Jonsson, I. G. 1990. "Wave-Current Interactions," *The Sea,* B. Le Méhauté and D. M. Hanes, eds., Part A, Vol 9, pp 65-120.

Kadono et al. 1986
Kadono, T., et al. 1986. "An Observation of Nonlinear Coastal Waves," *Proc. 33rd Japanese Coastal Engr. Conf.,* Vol 1, pp 149-153.

Keller 1948
Keller, J. B. 1948. "The Solitary Wave and Periodic Waves in Shallow Water," *Commun. Appl. Math.,* Vol 1, pp 323-339.

Keulegan and Patterson 1940
Keulegan, G. H. and Patterson, G. W. 1940. "Mathematical Theory of Irrotational Translation Waves," *Res. Jour. Natl. Bur. Stand.,* Vol 24, pp 47-101.

Kimura 1980
Kimura, A. 1980. "Statistical Properties of Random Wave Groups," *Proc. 1th Coastal Engr. Conf.,* Vol 2, pp 2955-2973.

Kinsman 1965
Kinsman, B. 1965. *Wind Waves,* Prentice-Hall, Englewood Cliffs, NJ.

Kitaigorodoskii et al. 1975
Kitaigorodoskii, S. A. et al. 1975. "On Phillips' Theory of equilibrium Range in the Spectra of Wind-Generated Gravity Waves," *Jour. Phys. Ocean.,* Vol 5, No. 3, pp 410-420.

Klopman 1990
Klopman, G. 1990. "A note on integral properties of periodic gravity waves int eh case of non-zero mean Eulerian velocity," *J. Fluid Mech,* Vol 211, pp 609-615.

Korteweg and de Vries 1895
Korteweg, D. J. and de Vries, G. 1895. "On the Change of Form of Long Waves Advancing in a Rectangular Canal, and on a New Type of Stationary Waves," *Phil. Mag.,* 5th Series, Vol 39, pp 422-443.

Kuo and Kuo 1974
Kuo, C. T., and Kuo, S. T. 1974. "Effect of Wave Breaking on Statistical Distribution of Wave Heights," *Proceedings of Civil Engineering Oceans,* pp 1211-1231.

Laitone 1960
Laitone, E. V. 1960. "The Second Approximation to Cnoidal and Solitary Waves," *Jour. Fluid Mech.,* Vol 9, pp 430-444.

Laitone 1962
Laitone, E. V. 1962. "Limiting Conditions for Cnoidal and Stokes Waves," *Journal of Geophysical Research,* Vol 67, pp 1555-1564.

Laitone 1965
Laitone, E. V. 1965. "Series Solutions for Shallow Water Waves," *Journal of Geophysical Research,* Vol 70, pp 995-998.

Lamb 1945
Lamb, H. 1945. *Hydrodynamics,* 6th ed., Dover, New York.

Le Méhauté 1976
Le Méhauté, B. 1976. *Introduction to Hydrodynamics and Water Waves,* Springer-Verlag, New York.

Le Méhauté and Webb 1964
Le Méhauté, B., and Webb, L. M. 1964. "Periodic Gravity Waves Over a Gentle Slope at a Third Order of Approximation," *Proc. 10th Coastal Engr. Conf.,* Vol 1, pp 23-40.

Leenknecht, Szuwalski, and Sherlock 1992
Leenknecht, D. A., Szuwalski, A. and Sherlock, A. R. 1992. "Automated Coastal Engineering System, User Guide and Technical Reference, Version 1.07," Coastal Engineering Research Center, U.S. Army Engineer Waterways Experiment Station, Vicksburg, MS 39180.

Liu 1971
Liu, P. C. 1971. "Normalized and Equilibrium Spectra of Wind-Wave in Lake Michigan," *Jour. Phys. Ocean.,* Vol 1 No. 4, pp 249-257.

Liu 1984
Liu, P.-L. F. 1984. "Wave-Current Interactions on a Slowly Varying Topography," *Journal of Geophysical Research,* Vol 88, pp 4421-4426.

Liu and Dalrymple 1978
Liu, P.-L. F., and Dalrymple, R. A. 1978. "Bottom Friction Stresses and Longshore Currents due to Waves with Large Angles of Incidence," *Jour. Marine Res.,* Vol 36, pp 357-375.

Longuet-Higgins 1952
Longuet-Higgins, M.S. 1952. "On the Statistical Distribution of the Wave Heights of Sea Waves," *Jour. Marine Res.,* Vol 11, pp 245-266.

Longuet-Higgins 1953
Longuet-Higgins, M. S. 1953. "Mass Transport in Waves," *Phil. Trans. Roy. Soc. London, Series A,* Vol 245, pp 535-581.

Longuet-Higgins 1957
Longuet-Higgins, M.S. 1957. "Statistical Analysis of a Random, Moving Surface," *Phil Trans. Roy. Soc. London, Series A,* Vol 249, pp 321-387.

Longuet-Higgins 1962
Longuet-Higgins, M.S. 1962. "The Distribution of Intervals Between Zeros of Random Function," *Phil. Trans. Roy. Soc. London,* Series A, Vol 254, pp 557-599.

Longuet-Higgins 1970
Longuet-Higgins, M. S. 1970. "Longshore Currents Generated by Obliquely Incident Sea Waves," *Journal of Geophysical Research,* Vol 75, pp 6778-6789.

Longuet-Higgins 1974
Longuet-Higgins, M. S. 1974. "On the Mass, Momentum, Energy, and Circulation of a Solitary Wave," *Proc. Roy. Soc. London, Series A,* Vol 337, pp 1-13.

Longuet-Higgins 1975a
Longuet-Higgins, M. S. 1975a. "Longshore Currents Generated by Obliquely Incident Sea Waves," *Journal of Geophysical Research,* Vol 75, pp 6778-6789.

Longuet-Higgins 1975b
Longuet-Higgins, M. S. 1975b. "On the Joint Distribution of the Periods and Amplitudes of Sea Waves," *Journal of Geophysical Research,* Vol 80, pp 2688-2694.

Longuet-Higgins 1976
Longuet-Higgins, M. S. 1976. "Recent Developments in the Study of Breaking Waves," *Proc. 15th Coastal Engr. Conf.,* Vol 1, pp 441-460.

Longuet-Higgins 1980
Longuet-Higgins, M. S. 1980. "On the Joint Distribution of Heights of Sea Waves: Some Effects of Nonlinearity and Finite Band Width," *Journal of Geophysical Research,* Vol 85, pp 1519-1523.

Longuet-Higgins 1983
Longuet-Higgins, M. S. 1983. "On the Joint Distribution of Wave Period and Amplitudes in a Random Wave Field," *Proc. Roy. Soc. London, Series A,* Vol 389, pp 241-258.

Longuet-Higgins 1984
Longuet-Higgins, M. S. 1984. "Statistical Properties of Wave Groups in a Random Sea State," *Phil. Trans. Roy. Soc. London, Series A,* Vol 310, pp 219-250.

Longuet-Higgins 1985
Longuet-Higgins, M. S. 1985. "A New Way to Calculate Steep Gravity Waves," *The Ocean Surface*, Y. Toba and H. Mitsuyasu, eds., Univ. of Tokyo Press, pp 1-15.

Longuet-Higgins and Fenton 1974
Longuet-Higgins, M. S., and Fenton, J. D. 1974. "On Mass, Momentum, Energy, and Calculation of a Solitary Wave," *Proc. Roy. Soc. London, Series A,* Vol 340, pp 471-493.

Longuet-Higgins and Fox 1977
Longuet-Higgins, M. S., and Fox, M. J. H. 1977. "Theory of the Almost Highest Wave: the Inner Solution," *Jour. Fluid Mech.,* Vol 80, pp 721-741.

Longuet-Higgins and Stewart 1962
Longuet-Higgins, M. S., and Stewart, R. W. 1962. "Radiation Stress and Mass Transport in Gravity Waves, with Application to 'Surf Beats'," *Journal of Fluid Mechanics,* Vol 13, No. 4, pp 481-504.

Longuet-Higgins and Stewart 1964
Longuet-Higgins, M. S., and Stewart, R. W. 1964. "Radiation Stresses in Water Waves: A Physical Discussion with Applications," *Deep Sea Research,* Vol 11, pp 529-562.

Mei 1991
Mei, C. C. 1991. *The Applied Dynamics of Ocean Surface Waves*, World Scientific Pub. Co., Teaneck, NJ.

Miche 1944
Miche, R. 1944. "Mouvements Ondulatoires des Mers en Profondeur Constante on Decroisante," *Annales des Ponts et Chaussees,* pp 25-78, 131-164, 270-292, 369-406.

Michell 1893
Michell, J. H. 1893. "On the Highest Wave in Water," *Phil. Mag.,* Vol 36, pp 430-435.

Miles 1979
Miles, J. W. 1979. "On the Korteweg de Vries Equation for Gradually Varying Channel," *Jour. Fluid Mech.,* Vol 91, pp 181-190.

Miles 1980
Miles, J. W. 1980. "Solitary Waves," *Annual Rev. Fluid Mech.,* Vol 12, pp 11-43.

Miles 1981
Miles, J. W. 1981. "The Korteweg de Vries Equation: A Historical Essay," *Jour. Fluid Mech.,* Vol 106, pp 131-147.

Milne-Thompson 1976
Milne-Thompson, L. M. 1976. *Theoretical Hydrodynamics*, The MacMillan Press Ltd., London.

Mitsuyasu 1971
Mitsuyasu, H. 1971. "On the Form of Fetch-Limited Wave Spectrum," *Coastal Engr. in Japan,* Vol 14, pp 7-14.

Mitsuyasu 1972
Mitsuyasu, H. 1972. "The One-Dimensional Wave Spectra at Limited Fetch," *Proc. 13th Coastal Engr. Conf.*, Vol 1, pp 289-306.

Mitsuyasu et al. 1975
Mitsuyasu, H. et al. 1975. "Observations of the Directional spectrum of Ocean Waves Using a Cloverleaf Buoy," *Jour. Phys. Ocean.*, Vol 5, pp 750-760.

Morison et al. 1950
Morison, J. R., et al. 1950. "The Forces Exerted by Surface Waves on Piles," *Petroleum Trans. AIME*, Vol 189, pp 149-157.

Munk 1949
Munk, W. H. 1949. "The Solitary Wave Theory and Its Application to Surf Problems," *Annals New York Acad. Sci.*, Vol 51, pp 376-423.

Myrhaug and Kjeldsen 1986
Myrhaug, D. and Kjeldsen, S. P. 1986. "Steepness and Asymmetry of Extreme Waves and the Highest Waves in Deep Water," *Intl. Jour. Ocean Engr.*, Vol 13, pp 549-568.

Nishimura et al. 1977
Nishimura, H., et al. 1977. "Higher Order Solutions of Stokes and Cnoidal Waves," *Jour. Faculty Engr. Univ. of Tokyo*, Series B, Vol 34, pp 267-293.

Ochi 1973
Ochi, M. K. 1973. "On Prediction of Extreme Values," *Jour. Ship Res.*, Vol 1, pp 29-37.

Ochi 1982
Ochi, M. K. 1982. "Stochastic Analysis and Probabilistic Prediction of Random Seas," *Advances in Hydrosci,*, V. T. Chow, ed., Academic Press, Vol 13, pp 217-375.

Ochi and Hubble 1976
Ochi, M. K. and Hubble, E. N. 1976. "Six Parameter Wave Spectra," *Proc. 15th Coastal Engr. Conf.*, Vol 1, pp 301-328.

Ottesen-Hansen 1980
Ottesen-Hansen, N.-E. 1980. "Correct Reproduction of Group-Induced Long Waves," *Proc. 17th Coastal Engr. Conf.*, Vol 1, pp 784-800.

Peregrine 1972
Peregrine, D. H. 1972. "Equations for Water Waves and the Approximation Behind Them," *Waves on Beaches and Resulting Sediment Transport*, R. E. Meyer, ed., Academic Press, pp 95-121.

Peregrine 1976
Peregrine, D. H. 1976. "Interaction of Water Waves and Currents," *Advances in Applied Mechanics*, Academic Press, New York, Vol 16, pp 9-117.

Phillips 1958
Phillips, O. M. 1958. "On the Generation of Waves by Turbulent Wind," *Jour. Fluid Mech.*, Vol 2, pp 417-445.

Phillips 1977
Phillips, O. M. 1977. *The Dynamics of the Upper Ocean*, 2nd ed., Cambridge University Press.

Pierson and Moskowitz 1964
Pierson, W. J., and Moskowitz, L. 1964. "A Proposed Spectral Form for Fully-Developed Wind Sea Based on the Similarity Law of S. A. Kitaigorodoskii," *Journal of Geophysical Research*, Vol 69, pp 5181-5203.

Price and Bishop 1974
Price, W. G., and Bishop, R. E. D. 1974. *Probabilistic Theory of Ship Dynamics*, Chapman and Hall, London.

Rayleigh 1876
Rayleigh, L. 1876. "On Waves," *Phil. Mag.,* Vol 1, pp 257-279.

Rice 1944-1945
Rice, S. O. 1944-1945. "Mathematical Analysis of Random Noise," *Bell System Tech. Jour.,* Vol 23, pp 282-332.; Vol 24, pp 45-156.

Rienecker and Fenton 1981
Rienecker, M. M. and Fenton, J. D. 1981. "A Fourier Representation Method for Steady Water Waves," *Jour. Fluid Mech.,* Vol 104, pp 119-137.

Russel and Osorio 1958
Russel, R. C. H., and Osorio, J. D. C. 1958. "An Experimental Investigation of Drift Profiles in a Closed Channel," *Proc. 6th Coastal Engr. Conf.,* Vol 1, pp 171-193.

Russell 1838
Russell, J. S. 1838. "Report of the Committee on Waves," *Rep. Meet. British Assoc. Adv. Sci 7ᵗʰ,* Liverpool, 1837, John Murray, London, pp 417-496.

Russell 1844
Russell, J. S. 1844. "Report on Waves," *14th Meeting Brit. Assoc. Adv. Sci.,* pp 311-390.

Sakai and Battjes 1980
Sakai, T., and Battjes, J. A. 1980. "Wave Shoaling Calculated from Cokelet's Theory," *Coastal Engr.,* Vol 4, pp 65-84.

Sand 1982
Sand, S. S. 1982. "Long Wave Problems in Laboratory Models," *ASCE Jour. Waterw., Port, Coastal and Ocean Engr.,* Vol 108, pp 492-503.

Sarpkaya and Isaacson 1981
Sarpkaya, T. and Isaacson, M. 1981. *Mechanics of Wave Forces on Offshore Structures*, Van Nostrand Reinhold Co., New York.

Schwartz 1974
Schwartz, L. W. 1974. "Computer Extension and Analytic Continuation of Stokes' Expansion for Gravity Waves," *Jour. Fluid Mech.,* Vol 62, pp 553-578.

Scott 1965
Scott, J. R. 1965. "A Sea Spectrum for Model Tests and Long-Term Ship Prediction," *Jour. Ship Res.*, Vol 9, pp 145-152.

Shore Protection Manual 1984
Shore Protection Manual. 1984. 4th ed., 2 Vol. U.S. Army Engineer Waterways Experiment Station, U.S. Government Printing Office, Washington, DC.

Silvester 1974
Silvester, R. 1974. *Coastal Engineering*, Vols I & II, Elsevier, Amsterdam.

Skjelbreia and Hendrickson 1961
Skjelbreia, L., and Hendrickson, J. 1961. "Fifth Order Gravity Wave Theory," *Proc. 7th Coastal Engr. Conf.*, Vol 1, pp 184-196.

Sobey 1990
Sobey, R. J. 1990. "Wave Theory Predictions of Crest Kinematics," *Water Wave Kinematics*, A. Torum and O. T. Gudmestad, eds., Kluwer Acad. Pub, The Netherlands, pp 215-231.

Sobey et al. 1987
Sobey, R. J., et al. 1987. "Application of Stokes, Cnoidal, and Fourier Wave Theories," *ASCE Jour. Waterw., Port, Coastal and Ocean Engr.*, Vol 113, pp 565-587.

Spanos 1983
Spanos, P. D. 1983. "ARMA Algorithms for Ocean Wave Modeling," *ASME Jour. Energy Resources Tech.*, Vol 105, pp 300-309.

St. Denis and Pierson 1953
St. Denis, M., and Pierson, W. J. 1953. "On the Motion of Ships in Confused Seas," *Trans. SNAME*, Vol 61, pp 280-357.

Stoker 1957
Stoker, J. J. 1957. *Water Waves, The Mathematical Theory with Applications*, Interscience, New York.

Stokes 1846
Stokes, G. G. 1846. "Report on Recent Research in Hydrodynamics," Mathematical and Physical Paper (1880), Vol 1, pp 167-187, Cambridge University Press.

Stokes 1847
Stokes, G. G. 1847. "On the Theory of Oscillatory Waves," *Trans. Camb. Phil. Soc.*, Vol 8, pp 441-455.

Stokes 1880
Stokes, G. G. 1880. *Math. Phys. Papers*, Vol 1, Camb. Univ. Press.

Su 1984
Su, M.-Y. 1984. "Characteristics of Extreme Wave Groups," *Oceans '84 Conf.*, Washington, DC.

Sverdrup and Munk 1947
Sverdrup, H. U., and Munk, W. H. 1947. *Wind, Sea and Swell: Theory of Relations for Forecasting*, U.S. Navy Hydro. Office, Publication No. 601.

Tayfun 1981
Tayfun, A. 1981. "Distribution of Crest-to-Trough Wave height," *ASCE Jour. Waterw., Port, Coastal and Ocean Engr.,* Vol 107, pp 149-158.

Tayfun 1983a
Tayfun, A. 1983a. "Effects of Spectrum Bandwidth on the Distribution of Wave Heights and Periods," *Intl. Jour. Ocean Engr.,* Vol 10, pp 107-118.

Tayfun 1983b
Tayfun, A. 1983b. "Nonlinear Effects on the Distribution of Crest-to-Trough Wave Heights," *Intl. Jour. Ocean Engr.,* Vol 10, pp 97-106.

Tayfun, Dalrymple, and Yang 1976
Tayfun, A., Dalrymple, R. A. and Yang, C. Y. 1976. "Random Wave-Current Interaction in Water of Varying Depth," *Ocean Engr.,* Vol 3, pp 403-420.

Thom 1961
Thom, H. C. S. 1961. "Distribution of Extreme Winds in the United States," *Trans ASCE,* Vol 126, Part II.

Thom 1973
Thom, H. C. S. 1973. "Distribution of Extreme Winds Over Ocean," and "Extreme Wave Height Distribution Over Oceans," *ASCE Jour. Waterw., Port, Coastal and Ocean Engr.,* Vol 99, pp 1-17 and 355-374.

Thompson 1977
Thompson, E. F. 1977. "Wave Climate at Selected Locations Along U.S. Coasts," Technical Report TR 77-1, U. S. Army Engineer Waterways Experiment Station, Vicksburg, MS.

Thompson and Vincent 1985
Thompson, E. F., and Vincent, C. L. 1985. "Significant wave Height for Shallow Water Design," *ASCE Jour. Waterw., Port, Coastal and Ocean Engr.,* Vol 111, pp 828-842.

Thornton and Guza 1983
Thornton, E. B., and Guza, R. T. 1983. "Transformation of Wave Height Distribution," *Journal of Geophysical Research,* Vol 88, No. C10, pp 5925-5938.

Tucker 1963
Tucker, M. J. 1963. "Analysis of Records of Sea Waves," *Proc. Inst. Civil Engrs.,* Vol 26, pp 305-316.

Tucker et al. 1984
Tucker, M. J., et al. 1984. "Numerical Simulation of a Random Sea: A Common Error and Its Effect Upon Wave Group Statistics," *Jour. Applied Ocean Res.,* Vol 6, pp 118-122.

Tung and Huang 1976
Tung, C. C. and Huang, N. E. 1976. "Interaction Between Waves and Currents and Their Influence on Fluid Forces," *Proc. BOSS'76,* Vol 1, pp 129-143.

Ursell 1953
Ursell, F. 1953. "The Long Wave Paradox in the Theory of Gravity Waves," *Proc. Camb. Philos. Soc.,* Vol 49, pp 685-694.

Van Dorn 1965

Van Dorn, W. G. 1965. "Source Mechanism of the Tsunami of March 28, 1964 in Alaska," *Proc. 9th Coastal Engr. Conf., ASCE,* pp 166-190.

Venezian 1980

Venezian, G. 1980. Discussion of "Direct Solution of Wave Dispersion Equation," by J.N. Hunt, *ASCE Jour. Waterw., Port, Coastal and Ocean Engr.,* Vol 106, pp 501-502.

Venezian and Demirbilek 1979

Venezian, G., and Demirbilek, Z. 1979. "Pade Approximants for Water Waves," Texas Engr. and Exp. Station Research Center Notes, College Station, Texas.

Vincent 1984

Vincent, C. L. 1984. "Shallow Water Waves: A Spectral Approach," *Proc. 19th Coastal Engr. Conf.,* Vol 1, pp 370-382.

Vincent 1985

Vincent, C. L. 1985. "Depth-Controlled Wave Height," *ASCE Jour. Waterw., Port, Coastal and Ocean Engr.,* Vol 111, pp 459-475.

Vincent and Hughes 1985

Vincent, C. L. and Hughes, S. A. 1985. "Wind Wave Growth in Shallow Water," *ASCE Jour. Waterw., Port, Coastal and Ocean Engr.,* Vol 111, pp 765-770.

Wang 1985

Wang, W. C. 1985. "Non-Gaussian Characteristics of Waves in Finite Water Depth," Rept. No.85-009, Univ. of Florida, Gainesville, FL.

Wehausen and Laitone 1960

Wehausen, J. V., and Laitone, E. V. 1960. "Surface Waves," *Handbuch der Physik,* S. Flugge, ed., Springer-Verlag, Berlin, Vol IX, pp 446-778.

Whitham 1967

Whitham, G. B. 1967. "Variational Methods and Applications to Water Waves," *Proc. Roy. Soc. London, Series A,* Vol 299, pp 6-25.

Whitham 1974

Whitham, G. B. 1974. *Linear and Nonlinear Waves,* Wiley Interscience, New York.

Wiegel 1954

Wiegel, R. L. 1954. *Gravity Waves, Tables of Functions*, Univ. of Calif., Council on Wave Res., The Engr. Found., Berkeley, CA.

Wiegel 1960

Wiegel, R. L. 1960. "A Presentation of Cnoidal Wave Theory for Practical Application," *Jour. Fluid Mech.,* Vol 7, pp 273-286.

Wiegel 1964

Wiegel, R. L. 1964. *Oceanographical Engineering*, Prentice-Hall, Englewood Cliffs, NJ.

Williams 1981
Williams, J. M. 1981. "Limiting Gravity Waves in Water of Finite Depth," *Phil. Trans. Roy. Soc. London, Series A,* Vol 302, pp 139-188.

Williams 1985
Williams, J. M. 1985. *Tables of Progressive Gravity Waves*, Pitman, Boston.

Wirsching 1986
Wirsching, P. 1986. "Reliability Methods in Mechanical and Structural Design," Lecture Notes on Short Course for Conoco R&D Dept., Ponca City, OK.

Wu and Liu 1984
Wu, C-S., and Liu, P.-L. F. 1984. "Effects of Nonlinear Inertial Forces on Nearshore Currents," *Coastal Engr.,* Vol 8, pp 15-32.

Wu and Thornton 1986
Wu, C-S., and Thornton, E. B. 1986. "Wave Numbers of Linear Progressive Waves," *ASCE Jour. Waterw., Port, Coastal and Ocean Engr.,* Vol 112, pp 536-540.

Yu 1952
Yu, Y.-Y. 1952. "Breaking of Waves by an Opposing Current," *Trans. Amer. Geophys. Union,* Vol 33, No.1, pp 39-41.

II-1-5. Definitions of Symbols

α	Dimensionless scaling parameter used in the JONSWAP spectrum for fetch-limited seas
α	Phillips' constant (= 0.0081) (Equation II-1-153)
α_x, α_z	Fluid particle accelerations [length/time^2]
γ	Peak enhancement factor used in the JONSWAP spectrum for fetch-limited seas
Γ	Gamma function
Δp	Difference in pressure at a point due to the presence of the solitary wave [force/length^2]
Δt	Sampling interval (Equation II-1-144) [time]
ϵ	Dimensionless pertubation expansion parameter
ϵ	Spectral bandwidth used in the probability distribution for waves (Equation II-1-150)
ϵ	Wave steepness (= H/L)
ζ	Vertical displacement of the water particle from its mean position (Equation II-1-27) [length]
η	Displacement of the water surface relative to the SWL [length]
$\eta(t)$	Sea state depicted in time series of the wave profile [length]
$\eta(x,t)$	Time series of the wave profile at a point x and time t (Equation II-1-175) [length]
$\overline{\eta}$	Mean or expected value of the sea state (Equation II-1-119) [length]
$\eta_{envelope}$	Envelope wave form of two or more superimposed wave trains (Equation II-1-48) [length]
η_{rms}	Root-mean-square surface elevation [length]
θ	Angle between the plane across which energy is being transmitted and the direction of wave advance [deg]
θ	Principal (central) direction for the spectrum measured counterclockwise from the principal wave direction [deg]
$\lambda_{1,2}$	Spectral shape parameters controlling the shape and the sharpness of the spectral peak of the Ochi-Hubble spectral model
μ_η	Mean or expected value of the sea state (Equation II-1-119) [length]
ν	Dimensionless Spectral width parameter
ξ	Horizontal displacement of the water particle from its mean position (Equation II-1-26) [length]

ρ	Mass density of water (salt water = 1,025 kg/m^3 or 2.0 slugs/ft^3; fresh water = 1,000kg/m^3 or 1.94 slugs/ft^3) [force-time^2/length^4]
ρ_η	Autocorrelation coefficient (Equation II-1-122)
σ_η	Standard deviation or square root of the variance
Φ	Velocity potential [length^2/time]
Φ	Weighing factor (Equation II-1-159)
Ψ	Stream function
ω	Wave angular or radian frequency (= 2π/T) [time^{-1}]
a	Wave amplitude [length]
A, B	Major- (horizontal) and minor- (vertical) ellipse semi-axis of wave particle motion (Equations II-1-34 and II-1-35) [length]. The lengths of A and B are measures of the horizontal and vertical displacements of the water particles (Figure II-1-4).
B_j	Dimensionless Fourier coefficients (Equation II-1-103)
C	Phase velocity or wave celerity (= L/T = ω/k) [length/time]
$C(s)$	Dimensionless constant satisfying the normalization condition
C_g	Wave group velocity [length/time]
cn	Jacobian elliptic cosine function
d	Water depth [length]
E	Total wave energy in one wavelength per unit crest width [length-force/length^2]
$E(\omega)$	Phillips' equilibrium range of the spectrum for a fully-developed sea in deep water (Equation II-1-153)
$E(f)$	One-dimensional spectrum or frequency energy spectrum or wave energy spectral density (Equation II-1-144)
$\overline{E}$	Total average wave energy per unit surface area or specific energy or energy density (Equation II-1-58) [length-force/length^2]
$\overline{E}_k$	Kinetic energy per unit length of wave crest for a linear wave (Equation II-1-53) [length-force/length^2]
$\overline{E}_p$	Potential energy per unit length of wave crest for a linear wave (Equation II-1-55) [length-force/length^2]
$E[\eta]$	Mean or expected value of the sea state (Equation II-1-119) [length]
F	Fetch length [length]
f_p	Frequency of the spectral peak used in the JONSWAP spectrum for fetch-limited seas [time^{-1}]

g	Gravitational acceleration [length/time^2]
$G(f,\theta)$	Dimensionless directional spreading function
H	Wave height [length]
$\overline{H}$	Mean wave height [length]
$H_{1/3}$	Significant wave height [length]
$H_{1/n}$	The average height of the largest $1/n$ of all waves in a record [length]
H_d	Design wave height [length]
H_j	Ordered individual wave heights in a record (Equation II-1-115) [length]
H_{max}	Maximum wave height [length]
H_{rms}	Root-mean-square of all measured wave heights [length]
H_s	Significant wave height [length]
k	Modulus of the elliptic integrals
k	Number of lags between waves in a sequence in a record (Equation II-1-169)
k	Wave number ($= 2\pi/L = 2\pi/CT$) [length^{-1}]
$K(k)$	Complete elliptic integral of the first kind
K_z	Pressure response factor (Equation II-1-43) [dimensionless]
L	Wave length [length]
M	Dimensionless parameter which is a function of H/d used in calculating water particle velocities for a solitary wave (Equations II-1-92 & II-1-93) (Figure II-1-17).
$m_{0,1,2,4}$	Moments of the wave spectrum
N	Dimensionless correction factor in determination of η from subsurface pressure (Equation II-1-46)
N	Dimensionless parameter which is a function of H/d used in calculating water particle velocities for a solitary wave (Equations II-1-92 & II-1-93) (Figure II-1-17).
N	Number of waves in a record
N_c	Number of crests in the wave record
N_z	Number of zero-upcrossings in the wave record
$-_0$	The subscript 0 denotes deepwater conditions
p	Pressure at any distance below the fluid surface [force/length^2]
P	Probability

$p(x)$ — Probability density

$P(x)$ — Probability distribution function - fraction of events that a particular event is not exceeded (Equation II-1-124)

$\overline{P}$ — Wave power or average energy flux per unit wave crest width transmitted across a vertical plane perpendicular to the direction of wave advance (Equation II-1-59) [length-force/time-length]

p_a — Atmospheric pressure [force/length^2]

p' — Total or absolute subsurface pressure -- includes dynamic, static, and atmospheric pressures (Equation II-1-39) [force/length^2]

$Q(x)$ — Probability of exceedence (Equation II-1-128)

Q_p — Spectral peakedness parameter proposed by Goda (1974) (Equation II-1-151)

R — Bernoulli constant (Equation II-1-102)

R — Cross-correlation coefficient - measures the degree of correlation between two signals (Equation II-1-123)

R_η — Autocorrelation or autocovariance function of the sea state (Equation II-1-121)

R_H — Correlation coefficient describing the correlation between wave heights as a function of μ and standard deviation σ (Equation II-1-169)

R_{HH} — Correlation coefficient of the wave envelope, relating wave height variation between successive wave heights (Equation II-1-173)

s — Dimensionless controlling parameter for the angular distribution that determines the peakedness of the directional spreading

T — Wave period [time]

$\overline{T}$ — Mean wave period [time]

$\overline{T_c}$ — Mean crest period [time]

$\overline{T_z}$ — Mean zero-upcrossing wave period [time]

T_c — Average wave period between two neighboring wave crests (Equation II-1-116) [time]

T_p — Wave period associated with the largest wave energy [time]

T_r — Sampling record length [time]

T_r — Total wave record length [time]

T_z — Zero-crossing wave period (Equation II-1-116) [time]

T_*^{max} — Most probable maximum wave period (Equation II-1-141) [time]

u — Fluid velocity (water particle velocity) in the x-direction [length/time]

U	Current speed [length/time]
U	Wind speed at the 10-m elevation above the sea surface [length/time]
$\overline{U}(z)$	Mass transport velocity (Equation II-1-69) [length/time]
u_{max}	Maximum fluid velocity in the horizontal direction [length/time]
U_P	Universal parameter for classification of wave theories
U_R	Dimensionless Ursell number (Equation II-1-67)
U_w	Wind speed at 19.5 m above mean sea level (Equation II-1-155) [length/time]
V	Volume of water within the wave above the still-water level per unit crest (Equation II-1-90) [length^3/length of crest]
w	Fluid velocity (water particle velocity) in the z-direction [length/time]
y_c	Vertical distance from seabed to the wave crest (Equation II-1-79) [length]
y_s	Vertical distance from seabed to the water surface (Equation II-1-77) [length]
y_t	Vertical distance from seabed to the wave trough [length]
z	Water depth below the SWL (Figure II-1-1) [length]

II-1-6. Acknowledgments

Authors of Chapter II-1, "Water Wave Mechanics:"

Zeki Demirbilek, Ph.D., Coastal and Hydraulics Laboratory (CHL), Engineer Research and Development Center, Vicksburg, Mississippi.
C. Linwood Vincent, Ph.D., Office of Naval Research, Arlington, Virginia.

Reviewers:

Robert A Dalrymple, Ph.D., Center for Applied Coastal Research, University of Delaware, Newark, Delaware.
Yoshimi Goda, Ph.D., Yokohama National University, Yokohama, Japan (emeritus).
Lee E. Harris, Ph.D., Department of Marine and Environmental Systems, Florida Institute of Technology, Melbourne, Florida.
Bernard LeMéhauté, Ph.D., University of Miami, Miami, Florida, (deceased).
Philip L.-F. Liu, Ph.D., School of Civil and Environmental Engineering, Cornell University, Ithaka, New York.
J. Richard Weggel, Ph.D., Dept. of Civil and Architectural Engineering, Drexel University, Philadelphia, Pennsylvania.
Robert L. Wiegel, University of California at Berkeley, Berkeley, California, (emeritus).

Table of Contents

Page

II-2-1. Meteorology II-2-1
   a. Introduction II-2-1
     (1) Background II-2-1
     (2) Organized scales of motion in the atmosphere II-2-1
     (3) Temporal variability of wind speeds II-2-3
   b. General structure of winds in the atmosphere II-2-5
   c. Winds in coastal and marine areas II-2-7
   d. Characteristics of the atmospheric boundary layer II-2-8
   e. Characteristics of near-surface winds II-2-9
   f. Estimating marine and coastal winds II-2-11
     (1) Wind estimates based on near-surface observations II-2-11
     (2) Wind estimates based on information from pressure fields and weather maps II-2-15
   g. Meteorological systems and characteristic waves II-2-23
   h. Winds in hurricanes II-2-27
   i. Step-by-step procedure for simplified estimate of winds for wave prediction II-2-34
     (1) Introduction II-2-34
     (2) Wind measurements II-2-34
     (3) Procedure for adjusting observed winds II-2-34
       (a) Level II-2-34
       (b) Duration II-2-34
       (c) Overland or overwater II-2-36
       (d) Stability II-2-36
     (4) Procedure for adjusting winds from synoptic weather charts II-2-36
       (a) Geostrophic wind speed II-2-36
       (b) Level and stability II-2-36
       (c) Duration II-2-36
     (5) Procedure for estimating fetch II-2-36

II-2-2. Wave Hindcasting and Forecasting II-2-37
   a. Introduction II-2-37
   b. Wave prediction in simple situations II-2-43
     (1) Assumptions in simplified wave predictions II-2-44
       (a) Deep water II-2-44
       (b) Wave growth with fetch II-2-44
       (c) Narrow fetches II-2-45
       (d) Shallow water II-2-45
     (2) Prediction of deepwater waves from nomograms II-2-46
     (3) Prediction of shallow-water waves II-2-47
   c. Parametric prediction of waves in hurricanes II-2-47

II-2-3. Coastal Wave Climates in the United States II-2-50
   a. Introduction II-2-50
   b. Atlantic coast II-2-52
   c. Gulf of Mexico II-2-56
   d. Pacific coast II-2-58
   e. Great lakes II-2-58

II-2-4. Additional Example Problem II-2-61

II-2-5. References II-2-63

II-2-6. Definitions of Symbols II-2-70

II-2-7. Acknowledgments II-2-72

List of Tables

Page

Table II-2-1. Ranges of Values for the Various Scales of Organized Atmospheric Motions II-2-2

Table II-2-2. Local Seas Generated by Various Meteorological Phenomena II-2-25

Table II-2-3. Wave Statistics in the Atlantic Ocean II-2-53

Table II-2-4. Wave Statistics in the Gulf of Mexico II-2-57

Table II-2-5. Wave Statistics in the Pacific Ocean II-2-59

Table II-2-6. Wave Statistics in the Great Lakes II-2-60

List of Figures

Page

Figure II-2-1. Ratio of wind speed of any duration U_t to the 1-hr wind speed U_{3600} II-2-4

Figure II-2-2. Duration of the fastest-mile wind speed as a function of wind speed U_f
(for open terrain conditions) II-2-5

Figure II-2-3. Equivalent duration for wave generation as a function of fetch and wind speed II-2-6

Figure II-2-4. Wind profile in atmospheric boundary layer II-2-7

Figure II-2-5. Coefficient of drag versus wind speed II-2-11

Figure II-2-6. Ratio of wind speed at any height to the wind speed at the 10-m height as a
function of measurement height for selected values of air-sea temperature
difference and wind speed: a) ΔT=+3°C; b) ΔT=0°C; c) ΔT=-3°C II-2-12

Figure II-2-7. Ratio R_L of wind speed over water U_W to wind speed over land U_L as a function
of wind speed over land U_L (after Resio and Vincent (1977)) II-2-14

Figure II-2-8. Amplification ratio R_T accounting for effects of air-sea temperature difference . . . II-2-15

Figure II-2-9. Surface synoptic chart for 0030Z, 27 October 1950 . . . . II-2-16

Figure II-2-10. Surface synoptic weather charts for the Halloween storm of 1991 . . . . II-2-17

Figure II-2-11. Key to plotted weather report II-2-18

Figure II-2-12. Geostrophic (free air) wind scale (after Bretschneider (1952)) II-2-20

Figure II-2-13. Ratio of wind speed at a 10-m level to wind speed at the top of the boundary
layer as a function of wind speed at the top of the boundary layer, for selected
values of air-sea temperature difference II-2-22

Figure II-2-14. Ratio of U_*/U_g as a function of U_g, for selected values of air-sea temperature
difference II-2-22

Figure II-2-15. Common wind direction conventions II-2-23

Figure II-2-16. Climatological variation in Holland's "B" factor (Holland 1980) II-2-30

Figure II-2-17. Relationship of estimated maximum wind speed in a hurricane at 10-m
elevation as a function of central pressure and forward speed of storm (based
on latitude of 30 deg, R_{max}=30 km, 15- to 30-min averaging period) II-2-32

Figure II-2-18. Definition of four radial angles relative to direction of storm movement II-2-32

Figure II-2-19. Horizontal distribution of wind speed along Radial 1 for a storm with forward velocity V_F of (a) 2.5 m/s; (b) 5 m/s; (c) 7.5 m/s II-2-33

Figure II-2-20. Logic diagram for determining wind speed for use in wave hindcasting and forecasting models II-2-35

Figure II-2-21. Phillips constant versus fetch scaled according to Kitaigorodskii. Small-fetch data are obtained from wind-wave tanks. Capillary-wave data were excluded where possible. (Hasselmann et al. 1973) II-2-42

Figure II-2-22. Definition of JONSWAP parameters for spectral shape II-2-43

Figure II-2-23. Fetch-limited wave heights II-2-46

Figure II-2-24. Fetch-limited wave periods II-2-47

Figure II-2-25. Duration-limited wave heights II-2-48

Figure II-2-26. Duration-limited wave periods II-2-49

Figure II-2-27. Maximum value of H_{mo} in a hurricane as a function of V_{max} and forward velocity of storm (Young 1987) II-2-50

Figure II-2-28. Values of $H_{mo}/H_{mo\,max}$ plotted relative to center of hurricane (0,0). II-2-51

Figure II-2-29. Reference locations for Tables II-2-3 through II-2-6 . . . . II-2-52

Chapter II-2
Meteorology and Wave Climate

II-2-1. Meteorology

a. Introduction.

(1) Background.

(a) A basic understanding of marine and coastal meteorology is an important component in coastal and offshore design and planning. Perhaps the most important meteorological consideration relates to the dominant role of winds in wave generation. However, many other meteorological processes (e.g., direct wind forces on structures, precipitation, wind-driven coastal currents and surges, the role of winds in dune formation, and atmospheric circulations of pollution and salt) are also important environmental factors to consider in man's interactions with nature in this sometimes fragile, sometimes harsh environment.

(b) The primary driving mechanisms for atmospheric motions are related either directly or indirectly to solar heating and the rotation of the earth. Vertical motions are typically driven by instabilities created by direct surface heating (e.g., air mass thunderstorms and land-sea breeze circulations), by advection of air into a region of different ambient air density, by topographic effects, or by compensatory motions related to mass conservation. Horizontal motions tend to be driven by gradients in near-surface air densities created by differential heating (for example north-south variations in incoming solar radiation, called insolation, and differences in the thermal response of ocean and continental areas), and by compensatory motions related to conservation of mass. The general structure and circulation of the earth's atmosphere is described in many excellent textbooks (Hess 1959)

(c) The rotation of the earth influences all motions in the earth's coordinate system. The net effect of the earth's rotation is to deflect all motion to the right in the Northern Hemisphere and to the left in the Southern Hemisphere. The strength of this deflection (termed Coriolis acceleration) is proportional to the sine of the latitude. Hence Coriolis effects are strongest in polar regions and vanish at the equator. Coriolis effects become significant when the trajectory of an individual fluid/gas particle moves over a distance of the same order as the Rossby radius of deformation, defined as

$$R_o = \frac{c}{f} \qquad\qquad (\text{II-2-1})$$

where

R_o = Rossby radius of deformation

f = Coriolis parameter defined as $1.458 \times 10^{-4} \sin \phi$, where ϕ is latitude (note f here is in $\sec^{-1}$)

c = characteristic velocity of the particle

For a velocity of 10 m/s at a latitude of 45 deg, R_o is about 100 km. This suggests that scales of motion with this velocity and with particle excursions of about 10 km and greater will begin to be significantly affected by Coriolis at this latitude.

(2) Organized scales of motion in the atmosphere.

(a) Table II-2-1 presents ranges of values for the various scales of organized atmospheric motions. This table should be regarded only as approximate spatial and temporal magnitudes of typical motions characteristic of these scales, and not as any specific limits of these scales. As can be seen in this table, the smallest scale of motion involves the transfer of momentum via molecular-scale interactions. This scale of motion is extremely ineffective for momentum transport within the earth's atmosphere and can usually be neglected at all but the slowest wind speeds and/or extremely small portions of some boundary layers. The next larger scale is that of turbulent momentum transfer. Turbulence is the primary transfer mechanism for momentum passing from the atmosphere into the sea; consequently, it is of extreme importance to most scientists and engineers. The next larger scale is that of organized convective motions. These motions are responsible for individual thunderstorm cells, usually associated with unstable air masses.

Table II-2-1
Ranges of Values for the Various Scales of Organized Atmospheric Motions

Transfer Mechanism	Typical Length Scale, meters	Typical Time Scale, sec
Molecular	$10^{-7} - 10^{-2}$	10^{-1}
Turbulent	$10^{-2} - 10^{3}$	10^{1}
Convective	$10^{3} - 10^{4}$	10^{3}
Meso-scale	$10^{4} - 10^{5}$	10^{4}
Synoptic-scale	$10^{5} - 10^{6}$	10^{5}
Large	$> 10^{6}$	10^{6}

(b) The next larger scale is termed the meso-scale. Meso-scale motions such as land-sea breeze circulations, coastal fronts, and katabatic winds (winds caused by cold air flowing down slopes due to gravitational acceleration) are important components of winds in near-coastal areas. Important organized meso-scale motions also exist in frontal regions of extratropical storms, within the spiral bands of tropical storms, and within tropical cloud clusters. An important distinction between meso-scale motions and smaller-scale motions is the relative importance of Coriolis accelerations. In meso-scale motions, the lengths of trajectories are sufficient to allow Coriolis effects to become important, whereas the trajectory lengths at smaller scales are too small to allow for significant Coriolis effects. Consequently, the first signs of trajectory curvature are found in meso-scale motions. For example, the land-breeze/sea-breeze system in most coastal areas of the United States does not simply blow from sea to land during the day and from land to sea at night. Instead, the wind direction tends to rotate clockwise throughout the day, with the largest rotation rates occurring during the transition periods when one system gives way to the next.

(c) The next larger scale of atmospheric motion is termed the synoptic scale. To many engineers and scientists, the synoptic scale is synonymous with the term storm scale, since the major storms in ocean areas occupy this niche in the hierarchy of scales. Storms that originate outside of tropical areas (extratropical storms) take their energy from horizontal instabilities created by spatial gradients in air density. Storms originating in tropical regions gain their energy from vertical fluxes of sensible and latent heat. Both the extratropical (or frontal) storms and tropical storms form closed or semi-closed trajectory motions around their circulation centers, due to the importance of Coriolis effects at this scale.

(d) The next larger scale of atmospheric motions is termed large scale. This scale of motion is more strongly influenced by thermodynamic factors than by dynamic factors. Persistent surface temperature differentials over large regions of the globe produce motions that can persist for very long time periods. Examples of such phenomena are found in subtropical high pressure systems, which are found in all oceanic areas and in seasonal monsoonal circulations developed in certain regions of the world.

(e) Scales of motion larger than large scale can be termed interannual scale, and beyond that, climatic scale. El Niño Southern Oscillation (ENSO) episodes, variations in year-to-year weather, changes in storm patterns and/or storm intensity, and long-term (secular) climatic variations are all examples of these longer-term scales of motion. The effects of these phenomena on engineering and planning considerations are very poorly understood at present. This is compounded by the fact that there does not even exist any real consensus among atmospheric scientists as to what mechanism or mechanisms control these variations. This may not diminish the importance of climatic variability but certainly detracts from the ability to treat it objectively. As better information is collected over longer time intervals, these scales of motion will be better understood.

(3) Temporal variability of wind speeds.

(a) Winds at any point on the earth represent a superposition of various atmospheric scales of motion, all interacting to produce local weather phenomena. Each scale plays a specific role in the transfer of momentum in the atmosphere. Due to the combination of different scales of motion, winds are rarely, if ever, constant for any prolonged interval of time. Because of this, it is important to recognize the averaging interval (explicit or implicit) of any data used in applications. For example, some winds represent "fastest mile" estimates, some winds represent averages over small, fixed time intervals (typically from 1 to 30 min), and some estimates (such as those derived from synoptic pressure fields) can even represent average winds over intervals of several hours. Design and planning considerations require different averages for different purposes. Individual gusts may contribute to the failure mode of some small structures or of certain structural elements on larger structures. For other structures, 1-min (or even longer) average wind speeds may be more related to critical structural forces.

(b) When dealing with wave generation in water bodies of differing sizes, different averaging intervals may also be appropriate. In small lakes and reservoirs or in riverine areas, a 1- to 5-min wind speed may be all that is required to attain a fetch-limited condition. In this case, the fastest 1- to 5-min wind speed will produce the largest waves, and thus be the appropriate choice for design and planning considerations. In large lakes and oceanic regions, the wave generation process tends to respond to average winds over a 15- to 30-min interval. Consequently, it is important in all applications to be aware of and use the proper averaging interval for all wind information.

(c) Figure II-2-1 shows the estimated ratio of winds of various durations to 1-hr average wind speeds. The proper application of Figure II-2-1 would be in converting extremal estimates of wind speeds from one averaging interval to another. For example, this graph shows that a 100-sec extreme wind speed is expected to be 1.2 times as high as a 1-hr extreme wind speed. This means that the highest average wind speed in 36 samples of 100-sec duration is expected to be 1.2 times higher than the average for all 36 samples added together.

(d) Occasionally, wind measurements are reported as fastest-mile wind speeds. The averaging time is the time required for the wind to travel a distance of 1 mile. The averaging time, which varies with wind speed, can be estimated from Figure II-2-2. Note that two axis are provided, for metric and English units.

(e) Figure II-2-3 shows the estimated time to achieve fetch-limited conditions as a function of wind speed and fetch length, based on the calculations of Resio and Vincent (1982). The proper averaging time for design and planning considerations varies dramatically as a function of these parameters. At first, it might not seem intuitive that the duration required to achieve fetch-limited conditions should be a function of wind speed; however, this comes about naturally due to the nonlinear coupling among waves in a wind-generated wave spectrum. The importance of nonlinear coupling is discussed further in the wave prediction section of this chapter. The examples are intended to illustrate the correct usage of figures and tables. Numerical values given in the solution of the examples were read from figures as approximate values or rounded off from the equations. Users need to use their own estimates and professional judgement when applying figures or equations to their particular engineering conditions or projects.

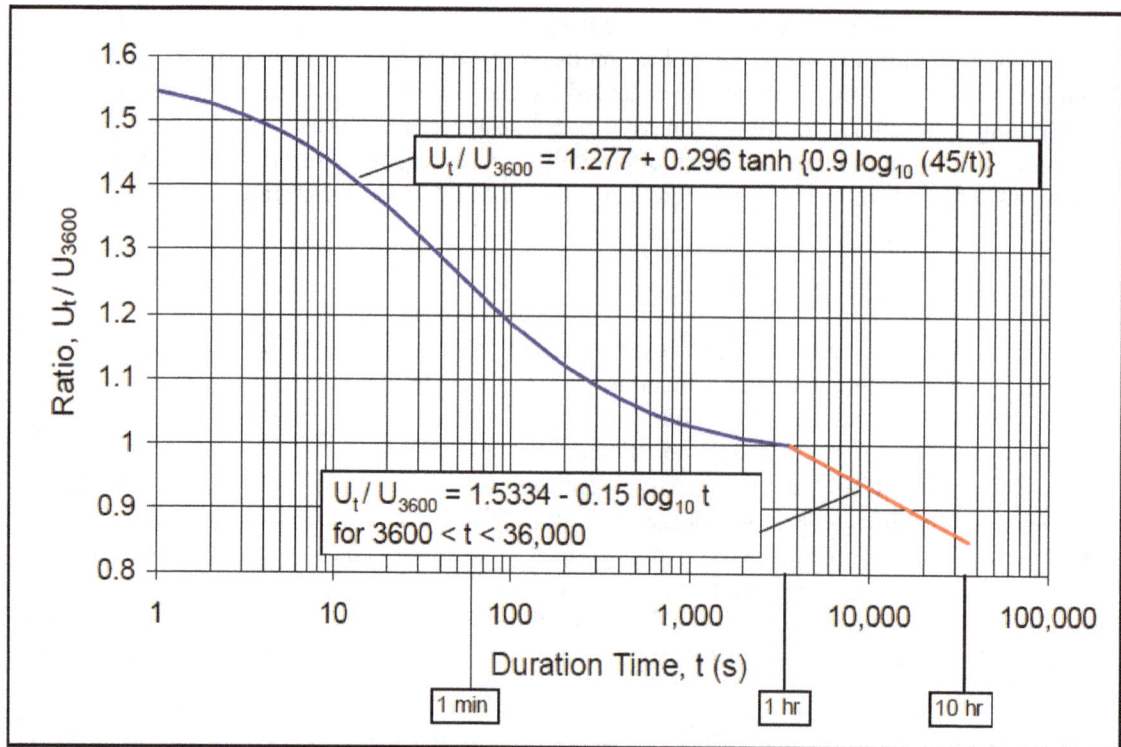

Figure II-2-1. Ratio of wind speed of any duration U_t to the 1-hr wind speed U_{3600}

EXAMPLE PROBLEM II-2-1

FIND:
    1-hr average winds for wave prediction

GIVEN:
    10-, 50-, and 100-year values of observed winds at a buoy located in the center of a large lake (U_{10} = 20.3 m/s, U_{50} = 24.8 m/s, U_{100} = 28.2 m/s). It is also known that the averaging interval for the buoy winds is 5 min.

SOLUTION:
    Using Figure II-2-1, the ratio of the fastest 5-min wind speed to the average 1-hr wind speed is approximately 1.09. Using this as a constant conversion factor, the 10-, 50-, and 100-year, 1-hr wind speeds are estimated as U'_{10}=18.6 m/s, U'_{50}=22.8 m/s, and U'_{100}=25.9 m/s.

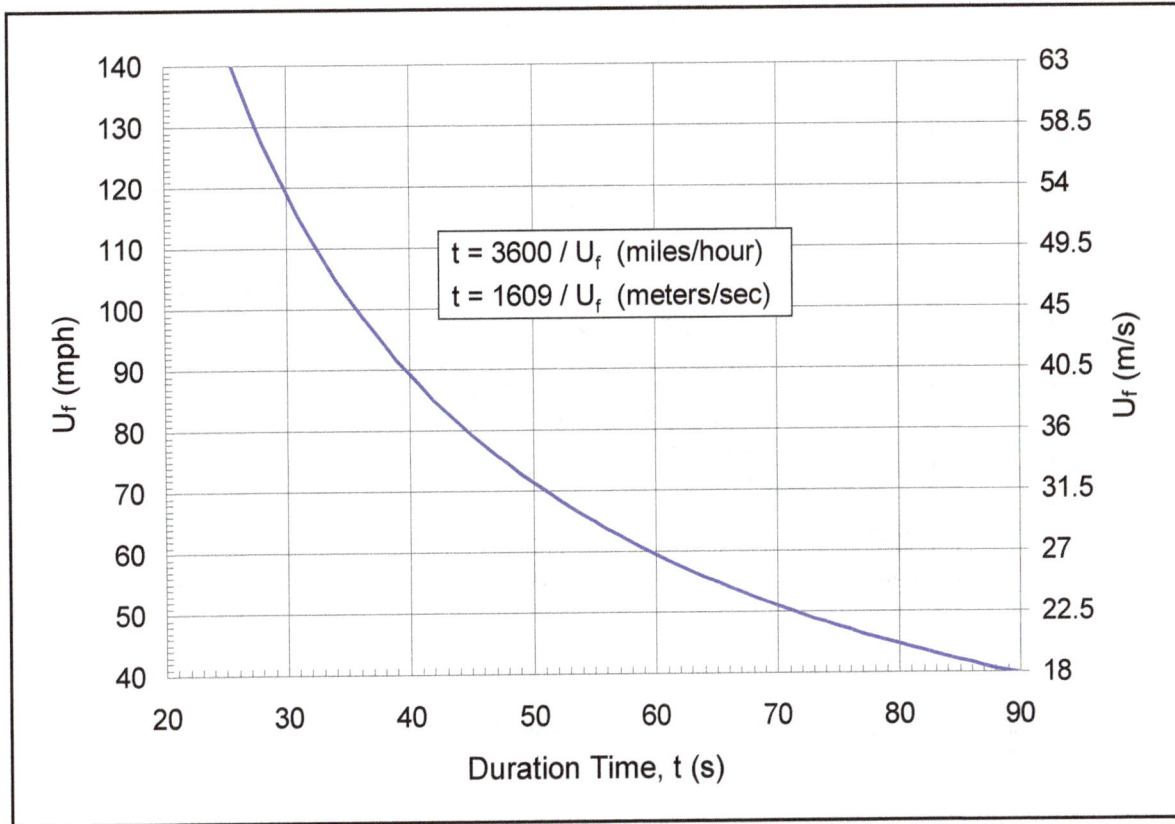

Figure II-2-2. Duration of the fastest-mile wind speed U_f as a function of wind speed (for open terrain conditions)

b. General structure of winds in the atmosphere.

(1) The earth's atmosphere extends to heights in excess of 100 km. Considerable layering in the vertical structure of the atmosphere occurs away from the earth's surface. The layering is primarily due to the absorption of specific bands of radiation in vertically localized regions. Absorbed radiation creates substantial warming in these regions which, in turn, produces inversion layers that inhibit local mixing. Processes essential to coastal engineering occur in the *troposphere*, which extends from the earth's surface up to an average altitude of 11 km. Most of the meteorological information used in estimating surface winds in marine areas falls within the troposphere. The lower portion of the troposphere is called the atmospheric or planetary boundary layer, within which winds are influenced by the presence of the earth's surface. The boundary layer typically reaches up to an altitude of 2 km or less.

(2) Figure II-2-4 shows an idealized relationship for an extended wind profile in a spatially homogeneous marine area (i.e. away from any land). The lowest portion is sometimes termed the constant stress layer, since there is essentially a constant flux of momentum through this layer. In this bottom layer, the time scale of momentum transfer is so short that there is little or no Coriolis effect; hence, the wind direction remains approximately constant. Above this layer is a region that is sometimes termed the Ekman layer. In this region, the influence of Coriolis becomes more pronounced and wind direction can vary significantly with

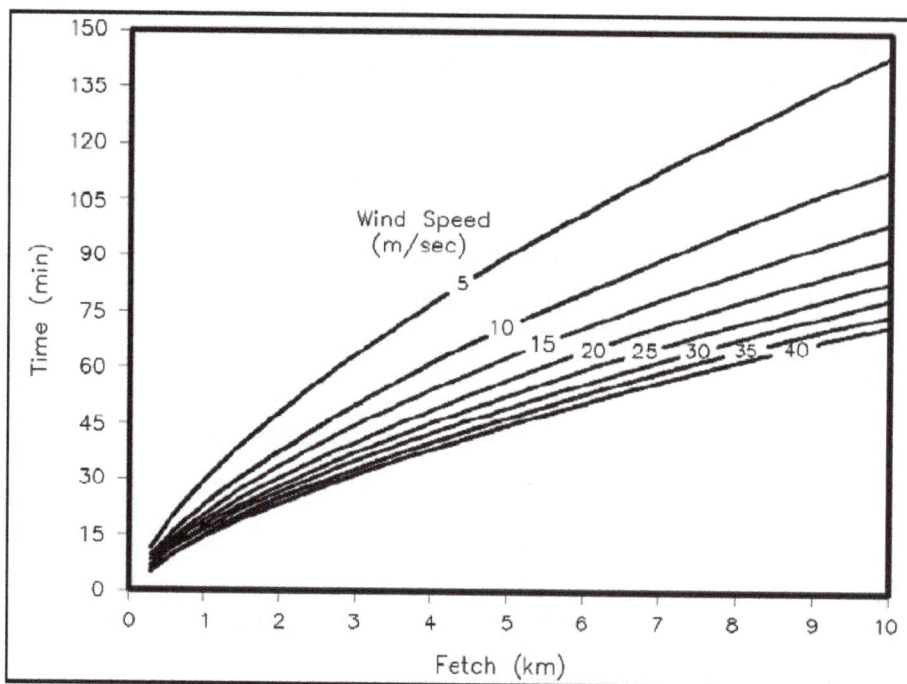

Figure II-2-3. Equivalent duration for wave generation as a function of fetch and wind speed

---

EXAMPLE PROBLEM II-2-2

FIND:
The appropriate 100-year wind speed for a basin with a fetch length of 10 km.

GIVEN:
A 100-year wind speed of 19.9 m/s, derived from 3-hr synoptic charts.

SOLUTION:
Figure II-2-3 requires knowledge of both wind speed and fetch distance; however, reasonable accuracy is gained by simply using the original wind speed and the appropriate fetch. In this case from Figure II-2-3, the appropriate wind-averaging interval is approximately 90 min. Using information from Figure II-2-1, the ratio of the highest 90-min wind speed to the highest 3-hr wind speed is given by the relationship

$$U_{5400}/U_{10800} = [-0.15 \log_{10}(5400)+1.5334]/[-0.15 \log_{10}(10800)+1.5334] = 0.9735/0.9284 = 1.048$$

Thus, the appropriate wind speed should be 1.048 times 19.9 m/sec, or 20.8 m/sec.

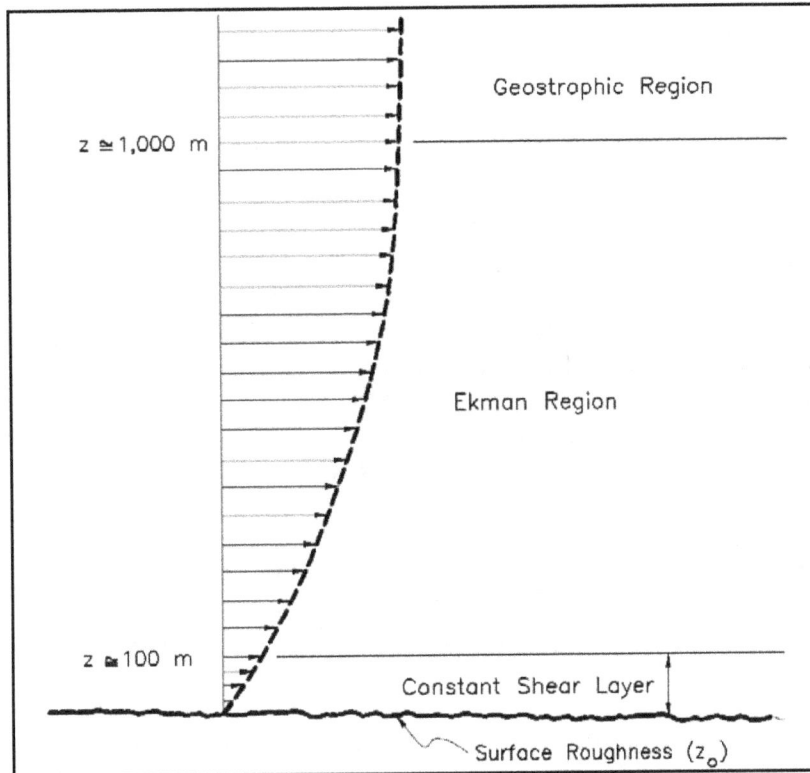

Figure II-2-4. Wind profile in atmospheric boundary layer

height. This results in wind directions at the top of the boundary layer which typically deviate about 10 to 15 deg to the right of near-surface wind directions over water and about 25 to 35 deg to the right of near-surface winds over land. Above the Ekman layer, the so-called geostrophic level is (asymptotically) approached. Winds in this level are assumed to be outside of the influence of the planetary surface; consequently, variations in winds above the Ekman layer are produced by different mechanisms than exist in the atmospheric boundary layer.

(3) Estimates of near-surface winds for wave prediction have historically been based primarily on two methods: direct interpolation/extrapolation/transformation of local near-surface measurements and transformation of surface winds from estimates of winds at the geostrophic level. The former method has mainly been applied to winds in coastal areas or to winds over large lakes. The latter method has been the main tool for estimating winds over large oceanic areas. A third method, termed "kinematic analysis" has received little attention in the engineering literature. All three of these methods will be discussed following a brief treatment of the general characteristics of winds within the atmospheric boundary layer.

c. *Winds in coastal and marine areas.*

(1) Background.

(a) Winds in marine and coastal areas are influenced by a wide range of factors operating at different space and time scales. Two potentially important local effects in the coastal zone, caused by the presence of land, are orographic effects and the sea breeze effect. Orographic effects are the deflection, channeling, or blocking of air flow by land forms such as mountains, cliffs, and high islands. A rule of thumb for blocking of low-level air flow perpendicular to a land barrier is given by the following:

$$\frac{U}{h_m} \begin{cases} < 0.1 \rightarrow blocked \\ > 0.1 \rightarrow no\ blocking \end{cases}$$ (II-2-2)

where

  1. U = wind speed

  h_m = height of the land barrier (in units consistent with U)

(b) An elevation of only 100 m will cause blocking of wind speeds less than about 10m/s, which includes most onshore winds (Overland 1992). The horizontal scale of these effects is on the order of 50 - 150 km. Another orographic effect called katabatic wind is caused by gravitational flow of cold air off higher ground such as a mountain pass. Since katabatic winds require cold air they are more frequent and strongest in high latitudes. These winds can have a significant impact on local coastal areas and are very site-specific (horizontal scale on the order of 25 km).

(c) Another local process, the sea breeze effect, is air flow caused by the differences in surface temperature and heat flux between land and water. Land temperatures change on a daily cycle while water temperatures remain relatively constant. This results in a sea breeze with a diurnal cycle. The on/offshore extent of the sea breeze is about 10 -20 km with wind speeds less than 10 m/s.

(d) Although understanding of atmospheric flows in complicated areas is still somewhat limited, considerable progress has been made in understanding and quantifying flow characteristics in simple, idealized situations. In particular, synoptic-scale winds in open-water areas (more than 20 km or so from land) are known to follow relatively straightforward relationships within the atmospheric boundary layer. The flow can be considered as a horizontally homogeneous, near-equilibrium boundary layer regime. As described in Tennekes (1973), Wyngaard (1973,1988), and Holt and Raman (1988), present-day boundary layer parameterizations appear to provide a relatively accurate depiction of flows within the homogeneous, near-equilibrium atmospheric boundary layers. Since these boundary-layer parameterizations have a substantial basis in physics, it is recommended that they be used in preference of older, less-verified methods.

  d. Characteristics of the atmospheric boundary layer.

(1) Since the 1960's, evidence from field and laboratory studies (Clarke 1970, Businger et al. 1971, Willis and Deardorff 1974, Smith 1988) and from theoretical arguments (Deardorff 1968, Tennekes 1973, Wyngaard 1973, 1988) have supported the existence of a self-similar flow regime within a homogeneous, near-equilibrium boundary layer in the atmosphere. In the absence of buoyancy effects (due to vertical gradients in potential temperature) and if no significant horizontal variations in density (baroclinic effects) exist, the atmospheric boundary layer can be considered as a neutral, barotropic flow. In this case, all flow characteristics can be shown to depend only on the speed of the flow at the upper edge of the boundary layer, roughness of the surface at the bottom of the boundary layer, and local latitude (because of the influence of the earth's rotation on the boundary-layer flow). Significantly for engineers and scientists, this theory predicts that wind speed at a fixed elevation above the surface cannot have a constant ratio of proportionality to wind speed at the top of the boundary layer.

(2) Deardorff (1968), Businger et al. (1971), and Wyngaard (1988) clearly established that flow characteristics within the atmospheric boundary layer are very much influenced by thermal stratification and horizontal density gradients (baroclinic effects). Thus, various relationships can exist between flows at the top of the boundary layer and near-surface flows. This additional level of complication is not negligible in many applications; therefore, stability effects should be included in wind estimates in important applications.

e. Characteristics of near-surface winds.

(1) Winds very close to a marine surface (within the constant-stress layer) generally follow some form of the "law-of-the-wall" for near-boundary flows. At wind speeds above about 5 m/s (at a 10-m reference level), turbulent transfers, rather than molecular processes, dominate air-sea interaction processes. Given a neutrally stable atmosphere, the wind speed close to the surface follows a logarithmic profile of the form

$$U_z = \frac{U_*}{k} \ln\left(\frac{z}{z_0}\right) \qquad\qquad (\text{II-2-3})$$

where

U_z = wind speed at height z above the surface

U_* = friction velocity

k = von Kármán's constant (approximately equal to 0.4)

z_0 = roughness height of the surface

(2) In this case, the rate of momentum transfer into a water column (of unit surface area) from the atmosphere can be written in the parametric form

$$\tau = \rho_a\, U_*^2$$
$$= \rho_a\, C_{Dz}\, U_z^2 \qquad\qquad (\text{II-2-4})$$

where

τ = wind stress

ρ_a = density of air

ρ_w = density of water

C_{Dz} = coefficient of drag for winds measured at level z

(3) The international standard reference height for winds is now taken to be 10 m above the surface. If winds are taken from this level, the z is usually dropped from the subscript notation and the momentum transfer is represented as

$$\tau = \rho_a\, C_D\, U^2 \qquad\qquad (\text{II-2-5})$$

where C_D specifically refers now to a 10-m reference level.

(4) Extensive evidence shows that the coefficient of drag over water depends on wind speed (Garratt 1977, Large and Pond 1981, Smith 1988).

(5) When surfaces (land or water) are significantly warmer or cooler than the overlying air, thermal stability effects tend to modify the logarithmic profile in Equation II-2-3. If the underlying surface is colder

than the air, the atmosphere becomes stably stratified and turbulent transfers are suppressed. If the surface is warmer than the air, the atmosphere becomes unstably stratified and turbulent transfers are enhanced. In this more general case, the form of the near-surface wind profile can be approximated as

$$U_z = \frac{U_*}{k} \left[\ln\left(\frac{z}{z_0}\right) - \phi\left(\frac{z}{L}\right) \right]$$

(II-2-6)

where

ϕ = universal similarity function characterizing the effects of thermal stratification

L = parameter with dimensions of length that represent the relative strength of thermal stratification effects (Obukov stability length)

(6) L is positive for stable stratification, negative for unstable stratification, and infinite for neutral stratification. Algebraic forms for ϕ and additional details on the specification of near-surface flow characteristics can be found in Resio and Vincent (1977), Hsu (1988), and the ACES Technical Reference (Leenhnecht, Szuwalski, and Sherlock 1992; Sec. 1-1).

(7) Transfer of momentum into water from the atmosphere can be markedly influenced by stability effects. For example, at the 10-m reference level, Equations II-2-4 through II-2-6 give

$$C_D = \left(\frac{U_*}{U}\right)^2$$
$$= \left[\frac{k}{\ln\left(\frac{z}{z_0}\right) - \phi\left(\frac{z}{L}\right)} \right]^2$$

(II-2-7)

(8) The system of equations representing the boundary layer is readily solved via a number of numerical techniques. However, a relationship between z_0 and U_* must also be specified.

(9) Since ϕ is negative for stable conditions and positive for unstable conditions, stratification clearly reduces the coefficient of drag for stable conditions and increases the coefficient of drag for unstable conditions (Figure II-2-5). Consequently, for the same wind speed at a reference level, the momentum transfer rate is lower in a stable atmosphere than in an unstable atmosphere.

(10) Studies by Hsu (1974); Geernaert, Katsaros, and Richter (1986); Huang et al. (1986); Janssen (1989, 1991); and Geernaert (1990) suggest that the coefficient of drag depends not only on wind speed but also on the stage of wave development. The physical mechanism responsible for this appears to be related to the phase speed of the waves in the vicinity of the spectral peak relative to the wind speed. At present, there does not appear to be sufficient information to establish this behavior definitively. Future studies may shed more light on these effects and their importance to marine and coastal winds.

Coefficient of Drag Vs. Wind Speed

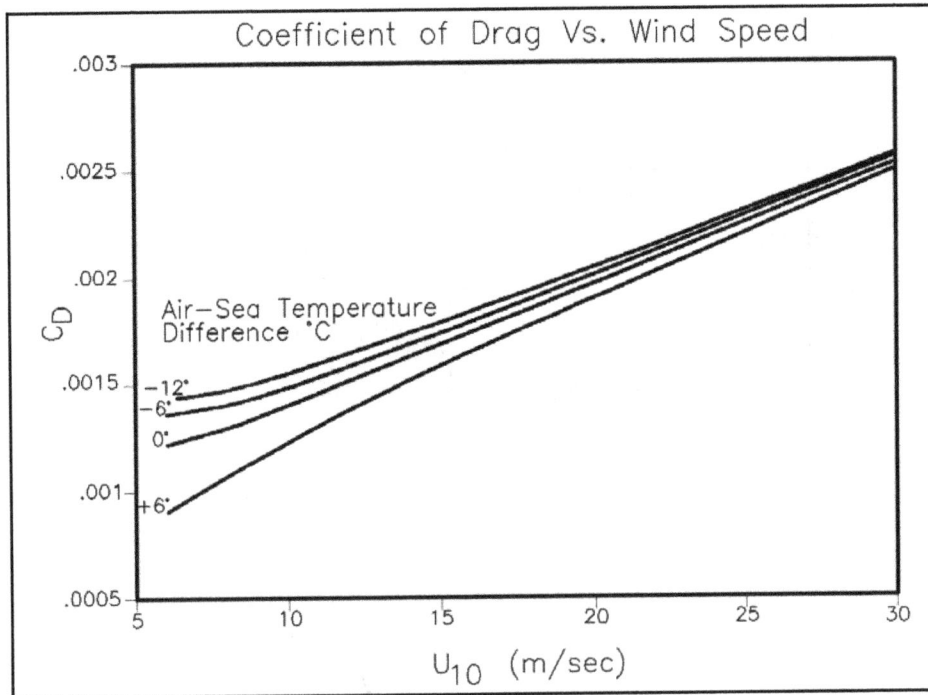

Figure II-2-5. Coefficient of drag versus wind speed

f.   Estimating marine and coastal winds.

(1)   Wind estimates based on near-surface observations. Three methods are commonly used to estimate surface marine wind fields. The first of these, estimation of winds from nearby measurements, has the appeal of simplicity and has been shown to work well for water bodies up through the size of the Great Lakes. To use this method, it is often necessary to transfer the measurements to different locations (e.g. from overland to overwater) and different elevations. Such complications necessitate consideration of the factors given below.

(a)   Elevation correction of wind speed. Often winds taken from observations of opportunity (ships, oil rigs, offshore structures, buoys, aircraft, etc.) do not coincide with the standard 10-m reference level. They must be converted to the 10-m reference level for predicting waves, currents, surges, and other wind-generated phenomena. Failure to do so can produce extremely large errors. For the case of winds taken in near-neutral conditions at a level near the 10-m level (within the elevation range of about 8-12 m), the "1/7" rule can be applied. This simple approximation is given as

$$U_{10} = U_z \left(\frac{10}{z} \right)^{\frac{1}{7}}$$

(II-2-9)

where z is measured in meters.

(b)   Elevation and stability corrections of wind speed. Figure II-2-6 provides a more comprehensive method to accomplish the above transformation, including both elevation and stability effects. The "1/7" rule is given as a special case. In Figure II-2-6, the ratio of the wind speed at any height to the wind speed

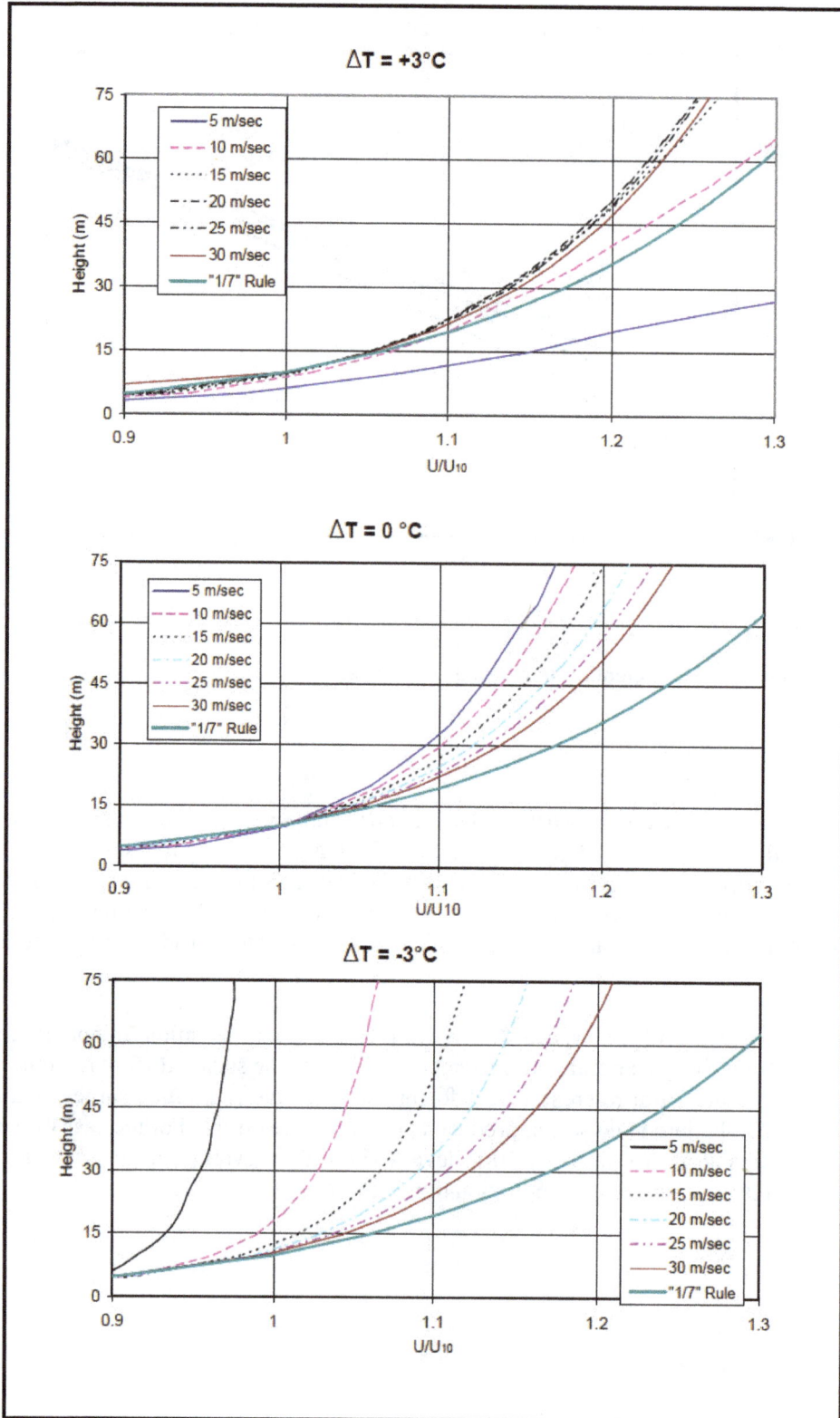

Figure II-2-6. Ratio of wind speed at any height to the wind speed at the 10-m height as a function of measurement height for selected values of air-sea temperature difference and wind speed: a) $\Delta T=+3°C$; b) $\Delta T=0°C$; c) $\Delta T=-3°C$. Plots generated with following conditions: duration of observed and final wind = 3 hrs; latitude = 30° N; fetch = 42 km; wind observation type - over water; fetch conditions - deep open water

at the 10-m height is given as a function of measurement height for selected values of air-sea temperature difference and wind speed. Air-sea temperature difference is defined as

$$\Delta T = T_a - T_s \qquad\qquad (\text{II-2-9})$$

where

ΔT = air-sea temperature difference, in deg C

T_a = air temperature, in deg C

T_s = water temperature, in deg C

As can be seen in Figure II-2-6, the "1/7" rule should not be used as a general method for transforming wind speeds from one level to another in marine areas. The ACES software package (Leenhnecht, Szuwalski, and Sherlock 1992) contains algorithms, based on planetary boundary layer physics, which compute the values shown in Figure II-2-6; so it is recommended that ACES be used if at all possible for individual situations.

EXAMPLE PROBLEM II-2-3

FIND:
    The estimated wind speed at a height of 10 m.

GIVEN:
    The wind speed at a height of 25 m is 20 m/s and the air-sea temperature difference is +3°C.

SOLUTION:
    From Figure II-2-6 (a), the ratio U/U_{10} is about 1.18 for a 20-m/s wind at a height of 25 m. So the estimated wind speed at a 10-m height U_{10} is equal to U at 25 m (20 m/s) divided by U/U_{10} (1.18), which gives U_{10} = 16.9 m/s.

(c) Simplified estimation of overwater wind speeds from land measurements. Due to the behavior of water roughness as a function of wind speed, the ratio of overwater winds at a fixed level to overland wind speeds at a fixed level is not constant, but varies nonlinearly as a function of wind speed. Figure II-2-7 provides guidance for the form of this variation. **The specific values shown in this figure are from a study of winds in the Great Lakes and care should be taken in applying them to other areas.** Figure II-2-8 indicates the expected variation with air-sea temperature difference (calculated with ACES). Although air-sea temperature difference can significantly affect light and moderate winds, it has only a small impact (5 percent or less) on high wind speeds typical of design. If at all possible, it is advisable to use locally collected data to respecify the exact form of Figures II-2-7 and II-2-8 for a particular project. One concern here would be the use of wind measurements from aboveground elevations that are markedly different from those used in the Resio and Vincent study (9.1 m or 30 ft).

(d) Wind speed variation with fetch. When winds pass over a discontinuity in roughness (e.g., a land-sea interface), an internal boundary layer is generated. The height of such a boundary layer forms a slope in the neighborhood of 1:30 in the downwind direction from the roughness discontinuity. This complication can make it difficult to use winds from certain locations at which winds from some directions fall within the

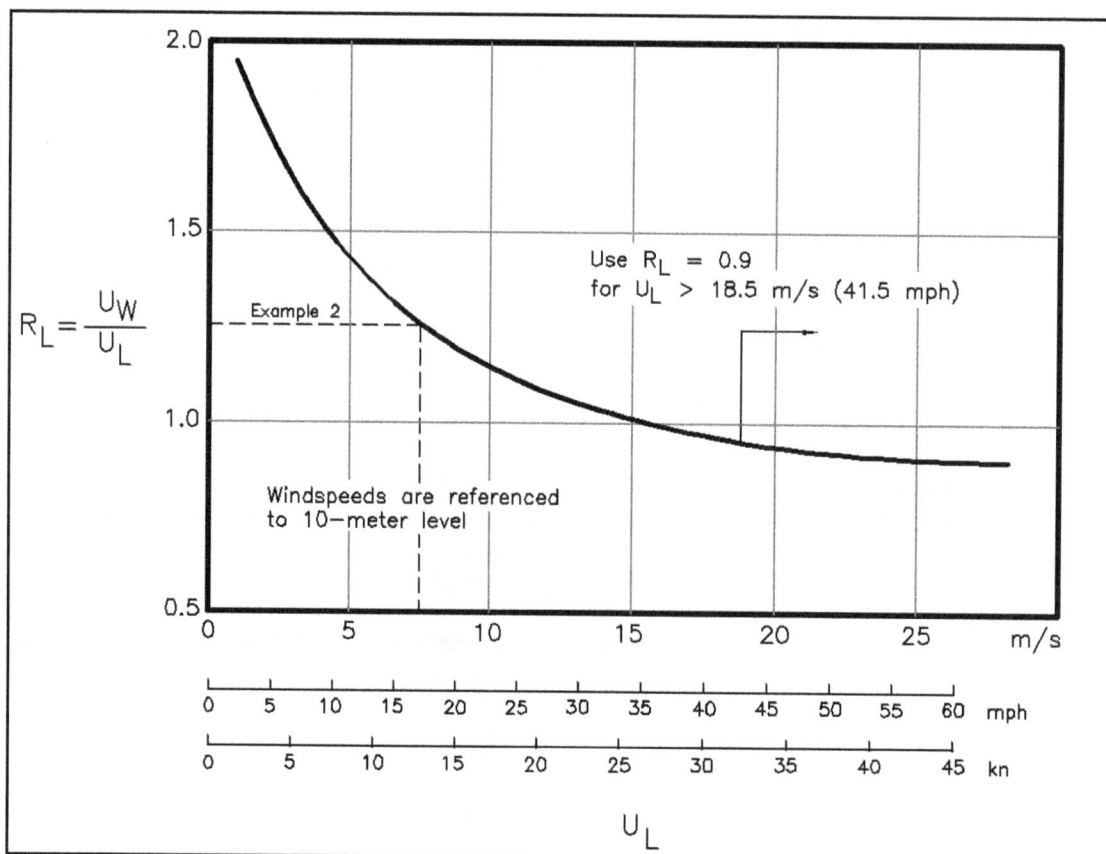

Figure II-2-7. Ratio R_L of windspeed over water U_W to windspeed over land U_L as a function of windspeed over land U_L (after Resio and Vincent (1977))

marine boundary layer and winds from other directions fall within a land boundary layer. In areas such as this, a land-to-sea transform (similar to that shown in Figures II-2-7 and II-2-8) can be used for all angles coming from the land. Depending on the distance to the water and the elevation of the measurement site, winds coming from the direction of open water may or may not still be representative of a marine boundary layer. Guidance for determining the effects of fetch on wind speed modifications can be found in Resio and Vincent (1977) and Smith (1983). These studies indicate that fetch effects wind speeds significantly only at locations within about 16 km (10 miles) of shore.

(e) Wind speed transition from land to water. The net effect of wind speed variation with fetch is to provide a smooth transition from the (generally lower) wind speed over land to the (generally higher) wind speed over water. Thus, wind speeds tend to increase with fetch over the first 10 miles or so after a transition from a land surface. The exact magnitude and characteristics of this transition depend on the roughness characteristics of the terrain and vegetation and on the stability of the air flow. A very simplistic approximation to this wind speed variation for the Resio and Vincent curves used here could be obtained by fitting a logarithmic curve to the asymptotic overland and overwater wind speed values. However, for most design and engineering purposes, it is probably adequate to simply use the long-fetch values with the recognition that they are somewhat conservative. The one situation that should cause some concern would be if overwater wind speed measurements are taken near the upwind end of a fetch. These winds could be considerably lower than wind speeds at the end of the fetch and underconservative values for wave conditions could result from the use of such (uncorrected) winds in a predictive scheme.

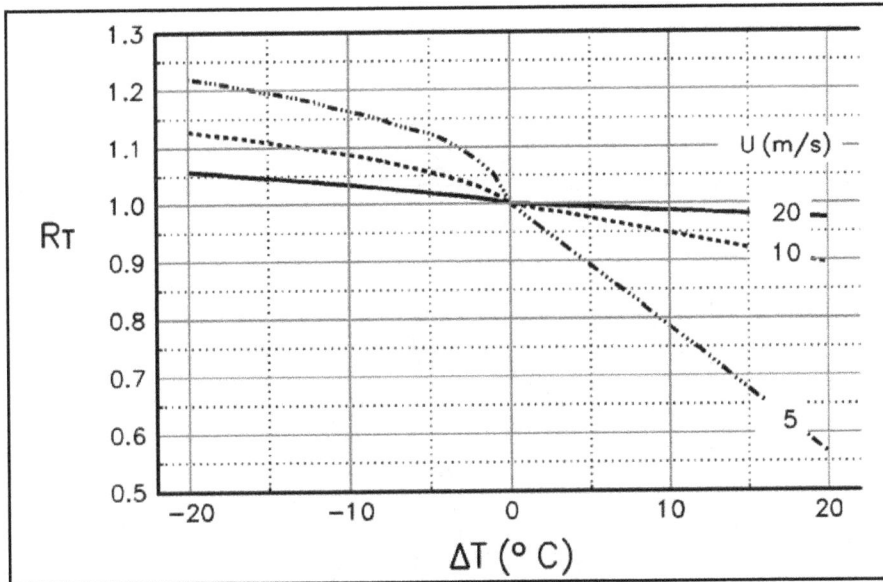

Figure II-2-8. Amplification R_T ratio of W_c (wind speed accounting for effects of air-sea temperature difference) to W_w (wind speed over water without temperature effects)

(f)   Empirical relationship. A rough empirical relationship between overwater wind speeds and land measurements is discussed in Part III-4-2-b. This highly simplified relationship is based on several restrictive assumptions including land measurements over flat, open terrain near the coast; and wind direction is within 45 deg of shore-normal. The approach may be helpful where wind measurements are available over both land and sea at a site, but the specific relationship of Equation III-4-12 is not recommended for general hydrodynamic applications.

(2)   Wind estimates based on information from pressure fields and weather maps. A primary driving force of synoptic-scale winds above the boundary layer is produced by horizontal pressure gradients. Figure II-2-9 is a simplified surface chart for the north Pacific Ocean. The area labeled L in the right center of the chart and the area labeled H in the lower left corner of the chart are low- and high-pressure areas. The pressures increase moving outward from L (isobars 972, 975, etc.) and decrease moving outward from H (isobars 1026, 1023, etc.). Synoptic-scale winds at latitudes above about 20 deg tend to blow parallel to the isobars, with the magnitude of the wind speed being inversely proportional to the spacing between the isobars. Scattered about the chart are small arrow shafts with a varying number of feathers. The direction of a shaft shows the direction of the wind, with each one-half feather representing a unit of 5 kt (2.5 m/s) in wind speed.

(a)   Figure II-2-10 shows a sequence of weather maps with isobars (lines of equal pressure) for the Halloween Storm of 1991. An intense extratropical storm (extratropical cyclone) is located off the coast of Nova Scotia. Other information available on this weather map besides observed wind speeds and directions includes air temperatures, cloud cover, precipitation, and many other parameters that may be of interest. Figure II-2-11 provides a key to decode the information.

(b)   Historical pressure charts are available for many oceanic areas back to the end of the 1800's. This is a valuable source of wind information when the pressure fields and available wind observations can be used to create marine wind fields. However, the approach for linking pressure fields to winds can be complex, as discussed in the following paragraphs.

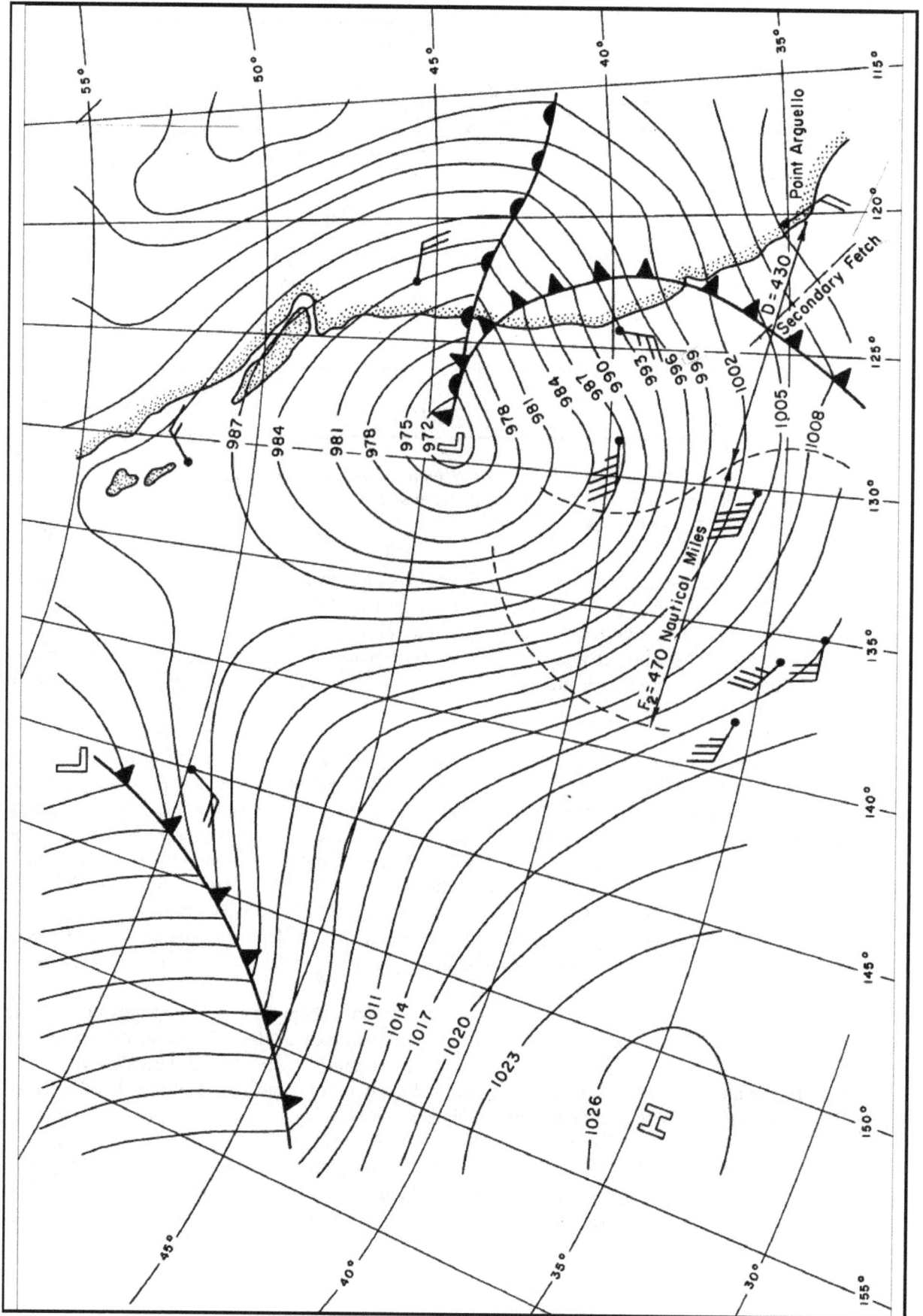

Figure II-2-9. Surface synoptic chart for 0030Z, 27 October 1950

(A) OCT. 28, 1500 UTC

(C) OCT. 30, 1500 UTC

(B) OCT. 29, 1800 UTC

(D) OCT. 31, 1200 UTC

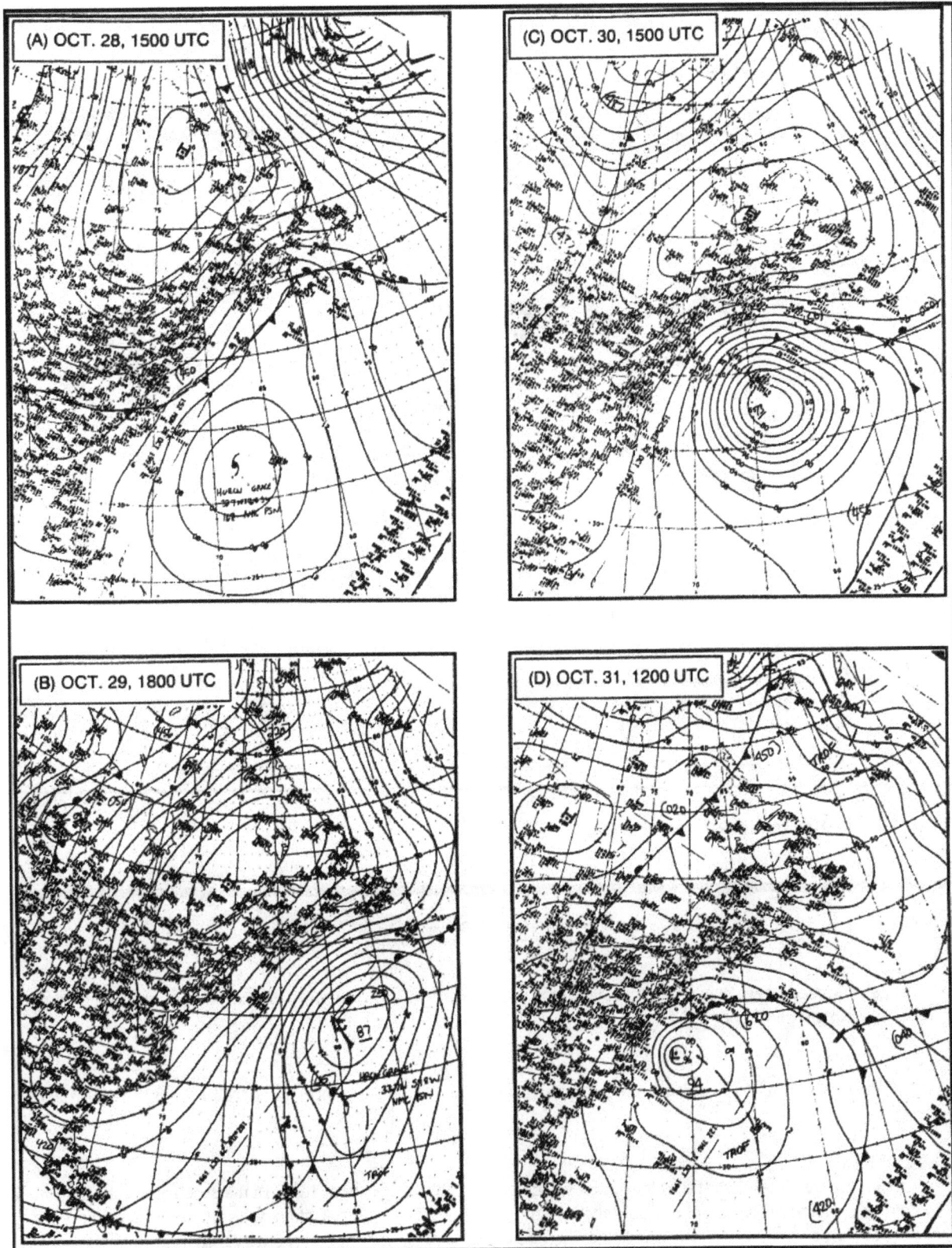

Figure II-2-10.   Surface synoptic weather charts for the Halloween storm of 1991

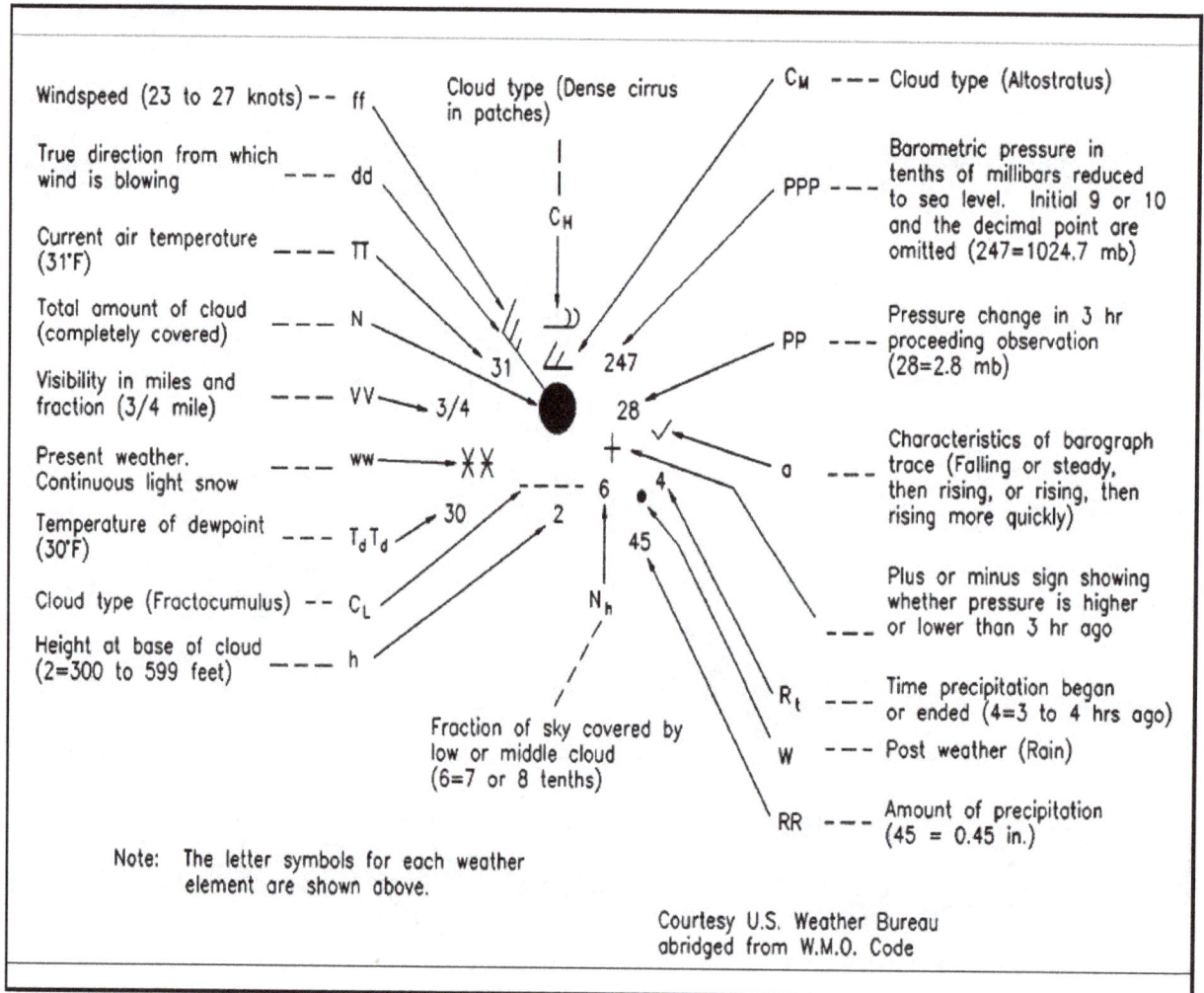

Figure II-2-11. Key to plotted weather report

EXAMPLE PROBLEM II-2-4

FIND:

The estimated overwater wind speed at a site over 10 miles from shore, given that the air-sea temperature difference is near zero ($\Delta T \approx 0$).

GIVEN:

A wind speed of 7.5 m/s at an airport location well inland (at the airport standard of 30 ft above ground elevation).

SOLUTION:

From Figure II-2-7 the ratio of overwater wind to overland wind is about 1.25. In the absence of information to calibrate a local relationship, multiply the 7.5-m/s wind speed by 1.25 to obtain an estimated overwater wind speed of 9.4 m/s. It should be recognized that the 90-percent confidence interval for this estimate is approximately 15 percent. It may be desirable to include this factor of conservatism in some calculations. However, at this short fetch, there is already conservatism due to the lack of consideration of wind speed variations with fetch.

(c)   Synoptic-scale winds in nonequatorial regions are usually close to a geostrophic balance, given that the isobars are nearly straight (i.e. the radius of curvature is large). For this balance to be valid, the flow must be steady state or very nearly steady state. Furthermore, frictional effects, advective effects, and horizontal and vertical mixing must all be negligible. In this case, the Navier-Stokes equation for atmospheric motions reduces to the geostrophic balance equation given by

$$U_g = \frac{1}{\rho_a f} \frac{dp}{dn}$$ (II-2-10)

where

U_g = geostrophic wind speed (located at the top of the atmospheric boundary layer)

dp/dn = gradient of atmospheric pressure orthogonal to the isobars

Wind direction at the geostrophic level is taken to be parallel to the local isobars. Hence, purely geostrophic winds in a large storm would move around the center of circulation, without converging on or diverging from the center.

(d)   Figure II-2-12 may be used for simple estimates of geostrophic wind speed. The distance between isobars on a chart is measured in degrees of latitude (an average spacing over a fetch is ordinarily used), and the latitude position of the fetch is determined. Using the spacing as ordinate and location as abscissa, the plotted, or interpolated, slant line at the intersection of these two values gives the geostrophic wind speed. For example, in Figure II-2-9, a chart with 3-mb isobar spacing, the average isobar spacing (measured normal to the isobars) over fetch F_2 located at 37 deg N. latitude, is 0.70 deg latitude. Scales on the bottom and left side of Figure II-2-12 are used to find a geostrophic wind of 34.5 m/s (67 kt).

(e)   If isobars exhibit significant curvature, centrifugal effects can become comparable or larger than Coriolis accelerations. In this situation, a simple geostrophic balance must be replaced by the more general gradient balance. The equation for this motion is

$$U_{gr} = \frac{1}{\rho_a f} \frac{dp}{dn} + \frac{U_{gr}^2}{f r_c}$$ (II-2-11)

where

U_{gr} = gradient wind speed

r_c = radius of curvature of the isobars

Winds near the centers of small extratropical storms and most tropical storms can be significantly affected and even at times dominated by centrifugal effects, so the more general gradient wind approximation is usually preferred to the geostrophic approximation. Gradient winds tend to form a small convergent angle (about 5° to 10°) relative to the isobars.

(f)   An additional complication results when the center of a storm is not stationary. In this case, the steady-state approximation used in both the geostrophic and gradient approximations must be modified to include non-steady-state effects. The additional wind component due to the changing pressure fields is

$$U_g = \frac{1}{\rho_a f} \frac{\Delta p}{\Delta n}$$

For T = 10° C

Δp = 3 mb and 4 mb

Δn = isobar spacing measured in degrees latitude

p = 1013.3 mb

ρ_a = 1.247 X 10⁻³ gm/cm³

f = Coriolis parameter = $2\omega \sin \emptyset$

where

ω = angular velocity of earth, 0.2625 rad/hr

Ø = latitude in degrees

GEOSTROPHIC WIND VELOCITY, U_g IN KNOTS

Left Scale – 3-mb Isobar Spacing
Right Scale – 4-mb Isobar Spacing

(after Bretschneider, 1952)

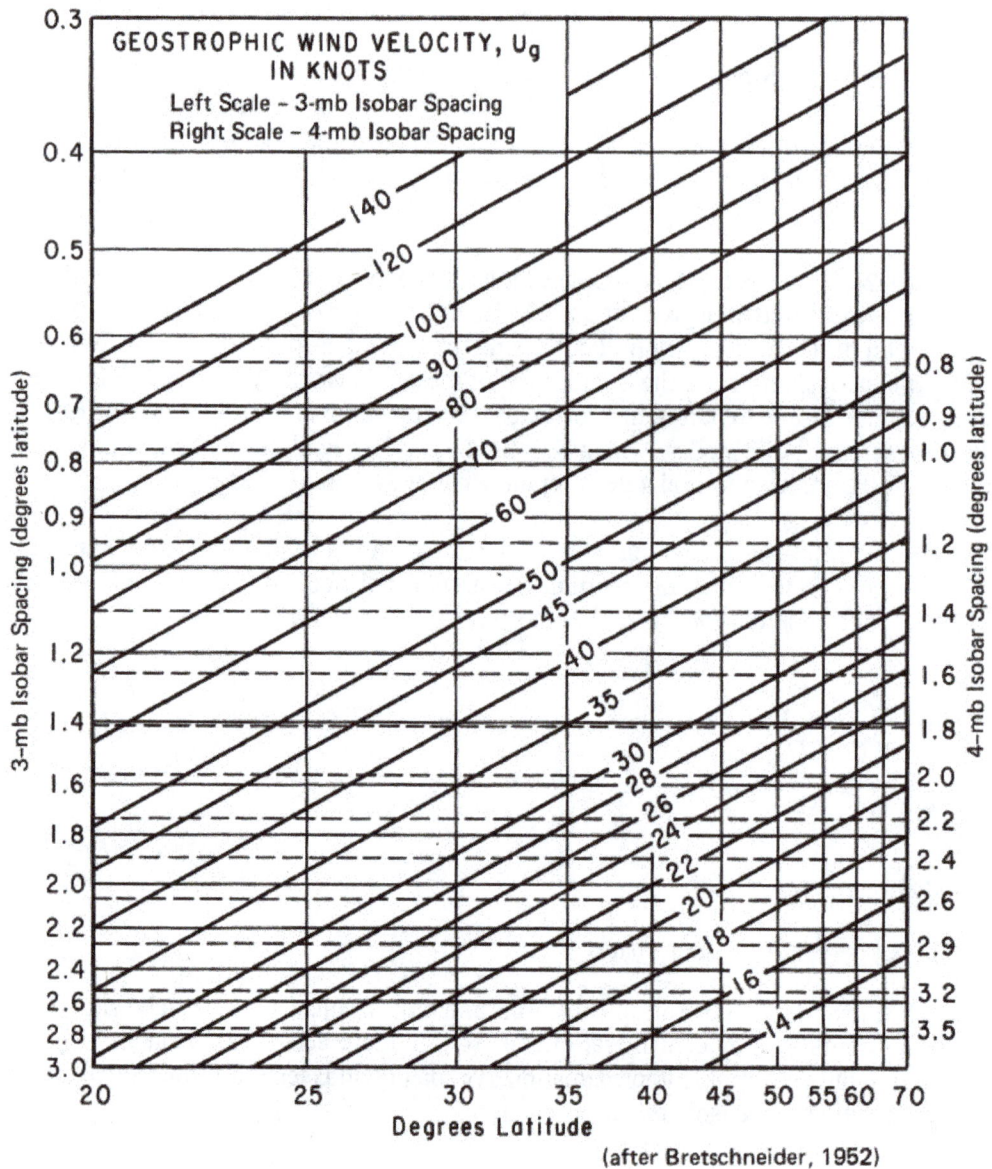

Figure II-2-12. Geostrophic (free air) wind scale (after Bretschneider (1952))

termed the isallobaric wind. In certain situations, the isallobaric wind can attain magnitudes nearly equal to those of geostrophic wind.

(g)   Due to the factors discussed above, winds at the geostrophic level can be quite complicated. Therefore, it is recommended that these calculations be performed with numerical computer codes rather than manual methods.

(h)   Once the wind vector is estimated at a level above the surface boundary layer, it is necessary to relate this wind estimate to wind conditions at the 10-m reference level. In some past studies, a constant proportionality was assumed between the wind speeds aloft and the 10-m wind speeds. Whereas this might suffice for a narrow range of wind speeds if the atmospheric boundary layer were near neutral and no horizontal temperature gradients existed, it is not a very accurate representation of the actual relationship between surface winds and winds aloft. Use of a single constant of proportionality to convert wind speeds at the top of the boundary layer to 10-m wind speeds is not recommended.

(i)   Over land, the height of the atmospheric boundary layer is usually controlled by a low-level inversion layer. This is typically not the case in marine areas where, in general, the height of boundary layer (in non-equatorial regions) is a function of the friction velocity at the surface and the Coriolis parameter, i.e.

$$h \approx \lambda \frac{U_*}{f} \qquad \qquad \text{(II-2-12)}$$

where

λ = dimensionless constant

(j)   Researchers have shown that, within the boundary layer, the wind profile depends on latitude (via the Coriolis parameter), surface roughness, geostrophic/gradient wind velocity, and density gradients in the vertical (stability effects) and horizontal (baroclinic effects). Over large water bodies, if the effects of wave development on surface roughness are neglected, the boundary-layer problem can be solved directly from specification of these factors. Figure II-2-13 shows the ratio of the wind at a 10-m level to the wind speed at the top of the boundary layer (denoted by the general term U_g here) as a function of wind speed at the top of the boundary layer, for selected values of air-sea temperature difference. Figure II-2-14 shows the ratio of friction velocity at the water's surface to the wind speed at the upper edge of the boundary layer as a function of these same parameters. It might be noted from Figure II-2-14 that a simple approximation for U_* in neutral stratification as a function of U_g is given by

$$U_* \approx 0.0275 \ U_g \qquad \qquad \text{(II-2-13)}$$

This approximation is accurate within 10 percent for the entire range of values shown in Figure II-2-14.

(k)   Measured wind directions are generally expressed in terms of azimuth angle from which winds come. This convention is known as a *meteorological coordinate system*. Sometimes (particularly in relation to winds calculated from synoptic information), a mathematical vector coordinate or *Cartesian coordinate system* is used (Figure II-2-15). Conversion from the vector Cartesian to meteorological convention is accomplished by

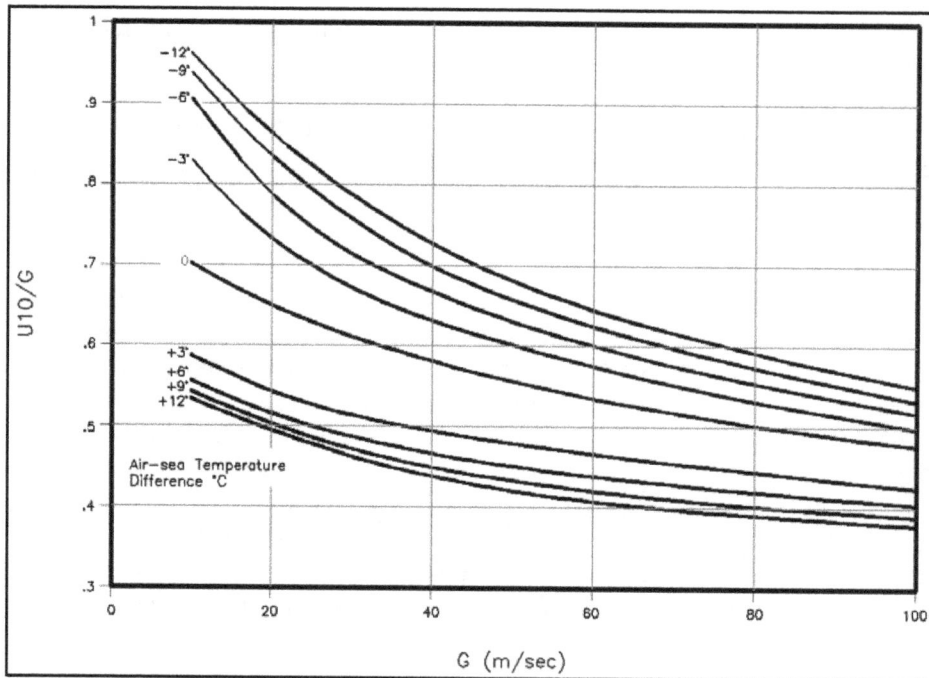

Figure II-2-13. Ratio of wind speed at a 10-m level to wind speed at the top of the boundary layer as a function of wind speed at the top of the boundary layer, for selected values of air-sea temperature difference

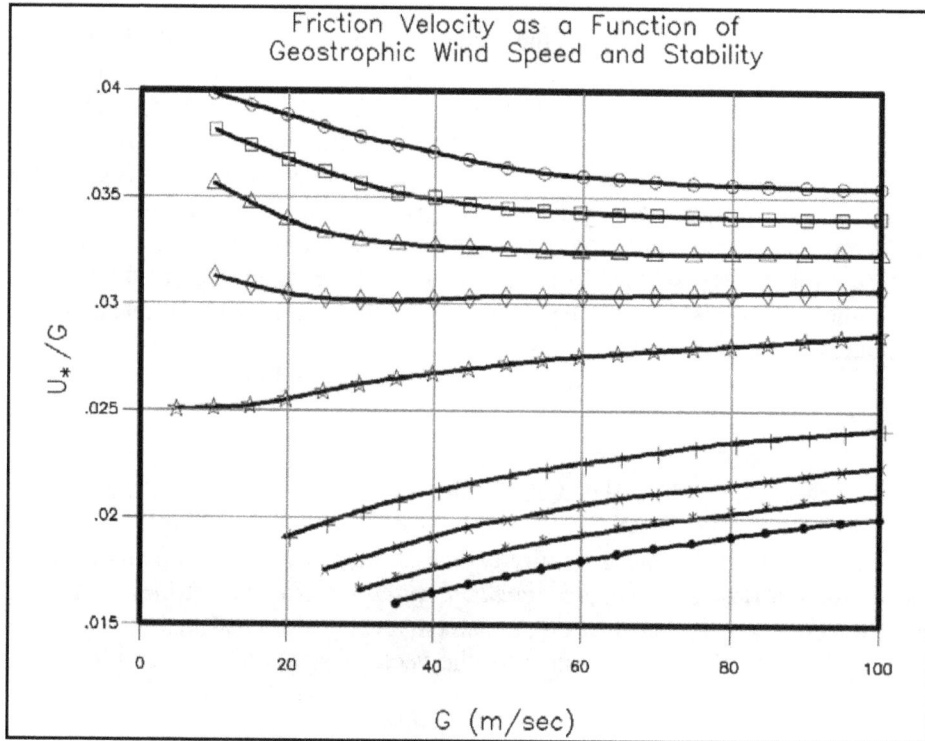

Figure II-2-14. Ratio of U*/Ug as a function of Ug for selected values of air-sea temperature difference

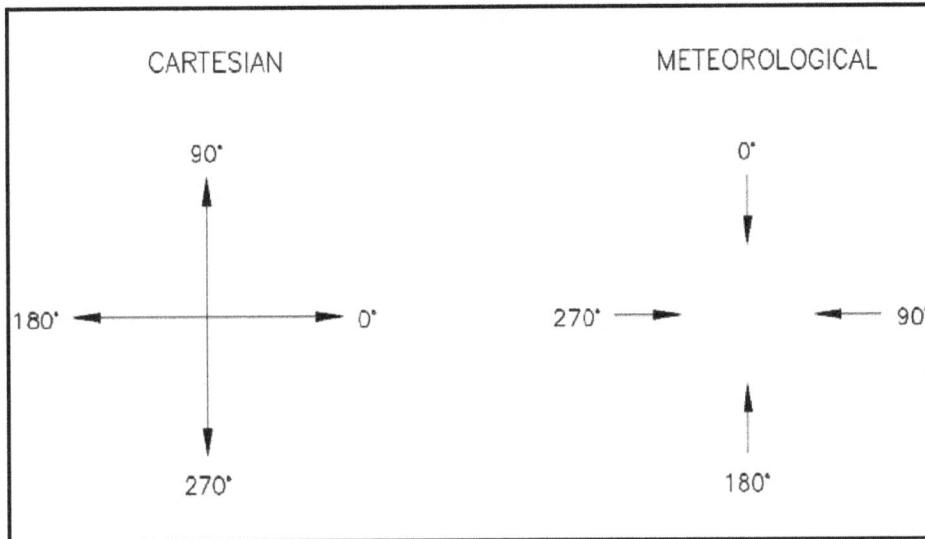

Figure II-2-15. Common wind direction conventions

$$\theta_{met} = 270 - \theta_{vec} \qquad\qquad (II\text{-}2\text{-}14)$$

where

θ_{met} = direction in standard meteorological terms

θ_{vec} = direction in a Cartesian coordinate system with the zero angle wind blowing toward the east

(l)   Wind estimates based on kinematic analyses of wind fields. In several careful studies, it has been shown that one method of obtaining very accurate wind fields is through the application of "kinematic analysis" (Cardone 1992). In this technique, a trained meteorological analyst uses available information from weather charts and other sources to construct detailed pressure fields and frontal positions. Using concepts of continuity along with this information, the analyst then constructs streamlines and isotachs over the entire analysis region. Unfortunately, this procedure is very labor-intensive; consequently, most analysts combine kinematic analyses of small subregions within their region with numerical estimates over the entire region. This method is sometimes referred to as a man-machine mix.

g.   *Meteorological systems and characteristic waves.* Many engineers and scientists working in marine areas do not have a firm understanding of wave conditions expected from different wind systems. Such an understanding is helpful not only for improving confidence in design conditions, but also for establishing guidelines for day-to-day operations. Two problems that can arise directly from this lack of experience are (1) specification of design conditions with a major meteorological component missing, and 2) underestimation of the wave generation potential of particular situations. An example of the former situation might be the neglect of extratropical waves in an area believed to be dominated by tropical storms. For example, in the southern part of the Bay of Campeche along the coast of Mexico, one might expect that hurricanes dominate the extreme wave climate. However, outbursts of cold air termed "northers" actually contribute to and even control some of the extreme wave climate in this region. An example of the second situation can be found in decisions to operate a boat or ship in a region where storm waves can endanger life and property. Table II-2-2 assists users of this manual in understanding such problems. Potentially threatening wind and wave conditions from various scales of the meteorological system are categorized.

EXAMPLE PROBLEM II-2-5

FIND:
The 10-m wind speed, the wind direction, and the coefficient of drag.

GIVEN:
A pressure gradient of 5 mb in 100 km, an air-sea temperature difference of -5° C (i.e. the water is warmer than the air, as is typical in autumn months), the latitude of the location of interest (equal to 45° N), and the geographic orientation of the isobars.

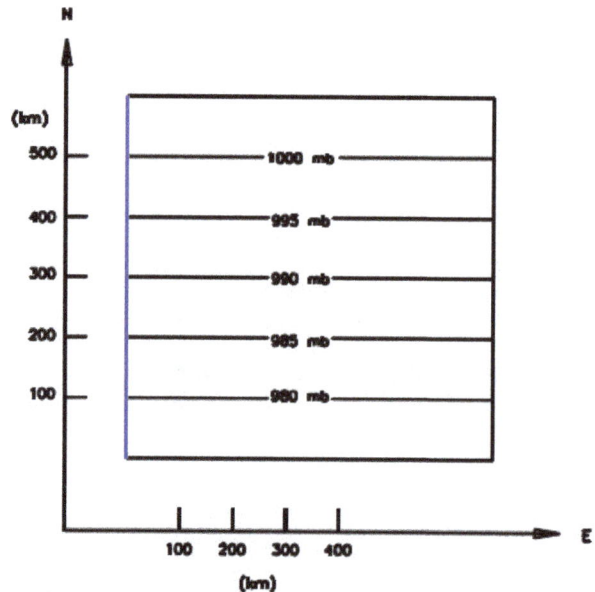

SOLUTION:
Option 1 - From Equation II-2-10, wind speed is calculated (in cgs units) as

$U_g = 1/(1.2 \times 10^{-3} \times 1.03 \times 10^{-4}) \times (dp/dn)$ (a)
$= 1/1.236 \times 10^{-7} \times (5 \times \underline{1000})/(100 \times 100000)$ (b)
$= 4045$ cm/sec   (c)
$= 40.45$ m/sec

The 1.2×10^{-3} factor in step (a) is air density in g/cm³.

The underlined 1000 factor in step (b) converts mb to dynes/cm². The 100000 factor in step (b) converts km to cm. From U_g and ΔT and Figure II-2-13

$U_{10}/U_g = 0.68$ and $U_{10} = U_g \times U_{10}/U_g = 40.45 \times 0.68 = 27.5$ m/s

From Figure II-2-5

$C_D = 0.0024$

Wind Direction: Parallel to isobars, counterclockwise circulation around low, therefore the direction is west

Option 2 - Use Figure II-2-12, though it requires pressure gradient information in a different form than given in this example.

Table II-2-2
Local Seas Generated by Various Meteorological Phenomena

Type of Wind System	Wave Characteristics	Characteristic Height and Period
Individual thunderstorm No significant horizontal rotation. Size, 1-10 km	Very steep waves. Waves can become relatively large if storm speed and group velocity of spectral peak are nearly equal. Can be a threat to some operations in open-ocean, coastal, and inland waters.	H 0.5 - 1.5 m T 1.5 - 3 sec
Supercell thunderstorms Begins to exhibit some rotation. Size, 5-20 km	Very steep waves. Waves can become relatively large if storm speed and group velocity of spectral peak are nearly equal. Can pose a serious threat to some operations in open-ocean, coastal, and inland waters.	H 2 - 3 m T 3 - 6 sec
Sea breeze Thermally driven near-coast winds. Size, 10-100 km	Waves of intermediate steepness. Can modify local wave conditions when superposed on synoptic systems. Can affect some coastal operations.	H 0.5 - 1.5 m T 3 - 5 sec
Coastal fronts Results from juxtaposition of cold air and warm water. Size, 10 km across and 100 km long	Can modify local wave conditions near coasts. Minimal effects on wave conditions due to orientation of winds and fetches.	H 0.5 - 1.0 m T 3 - 4 sec
Lee waves "Spin-off" eddies due to interactions between synoptic winds and coastal topography Size, 10's of km	Generates waves that can deviate significantly in direction from synoptic conditions. Can affect coastal wave climates.	H 0.5 - 1.5 m T 2 - 5 sec
Frontal squall lines Organized lines of thunderstorms moving within a frontal area. Size, 100's of km long and 10 km across	Can create severe hazards to coastal and offshore operations. Can generate extreme wave conditions for inland waters. Waves can become quite large if frontal area becomes stationary or if rate of frontal movement matches wave velocity of spectral peak. Can create significant addition to existing synoptic-scale waves.	H 1 - 5 m T 4 - 7 sec

(Sheet 1 of 3)

Table II-2-2 (Continued)

Mesoscale Convective Complex (MCC) Large, almost circular system of thunderstorms with rotation around a central point (2-3 form in the U.S. per year). Size, 100-400 km in diameter	Important in interior regions of U.S. Can generate extreme waves for short-fetch and intermediate-fetch inland areas.	H fetch-limited T fetch-limited U ≈ 20 m/s
Tropical depression Weakly circulating tropical system with winds under 45 mph.	Squall lines superposed on background winds can produce confused, steep waves.	H 1 - 4 m T 4 - 8 sec
Tropical storm Circulating tropical system with winds over 45 mph and less than 75 mph.	Very steep seas. Highest waves in squall lines.	H 5 - 8 m T 5 - 9 sec
Hurricane Intense circulating storm of tropical origin with wind speeds over 75 mph. Shape is usually roughly circular.	Can produce large wave heights. Directions near storm center are very short-crested and confused. Highest waves are typically found in the right rear quadrant of a storm. Wave conditions are primarily affected by storm intensity, size, and forward speed, and in weaker storms by interactions with other synoptic scale and large-scale features.	Saffir Simpson Hurricane Scale SS H(m) T(sec) 1 4-8 7-11 2 6-10 9-12 3 8-12 11-13 4 10-14 12-15 5 12-17 13-17 (see Table IV-1-4)
Extratropical cyclones Low pressure system formed outside of tropics. Shapes are variable for weak and moderate strength storms, with intense storms tending to be elliptical or circular.	Extreme waves in most open-ocean areas north of 35° are produced by these systems. Large waves tend to lie in region of storm with winds parallel to direction of storm movement. Predominant source of swell for most U.S. east coast and west coast areas.	Weak: H 3-5m T 5-10 sec Moderate: H 5-8m T 9-13 sec Intense: H 8-12m T 12-17sec Extreme: H 13-18m T 15-20sec
Migratory highs Slowly moving high-pressure systems.	Produce moderate storm conditions along U.S. east coast south of 30° latitude when pressure gradients become steep.	H 1 - 4 m T 4 - 10 sec
Stationary highs Permanent systems located in subtropical ocean areas. Southern portions constitute the trade winds.	Produce low swell-like waves due to long fetches. Can interact with synoptic-scale and large-scale weather systems to produce moderately intense wave generation. Very persistent wave regime.	H 1 - 3 m T 5 - 10 sec

(Sheet 2 of 3)

Table II-2-2 (Concluded)

<u>Monsoonal winds</u> Biannual outbursts of air from continental land masses.	Episodic wave generation can generate large wave conditions. Very important in the Indian Ocean, part of the Gulf of Mexico, and some U.S. east coast areas.	H 4 - 7 m T 6 - 11 sec
<u>Long-wave generation</u>	Long waves can be generated by moving pressure/wind anomalies (such as can be associated with fronts and squall lines) and can resonate with long waves if the speed of frontal or squall line motion is approximately $\sqrt{g\,d}$. Examples of this phenomenon have been linked to inundations of piers and beach areas in Lake Michigan and Daytona Beach in recent years.	
<u>Gap winds</u> Wind acceleration due to local topographic funneling.	These winds may be extremely important in generating waves in many U.S. west coast areas not exposed to open-ocean waves.	U ≈ 40 m/s

(Sheet 3 of 3)

h.   Winds in hurricanes.

(1)   In tropical and in some subtropical areas, organized cloud clusters form in response to perturbations in the regional flow. If a cloud cluster forms in an area sufficiently removed from the Equator, then Coriolis accelerations are not negligible and an organized, closed circulation can form. A tropical system with a developed circulation but with wind speeds less than 17.4 m/s (39 mph) is termed a tropical depression. Given that conditions are favorable for continued development (basically warm surface waters, little or no wind shear, and a high pressure area aloft), this circulation can intensify to the point where sustained wind speeds exceed 17.4 m/s, at which time it is termed a tropical storm. If development continues to the point where the maximum sustained wind speed equals or exceeds 33.5 m/s (75 mph), the storm is termed a hurricane. If such a storm forms west of the international date line, it is called a typhoon. In this section, the generic term hurricane includes hurricanes and typhoons, since the primary distinction between them is their point of origin. Tropical storms will also follow some of the wind models given in this section, but since these storms are weaker, they tend to be more poorly organized.

(2)   Although it might be theoretically feasible to model a hurricane with a primitive equation approach (i.e. to solve the coupled dynamic and thermodynamic equations directly), information to drive such a model is generally lacking and the roles of all of the interacting elements within a hurricane are not well-known. Consequently, practical hurricane wind models for most applications are driven by a set of parameters that characterize the size, shape, rate of movement, and intensity of the storm, along with some parametric representation of the large-scale flow in which the hurricane is imbedded. Myers (1954); Collins and Viehmann (1971); Schwerdt, Ho, and Watkins (1979); Holland (1980); and Bretschneider (1990) all describe and justify various parametric approaches to wind-field specification in tropical storms. Cardone, Greenwood, and Greenwood (1992) use a modified form of Chow's (1971) moving vortex model to specify winds with a gridded numerical model. However, since this numerical solution is driven only by a small set of parameters and assumes steady-state conditions, it produces results that are similar in form to those of parametric models (Cooper 1988). Cardone et al. (1994) and Thompson and Cardone (1996) describe a more general model version that can approximate irregularities in the radial wind profile such as the double maxima observed in some hurricanes.

(3)   All of the above models have been shown to work relatively well in applications; however, the Holland (1980) model appears to provide a better fit to observed wind fields in early stages of rapidly

Meteorology and Wave Climate

developing storms and appears to work as well as other models in mature storms. Consequently, this model will be described in some detail here. In presently available hurricane models, wind fields are assumed to have no memory and thus can be determined by only a small set of parameters at a given instant.

(4) In the Holland model, hurricane pressure profiles are normalized via the relationship

$$\beta = \frac{p - p_c}{p_n - p_c} \qquad\qquad \text{(II-2-15)}$$

where

p = pressure at radius r

r = arbitrary radius

p_c = central pressure in the storm

p_n = ambient pressure at the periphery of the storm

(5) Holland showed that the family of β-curves for a number of storms resembled a family of rectangular hyperbolas and could be represented as

$$r^B \ln (\beta^{-1}) = A$$

or

$$\beta^{-1} = \exp\left(\frac{A}{r^B}\right) \qquad\qquad \text{(II-2-16)}$$

or

$$\beta = \exp\left(\frac{-A}{r^B}\right)$$

A = scaling parameter with units of length

B = dimensionless parameter that controls the peakedness of the wind speed distribution

(6) This leads to a representation for the pressure profile as

$$p = p_c + (p_n - p_c) \exp\left(\frac{-A}{r^B}\right) \qquad\qquad \text{(II-2-17)}$$

which then leads to a gradient wind approximation of the form

$$U_{gr} = \left[\frac{AB(p_n - p_c)\exp\left(\frac{-A}{r^B}\right)}{\rho_a r^B} + \frac{r^2 f^2}{4}\right]^{\frac{1}{2}} - \frac{rf}{2} \qquad\qquad \text{(II-2-18)}$$

where

U_{gr} = gradient approximation to the wind speed

(7)   In the intense portion of the storm, Equation II-2-18 reduces to a cyclostrophic approximation given by

$$U_c = \left[\frac{AB(p_n - p_c) \exp\left(\frac{-A}{r^B}\right)}{\rho_a r^B} \right]^{\frac{1}{2}}$$ (II-2-19)

where

U_c = cyclostrophic approximation to the wind speed

which yields explicit forms for the radius to maximum winds as

$$R_{\text{max}} = A^{\frac{1}{B}}$$ (II-2-20)

where

R_{max} = distance from the center of the storm circulation to the location of maximum wind speed

(8)   The maximum wind speed can then be approximated as

$$U_{\text{max}} = \left(\frac{B}{\rho_a e} \right)^{\frac{1}{2}} (p_n - p_c)^{\frac{1}{2}}$$ (II-2-21)

where

U_{max} = maximum velocity in the storm

e = base of natural logarithms, 2.718

(9)   Rosendal and Shaw (1982) showed that pressure profiles and wind estimates from the Holland model appeared to fit observed typhoon characteristics in the central North Pacific. If B is equal to 1 in this model, the pressure profile and wind characteristics become similar to results of Myers (1954); Collins and Viehmann (1971); Schwerdt, Ho, and Watkins (1979); and Cardone, Greenwood, and Greenwood (1992). In the case of the Cardone, Greenwood, and Greenwood model, this similarity would exist only for the case of a storm with no significant background pressure gradient.

(10) Holland argues that B=1 is actually the lower limit for B and that, in most storms, the value is likely to be more in the range of 1.5 to 2.5. As shown in Figure II-2-16, this argument is supported by the data from Atkinson and Holliday (1977) and Dvorak (1975) taken from studies of Pacific typhoons. The effect of a higher value of B is to produce a more peaked wind distribution in the Holland model than exists in models with B set to a value of 1. According to Holland (1980), use of a wind field model with B=1 will underestimate winds in many tropical storms. In applications, the choices of A and B can either be based on

the best two-parameter fit to observed pressure profiles or on the combination of an R_{max} value with the data shown in Figure II-2-16. It is worth noting here that the Holland model is similar to several other parametric models, except that it uses two parameters rather than one in describing the shape of the wind profile. This second parameter allows the Holland model to represent a range of peakedness rather than only a single peakedness in applications.

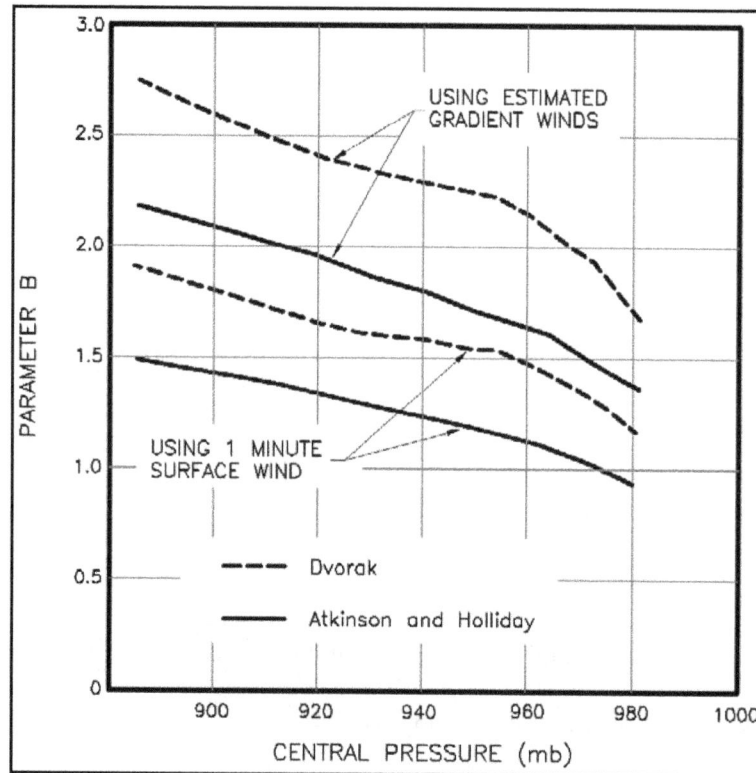

Figure II-2-16. Climatological variation in Holland's "B" factor (Holland 1980)

(11) As a final element in application of the Holland wind model, it is necessary to consider the effects of storm movement on the surface wind field. Since a hurricane moves most of its mass along with it (unlike an extratropical storm), this step is a necessary adjustment to the storm wind field and can create a marked asymmetry in the storm wind field, particularly for the case of weak or moderate storms. Hughes' (1952) composite wind fields from moving hurricanes indicated that the highest wind speeds occurred in the right rear quadrant of the storm. This supports the interpretation that the total wind in a hurricane can be obtained by adding a wind vector for storm motion to the estimated winds for a stationary storm. On the other hand, Chow's (1971) numerical results suggest that winds in the front right and front left quadrants are more likely to contain the maximum wind speeds in a moving hurricane. These contradictory results have made it difficult to treat the effects of storm movement of surface wind fields in a completely satisfactory fashion. Various researchers have either ignored the problem or suggested that, at least in simple parametric models, the effects of storm movement can be adequately approximated by adding a constant vector representative of the forward storm motion to the estimated wind for a stationary storm. In light of the overall lack of definitive information on this topic, the latter approach is considered sufficient.

(12) At this point, it should be stressed that Equations II-2-18, 19, and 21 and superposition of the storm motion vector are only applicable to winds above the surface boundary layer. In order to convert these winds to winds at a 10-m reference level, it is necessary to apply a model of the type described in Part II-2-1-c-(3)-

(b). As shown in that section, it is not advisable to use a constant ratio between winds at the top of the boundary layer and winds at a 10-m level. If a complete wind field is required for a particular application it is recommended to use a planetary-boundary-layer (PBL) model combined with either a moving vortex formulation or a numerical version of a parametric model.

(13) To provide some guidance regarding maximum sustained wind speeds at a 10-m reference level, Figure II-2-17 shows representative curves of maximum sustained wind speed versus central pressure for selected values of forward storm movement. It should be noted that maximum winds at the top of the boundary layer are relatively independent of latitude, since the wind balance equation is dominated by the cyclostrophic term; however, there is a weak dependence on latitude through the boundary-layer scaling, which is latitude-dependent. This dependence and dependence of the maximum wind speed on the radius to maximum wind were both found to be rather small; consequently, only fixed values of latitude and R_{max} have been treated here. From the methods used in deriving these estimates, winds given here can be regarded as typical values for about a 15- to 30-min averaging period. Thus, winds from this model are appropriate for use in wave models and surge models, but must be transformed to shorter averaging times for most structural applications.

(14) Values for wind speeds in Figure II-2-17 may appear low to people who recall reports of maximum wind speeds for many hurricanes in the range of 130-160 mph (about 58-72 m/s). First, it should be recognized that very few good measurements of hurricane wind speed exist today. Where such measurements exist, they give support to the values presented in Figure II-2-17. Second, the values reported as sustained wind speeds often come from airplane measurements, so they tend to be considerably higher than corresponding values at 10 m. Third, winds at airports and other land stations often use only a 1-min averaging time in their wind speed measurements. These winds are subsequently reported as sustained wind speeds. An idea of the magnitude that some of these effects can have on wind estimates may be gained via the following example. The central pressure of Hurricane Camille as it moved onshore at a speed of about 6 m/s in 1969 was about 912 mb. From Figure II-2-17, the 15- to 30-min average wind speed is estimated to be 52.5 m/s. Converting this to a 1-min wind speed in miles per hour yields approximately 150 mph, which is in very reasonable agreement with the measured and estimated winds in this storm. It is important to recognize though that these higher wind speeds are not appropriate for applications in surge and wave models.

(15) Figures II-2-18 and II-2-19 are examples of the output from the hurricane model presented here. Figure II-2-18 shows the four radials. Figure II-2-19 shows wind speed along Radials 1 and 3, as a function of dimensionless distance along the radial (r/R_{max}) for a central pressure p_c of 930 mb and forward speeds of 2.5 m/s, 5.0 m/s, and 7.5 m/s. The inflow angle along these radii (not shown) can be quite variable. The behavior of this angle is a function of several factors and is still the subject of some debate.

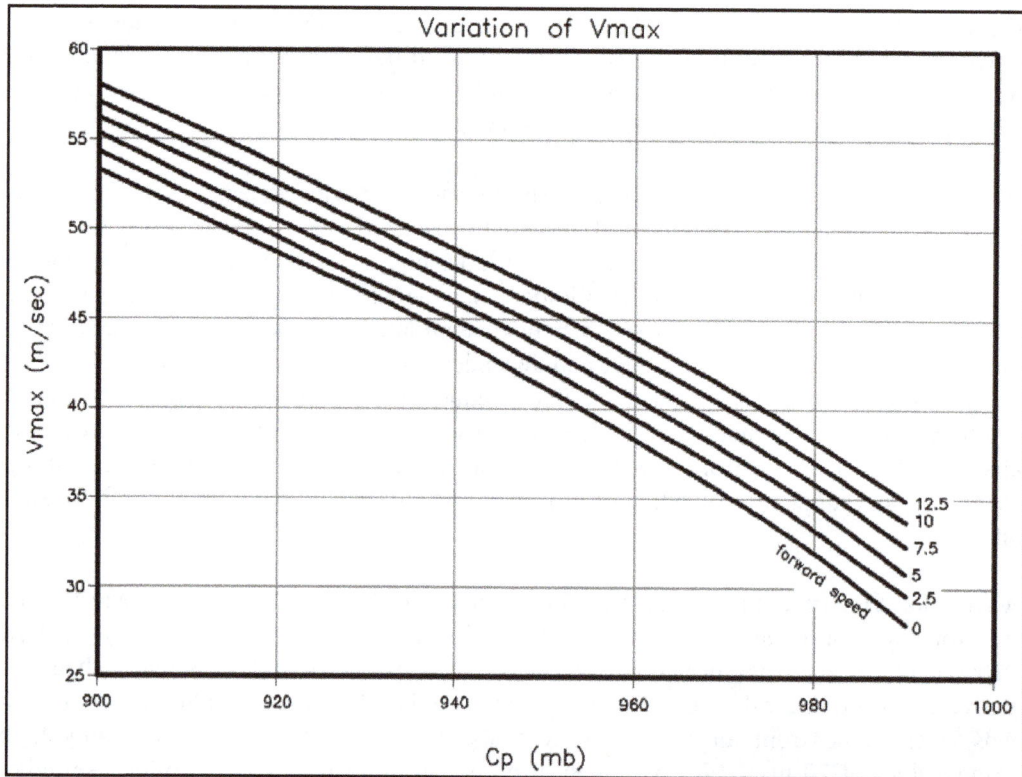

Figure II-2-17.   Relationship of estimated maximum wind speed in a hurricane at 10-m elevation as a function of central pressure and forward speed of storm (based on latitude of 30 deg, R_{max}=30 km, 15- to 30-min averaging period)

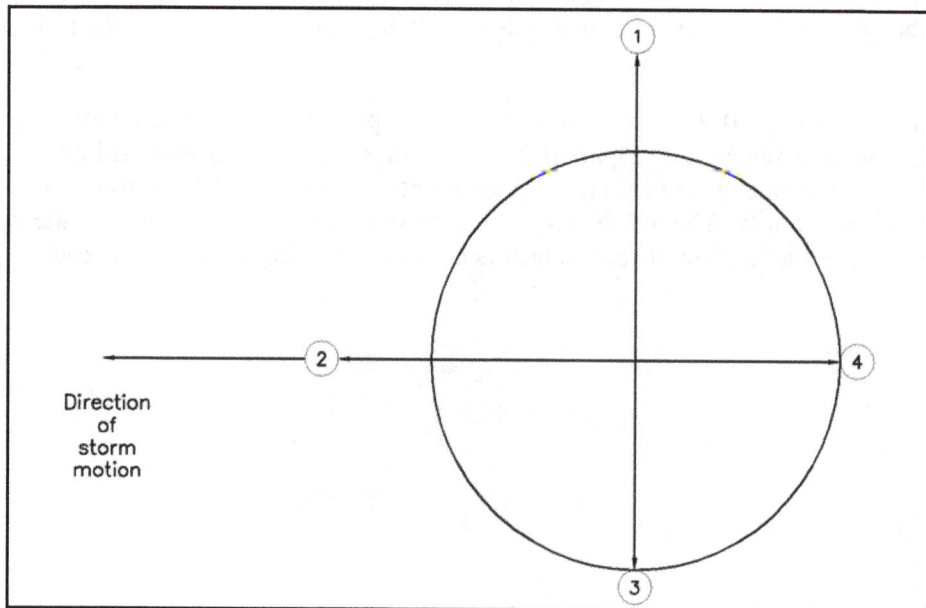

Figure II-2-18.   Definition of four radial angles relative to direction of storm movement

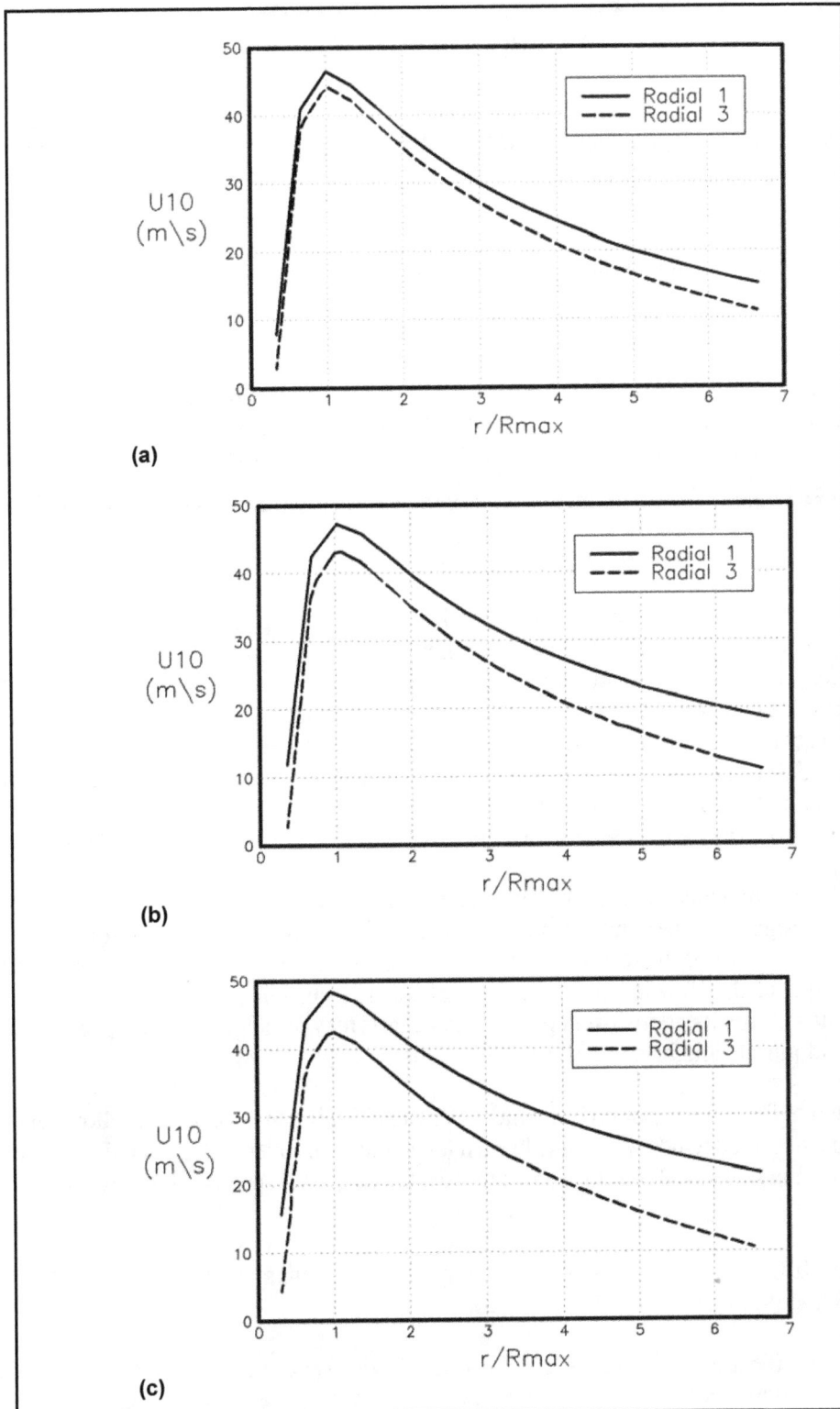

Figure II-2-19. Horizontal distribution of wind speed along Radial 1 for a storm with forward velocity V_F of (a) 2.5 m/s; (b) 5 m/s; (c) 7.5 m/s

---

EXAMPLE PROBLEM II-2-6

FIND:
The expected maximum sustained wind speed for this storm for surge and/or wave prediction and the maximum 1-min wind speed.

GIVEN:
A hurricane located at a latitude of 28° with a central pressure of 935 mb and a forward velocity of 10 m/s.

SOLUTION:
Using Figure II-2-17, the maximum wind speed in a moving storm with the parameters given here is approximately 47.3 m/s for a 15- to 30-min average at the 10-m level. From Figure II-2-1, the ratio of a 30-min wind (chosen here to give a conservative approximation) to a 1-min wind is approximately 1.23. Multiplying this factor times 47.3 yields a 1-min wind speed of 58.2 m/s (130 mph).

---

i.   Step-by-step procedure for simplified estimate of winds for wave prediction.

(1)   Introduction. This section presents simplified, step-by-step methods for estimating winds to be used in wave prediction. The methods include the key assumption that wind fields are well-organized and can be adequately represented as an average wind speed and direction over the entire fetch. Most engineers can conveniently use computer-based wind estimation tools such as ACES, and such tools should be used in preference to the corresponding methods in this section. The simplified methods provide an approximation to the processes described earlier in this chapter. The methods embody graphs presented earlier, some of which were generated with ACES. The simplified methods are particularly useful when quick, low-cost estimates are needed. They are reasonably accurate for simple situations where local effects are small.

(2)   Wind measurements. Winds can be estimated using direct measurements or synoptic weather charts. For preliminary design, extreme winds derived from regional records may also be useful (Part II-9-6). Actual wind records from the site of interest are preferred so that local effects such as orographic influences and sea breeze are included. If wind measurements at the site are not available and cannot be collected, measurements at a nearby location or synoptic weather charts may be helpful. Wind speeds must be properly adjusted to avoid introducing bias into wave predictions.

(3)   Procedure for adjusting observed winds. When ACES is unavailable, the following procedure can be used to adjust observed winds with some known level, location (over water or land), and averaging time. A logic diagram (Figure II-2-20) outlines the steps in adjusting wind speeds for application in wave growth models.

(a)   Level. If the wind speed is observed at any level other than 10 m, it should be adjusted to 10 m using Figure II-2-6 (see Example Problem II-2-3).

(b)   Duration. If extreme winds are being considered, wind speed should be adjusted from the averaging time of the observation (fastest mile, 5-min average, 10-min average, etc.) to an averaging time appropriate for wave prediction using Figure II-2-1 (see Example Problem II-2-1). Typically several different averaging times should be considered for wave prediction to ensure that the maximum wave growth scenario has been identified. When the fetch is limited, Figure II-2-3 can be used to estimate the maximum averaging time to

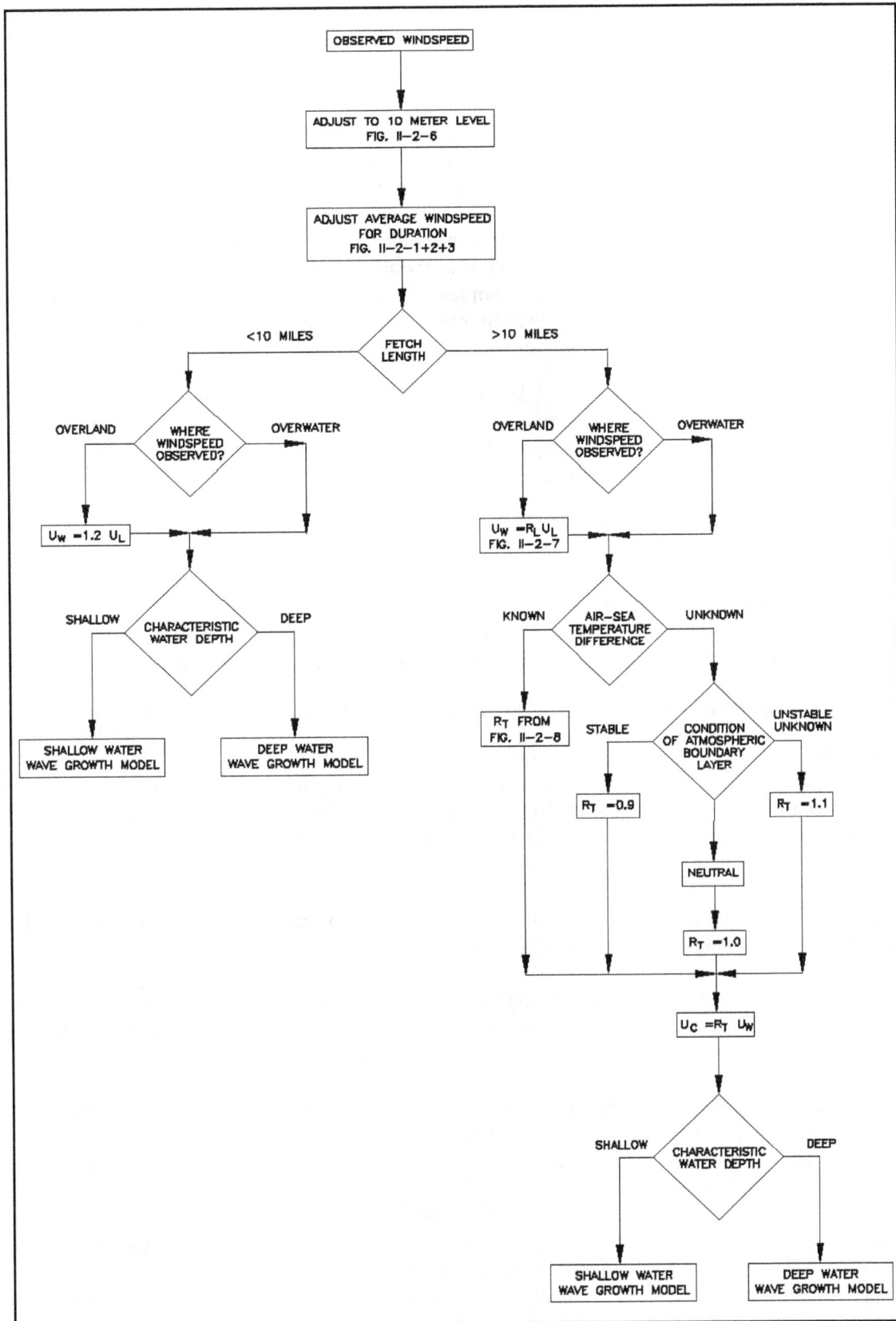

Figure II-2-20. Logic diagram for determining wind speed for use in wave hindcasting and forecasting models

be considered. When the observed wind is given in terms of the fastest mile, Figure II-2-2 can be used to convert to an equivalent averaging time.

(c) Overland or overwater. When the observation was collected overwater (within the marine boundary layer), this adjustment is not needed. When the observation was collected overland and the fetch is long enough for full development of a marine boundary layer (longer than about 16 km or 10 miles), the observed wind speed should be adjusted to an overwater wind speed using Figure II-2-7 (see Example Problem II-2-4). Otherwise (for overland winds and fetches less than 16 km), wave growth occurs in a transitional atmospheric boundary layer, which has not fully adjusted to the overwater regime. In this case, wind speeds observed overland must be increased to better represent overwater wind speeds. A factor of 1.2 is suggested here, but no simple method can accurately represent this complex case. In relation to all of these adjustments, the term overland implies a measurement site that is predominantly characterized as inland. If a measurement site is directly adjacent to the water body, it may, for some wind directions, be equivalent to overwater.

(d) Stability. For fetches longer than 16 km, an adjustment for stability of the boundary layer may also be needed. If the air-sea temperature difference is known, Figure II-2-8 can be used to make the adjustment. When only general knowledge of the condition of the atmospheric boundary layer is available, it should be categorized as stable, neutral, or unstable according to the following:

Stable - when the air is warmer than the water, the water cools air just above it and decreases mixing in the air column ($R_T = 0.9$).

Neutral - when the air and water have the same temperature, the water temperature does not affect mixing in the air column ($R_T = 1.0$).

Unstable - when the air is colder than the water, the water warms the air, causing air near the water surface to rise, increasing mixing in the air column ($R_T = 1.1$).

When the boundary layer condition is unknown, an unstable condition, $R_T = 1.1$, should be assumed.

(4) Procedure for adjusting winds from synoptic weather charts. As discussed earlier, synoptic weather charts are maps depicting isobars at sea level. The free air, or geostrophic, wind speed is estimated from these sea level pressure charts. Adjustments or corrections are then made to the geostrophic wind speed. Pressure chart estimations should be used only for large areas, and the estimated values should be compared with observations, if possible, to verify their accuracy.

(a) Geostrophic wind speed. To estimate geostrophic wind speed, Equation II-2-10 or Figure II-2-12 should be used (see Example Problem II-2-5).

(b) Level and stability. Wind speed at the 10-m level should be estimated from the geostrophic wind speed using Figure II-2-13. The resulting speed should then be adjusted for stability effects as needed using Figure II-2-8.

(c) Duration. Wind duration estimates are also needed. Since synoptic weather charts are prepared only at 6-hr intervals, it may be necessary to use interpolation to determine duration. Linear interpolation is adequate for most cases. Interpolation should not be used if short-duration phenomena, such as frontal passages or thunderstorms, are present.

(5) Procedure for estimating fetch. Fetch is defined as a region in which the wind speed and direction are reasonably constant. Fetch should be defined so that wind direction variations do not exceed 15 deg and wind speed variations do not exceed 2.5 m/s (5 knots) from the mean. A coastline upwind from the point of interest always limits the fetch. An upwind limit to the fetch may also be provided by curvature, or spreading, of the isobars or by a definite shift in wind direction. Frequently the discontinuity at a weather front will limit fetch.

II-2-2. Wave Hindcasting and Forecasting

    a.   Introduction.

    (1)   The theory of wave generation has had a long and rich history. Beginning with some of the classic works of Kelvin (1887) and Helmholtz (1888) in the 1800's, many scientists, engineers, and mathematicians have addressed various forms of water wave motions and interactions with the wind. In the early 1900's, the work of Jeffreys (1924, 1925) hypothesized that waves created a "sheltering effect" and hence created a positive feedback mechanism for transfer of momentum into the wave field from the wind. However, it was not until World War II that organized wave predictions began in earnest. During the 1940's, large bodies of wave observations were collated and the bases for empirical wave predictions were formulated. Sverdrup and Munk (1947, 1951) presented the first documented relationships among various wave-generation parameters and resulting wave conditions. Bretschneider (1952) revised these relationships based on additional evidence; methods derived from these exemplary pioneer works are still in active use today.

    (2)   The basic tenet of the empirical prediction method is that interrelationships among dimensionless wave parameters will be governed by universal laws. Perhaps the most fundamental of these laws is the fetch-growth law. Given a constant wind speed and direction over a fixed fetch, it is expected that waves will reach a stationary fetch-limited state of development. In this situation, wave heights will remain constant (in a statistical sense) through time but will vary along the fetch. If dimensionless wave height is taken as

$$\hat{H} = \frac{gH}{u_*^2} \qquad\qquad (\text{II-2-22})$$

where

    H = characteristic wave height, originally taken as the significant wave height but more recently taken as the energy-based wave height H_{m0}

    u_* = friction velocity

and dimensionless fetch is defined as

$$\hat{X} = \frac{gX}{u_*^2} \qquad\qquad (\text{II-2-23})$$

where

    X = straight line distance over which the wind blows

then idealized, fetch-limited wave heights are expected to follow a relationship of the form

$$\hat{H} = \lambda_1 \hat{X}^{m_1} \qquad\qquad (\text{II-2-24})$$

where

    λ_1 = dimensionless coefficient

    m_1 = dimensionless exponent

(3) If dimensionless wave frequency (defined simply as one over the spectral peak wave period) defined as

$$\hat{f}_p = \frac{u_* \, f_p}{g} \tag{II-2-25}$$

where

f_p = frequency of the spectral peak

then a stationary wave field also implies a fixed relationship between wave frequency and fetch of the form

$$\hat{f}_p = \lambda_2 \, \hat{X}^{m_2} \tag{II-2-26}$$

where

λ_2 and m_2 are more empirical coefficients.

(4) Since u_* scales the effective rate of momentum transfer from the atmosphere into the waves, all empirical coefficients in these wave generation laws are expected to be universal values. Unfortunately, there is still some ambiguity in these values; however, in lieu of any demonstrated improvements over values from the *Shore Protection Manual* (1984), those values for fetch-limited wave growth will be adopted here.

(5) From basic conservation laws and the dispersion relationship, it is anticipated that any law governing the rate of growth of waves along a fetch will also form a unique constraint on the rate of growth of waves through time. If we define dimensionless time as

$$\hat{t} = \frac{g \, t}{u_*} \tag{II-2-27}$$

where

t = time

additional relationships governing the duration-growth of waves will be

$$\hat{H} = \lambda_3 \, \hat{t}^{m_3} \tag{II-2-28}$$

and

$$\hat{f}_p = \lambda_4 \, \hat{t}^{m_4} \tag{II-2-29}$$

where

λ_4 and m_4 are more "universal" coefficients to be determined empirically.

(6) The form of Equations II-2-26 and II-2-27 imply that waves will continue to grow as long as fetch and time continue to increase. This concept was observed to be incorrect in the early compendiums of data

(Sverdrup and Munk 1947, Bretschneider 1952), which suggested that a "fully developed" wave height would evolve under the action of the wind. Available data indicated that this fully developed wave height could be represented as

$$H_\infty = \frac{\lambda_5 \, u^2}{g}$$ (II-2-30)

where

H_∞ = fully developed wave height

λ_5 = dimensionless coefficient (approximately equal to 0.27)

u = wind speed

Wave heights defined by Equation II-2-30 are usually taken as representing an upper limit to wave growth for any wind speed.

(7) In the 1950's, researchers began to recognize that the wave generation process was best described as a spectral phenomenon (e.g. Pierson, Neumann, and James (1955)). Theoreticians then began to reexamine their ideas on the wave-generation process, with regard to how a turbulent wind field could interact with a random sea surface. Following along these lines, Phillips (1958) and Miles (1957) advanced two theories that formed the cornerstone of the understanding of wave generation physics for many years. Phillips' concept involved the resonant interactions of turbulent pressure fluctuations with waves propagating at the same speed. Miles' concept centered on the mean flux of momentum from a "matched layer" above the wave field into waves travelling at the same speed. Phillips' theory predicted linear wave growth and was believed to control the early stages of wave growth. Miles' theory predicted an exponential growth and was believed to control the major portion of wave growth observed in nature. Direct measurements of the Phillips' resonance mechanism indicated that the measured turbulent fluctuations were too small by about an order of magnitude to explain the observed early growth in waves; however, it was still adopted as a plausible concept. Subsequent field efforts by Snyder and Cox (1966) and Snyder et al. (1981) have supported at least the functional form of Miles' theory for the transfer of energy into the wave field from winds.

(8) From basic concepts of energy conservation and the fact that waves do attain limiting fully developed wave heights, it is obvious that wave generation physics cannot consist of only wind source terms. There must be some physical mechanism or mechanisms that leads to a balance of wave growth and dissipation for the case of fully developed conditions. Phillips (1958) postulated that one such mechanism in waves would be wave breaking. Based on dimensional considerations and the knowledge that wave breaking has a very strong local effect on waves, Phillips argued that energy densities within a spectrum would always have a universal limiting value given by

$$E(f) = \frac{\alpha \, g^2 f^{-5}}{(2\pi)^4}$$ (II-2-31)

where E(f) is the spectral energy density in units of length squared per hertz and α was understood to be a universal (dimensionless) constant approximately equal to 0.0081. It should be noted here that energy densities in this equation are proportional to f^5 (as can be deduced from dimensional arguments) and that they are independent of wind speed. Phillips hypothesized that local wave breaking would be so strong that wind effects could not affect this universal level. In this context, a saturated region of spectral energy densities is assumed to exist in some region from near the spectral peak to frequencies sufficiently high that viscous effects would begin to be significant. This region of saturated energy densities is termed the equilibrium range of the spectrum.

(9) Kitaigorodskii (1962) extended the similarity arguments of Phillips to distinct regions throughout the entire spectrum where different mechanisms might be of dominant importance. Pierson and Moskowitz (1964) followed the dimensional arguments of Phillips and supplemented these arguments, with relationships derived from measurements at sea. They extended the form of Phillips spectrum to the classical Pierson-Moskowitz spectrum

$$E(f) = \frac{\alpha \ g^2 \ f^{-5}}{(2\pi)^4} \ \exp\left[-0.74\left(\frac{f}{f_u}\right)^{-4}\right] \tag{II-2-32}$$

where

f_u = limiting frequency for a fully developed wave spectrum (assumed to be a function only of wind speed)

(10) Based on these concepts of spectral wave growth due to wind inputs via Miles-Phillips mechanisms and a universal limiting form for spectral densities, first-generation (1G) wave models in the United States were born (Inoue 1967, Bunting 1970). It should be pointed out here that the first model of this type was actually developed in France (Gelci, Cazale, and Vassel 1957); however, that model did not incorporate the limiting Pierson-Moskowitz spectral form as did models in the United States. In these models, it was recognized that waves in nature are not only made up of an infinite (continuous) sum of infinitesimal wave components at different frequencies but that each frequency component is made up of an infinite (continuous) sum of wave components travelling in different directions. Thus, when waves travel outward from a storm, a single "wave train" moving in one direction does not emerge. Instead, directional wave spectra spread out in different directions and disperse due to differing group velocities associated with different frequencies. This behavior cannot be modeled properly in parametric (significant wave height) models and understanding of this behavior formed the basic motivation to model all wave components in a spectrum individually. The term discrete-spectral model has since been employed to describe models that include calculations of each separate (frequency-direction) wave component. The equation governing the energy balance in such models is sometimes termed the radiative transfer equation and can be written as

$$\frac{\partial E(f,\theta,x,y,t)}{\partial t} = -c_G \ \vec{\nabla} \ E(f,\theta,x,y,t) + \sum_{k=1}^{K} S(f,\theta,x,y,t)_k \tag{II-2-33}$$

where

$E(f, \theta, x, y, t)$ = spectral energy density as a function of frequency (f), propagation direction (θ), two horizontal spatial coordinates (x and y) and time (t)

$S(f, \theta, x, y, t)_k$ = the kth source term, which exists in the same five dimensions as the energy density

The first term on the right side of this equation represents the effects of wave propagation on the wave field. The second term represents the effects of all processes that add energy to or remove energy from a particular frequency and direction component at a fixed point at a given time.

(11) In the late 1960's evidence of spectral behavior began to emerge which suggested that the equilibrium range in wave spectra did not have a universal value for α. Instead, it was observed that α varied

as a function of nondimensional fetch (Mitsuyasu 1968). This presented a problem to the "first-generation" interpretation of wave generation physics, since it implied that energies within the equilibrium range are not controlled by wave breaking. Fortunately, a theoretical foundation already existed to help explain this discrepancy. This foundation had been established in 1961 in an exceptional theoretical formulation by Klaus Hasselmann in Germany. In this formulation, Hasselmann, using relatively minimal assumptions, showed that waves in nature should interact with each other in such a way as to spread energy throughout a spectrum. This theory of wave-wave interactions predicted that energy near the spectral peak region should be spread to regions on either side of the spectral peak.

(12) Hasselmann et al. (1973) collected an extensive data set in the Joint North Sea Wave Project (JONSWAP). Careful analysis of these data confirmed the earlier findings of Mitsuyasu and revealed a clear relationship between Phillips' α and nondimensional fetch (Figure II-2-21). This finding and certain other spectral phenomena, such as the tendency of wave spectra to be more peaked than the Pierson-Moskowitz spectrum during active generation, could not be explained in terms of "first-generation" concepts; however, they could be explained in terms of a nonlinear interaction among wave components. This pointed out the necessity of incorporating wave-wave interactions into wave prediction models, and led to the development of second-generation (2G) wave models. The modified spectral shape which came out of the JONSWAP experiment has come to bear the name of that experiment; hence we now have the JONSWAP spectrum, which can be written as

$$E(f) = \frac{\alpha g^2}{(2\pi)^4 f^5} \exp\left[-1.25\left(\frac{f}{f_p}\right)^{-4}\right] \gamma^{\exp\left[-\frac{\left(\frac{f}{f_p}-1\right)^2}{2\sigma^2}\right]} \qquad \text{(II-2-34)}$$

where

α = equilibrium coefficient

σ = dimensionless spectral width parameter, with value σ_a for $f < f_p$ and value σ_b for $f \geq f_p$

γ = peakedness parameter

The average values of the σ and γ parameters in the JONSWAP data set were found to be $\gamma = 3.3$, $\sigma_a = 0.07$, and $\sigma_b = 0.09$. Figure II-2-22 compares this spectrum to the Pierson-Moskowitz spectrum.

(13) Early second-generation models (Barnett 1968, Resio 1981) followed an f^5 equilibrium-range formulation since prior research had been formulated with that spectral form. Toba (1978) was the first researcher to present data suggesting that the equilibrium range in spectra might be better fit by an f^4 dependence. Following his work, Forristall et al. (1978); Kahma (1981); and Donelan, Hamilton, and Hu (1982) all presented evidence from independent field measurements supporting the tendency of equilibrium ranges to follow an f^4 dependence. Kitaigorodskii (1983); Resio (1987,1988); and Resio and Perrie (1989) have all presented theoretical analyses showing how this behavior can be explained by the nature of nonlinear fluxes of energy through a spectrum. Subsequently, Resio and Perrie (1989) determined that, although certain spectral growth characteristics were somewhat different between the f^4 and f^5 formulations,

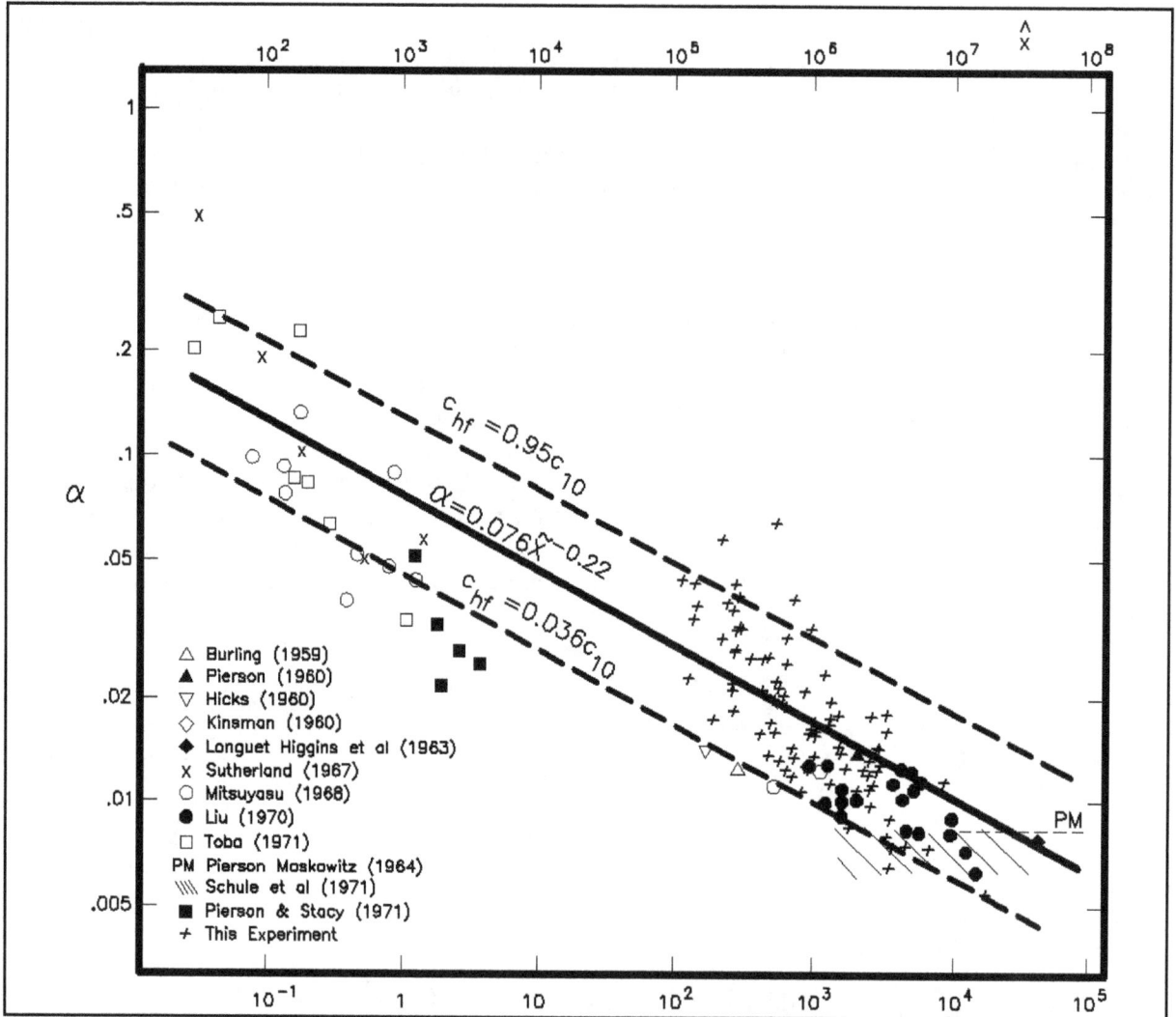

Figure II-2-21. Phillips' constant versus fetch scaled according to Kitaigorodskii. Small-fetch data are obtained from wind-wave tanks. Capillary-wave data were excluded where possible (Hasselmann et al. 1973)

the basic energy-growth equations were quite similar for the two formulations. The f^4 formulation is incorporated into CERC's WAVAD model, used in its hindcast studies.

(14) Since the early 1980's, a new class of wave model has come into existence (Hasselmann et al. 1985). This new class of wave model has been termed a third-generation wave model (3G). The distinction between second-generation and third-generation wave models is the method of solution used in these models. Second-generation wave models combine relatively broad-scale parameterizations of the nonlinear wave-wave interaction source term combined with constraints on the overall spectral shape to simulate wave growth. Third-generation models use a more detailed parameterization of the nonlinear wave-wave interaction source terms and relax most of the constraints on spectral shape in simulating wave growth. Various third-generation models are used around the world today; however, the third-generation model is probably the WAM model.

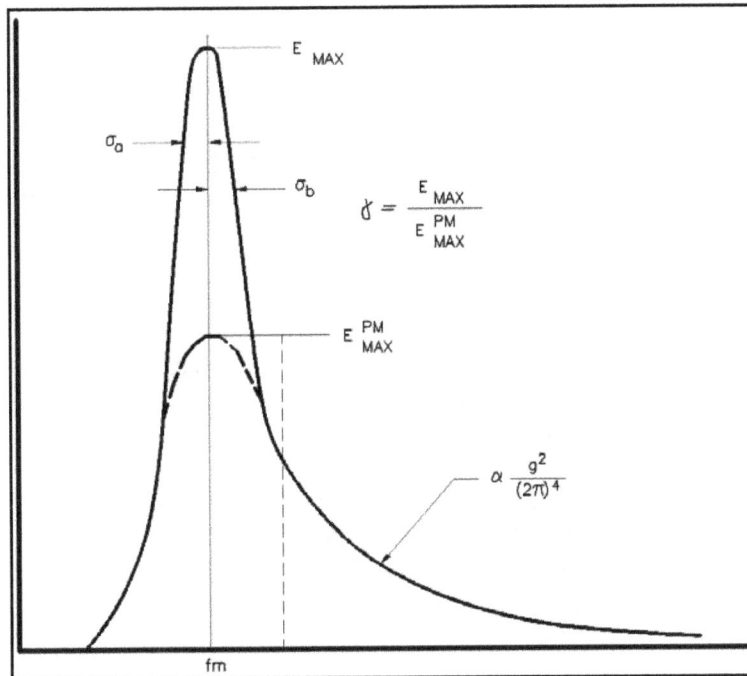

$$\gamma = \frac{E_{MAX}}{E_{MAX}^{PM}}$$

$$\alpha \frac{g^2}{(2\pi)^4}$$

Figure II-2-22. Definition of JONSWAP parameters for spectral shape

(15) Part of the motivation to use third-generation models is related to the hope that future simulations of directional spectra can be made more accurate via the direct solution of the detailed source-term balance. This is expected to be particularly important in complex wave generation scenarios where second-generation models might not be able to handle the general source term balance. However, recent research by Van Vledder and Holthuisen (1993) has demonstrated rather convincingly that the "detailed balance" equations in the WAM (WAMDI Group 1988)model at this time still cannot accurately simulate waves in rapidly turning winds. Hence, there remains much work to be done in this area before the performance of third-generation models can be considered totally satisfactory.

(16) First-generation models that have been modified to allow the Phillips equilibrium coefficient to vary dynamically (Cardone 1992) , second-generation models (Resio 1981; NORSWAM, Hubertz 1992), and third-generation models (Hasselmann et al. 1985) have all been shown to produce very good predictions and hindcasts of wave conditions for a wide range of meteorological situations. These models are recommended in developing wave conditions for design and planning situations having serious economic or safety implications, and should be properly verified with local wave data, wherever feasible. This is not meant to imply that wave models can supplant wave measurements, but rather that in most circumstances, these models should be used instead of parametric models.

b. Wave prediction in simple situations. In some situations it is desirable to estimate wave conditions for preliminary considerations in project designs or even for final design in cases where total project costs are minimal. In the past, nomograms have played an important role in providing such wave information. However, with today's proliferation of user-friendly computer software such as the ACES Program, reliance on nomograms is discouraged. ACES will assist a user in his or her calculations, will facilitate most applications, and will help avoid most potential pitfalls related to misuse of wave prediction schemes. In spite of this warning and advice to use ACES, conventional prediction methods will be discussed here in order to provide such information for appropriate applications.

(1) Assumptions in simplified wave predictions.

(a) Deep water. There are three situations in which simplified wave predictions can provide accurate estimates of wave conditions. The first of these occurs when a wind blows, with essentially constant direction, over a fetch for sufficient time to achieve steady-state, fetch-limited values. The second idealized situation occurs when a wind increases very quickly through time in an area removed from any close boundaries. In this situation, the wave growth can be termed duration-limited. It should be recognized that this condition is <u>rarely</u> met in nature; consequently, this prediction technique should only be used with great caution. Open-ocean winds rarely can be categorized in such a manner to permit a simple duration-growth scenario. The third situation that may be treated via simplified prediction methods is that of a fully developed wave height. Knowledge of the fully developed wave height can provide valuable upper limits for some design considerations; however, open-ocean waves rarely attain a limiting wave height for wind speeds above 50 knots or so. Equation II-2-30 provides an easy means to estimate this limiting wave height.

(b) Wave growth with fetch. In this section, SI units should be used in formulas and figures. Figure II-2-3 shows the time required to accomplish fetch-limited wave development for short fetches. The general equation for this can be derived by combining the JONSWAP growth law for peak frequency, an equation for the fully developed frequency, and the assumption that a local wave field propagates at a group velocity approximately equal to 0.85 times the group velocity of the spectral peak. This factor accounts for both frequency distribution of energy in a JONSWAP spectrum and angular spreading. This yields

$$t_{x,u} = 77.23 \frac{X^{0.67}}{u^{0.34} g^{0.33}} \tag{II-2-35}$$

where

$t_{x,u}$ = time required for waves crossing a fetch of length x under a wind of velocity u to become fetch-limited

Equation II-2-35 can be used to determine whether or not waves in a particular situation can be categorized as fetch-limited.

The equations governing wave growth with fetch are

$$\frac{gH_{m_0}}{u_*^2} = 4.13 \times 10^{-2} * \left(\frac{gX}{u_*^2} \right)^{\frac{1}{2}}$$

and

$$\frac{gT_p}{u_*} = 0.751 \left(\frac{gX}{u_*^2} \right)^{\frac{1}{3}} \tag{II-2-36}$$

$$C_D = \frac{u_*^2}{U_{10}^2}$$

$$C_D = 0.001(1.1 + 0.035\, U_{10})$$

where

X = straight line fetch distance over which the wind blows (units of m)

H_{m0} = energy-based significant wave height (m)

C_D = drag coefficient

U_{10} = wind speed at 10 m elevation (m/sec)

u_* = friction velocity (m/sec)

See Demirbilek, Bratos, and Thompson (1993) for more details.

Fully developed wave conditions in these equations are given by

$$\frac{gH_{m_*}}{u_*^2} = 2.115 \times 10^2$$

$$and$$

$$\frac{gT_p}{u_*} = 2.398 \times 10^2$$

(II-2-37)

Equations governing wave growth with wind duration can be obtained by converting duration into an equivalent fetch given by

$$\frac{gX}{u_*^2} = 5.23 \times 10^{-3} \left(\frac{gt}{u_*} \right)^{\frac{3}{2}}$$

(II-2-38)

where t in this equation is the wind duration. The fetch estimated from this equation can then be substituted into the fetch-growth equations to obtain duration-limited estimates of wave height and period. An example demonstrating these procedures is provided at the end of this chapter.

(c) Narrow fetches. Early wave prediction nomograms included modifications to predicted wave conditions based on a sort of aspect ratio for a fetch area, based on the ratio of fetch width to fetch length. Subsequent investigations (Resio and Vincent 1979) suggested that wave conditions in fetch areas were actually relatively insensitive to the width of a fetch; consequently, it is recommended here that fetch width not be used to estimate an effective fetch for use in nomograms or the ACES Program. Instead, it is recommended that the straightline fetch be used to define fetch length for applications.

(d) Shallow water. Many studies suggest that water depth acts to modify wave growth. Bottom friction and percolation (Putnam 1949, Putnam and Johnson 1949, Bretschneider and Reid 1953) have been postulated as significant processes that diminish wave heights in shallow water; however, recent studies in shallow water (Jensen 1993) indicate that fetch-limited wave growth in shallow water appears to follow growth laws that are quite close to deepwater wave growth for the same wind speeds, up to a point where an asymptotic depth-dependent wave height is attained. In light of this evidence, it seems prudent to disregard bottom friction effects on wave growth in shallow water. Also, evidence from Bouws et al. (1985) indicates that wave spectra in shallow water do not appear to have a noticeable dependence on variations in bottom sediments. Consequently, it is recommended that deepwater wave growth formulae be used for all depths, with the constraint that no wave period can grow past a limiting value as shown by Vincent (1985). This limiting wave period is simply approximated by the relationship

$$T_p \approx 9.78 \left(\frac{d}{g} \right)^{\frac{1}{2}}$$

(II-2-39)

In cases with extreme amounts of material in the water column (for example sediment, vegetation, man-made structures, etc.), it is likely that the dissipation rate of wave energy will become very large. In such cases, Camfield's work (1977) may be used as a guideline for estimating frictional effects on wave growth and dissipation; however, it should be recognized that little experimental evidence exists to confirm the exact values of these dissipation rates.

(2) Prediction of deepwater waves from nomograms. Figures II-2-23 through II-2-26 are wave prediction nomograms under fetch-limited and duration-limited conditions. The curves in these nomograms are based on Equations II-2-30 and II-2-36 through II-2-38 presented previously in this section. The asymptotic upper limits in both cases provide information on the fully developed wave heights as a function of wind speed. The same information can be obtained more expediently via the ACES Program.

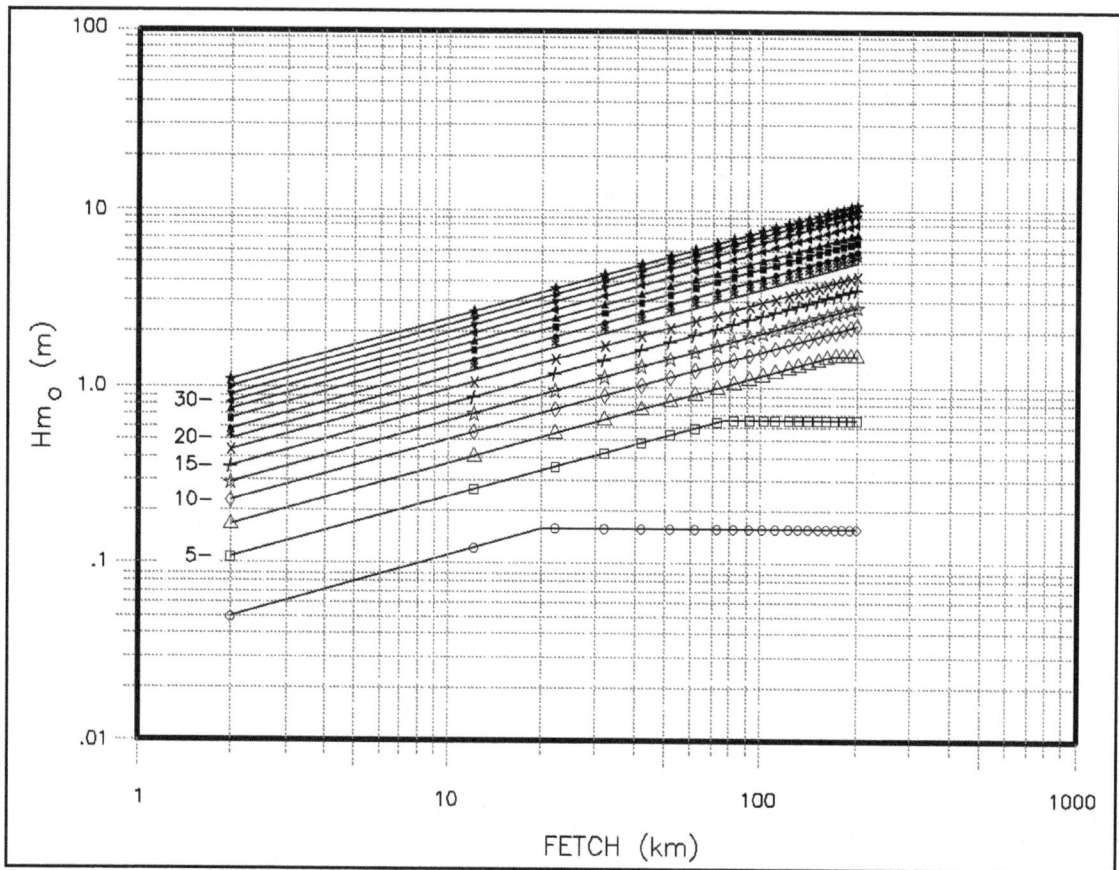

Figure II-2-23.   Fetch-limited wave heights

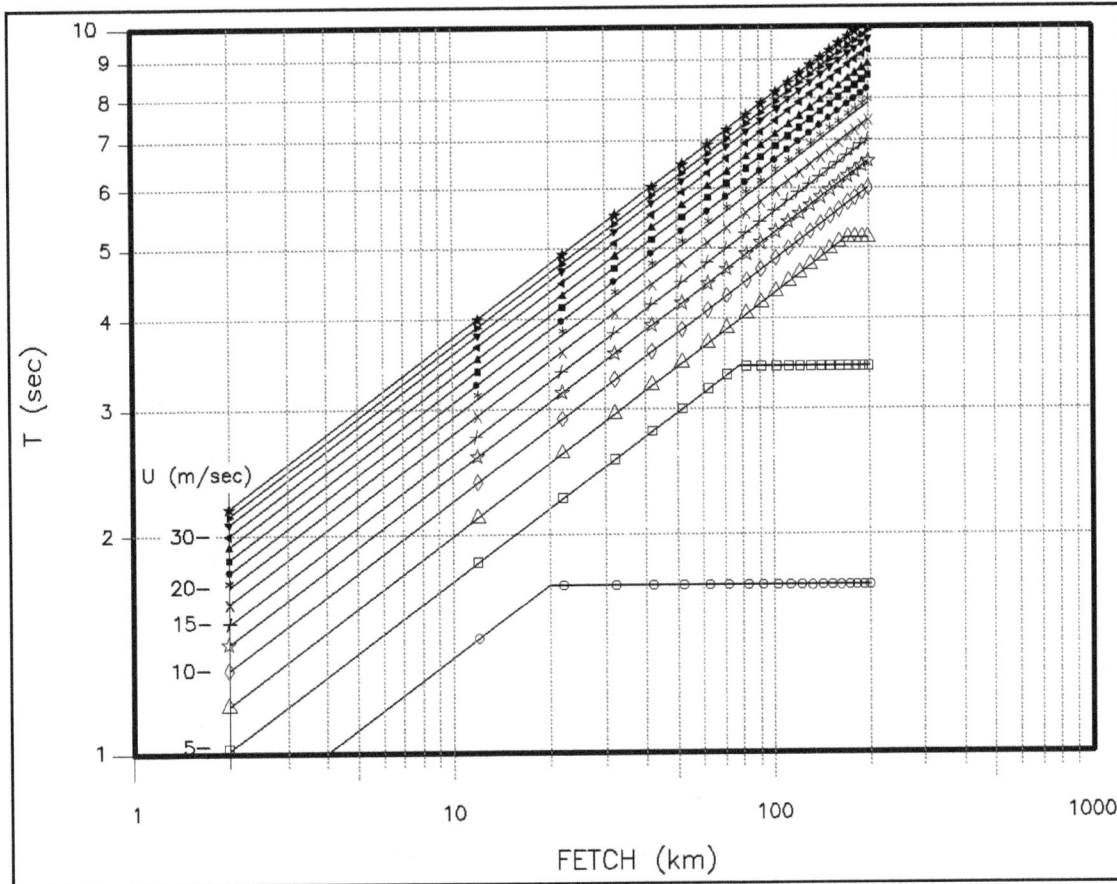

Figure II-2-24.   Fetch-limited wave periods (wind speeds are plotted in increments of 2.5 m/s)

(3)   Prediction of shallow-water waves. Rather than providing separate nomograms for shallow-water wave generation, the following procedure is recommended for estimating waves in shallow basins:

- Determine the straight-line fetch and over-water wind speed.

- Using the fetch and wind speed from (1), estimate the wave height and period from the deepwater nomograms.

- Compare the predicted peak wave period from (2) to the shallow-water limit given in Equation II-2-39. If that wave period is greater than the limiting value, then reduce the predicted wave period to this value. The wave height may be found by noting the dimensionless fetch associated with the limiting wave period and substituting this fetch for the actual fetch in the wave growth calculation.

- If the predicted wave period is less than the limiting value, then retain the deepwater values from (2).

- If wave height exceeds 0.6 times the depth, wave height should be limited to 0.6 times the depth.

c.   Parametric prediction of waves in hurricanes.

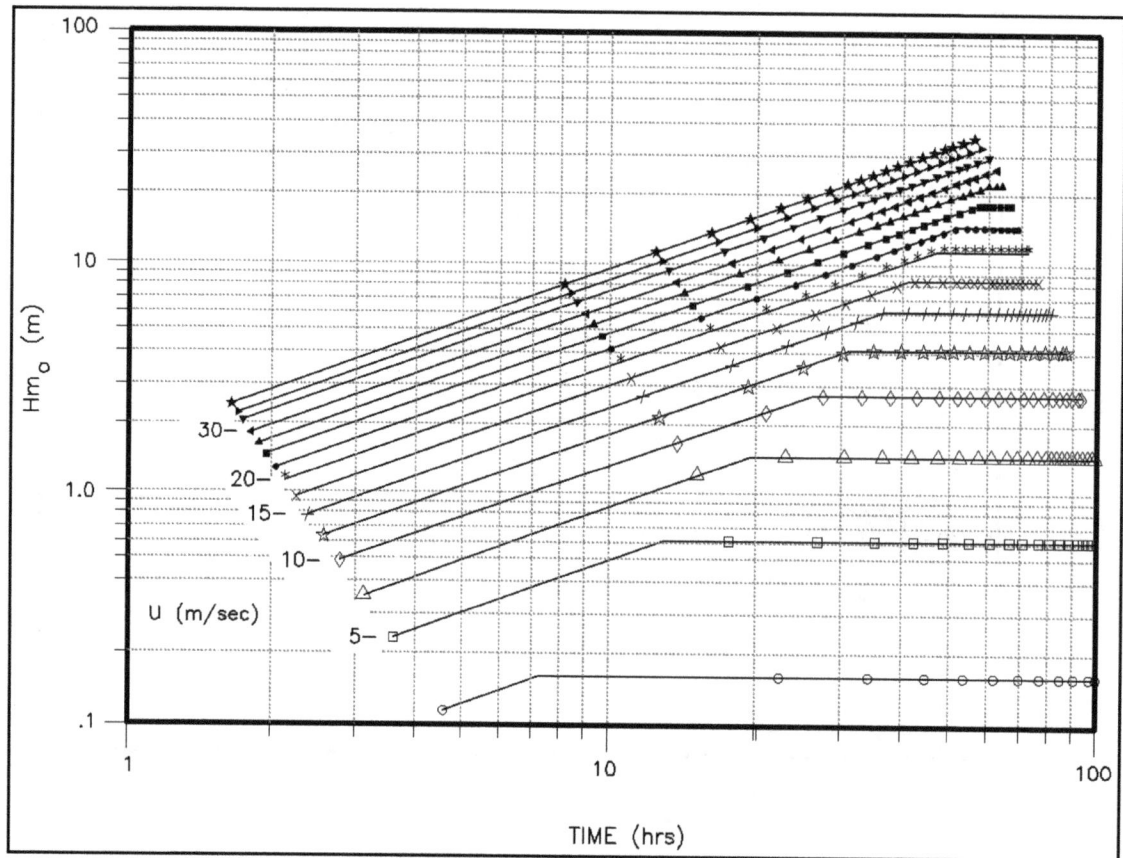

Figure II-2-25. Duration-limited wave heights (wind speeds are plotted in increments of 2.5 m/s)

(1) As shown Table II-2-2, waves from tropical storms, hurricanes, and typhoons represent a dominant threat to coastal and offshore structures and activities in many areas of the world. In this section, the generic term "hurricane" refers to all of these classes of storms. As pointed out previously in this chapter, the only distinction between tropical storms and hurricanes/typhoons is storm intensity (and somewhat the storm's degree of organization). The only distinction between hurricanes and typhoons is the point of origin of the storm.

(2) Spectral models have been shown to provide accurate estimates of hurricane wave conditions, when driven by good wind field information (Ward, Evans, and Pompa 1977; Corson et al. 1982; Cardone 1992; Hubertz 1992). Numerical spectral models can be run on most available PC's today, so there is little motivation not to use such models in any application with significant economic and/or safety implications. However, certain situations remain in which a parametric hurricane wave model may still play an important role in offshore and coastal applications. Therefore, some documentation of parametric models is still included in this manual.

(3) In general, parametric prediction methods tend to work well when applied to phenomena that have little or no dependence on previous states (i.e. systems with little or no memory). A good example of such a physical system is a hurricane wind field. It has been demonstrated (Ward et al. 1977) that hurricane wind fields can be well-represented by a small number of parameters, since winds in a hurricane tend always to remain very close to a dynamic balance with certain driving mechanisms. On the other hand, waves depend not only on the present wind field but also on earlier wind fields, bathymetric effects, pre-existing waves from other wind systems, and in general on the entire wave-generation process over the last to 12 to 24 hr.

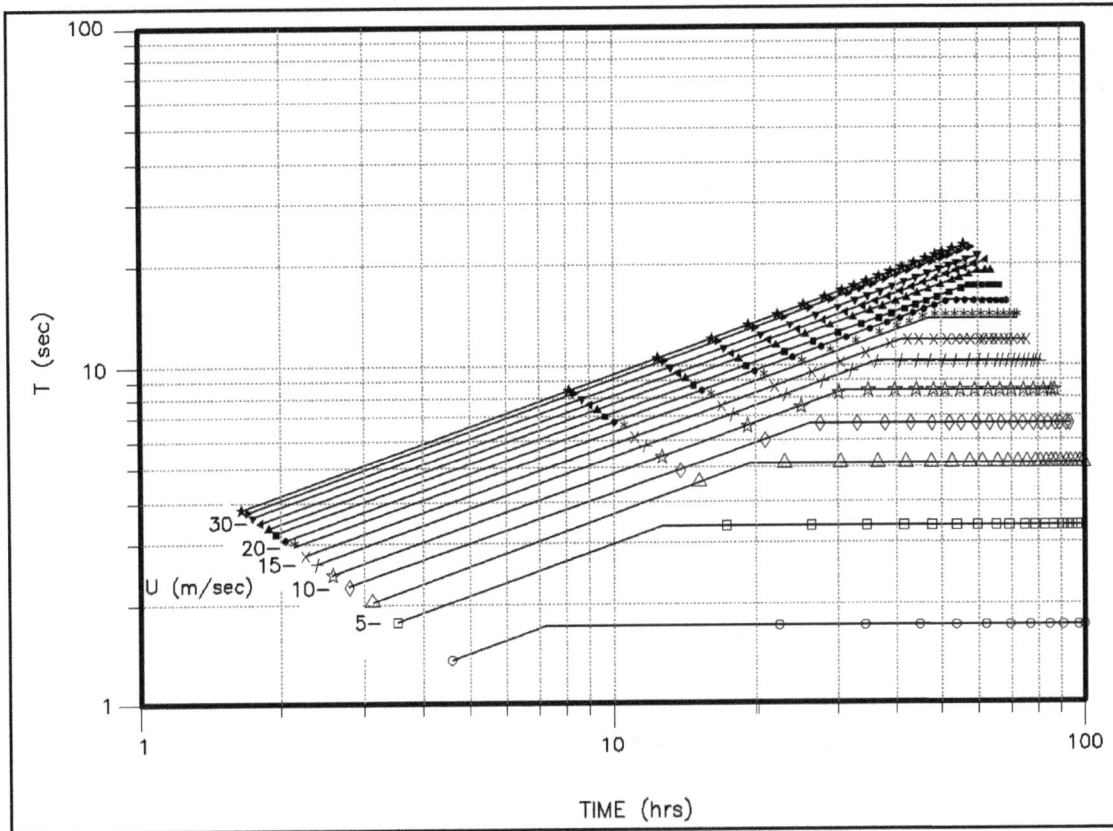

Figure II-2-26.   Duration-limited wave periods

Thus, parametric models do not work well for all hurricanes, but do provide accurate results when the following criteria are met for an interval of about 12 to 18 hr prior to the application of a parametric model:

● Hurricane intensity (maximum velocity) is relatively constant.

● Hurricane track is relatively straight.

● Hurricane forward speed is relatively constant.

● Hurricane is not affected by land or bathymetric effects.

● No strong secondary wind and/or wave systems affect conditions in the area of interest.

(4)   In certain situations, where there is a lack of detail on the actual characteristics of a hurricane (such as in hurricane forecasts, older historical storms, hurricanes in some regions of the world where meteorological data are sparse), parametric models may provide accuracies equal to those of spectral models, provided that land effects and bathymetric effects are minimal. However, even when these criteria are met, situations where secondary wind and/or wave systems can seriously affect wave conditions in an area should be avoided. Examples of this occur when large-scale pressure gradients (monsoonal or extratropical) significantly affect the shape and/or wind distribution of a hurricane. Winds and waves in such a storm will not be distributed in a manner consistent with the assumptions made in this section.

(5)   Young (1987) developed a parametric wave model based on results from simulations with a numerical spectral model. His results show that there is a strong dependence of wave height on the relative values of maximum wind speed and forward storm velocity (Figure II-2-27). These results can be used to estimate the maximum value of H_{m0} in a hurricane. The distribution of wave heights within a hurricane is also affected by the ratio of maximum wind speed to forward storm velocity; however, in an effort to simplify applications here, only one chart is presented (Figure II-2-28). This chart is characteristic of storms with strong winds (maximum wind speed greater than 40 m/sec) and slow-to-moderate forward velocities (V_f less than 12 m/sec).

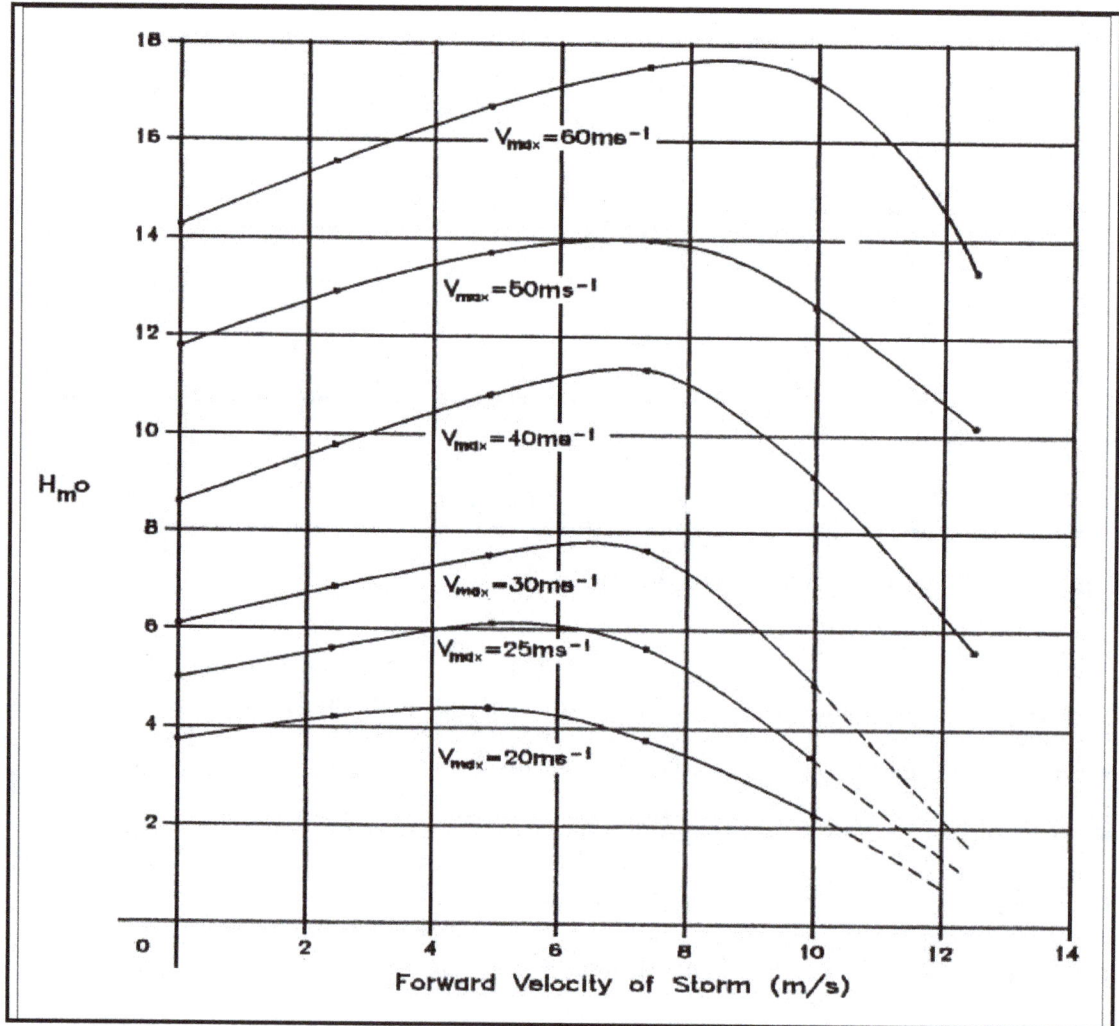

Figure II-2-27.   Maximum value of H_{mo} in a hurricane as a function of V_{max} and forward velocity of storm (Young 1987)

II-2-3. Coastal Wave Climates in the United States.

a.   Introduction.

(1)   Coastal wave climates around U.S. coastlines are extremely varied. Past studies such as that by Thompson (1977) have relied primarily on measured wave conditions in coastal areas to specify nearshore wave climates. However, we now know that coastal wave heights can vary markedly as a function of distance

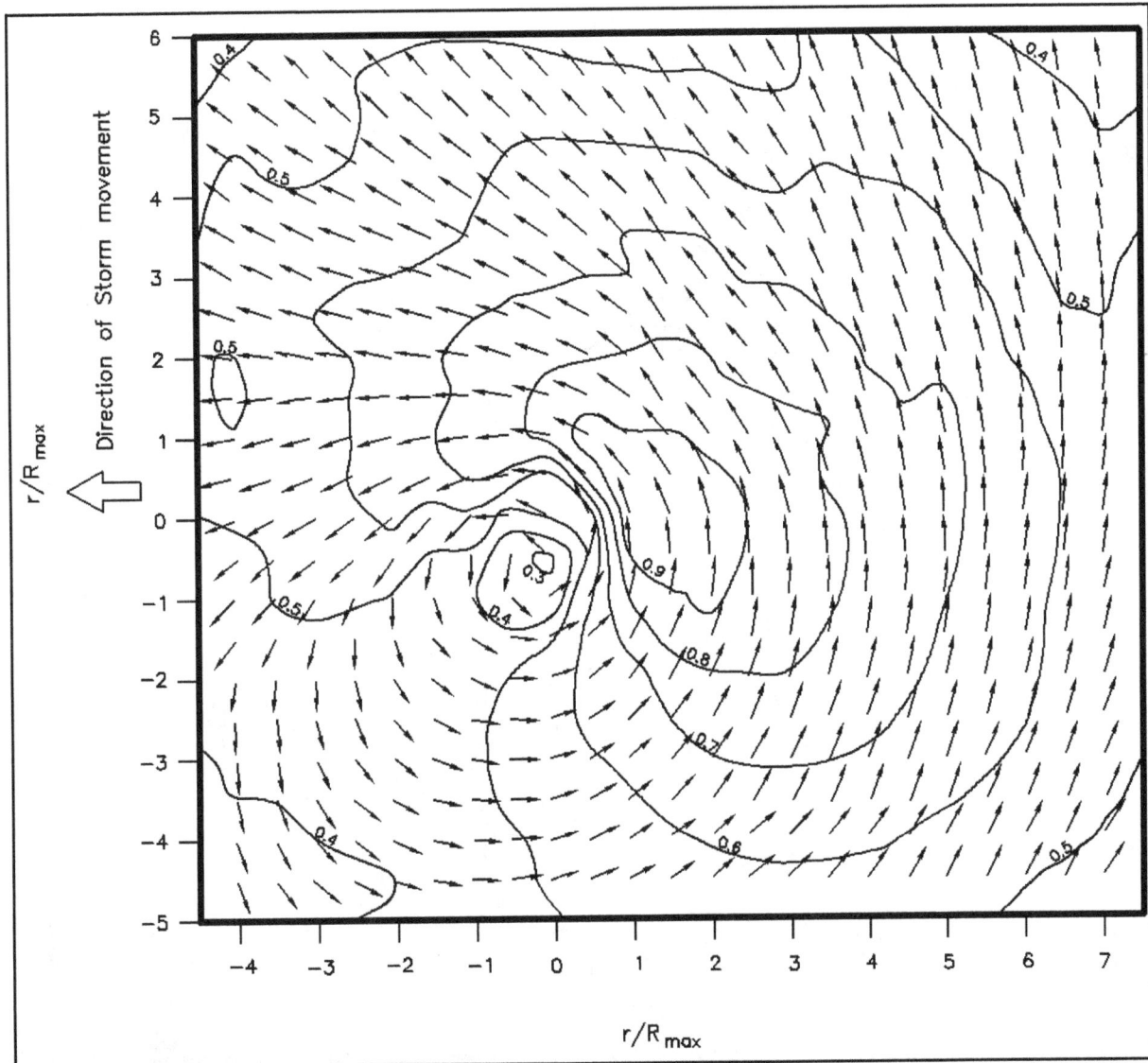

Figure II-2-28. Values of $H_{mo}/H_{mo\ max}$ plotted relative to center of hurricane (0,0)

offshore, degree of coastal sheltering, and various wave transformation factors. This means that measured waves in nearshore areas represent site-specific data. Also, even though measurements in U.S. waters have proliferated, they still do not offer comprehensive coverage. Because of these inherent difficulties in using measurements for a national climatology, hindcast information is used in this section to describe a general coastal wave climate. This is not meant to be interpreted that such models produce information that is as accurate as wave gauges or in any other way superior to wave measurements; but merely that they represent a consistent, comprehensive database for examining regional variations. In the near future, data assimilation methods will combine measurements and hindcasts into a unified database.

(2) In this section, typical wave conditions and storm waves for each of four general coastal areas will be described, along with some of the important meteorological systems that produce these waves. The areas covered here include all coastal areas within the United States, except for Alaska and Hawaii. The wave information presented in Tables II-2-3 through II-2-6 is based on numerical hindcast data provided by CERC's Wave Information Study (WIS). WIS is a multi-year study to develop wave climates for U.S. coastal regions. This information is not yet available for Alaskan and Hawaiian coastal areas; thus, these areas are

omitted in the presentations shown here. It should be noted that this information is very generalized. Waves at a specific site can vary from these estimates due to many site-specific factors, such as: variations in exposure to waves from different directions (primarily related to offshore islands and coastal orientation), bathymetric effects (refraction, shoaling, wave breaking, diffraction, etc.), interactions with currents near inlets or river mouths, and variations in fetches for wave generation.

(3) Figure II-2-29 provides the locations of reference sites along U.S. coastlines that will be used in subsequent parts of this section. A nominal depth of 20 m is assumed for these sites.

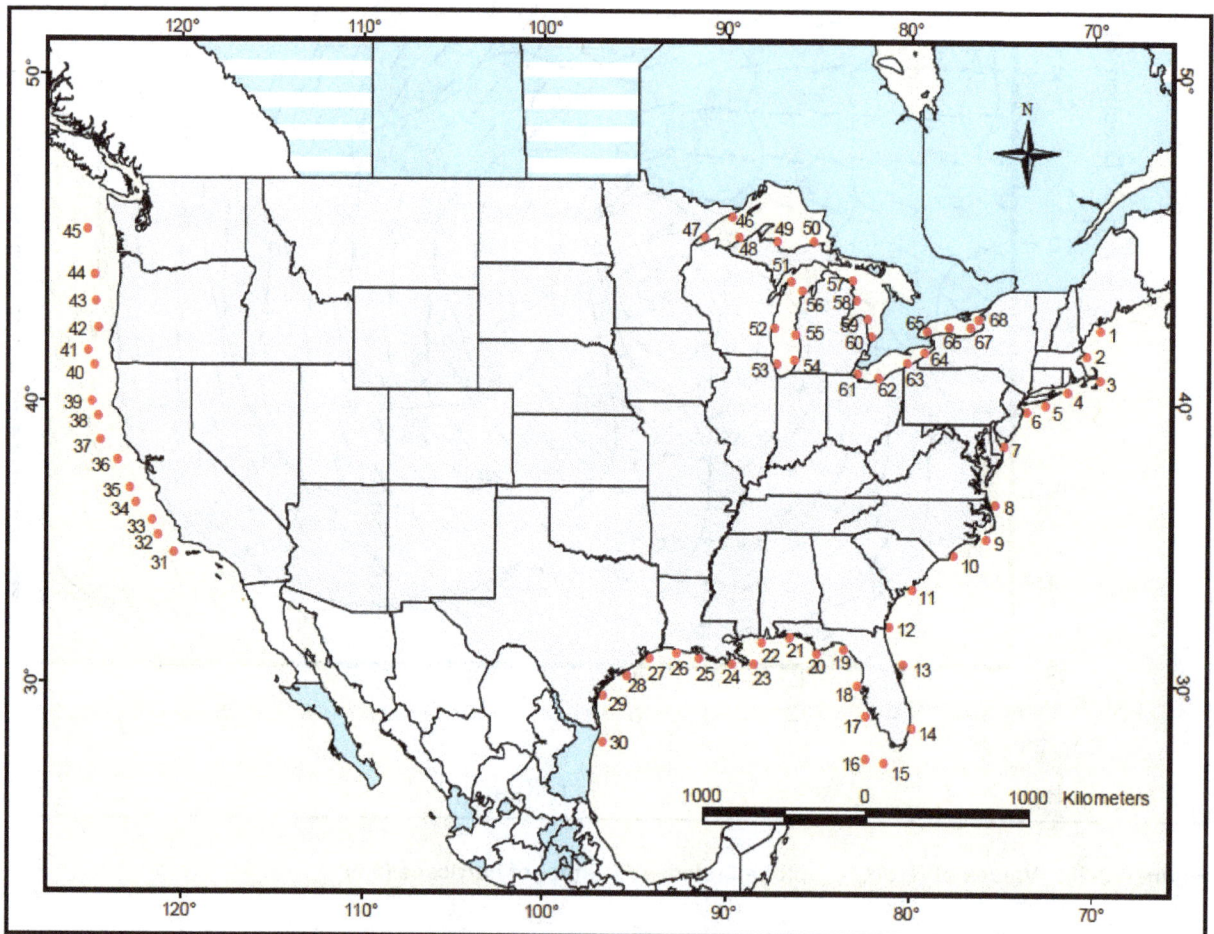

Figure II-2-29.   Reference locations for Tables II-2-3 through II-2-6

b.   *Atlantic coast.*

(1) Table II-2-3 provides wave information for the Atlantic coast. Mean wave heights are fairly consistent along the entire Atlantic coast (0.7 to 1.3 m); however, the overall distribution suggests a subtle multi-peak pattern with maxima at Cape Cod (1.3 m) and Cape Hatteras (1.2 m) and possibly a third peak in the vicinity of Cape Canaveral (1.1 m). These peaks are superimposed on a pattern of slight overall decreasing wave heights as one moves from north to south. Mean wave periods exhibit a relatively high degree of consistency along the entire Atlantic coast, varying only between 6.4 and 7.4 sec, except along the extreme southern part of Florida. The modal direction of the waves is taken here as the 22.5-deg direction class with the highest probability and appears to be primarily a function of coastal exposure.

Table II-2-3
Wave Statistics in the Atlantic Ocean

Map ID	WIS Station	Lat (deg)	Long (deg)	Depth (m)	H_s (m)	T_d (sec)	H_d (deg)	H_s90 (m)	T_d90 (sec)	H_d90 (deg)	H_s 5yr (m)	T_d 5yr (sec)	H_d 5yr (deg)
1	au2100	43.50	-69.75	110	1.4	7.5	160	2.6	8.8	165	11.6	13.8	131
2	au2095	42.50	-70.50	55	1.2	7.2	80	2.3	8.2	79	9.5	14.8	80
3	au2090	41.50	-69.75	18	1.5	7.8	118	2.7	9.2	113	8.4	14.0	52
4	au2082	41.00	-71.50	46	1.4	7.2	167	2.5	8.3	169	8.7	12.6	159
5	au2077	40.50	-72.75	37	1.4	7.2	162	2.5	8.1	170	8.2	12.7	147
6	au2072	40.25	-73.75	27	1.2	7.1	132	2.2	7.9	128	7.2	12.8	129
7	au2066	38.75	-75.00	18	1.0	7.2	124	1.9	7.9	126	5.7	12.8	124
8	au2056	36.25	-75.50	27	1.2	8.1	59	2.3	8.8	46	7.7	13.7	70
9	au2049	34.75	-76.00	35	1.3	7.5	139	2.3	8.1	142	8.1	14.0	129
10	au2042	34.00	-77.75	9	1.1	8.1	124	2.0	8.8	124	6.0	15.2	126
11	au2035	32.50	-80.00	9	1.0	7.5	122	1.7	8.2	122	5.1	14.2	132
12	au2028	30.75	-81.25	11	1.1	8.9	94	1.8	9.3	91	5.7	15.7	95
13	au2021	29.00	-80.50	18	1.2	9.4	65	2.1	10.0	61	6.5	15.4	65
14	au2009	26.00	-80.00	220	0.9	7.3	62	1.8	7.6	64	6.1	10.8	44

For location, refer to Figure II-2-29. Statistics computed from time-series hindcast covering 1976 – 1995 period.

Meteorology and Wave Climate

EXAMPLE PROBLEM II-2-7

FIND:

The significant wave height at the end of this fetch, assuming that the duration of the wind is sufficient to generate fetch-limited waves (from Figure II-2-3, this is found to be greater than about 1.25 hr).

GIVEN:

A constant wind speed of 15 m/sec over a fetch of 10 km in a basin with a constant depth of 3 m. (Note: as pointed out in the previous section on winds, wind speeds tend to increase with fetch over a fetch of this size, so care should be taken in estimating this wind speed)

SOLUTION:

OPTION 1 - Use ACES

OPTION 2 - From Figure II-2-24 the fetch-limited peak wave period is about 2.7 sec, from Equation II-2-39, the limiting wave period in 3 m is 5.4 sec; therefore, $T_p = 2.7$ sec and $H_{m0} = 1.0$ m (deepwater values).

EXAMPLE PROBLEM II-2-8

FIND:

The significant wave height at the end of this fetch.

GIVEN:

A constant wind speed of 25 m/sec over a fetch of 50 km in a basin with a constant depth of 1.6 m.

SOLUTION:

OPTION 1 - Use ACES

OPTION 2 - From Figure II-2-24, the fetch-limited peak wave period is about 5.8 sec, from Equation II-2-39, the limiting wave period in 1.6 m is 4.0 sec; therefore, the waves stopped growing at this limit. This corresponds to a fetch of 20 km at this wind speed; thus, the final values of T_p and H_{m0} are 4.0 sec and 2.1 m (using the 20-km fetch and 25-m/sec wind speed in Figure II-2-23). However, this value exceeds 0.6 times the depth, so the final answer should be 0.8 m. The wave height is limited in this example to be half the water depth. In shallow depths, this is a reasonable approximation.

(2)   These results appear consistent with the mean storminess expected in these Atlantic coastal regions. In the northern portion of the Atlantic coast, the primary source of large waves is migratory extratropical cyclones. Between storm intervals in this region, waves come primarily from swell propagating from storms moving away from the coast. Due to this direction of storm movement, the swell from these storms is usually not very large (less than 2 m). As one moves southward past Cape Hatteras, waves from high-pressure systems (both migratory and semipermanent) begin to become dominant in the wave population. Once south of Jacksonville, the wave climate is typically dominated by easterly winds from high pressure systems, with a secondary source of swell from northeasters. Farther south, as one approaches Miami, the Bahamas provide considerable shelter for waves approaching from the east. In coastal areas without significant swell, sea breeze winds can play a significant role in producing coastal waves during afternoon periods. This situation occurs over much of the U.S. east coast during intervals of the year.

(3)   The 90th percentile wave heights can be considered as representative of typical large wave conditions. As can be seen here, this wave height varies from 1.9 to 2.4 m along the New England region down to 1.4 to 1.9 m along the Florida coast. As was seen in the distribution of mean wave heights, the overall pattern appears to have maxima at Cape Cod (2.4 m), Cape Hatteras (2.1 m), and Cape Canaveral (1.9 m). The associated periods are very consistent along most of the Atlantic coast (8.5 to 9.9 sec) except for the southern half of Florida, where the periods are somewhat lower (6.2 to 7.7 sec). Directions of the 90th-percentile wave reflect the general coastal orientation.

(4)   Extreme waves along the Atlantic coast are often produced by both intense extratropical storms and tropical storms. Table II-2-3 does not provide any information that extends into the return period domain dominated by tropical storms; consequently, this table can be regarded as actually providing information only on extratropical storms. Since this table is not intended to be used directly for any coastal design considerations, information on large-return-period storms is specifically excluded.

(5)   The 5-year wave heights presented in Table II-2-3 can be considered as representing typical large storms that might affect short-term projects (beach nourishment, dredging operations, sand bypassing, etc.). Values of the 5-year wave height range from generally greater than 6 m north of Long Island to only 4.2 m in the Florida Keys. Again, north to south decreasing maxima appear in the regions of Cape Cod (6.7 m), Cape Hatteras (5.9 m), and Cape Canaveral (4.9 m). Associated wave periods are generally in the range of 11 to 13 sec, except for the Florida Keys site, where this period is only 9.5 sec.

(6)   Various types of extratropical storms have wreaked havoc along different coastal areas. These storms range from "bombs" (small, intense, rapidly developing storms) to large almost-stationary storms (developing typically after a change in the large-scale global circulation). Bombs produce higher wind speeds (sustained winds can exceed 70 knots) but due to fetch and duration considerations, the larger, slower-moving storms produce larger wave heights (a measured H_{m0} greater than 17 m south of Nova Scotia in the Halloween Storm). Other examples of classic storms along the U.S. east coast include the Ash Wednesday Storm of 1962 (affecting mainly the mid-Atlantic region), the Blizzard of 1978 (affecting mainly the northeastern states), and the Storm of March 1993, which affected most of the U.S east coast. This last storm has been called the "Storm of the Century" by some; however, it is by no means the worst storm in terms of waves for most areas along the east coast in this century. In fact, along much of the Atlantic coast, the wind direction was toward offshore; consequently, there was almost no wave action at the coast in many locations. Farther offshore the situation was considerably different and many ships and boats were lost.

(7)   Hurricanes can also produce extreme wave conditions along the coast. Particularly at the coast itself where storm surges of 10 ft or more can accompany waves, hurricane waves represent an extreme threat to both life and property. An excellent source of hurricane information is the HURDAT file at the National Climatic Center in Asheville, NC. This file (available on magnetic tape or PC diskette format) contains storm tracks, maximum wind speeds, central pressures, and other parameters of interest for all hurricanes affecting

the United States since 1876. The effects of Hurricanes Hugo in 1988 and Andrew in 1992 have shown the tremendous potential for coastal destruction that can accompany these storm systems in southern reaches of the Atlantic coast. The effects of the Hurricane of 1933 in New England and Hurricane Bob in 1990 show that even farther north, the risk of hurricanes cannot be neglected.

    c.   Gulf of Mexico.

    (1)   Table II-2-4 shows the same information for the U.S. Gulf coast as was given in Table II-2-3 for the Atlantic coast. Mean wave heights for this coast are often considered to be considerably lower than those on the Atlantic coast; however, as can be seen in this table, this is not evident in the wave data. In fact, mean wave heights near Brownsville are larger than anywhere on the Atlantic coast. The reason for this is that the mean wind direction in this location is directed toward land, whereas, along the Atlantic coast, the mean wind direction is directed away from land except for areas south of Jacksonville, FL. Mean wave heights generally decrease eastward to the Appalachicola area and then remain fairly constant southward to the Florida Keys.

    (2)   Many of the larger waves in the Gulf of Mexico are generated by storms that are centered well to the north over land. Thus, large waves can be experienced at offshore sites even when conditions along the coast are quite calm. Typical day-to-day wave conditions in many coastal areas are produced by a combination of relatively small synoptic-scale winds and sea-breeze circulations. As noted in Table II-2-2 in this section, these waves are rarely very large. At times, the Gulf of Mexico comes under the influence of large-scale high pressure systems, with winds blowing from east to west across much of the Gulf. These winds are primarily responsible for the higher wave conditions in the western Gulf. Due to the lack of strong storms centered within the Gulf, there is little or no swell reaching Gulf shorelines, with the notable exception being swell from remote tropical systems. Consequently, except for the extreme western Gulf of Mexico, mean wave periods tend to be somewhat smaller than those along the Atlantic coast (4 to 6 sec).

    (3)   The 90th percentile wave heights indicate that typical large wave conditions along the coast are only about 50 percent larger than the mean wave heights (compared to about a 100-percent factor for the Atlantic coast). This is consistent with the idea that the Gulf of Mexico is, in fact, a calmer basin than the Atlantic. These wave heights in the Gulf vary from a maximum of 1.5 m near Brownsville to 1.2 m along Florida's west coast. Associated wave periods range from 6 to 8 sec.

    (4)   Values of the 5-year wave heights in the Gulf of Mexico vary from 3.2 m along the west coast of Florida to 4.6 m near Brownsville. Associated wave periods vary between 9 and 10.5 sec. Some of the higher non-tropical waves in the Gulf of Mexico are generated by wind systems called "Northers." Since these winds blow out of the north, they typically do not create problems at the coast itself, but can produce large waves at offshore sites. Occasionally an extratropical cyclone will develop within the Gulf. One example, the intense storm of 10-13 March, 1993 (the so-called "Storm of the Century"), produced high surges and large waves along extensive portions of Florida's west coast. Damages and loss of life from this storm demonstrated that, although rare, strong extratropical storms can still be a threat to some Gulf coastal areas.

    (5)   The primary source of extreme waves in the Gulf of Mexico is hurricanes. Hurricanes Betsy (1965), Camille (1969), Carmen (1975), Frederick (1979), Alicia (1985), and Andrew (1992) have clearly shown the devastating potential of these storms in the Gulf of Mexico. Even though shallow-water effects may diminish coastal wave heights from the values listed in Table II-2-2, wave conditions are still sufficient to control design and planning considerations for most coastal and offshore structures/facilities in the Gulf.

Table II-2-4
Wave Statistics in the Gulf of Mexico

Map ID	WIS Station	Lat (deg)	Long (deg)	Depth (m)	Hs (m)	Td (sec)	Hd (deg)	Hs90 (m)	Td90 (sec)	Hd90 (deg)	Hs 5yr (m)	Td 5yr (sec)	Hd 5yr (deg)
15	gu1002	24.25	-81.50	451	0.8	4.4	83	1.5	5.7	77	4.8	9.9	140
16	gu1006	24.50	-82.50	14	0.7	4.3	48	1.4	5.8	37	3.9	9.3	202
17	gu1017	26.50	-82.50	18	0.4	3.9	270	0.9	5.1	303	4.5	10.5	219
18	gu1023	28.00	-83.00	11	0.4	3.9	264	0.8	5.2	274	4.6	12.2	244
19	gu1030	29.75	-83.75	5	0.4	3.8	224	0.8	4.8	252	3.1	10.6	210
20	gu1036	29.50	-85.25	19	0.7	4.2	170	1.3	5.4	113	5.5	11.8	183
21	gu1042	30.25	-86.75	14	0.6	4.2	153	1.1	5.3	149	5.5	11.9	174
22	gu1048	30.00	-88.25	29	0.7	4.4	135	1.3	5.7	135	5.6	10.6	128
23	gu1054	29.00	-88.75	209	0.9	4.9	98	1.8	6.6	89	6.8	11.7	136
24	gu1059	29.00	-90.00	25	0.8	4.7	138	1.5	6.0	138	5.4	11.4	167
25	gu1066	29.25	-91.75	7	0.7	4.6	162	1.3	6.0	164	4.0	11.3	190
26	gu1071	29.50	-93.00	12	0.8	4.7	150	1.4	6.1	153	4.2	11.1	167
27	gu1077	29.25	-94.50	15	0.8	5.1	140	1.5	6.5	142	4.5	11.9	145
28	gu1082	28.50	-95.75	20	0.9	5.4	137	1.5	6.8	143	4.5	11.7	132
29	gu1088	27.50	-97.00	27	0.9	5.8	119	1.5	7.0	123	5.3	11.2	108
30	gu1102	25.25	-97.00	48	1.1	6.0	100	1.7	7.2	102	5.8	9.9	73

For location, refer to Figure II-2-29. Statistics computed from time-series hindcast covering 1976 - 1995 period.

d.   Pacific coast.

(1)   Table II-2-5 provides information for the Pacific coast that is comparable to that presented in Tables II-3-3 and II-2-4 for the Atlantic and Gulf of Mexico coasts, respectively. The Pacific coast is very different from the east coast in that wave-producing storms within the Pacific Ocean are travelling toward this coast. This means that the west coast typically has a much richer source of swell waves tha do other U.S. coastal areas. As can be seen by comparison to the Atlantic coast results (Table II-2-3), this results in higher wave conditions along the Pacific coast, with mean wave heights ranging from 2.5 m near the Mexican border to 3.2 m near the Canadian border. This difference is also reflected in the mean periods along these coasts, which vary from 9.6 to 12.1 sec. During (Northern Hemisphere) summer months, storm tracks usually move far to the north and storms are less intense. Consequently, swell from mid-latitude storms in the Northern Hemisphere diminish in size and frequency, allowing swell from tropical storms spawned off the west coast of Mexico and from large winter storms in the Southern Hemisphere to become important elements in the summer wave climate.

(2)   Typical winter storm tracks move storm centers inland in the region from northern California to the Canadian border. Hence, large waves in these regions frequently come in the form of local seas. South of San Francisco, local storms strike the coast with less frequency; thus, many of the large waves in this area arrive in the form of swell. Many notable exceptions to this general rule of thumb can be found in the late 1970's and 1980's, however. In particular, the storm of January 1989 moved across the California coast in the vicinity of Los Angeles and caused much damage to southern California coastal areas.

(3)   The 90th percentile wave heights along the Pacific coast are about twice their Atlantic coast counterparts. In the southern California region, these values are typically in the 3.9- to 4.2-m range. As one moves northward, the 90th percentile wave height increases to a maximum of about 5.4 m along the coast of Washington. Periods associated with these waves tend to be quite long, ranging between 11 and 14 sec.

(4)   The 5-year wave heights in the southern California region are comparable to those found along the New England coast on the Atlantic (6.8-6.9 m compared to 6.7 m). However, associated periods are considerably longer (16.8 sec compared to 12-13 sec). As one moves northward, these wave heights increase to levels greater than 10 m along much of the coast north of the California-Oregon border. Periods of these large waves tend to fall in the 14- to 16-sec range.

(5)   Although many studies have dismissed the importance of tropical storms to the extreme wave climate along the Pacific coast, at least one tropical storm has moved into the Los Angeles basin during this century, suggesting that this threat is not negligible. Given the curvature of the coast and the water temperatures north of Point Conception, it is unlikely that tropical storms can produce a significant threat at coastal sites north of this point; however, south of this point it is important to include tropical storms in any design and planning considerations.

e.   Great Lakes.

(1)   Table II-2-6 provides comparable information for the Great Lakes as provided for previous coastal areas in Tables II-2-3 through II-2-5. Wave conditions within the Great Lakes are strongly influenced by fetches aligned with the dominant directions of storm winds. These winds are mainly produced by various extratropical storms moving across the Great Lakes region. Table II-2-6 compares the largest 50-year (return period) wave heights for each lake. Since strong storms are not very frequent in late spring through early autumn, this interval is usually relatively calm along most shorelines. During the period from mid-autumn until ice effects on the lakes reduce the wave generation potential, the largest waves are generated. Again

Meteorology and Wave Climate

Table II-2-5
Wave Statistics in the Pacific Ocean

Map ID	WIS Station	Lat (deg)	Long (deg)	Depth (m)	Hs (m)	Td (sec)	Hd (deg)	Hs90 (m)	Td90 (sec)	Hd90 (deg)	Hs 5yr (m)	Td 5yr (sec)	Hd 5yr (deg)
31	p2008	34.01	-120.92	1650	2.7	10.3	293	4.1	12.5	290	7.0	15.4	285
32	p2011	34.82	-121.82	2122	2.8	10.3	294	4.2	12.5	291	7.5	15.7	283
33	p2013	35.41	-122.16	2944	2.8	10.3	292	4.2	12.5	287	7.8	15.8	279
34	p2016	36.22	-123.06	3365	2.9	10.4	295	4.3	12.5	290	7.9	15.1	264
35	p2018	36.82	-123.41	3387	3.0	10.5	294	4.4	12.6	289	8.2	14.8	250
36	p2022	38.04	-124.11	3655	3.1	10.8	292	4.6	12.9	285	8.5	14.6	245
37	p2025	38.85	-125.05	3435	3.2	10.7	287	4.9	12.7	281	9.3	14.3	242
38	p2028	39.87	-125.17	2471	3.2	10.7	286	4.8	12.7	278	9.4	13.8	235
39	p2030	40.49	-125.53	2702	3.2	10.7	284	4.9	12.6	275	9.7	14.0	219
40	p2034	41.93	-125.39	3151	3.2	10.7	274	5.2	12.9	265	10.4	14.1	235
41	p2036	42.55	-125.74	2993	3.3	10.7	271	5.4	12.9	262	10.8	13.6	224
42	p2038	43.37	-125.21	1783	3.3	10.9	271	5.4	13.0	263	10.4	14.1	238
43	p2041	44.41	-125.29	2227	3.0	10.9	271	4.9	13.0	264	9.2	14.0	219
44	p2044	45.45	-125.36	2033	3.0	10.9	266	4.8	12.9	259	9.5	14.3	213
45	p2049	47.14	-125.77	1630	3.2	10.7	257	5.4	12.7	250	10.5	14.4	224

For location, refer to Figure II-2-29. Statistics computed from time-series hindcast covering 1956 - 1975 period.

Meteorology and Wave Climate

Table II-2-6
Wave Statistics in the Great Lakes

Map ID	WIS Station	Lat (deg)	Long (deg)	Depth (m)	H_s (m)	T_d (sec)	H_d (deg)	$H_s 90$ (m)	$T_d 90$ (sec)	$H_d 90$ (deg)	$H_s 5yr$ (m)	$T_d 5yr$ (sec)	$H_d 5yr$ (deg)
46	s0004	47.67	-90.07	174	0.8	3.9	255	1.6	5.2	279	7.4	11.1	88
47	s0014	46.95	-91.57	154	0.7	3.6	27	1.3	4.6	325	6.0	10.2	56
48	s0028	46.95	-89.63	101	0.8	4.0	318	1.8	5.7	313	5.9	9.7	17
49	s0046	46.80	-87.50	55	1.0	4.2	2	2.0	5.7	351	7.7	10.2	10
50	s0055	46.80	-85.57	119	1.1	4.6	322	2.2	6.3	328	7.0	10.3	310
51	mu0025	45.27	-86.75	55	0.9	4.4	172	1.7	5.9	164	4.9	8.4	194
52	mu0013	43.55	-87.67	64	0.8	4.3	98	1.6	5.6	109	4.6	8.4	54
53	mu0003	42.10	-87.55	24	0.9	4.1	22	1.6	5.5	23	5.4	9.3	11
54	mu0058	42.22	-86.58	36	1.0	4.5	308	1.8	6.2	302	5.1	8.8	342
55	mu0051	43.23	-86.50	73	1.0	4.4	256	1.9	6.3	259	5.3	8.6	238
56	mu0039	44.95	-86.17	36	0.9	4.4	286	1.8	6.0	291	5.2	9.0	222
57	h0020	45.35	-83.32	30	0.8	4.0	5	1.7	5.3	360	5.8	9.1	83
58	h0015	44.63	-83.12	20	0.9	4.2	83	1.9	5.6	301	6.7	9.4	52
59	h0007	43.90	-82.52	30	0.8	4.2	346	1.7	5.6	335	6.1	9.4	12
60	h0002	43.18	-82.32	14	0.9	4.0	334	1.7	5.2	322	6.8	9.8	356
61 [1]	e0002	41.73	-83.08	9	0.6	3.5	285	1.1	4.4	299	2.4	5.5	14
62	e0009	41.58	-81.90	17	0.9	3.9	280	1.7	5.3	292	4.0	7.6	339
63	e0017	42.15	-80.35	20	1.0	4.3	263	2.0	5.9	272	4.8	8.5	271
64	e0022	42.58	-79.37	30	0.9	4.0	253	1.8	5.5	255	5.2	8.8	252
65 [2]	o0037	43.45	-79.25	128	0.4	3.0	271	0.9	3.9	273	3.5	6.6	236
66	o0040	43.60	-78.05	185	0.4	3.2	262	1.0	4.5	266	4.8	8.2	260
67	o0043	43.60	-76.85	174	0.4	3.3	266	1.0	4.7	266	5.5	9.4	265
68	o0020	43.88	-76.45	24	0.4	3.2	243	0.9	4.2	247	5.1	8.7	243

For location, refer to Figure II-2-29. Statistics computed from time-series hindcast covering 1956 - 1987 period.
[1] Station e0002 is sheltered from the east by islands.
[2] Station o0037 reflects west-to-east weather patterns.

in the spring, after the ice has thawed, large waves (although usually significantly smaller than waves in autumn) can be generated and can affect coastal areas.

(2)   One of the issues of critical concern in the Great Lakes is that of mean lake level. These levels have fluctuated considerably through recorded history in response to periods of low and high precipitation in the general geographic area.   Critical design criteria for many Great Lakes coastal areas are defined by the superposition of high wave conditions (generated by extratropical storms) on top of high mean lake levels and storm surges.

II-2-4. Additional Example Problem

This problem demonstrates use of assumptions in simplified wave predictions.

EXAMPLE PROBLEM II-2-9

FIND: The significant wave height and spectral peak wave period generated by a mean wind speed of 30 m/s over a fetch of 50 km. (Work the problem in metric units.)

SOLUTION:

Step 1. Check required wind duration. Given that x is the fetch in meters, g is the acceleration due to gravity in meters/second-squared, u_{10} is the wind speed in meters/second, we have

$$t_{x,u} = 77.23 \frac{x^{0.67}}{u_{10}^{0.34} g^{0.33}} = 77.23 \frac{(50,000)^{0.67}}{30^{0.34} 9.82^{0.33}} = 16,087 \sec = 4.47 \text{hrs}$$

If the wind duration is equal to or longer than this than a fetch-limited situation exists.

Step 2. Estimate friction velocity. First, estimate the coefficient of drag as

$$C_d \approx 0.001(1.1 + 0.035 u_{10});$$

Then, estimate the friction velocity as

$$u_* = \sqrt{C_d} u_{10} = \sqrt{0.00215} \times 30 = 1.39 m/\sec$$

Step 3. Estimate Significant Wave Height. Estimate nondimensional fetch as

$$\hat{x} = \frac{gx}{u_*^2} = 9.82 \times 50,000 / (1.39)^2 = 2.54 \times 10^5$$

Estimate nondimensional wave height as

$$\hat{H}_{m0} = \lambda_1 \hat{x}^{m_1};$$

$$\lambda_1 = 0.0413;$$

$$m_1 = \frac{1}{2};$$

$$\hat{H}_{m0} = 0.0413 \times (2.54 \times 10^5)^{\frac{1}{2}} = 20.8;$$

$$H_{m0} = \hat{H}_{m0} \times \frac{u_*^2}{g} = 20.8 \times \frac{(1.39)^2}{9.82} = 4.1m$$

Step 4. Estimate Spectral Peak Period. Since we already have calculated the nondimensional fetch in Step 3, we can proceed to estimate the nondimensional spectral peak period:

$$\hat{T}_p = \frac{gT_p}{u_*} = \lambda_2 \hat{x}^{m_2}$$

$$\lambda_2 = 0.751$$

$$m_2 = \frac{1}{3}$$

$$\hat{T}_p = 0.751 \times (2.54 \times 10^5)^{\frac{1}{3}} = 47.5$$

$$T_p = \hat{T}_p \frac{u_*}{g} = \frac{47.5 \times 1.39}{9.82} = 6.7 \sec$$

II-2-5. References

Atkinson and Holliday 1977
Atkinson, G. D., and Holliday, C. R. 1977. "Tropical Cyclone Minimum Sea Level Pressure/Maximum Sustained Wind Relationship for the Western Northern Pacific," *Mon. Wea. Rev.,* Vol 105, pp 421-427.

Barnett 1968
Barnett, T. P. 1968. "On the Generation, Dissipation, and Prediction of Ocean Wind Waves," *J. Geophys. Res.,* Vol 2, pp 531-534.

Bouws, Gunther, and Vincent 1985
Bouws, E., Gunther, H., and Vincent, C. L. 1985. "Similarity of Wind Wave Spectrum in Finite-Depth Water, Part I: Spectral Form," *J. Geophys. Res.,* Vol 85, No. C3, pp 1524-1530.

Bretschneider 1952
Bretschneider, C. 1952. "Revised Wave Forecasting Relationships," *Proceedings of the 2nd Coastal Engineering Conference,* American Society of Civil Engineers, pp 1-5.

Bretschneider 1990
Bretschneider, C. 1990. "Tropical Cyclones," *Handbook of Coastal and Ocean Engineering,* Vol 1, J. B. Herbich, ed., pp 249-270.

Bretschneider and Reid 1953
Bretschneider, C., and Reid, R. O. 1953. "Change in Wave Height Due to Bottom Friction, Percolation and Refraction," *34th Annual Meeting of American Geophysical Union.*

Bunting 1970
Bunting, D. C. 1970. "Evaluating Forecasts of Ocean-Wave Spectra," *Journal of Geophysical Research,* Vol 75, No. 21, pp 4131-4143.

Businger et al. 1971
Businger, J. A., Wyngaard, J. C., Izumi, Y., and Bradley, E. F. 1971. "Flux-Profile Relationships in the Atmoshperic Surface Layer," *J. Atmos. Sci.,* Vol 25, pp 1021-1025.

Camfield 1977
Camfield, F. E. 1977. "Wind-Wave Propagation Over Flooded, Vegetated Land," Technical Paper No. 77-12, Coastal Engineering Research Center, U.S. Army Engineer Waterways Experiment Station, Vicksburg, MS.

Cardone 1992
Cardone, V. J. 1992. "On the Structure of the Marine Surface Wind Field," *3rd International Workshop of Wave Hindcasting and Forecasting,* Montreal, Quebec, May 19-22, pp 54-66.

Cardone, Cox, Greenwood, and Thompson 1994
Cardone, V. J., Cox, A. T., Greenwood, J. A., and Thompson, E. F. 1994. "Upgrade of the Tropical Cyclone Surface Wind Field Model," Miscellaneous Paper CERC-94-14, U.S. Army Engineer Waterways Experiment Station, Vicksburg, MS.

Cardone, Greenwood, and Greenwood 1992
Cardone, V. J., Greenwood, C. V., and Greenwood, J. A. 1992. "Unified Program for the Specification of Hurricane Boundary Layer Winds over Surfaces of Specified Roughness," Contract Report CERC-92-1, U.S. Army Engineer Waterways Experiment Station, Vicksburg, MS.

Chow 1971
Chow, S. 1971. "A Study of the Wind Field in the Planetary Boundary Layer of a Moving Tropical Cyclone," M. S. thesis, New York University.

Clarke 1970
Clarke, R. H. 1970. "Observational Studies of the Atmospheric Boundary Layer," *Quart. J. Roy. Meteor. Soc,* Vol 96, pp 91-114.

Collins and Viehmann 1971
Collins, J. I., and Viehmann, M. J. 1971. "A Simplified Model for Hurricane Wind Fields," Paper 1346, *Offshore Technology Conference,* Houston, TX.

Cooper 1988
Cooper, C. K. 1988. Parametric Models of Hurricane-Generated Winds, Waves, and Currents in Deep Water," *Proc. Offshore Tech. Conf.*, Paper 5738, Houston, TX.

Corson et al. 1982
Corson, W. D., Resio, D. T., Brooks, R. M., Ebersole, B. A., Jensen, R. E., Ragsdale, D. S., and Tracy, B. A. 1982. "Atlantic Coast Hindcast Phase II, Significant Wave Information," WIS Report 6, U.S. Army Engineer Waterways Experiment Station, Vicksburg, MS.

Deardorff 1968
Deardorff, J. W. 1968. "Dependence of Air-Sea Transfer Coefficients on Bulk Stability," *J. Geophys. Res,* Vol 73, pp 2549-2557.

Demirbilek, Bratos, and Thompson 1993
Demirbilek, Z., Bratos, S. M., and Thompson, E. F. 1993. "Wind Products for Use in Coastal Wave and Surge Models,"Miscellaneous Paper CERC-93-7, U.S. Army Engineer Waterways Experiment Station, Vicksburg, MS.

Donelan, Hamilton, and Hu 1982
Donelan, M. A., Hamilton, J., and Hu, W. H. 1982. "Directions Spectra of Wind-Generated Waves," Unpublished manuscript, Canada Centre for Inland Waters.

Dvorak 1975
Dvorak, V. F. 1975. "Tropical Cyclone Intensity Analysis and Forecasting from Satellite Imagery," *Mon. Wea. Rev.,* Vol 105, pp 369-375.

Forristall, Ward, Cardone, and Borgman, 1978
Forristall, G. Z., Ward, E. G., Cardone, V. J., and Borgman, L. E. 1978. "The Directional Spectra and Kinematics of Surface Gravity Waves in Tropical Storm Delia," *J. Pmys. Oceanogr.,* Vol 8, pp 888-909.

Garratt 1977
Garratt, J. R. 1977. "Review of Drag Coefficients Over Oceans and Continents," *Mon. Wea. Rev.,* Vol 105, pp 915-929.

Geernaert 1990
Geernaert, G. L. 1990. " Bulk Parameterizations for the Wind Stress and Heat Fluxes," *Surface and Fluxes: Theory and Remote Sensing; Vol 1: Current Theory.* G. L. Geernaert and W. J. Plant, ed., Kluwer Academic Publisher, Worwell, MA, pp 91-172.

Geernaert, Katsaros, and Richter 1986
Geernaert, G. L., Katsaros, K. B., and Richter, K. 1986. "Variation of the Drag Coefficient and its Dependence on Sea State," *J. Geophys. Res.,* Vol 91, pp 7667-7679.

Gelci, Cazale, and Vassel 1957
Gelci, R., Cazale, H., and Vassel, J. 1957. "Prevision de la Houle," *La Methode des Densites Spectroangularies, Bull. Infor.,* Comite Central Oceangr. d' Etude Cotes, Vol 9, pp 416-435.

Hasselmann et al. 1973
Hasselmann, K., Barnett, T. P., Bouws, E., Carlson, H., Cartwright, D. E., Enke K., Weing, J. A., Gienapp, H., Hasselmann, D. E., Kruseman, P., Meerburg, A., Muller, P., Olbers, K. J., Richter, K., Sell, W., and Walden, W. H. 1973. "Measurements of Wind-Wave Growth and Swell Decay During the Joint North Sea Wave Project (JONSWAP)," *Deutsche Hydrograph, Zeit., Erganzung-self Reihe,* A 8(12).

Hasselmann, Hasselmann, Allender, and Barnett 1985
Hasselmann, S., Hasselmann, K., Allender, J. H., and Barnett, T. P. 1985. "Computations and Parameterizations of Nonlinear Energy Transfer in a Gravity-wave Spectrum; Part II: Parameterization of Nonlinear Transfer for Application in Wave Models," *J. Phys. Oceanogr.,* Vol 15, pp 1378-1391.

Helmholtz 1888
Helmholtz, H. 1888. "Uber Atmospharische Bewwgungen," *S. Ber. Preuss. Akad. Wiss. Berlin, Mathem. Physik Kl.*

Hess 1959
Hess, S. L. 1959. *Introduction to Theoretical Meteorology.* Holt, New York.

Holland 1980
Holland, G. J. 1980. "An Analytic Model of the Wind and Pressure Profiles in Hurricanes," *Mon. Wea. Rev.,* Vol 108, pp 1212-1218.

Holt and Raman 1988
Holt, T., and Raman, S. 1988. "A Review and Comparative Evaluation of Multilevel Boundary Layer Parameterizations for First-Order and Turbulent Kinetic Energy Closure Schemes," *Rev. Geophys.,* Vol 26, pp 761-780.

Hsu 1974
Hsu, S. A. 1974. "A Dynamic Roughness Equation and its Application to Wind Stress Determination at the Air-Sea Interface," *J. Phys. Oceanogr.,* Vol 4, pp 116-120.

Hsu 1988
Hsu S.A. 1988. *Coastal Meteorology,* Academic Press, New York.

Huang et al 1986
Huang, N. E., Bliven, L. F., Long, S. R., and DeLeonibus, P. S. 1986. "A Study of the Relationship Among Wind Speed, Sea State, and the Drag Coefficient for a Developing Wave Field," *J. Geophys. Res.,* Vol 91, No. C6, pp 7733-7742.

Hughes 1952
Hughes L. A. 1952. "On the Low-Level Wind Structure of Tropical Storms," *J. of Meteorol.,* Vol 9, pp 422-428.

Hubertz 1992
Hubertz, J. M. 1992. "User's Guide to the Wave Information Studies (WIS) Wave Model, Version 2.0," WIS Report 27, U.S. Army Engineer Waterways Experiment Station, Vicksburg, MS.

Hydraulics Research Station 1977
Hydraulics Research Station. 1977. "Numerical Wave Climate Study for the North Sea (NORSWAM)," Report EX 775, Wallingford, England.

Inoue 1967
Inoue, T. 1967. "On the Growth of the Spectrum of a Wind Generated Sea According to a Modified Miles-Phillips Mechanism and Its Application to Wave Forecasting," Geophysical Sciences Laboratory Report No. TR-67-5, Department of Meteorology and Oceanography, New York University.

Janssen 1989
Janssen, P. A. E. M. 1989. "BAR Wave-Induced Stress and the Drag of Air Flow Over Sea Waves," *J. Phys. Oceanogr.,* Vol 19, pp 745-754.

Janssen 1991
Janssen, P. A. E. M. 1991. "Quasi-linear Theory of Wind wave generation Applied to Wave Forecasting," *J. Phys. Oceanogr.,* Vol 21, pp 745-754.

Jeffreys 1924
Jeffreys, H. 1924. "On the Formation of Waves by Wind," *Proc. Roy. Soc. Lond.,* Vol 107, pp 189-206.

Jeffreys 1925
Jeffreys, H. 1925. "On the Formation of Waves by Wind," *Proc. Roy. Soc. Lond.,* Ser. A., Vol 110, pp 341-347.

Kahma 1981
Kahma, K. K. 1981. "A Study of the Growth of the Wave Spectrum with Fetch," *Journal of Physical Oceanography,* Vol 11, Nov., pp 1503-15.

Kelvin 1887
Kelvin, Lord, 1887. "On the Waves Produced by a Single Impulse in Water of Any Depth or in a Dispersive Medium," *Mathematical and Physical Papers,* Vol IV, London, Cambridge University Press, 1910, pp 303-306.

Kitaigorodskii 1962
Kitaigorodskii, S. A. 1962. "Application of the Theory of Similarity to the Analysis of Wind Generated Wave Motion as a Stochastic Process," *Bull. Acad. Sci.,* USSR Ser. Geophys., Vol 1, No.1, pp 105-117.

Kitaigorodskii 1983
Kitaigorodskii, S. A. 1983. "On the Theory of the Equilibrium Range in the Spectrum of Wind-Generated Gravity Waves," *J. Phys. Oceanogr.,* Vol 13, pp 816-827.

Large and Pond 1981
Large, W. G., and Pond, S. 1981. "Open Ocean Momentum Flux Measurements in Moderate to Strong Winds," *J. Phys. Oceanogr.,* Vol 11, pp 324-336.

Leenhnecht, Szuwalski, and Sherlock 1992
Leenhnecht, D. A., Szuwalski, A., and Sherlock, A. R. 1992. "Automated Coastal Engineering System Technical Reference, Version 1.07, Coastal Engineering Research Center, Waterways Experiment Station, Vicksburg, MS.

Miles 1957
Miles, J. W. 1957. "On the Generation of Surface Waves by Shear Flows," *Journal of Fluid Mechanics*, Vol 3, pp185-204.

Mitsuyasu 1968
Mitsuyasu, H. 1968. "On the Growth of the Spectrum of Wind-Generated Waves (I).," Reports of the Research Institute of Applied Mechanics, Kyushu University, Fukuoka, Japan, Vol 16, No. 55, pp 459-482.

Myers 1954
Myers, V. A. 1954. "Characteristics of United States Hurricanes Pertinent to Levee Design for Lake Okeechobee, Florida," Hydromet. Rep. No. 32., U.S. Weather Bureau, Washington, DC.

Phillips 1958
Phillips, O. M. 1958. "The Equilibrium Range in the Spectrum of Wind-Generated Waves," *Journal of Fluid Mechanics*, Vol 4, pp 426-434.

Pierson and Moskowitz 1964
Pierson, W. J., and Moskowitz, L. 1964. "A Proposed Spectral Form for Fully Developed Wind Seas Based in the Similarity Theory of S. A. Kitiagorodskii," *J Geophys. Res.,* Vol 9, pp 5181-5190.

Pierson, Neuman, and James 1955
Pierson, W. J., Neuman, G., and James, R. W. 1955. "Observing and Forecasting Oceanwaves by Means of Wave Spectra and Statistics," U.S. Navy Hydrographic Office Pub. No. 60.

Putnam 1949
Putnam, J. A. 1949. "Loss of Wave Energy Due to Percolation in a Permeable Sea Bottom," *Transactions of the American Geophysical Union*, Vol 30, No. 3, pp 349-357.

Putnam and Johnson 1949
Putnam, J. A., and Johnson, J. W. 1949. "The Dissipation of Wave Energy by Bottom Friction," *Transactions of the American Geophysical Union*, Vol 30, No. 1, pp 67-74.

Resio 1981
Resio, D. T. 1981. "The Estimation of a Wind Wave Spectrum in a Discrete Spectral Model," *J. Phys. Oceanogr.,* Vol 11, pp 510-525.

Resio 1987
Resio, D. T. 1987. "Shallow Water Waves; Part I: Theory," *J. Waterway, Port, Coastal and Ocean Eng.,* Vol 113, pp 264-281.

Resio 1988
Resio, D. T. 1988. "Shallow Water Waves; Part II: Data Comparisons," *J. Waterway, Port, Coastal and Ocean Eng.,* Vol 114, pp 50-65.

Resio and Perrie 1989
Resio, D. T., and Perrie, W. 1989. "Implications d and f Equilibrium Range for Wind-Generated Waves," *J. Phys. Oceanogr.,* Vol 19, pp 193-204.

Resio and Vincent 1977
Resio, D. T., and Vincent, C. L. 1977. "Estimation of Winds Over the Great Lakes," *J. Waterways Harbors and Coastal Div.,* American Society of Civil Engineers, Vol 102, pp 263-282.

Resio and Vincent 1979
Resio, D. T., and Vincent, C. L. 1979. "A Comparison of Various Numerical Wave Prediction Techniques," *Proceedings of the 11th Annual Ocean Technology Conference,* Houston, TX, p 2471.

Resio and Vincent 1982
Resio, D. T., and Vincent, C. L. 1982. "A Comparison of Various Numerical Wave Prediction Techniques," *Proc. 11th Annual Offshore Technology Conf.,* Houston, TX, pp 2471-2485.

Rosendal and Shaw 1982
Rosendal, H., and Shaw, S. L. 1982. "Relationship of Maximum Sustained Wind to Minimum Sea Level Pressure in Central North Pacific Tropical Cyclones," NOAA Tech. Memo. NWSTM PR24.

Schwerdt et. al 1979
Schwerdt, R. W., Ho, F. P., and Watkins, R. R. 1979. "Meteorological Criteria for Standard Project Hurricane and Probable Maximum Hurricane Windfields, Gulf and East Coasts of the United States," Tech. Rep. NOAA-TR-NWS-23, National Oceanic and Atmospheric Administration.

Shore Protection Manual 1984
Shore Protection Manual. 1984. 4th ed., 2 Vol, U.S. Army Engineer Waterways Experiment Station, U.S. Government Printing Office, Washington, DC.

Smith 1988
Smith, S. D. 1988. "Coefficients for Sea Surface Wind Stress, Heat Flux, and Wind Profiles as a Function of Wind Speed and Temperature," *J. Geophys. Res.,* Vol 93, pp 467-47.

Snyder and Cox 1966
Snyder, R. L., and Cox, C. S. 1966. "A Field Study of the Wind Generation of Ocean Waves," *J. Mar. Res.,* Vol 24, p 141.

Snyder, Dobson, Elliott, and Long 1981
Snyder, R., Dobson, F. W., Elliott, J. A., and Long, R. B. 1981. "Array Measurements of Atmospheric Pressure Fluctuations Above Surface Gravity Waves," *Journal of Fluid Mechanics,* Vol 102, pp 1-59.

Sverdrup and Munk 1947
Sverdrup, H. U., and Munk, W. H. 1947. "Wind, Sea, and Swell: Theory of Relations for Forecasting." Pub. No. 601, U.S. Navy Hydrographic Office, Washington, DC.

Szabados 1982
Szabados, M. W. 1982. "Intercomparsion of the Offshore Wave Measurements During ARSLOE," *Proceedings OCEANS 82 Conference*, pp 876-881.

Tennekes 1973
Tennekes, H. 1973. "Similarity Laws and Scale Relations in Planetary Boundary Layers," *Workshop on Micrometeorology*, D. A. Haugen, ed., American Meteorology Society, pp 177-216.

Thompson 1977
Thompson, E. F. 1977. "Wave Climate at Selected Locations Along U.S. Coasts," TR 77-1, Coastal Engineering Research Center, U.S. Army Engineer Waterways Experiment Station, Vicksburg, MS.

Thompson and Cardone 1996
Thompson, E. F., and Cardone, V. J. 1996. "Practical Modeling of Hurricane Surface Wind Fields," *Journal of Waterway, Port, Coastal, and Ocean Engineering*, Vol 122, No. 4, pp 195-205.

Toba 1978
Toba, Y. 1978. "Stochastic Form of the Growth of Wind Waves in a Single Parameter Representation with Physical Implications," *Journal of Physical Oceanography*, Vol 8, pp 494-507.

Van Vledder and Holthuijsen 1993
Van Vledder, G. P., and Holthuijsen, L. H. 1993. "The Directional Response of Ocean Waves to Turning Winds," *J. Phys. Oceanogr.*, Vol 23, pp 177-192.

Vincent 1985
Vincent, C. L. 1985. "Depth-Controlled Wave Height," *J. Waterway, Port, Coastal and Ocean Eng.*, Vol 111, No. 3, pp 459-475.

WAMDI Group 1988
WAMDI Group. 1988. "The WAM Model - a Third Generation Wave Prediction Model," *Journal of Physical Oceanography*, Vol 18, 1775-1810.

Ward, Evans, and Pompa, 1977
Ward, E. G., Evans, D. J., and Pompa, J. A. 1977. "Extreme Wave Heights Along the Atlantic Coast of the United States. *Offshore Technology Conference*, OTC 2846, pp 315-324.

Willis and Deardorff 1974
Willis, G. E., and Deardorff, J. W. 1974. "A Laboratory Model of The Unstable Planetary Boundary Layer," *J. Atmos. Sci.*, Vol 31, pp 1297-1307.

Wyngaard 1973
Wyngaard, J. C. 1973. "On Surface-Layer Turbulence," *Workshop on Micrometeorology*, D. A. Haugen, ed., American Meteorology Society, Boston, pp 101-149.

Wyngaard 1988
Wyngaard, J. C. 1988. "Structure of the PBL," *Lectures on Air Pollution Modeling.* A. Venkatram and J. Wyngaard, ed., American Meteorological Society, Boston.

Young 1987
Young, I. R. 1987. "Validation of the Spectral Wave Model ADFA1," *Res. Rep. 17, Dep. Civ. Eng., Augt.* Defense Force Acad., Canberra, Australia.

II-2-6. Definitions of Symbols

α	Equilibrium coefficient
γ	Peak enhancement factor used in the JONSWAP spectrum for fetch-limited seas
ΔT	Air-sea temperature difference [deg °C]
θ_{met}	Measured wind direction in standard meteorological terms (Equation II-2-14) [deg]
θ_{vec}	Measured wind direction in a Cartesian system with the zero angle wind blowing toward the east (Equation II-2-14) [deg]
λ	Dimensionless constant in determining the height of the atmospheric boundary layer (Equation II-2-12)
λ_{1-5}	Dimensionless empirical coefficients used in empirical wave predictions
ρ_a	Mass density of air [force-time^2/length^4]
ρ_w	Mass density of water (salt water = 1,025 kg/m^3 or 2.0 slugs/ft^3; fresh water = 1,000 kg/m^3 or 1.94 slugs/ft^3) [force-time^2/length^4]
σ	Dimensionless spectral width parameter
τ	Wind stress [force/length^2]
φ	Dimensionless universal function characterizing the effects of thermal stratification
ω	Angular velocity of the earth (= 0.2625 rad/hr = 7.292x10^{-5} red/sec)
A	Scaling parameter in the Holland wind model [length]
B	Dimensionless parameter that controls the peakedness of the wind speed distribution in the Holland wind model
c	Particle velocity [length/time]
C_D, c_d	Coefficient of drag for winds measured at 10-m [dimensionless]
C_{Dz}	Coefficient of drag for winds measured at level z [dimensionless]
e	Base of natural logarithms (= 2.718)
$E(f)$	Spectral energy density [length/hertz]
f	Coriolis parameter (= $2\,\omega \sin \phi$ = 1.458 x 10^{-4} sin ϕ), where ϕ is geographical latitude [sec^{-1}]. Also, f = frequency [Hz] = $\dfrac{1}{\text{Period [sec]}}$
f_p	Peak frequency of the spectral peak
f_u	Limiting frequency for a fully developed wave spectrum (Equation II-2-32)
g	Gravitational acceleration [length/time^2]

h Height of the boundary layer (Equation II-2-12) [length]

$\dot{H}$ Dimensionless wave height (Equation II-2-22)

h_m Height of the land barrier [length]

H_{m0} Energy-based significant wave height [length]

H_{stable} Stable wave height (Equation II-4-14) [length]

H_∞ Fully developed wave height (Equation II-2-30) [length]

k Dimensionless von Kármán's constant (approximately equal to 0.4). Also, $k =$ wave number [length^{-1}] defined as $= \dfrac{2\pi}{L}$ where L = wave length [length]

L Parameter that represents the relative strength of thermal stratification effects [length]

m_{1-5} Dimensionless empirical exponents used in empirical wave predictions

$\cdot_0$ The subscript 0 denotes deepwater conditions

p Pressure at radius r of a storm [force/length^2]

p_c Central pressure in the storm [force/length^2]

p_n Ambient pressure at the periphery of the storm [force/length^2]

r Arbitrary radius [length]

r_c Radius of curvature of the isobars [length]

R_L Ratio of over water windspeed, U_W to over land windspeed, U_L as a function of over land windspeed (Figure II-2-7)

R_{max} Distance from the center of the storm circulation to the location of maximum wind speed (Equation II-2-20) [length]

R_O Rossby radius of deformation (Equation II-2-1) [length]

R_T Amplification ratio (Figure II-2-8), ratio of wind speed accounting for effects of air-sea temperature difference to wind speed over water without temperature effects

t Duration [time]

T_a Air temperature [deg C]

T_p Limiting wave period (Equation II-2-39) [time]

T_s Water temperature [deg C]

$t_{x,u}$ Time required for waves crossing a fetch (Equation II-2-35) [time]

u Wind speed [length/time]

U'_t Estimated wind speed of any duration [length/time]

U_c	Cyclostrophic approximation to the wind speed [length/time]
U_f	Fastest mile wind speed [length/time]
U_g	Geostrophic wind speed (Equation II-2-10) [length/time]
U_{gr}	Gradient wind speed (Equations II-2-11 and II-2-18) [length/time]
U_L	Wind speed over land [length/time]
U_{max}	Maximum velocity in the storm (Equation II-2-21) [length/time]
U_t	Wind speed of any duration [length/time]
U_W	Wind speed over water [length/time]
U_z	Wind speed at height z above the surface (Equation II-2-3) [length/time]
u_*	Wind friction velocity [length/time]
U_*	Wind friction velocity [length/time]
W_C	Wind speed accounting for effects of air-sea temperature difference [length/time]
W_W	Wind speed over water without temperature effects [length/time]
X	Straight line distance over which the wind blows [length]
z_0	Roughness height of the surface [length]

II-2-7. Acknowledgments

Authors of Chapter II-2, "Meteorology and Wave Climate:"

Donald T. Resio, Ph.D., Coastal and Hydraulics Laboratory (CHL), Engineer Research and Development Center, Vicksburg, Mississippi.
Steven M. Bratos, U.S. Army Engineer District, Jacksonville, Jacksonville, Florida.
Edward F. Thompson, Ph.D., CHL.

Reviewers:

Lee E. Harris, Ph.D., Department of Marine and Environmental Systems, Florida Institute of Technology, Melbourne, Florida.
Robert O. Reid, Ph.D., Texas A. & M. University, College Station, Texas (emeritus).
J. Richard Weggel, Ph.D., Dept. of Civil and Architectural Engineering, Drexel University, Philadelphia, Pennsylvania.
Zeki Demirbilek, Ph.D., CHL.
H. Lee Butler, CHL (retired)

Table of Contents

Page

II-3-1. Introduction II-3-1
   a. Background II-3-1
   b. Practical limitations II-3-1
   c. Importance of water level II-3-2
   d. Role of gauging II-3-3
   e. Physical modeling II-3-3

II-3-2. Principles of Wave Transformation II-3-4
   a. Introduction II-3-4
   b. Wave transformation equation II-3-4
   c. Types of wave transformation II-3-5

II-3-3. Refraction and Shoaling II-3-6
   a. Wave rays II-3-6
   b. Straight and parallel contours II-3-7
   c. Realistic bathymetry II-3-11
   d. Problems in ray approach II-3-15
   e. Wave diffraction II-3-16
   f. Reflection II-3-16
   g. Refraction and shoaling of wave spectra II-3-17
   h. Alternate formulations II-3-17
      (1) Mild slope equation II-3-17
      (2) Boussinesq equations II-3-17

II-3-4. Transformation of Irregular Waves II-3-18

II-3-5. Advanced Propagation Methods II-3-19
   a. Introduction II-3-19
   b. RCPWAVE II-3-20
      (1) Introduction II-3-20
      (2) Examples of RCPWAVE results II-3-21
      (3) Data requirements for RCPWAVE II-3-24
   c. REFDIF II-3-24
      (1) Introduction II-3-24
      (2) Wave breaking II-3-25
      (3) Wave damping mechanisms II-3-25
      (4) Wave nonlinearity II-3-25
      (5) Numerical noise filter II-3-25
      (6) Examples of REF/DIF1 results laboratory verification II-3-26
      (7) Data requirements for REFDIF II-3-26
   d. STWAVE II-3-26

(1) Introduction II-3-26
(2) Examples of STWAVE results II-3-29
    (a) Spectral versus monochromatic calculations II-3-29
    (b) Effects of coupled source terms II-3-29
    (c) Wind effects II-3-29
(3) Data requirements for STWAVE II-3-30
  e. Limitations II-3-30

II-3-6. Guidance for Performing Wave Transformation Studies II-3-31
  a. Introduction II-3-31
  b. Problem formulation II-3-32
  c. Site analysis II-3-32
  d. Selection of input data site II-3-33
  e. Selection of wave transformation method II-3-33
  f. Calibration/verification II-3-33
  g. Post-processing II-3-34

II-3-7. References II-3-34

II-3-8. Definitions of Symbols II-3-40

II-3-9. Acknowledgments II-3-41

List of Tables

Page

Table II-3-1.   Example Problem II-3-1 Refraction and Shoaling Results II-3-14

Table II-3-2.   Guidance for Selection of Wave Transformation Methods II-3-33

List of Figures

Page

Figure II-3-1.   Waves propagating through shallow water influenced by the underlying bathymetry and currents II-3-2

Figure II-3-2.   Amplification of wave height behind a shoal for waves with different spreads of energy in frequency and direction II-3-3

Figure II-3-3.   Straight shore with all depth contours evenly spaced and parallel to the shoreline II-3-7

Figure II-3-4.   Idealized plots of wave rays II-3-8

Figure II-3-5.   Wave-height variation along a wave ray II-3-10

Figure II-3-6.   Solution nomogram II-3-12

Figure II-3-7.   Highly regular bathymetry but undulatory contours II-3-16

Figure II-3-8.   Typical RCPWAVE application, bathymetry II-3-22

Figure II-3-9.   Typical RCPWAVE application, wave height II-3-23

Figure II-3-10.   Bathymetry input to REF/DIF1 for a simulation of wave propagations at Revere Beach, MA II-3-27

Figure II-3-11.   Wave heights calculated by REF/DIF 1 . . . . II-3-28

Figure II-3-12.   Spectral model results compared to laboratory measurements for broad directional spectrum II-3-30

Figure II-3-13.   STWAVE results for a 1:30 sloping beach II-3-31

Figure II-3-14.   STWAVE results for CHL's Field Research Facility at Duck, NC II-3-32

Chapter II-3
Estimation of Nearshore Waves

II-3-1. Introduction

a. Background.

(1) Coastal engineering considers problems near the shoreline normally in water depths of less than 20 m. Project designs usually require knowledge of the wave field over an area of 1-10 km² in which the depth may vary significantly. Additionally, study of shoreline change and beach protection frequently requires analysis of coastal processes over entire littoral cells, which may span 10-100 km in length. Wave data are generally not available at the site or depths required. Often a coastal engineer will find that data have been collected or hindcast at sites offshore in deeper water or nearby in similar water depths. This chapter provides procedures for transforming waves from offshore or nearby locations to nearshore locations needed by the engineer.

(2) Understanding the processes that affect coastal waves is essential to coastal engineering. Waves propagating through shallow water are strongly influenced by the underlying bathymetry and currents (Figure II-3-1). A sloping or undulating bottom, or a bottom characterized by shoals or underwater canyons, can cause large changes in wave height and direction of travel. Shoals can focus waves, in some cases more than doubling wave height behind the shoal. Other bathymetric features can reduce wave heights. The magnitude of these changes is particularly sensitive to wave period and direction and how the wave energy is spread in frequency and direction (Figure II-3-2). In addition, wave interaction with the bottom can cause wave attenuation. The influence of bathymetry on local wave conditions cannot be overstated as a critical factor in coastal engineering design.

(3) Wave height is often the most significant factor influencing a project. Designing with a wave height that is overly conservative can greatly increase the cost of a project and may make it uneconomical. Conversely, underestimating wave height could result in catastrophic failure of a project or significant maintenance costs. Approaches for transforming waves are numerous and differ in complexity and accuracy. Consequently transformation studies require careful analysis. They are but one part of selecting project design criteria, which will be treated in Part II-9.

(4) Wave transformation across irregular bathymetry is complex. Simplifying assumptions admit valid and useful approximations for estimating nearshore waves. After this introduction, a basic principles section provides an overview of the theoretical basis for wave transformation analyses, followed by development of a simple method for refraction and shoaling estimates. Transformation of irregular waves is then discussed. Next, advanced wave transformation models currently used by the Corps of Engineers are discussed. A final section provides guidance on selecting the approach used in calculating wave transformation. This chapter is primarily directed at open coast wave problems excluding structures such as breakwaters or jetties. Analyses involving structures are provided in Part II-7.

b. Practical limitations.

(1) The purpose of this chapter is to provide methods for estimating waves at one site given information at another. The assumption made is that *the wave information used as input to the analysis is characteristic of the waves that would propagate to the site.* In each case, the engineer should assure that there is no limitation of fetch, sheltering of waves, or oddness of bathymetry that would make selection of the input site inappropriate.

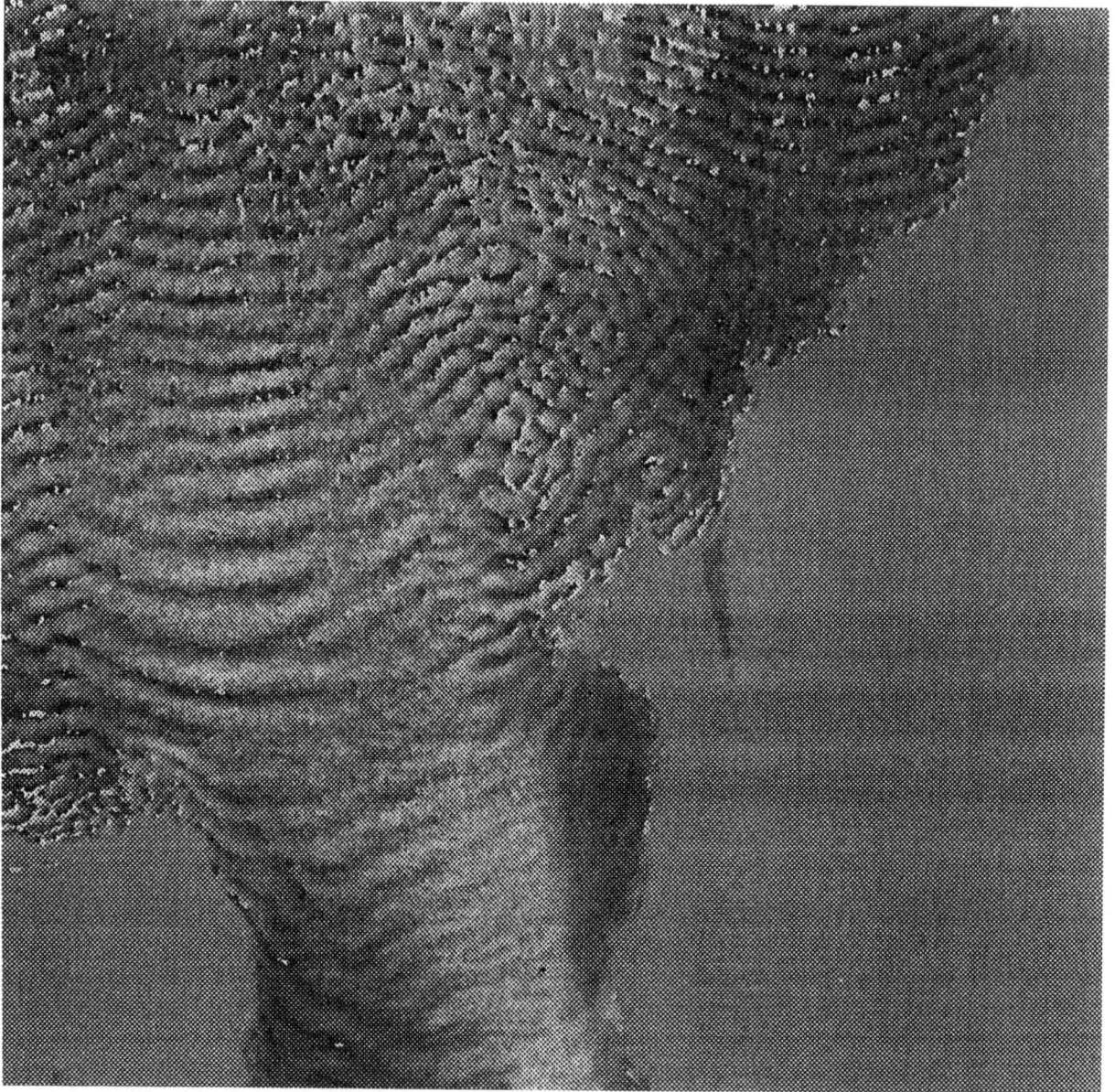

Figure II-3-1. Waves propagating through shallow water influenced by the underlying bathymetry and currents

(2) For most of the open U.S. coastline, Wave Information Study data or data from gauges provide adequate spacing of sites along the coast to give estimates of the wave climate that can be used as input to nearshore transformation studies. In other places or for simulation of a specific event, a special hindcast of the deepwater wave climate may be required to provide input for a transformation analysis.

c. *Importance of water level.* Near the coast, variable water depths can produce major variations in wave conditions over short distances. The important physical parameter is the depth of the water on which the surface waves are traveling. In nature, water depth is not a constant: it varies with tide stage, hurricane or extratropical storm surge, or for a variety of other reasons (Part II-5). These variations in water level influence wave breaking. *Hence, any study of wave transformation must account for expected water levels for the site and the situation of interest.*

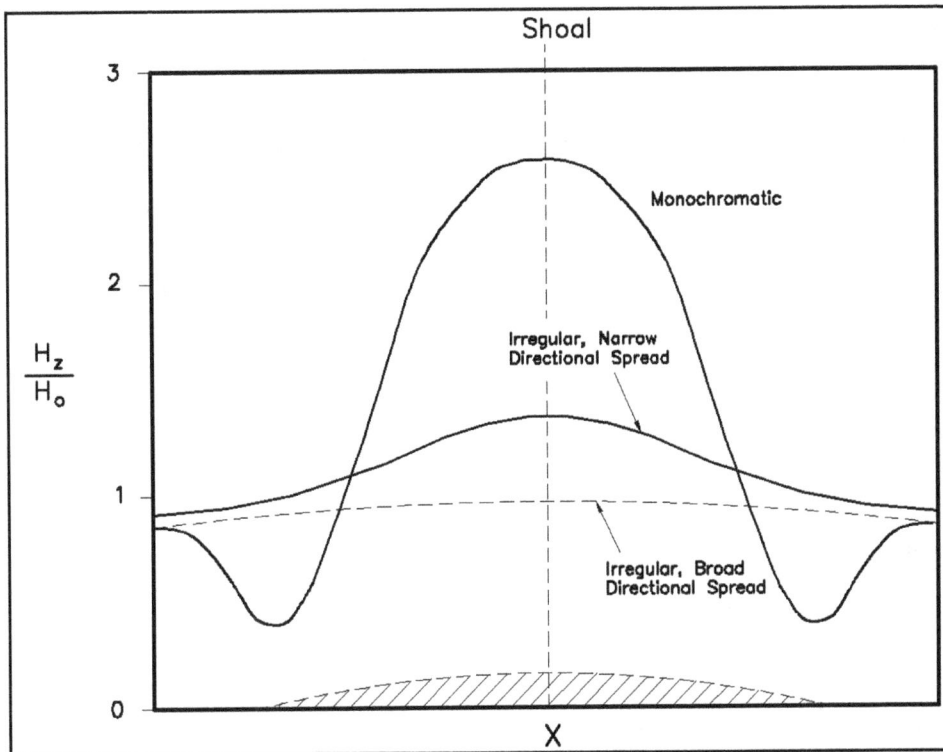

Figure II-3-2. Amplification of wave height behind a shoal for waves with different spreads of energy in frequency and direction

d. Role of gauging. The procedures described here are needed because long-term site-specific data often do not exist. If time and funding are available, a short-term gauging program should be considered. A gauging program can help in two ways:

(1) It may provide a simple statistically based transformation procedure.

(2) It can be used to validate/calibrate a numerical model as a transformation procedure for the project.

Even a few months of gauge data can be a significant complement to any wave-transformation analysis. Short-term gauging is generally not useful in providing, by itself, a design-wave height.

e. Physical modeling. This chapter emphasizes calculation procedures for estimating nearshore waves. However, some sites are so complicated that a physical model of the site may be required to determine the wave conditions. Physical modeling is a well-established procedure for analysis of wave propagation and breaking effects and is particularly useful in analysis of the effects of structures on the wave field. Physical modeling is not useful for evaluating bottom friction or percolation effects or inclusion of wind inputs. Because of scaling limitations and costs, physical models are generally used for small areas (a few square kilometers or less). If strong currents transverse to a wave field are present, such as at a tidal inlet, a physical model may be the only dependable method for estimating the wave field.

II-3-2. Principles of Wave Transformation

a. Introduction.

(1) In this section, the scientific principles governing the transformation of waves from deep water to shallow will be presented in sufficient detail to highlight critical assumptions and simplifications. Unfortunately, the problem is so complex that detailed computations require use of complicated numerical models whose background and implementation are beyond the scope of the Coastal Engineering Manual. This chapter provides the principles of wave transformation, a simplified approach, and an introduction to three numerical models used by the Corps of Engineers.

(2) Processes that can affect a wave as it propagates from deep into shallow water include:

(a) Refraction.

(b) Shoaling.

(c) Diffraction.

(d) Dissipation due to friction.

(e) Dissipation due to percolation.

(f) Breaking.

(g) Additional growth due to the wind.

(h) Wave-current interaction.

(i) Wave-wave interactions

The first three effects are *propagation* effects because they result from convergence or divergence of waves caused by the shape of the bottom topography, which influences the direction of wave travel and causes wave energy to be concentrated or spread out. Diffraction also occurs due to structures that interrupt wave propagation. The second three effects are *sink* mechanisms because they remove energy from the wave field through dissipation. The wind is a *source* mechanism because it represents the addition of wave energy if wind is present. The presence of a large-scale current field can affect wave propagation and dissipation. Wave-wave interactions result from nonlinear coupling of wave components and result in transfer of energy from some waves to others. The procedures presented will stop just seaward of the surf zone, which is treated in Part II-4, "Surf Zone Hydrodynamics."

b. Wave transformation equation.

(1) The general problem of wave transformation will be introduced in terms of the concept of directional wave spectra discussed in Part II-1 and II-2. Adopting the notation of Part II-2, consider a directional spectrum $E(x,y,t,f,\theta)$ where f,θ represents a particular frequency-direction component, x,y represents a location in geographic space, and t represents time. The waves are propagating over a region with varying water depths with no current. Water level will not be time-dependent in the following analyses. Structures are not considered. The general equation used to estimate wave transformation is the radiative transfer equation introduced in Part II-2.

$$\underset{A}{\frac{\partial E(x,y,t,f,\theta)}{\partial t}} + \underset{B}{\nabla \cdot [C_g(x,y,f) \; E(x,y,t,f,\theta)]} = \underset{C}{S_w} + \underset{D}{S_n} + \underset{E}{S_D} + \underset{F}{S_F} + \underset{G}{S_P}$$

(II-3-1)

(2) Although multidimensional, this equation is fundamentally simple. Term A represents the temporal rate of change of the spectrum, term B represents the propagation of wave energy, term C represents inputs from the wind, term D represents the redistribution of wave energy between different wave components that arise from nonlinearities of the waves, term E represents dissipation due to breaking, term F represents losses due to bottom friction, and term G represents losses due to percolation. Many different algebraic forms have been suggested for the various S_i; three references that provide examples are WAMDI (1988), Sobey and Young (1986), and Young (1988). Since they are complicated and cannot be used in manual computations, their algebraic form is not provided here. More detailed discussion of spectral wave mechanics may be found in Leblond and Mysak (1978), Hasselmann (1962, 1963a, 1963b), Hasselmann et al. (1973), Barnett (1968), Phillips (1977), Resio (1981), WAMDI (1988), and in Parts II-1 and II-2.

(3) Surface wave motions produce a velocity field that extends to some depth in the water column. This depth for a deepwater wave is $L/2$ where L is the deepwater wave length. If the water depth is less than $L/2$, the motion extends to the bottom. In cases where the wave motion interacts with the bottom, several physical changes occur as shown in Part II-1: the celerity C and group velocity C_g are changed, as is the wavelength. If the waves are propagating in a region in which the depths are variable (and sufficiently shallow so that the wave interacts with the bottom), the changes in wave speed change the direction of wave travel and change the amplitude of the wave (*refraction and shoaling*). If the patterns of wave propagation lead to strong focusing of waves, wave energy may be radiated away from the convergence by diffraction (Penny and Price 1944; Berkhoff 1972). The interaction of the wave with the bottom produces a boundary layer, which will result in the loss of wave energy to the bed due to bottom friction (Term F) resulting from bottom materials and bed forms (Bagnold 1946). If the bed is reasonably porous, the pressure field associated with the passing wave can induce flow into and out of the bed (Bretschneider and Reid 1953), resulting in energy losses due to percolation (Term G). If the bed is muddy or visco-elastic other losses may occur (Forristall and Reece 1985). Typically, only one of the bottom loss mechanisms is dominant at one locality although in a large, complicated area a variety of bottom types may exist with differing mechanisms important at different sites along the path of wave travel. However, the bottom-loss terms are often not applied because inadequate information is available on bottom-material composition to allow their proper use.

(4) Wind input, interwave transfers, and breaking follow the principles outlined in Part II-2, though modified due to depth effects. Of the three, wave breaking is most affected by depth. If shoals exist, depth-induced breaking may be significant even though it is outside of the surf zone. Surf zone wave breaking is treated in Part II-4. The effect of sporadic breaking of large waves on shoals or other depth-related features outside the surf zone is not negligible in high sea states. Even in deep water, waves break through whitecapping or oversteepening due to superposition of large waves. The interaction of waves and an underlying current can result in refraction of the waves and wave breaking (Jonsson 1978; Peregrine 1976).

c. *Types of wave transformation.*

(1) Three classic cases of wave transformation describe most situations found in coastal engineering:

(a) A large storm generates deepwater waves that propagate across shallower water while the waves continue to grow due to wind.

(b) A large storm generates winds in an area remote from the site of interest and as waves cross shallower water with negligible wind, they propagate to the site as swell.

(c) Wind blows over an area of shallow water generating waves that grow so large as to interact with the bottom (no propagation of waves from deeper water into the site).

(2) All cases are important, but the first and third are relatively complex and require a numerical model for reasonable treatment. The second case, swell propagating across a shallow region, is a classic building block that has served as a basis for many coastal engineering studies. Often the swell is approximated by a monochromatic wave, and simple refraction and shoaling methods are used to make nearshore-wave estimates. Since the process of refraction and shoaling is important in coastal engineering, the next section is devoted to deriving some simple approaches to illustrate the need for more complex approaches.

(3) Often it is necessary for engineers to make a steady-state assumption: i.e., wave properties along the outer boundary of the region of interest and other external forcing are assumed not to vary with time. This is appropriate if the rate of variation of the wave field in time is very slow compared to the time required for the waves to pass from the outer boundary to the shore. If this is not the case, then a time-dependent model is required. Cases (a) and (c) would more typically require a time-dependent model. Time-dependent models are not discussed here due to their complexity. Examples are described by Resio (1981), Jensen et al. (1987), WAMDI (1988), Young (1988), SWAMP Group (1985), SWIM Group (1985), and Demirbilek and Webster (1992a,b).

II-3-3. Refraction and Shoaling

In order to understand wave refraction and shoaling, consider the case of a steady-state, monochromatic (and thereby long-crested) wave propagating across a region in which there is a straight shoreline with all depth contours evenly spaced and parallel to the shoreline (Figure II-3-3). In addition, no current is present. If a wave crest initially has some angle of approach to the shore other than 0 deg, part of the wave (point A) will be in shallower water than another part (point B). Because the depth at A, h_A, is less than the depth at B, h_B, the speed of the wave at A will be slower than that at B because

$$C_A = \frac{g}{\omega} \tanh k h_A < \frac{g}{\omega} \tanh k h_B = C_B \qquad \text{(II-3-2)}$$

The speed differential along the wave crest causes the crest to turn more parallel to shore. The propagation problem becomes one of plotting the direction of wave approach and calculating its height as the wave propagates from deep to shallow water. For the case of monochromatic waves, wave period remains constant (Part II-1). In the case of an irregular wave train, the transformation process may affect waves at each frequency differently; consequently, the peak period of the wave field may shift.

a. Wave rays.

(1) The wave-propagation problem can often be readily visualized by construction of wave rays. If a point on a wave crest is selected and a wave crest orthogonal is drawn, the path traced out by the orthogonal as the wave crest propagates onshore is called a ray. Hence, a group of wave rays map the path of travel of the wave crest. For simple bathymetry, a group of rays can be constructed by hand to show the wave transformation, although it is a tedious procedure. Graphical computer programs also exist to automate this process (Harrison and Wilson 1964, Dobson 1967, Noda et al. 1974), but to a large degree such approaches have been superseded by the numerical methods discussed in Part II, Section 3-5. Refraction and shoaling analyses typically try to specify the wave height and direction along a ray.

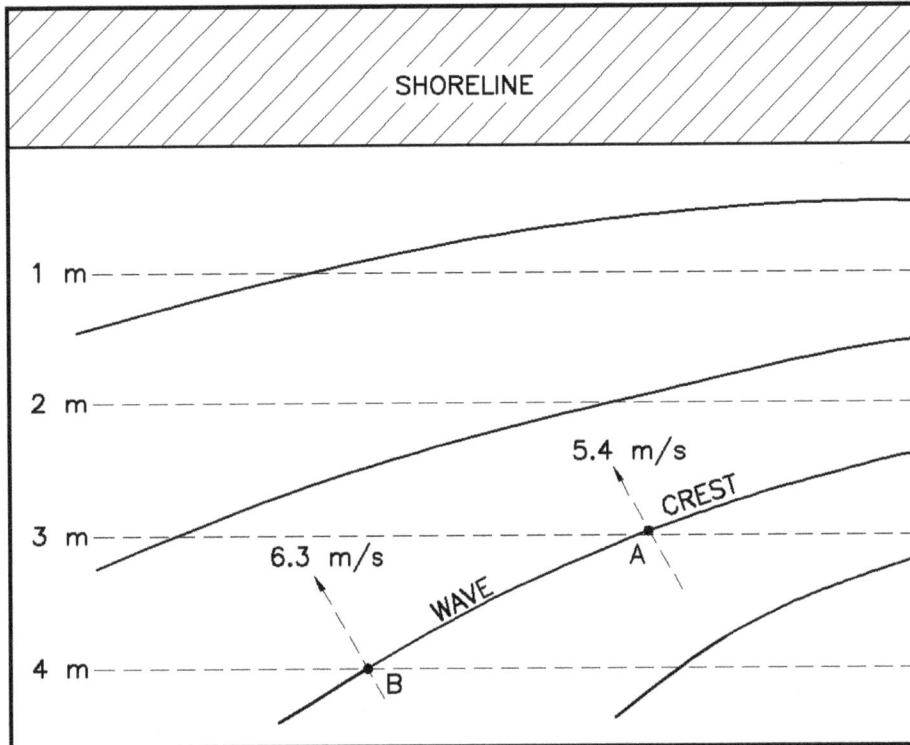

Figure II-3-3. Straight shore with all depth contours evenly spaced and parallel to the shoreline

(2) Figure II-3-4 provides idealized plots of wave rays for several typical types of bathymetry. Simple parallel contours tend to reduce the energy of waves inshore if they approach at an angle. Shoals tend to focus rays onto the shoals and spread energy out to either side. Canyons tend to focus energy to either side and reduce energy over the head of the canyon. The amount of reduction or amplification will depend not only on bathymetry, but on the initial angle of approach and period of the waves. For natural sea states that have energy spread over a range of frequencies and directions, reduction and amplification are also dependent upon the directional spread of energy (Vincent and Briggs 1989).

(3) Refraction and shoaling have been derived and treated widely. The following presentation follows that of Dean and Dalrymple (1991) very closely. Other explanations are provided in Ippen (1966), the *Shore Protection Manual* (1984), and Herbich (1990).

b. Straight and parallel contours.

(1) First, the equation for specifying how wave angle changes along the ray is developed, followed by the equation for wave height. The derivation is only for parallel and straight contours with no currents present. The x-component of the coordinate system will be taken to be orthogonal to the shoreline; the y-coordinate is taken to be shore-parallel. The straight and shore-parallel contours assumption will imply that any derivative in the y-direction is zero because dh/dy is zero.

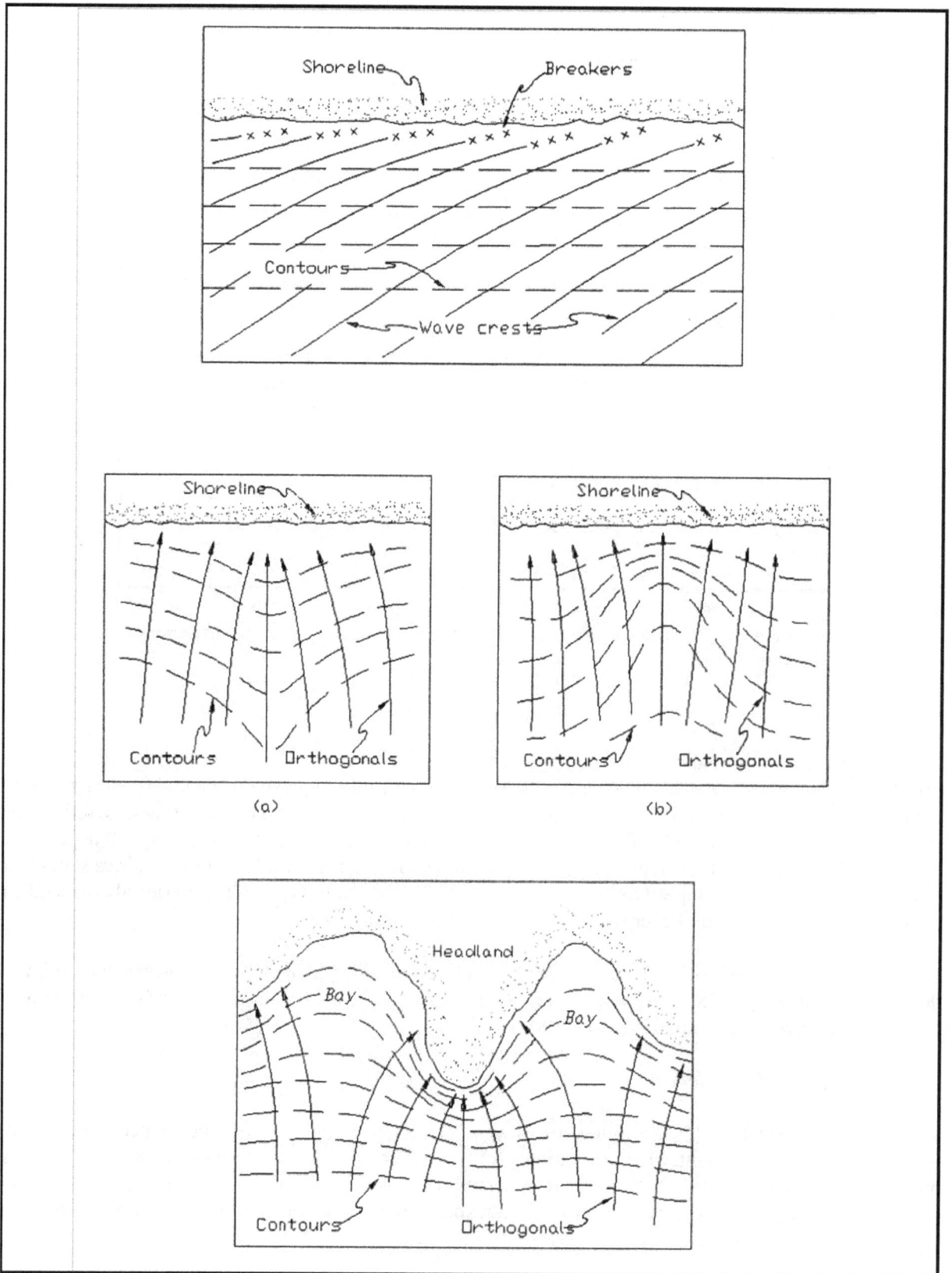

Figure II-3-4.   Idealized plots of wave rays

(2) For a monochromatic wave, the wave phase function

$$\Omega\ (x,y,t)\ =\ (k\ \cos\theta\ +\ k\ \sin\theta\ -\ \omega t) \tag{II-3-3}$$

can be used to define the wave number vector $\vec{k}$ by

$$\vec{k}\ =\ \nabla\ \Omega \tag{II-3-4}$$

(3) Since $\vec{k}$ is a vector, one can take the curl of $\vec{k}$

$$\nabla \times \vec{k}\ =\ 0 \tag{II-3-5}$$

which is zero because $\vec{k}$ by definition is the gradient of a scaler and the curl of a gradient is zero.

(4) Substituting the components of $\vec{k}$, Equation II-3-5 yields

$$\frac{\partial (k\ \sin\theta)}{\partial x}\ -\ \frac{\partial (k\ \cos\theta)}{\partial y}\ =\ 0 \tag{II-3-6}$$

(5) Since the problem is defined to have straight and parallel contours, derivatives in the y direction are zero and using the dispersion relation linking k and C (and noting that $k = 2\pi/CT$ and wave period is constant) Equation II-3-6 simplifies to

$$\frac{d}{dx}\left(\frac{\sin\theta}{C} \right)\ =\ 0 \tag{II-3-7}$$

or

$$\frac{\sin\ \theta}{C}\ =\ constant \tag{II-3-8}$$

(6) Let C_0 be the deepwater celerity of the wave. In deep water, $\sin\ (\theta_0)/c_0$ is known if the angle of the wave is known, so Equation II-3-8 yields

$$\frac{\sin\ \theta}{C}\ =\ \frac{\sin\ \theta_0}{C_0} \tag{II-3-9}$$

along a ray. This identity is the equivalent of Snell's law in optics. The equation can be readily solved by starting with a point on the wave crest in deep water and incrementally estimating the change in C because of changes in depth. The direction $\ominus$ of wave travel is then estimated plotting the path traced by the ray. The size of increment is selected to provide a smooth estimate of the ray.

(7) The wave-height variation along the ray can be estimated by considering two rays closely spaced together (Figure II-3-5). In deep water, the energy flux (EC_n), which is also EC_g across the wave crest distance b_0 can be estimated by $(ECn)_0 b_0$. Considering a location a short distance along the ray, the energy flux is $(ECn)_1 b_1$. Since the rays are orthogonal to the wave crest, there should be no transfer of energy across the rays and conservation principles give

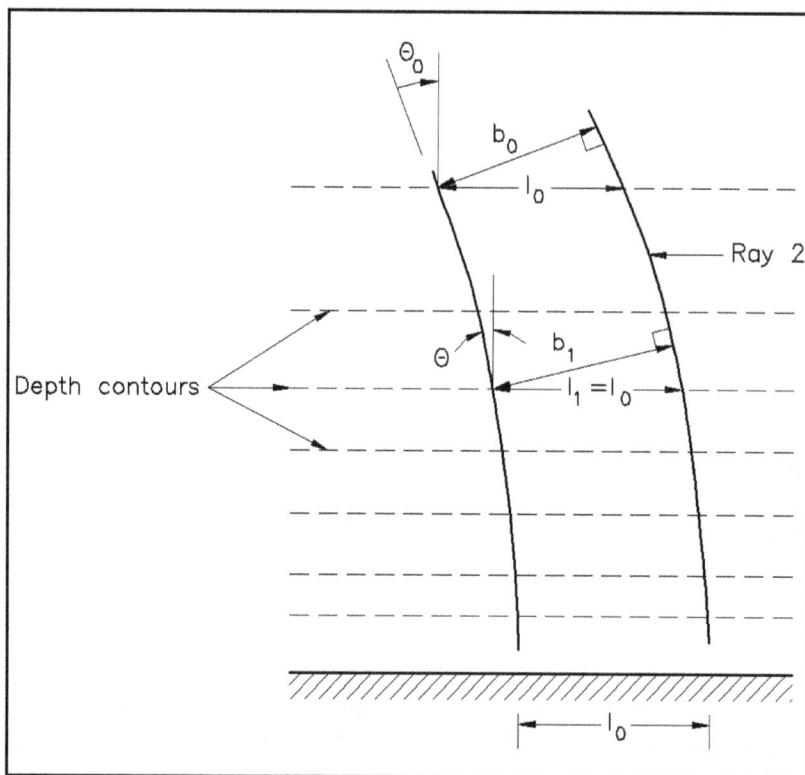

Figure II-3-5.   Wave-height variation along a wave ray

$$(ECn)_0 b_0 = (ECn)_1 b_1 \tag{II-3-10}$$

(8) From Part II-1, the height and energy of a monochromatic wave are given by

$$E = \frac{1}{8} \rho g H^2 \tag{II-3-11}$$

and the wave height at location 1 is thus related to the wave height in deep water by

$$H_1 = H_0 \sqrt{\frac{C_{g_0}}{C_{g_1}}} \sqrt{\frac{b_0}{b_1}} \tag{II-3-12}$$

(9) This equation is usually written as

$$H_1 = H_0 \, K_s \, K_r \tag{II-3-13}$$

where K_s is called the shoaling coefficient and K_r is the refraction coefficient. From the case of simple, straight, and parallel contours, the value at b_1 can be found from b_0

$$K_r = \left(\frac{b_o}{b_1} \right)^{\frac{1}{2}} = \left(\frac{\cos \theta_0}{\cos \theta_1} \right)^{\frac{1}{2}} = \left(\frac{1 - \sin^2 \theta_0}{1 - \sin^2 \theta_1} \right)^{\frac{1}{4}} \tag{II-3-14}$$

by noting that ray 2 is essentially ray 1 shifted downcoast. For straight and parallel contours, Figure II-3-6 is a solution nomogram. This is automated in the ACES program (Leenknecht, Szuwalski, and Sherlock 1992) and the program NMLONG (Kraus 1991). Figure II-3-6 provides the local wave angle K_R and $K_R K_S$ in terms of initial deepwater wave angle and d/gT^2. Although the bathymetry of most coasts is more complicated than this, these procedures provide a quick way of estimating approximate wave approach angles.

c. Realistic bathymetry.

(1) The previous discussion was for the case of straight and parallel contours. If the topography has variations in the y direction, then the full equation must be used. Dean and Dalrymple (1991) show the derivation in detail for ray theory in this case. Basically, the (x,y) coordinate system is transformed to (s,n) coordinates where s is a coordinate along a ray and n is a coordinate orthogonal to it. Algebraically, the equation for wave angle can be derived in the ray-based coordinate system

$$\frac{\partial \theta}{\partial s} = \frac{1}{k}\frac{\partial k}{\partial n} = -\frac{1}{C}\frac{\partial C}{\partial n} \tag{II-3-15}$$

and the ray path defined by

$$\frac{ds}{dt} = C \tag{II-3-16}$$

$$\frac{dx}{dt} = C \cos\theta \tag{II-3-17}$$

$$\frac{dy}{dt} = C \sin\theta \tag{II-3-18}$$

(2) Equation II-3-15 represents the discussion at the beginning of this section; the rate at which the wave turns depends upon the local gradient in wave speed along the wave crest. Munk and Arthur's computation for the refraction coefficient is more complicated: defining

$$K_r = \left(\frac{1}{\beta}\right)^{\frac{1}{2}} \tag{II-3-19}$$

where $\beta = b/b_0$ then

$$\frac{d^2\beta}{ds^2} + p\,\frac{d\beta}{ds} + q\beta = 0 \tag{II-3-20}$$

with

$$p(s) = -\frac{\cos\theta}{C}\frac{\partial C}{\partial x} - \frac{\sin\theta}{C}\frac{\partial C}{\partial y} \tag{II-3-21}$$

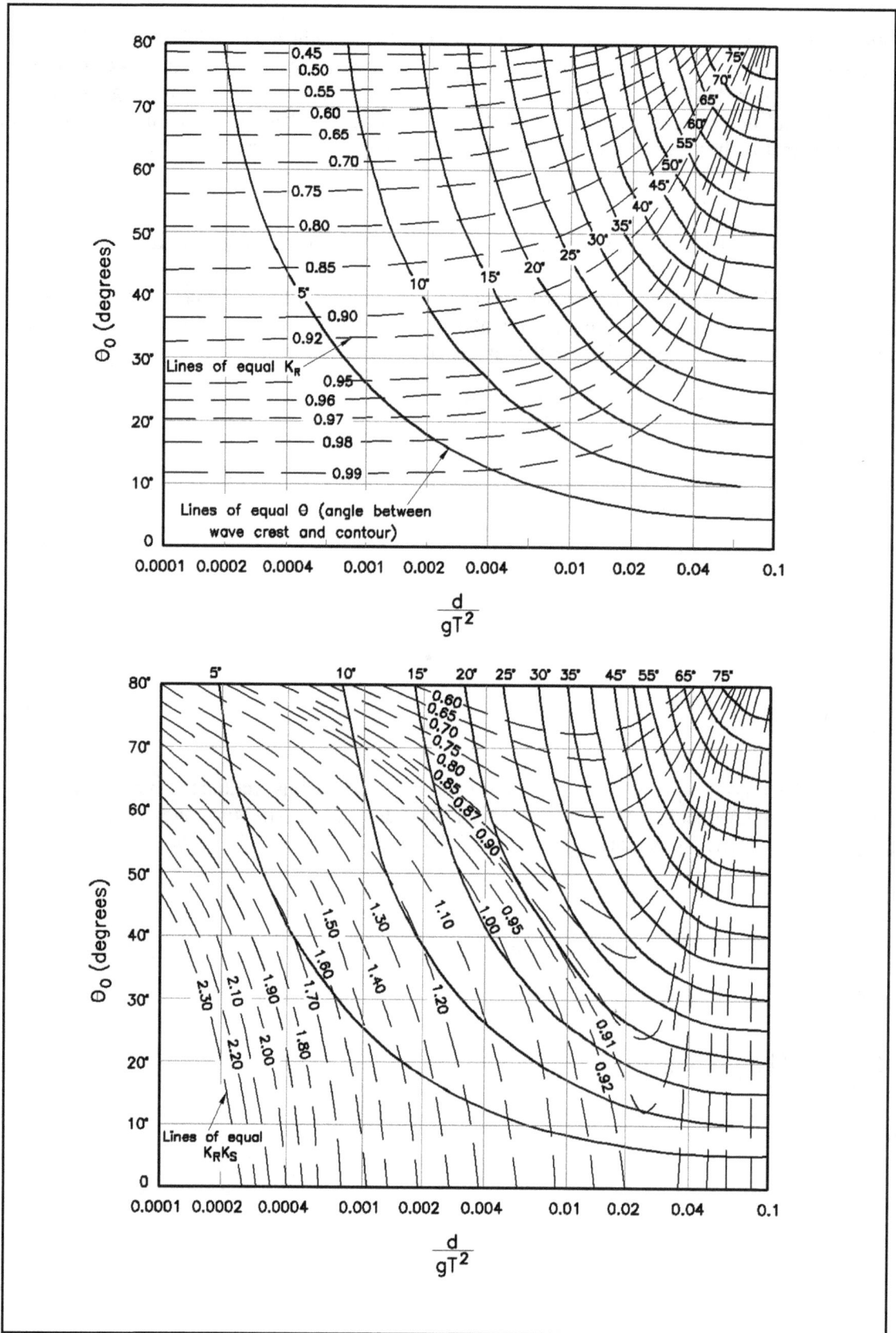

Figure II-3-6. Solution nomogram

EXAMPLE PROBLEM II-3-1

FIND:

Wave height H and angle θ at water depths of 200, 100, 90, 80, 70, 60, 50, 40, 30, 20, 10, 16, 14, 12, 10, 8, 6, and 4 m for deepwater wave angles of 0°, 15°, and 45°.

GIVEN:

A wave 1 m high and 15-sec period in 500 m of water, with a plane, sloping beach.

SOLUTION:

Routine solutions for a plane beach can be obtained using the ACES wave transformation code, by direct calculation, or graphically using Figure II-3-6.

Table II-3-1 provides the results obtained by directly using the ACES code. On a personal computer with a 486-level microprocessor, the results may be obtained in seconds.

For a wave with a depth of 10 m and an initial wave angle of 45 deg, wave height and angle are calculated as follows:

Since the deepwater wave length of a 15-sec wave is

$$L_0 = 1.56 \ T^2 = 1.56 \ (15)^2 = 351 \ m$$

and since 500 m is greater than $L_0/2$, the given initial wave is a deepwater condition. The wave length of the wave in 10 m must be estimated from

$$L = \frac{g \ T^2}{2 \ \pi} \ \tanh \left(\frac{2 \ \pi \ d}{L} \right)$$

and is 144 m (see Problem II-1-1).

The shoaling coefficient K_s can be estimated from

$$K_s = \left(\frac{C_{g0}}{C_{gl}} \right)^{\frac{1}{2}}$$

In deep water C_{g0} for a 15-sec wave is

$$\frac{1}{2} \ C_0 = \frac{1}{2} \ (1.56 \ T) = \frac{23.4}{2} = 11.7 \ \text{m/s}$$

The group velocity is given by

$$C_g = nC = \frac{1}{2} \left(1 + \frac{4\pi d/L}{\sinh \ (4\pi d/L)} \right) \frac{gT}{2\pi} \ \tanh \left(\frac{2\pi d}{L} \right)$$

Substitution of $d = 10$ m, $L = 144$ m, $T = 15$ sec, and $g = 9.8$ m/sec^2 yields 9.05 m/s.

$$K_s = \left(\frac{11.70}{9.05} \right)^{\frac{1}{2}} = 1.14$$

Solution for K_r involves

$$K_r = \left(\frac{1 \ - \ \sin^2\theta_0}{1 \ - \ \sin^2\theta_1} \right)^{\frac{1}{4}}$$

(Continued)

Example Problem II-3-1 (Concluded)

In deep water, θ is 45 deg. From Equation II-3-9,

$$\sin \theta = \frac{C_1 \sin \theta_0}{C_0}$$

In deep water $C_0 = 1.56T = 23.4$ m/s. In 10 m of water, $C_1 = L_1/T = 144$ m/$15_s = 9.60$ m/s.

$$\sin \theta = \frac{9.6 \sin (45^0)}{23.4} = \frac{9.6 (7.07)}{23.4} = 0.29$$

$$K_r = \left(\frac{1 - \sin^2\theta_0}{1 - \sin^2\theta} \right)^{\frac{1}{4}} = \left(\frac{1 - (0.707)^2}{1 - (0.29)^2} \right)^{\frac{1}{4}} = \left(\frac{0.50}{0.91} \right)^{\frac{1}{4}} = 0.86$$

Therefore: $H_i = H_0 K_s K_r = 1(1.14)(0.86) = 0.98$ m.

The angle of approach is arc sin (sinθ) = 16.8⁰. Thus, the l-m, 15-sec wave has changed 2 percent in height by 28.2 deg in angle of approach.

The largest differences caused by refraction and shoaling will be seen at the shallowest depths. From Table II-3-1 at the 4-m depth, the wave height for a 45-deg initial angle is 1.18 m compared to 1.39 m for a wave with initial angle of 0 deg. If the initial angle had been 70 deg, $K_r K_s$ would be about 0.8.

Table II-3-1
Example Problem II-3-1 Refraction and Shoaling Results

Depth	$\theta_0 = 0^\circ$		$\theta_0 = 15^\circ$		$\theta_0 = 45^\circ$	
	θ	H	θ	H	θ	H
500	0	1.00	15.0	1.00	45.0	1.00
400	0	1.00	15.0	1.00	45.0	1.00
300	0	1.00	15.0	1.00	45.0	1.00
200	0	1.00	15.0	1.00	45.0	1.00
100	0	0.94	14.3	0.94	42.4	0.92
90	0	0.93	14.0	0.93	41.2	0.91
80	0	0.93	13.7	0.92	30.4	0.89
70	0	0.92	13.2	0.91	38.9	0.88
60	0	0.91	12.7	0.91	37.0	0.86
50	0	0.91	12.0	0.91	34.5	0.85
40	0	0.92	11.1	0.92	31.8	0.84
30	0	0.95	9.9	0.94	28.1	0.85
20	0	1.00	8.4	0.99	23.4	0.88
18	0	1.02	8.0	1.01	22.3	0.89
16	0	1.04	7.8	1.03	21.1	0.91
14	0	1.07	7.1	1.05	19.8	0.92
12	0	1.10	6.6	1.08	18.4	0.95
10	0	1.14	6.1	1.12	16.8	0.98
8	0	1.19	5.5	1.17	15.1	1.02
6	0	1.27	4.8	1.25	13.15	1.08
4	0	1.39	3.9	1.37	10.8	1.18

and

$$q(s) = \frac{\sin^2\theta}{C}\frac{\partial^2 C}{\partial x^2} - 2\frac{\sin\theta\,\cos\theta}{C}\frac{\partial^2 C}{\partial x\,\partial y} + \cos^2\theta\frac{\partial^2 C}{\partial y^2} \tag{II-3-22}$$

(3) These equations are solved for a set of rays for each wave component of interest (typically combinations of periods and directions). Since this analysis is linear, often a unit wave height is applied for the offshore wave height, which yields a series of refraction and shoaling coefficients at sites of interest. Then the wave transformation for any non-unit initial wave height is obtained by multiplication. This is permissible as long as wave breaking does not occur along a wave ray.

d. Problems in ray approach.

(1) Estimating wave propagation patterns with wave rays is intuitively and visually satisfying, and often very useful. The engineer obtains a good picture of how a wave propagates to a site. However, the procedure has several drawbacks when applied to even mildly irregular bathymetry. One problem is ray convergence/crossing; another is bathymetry inadequacy on ray paths.

(2) An example calculation from Noda et al. (1974) illustrates the basic problem. Bathymetry is highly regular, but has undulatory contours (Figure II-3-7). From the ray pattern, convergence and divergence of adjacent rays are apparent as the waves sweep over the undulations in bathymetry. However, in shallow water near the shore, the rays are sufficiently perturbed by the bathymetry that several converge, with the ray spacing going to zero (in some ray programs the rays actually are computed to cross). Remembering the conservation of wave energy argument used to define the refraction coefficient, the flux across an orthogonal between the rays remains constant. As the spacing between rays approaches zero, the energy flux becomes infinite. Practically, if strong wave convergence occurs, breaking either due to depth constraints (Part II-4) or steepness constraints (Part II-1) naturally limits the wave height. However, situations which generate strong gradients or discontinuities in wave height along a wave crest give rise to *diffraction* effects, which can reduce the wave height and keep it below the breaking value.

(3) The second problem with ray theory is the sensitivity of the wave ray calculations. In most locations, the bathymetry is not well-known. Discretizations of the bathymetry can produce sharper local gradients in the computational depth field than may exist locally or, conversely, may reduce local gradients. Most wave ray calculation schemes calculate each wave ray uncoupled from all others. Ray paths are very sensitive to gradients in bathymetry. The smoothing algorithms that are used to numerically compute the required derivatives can alter the ray field significantly. Since the ray calculations are uncoupled, adjacent rays may take radically different paths due to how the bathymetry was discretized or smoothed. Also, if the ray calculations were started at slightly different spatial locations, the resulting patterns may be significantly different for the same reason. *In the cases where ray patterns are unstable with respect to perturbations of initial positions or where adjacent rays show unusual divergences and crossings, the coastal engineer must carefully assess whether the propagation is indeed that unusual (in which case ray theory results may not be accurate) or decide that more careful analysis of the bathymetry and smoothing is needed.*

(4) Wave propagation discussion has centered on the concept of waves traveling from deep water to shallow. At some locations, the bathymetry is such that waves propagating from offshore towards a beach may initially propagate from deeper to shallow water, then propagate across a zone where the water becomes deeper again. In the region where the wave is propagating at an angle to the progressively shallower depths, the process of refraction previously described occurs: the waves turn more shore-normal. Once the depth gradient reverses, the wave turns in the opposite direction (because of the reversed depth gradient, the part

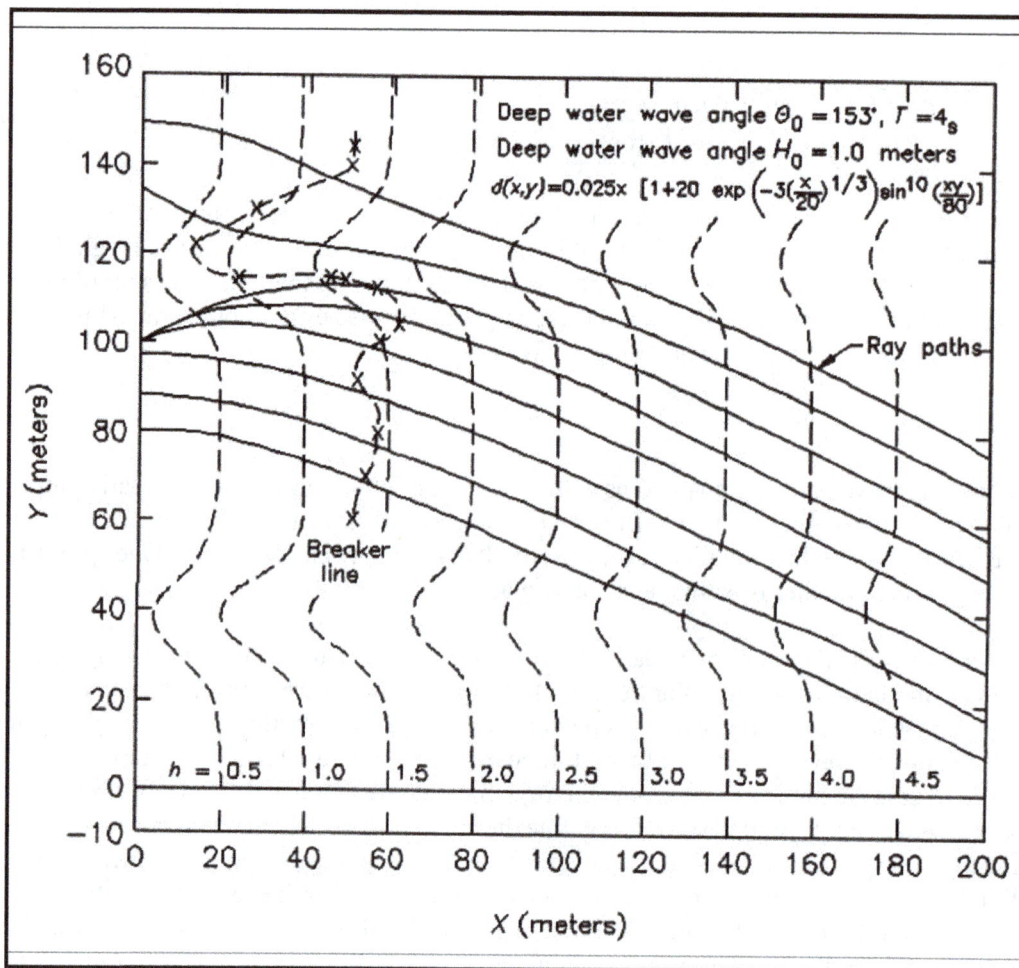

Figure II-3-7. Highly regular bathymetry but undulatory contours

of the wave crest nearer shore is in deeper water than the part of the wave offshore and is hence moving faster). If the combination of wave angle of approach to the bathymetry and wave period is correct, the wave will turn back and go offshore, giving the appearance of a reflected wave. Calculation of wave heights where waves crook or bend backwards is not treated here.

  e Wave diffraction. Wave diffraction is a process of wave propagation that can be as important as refraction and shoaling. The classical introduction to diffraction treats a wave propagating past the tip of a breakwater. Since diffraction theory is most often applied to the interaction of waves with harbor structures, derivation of wave diffraction is deferred until Part II-7 (Harbor Hydrodynamics). Any process that produces an abrupt or very large gradient in wave height along a wave crest also produces diffracted waves that tend to move energy away from higher waves to the area of lower waves. So initial wave energy is reduced as diffracted waves are produced. However, if the rate of convergence is too great, the wave may still break.

  f. Reflection. Waves that propagate into a solid object such as a breakwater, a seawall, a cliff, or a sloping beach may reflect. In the case of a vertical, hard structure, the fraction of wave energy reflected can be large. For permeable structures or gentle slopes, the reflection will be much less. For nearshore wave propagation problems, reflections are usually ignored because the reflected wave may often be less than 10 percent of the incident wave.

Estimation of Nearshore Waves

g. *Refraction and shoaling of wave spectra.* The previous discussion of refraction and shoaling was for a single wave component. However, wave propagation was introduced in Equation II-3-1 through the concept of spectral components. In principle, refraction and shoaling of a wave field in terms of its spectral components simply requires computing the refraction and shoaling coefficient for each frequency-direction (f,θ) component and computing the transformed sum:

$$E(f,\theta) = \sum K_s^2 (f,\theta) \quad K_r^2 (f,\theta) \quad E_o (f,\theta) \quad \Delta f \, \Delta\theta \tag{II-3-23}$$

where $E_o(f,\theta)$ is the offshore directional spectrum. This is possible as long as no breaking or other loss or gain occurs along the propagation path of the individual waves. If it does occur, most advanced spectral models compute the wave transformation locally. In this approach, the area of interest is covered by a discrete series of computation points and the ray path for each (f,θ) component in the spectrum is computed for each grid point only by tracing the ray back to the grid cell boundary defined by adjacent grid points. This approximation, called backward ray tracing, is adequate as long as the wave energy and bathymetry vary smoothly and gently over the domain.

h. *Alternate formulations.*

(1) Mild slope equation. The refraction and shoaling analyses presented above were based on linear wave theory and a ray approach equivalent to geometrical optics. This works well for simple cases, but once the bathymetry becomes even moderately undulatory, the ray approach runs into difficulty. Berkhoff (1972) formulated a more advanced approach for wave propagation that includes refraction, shoaling, and diffraction simultaneously and can incorporate structures. Berkhoff developed what is termed the mild-slope equation given by

$$\nabla (CC_g\Phi) + \omega^2 \left(\frac{C_g}{C} \right) \Phi = 0 \tag{II-3-24}$$

with

$$\varphi(x,y,z) = \Phi \, \frac{\cosh \, k(h+z)}{\cosh \, kh} \tag{II-3-25}$$

where

$$\nabla = \left(\frac{\partial}{\partial x_i} \, , \, \frac{\partial}{\partial y_j} \right) \tag{II-3-26}$$

which provides a solution φ for amplitude and phase of the waves in the horizontal plane. To obtain the equation, Berkhoff assumed that the bottom slope was mild (no abrupt steps, shoals, or trenches). Often slopes of interest violate this assumption, but the models based on the mild-slope equation perform better than the ray approach. Many approaches have been taken to computationally solve this equation. Berkhoff's approach solves the velocity potential of the wave in the horizontal, which can require 5-10 computational grid points per wave length. This is impractical for many cases. Another approach, developed by Radder (1979), is to use a parabolic approximation, which is far more computationally efficient (but subsequently adds more limitations).

(2) Boussinesq equations. Another approach for wave propagation problems close to the coast and in harbors is the use of vertically integrated shallow-water equations in which a Boussinesq (Part II-1) approximation has been made. The numerical models (e.g., Abbott, Peterson, and Skovgaard 1978) resulting from this approach require 10-20 grid points per wave length but have the advantage of being time-dependent

so that the pattern of wave propagation can be directly visualized. Wave crests evolve during the shoaling process to have nonsinusoidal shapes characteristic of shallow-water waves. Currents may be applied directly. Wave breaking, however, is simulated empirically.

II-3-4. Transformation of Irregular Waves

a. The preceding discussion emphasized the refraction and shoaling of monochromatic waves. When this process is applied to an initial significant wave height and period, it is called a *significant wave analysis.* For many conditions where propagation is the dominant factor (as opposed to additional wave growth or bottom dissipation) the significant wave analysis provides a reasonable and generally conservative approximation. The significant wave analysis may be inadequate when wave conditions have spectra characterized by wide directional spreads or broad (frequency) spectral widths or multiple spectral peaks. Cases where the significant wave analysis is adequate primarily involve narrow band swell. This section outlines differences that may be expected between the application of significant wave analyses and application of an irregular wave approach.

b. Rrefraction and shoaling for monochromatic waves may be applied to the individual frequency and direction components of the spectrum of an irregular wave system. Two factors become important: directional spreading and spectral wave mechanics. Directional spreading is important whenever it is present. Spectral wave dynamics are most important in high-energy, high-steepness wave cases, and negligible for low-energy, low-steepness cases.

c. Directional spreading is important for two reasons. First, laboratory tests (Berkhoff, Boij, and Radder 1982; Vincent and Briggs 1989) with unidirectional waves indicate that shoals concentrate wave energy immediately behind the shoal and reduce it on the flanks. The increase behind the shoal can be nearly 250 percent of the initial wave height; the reduction to either side can be about 50 percent. However, laboratory tests with wave spectra having significant directional spread (Vincent and Briggs 1989) show only a 110- to 140-percent increase behind the shoal and only a 10- to 15-percent reduction on the sides. Numerical models incorporating directional spread also replicate this (Panchang et al. 1990). In the case with directional spread, the shoal focusses each frequency-direction component at a different location behind the shoal rather than at one spot as in the unidirectional case. Consequently some of the high- and low-energy regions overlap and cancel each other out. Secondly, if the mean angle of wave approach is not directly onshore, one consequence of directional spreading is that some fraction of the wave energy is heading parallel to shore or offshore. In the case of a wave system with symmetric directional spread (i.e., 50 percent to the left and right of the mean direction), if the mean direction were parallel to a straight shoreline (and the measurement were made in deep water), half of the energy would be moving in directions that could not refract towards shore. So even for angles up to 30 deg offshore-parallel, significant amounts of energy are not propagating shoreward. In a significant wave analysis, all the energy would propagate shoreward. If the shoreline, fetch, or bathymetry is complicated, the fraction of energy that propagates towards shore is more difficult to define.

d. Spectral dynamics arise because waves of different lengths and steepness are propagating through and with other waves. According to Equation II-3-1, these waves can exchange energy between each other (nonlinear transfers) and superposition of waves can lead to dissipation due to breaking. Analysis of thousands of wave records (Bouws et al. 1987; Bouws, Gunther, and Vincent 1985; Miller and Vincent 1990) indicates that higher energy wind sea spectra achieve a characteristic shape that is different from that obtained simply by shoaling. As a result, the energy level for shoaling irregular wave tends to be less than that predicted from linear monochromatic shoaling of the wave components, especially near the surf zone. Smith and Vincent (1992) also indicate that the shoaling and breaking of irregular waves with two spectral peaks can substantially differ from the monochromatic (and even single peak spectral) case. Moreover, the wave spectrum after refraction and shoaling can have a substantially different peak period. Although a satisfactory explanation of these phenomena is not available, their impact is significant, with differences up to 30-40 percent from the significant wave approach.

e. Treatment of spectral wave mechanics in any detail requires use of a numerical model. However, in using a significant wave approach, it can generally be assumed that:

(1) It may overestimate wave focussing effects.

(2) Careful estimates of the fraction of wave energy heading shoreward should be made for oblique angles cases.

(3) Shoaling calculations may overestimate wave heights in high energy conditions.

f. Shifts in wave period may also occur. As a result, significant wave analysis tends to be conservative; this may be why it has been an acceptable approach for design. However, for cost-sensitive projects, a more complicated approach may be warranted.

g. The following precautions are suggested. In a significant wave analysis, if regions of highly focussed wave energy occur with corresponding lobes of low energy, the regions of low energy should be carefully considered. In the field, natural wave systems generally have significant directional spread, so calculated values in the low energy lobes may significantly underestimate wave heights. In cases where irregular waves are modeled spectrally, typically only the wave height H_s is estimated. In shallow water, larger waves do occur ($H_{1/10}$, etc.) and combinations of individual wave height, period, and bottom depth can result in individual waves or groups of waves significantly larger than H_s (see Part II-2).

II-3-5. Advanced Propagation Methods

a. Introduction.

(1) As indicated in Part II-3-2, as waves propagate, they may continue to grow due to the continued action of the wind or may lose energy due to breaking, bottom friction, or percolation. These effects cannot be realistically incorporated through manual calculations. The preceding discussion indicates that computations involving rays are tedious by hand and subject to many inaccuracies. Advances have been made in computing wave transformation; they were briefly indicated in the preceding sections. Many of these procedures may run efficiently on a personal computer or a work station and do not require a large mainframe or supercomputer. Hence they can be applied readily by most engineers (ACES 1992).

(2) This section describes three computer programs that are available and in use by the Corps of Engineers. Each program is briefly described and a reference indicates where the program can be obtained. Each program is complicated and requires some effort to use properly. A short description is provided here to indicate to the engineer the potentials of these codes. The three have been selected to provide a cross section of the types of technology available. Other computer programs can be obtained and may be as suitable for use as those described here.

(3) Examples of technology available to practicing engineers is provided. **The Corps of Engineers does not endorse the codes discussed or certify their accuracy.** Indeed, the suitability and accuracy of any of these codes depend upon the problem under study and the way in which the code is applied. **With the exception of very simple bathymetry, it is recommended that nearshore wave transformation studies use a numerical code capable of handling the complexities required.** However, the particular numerical approach selected depends upon the problem.

(4) The three models discussed below are all steady-state models. Time-dependent, shallow-water models are available (Jensen et al. 1987; Demirbilek and Webster 1992a, 1992b). They are not discussed here because they require extensive sets of meteorological data and cannot be easily applied. The basic characteristics of the three models discussed are as follows:

(a) (RCPWAVE) RCPWAVE is a steady-state, linear-wave model based on the mild-slope equation and includes wave breaking. It is applicable for open coast areas without structures. It is basically a monochromatic-wave approach.

(b) (REFDIF1) REFDIF is a steady-state model based on the parabolic approximation solution to the mild-slope equation. The model includes wave breaking, wave damping, and some nonlinear effects. Although primarily used as a monochromatic wave model, a spectral version is available. The model can simulate aspects of propagation associated with simple currents and can include structures.

(c) (STWAVE) STWAVE is a steady-state, linear wave model that computes the evolution of the directional spectrum over space (Equation II-3-1). The model includes breaking, bottom friction, percolation, and wind input and solves for the nonlinear transfers of energy within the wave spectrum. It has two modes for handling diffraction of wave energy and the computational domain may include simple structures. The models can handle aspects of propagation associated with simple currents.

(5) The three models are theoretically complicated and computationally demanding. All can be effectively used on a powerful PC-type computer or work station. Each model has considerable strengths and each can be an appropriate choice for wave transformation. However, none can be considered universally applicable and the results from all can be inaccurate if the assumptions made in model development are significantly violated. Users of any of the models must become thoroughly familiar with the model, its assumptions, and limitations.

b. RCPWAVE.

(1) Introduction.

(a) The RCPWAVE model (Ebersole 1985; Ebersole, Cialone, and Prater 1986) was developed in the early 1980's as an engineering tool for calculating the properties of waves as they propagate into shallow water and eventually break. The theoretical basis for the model (linear-wave theory) and the types of information generated by the model (wave height, period, and direction as a function of location) are consistent with current theories and equations used by the engineering community to calculate potential longshore sand-transport rates and shoreline and beach change. The model was designed to operate efficiently for coastal regions that may be tens of kilometers in length, and to overcome deficiencies of previously developed refraction models that could be applied on a regional scale. The wave ray refraction models of Harrison and Wilson (1964), Dobson (1967), Noda et al. (1974), and others "failed" in regions of strong wave convergence and divergence (i.e., highly irregular bathymetry), leaving users with no wave solutions and little guidance for interpreting results in these regions. Berkhoff (1972, 1976) derived an elliptic equation that approximately represented the complete transformation process for linear waves over arbitrary bathymetry, where the bathymetry was only constrained to have mild slopes. Numerical solution of this equation requires discretization of the spatial domain and subsequent computations with grid resolutions that are a fraction of the wave lengths being considered (typically one tenth or smaller). This requirement limits the utility of the approach for large regions of coastline.

(b) RCPWAVE is based on the mild-slope equation. An assumed form for the velocity potential associated with only the forward scattered wave field is used with the mild slope equation to develop two equations, one describing the conservation of wave energy (assuming a constant wave frequency) and the other defining the magnitude of the wave phase gradient as a function of the wave number and the spatial variability in wave amplitude. The finite difference forms of these two equations and a third equation defining the irrotationality of the wave phase function gradient are solved for the wave height, direction, and phase gradient. A forward-marching (in the direction of wave propagation) solution scheme is used, and solutions are obtained at the center of each rectangular grid cell of the discretized model domain. RCPWAVE differs from the complete solution of the mild-slope equation by neglecting reflections of waves by structures and bathymetry. This allows RCPWAVE to use a grid resolution only sufficient to resolve bathymetric gradients, permitting RCPWAVE to cover larger spatial regions than those usually covered in full solutions to the mild-slope equation.

(c) Wave breaking is also treated in the model. The occurrence of wave breaking is first considered by comparing the computed local wave height with a limiting wave height calculated using the method of Weggel (1972). If the computed wave height exceeds the limiting value, then energy is dissipated according to the breaker decay model of Dally, Dean, and Dalrymple (1984). When the calculated local wave height falls below the stable value proposed by Dally, Dean, and Dalrymple, energy dissipation is again set to zero. The influence of wave height variability on wave phase is neglected in the surf zone. Details of the RCPWAVE model derivation and solution scheme can be found in Ebersole, Cialone and Prater (1986).

(2) Examples of RCPWAVE results.

(a) Figures II-3-8 and II-3-9 show results from a typical application of RCPWAVE. The model domain is the region offshore of Homer Spit, Alaska. The section of coast being considered is approximately 33 km in length. A rectangular grid mesh was constructed within the domain, with grid resolution of approximately 130 m in the on-offshore direction and 250 m in the alongshore direction. Figure II-3-8 shows bathymetric contours in the model domain. The nearshore region is characterized by a fairly broad shelf with depths of 20 m or less, and offshore the depths increase to 200 ft and greater in the lower right-hand corner of the domain. The shallow-water region is characterized by irregular contours, with extensive shoals at locations A and B in Figure II-3-8.

(b) Figure II-3-9 shows the wave height field (shaded contours) throughout the model domain for an incident deepwater wave with the characteristics shown. Note the areas of wave convergence and divergence and the resulting variation in wave height observed along the coast (darker shades indicate convergence and lighter shades indicate divergence). The shoals cause wave convergence, which is evidenced by zones of higher wave height in the lee of the shoals. Wave heights are lower in divergent zones that are created as waves attempt to align their propagation direction to be perpendicular to bathymetric contours and propagate toward the shoals. A plot of wave direction vectors would also indicate zones of wave convergence and divergence. Also note the position of the breaker line (indicated by the seawardmost pattern of dots), as it follows the shallower bathymetric contours. Shoals can cause the focussed incident wave to break at a greater distance from shore than elsewhere in the region. Information that can be obtained using the model includes wave height and direction variability at many locations along the coast under different incident wave conditions, variability of inshore wave conditions with changing water levels, and zones of potentially high and low longshore sand transport.

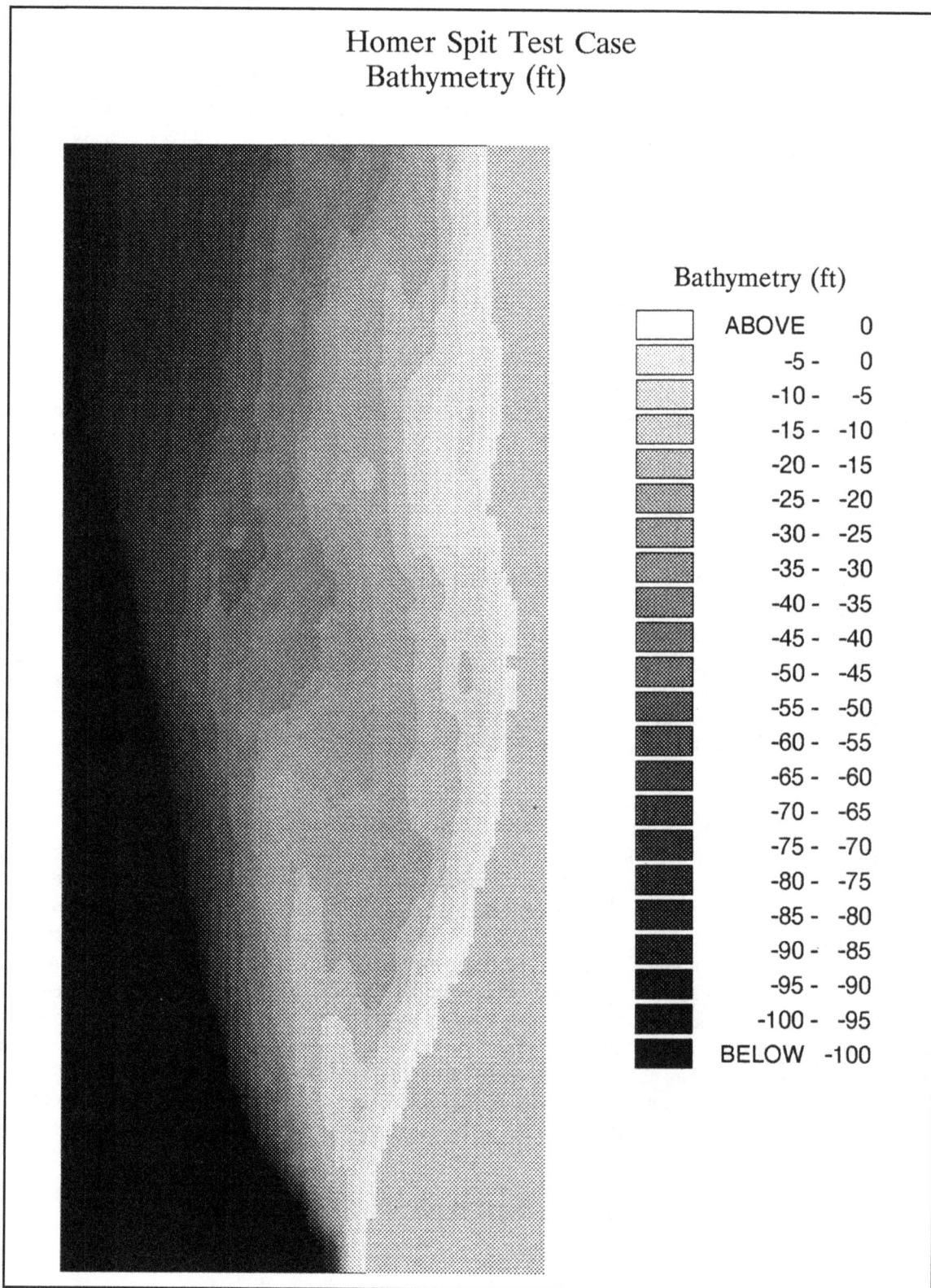

Figure II-3-8.   Typical RCPWAVE application, bathymetry

Estimation of Nearshore Waves

Homer Spit Test Case
Wave Height (ft)

19931211 1700   Incident Cond T: 10.0 Ho: 8.0 THo: -45.0

Wave Height (ft)

	ABOVE 10.5
	10.0 - 10.5
	9.5 - 10.0
	9.0 - 9.5
	8.5 - 9.0
	8.0 - 8.5
	7.5 - 8.0
	7.0 - 7.5
	6.5 - 7.0
	6.0 - 6.5
	5.5 - 6.0
	5.0 - 5.5
	4.5 - 5.0
	4.0 - 4.5
	3.5 - 4.0
	3.0 - 3.5
	2.5 - 3.0
	2.0 - 2.5
	1.5 - 2.0
	1.0 - 1.5
	0.5 - 1.0
	0.0 - 0.5
	BELOW 0.0

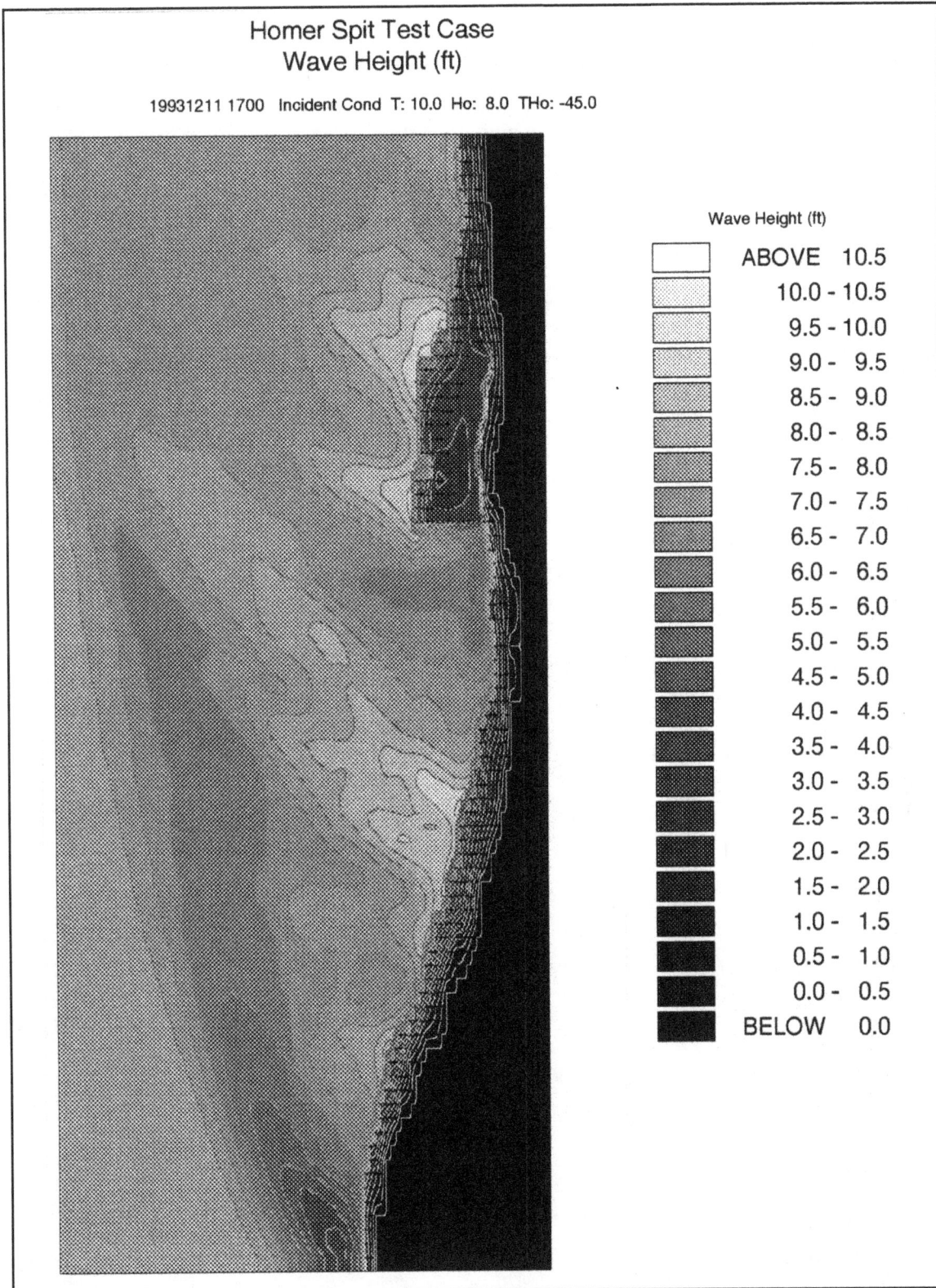

Figure II-3-9.   Typical RCPWAVE application, wave height

Estimation of Nearshore Waves

(3) Data requirements for RCPWAVE.

(a) Primary input to the RCPWAVE model includes the following: parameters describing the domain to be modeled, such as the number of computational grid cells in each direction and the cell dimensions; definition of the water depth at each cell; and definition of the incident wave height, period, and direction along the offshore domain boundary for each wave condition to be simulated. Model output includes wave height, period, and direction at each cell of the computational domain, and an indication of whether or not the wave is calculated to be a broken wave.

(b) Typically the first step in the model application process is to discretize the model domain into a rectangular mesh. The grid mesh that is created can be overlaid on a bathymetric chart, assuming the grid and chart are plotted to the same horizontal scale, and depths at each cell can be digitized for use as model input. A constant correction to the depths, representing a datum change or a specific water level change, can be included in the input data set. An arbitrary number of wave conditions, each defined by a unique combination of height, period and direction, can be simulated. Wave conditions to be simulated are usually defined after a statistical analysis of the wave climate in the region being studied. Bathymetry specification and incident wave conditions comprise the bulk of the effort to create the input data set.

c. *REFDIF*.

(1) Introduction. The model REF/DIF 1, which has been developed for practical application, is based on the mild-slope, wave-current model equation developed by Kirby (1984), which may be written as

$$\frac{D^2\varphi}{Dt^2} + \nabla \cdot U \frac{D\varphi}{Dt} - \nabla \cdot (CC_g \nabla \varphi) + (\sigma^2 - k^2 CC_g)\,\varphi = 0 \qquad \text{(II-3-27)}$$

where φ is the velocity potential at the free surface, and where

$$\frac{D}{Dt} = \frac{\partial}{\partial t} + U \cdot \nabla \qquad \text{(II-3-28)}$$

$$\nabla = \left(\frac{\partial}{\partial x} \, , \, \frac{\partial}{\partial y} \right) \qquad \text{(II-3-29)}$$

$$U = (U(x,y) \, , \, V(x,y)) = \text{ambient current vector} \qquad \text{(II-3-30)}$$

$$\sigma = \omega - k \cdot U \qquad \text{(II-3-31)}$$

$$C = \frac{\sigma}{k} \qquad \text{(II-3-32)}$$

$$C_g = \frac{\partial \sigma}{\partial k} \qquad \text{(II-3-33)}$$

$$\sigma^2 = gk \tanh (kh) \qquad \text{(II-3-34)}$$

Several additional features are included in the model in order to increase its range of application and accuracy.

(2) Wave breaking. The model tests whether the local wave height has exceeded a fixed threshold, which is set at $h/d = 0.78$. For local wave heights exceeding this value, a breaking wave energy flux decay model is started in order to remove energy from the wave train. The model used is described in Dally, Dean, and Dalrymple (1985). The reader is referred to Kirby and Dalrymple (1986a) for further details.

(3) Wave damping mechanisms.

(a) In addition to the strong wave breaking mechanism described above, REF/DIF 1 also provides the user with three selectable bottom damping mechanisms. These are: laminar bottom boundary layer damping, sand-bed percolation damping, and turbulent bottom boundary layer damping.

(b) At present, no laboratory or field data sets clearly point to the need for including bottom damping effects in model simulations. Laboratory experiments usually include too short a propagation distance for damping effects to accumulate significantly. In the field, damping due to bottom effects may be balanced or overshadowed by wave growth resulting from wind-wave interaction, and so one should not be considered in the absence of the other. At present, it is recommended that these user-selectable damping mechanisms not be included in model simulations.

(4) Wave nonlinearity.

(a) Wave nonlinearity has a strong effect on the phase speed of waves and thus can significantly modify both refraction and diffraction effects. For example, waves shoaling on a plane beach refract more slowly than predicted by linear theory, since the increase in wave height with decreasing water depth speeds up the waves, in opposition to the direct, linear-theory effect decreasing depth, which slows them. Diffraction effects are typically enhanced. Phase speed is greater in a high-amplitude, illuminated area than in a low-amplitude, shadowed area; this causes refractive bending of waves into the shadow area, causing an increase in wave height in the shadow zone relative to the predictions of linear theory.

(b) REF/DIF 1, designed to predict the propagation of a monochromatic wave in intermediate water depth, includes the effects of nonlinearity as predicted by third-order Stokes wave theory (Kirby and Dalrymple 1983). Since the model is often used to predict wave-height distributions into the surf zone and up to dry land boundaries, the model must also be corrected to avoid the singularities arising from the invalidity of Stokes theory in shallow water. In order to provide a smooth correction to the model results in the shallow-water limit, Kirby and Dalrymple (1986b) provided an algorithm that gives a smooth patch between Stokes theory and an empirical modification to linear theory developed by Hedges (1976). The approximate theory does not cause any degradation in solution accuracy in comparison to the Stokes theory alone for intermediate depth experiments; see Kirby and Dalrymple (1986b) for relevant documentation.

(5) Numerical noise filter. Higher-order forms of the parabolic approximation have the undesirable effect of allowing high-wave number noise (i.e., noise with rapid lateral variation) to propagate rapidly across the computational grid. This effect has been described in detail by Kirby (1986a), and is usually found in association with the start of surf zones around complicated planforms such as island shores. The resulting noise component may be damped by the application of various types of smoothing filters. The three-point moving average filter described by Kirby (1986a) has been found to be heavy-handed in practical applications, and has been replaced in present versions of the REF/DIF 1 model by a damping filter included in the governing differential equation, whose effect is centered around the lateral wave number, which spread rapidly in the undamped model. A full description of the damping method and a range of tests may be found in Kirby (1993).

(6) Examples of REF/DIF 1 results laboratory verification. REF/DIF 1 (and the parabolic approximation model in general) are capable of providing a detailed picture of the water surface in the region of study if the

grid resolution is sufficiently high. This picture includes the geometry of crests and troughs as well as the location of regions of high or low wave height resulting from short-crestedness of the wave field. Since irregular waves in the field usually lead to a fairly smooth spatial variation in wave height estimates (after statistical averaging), a more stringent test of model accuracy is provided by comparison to laboratory tests with monochromatic waves. Parabolic models have been tested against data of this type in a number of studies, including Berkhoff, Booij, and Radder (1982); Tsay and Liu (1982); Kirby and Dalrymple (1984), Panchang et al. (1990), and Demirbilek (1994). The results showed that the higher-order parabolic approximation, together with nonlinear correction to the wave phase speed, can correctly predict the distribution of wave heights and nodal points in the evolving wave field. Figure II-3-10 shows the bathymetry input to REF/DIF1 for a simulation of wave propagations at Revere Beach, MA. Figure II-3-11 shows the wave heights calculated by the models.

(7) Data requirements for REFDIF.

(a) REF/DIF 1 computes a grid-based wave evolution over an arbitrary bathymetry and current field. To run the model, the user must provide, at minimum, an array of depth values h on a grid with regular spacing in x and y. The model always assumes that x is the preferred direction, or the direction in which the computation marches. No provision is made at present for relating the model coordinate system to a global coordinate system. If the user wishes to include the effects of tidal currents in the model study, then arrays of velocity components U and V must also be provided for the same regular grid used to specify h values. This information establishes the geometry for the model run.

(b) The user must also specify the form of the wave train at the offshore boundary. This may be done by specifying a combination of one or more monochromatic waves at the offshore boundary, or the offshore wave field may be specified at the first grid row by means of input data. The user's manual provided in Kirby and Dalrymple (1992) should be consulted for more details about the input data.

(c) The model provides the user with a grid of computed wave heights and directions on the same geometric grid used for input. In addition, the complex amplitude values are provided and may be used to reconstruct plots of the computed wave field, if these are desired and if the grid resolution is fine enough to permit it. For larger-scaled model areas, this last step is often not feasible, as it requires 5 to 6 grid points per modeled wave length in the input bathymetry grid. A version of REF/DIF capable of simulating wave spectra has recently been released.

d. STWAVE.

(1) Introduction.

(a) STWAVE is a steady-state spectral model for predicting wave conditions in coastal areas. It solves the complete radiative transfer equation (Equation II-3-1) including both propagation effects (refraction, shoaling, diffraction, and wave-current interactions) and source-term effects (wave breaking, wind inputs, and nonlinear wave-wave interactions). STWAVE was developed under the premise that waves in nature should be treated as nonlinearly interacting stochastic wave components rather than as deterministic nonlinear waves. This is particularly relevant when dealing with wave transformations over distances of hundreds of thousands of wavelengths (typical of many coastal wave transformation studies). At much shorter distances a deterministic, long-crested approximation can provide an appropriate framework for understanding and interpreting wave behavior. At longer distances, theoretical and empirical evidence

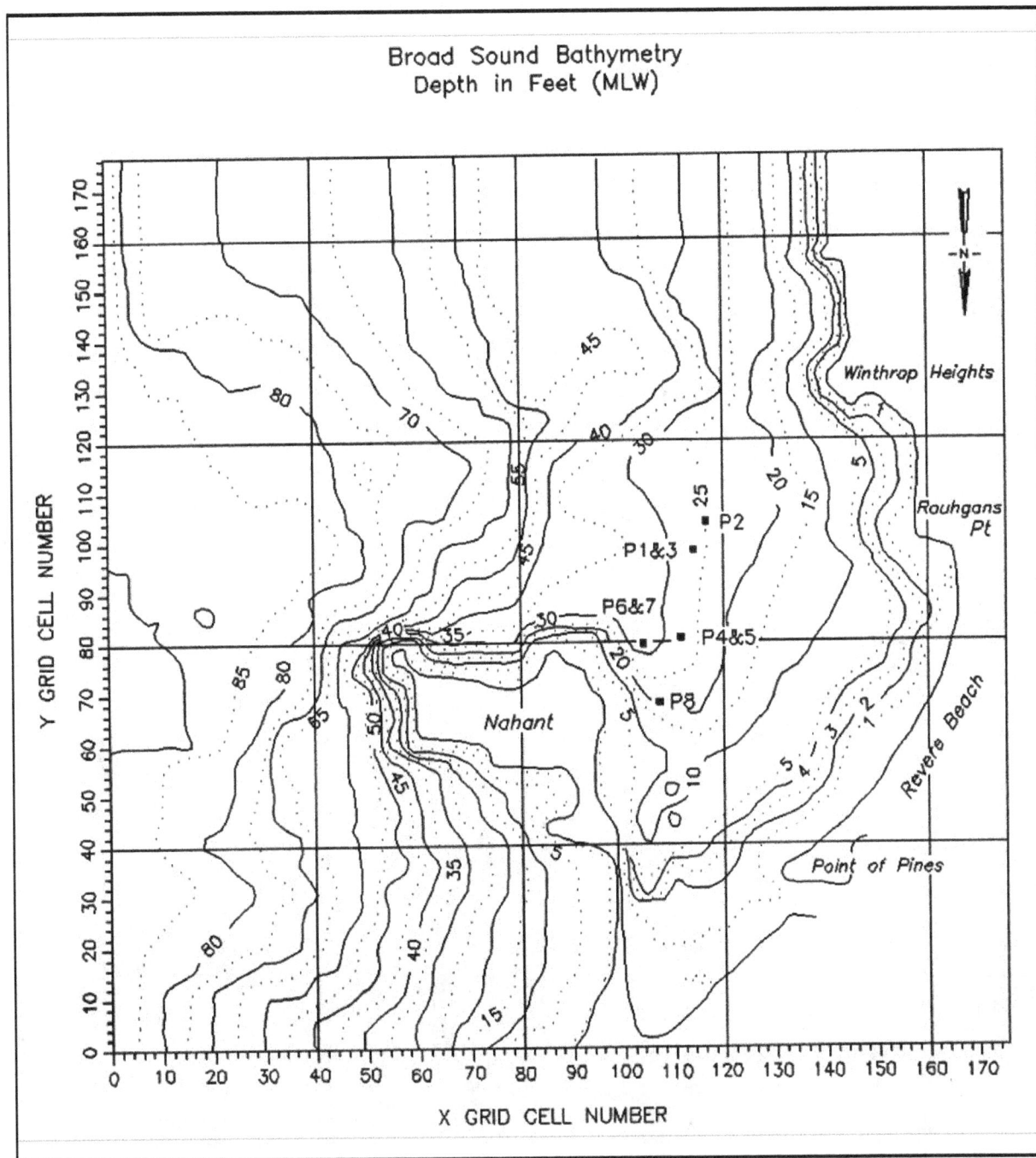

Figure II-3-10. Bathymetry input to REF/DIF1 for a simulation of wave propagations at Revere Beach, MA

strongly supports a stochastic approximation for wave phenomena (West 1981). Over small distances, near discontinuities in a wave field (such as breakwaters), STWAVE can incorporate wave phase information into its solution; otherwise, it uses a random-phase approximation for its diffraction and combined refraction-diffraction (CRD) calculations. Theoretical details of STWAVE can be found in Resio (1993).

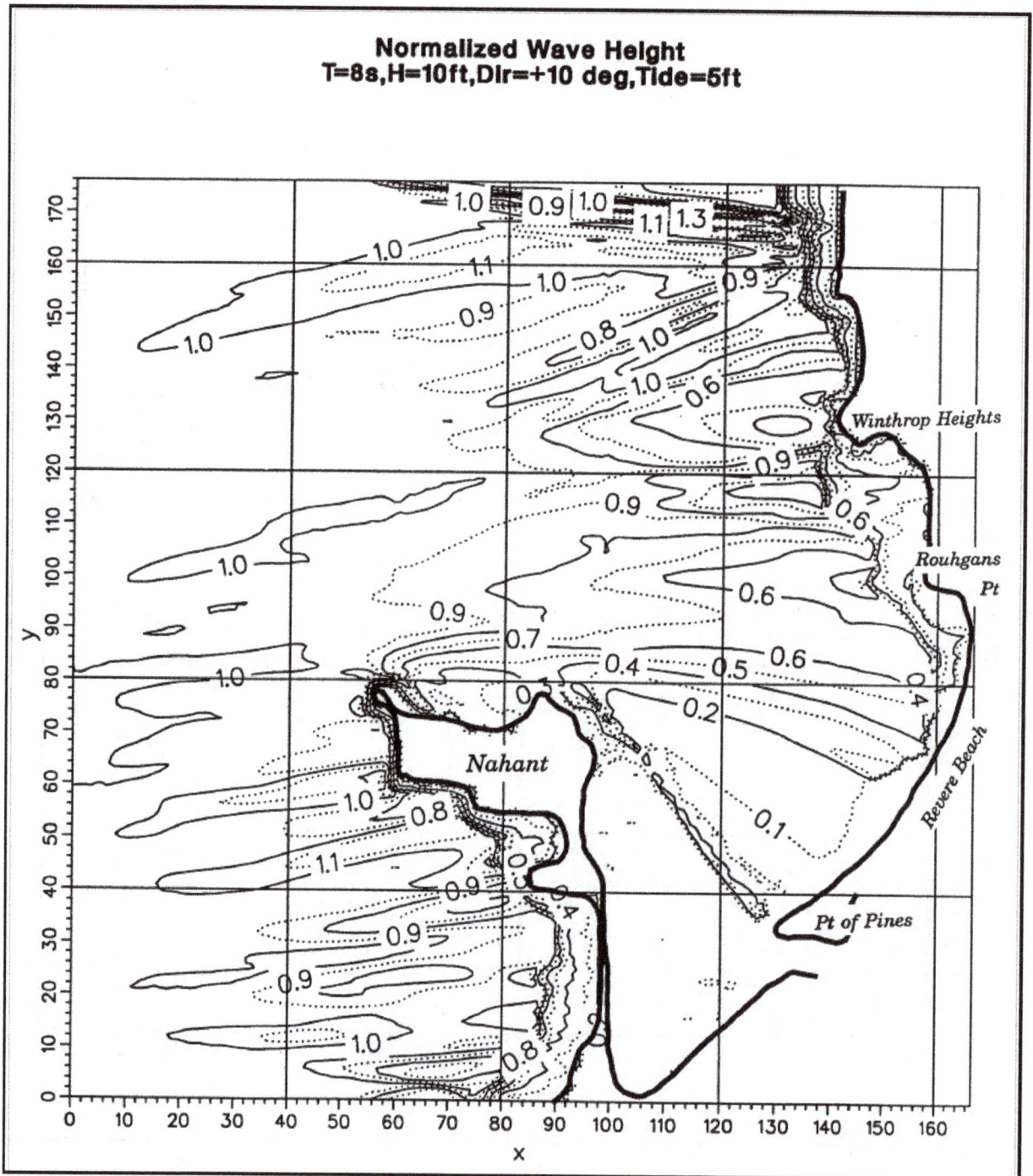

Figure II-3-11. Wave heights calculated by REF/DIF 1

(b) The following two assumptions have been inherent in essentially all previous steady-state models for predicting nearshore wave transformations:

- Predictions based on unidirectional, monochromatic wave theories can provide solutions that are equivalent to the behavior of naturally occurring directional wave spectra.

● Nearshore transformations are dominated by conservative processes (refraction, shoaling, and diffraction) and hence nonconservative effects (energy sinks and sources) can be neglected as a first approximation.

(c) A corollary to the first assumption above is that increased accuracy in deterministic propagation estimates translates into commensurate increases in accuracy in real-world applications. Unfortunately, laboratory studies by Thompson and Vincent (1984) and Vincent and Briggs (1989) have clearly demonstrated that the first assumption is not valid unless the wave field is narrow-banded in both frequency and direction. Thus, for most coastal wave predictions to be accurate, they must solve all wave components and not just a hypothetical "dominant" component. This presents significant problems for wave models that solve only one wave component at a time, since wave energy traveling in one direction can be "scattered" into another direction via diffraction. Hence, diffraction causes wave components in a spectrum to interact and attempts to solve the CRD equation on a component-by-component basis have difficulty properly accounting for this effect. STWAVE overcomes this problem by using a piecewise solution method that simulates the propagation of all wave components simultaneously.

(d) Returning to the second assumption above, field and laboratory data presented in Bouws, Gunther, and Vincent (1985) and Resio (1988) show that nonconservative effects, rather than conservative propagation effects, dominate wave transformations in many coastal areas, particularly during storm conditions. Moreover, the form of many of the source terms affecting shallow-water wave transformations is such that they depend on energy content within the entire wave spectrum. Methods that solve for each component of the spectrum independently cannot provide suitable estimates of coupled source terms. STWAVE is formulated in a manner that permits straightforward solution of these processes.

(2) Examples of STWAVE results. The following comparisons are intended to demonstrate the importance of various terms in coastal wave transformations and the ability of STWAVE to handle these terms.

(a) Spectral versus monochromatic calculations. Figure II-3-12 compares predicted wave heights behind a shoal using STWAVE, for a unidirectional, monochromatic wave and for a JONSWAP spectrum with a spectral peak frequency of 0.1 Hz and a $\cos^4$ angular distribution of energy. Monochromatic calculations from the laboratory study of Vincent and Briggs (1989), while mathematically accurate, do not reasonably represent propagation effects in a wave spectrum with natural frequency and direction energy spreads.

(b) Effects of coupled source terms. Figure II-3-13 compares spectral transformation over 1:30, 1:100, and 1:500 slopes for the same JONSWAP spectrum as above with a mean approach angle to the coast of 30 deg, for the case of no source terms and for the case of wave breaking and nonlinear wave-wave interaction source terms included. This comparison suggests that CRD effects account for only about 5 percent of the total energy variations in coastal waves passing over moderate to shallow slopes. This finding is consistent with those of Resio (1988) and helps to explain why nearshore wave spectra tend strongly toward self-similar forms during local storms (Bouws, Gunther, and Vincent 1985; Resio 1987; Miller and Vincent 1990).

(c) Wind effects. Figure II-3-14 shows the differences in wave transformations with and without a 20-m/sec onshore wind over an offshore profile typical of the U.S. east coast. In this example, waves at the seaward boundary are set to the same JONSWAP spectrum as Examples 1 and 2. These results show a marked difference between the two cases. This difference is consistent with theoretically expected wind input and indicates that, particularly during storm conditions, neglecting wind input can lead to significant misestimations of wave conditions.

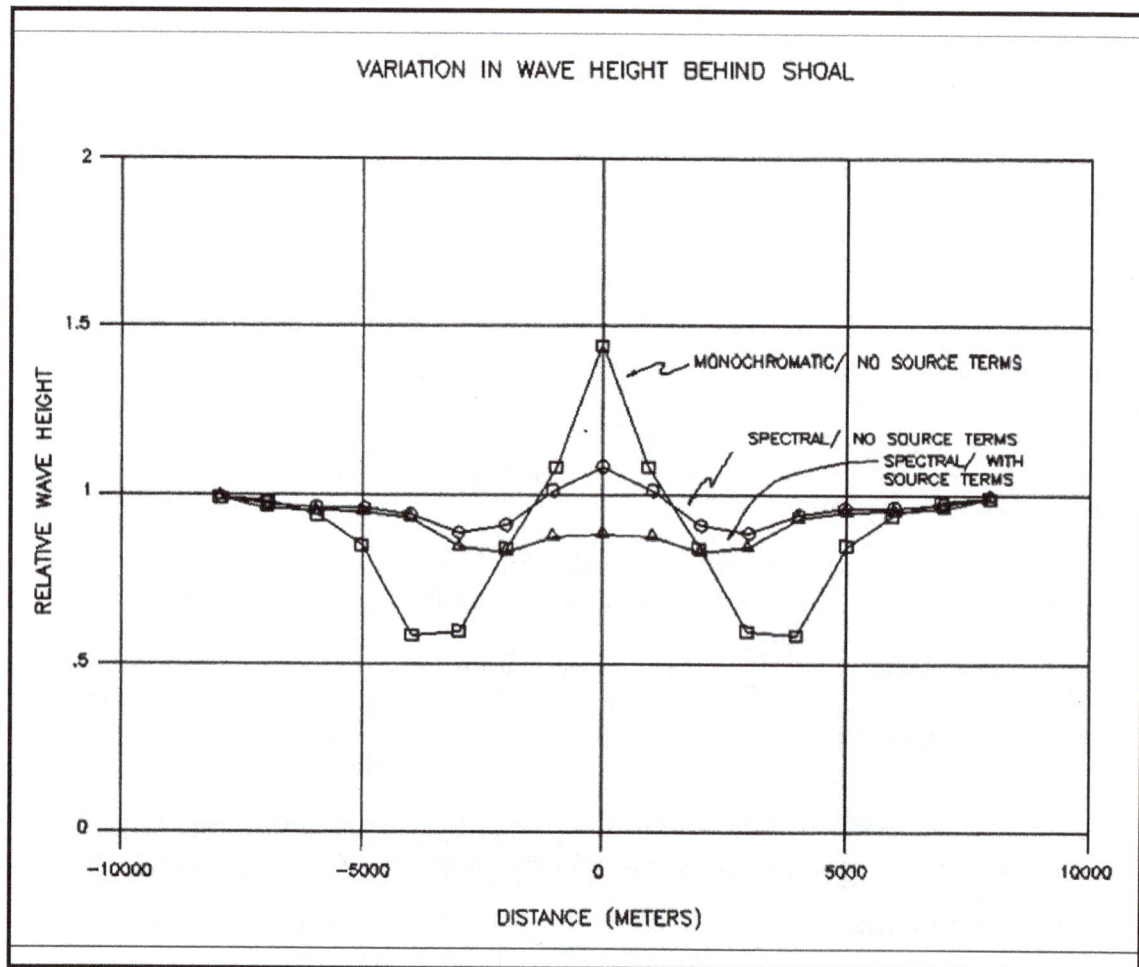

Figure II-3-12. Spectral model results compared to laboratory measurements for broad directional spectrum

(3) Data requirements for STWAVE. In STWAVE, a square grid mesh covers the computational domain. Water depth must be supplied at each node. If currents are included, a current must be supplied at each point. Wave characteristics are computed at each of these grid points. The model requires an input directional spectrum for the outer boundary and information about wind speed and direction and bottom friction coefficients.

   e.   *Limitations.*

(1) Each model has natural limitations reflecting its theoretical basis. The references provided discuss these in some detail. If strictly interpreted, each model has a narrow range over which it is valid. Almost all of these models are regularly used to simulate conditions outside a strict interpretation of limits, with the results often effectively accurate. Considerable judgement and experience are required to determine if the simulation is valid.

(2) The following limitations indicate where the model may or may not be useful. RCPWAVE may be inaccurate for waves crossing behind shoals, or in the vicinity of structures. Wave approach directions should not be too oblique relative to the offshore boundary. REF/DIF1 can allow for some structures and islands but again should not use waves with highly oblique wave angles (In both RCPWAVE and REF/DIF1,

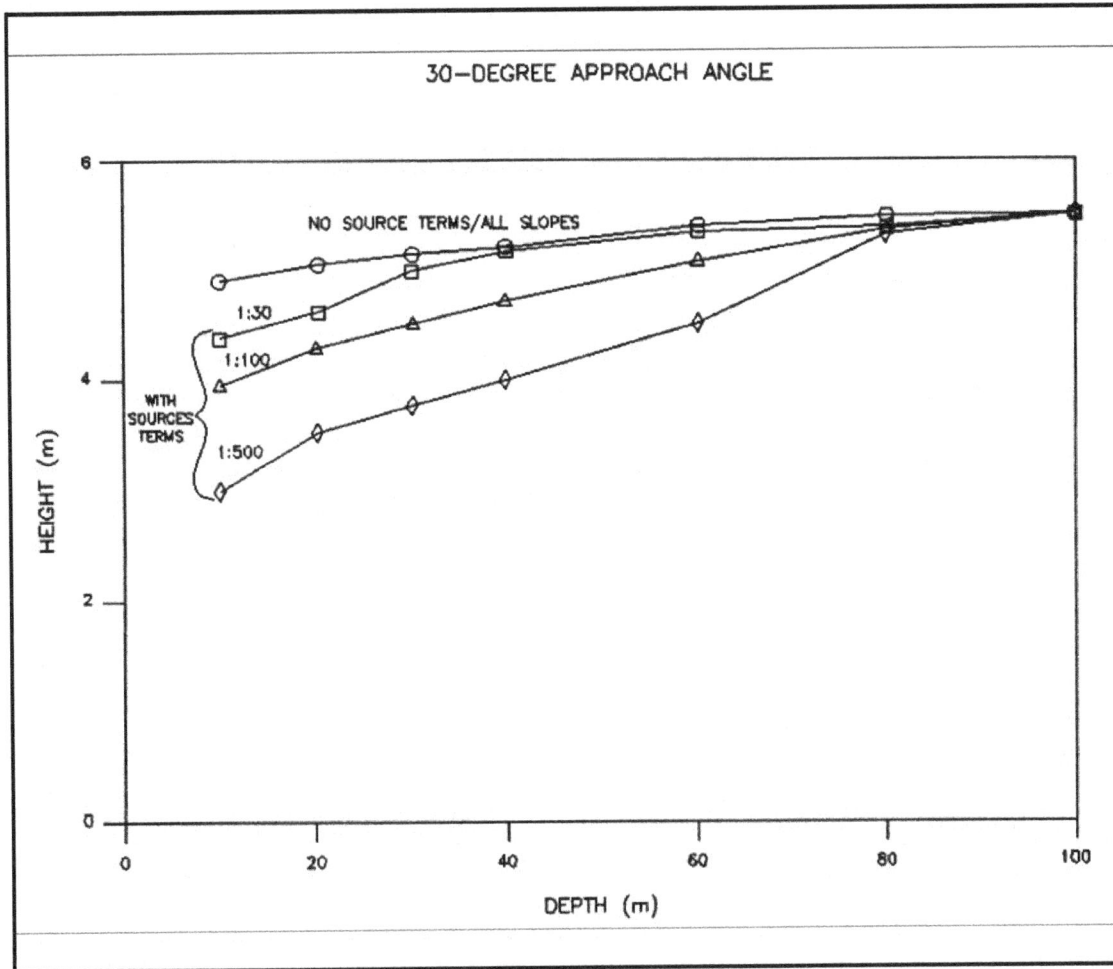

Figure II-3-13.   STWAVE results for a 1:30 sloping beach

the oblique angle dilemma often can be resolved by using a different grid). STWAVE may underrepresent
wave focussing for very narrow swell.

II-3-6. Guidance for Performing Wave Transformation Studies

a. Introduction.

(1) The preceding parts of this chapter provide the engineer with an understanding and some techniques
for taking a wave condition offshore of a project or nearby and transforming it to the site of interest. In
practice, an engineer will typically consider a suite of wave conditions perhaps representing different storms,
different seasonal characteristics, and different water levels (particularly in shallow water or at the beach if
there is a high tide or storm surge to be considered). Selection of the conditions for project design studies
is a very important component of any coastal engineering study and Part II-2 and Part II-3 both treat this
problem.

(2) Transformation analyses are needed because there is often a lack of site-specific data. In some
instances, a cursory transformation analysis may be required to help decide whether an offshore or nearby
site is adequate for determining offshore boundary conditions. Typically this may be approached by setting
up one of the transformation procedures described and running a small set of wave conditions that might span

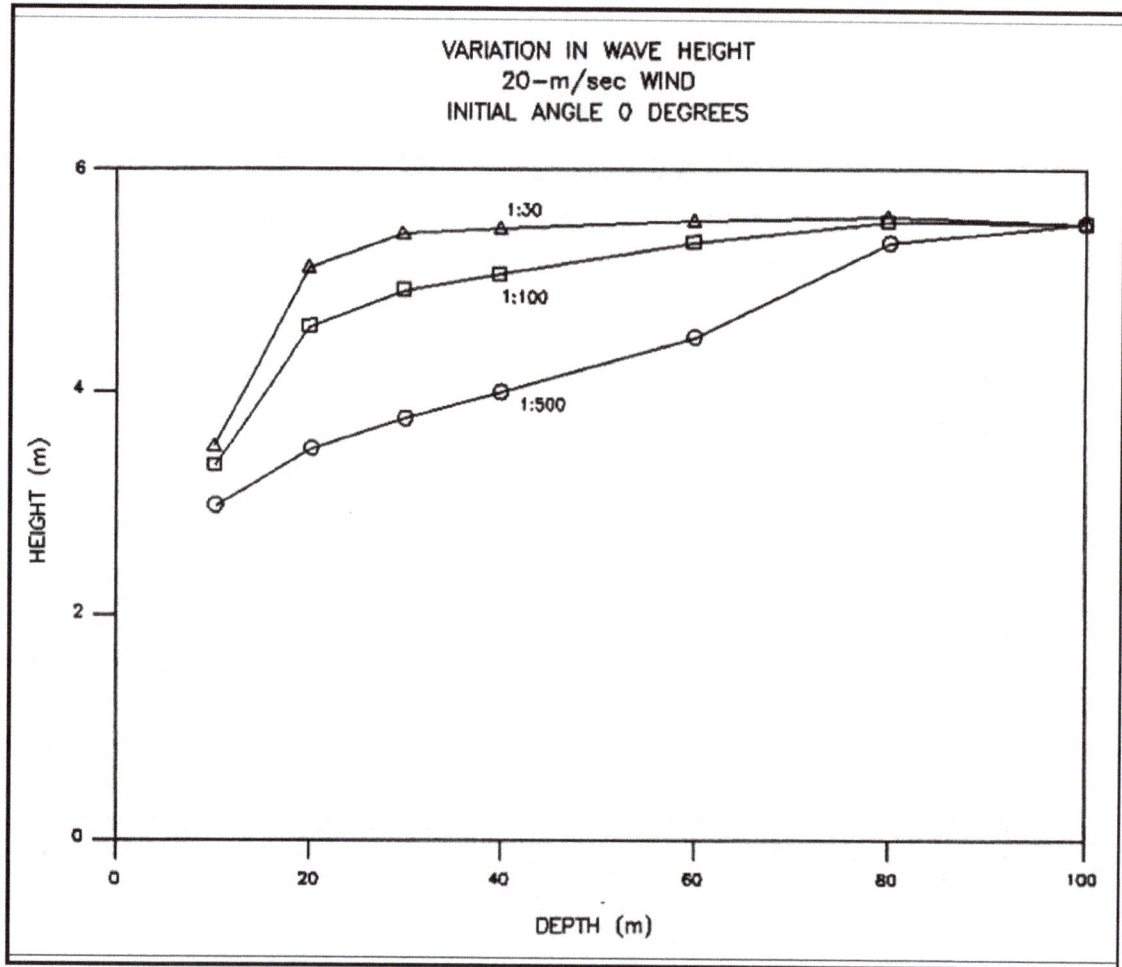

Figure II-3-14.   STWAVE results for CHL's Field Research Facility at Duck, NC

the final set to be studied. As an example, waves of a certain period or direction offshore may not propagate to the site, and the engineer can thus ignore such wave conditions in a more detailed study.

(3) Some of the decisions and actions an engineer will need to make in performing a wave transformation analysis follow.

b. Problem formulation. At the initiation of the study, the engineer should clearly understand what wave information must be produced for the site, how it will be used, and the accuracy required. The engineer should gather all pertinent bathymetry data, water level data, and nearby wave data. Aerial photography of the site can be very useful by providing the engineer with indications of wave propagation patterns, areas of offshore breaking, etc., that a transformation procedure should properly simulate. Short-term gauge records can be used in checking the procedure. Again, a short-term gauging program is desirable.

c. Site analysis. The physical characteristics of the site and any ancillary information should be carefully scrutinized so that the engineer can understand how irregular the bathymetry is, the presence of significant currents, shoals, canyons, islands, structures, etc., that would be important in selecting the offshore or nearby site for a source of data input, for selecting the transformation procedure used, and in understanding what problems may arise in the analysis. Usually this type of knowledge is gained through experience, and a consultant may be required to assist someone unexperienced in such analyses. If time permits, one of the

Estimation of Nearshore Waves

advanced models could be set up and run in an exploratory mode to help the engineer understand possible problems.

d. Selection of input data site. Based on project formulation and site analysis, offshore/nearby sites are evaluated in terms of any feature that would preclude their use (see Part II-3-1d). In particular, the use of nearby sites in similar depths of water must be evaluated in terms of whether waves reaching the site have broken. As an example, if waves at a nearshore site have propagated over a shoal where breaking can occur, there is no way to "unbreak" the waves. So they cannot be used to eliminate offshore wave conditions. In general the offshore data site will need as a minimum information on **wave height, period, and direction.** If adequate data are not available, methods for hindcasting, as described in Part II-2, may be used to simulate the information required. The methods of Parts II-2 and 3 should be used to develop the wave information to be transformed.

e. Selection of wave transformation method. Table II-3-2 provides guidance on the applicability of the various methods described in this chapter. It does not provide guidelines for all cases. With some skill, the models described can be pushed somewhat beyond their inherent limitations (but such results must be carefully scrutinized and used conservatively). In very complicated cases or in cases in which a time-dependent model is required, use of an expert consultant to provide assistance is recommended. In some complicated cases, a physical model may be required.

Table II-3-2
Guidance for Selection of Wave Transformation Methods

Case	Fig.II-3-6 or ACES	NMLONG	RCPWAVE	REFDIF1	STWAVE
Planar topography (no shoals, etc)	yes	yes	yes	yes	yes
Highly Irregular Bathymetry					
Swell, no structures	no	no	yes	yes	yes
Swell, structures	no	no	no	yes	yes
Complicated directional Spectra, but narrow frequency spectra	no	no	no	yes	yes
High winds or broad band frequency spectra	no	no	no	no	yes
Irregular Bathymetry, High resolution Computations Near Structure					
Swell	no	no	no	yes	no

f. Calibration/verification. After the method is set up, it is important to check the calculations with observations if at all possible. If measured wave data are not available, then aerial photographs can be helpful in deciding if the model reproduces observed wave patterns. If no wave data or photographs are available, the method should be applied to a range of heights, periods, and directions and the results should be carefully scrutinized for odd or unstable results. If the calculations are overly sensitive to small variations in input data, a careful decision should be made as to whether the technique should be applied. A physical model may be appropriate in situations with very irregular bathymetry, complicated or multiple structures, reefs, and where currents are important.

g. Post-processing.

(1) Plotted results should be carefully examined for any signs of computational instability. These typically are unreasonable variations in height or direction over short distances.

(2) The techniques provided in this chapter, if used carefully by an experienced engineer, can provide very useful information in a wide range of cases. However, there are some cases where they simply will not work. Anyone who applies these techniques should understand the limitations of the techniques, and be versed in understanding when they have been used inappropriately. The user should be aware that the models can provide realistic-looking answers that unfortunately are just wrong.

II-3-7. References

Abbott, Peterson, and Skovgaard 1978
Abbot, M. B., Peterson, H. M., and Skovgaard, O. 1978. "On the Numerical Modelling of Short Waves in Shallow Water," *Journal Hydraulic Research*, Vol 16, No. 3, pp 173-203.

Bagnold 1946
Bagnold, R. A. 1946. "Motion of Waves in Shallow Water - Interaction Between Waves and Sand Bottoms," *Royal Society of London*, Vol 187, pp 1-18.

Barnett 1968
Barnett, T. P. 1968. "On the Generation, Dissipation, and Prediction of Wind Waves," *Journal of Geophysical Research,* Vol 73, pp 6879-6885.

Berkhoff 1972
Berkhoff, J. C. W. 1972. "Computation of Combined Refraction - Diffraction," *Proceedings, 13th International Conference on Coastal Engineering,* American Society of Civil Engineers, Vol 1, pp 471-490.

Berkhoff 1976
Berkhoff, J. C. W. 1976. "Mathematical Models for Simple Harmonic Linear Water Waves, Wave Diffraction and Refraction," Publication 1963, Delft Hydraulics Laboratory, Delft, The Netherlands, April.

Berkhoff, Booij, and Radder 1982.
Berkhoff, J. C., Booij, N., and Radder, A. C. 1982. "Verification of Numerical Wave Propagation Models for Simple Harmonic Linear Water Waves," *Coastal Engineering,* Vol 6, pp 255-279.

Bouws et al. 1987
Bouws et al. 1987. "Similarity of the Wind Wave Spectrum in Finite Depth Water; Part 2: Statistical Relations Between Shape and Growth Stage," *Deutsche Hydogr. Z.,* Vol 40, pp 1-24.

Bouws, Gunther, and Vincent 1985
Bouws, E, Gunther, H., and Vincent, C. L. 1985. "Similarity of Wind Wave Spectrum in Finite-Depth Water, Part I - Spectral Form," *Journal of Geophysical Research*, Vol 85, No. C3, pp 1524-1530.

Bretschneider and Reid 1953
Bretschneider, C. L., and Reid, R. O. 1953. "Change in Wave Height Due to Bottom Friction, Percolation and Refraction," *34th Annual Meeting of American Geophysical Union.*

Dally, Dean, and Dalrymple 1984
Dally, W. R., Dean, R. G., and Dalrymple, R. A. 1984. "Modeling Wave Transformation in the Surf Zone," Miscellaneous Paper CERC-84-8, U.S. Army Engineer Waterways Experiment Station, Vicksburg, MS.

Dean and Dalrymple 1991
Dean, R. G., and Dalrymple, R. A. 1991. *Water Wave Mechanics for Engineers and Scientists*, World Scientific Pub. Co., Teaneck, NJ.

Demirbilek 1994
Demirbilek, Z. 1994. "Comparison of REFDIF Model and CERC Elliptical Shoal Laboratory Data," Memorandum for Record, dated 18 June 1994, for the Coastal Inlets Research Program, Coastal Engineering Research Center, Vicksburg, MS.

Demirbilek and Webster 1992a
Demirbilek, Z., and Webster, W. C. 1992a. "Application of the Green-Naghdi Theory of Fluid Sheets to Shallow-Water Wave Problems; Report 1: Model Development," Technical Report CERC-92-11, U.S. Army Engineer Waterways Experiment Station, Vicksburg, MS.

Demirbilek and Webster 1992b
Demirbilek, Z., and Webster, W.C. 1992b. "User's Manual and Examples for GNWAVE, Final Report," Technical Report CERC-92-13, U.S. Army Engineer Waterways Experiment Station, Vicksburg, MS.

Dobson 1967
Dobson, R. S. 1967. "Some Applications of a Digital Computer to Hydraulic Engineering Problems," Technical Report No. 80, Department of Civil Engineering, Stanford University, Stanford, CA.

Ebersole 1985
Ebersole, B. A. 1985. "Refraction-Diffraction Model for Linear Water Waves," *Journal of Waterways, Port, Coastal, and Ocean Engineering*, American Society of Civil Engineers, Vol III, No. WW6, pp 939-953.

Ebersole, Cialone, and Prater 1986
Ebersole, B. A., Cialone, M. A., and Prater, M. D. 1986. "RCPWAVE - A Linear Wave Propagation Model for Engineering Use," Technical Report CERC-86-4, U.S. Army Engineer Waterways Experiment Station, Vicksburg, MS.

Forristall and Reece 1985
Forristall, G. Z., and Reece, A. M. 1985. "Measurements of Wave Attenuation Due to a Soft Bottom: The SWAMP Experiment," *Journal of Geophysical Research*, Vol 90, No. C2, pp 3367-3380.

Harrison and Wilson 1964
Harrison, W., and Wilson, W. S. 1964. "Development of a Method for Numerical Calculation of Wave Refraction," Technical Memorandum No. 6, Coastal Engineering Research Center, U.S. Army Engineer Waterways Experiment Station, Vicksburg, MS.

Hasselmann 1962
Hasselmann, K. 1962. "On the Non-Linear Energy Transfer In A Gravity-Wave Spectrum - General Theory," *J. Fluid Mech.*, Vol 12, pp 481-500.

Hasselmann 1963a
Hasselmann, K. 1963a. "On the Non-Linear Energy Transfer In A Gravity-Wave Spectrum - Part 2," *J. Fluid Mech.*, Vol 15, pp 273-281.

Hasselmann 1963b
Hasselmann, K. 1963b. "On the Non-Linear Energy Transfer In A Gravity-Wave Spectrum - Part 3," *J. Fluid Mech.*, Vol 15, pp 385 - 398.

Hasselmann et al. 1973
Hasselmann et al. 1973. "Measurements of Wind-Wave Growth and Swell Decay During the Joint North Sea Wave Project (JONSWAP)," *Deutsche Hydrograph. Zeit.,* Erganzungsheft Reihe A (8^0), No. 12.

Hedges 1976
Hedges, T. S. 1976. "An Empirical Modification to Linear Wave Theory," *Proc. Inst. Civ. Engrs.*, Vol 61, pp 575-579.

Jensen et al. 1987
Jensen, R. E., et al. 1987. "SHALWV - Hurricane Wave Modeling and Verification; User's Guide to SHALWV: Numerical Model for Simulation of Shallow-Water Wave Growth, Propagation, and Decay," Instruction Report CERC-86-2, No. 2, U.S. Army Engineer Waterways Experiment Station, Coastal Engineering Research Center, Vicksburg, MS.

Jonsson 1978
Jonsson, I. G. 1978. "Combinations of Waves and Currents," *Stability of Tidal Inlets*, Per Bruun, ed., Amsterdam, Elsenir, pp 162-203.

Kirby 1984
Kirby, J. T. 1984. "A Note on Linear Surface Wave-Current Interaction Over Slowly Varying Topography," *Journal of Geophysical Research*, Vol 89, pp 745-747.

Kirby 1986a
Kirby, J. T. 1986a. "Higher-Order Approximations in the Parabolic Equation Method for Water Waves," *Journal of Geophysical Research*, Vol 91, pp 933-952.

Kirby and Dalrymple 1983
Kirby, J. T., and Dalrymple, R. A. 1983. "A Parabolic Equation for the Combined Refraction-Diffraction of Stokes Waves by Mildly Varying Topography," *J. Fluid Mech.*, Vol 136, pp 453-466.

Kirby and Dalrymple 1984
Kirby, J. T., and Dalrymple, R. A. 1984. "Verification of a Parabolic Equation for Propagation of Weakly-Nonlinear Waves," *Coastal Engineering*, Vol 8, pp 219-232.

Kirby and Dalrymple 1986a
Kirby, J. T., and Dalrymple, R. A. 1986a. "Modeling Waves in Surfzones and Around Islands," *J. Waterways, Port, Coastal and Ocean Engrg.*, Vol 112, pp 78 -93.

Kirby and Dalrymple 1986b
Kirby, J. T., and Dalrymple, R. A. 1986b. "An Approximate Model for Nonlinear Dispersion in Monochromatic Wave Propagation Models," *Coastal Engineering*, Vol 9, pp 545-561.

Kirby and Dalrymple 1992
Kirby, J. T., and Dalrymple, R. A. 1992. "Combined Refraction/Diffraction Model REF/DIF 1, Version 2.4.; Documentation and User's Manual," Research Report CACR-92-04, Center for Applied Coastal Research, University of Delaware.

Kraus 1991
Kraus, N. C. 1991. "NMLONG: Numerical Model for Simulating the Longshore Current," U.S. Army Engineer Waterways Experiment Station, Vicksburg, MS.

Leblond and Mysak 1978
Leblond, P. H., and Mysak, L. A. 1978. *"Waves in the Ocean,"* Elsevier Scientific Pub. Co., Amsterdam-Oxford, NY.

Leenknecht, Szuwalski, and Sherlock 1992
Leenknecht, D. A., Szuwalski, A., and Sherlock, A. R. 1992. "Automated Coastal Engineering System; User Guide and Technical Reference, Version 1.07," U.S. Army Engineer Waterways Experiment Station, Vicksburg, MS.

Miller and Vincent 1990
Miller, H. C., and Vincent, C.L. 1990. "FRF Spectrum: TMA with Kitaigoredskii's F^4 Scaling," *Journal of Waterway, Port, Coastal, and Ocean Engineering,* Vol 116, No. 1, pp 57-78.

Noda 1974
Noda, E. K. 1974. "Wave-Induced Nearshore Circulation," *Journal of Geophysical Research,* Vol 79, No. 27, pp 4097-4106.

Noda et al. 1974
Noda, E. K., et al. 1974. "Nearshore Circulations Under Sea Breeze Conditions and Wave-Current Interactions in the Surf Zone," Technical Report No. 4, Tetra Tech, Inc., Pasadena, CA.

Panchang et al. 1990
Panchang, V. G., Pearce, B. R., Wei, G., and Briggs, M. J. 1990. "Numerical Simulation of Irregular Wave Propagation Over Shoal," *J. Waterway, Port, Coastal, and Ocean Engrg.,* Vol 116, pp 324-340.

Penny and Price 1944
Penny, W. G., and Price, A. T. 1944. "Diffraction of Waves by Breakwaters," Directorate of Miscellaneous Weapons Development History No. 20 - Artificial Harbors, Sec 3D.

Peregrine 1976
Peregrine, D. H. 1976. "Interaction of Water Waves and Currents," *Advances in Applied Mechanics,* Academic Press, New York, Vol 16, pp 9-117.

Phillips 1977
Phillips, O. M. 1977. *The Dynamics of The Upper Ocean,* 2nd ed., Cambridge University Press.

Radder 1979
Radder, A. C. 1979. "On the Parabolic Equation Method for Water-Wave Propagation," *Journal of Fluid Mechanics,* Vol 95, Part 1, pp 159-176.

Resio 1981
Resio, D. T. 1981. "The Estimation of Wind-Wave Generation in a Discrete Spectral Model," *Journal of Geophysical Research*, Vol 11, pp 510 - 525.

Resio 1987
Resio, D. T. 1987. "Shallow Water Waves. I - Theory," *J. Waterway, Port, Coastal and Ocean Engr.*, American Society of Civil Engineers, Vol 113, pp 264-281.

Resio 1988
Resio, D. T. 1988. "Shallow Water Waves II - Data Comparisons," *J. Waterway, Port, Coastal and Ocean Engr.*, Vol 114, pp 50-65.

Resio 1993
Resio, D. T. 1993. "STWAVE: Wave Propagation Simulation Theory, Testing and Application," Department of Oceanography, Ocean Engineering, and Environmental Science, Florida Institute of Technology.

Shore Protection Manual 1984
Shore Protection Manual. 1984. 4th ed., 2 Vol, U. S. Army Engineer Waterways Experiment Station, U.S. Government Printing Office, Washington, DC.

Smith and Vincent 1992
Smith, J. M., and Vincent, C. L. 1992. "Shoaling and Decay of Two Wave Trains on a Beach," *J. Waterways, Port, Coastal and Ocean Engr.*, Vol 118, No. 5, pp 517-533.

Sobey and Young 1986
Sobey, R. J., and Young, I. R. 1986. "Hurricane Wind Waves - A Discrete Spectral Model," *J. Waterway, Port, Coastal and Ocean Engineering*, Vol 112, No. 3, pp 370-389.

SWAMP Group 1985
SWAMP Group. 1985. *"The Sea Wave Modeling Project (SWAMP): Principal Results and Conclusion,"* Plenum, New York, p 256.

SWIM Group 1985
SWIM Group. 1985. "A Shallow Water Intercomparison of Three Numerical Prediction Models (SWIM)," *Quarterly Journal of the Royal Meteorological Society*, Vol 111, pp 1087-1112.

Thompson and Vincent 1984
Thompson, E. F., and Vincent, C. L. 1984. "Shallow-Water Wave Height Parameters," *J. Waterway, Port, Coastal and Ocean Engr.*, American Society of Civil Engineers, Vol 110, No. 2, pp 293-299.

Tsay and Liu 1982
Tsay, T. K., and Liu, P. L. -F. 1982. "Numerical Solution of Water-Wave Refraction and Diffraction Problems in the Parabolic Approximation," *Journal of Geophysical Research*, Vol 87, pp 7932-7940.

Vincent and Briggs 1989
Vincent, C. L., and Briggs, M. J. 1989. "Refraction-Diffraction of Irregular Waves Over A Mound," *J. Waterway, Port, Coastal and Ocean Engrg*, American Society of Civil Engineers, Vol 115, No. 2, pp 269-284.

WAMDI 1988
WAMDI. 1988. "The WAM Model - A Third Generation Ocean Wave Prediction Model," *Journal of Physical Oceanography*, Vol 18, pp 1775-1810.

Weggel 1972
Weggel, J. R. 1972. "Maximum Breaker Height," *Journal of the Waterways, Harbors, and Coastal Engineering Division*, Vol 98, No. WW4, pp 529-548.

West 1981
West, B. J. 1981. "On the Simpler Aspects of Nonlinear Fluctuating Gravity Wave (Weak Interaction Theory)," *Lecture Notes in Physics 146*, Springer-Verlag, New York.

Young 1988
Young, L. R. 1988. "A Shallow Water Spectral Wave Model," *Journal of Geophysical Research*, Vol 93, No. C5, pp 5113-5129.

II-3-8. Definitions of Symbols

$\vec{k}$	Wave number vector (Equation II-3-4)
θ	Angle between the plane across which energy is being transmitted and the direction of wave advance [deg]
ρ	Mass density of water (salt water = 1,025 kg/m^3 or 2.0 slugs/ft^3; fresh water = 1,000kg/m^3 or 1.94 slugs/ft^3) [force-time^2/length^4]
φ	Velocity potential at the free surface [length^2/time]
ω	Wave angular or radian frequency (= $2\pi/T$) [time^{-1}]
Ω	Wave phase function (Equation II-3-3)
C	Wave celerity [length/time]
C_g	Wave group velocity [length/time]
d	Water depth [length]
E	Total wave energy in one wavelength per unit crest width [length-force/length^2]
$E(x,y,t,f,\theta)$	Directional spectrum where x,y represents a location in geographic space, t represents time, and f,θ represents a particular frequency-direction component
h	Water depth [length]
H	Wave height [length]
k	Wave number (= $2\pi/L = 2\pi/CT$) [length^{-1}]
K_r	Refraction coefficient [dimensionless]
K_s	Shoaling coefficient [dimensionless]
L	Wave length [length]
$-_0$	The subscript 0 denotes deepwater conditions
T	Wave period [time]
U	Ambient current vector (Equation II-3-30) [length/time]
φ	Velocity potential at the free surface [length^2/time]

Estimation of Nearshore Waves

II-3-9. Acknowledgments

Authors of Chapter II-3, "Estimation of Nearshore Waves:"

C. Linwood Vincent, Ph.D., Office of Naval Research, Arlington, Virginia.
Zeki Demirbilek, Ph.D., Coastal and Hydraulics Laboratory, Engineer Research and Development Center, Vicksburg, Mississippi.
J. Richard Weggel, Ph.D., Dept. of Civil and Architectural Engineering, Drexel University, Philadelphia, Pennsylvania.

Reviewers:

Robert A Dalrymple, Ph.D., Center for Applied Coastal Research, University of Delaware, Newark, Delaware.
Lee E. Harris, Ph.D., Department of Marine and Environmental Systems, Florida Institute of Technology, Melbourne, Florida.
Philip L.-F. Liu, Ph.D., School of Civil and Environmental Engineering, Cornell University, Ithaca, New York.

Table of Contents

Page

II-4-1. Introduction II-4-1

II-4-2. Surf Zone Waves II-4-1
    a. Incipient wave breaking II-4-1
        (1) Breaker type II-4-1
        (2) Breaker criteria II-4-3
        (3) Regular waves II-4-3
        (4) Irregular waves II-4-4
    b. Wave transformation in the surf zone II-4-5
        (1) Similarity method II-4-5
        (2) Energy flux method II-4-6
        (3) Irregular waves II-4-8
        (4) Waves over reefs II-4-11

II-4-3. Wave Setup II-4-12

II-4-4. Wave Runup on Beaches II-4-14
    a. Regular waves II-4-17
    b. Irregular waves II-4-18

II-4-5. Infragravity Waves II-4-19

II-4-6. Nearshore Currents II-4-20
    a. Introduction II-4-20
    b. Longshore current II-4-22
    c. Cross-shore current II-4-25
    d. Rip currents II-4-26

II-4-7. References II-4-27

II-4-8. Definitions of Symbols II-4-37

II-4-9. Acknowledgments II-4-40

List of Figures

Page

Figure II-4-1.   Breaker types II-4-2

Figure II-4-2.   Breaker depth index as a function of $H_b/(gT^2)$ (Weggel 1972) II-4-5

Figure II-4-3.   Change in wave profile shape from outside the surf zone (a,b) to inside the
surf zone (c,d). Measurements from Duck, North Carolina (Ebersole 1987) II-4-7

Figure II-4-4.   Transformation of H_{rms} with depth based on the initial wave approach and
the Dally, Dean, and Dalrymple (1985) model II-4-9

Figure II-4-5.   NMLONG simulation of wave height transformation (Leadbetter Beach,
Santa Barbara, California, 3 Feb 1980 (Thornton and Guza 1986)) II-4-10

Figure II-4-6.   Shallow-water transformation of wave spectra (solid line - incident, d = 3.0m;
dotted line - incident breaking zone, d = 1.7m; dashed line - surf zone,
d = 1.4m) II-4-11

Figure II-4-7.   Definition sketch for wave setup II-4-13

Figure II-4-8.   Irregular wave setup for plane slope of 1/100 . . . . II-4-15

Figure II-4-9.   Irregular wave setup for plane slope of 1/30 . . . . II-4-15

Figure II-4-10.   Example problem II-4-2 . . . . II-4-16

Figure II-4-11.   Definition sketch for wave runup II-4-17

Figure II-4-12.   Definition of runup as local maximum in elevation II-4-17

Figure II-4-13.   Measured cross-shore and longshore flow velocities II-4-21

Figure II-4-14.   Nearshore circulation systems II-4-22

Figure II-4-15.   Longshore current profiles (solid line - no lateral mixing; dashed lines - with
lateral mixing) II-4-24

Figure II-4-16.   Equation II-4-37 compared with field and laboratory data (Komar 1979) II-4-25

Figure II-4-17.   NMLONG simulation of longshore current (Leadbetter Beach, Santa Barbara,
California, 3 Feb 1989 (Thornton and Guza 1986)) II-4-26

Figure II-4-18.   Field measurement of cross-shore flow on a barred profile (Duck,
North Carolina, October 1990) II-4-27

Chapter II-4
Surf Zone Hydrodynamics

II-4-1. Introduction

a. Waves approaching the coast increase in steepness as water depth decreases. When the wave steepness reaches a limiting value, the wave breaks, dissipating energy and inducing nearshore currents and an increase in mean water level. Waves break in a water depth approximately equal to the wave height. The *surf zone* is the region extending from the seaward boundary of wave breaking to the limit of wave uprush. Within the surf zone, wave breaking is the dominant hydrodynamic process.

b. The purpose of this chapter is to describe shallow-water wave breaking and associated hydrodynamic processes of wave setup and setdown, wave runup, and nearshore currents. The surf zone is the most dynamic coastal region with sediment transport and bathymetry change driven by breaking waves and nearshore currents. Surf zone wave transformation, water level, and nearshore currents must be calculated to estimate potential storm damage (flooding and wave damage), calculate shoreline evolution and cross-shore beach profile change, and design coastal structures (jetties, groins, seawalls) and beach fills.

II-4-2. Surf Zone Waves

The previous chapter described the transformation of waves from deep to shallow depths (including refraction, shoaling, and diffraction), up to wave breaking. This section covers incipient wave breaking and the transformation of wave height through the surf zone.

a. Incipient wave breaking. As a wave approaches a beach, its length L decreases and its height H may increase, causing the wave steepness H/L to increase. Waves break as they reach a limiting steepness, which is a function of the relative depth d/L and the beach slope $\tan \beta$. Wave breaking parameters, both qualitative and quantitative, are needed in a wide variety of coastal engineering applications.

(1) Breaker type.

(a) *Breaker type* refers to the form of the wave at breaking. Wave breaking may be classified in four types (Galvin 1968): as spilling, plunging, collapsing, and surging (Figure II-4-1). In *spilling breakers*, the wave crest becomes unstable and cascades down the shoreward face of the wave producing a foamy water surface. In *plunging breakers*, the crest curls over the shoreward face of the wave and falls into the base of the wave, resulting in a high splash. In *collapsing breakers* the crest remains unbroken while the lower part of the shoreward face steepens and then falls, producing an irregular turbulent water surface. In *surging breakers*, the crest remains unbroken and the front face of the wave advances up the beach with minor breaking.

(b) Breaker type may be correlated to the surf similarity parameter ξ_o, defined as

$$\xi_o = \tan\beta \left(\frac{H_o}{L_o} \right)^{-\frac{1}{2}}$$

(II-4-1)

where the subscript o denotes the deepwater condition (Galvin 1968, Battjes 1974). On a uniformly sloping beach, breaker type is estimated by

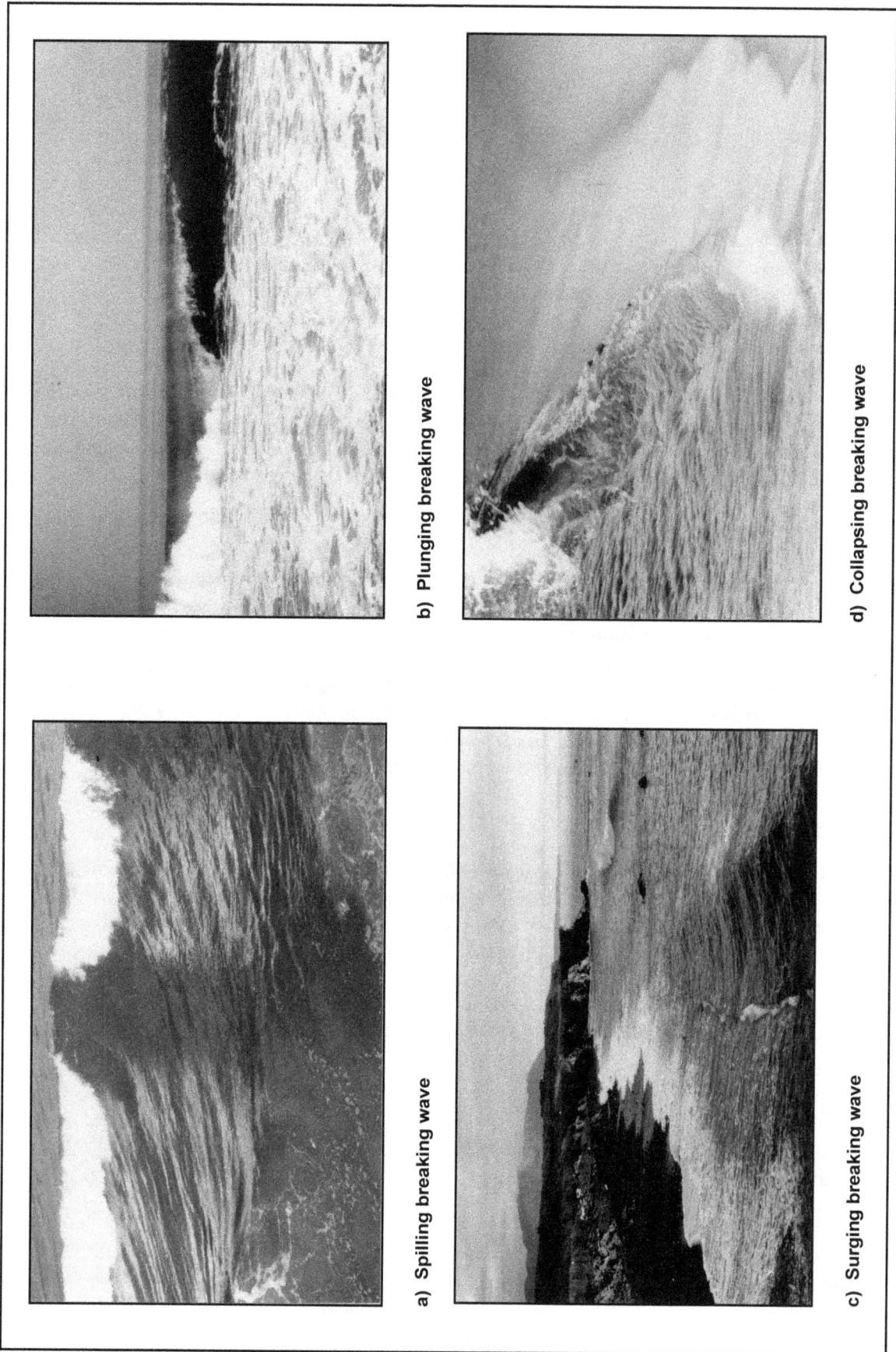

a) Spilling breaking wave

b) Plunging breaking wave

c) Surging breaking wave

d) Collapsing breaking wave

Figure II-4-1. Breaker types

Surging/collapsing $\quad \xi_o > 3.3$

Plunging $\quad 0.5 < \xi_o < 3.3$ $\qquad\qquad\qquad\qquad$ (II-4-2)

Spilling $\quad \xi_o < 0.5$

(c) As expressed in Equation II-4-2, spilling breakers tend to occur for high-steepness waves on gently sloping beaches. Plunging breakers occur on steeper beaches with intermediately steep waves, and surging and collapsing breakers occur for low steepness waves on steep beaches. Extremely low steepness waves may not break, but instead reflect from the beach, forming a standing wave (see Part II-3 for discussion of reflection and Part II-5 for discussion of tsunamis).

(d) Spilling breakers differ little in fluid motion from unbroken waves (Divoky, Le Méhauté, and Lin 1970) and generate less turbulence near the bottom and thus tend to be less effective in suspending sediment than plunging or collapsing breakers. The most intense local fluid motions are produced by a plunging breaker. As it breaks, the crest of the plunging wave acts as a free-falling jet that may scour a trough into the bottom. The transition from one breaker type to another is gradual and without distinct dividing lines. Direction and magnitude of the local wind can affect breaker type. Douglass (1990) showed that onshore winds cause waves to break in deeper depths and spill, whereas offshore winds cause waves to break in shallower depths and plunge.

(2) Breaker criteria. Many studies have been performed to develop relationships to predict the wave height at incipient breaking H_b. The term *breaker index* is used to describe nondimensional breaker height. Two common indices are the *breaker depth* index

$$\gamma_b = \frac{H_b}{d_b} \qquad\qquad\qquad\qquad\qquad\qquad\text{(II-4-3)}$$

in which d_b is the depth at breaking, and the *breaker height* index

$$\Omega_b = \frac{H_b}{H_o} \qquad\qquad\qquad\qquad\qquad\qquad\text{(II-4-4)}$$

Incipient breaking can be defined several ways (Singamsetti and Wind 1980). The most common definition is the point that wave height is maximum. Other definitions are the point where the front face of the wave becomes vertical (plunging breakers) and the point just prior to appearance of foam on the wave crest (spilling breakers). Commonly used expressions for calculating breaker indices follow.

(3) Regular waves.

(a) Early studies on breaker indices were conducted using solitary waves. McCowan (1891) theoretically determined the breaker depth index as $\gamma_b = 0.78$ for a solitary wave traveling over a horizontal bottom. This value is commonly used in engineering practice as a first estimate of the breaker index. Munk (1949) derived the expression $\Omega_b = 0.3(H_o/L_o)^{-1/3}$ for the breaker height index of a solitary wave. Subsequent studies, based on periodic waves, by Iversen (1952), Goda (1970), Weggel (1972), Singamsetti and Wind (1980), Sunamura (1980), Smith and Kraus (1991), and others have established that the breaker indices depend on beach slope and incident wave steepness.

(b) From laboratory data on monochromatic waves breaking on smooth, plane slopes, Weggel (1972) derived the following expression for the breaker depth index

$$\gamma_b = b - a \frac{H_b}{g\,T^2} \tag{II-4-5}$$

for $\tan\beta \leq 0.1$ and $H_o'/L_o \leq 0.06$, where T is wave period, g is gravitational acceleration, and H_o' is equivalent unrefracted deepwater wave height. The parameters a and b are empirically determined functions of beach slope, given by

$$a = 43.8 \left(1 - e^{-19\tan\beta}\right) \tag{II-4-6}$$

and

$$b = \frac{1.56}{\left(1 + e^{-19\,5\tan\beta}\right)} \tag{II-4-7}$$

(c) The breaking wave height H_b is contained on both sides of Equation II-4-5, so the equation must be solved iteratively. Figure II-4-2 shows how the breaker depth index depends on wave steepness and bottom slope. For low steepness waves, the breaker index (Equation II-4-5) is bounded by the theoretical value of 0.78, as the beach slope approaches zero, and twice the theoretical value (sum of the incident and perfectly reflected component), or 1.56, as the beach slope approaches infinity. For nonuniform beach slopes, the average bottom slope from the break point to a point one wavelength offshore should be used.

(d) Komar and Gaughan (1973) derived a semi-empirical relationship for the breaker height index from linear wave theory

$$\Omega_b = 0.56 \left(\frac{H_o'}{L_o}\right)^{-\frac{1}{5}} \tag{II-4-8}$$

(e) The coefficient 0.56 was determined empirically from laboratory and field data.

(4) Irregular waves. In irregular seas (see Part II-1 for a general discussion of irregular waves), incipient breaking may occur over a wide zone as individual waves of different heights and periods reach their steepness limits. In the *saturated* breaking zone for irregular waves (the zone where essentially all waves are breaking), wave height may be related to the local depth d as

$$H_{rms,b} = 0.42\,d \tag{II-4-9}$$

for root-mean-square (rms) wave height (Thornton and Guza 1983) or, approximately,

$$H_{mo,b} = 0.6\,d \tag{II-4-10}$$

for zero-moment wave height (see Part II-1). Some variability in $H_{rms,b}$ and $H_{mo,b}$ with wave steepness and beach slope is expected; however, no definitive study has been performed. The numerical spectral wave transformation model STWAVE (Smith et al. 2001) uses a modified Miche Criterion (Miche 1951).

$$H_{mo,b} = 0.1\,L\,\tan h\,kd \tag{II-4-11}$$

to represent both depth- and steepness-induced wave breaking.

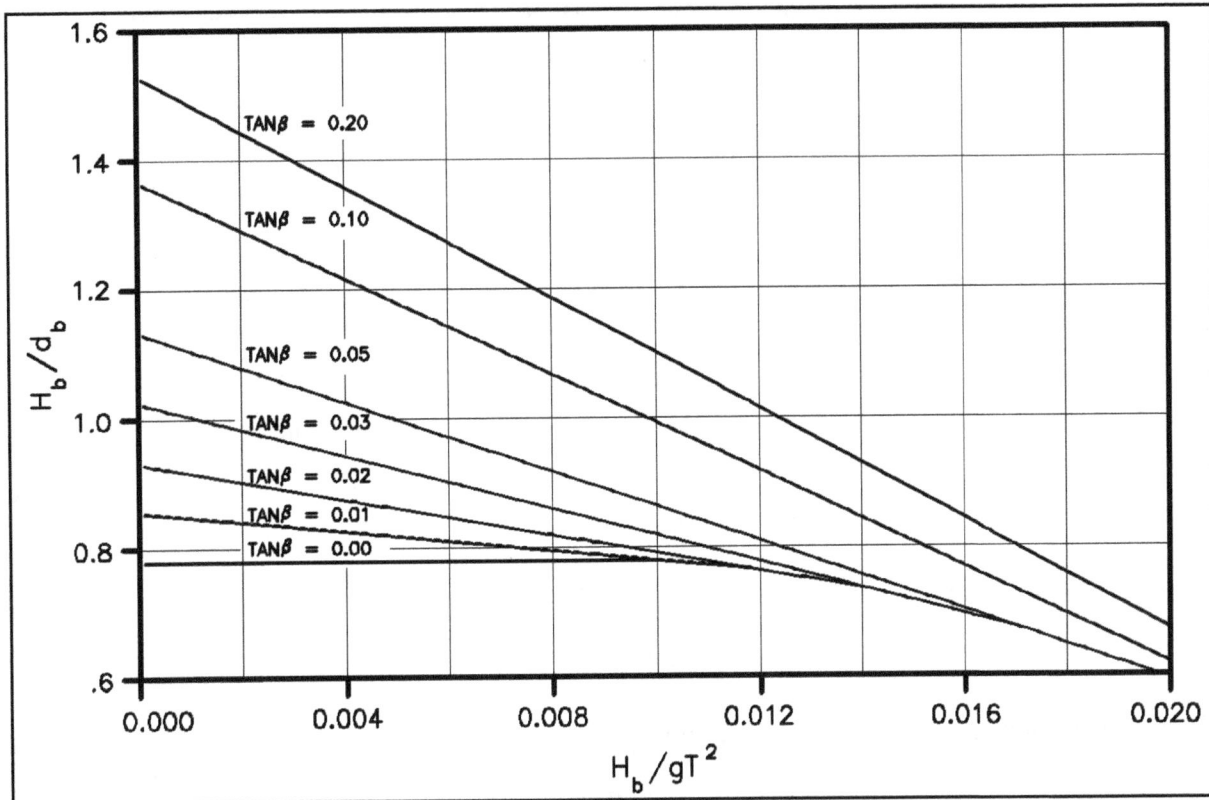

Figure II-4-2.   Breaker depth index as a function of $H_b/(gT^2)$ (Weggel 1972)

  b. *Wave transformation in the surf zone.* Following incipient wave breaking, the wave shape changes rapidly to resemble a bore (Svendsen 1984). The wave profile becomes sawtooth in shape with the leading edge of the wave crest becoming nearly vertical (Figure II-4-3). The wave may continue to dissipate energy to the shoreline or, if the water depth again increases as in the case of a barred beach profile, the wave may cease breaking, re-form, and break again on the shore. The transformation of wave height through the surf zone impacts wave setup, runup, nearshore currents, and sediment transport.

  (1) Similarity method. The simplest method for predicting wave height through the surf zone, an extension of Equation II-4-3 shoreward of incipient breaking conditions, is to assume a constant height-to-depth ratio from the break point to shore

$$H_b = \gamma_b \, d_b \tag{II-4-12}$$

This method, also referred to as saturated breaking, has been used successfully by Longuet-Higgins and Stewart (1963) to calculate setup, and by Bowen (1969a), Longuet-Higgins (1970a,b), and Thornton (1970) to calculate longshore currents. The similarity method is applicable only for monotonically decreasing water depth through the surf zone and gives best results for a beach slope of approximately 1/30. On steeper slopes, Equation II-4-12 tends to underestimate the wave height. On gentler slopes or barred topography, it tends to overestimate the wave height. Equation II-4-12 is based on the assumption that wave height is zero at the mean shoreline (see Part II-4-3 for discussion of mean versus still-water shoreline). Camfield (1991) shows that a conservative estimate of wave height at the still-water shoreline is $0.20 \, H_b$ for $0.01 \le \tan \beta \le 0.1$.

EXAMPLE PROBLEM II-4-1

FIND:
Wave height and water depth at incipient breaking.

GIVEN:
A beach with a 1 on 100 slope, deepwater wave height $H_o = 2$ m, and period $T = 10$ sec. Assume that a refraction analysis (Part II-3) gives a refraction coefficient $K_R = 1.05$ at the point where breaking is expected to occur.

SOLUTION:
The equivalent unrefracted deepwater wave height H_o' can be found from the refraction coefficient (see Part II-3, Equation II-3-14)

$$H_o' = K_R H_o = 1.05 \, (2.0) = 2.1 \text{ m}$$

and the deepwater wavelength L_o is given by (Part II-1)

$$L_o = g \, T^2/(2\pi) = 9.81 \, (10^2)/(2\pi) = 156 \text{ m}$$

Estimate the breaker height from Equation II-4-8

$$\Omega_b = 0.56 \, (H_o'/L_o)^{-1/5} = 0.56 \, (2.1/156.)^{-1/5} = 1.3$$
$$H_b \text{ (estimated)} = \Omega_b \, H_o' = 2.7 \text{ m}$$

From Equations II-4-6 and II-4-7, determine a and b used in Equation II-4-5, $\tan \beta = 1/100$

$$a = 43.8(1 - e^{-19 \, (1/100)}) = 7.58$$
$$b = 1.56 \, / \, (1 + e^{-19.5 \, (1/100)}) = 0.86$$

$$\gamma_b = b - a \, H_b \, / \, (gT^2) = 0.86 - 7.58 \, (2.7)/(9.81 \, 10^2) = 0.84$$
$$d_b = H_b \, / \, \gamma_b = 2.7/0.84 = 3.2 \text{ m}$$

Breaker height is approximately 2.7 m and breaker depth is 3.2 m. The initial value selected for the refraction coefficient would now be checked to determine if it is correct for the actual breaker location. If necessary, a corrected refraction coefficient should be used to recompute breaker height and depth.

(2) Energy flux method.

(a) A more general method for predicting wave height through the surf zone for a long, straight coast is to solve the steady-state energy balance equation

$$\frac{d(EC_g)}{dx} = -\delta \qquad\qquad\qquad \text{(II-4-13)}$$

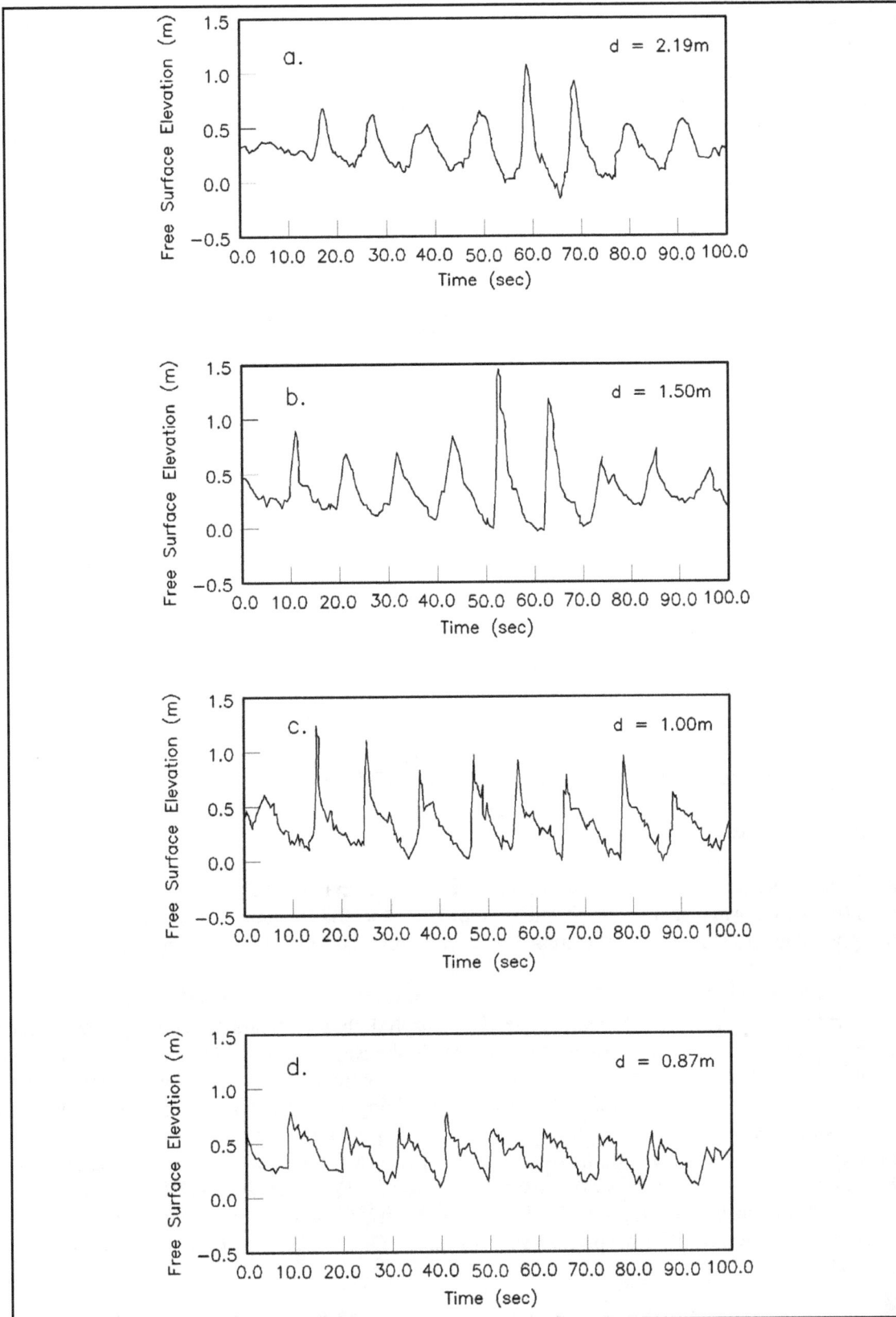

Figure II-4-3.   Change in wave profile shape from outside the surf zone (a,b) to inside the surf zone (c,d). Measurements from Duck, NC (Ebersole 1987)

where E is the wave energy per unit surface area, C_g is the wave group speed, and δ is the energy dissipation rate per unit surface area due to wave breaking. The wave energy flux EC_g may be specified from linear or higher order wave theory. Le Méhauté (1962) approximated a breaking wave as a hydraulic jump and substituted the dissipation of a hydraulic jump for δ in Equation II-4-13 (see also Divoky, Le Méhauté, and Lin 1970; Hwang and Divoky 1970; Svendsen, Madsen, and Hansen 1978).

(b) Dally, Dean, and Dalrymple (1985) modeled the dissipation rate as

$$\delta = \frac{\kappa}{d}(EC_g - EC_{g,s})$$ (II-4-14)

where κ is an empirical decay coefficient, found to have the value 0.15, and $EC_{g,s}$ is the energy flux associated with a *stable* wave height

$$H_{stable} = \Gamma d$$ (II-4-15)

(c) The quantity Γ is an empirical coefficient with a value of approximately 0.4. The stable wave height is the height at which a wave stops breaking and re-forms. As indicated, this approach is based on the assumption that energy dissipation is proportional to the difference between local energy flux and stable energy flux. Applying linear, shallow-water theory, the Dally, Dean, and Dalrymple model reduces to

$$\frac{d(H^2 d^{\frac{1}{2}})}{dx} = -\frac{\kappa}{d}\left(H^2 d^{\frac{1}{2}} - \Gamma^2 d^{\frac{5}{2}}\right) \quad \text{for} \quad H > H_{stable}$$

$$= 0 \quad \text{for} \quad H < H_{stable}$$ (II-4-16)

This approach has been successful in modeling wave transformation over irregular beach profiles, including bars (e.g., Ebersole (1987), Larson and Kraus (1991), Dally (1992)).

(3) Irregular waves.

(a) Transformation of irregular waves through the surf zone may be analyzed or modeled with either a statistical (individual wave or wave height distribution) or a spectral (parametric spectral shape) approach. Part II-1 gives background on wave statistics, wave height distributions, and parametric spectral shapes.

(b) The most straightforward statistical approach is transformation of individual waves through the surf zone. Individual waves seaward of breaking may be measured directly, randomly chosen from a Rayleigh distribution, or chosen to represent wave height classes in the Rayleigh distribution. Then the individual waves are independently transformed through the surf zone using Equation II-4-13. Wave height distribution can be calculated at any point across the surf zone by recombining individual wave heights into a distribution to calculate wave height statistics (e.g., $H_{1/10}$, $H_{1/3}$, H_{rms}). This method does not make a priori assumptions about wave height distribution in the surf zone. The individual wave method has been applied and verified with field measurements by Dally (1990), Larson and Kraus (1991), and Dally (1992). Figure II-4-4 shows the nearshore transformation of H_{rms} with depth based on the individual wave approach and the Dally, Dean, and Dalrymple (1985) model for deepwater wave steepness (H_{rmso}/L_o) of 0.005 to 0.05 and plane beach slopes of 1/100 and 1/30.

(c) A numerical model called NMLONG (Numerical Model of the LONGshore current) (Larson and Kraus 1991) calculates wave breaking and decay by the individual wave approach applying the Dally, Dean,

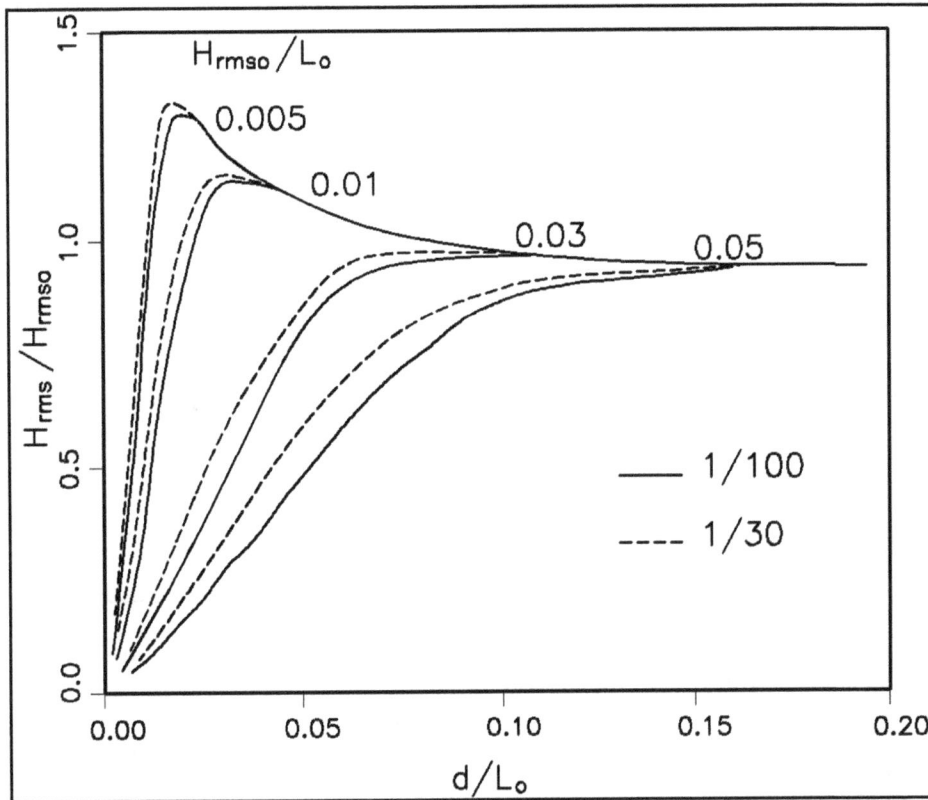

Figure II-4-4. Transformation of H_{rms} with depth based on the individual wave approach and the Dally, Dean, and Dalrymple (1985) model

and Dalrymple (1985) wave decay model (monochromatic or irregular waves). The main assumption underlying the model is uniformity of waves and bathymetry alongshore, but the beach profile can be irregular across the shore (e.g., longshore bars and nonuniform slopes). NMLONG uses a single wave period and direction and applies a Rayleigh distribution wave heights outside the surf zone. The model runs on a personal computer and has a convenient graphical interface. NMLONG calculates both wave transformation and longshore current (which will be discussed in a later section) for arbitrary offshore (input) wave conditions and provides a plot of results. Figure II-4-5 gives an example NMLONG calculation and a comparison of wave breaking field measurements reported by Thornton and Guza (1986).

(d) A second statistical approach is based on assuming a wave height distribution in the surf zone. The Rayleigh distribution is a reliable measure of the wave height distribution in deep water and at finite depths. In the surf zone, depth-induced breaking acts to limit the highest waves in the distribution, contrary to the Rayleigh distribution, which is unbounded. The surf zone wave height distribution has generally been represented as a truncated Rayleigh distribution (e.g., Collins (1970), Battjes (1972), Kuo and Kuo (1974), Goda 1975). Battjes and Janssen (1978) and Thornton and Guza (1983) base the distribution of wave heights at any point in the surf zone on a Rayleigh distribution or a truncated Rayleigh distribution (truncated above a maximum wave height for the given water depth). A percentage of waves in the distribution is designated as broken, and energy dissipation from these broken waves is calculated from Equation II-4-13 through a model of dissipation similar to a periodic bore. Battjes and Janssen (1978) define the energy dissipation as

$$\delta = 0.25 \rho g Q_b f_m (H_{max})^2$$

(II-4-17)

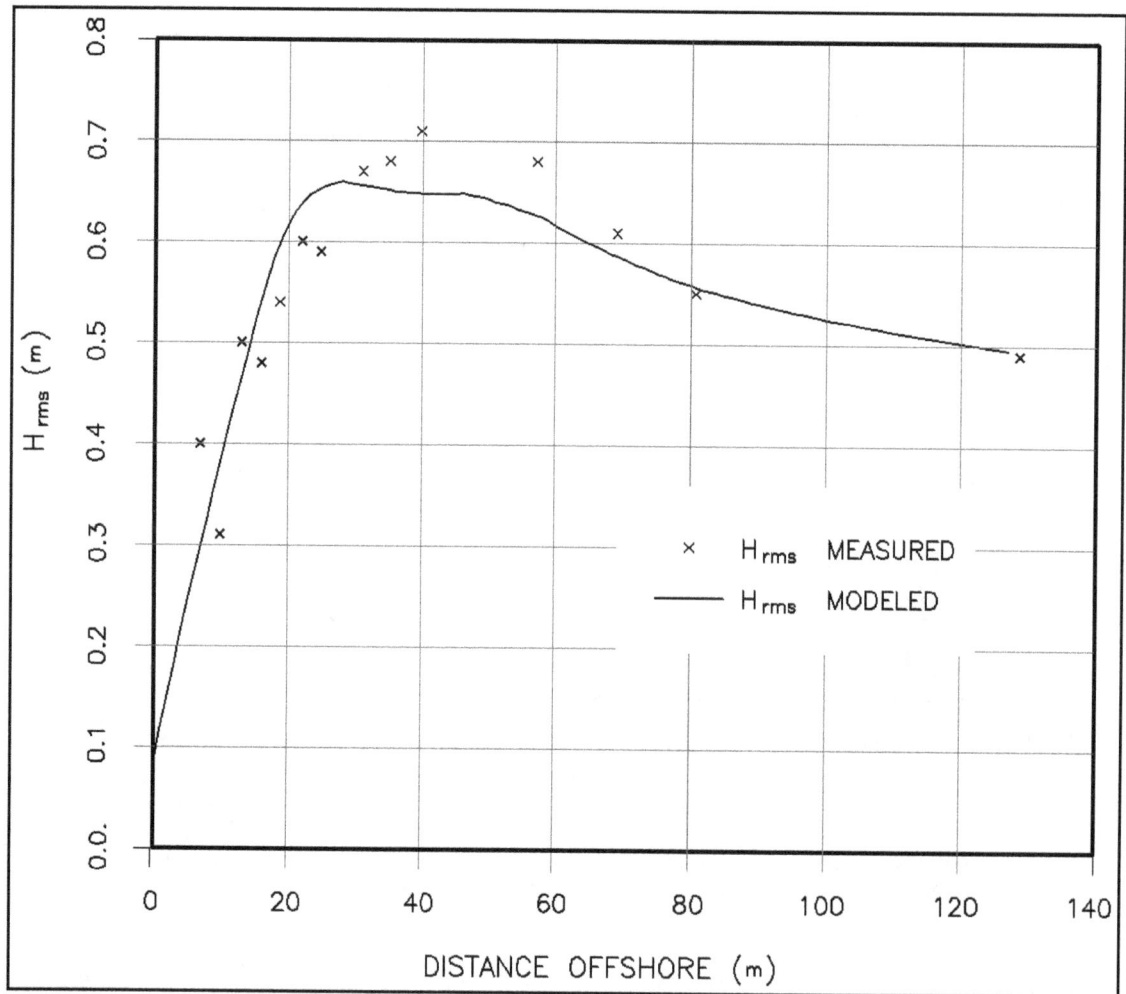

Figure II-4-5. NMLONG simulation of wave height transformation (Leadbetter Beach, Santa Barbara, California, 3 Feb 1980 (Thornton and Guza 1986))

where Q_b is the percentage of waves breaking, f_m is the mean wave frequency, and the maximum wave height is based on the Miche (1951) criterion

$$H_{max} = 0.14L \tanh(kd) \tag{II-4-18}$$

where k is wave number. Battjes and Janssen base the percentage of waves breaking on a Rayleigh distribution truncated at H_{max}. Baldock et al. (1998) show improved results and reduced computational time by basing Q_b on the full Rayleigh distribution (Smith 2001). Goda (2002) documented that although the wave height distribution in the midsurf zone is narrower than the Rayleigh distribution, in the outer surf zone and near the shoreline the distribution is nearly Rayleigh. This method has been validated with laboratory and field data (e.g., Battjes and Janssen 1978; Thornton and Guza 1983) and implemented in numerical models (e.g., Booij 1999). Specification of the maximum wave height in terms of the Miche criterion (Equation II-4-18) has the advantage of providing reasonable results for steepness-limited breaking (e.g., waves breaking on a current) as well as depth-limited breaking (Smith et al. 1997).

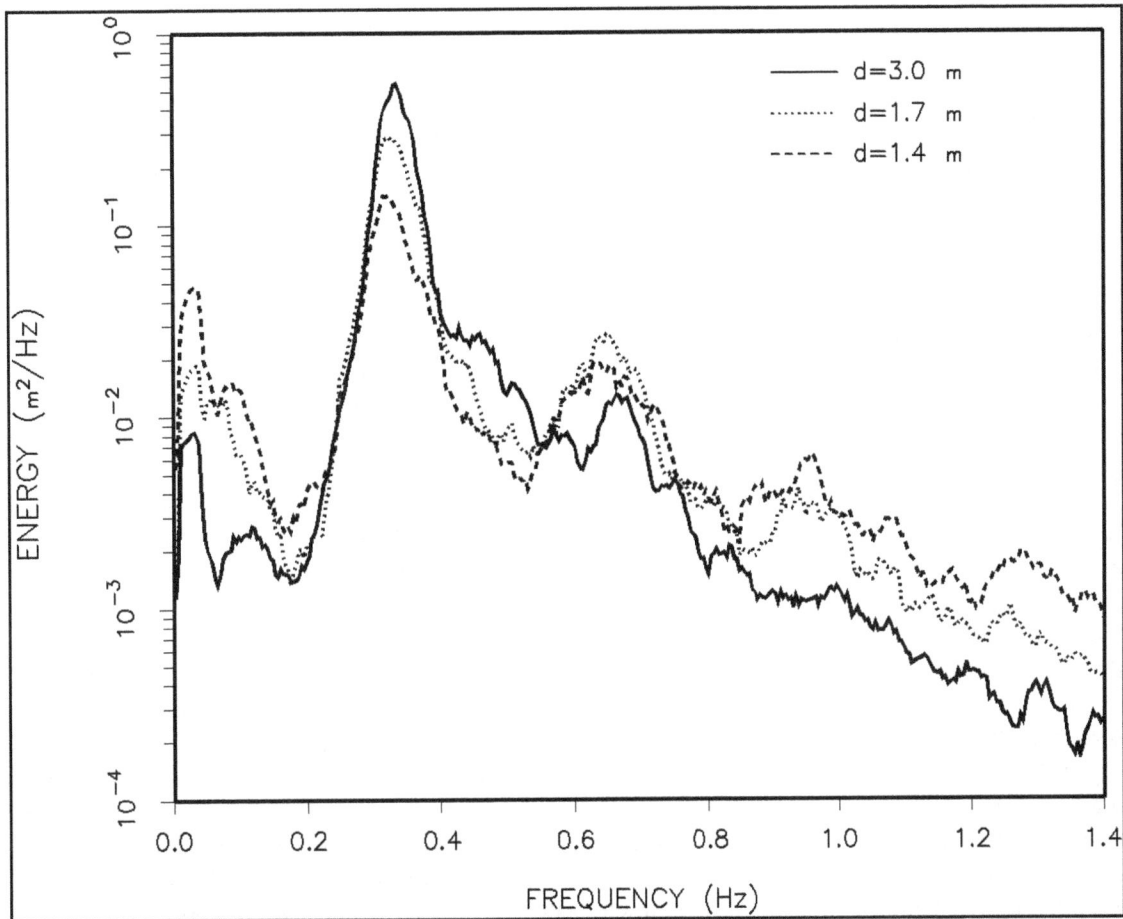

Figure II-4-6. Shallow-water transformation of wave spectra (solid line - incident, d = 3.0m; dotted line - incident breaking zone, d = 1.7m; dashed line - surf zone, d = 1.4m)

(e) In shallow water, the shape of the wave spectrum is influenced by nonlinear transfers of wave energy from the peak frequency to higher frequencies and lower frequencies (Freilich and Guza 1984; Freilich, Guza, and Elgar 1990). Near incipient breaking higher harmonics (energy peaks at integer multiples of the peak frequency) appear in the spectrum as well as a general increase in the energy level above the peak frequency as illustrated in Figure II-4-6. Low-frequency energy peaks (subharmonics) are also generated in the surf (Figure II-4-6, also see Part II-4-5). Figure II-4-6 shows three wave spectra measured in a large wave flume with a sloping sand beach. The solid curve is the incident spectrum (d = 3.0 m), the dotted curve is the spectrum at the zone of incipient breaking (d = 1.7 m), and the dashed curve is within the surf zone (d = 1.4 m). Presently, no formulation is available for the dissipation rate based on spectral parameters for use in Equation II-4-13. Therefore, the energy in the spectrum is often limited using the similarity method. Smith and Vincent (2002) found that in the inner surf zone, wave spectra evolve to a similar, single-peaked shape regardless of the complexity of the shape outside the surf zone (e.g., multipeaked spectra evolve to a single peak). It is postulated that the spectral shape evolves from the strong nonlinear interactions in the surf zone.

(4) Waves over reefs. Many tropical coastal regions are fronted by coral reefs. These reefs offer protection to the coast because waves break on the reefs, so the waves reaching the shore are less energetic. Reefs typically have steep seaward slopes with broad, flat reef tops and a deeper lagoon shoreward of the reef. Transformation of waves across steep reef faces and nearly flat reef tops cannot be modeled by simple wave breaking relationships such as Equation II-4-12. Generally, waves refract and shoal on the steep reef face,

break, and then reform on the reef flat. Irregular transformation models based on Equation II-4-13 give reasonable results for reef applications (Young 1989), even though assumptions of gentle slopes are violated at the reef face. Wave reflection from coral reefs has been shown to be surprisingly low (Young 1989; Hardy and Young 1991). Although the dominant dissipation mechanism is depth-limited wave breaking, inclusion of an additional wave dissipation term in Equation II-4-13 to represent bottom friction on rough coral improves wave estimates. General guidance on reef bottom friction coefficients is not available, site-specific field measurements are recommended to estimate bottom friction coefficients.

(5) Advanced modeling of surf zone waves. Numerical models based on the Boussinesq equations have been extended to the surf zone by empirically implementing breaking. In time-domain Boussinesq models, a surface roller (Schäffer et al. 1993) or a variable eddy viscosity (Nwogu 1996; Kennedy et al. 2000) is used to represent breaking induced mixing and energy dissipation. Incipient breaking for individual waves is initiated based on velocity at the wave crest or slope of the water surface. These models accurately represent the time-varying, nonlinear wave profile (including vertical and horizontal wave asymmetry) and depth-averaged current. Boussinesq models also include the generation of low-frequency waves in the surf zone (surf beat and shear waves) (e.g., Madsen, Sprengen, and Schäffer 1997; Kirby and Chen 2002). Wave runup on beaches and interaction with coastal structures are also included in some models. Although Boussinesq models are computationally intensive, they are now being used for many engineering applications (e.g., Nwogu and Demirbilek 2002). The one-dimensional nonlinear shallow-water equations have also been used to calculate time-domain irregular wave transformation in the surf zone (Kobayashi and Wurjanto 1992). This approach has been successful in predicting the oscillatory and steady fluid motions in the surf and swash zones (Raubenheimer et al. 1994). Reynolds Averaged Navier Stokes (e.g., Lin and Liu 1998) and Large Eddy Simulation (Watanabe and Saeki 1999; Christensen and Deigaard 2001) models have been developed to study the turbulent 3-D flow fields generated by breaking waves. These models can represent obliquely descending eddies generated by breaking waves (Nadaoka, Hino, and Koyano 1989) which increase the turbulent intensity, eddy viscosity, and near-bottom shear stress (Okayasu et al. 2002). Results from these models may help explain the difference in sediment transport patterns under plunging and spilling breakers (Wang, Smith, and Ebersole 2002). These detailed large-scale turbulence models are still research tools requiring large computational resources for short simulations. However, results from the models are providing insights to surf zone turbulent processes that are difficult to measure in the laboratory or field.

II-4-3. Wave Setup

a. Wave setup is the superelevation of mean water level caused by wave action (additional changes in water level may include wind setup or tide, see Part II-5). Total water depth is a sum of still-water depth and setup

$$d = h + \overline{\eta} \qquad \text{(II-4-19)}$$

where

h = still-water depth

$\overline{\eta}$ = mean water surface elevation about still-water level

b. Wave setup balances the gradient in the cross-shore directed radiation stress, i.e., the pressure gradient of the mean sloping water surface balances the gradient of the incoming momentum. Derivation of radiation stress is given in Part II-1.

c. Mean water level is governed by the cross-shore balance of momentum

$$\frac{d\overline{\eta}}{dx} = -\frac{1}{\rho g d}\frac{dS_{xx}}{dx} \tag{II-4-20}$$

where S_{xx} is the cross-shore component of the cross-shore directed radiation stress, for longshore homogeneous waves and bathymetry (see Equations II-4-34 through II-4-36 for general equations). Radiation stress both raises and lowers (*setdown*) the mean water level across shore in the nearshore region (Figure II-4-7).

d. Seaward of the breaker zone, Longuet-Higgins and Stewart (1963) obtained setdown for regular waves from the integration of Equation II-4-20 as

$$\overline{\eta} = -\frac{1}{8}\frac{H^2\frac{2\pi}{L}}{\sinh\left(\frac{4\pi}{L}d\right)} \tag{II-4-21}$$

assuming linear wave theory, normally incident waves, and $\overline{\eta} = 0$ in deep water. The maximum lowering of the water level, setdown, occurs near the break point $\overline{\eta}_b$.

e. In the surf zone, $\overline{\eta}$ increases between the break point and the shoreline (Figure II-4-7). The gradient, assuming linear theory ($S_{xx} = 3/16\,\rho\,g\,H^2$), is given by

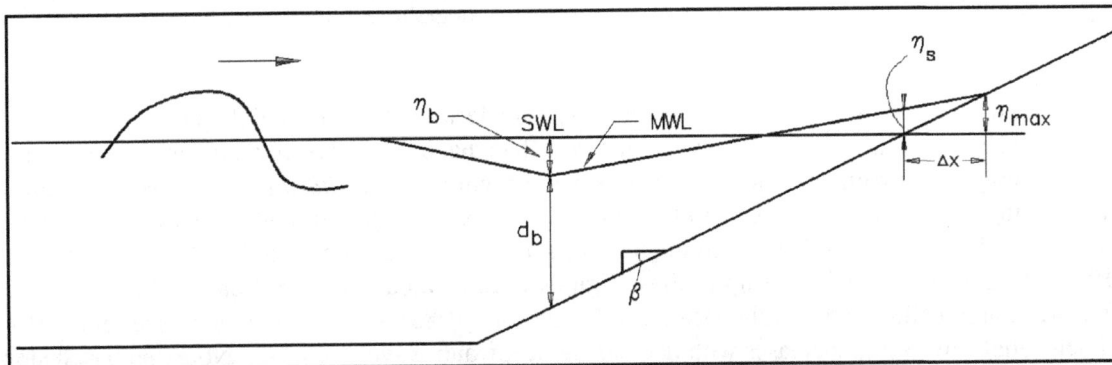

Figure II-4-7. Definition sketch for wave setup

$$\frac{d\overline{\eta}}{dx} = -\frac{3}{16}\frac{1}{h+\overline{\eta}}\frac{d(H^2)}{dx} \tag{II-4-22}$$

where the shallow-water value of $S_{xx} = 3/16\,\rho\,g\,d\,H^2$ has been substituted into Equation II-4-20. The value of $\overline{\eta}$ depends on wave decay through the surf zone. Applying the saturated breaker assumption of linear wave height decay on a plane beach, Equation II-4-22 reduces to

$$\frac{d\overline{\eta}}{dx} = \frac{1}{1+\frac{8}{3\gamma_b^2}}\tan\beta \tag{II-4-23}$$

f. Combining Equations II-4-21 and II-4-23, setup at the still-water shoreline $\overline{\eta}_s$ is given by

$$\overline{\eta}_s = \overline{\eta}_b + \left[\frac{1}{1 + \dfrac{8}{3\gamma_b^2}}\right] h_b \qquad\qquad\text{(II-4-24)}$$

g. The first term in Equation II-4-24 is setdown at the break point and the second term is setup across the surf zone. The setup increases linearly through the surf zone for a plane beach. For a breaker depth index of 0.8, $\overline{\eta}_s \approx 0.15\,d_b$. Note that, for higher breaking waves, d_b will be greater and thus setup will be greater. Equation II-4-24 gives setup at the still-water shoreline; to calculate maximum setup and position of the mean shoreline, the point of intersection between the setup and beach slope must be found. This can be done by trial and error, or, for a plane beach, estimated as

$$\Delta x = \frac{\overline{\eta}_s}{\tan\beta - \dfrac{d\overline{\eta}}{dx}} \qquad\qquad\text{(II-4-25)}$$

$$\overline{\eta}_{\max} = \overline{\eta}_s + \frac{d\overline{\eta}}{dx}\,\Delta x$$

where Δx is the shoreward displacement of the shoreline and $\overline{\eta}_{max}$ is the setup at the mean shoreline.

h. Wave setup and the variation of setup with distance on irregular (non-planar) beach profiles can be calculated based on Equations II-4-21 and II-4-22 (e.g., McDougal and Hudspeth 1983, Larson and Kraus 1991). NMLONG calculates mean water level across the nearshore under the assumptions previously discussed.

i. Setup for irregular waves should be calculated from decay of the wave height parameter H_{rms}. Wave setup produced by irregular waves is somewhat different than that produced by regular waves (Equation II-4-22) because long waves with periods of 30 sec to several minutes, called *infragravity waves*, may produce a slowly varying mean water level. See Part II-4-5 for discussion of magnitude and generation of infragravity waves. Figures II-4-8 and II-4-9 show irregular wave setup, nondimensionalized by H_{rmso}, for plane slopes of 1/100 and 1/30, respectively. Setup in these figures is calculated from the decay of H_{rms} given by the irregular wave application of the Dally, Dean, and Dalrymple (1985) wave decay model (see Figure II-4-4). Nondimensional wave setup increases with decreasing deepwater wave steepness. Note that beach slope is predicted to have a relatively small influence on setup for irregular waves.

II-4-4. Wave Runup on Beaches

Runup is the maximum elevation of wave uprush above still-water level (Figure II-4-11). Wave uprush consists of two components: superelevation of the mean water level due to wave action (setup) and fluctuations about that mean (*swash*). Runup, R, is defined in Figure II-4-12 as a local maximum or peak in the instantaneous water elevation, η, at the shoreline. The upper limit of runup is an important parameter for determining the active portion of the beach profile.

At present, theoretical approaches for calculating runup on beaches are not viable for coastal design. Difficulties inherent in runup prediction include nonlinear wave transformation, wave reflection, three-dimensional effects (bathymetry, infragravity waves), porosity, roughness, permeability, and groundwater elevation. Wave runup on structures is discussed in Chapter VI-2.

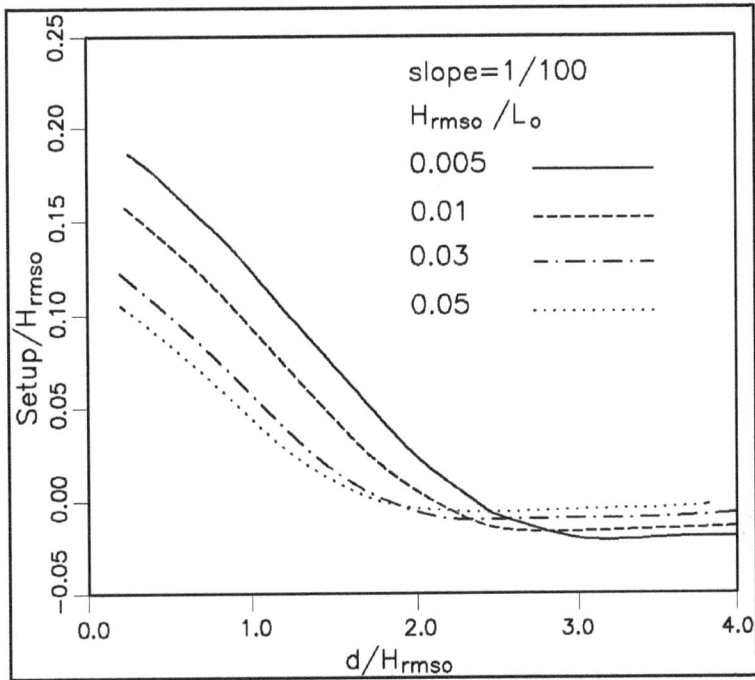

Figure II-4-8.   Irregular wave setup for plane slope of 1/100

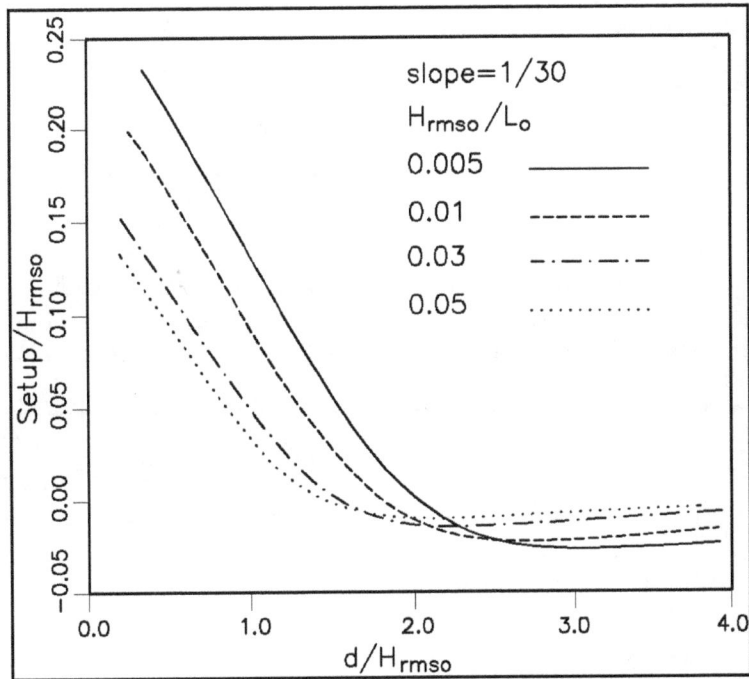

Figure II-4-9.   Irregular wave setup for plane slope of 1/30

EXAMPLE PROBLEM II-4-2

FIND:

Setup across the surf zone.

GIVEN:

A plane beach having a 1 on 100 slope, and normally incident waves with deepwater height of 2 m and period of 10 sec (see Example Problem II-4-1).

SOLUTION:

The incipient breaker height and depth were determined in Example Problem II-4-1 as 2.7 m and 3.2 m, respectively. The breaker index is 0.84, based on Equation II-4-5.

Setdown at the breaker point is determined from Equation II-4-21. At breaking, Equation II-4-21 simplifies to $\overline{\eta}_b = -1/16\, \gamma_b^2\, d_b$, (sinh $2\pi d/L \approx 2\pi d/L$, and $H_b = \gamma_b\, d_b$), thus

$$\overline{\eta}_b = -1/16\,(0.84)^2\,(3.2) = -0.14 \text{ m}$$

Setup at the still-water shoreline is determined from Equation II-4-24

$$\overline{\eta}_s = -0.14 + (3.2 + 0.14) + 1/(1 + 8/(3\,(.84)^2)) = 0.56 \text{ m}$$

The gradient in the setup is determined from Equation II-4-23 as

$$d\overline{\eta}/dx = 1/(1 + 8/(3\,(0.84)^2))(1/100) = 0.0021$$

and from Equation II-4-25, $\Delta x = (0.56)/(1/100 - 0.0021) = 70.9$ m, and

$$\overline{\eta}_{max} = 0.56 + 0.0021(64.6) = 0.65 \text{ m}$$

For the simplified case of a plane beach with the assumption of linear wave height decay, the gradient in the setup is constant through the surf zone. Setup may be calculated anywhere in the surf zone from the relation $\overline{\eta} = \overline{\eta}_b + (d\overline{\eta}/dx)(x_b - x)$, where x_b is the surf zone width and $x = 0$ at the shoreline (x is positive offshore).

x, m	h, m	η, m
334	3.3	-0.14
167	1.7	0.21
0	0.0	0.56
-71	-0.7	0.71

Setdown at breaking is -0.14 m, net setup at the still-water shoreline is 0.56 m, the gradient in the setup is 0.0021 m/m, the mean shoreline is located 71 m shoreward of the still-water shoreline, and maximum setup is 0.71 m (Figure II-4-10).

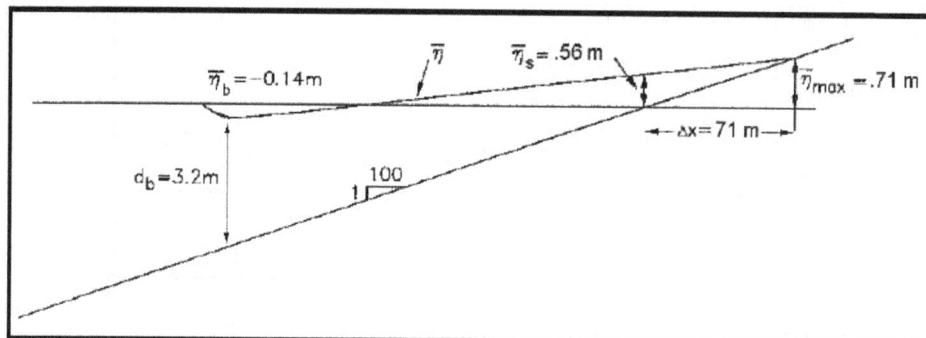

Figure II-4-10. Example problem II-4-2

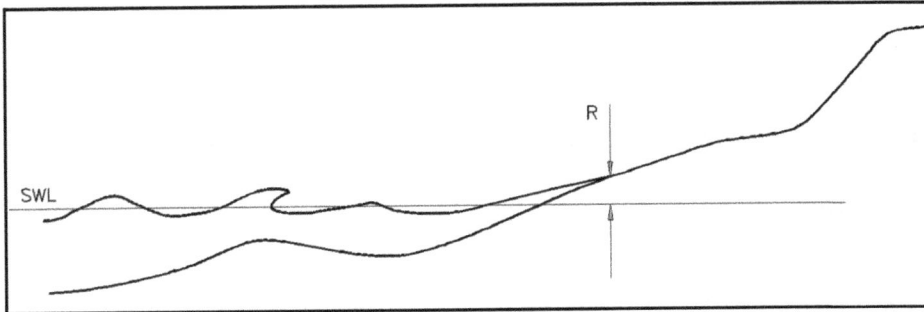

Figure II-4-11.   Definition sketch for wave runup

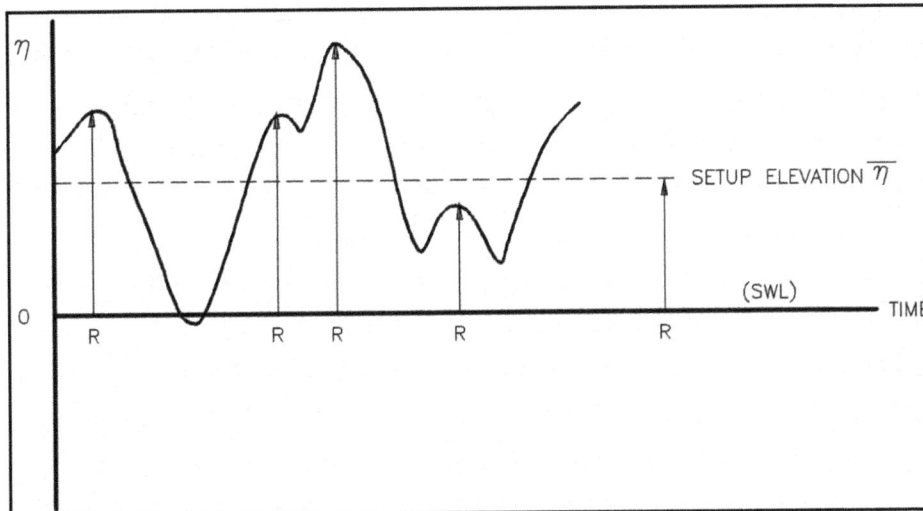

Figure II-4-12.   Definition of runup as local maximum in elevation

a.   Regular waves.

(1) For breaking waves, Hunt (1959) empirically determined runup as a function of beach slope, incident wave height, and wave steepness based on laboratory data. Hunt's formula, given in nondimensional form (Battjes 1974), is

$$\frac{R}{H_o} = \xi_o \quad for \quad 0.1 < \xi_o < 2.3 \tag{II-4-26}$$

for uniform, smooth, impermeable slopes, where ξ_o is the surf similarity parameter defined in Equation II-4-1. Walton et al. (1989) modified Equation II-4-26 to extend the application to steep slopes by replacing $\tan \beta$ in the surf similarity parameter, which becomes infinite as β approaches $\pi/2$, with $\sin \beta$. The modified Hunt formula was verified with laboratory data from Saville (1956) and Savage (1958) for slopes of 1/10 to vertical.

(2) The nonbreaking upper limit of runup on a uniform slope is given by

$$\frac{R}{H_o} = (2\pi)^{\frac{1}{2}} \left(\frac{\pi}{2\beta}\right)^{\frac{1}{4}}$$

(II-4-27)

based on criteria developed by Miche (1951) and Keller (1961) (Walton et al. 1989).

    b. Irregular waves.

    (1) Irregular wave runup has also been found to be a function of the surf similarity parameter (Holman and Sallenger 1985, Mase 1989, Nielsen and Hanslow 1991), but differs from regular wave runup due to the interaction between individual runup bores. Uprush may be halted by a large backrush from the previous wave or uprush may be overtaken by a subsequent large bore. The ratio of the number of runup crests to the number of incident waves increases with increased surf similarity parameter (ratios range from 0.2 to 1.0 for ξ_o of 0.15 to 3.0) (Mase 1989, Holman 1986). Thus, low-frequency (infragravity) energy dominates runup for low values of ξ_o. See Section II-4-5 for a discussion of infragravity waves.

    (2) Mase (1989) presents predictive equations for irregular runup on plane, impermeable beaches (slopes 1/5 to 1/30) based on laboratory data. Mase's expressions for the maximum runup (R_{max}), the runup exceeded by 2 percent of the runup crests ($R_{2\%}$), the average of the highest 1/10 of the runups ($R_{1/10}$), the average of the highest 1/3 of the runups ($R_{1/3}$), and the mean runup ($\bar{R}$) are given by

$$\frac{R_{max}}{H_o} = 2.32 \; \xi_o^{0.77}$$

(II-4-28)

$$\frac{R_{2\%}}{H_o} = 1.86 \; \xi_o^{0.71}$$

(II-4-29)

$$\frac{R_{1/10}}{H_o} = 1.70 \; \xi_o^{0.71}$$

(II-4-30)

$$\frac{R_{1/3}}{H_o} = 1.38 \; \xi_o^{0.70}$$

(II-4-31)

$$\frac{\bar{R}}{H_o} = 0.88 \; \xi_o^{0.69}$$

(II-4-32)

for $1/30 \le \tan \beta \le 1/5$ and $H_o/L_o \ge 0.007$, where H_o is the significant deepwater wave height and ξ_o is calculated from the deepwater significant wave height and length. The appropriate slope for natural beaches is the slope of the beach face (Holman 1986, Mase 1989). Wave setup is included in Equations II-4-28 through II-4-32. The effects of tide and wind setup must be calculated independently. Walton (1992) extended Mase's (1989) analysis to predict runup statistics for any percent exceedence under the assumption that runup follows the Rayleigh probability distribution.

    (3) Field measurements of runup (Holman 1986, Nielsen and Hanslow 1991) are consistently lower than predictions by Equations II-4-28 through II-4-32. Equation II-4-29 overpredicts the best fit to $R_{2\%}$ by a factor of two for Holman's data (with the slope defined as the beach face slope), but is roughly an upper envelope of the data scatter. Differences between laboratory and field results (porosity, permeability, nonuniform slope, wave reformation across bar-trough bathymetry, wave directionality) have not been quantified. Mase (1989) found that wave groupiness (see Part II-1 for a discussion of wave groups) had little impact on runup for gentle slopes.

EXAMPLE PROBLEM II-4-3

FIND:
Maximum and significant runup.

GIVEN:
A plane beach having a 1 on 80 slope, and normally incident waves with deepwater height of 4.0 m and period of 9 sec.

SOLUTION:
Calculation of runup requires determining deepwater wavelength

$$L_o = g \ T^2/(2\pi) = 9.81 \ (9^2)/(2\pi) = 126 \text{ m}$$

and, from Equation II-4-1, the surf similarity parameter

$$\xi_o = \tan \beta \ (H_o/L_o)^{-1/2} = (1/80) \ (4.0/126.)^{-1/2} = 0.070$$

Maximum runup is calculated from Equation II-4-28

$$R_{max} = 2.32 \ H_o \ \xi_o^{0.77} = 2.32 \ (4.0)(0.070)^{0.77} = 1.2 \text{ m}$$

Significant runup is calculated from Equation II-4-31

$$R_{1/3} = 1.38 \ H_o \ \xi_o^{0.70} = 1.38 \ (4.0)(0.070)^{0.70} = 0.86 \text{ m}$$

Maximum runup is 1.2 m and significant runup is 0.86 m.

II-4-5. Infragravity Waves

a. Long wave motions with periods of 30 sec to several minutes often contribute a substantial portion of the surf zone energy. These motions are termed *infragravity waves*. Swash at wind wave frequencies (period of 1-20 sec) dominates on reflective beaches (steep beach slopes, typically with plunging or surging breakers), and infragravity frequency swash dominates on dissipative beaches (gentle beach slopes, typically with spilling breakers) (see Wright and Short (1984) for description of dissipative versus reflective beach types).

b. Infragravity waves fall into three categories: a) bounded long waves, b) edge waves, and c) leaky waves. Bounded long waves are generated by gradients in radiation stress found in wave groups, causing a lowering of the mean water level under high waves and a raising under low waves (Longuet-Higgins and Stewart 1962). The bounded wave travels at the group speed of the wind waves, hence is bound to the wave group. Edge waves are freely propagating long waves which reflect from the shoreline and are trapped along shore by refraction. Long waves may be progressive or stand along the shore. Edge waves travel alongshore with an antinode at the shoreline, and the amplitude decays exponentially offshore. Leaky waves are also freely propagating long waves or standing waves, but they reflect from the shoreline to deep water and are not trapped by the bathymetry. Proposed generation mechanisms for the freely propagating long waves include time-varying break point of groupy waves (Symonds, Huntley, and Bowen 1982), release of bounded waves through wave breaking (Longuet-Higgins and Stewart 1964), and nonlinear wave-wave interactions (Gallagher 1971).

c. Field studies have clearly identified bounded long waves and edge waves in the nearshore (see discussion by Oltman-Shay and Hathaway (1989)). The relative amount of infragravity energy and incident wind wave energy is a function of the surf similarity parameter (Holman and Sallenger 1985, Holman 1986), with infragravity energy dominating for low values of the surf similarity parameters ($\xi_o < 1.5$). For low values, the energy spectrum at incident frequencies is generally saturated (the spectral energy density is independent of the offshore wave height, due to wave breaking), but at infragravity frequencies, the energy density increases linearly with increasing offshore wave height (Guza and Thornton 1982, Mase 1988). Storm conditions with high steepness waves tend to have low-valued surf similarity parameters, so infragravity waves are prevalent in storms. Velocities and runup heights associated with infragravity waves have strong implications for nearshore sediment transport, beach morphology evolution, structural stability, harbor oscillation, and energy transmission through structures, as well as amplification or damping of infragravity waves by the local morphology or structure configuration. Presently, practical questions of how to predict infragravity waves and design for their effects have not been answered.

II-4-6. Nearshore Currents

a. Introduction.

(1) The current in the surf zone is composed of motions at many scales, forced by several processes. Schematically, the total current u can be expressed as a superposition of these interrelated components

$$u = u_w + u_t + u_a + u_o + u_i \tag{II-4-33}$$

where u_w is the steady current driven by breaking waves, u_t is the tidal current, u_a is the wind-driven current, and u_o and u_i are the oscillatory flows due to wind waves and infragravity waves. Figure II-4-13 shows longshore and cross-shore currents measured in the surf zone at the Field Research Facility in Duck, NC. The mean value of the current in the figure is the steady current driven by breaking waves and wind, the long period oscillation is due to infragravity waves, and the short-period oscillation is the wind-wave orbital motion.

(2) Currents generated by the breaking of obliquely incident wind waves generally dominate in and near the surf zone on open coasts. Strong local winds can also drive significant nearshore currents (Hubertz 1986). Wave- and wind-driven currents are important in the transport and dispersal of sediment and pollutants in the nearshore. These currents also transport sediments mobilized by waves. Tidal currents, which may dominate in bays, estuaries, and coastal inlets, are discussed in Parts II-5 and II-7.

(3) Figure II-4-14 shows typical nearshore current patterns: a) an alongshore system (occurring under oblique wave approach), b) a symmetric cellular system, with longshore currents contributing equally to seaward-flowing rip currents (occurring under shore-normal wave approach), and c) an asymmetric cellular system, with longshore currents contributing unequally to rip currents (Harris 1969). The beach topography is often molded by the current pattern, but the current pattern also responds to the topography.

(4) Nearshore currents are calculated from the equations of momentum (Equations II-4-34 and II-4-35) and continuity (Equation II-4-36):

$$U\frac{\partial U}{\partial x} + V\frac{\partial U}{\partial y} = -g\frac{\partial \overline{\eta}}{\partial x} + F_{bx} + L_x + R_{bx} + R_{sx} \tag{II-4-34}$$

$$U\frac{\partial V}{\partial x} + V\frac{\partial V}{\partial y} = -g\frac{\partial \overline{\eta}}{\partial y} + F_{by} + L_y + R_{by} + R_{sy} \tag{II-4-35}$$

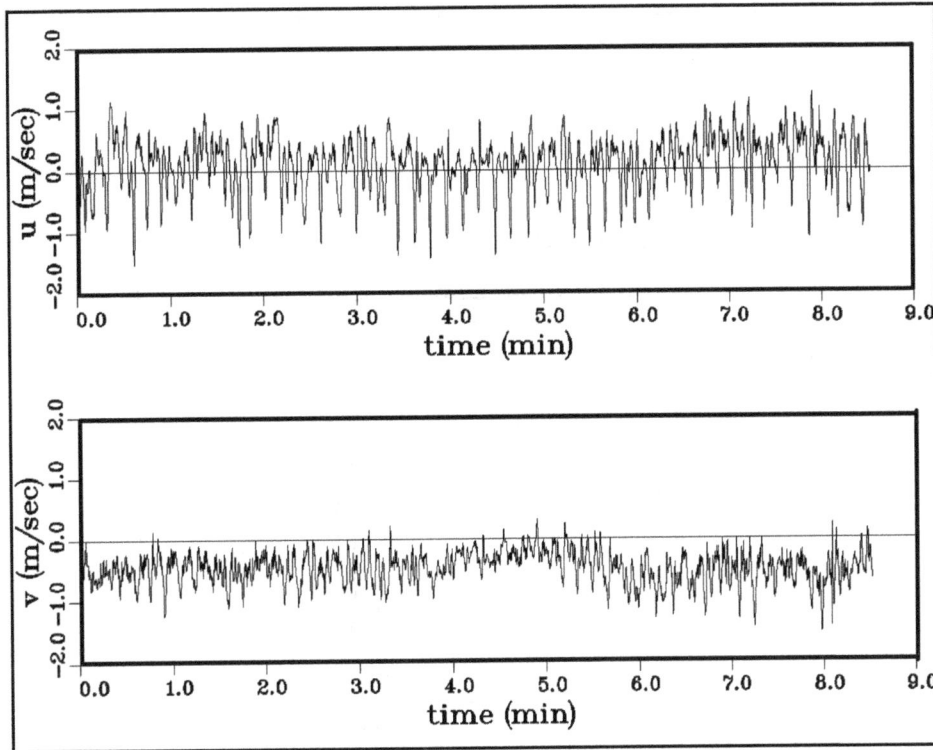

Figure II-4-13. Measured cross-shore and longshore flow velocities

$$\frac{\partial(Ud)}{\partial x} + \frac{\partial(Vd)}{\partial y} = 0 \tag{II-4-36}$$

where

U = time- and depth-averaged cross-shore current

V = time- and depth-averaged longshore current

F_{bx}, F_{by} = cross-shore and longshore components of bottom friction

L_x, L_y = cross-shore and longshore components of lateral mixing

R_{bx}, R_{by} = cross-shore and longshore components of wave forcing

R_{sx}, R_{sy} = cross-shore and longshore components of wind forcing

(5) These equations include wave and wind forcing, pressure gradients due to mean water level varia-tions, bottom friction due to waves and currents, and lateral mixing of the current. The primary driving force is the momentum flux of breaking waves (radiation stress), which induces currents in both the longshore and cross-shore directions. Radiation stress is proportional to wave height squared, so the forcing that generates currents is greatest in regions of steep wave height decay gradients. Bottom friction is the resisting force to the currents. Bottom roughness and wave and current velocities determine bottom friction. Lateral mixing is the exchange of momentum caused by turbulent eddies which tend to "spread out" the effect of wave forcing beyond the region of steep gradients in wave decay. Longshore, cross-shore, and rip current components of nearshore circulation are discussed in the following sections.

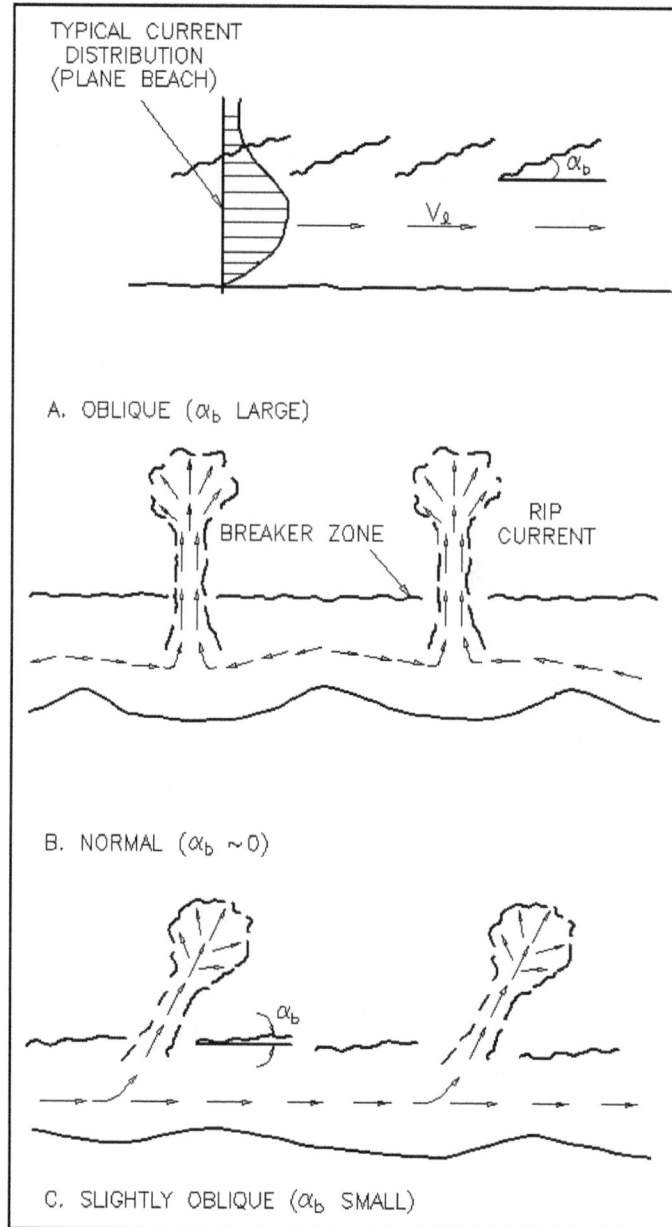

TYPICAL CURRENT
DISTRIBUTION
(PLANE BEACH)

V_ℓ

α_b

A. OBLIQUE (α_b LARGE)

BREAKER ZONE

RIP
CURRENT

B. NORMAL ($\alpha_b \sim 0$)

α_b

C. SLIGHTLY OBLIQUE (α_b SMALL)

Figure II-4-14.   Nearshore circulation systems

b.   Longshore current.

(1) Wave- and wind-induced longshore currents flow parallel to the shoreline and are strongest in the surf zone, decaying rapidly seaward of the breakers. These currents are generated by gradients in momentum flux (radiation stress) due to the decay of obliquely incident waves and the longshore component of the wind. Typically, longshore currents have mean values of 0.3 m/sec or less, but values exceeding 1 m/sec can occur in storms. The velocities are relatively constant over depth (Visser 1991).

(2) The concept of radiation stress was first applied to the generation of longshore currents by Bowen (1969a), Longuet-Higgins (1970a,b), and Thornton (1970). These studies were based on the assumptions of longshore homogeneity (Figure II-4-14a) and no wind forcing, reducing Equation II-4-35 to a balance

between the wave forcing, bottom friction, and lateral mixing. The wave driving force for the longshore current is the cross-shore gradient in the radiation stress component S_{xy},

$$R_{by} = -\frac{1}{\rho d}\frac{\partial S_{xy}}{\partial x}$$ (II-4-37)

where, using linear wave theory,

$$S_{xy} = \frac{n}{8}\rho g H^2 \cos\alpha \sin\alpha$$ (II-4-38)

where n is the ratio of wave group speed and phase speed. The variables determining wave-induced longshore current, as seen in the driving force given in Equations II-4-37 and II-4-38, are the angle between the wave crest and bottom contours, and wave height. Wave height affects not only longshore velocity, but also the total volume rate of flow by determining the width of the surf zone.

(3) A simple analytical solution for the wave-induced longshore current was given by Longuet-Higgins (1970a,b) under the assumptions of longshore homogeneity in bathymetry and wave height, linear wave theory, small breaking wave angle, uniformly sloping beach, no lateral mixing, and saturated wave breaking ($H = \gamma_b\, d$) through the surf zone. Under these assumptions, the longshore current in the surf zone is given by:

$$V = \frac{5\pi}{16}\frac{\tan\beta^*}{C_f}\gamma_b\sqrt{gd}\sin\alpha\cos\alpha$$ (II-4-39)

where

$\quad V$ = longshore current speed

$\tan\beta^*$ = beach slope modified for wave setup = $\tan\beta/(1+(3\gamma_b^2/8))$

$\quad C_f$ = bottom friction coefficient

$\quad \alpha$ = wave crest angle relative to the bottom contours

(4) The modified beach slope $\tan\beta^*$ accounts for the change in water depth produced by wave setup. The bottom friction coefficient C_f has typical values in the range 0.005 to 0.01, but is dependent on bottom roughness. This parameter is often used to calibrate the predictive equation, if measurements are available. The cross-shore distribution of the longshore current given by Equation II-4-39 is triangular in shape with a maximum at the breaker line and zero at the shoreline (Figure II-4-15) and seaward of the breaker line. Inclusion of lateral mixing smooths the current profile as shown by the dotted lines in Figure II-4-15. The parameter V_o in Figure II-4-15 is the maximum current for the case without lateral mixing, and it is used to nondimensionalize the longshore current.

(5) Komar and Inman (1970) obtained an expression for the longshore current at the mid-surf zone V_{mid} based on relationships for evaluating longshore sand transport rates which is given by Komar (1979):

$$V_{mid} = 1.17\sqrt{g\,H_{rms,b}}\sin\alpha_b\cos\alpha_b$$ (II-4-40)

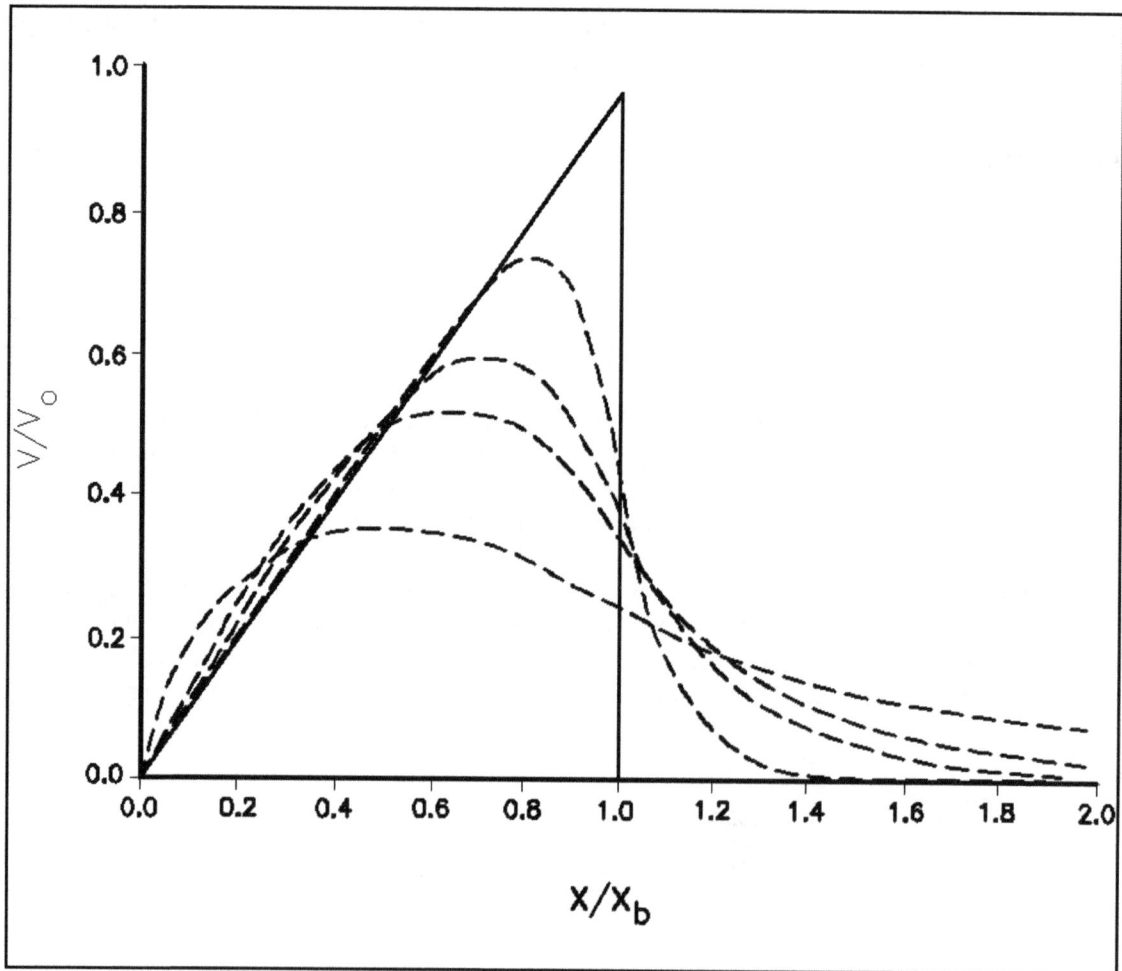

Figure II-4-15. Longshore current profiles (solid line - no lateral mixing; dashed lines - with lateral mixing)

(6) Equation II-4-40 shows good agreement with available longshore current data (Figure II-4-16). Although Equations II-4-39 and II-4-40 are similar in form, Equation II-4-4037 is independent of beach slope, which implies that $\tan \beta / C_f$ is constant in Equation II-4-39. The interdependence of $\tan \beta$ and C_f may result from the direct relationship of both parameters to grain size or an apparent dependence due to beach-slope effects on mixing (which is not included in Equation II-4-39) (Komar 1979, Huntley 1976, Komar and Oltman-Shay 1990).

(7) Longshore current, eliminating many simplifying assumptions used in Equation II-4-39, is solved numerically by the model NMLONG (Larson and Kraus 1991) for longshore-homogenous applications. NMLONG, which was briefly discussed for the simulation of breaking waves, calculates wave and wind-induced longshore current, wave and wind-induced setup, and wave height across the shore. Figure II-4-17 gives an example NMLONG calculation and comparison to field measurements of wave breaking and longshore current reported by Thornton and Guza (1986). The two-dimensional equations (Equations II-4-34 through II-4-36) are solved numerically by Noda (1974), Birkemeier and Dalrymple (1975), Ebersole and Dalrymple (1980), Vemulakonda (1984), and Wind and Vreugdenhil (1986).

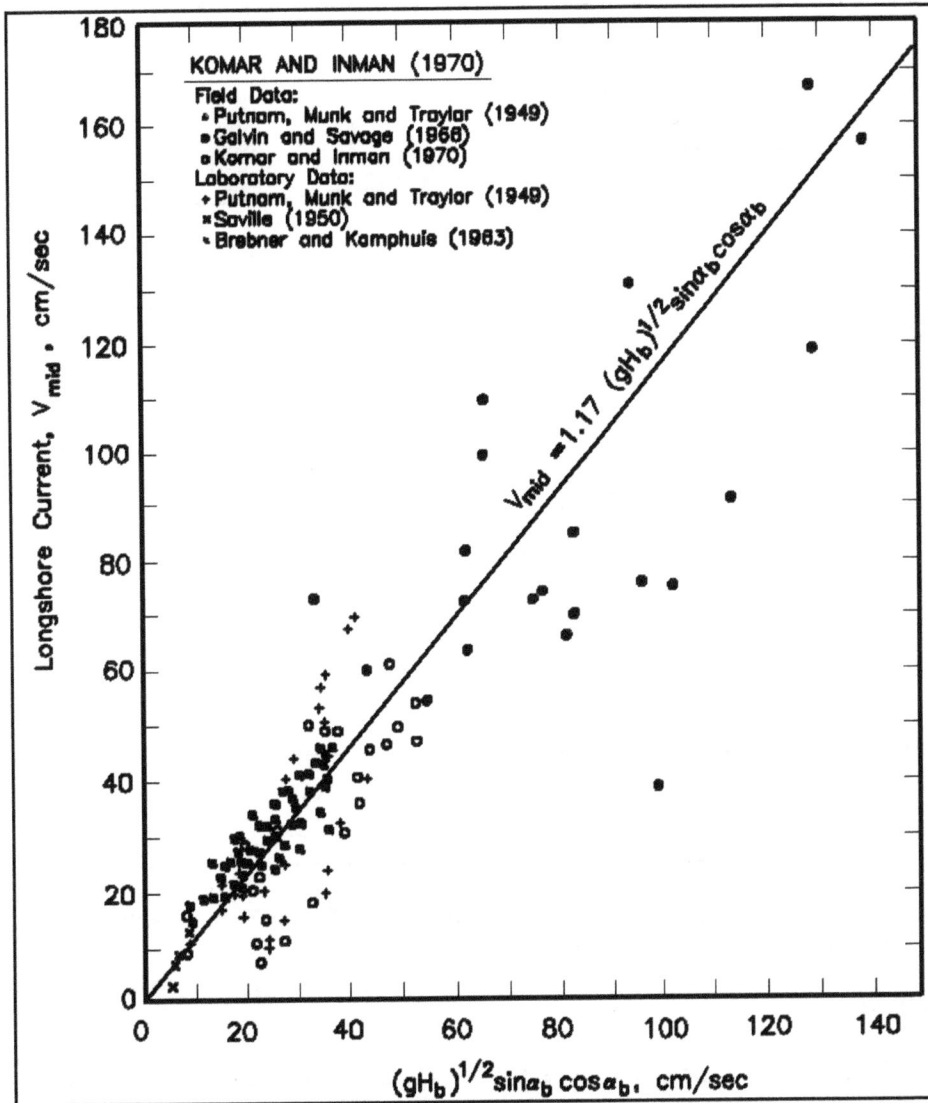

Figure II-4-16. Equation II-4-37 compared with field and laboratory data (Komar 1979)

  c. Cross-shore current. Unlike longshore currents, the cross-shore current is not constant over depth. The mass transport carried toward the beach due to waves (see Part II-1) is concentrated between the wave trough and crest elevations. Because there is no net mass flux through the beach, the wave-induced mass transport above the trough is largely balanced by a reverse flow or *undertow* below the trough. Figure II-4-18 shows field measurements of the cross-shore flow below trough level on a barred profile. The undertow current may be relatively strong, generally 8-10 percent of $\sqrt{gd}$ near the bottom. The vertical profile of the undertow is determined as a balance between radiation stresses, the pressure gradient from the sloping mean water surface, and vertical mixing. The first quantitative analysis of undertow was given by Dyhr-Nielsen and Sorensen (1970). The undertow profile is solved by Dally and Dean (1984), Hansen and Svendsen (1984), Stive and Wind (1986), and Svendsen, Schäffer, and Hansen (1987).

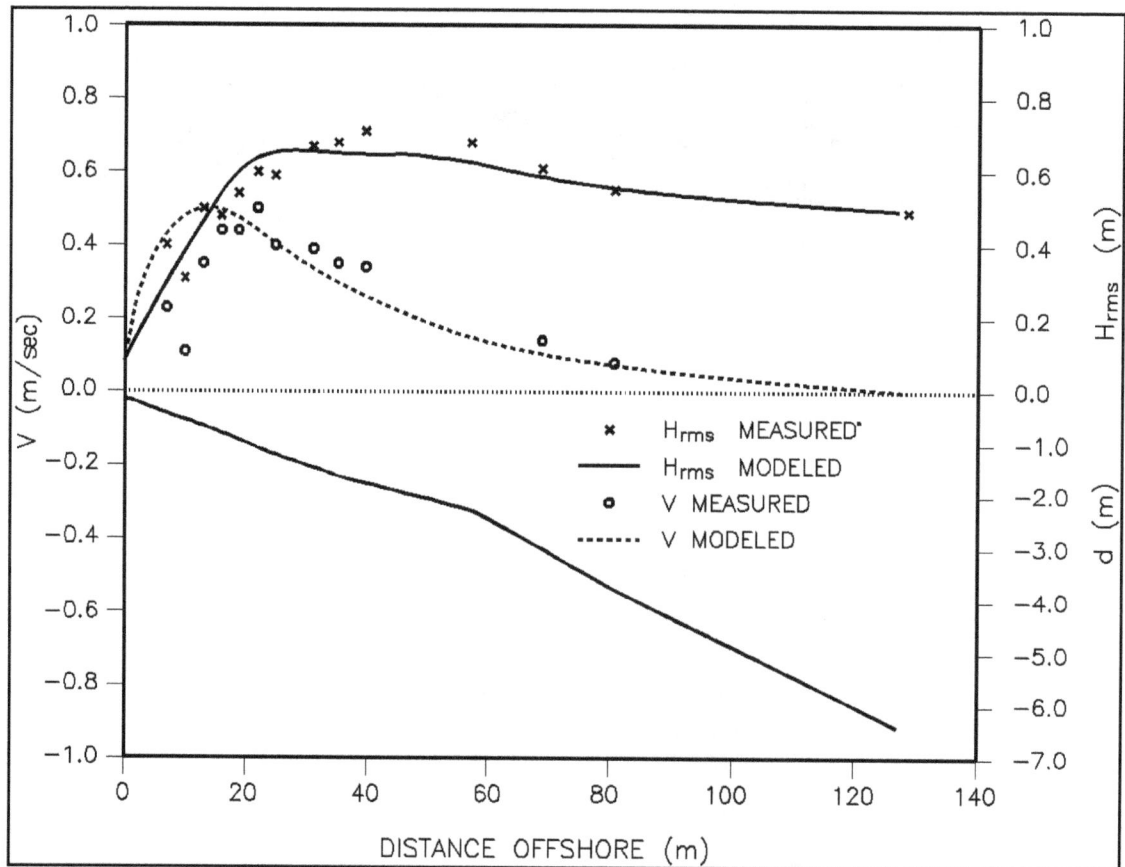

Figure II-4-17. NMLONG simulation of longshore current (Leadbetter Beach, Santa Barbara, California, 3 Feb 1989 (Thornton and Guza 1986))

d. Rip currents.

(1) The previous sections on longshore current, cross-shore current, and wave setup focused on processes that are two-dimensional, with waves, currents, and water levels changing only in the cross-shore and vertical directions, but homogeneous alongshore. *Rip currents*, strong, narrow currents that flow seaward from the surf zone, are features of highly three-dimensional current patterns. Rip currents are fed by longshore-directed surf zone currents, which increase from zero between two neighboring rips, to a maximum just before turning seaward to form a rip current. Rip currents often occur periodically along the beach, forming circulation cells (Figure II-4-14b,c). High offshore-directed flows in rip currents can cause scour of the bottom and be a hazard for swimmers.

(2) Rip currents and cell circulation can be generated by longshore variations in wave setup. Breaking wave height and wave setup are directly related; thus ,a longshore variation in wave height causes a longshore variation in setup. The longshore gradient in setup generates longshore flows from the position of highest waves and setup toward the position of the lowest waves and setup (Bowen 1969b). This effect can be seen in the term $\partial\overline{\eta}/\partial y$ in the longshore momentum equation (Equation II-4-35). The longshore variation in wave setup may be caused by convergence or divergence of waves transforming across bottom topography (Sonu 1972, Noda 1974) or the sheltering effect of headlands, jetties, or detached breakwaters (Gourley 1974, 1976; Sasaki 1975; and Mei and Liu 1977). Edge waves can interact with incident waves to produce a regular variation in the breaker height alongshore, and thus generate regularly spaced rip currents (Bowen 1969b, Bowen and Inman 1969). Interaction of two intersecting wave trains can similarly generate regularly spaced rip currents (Dalrymple 1975).

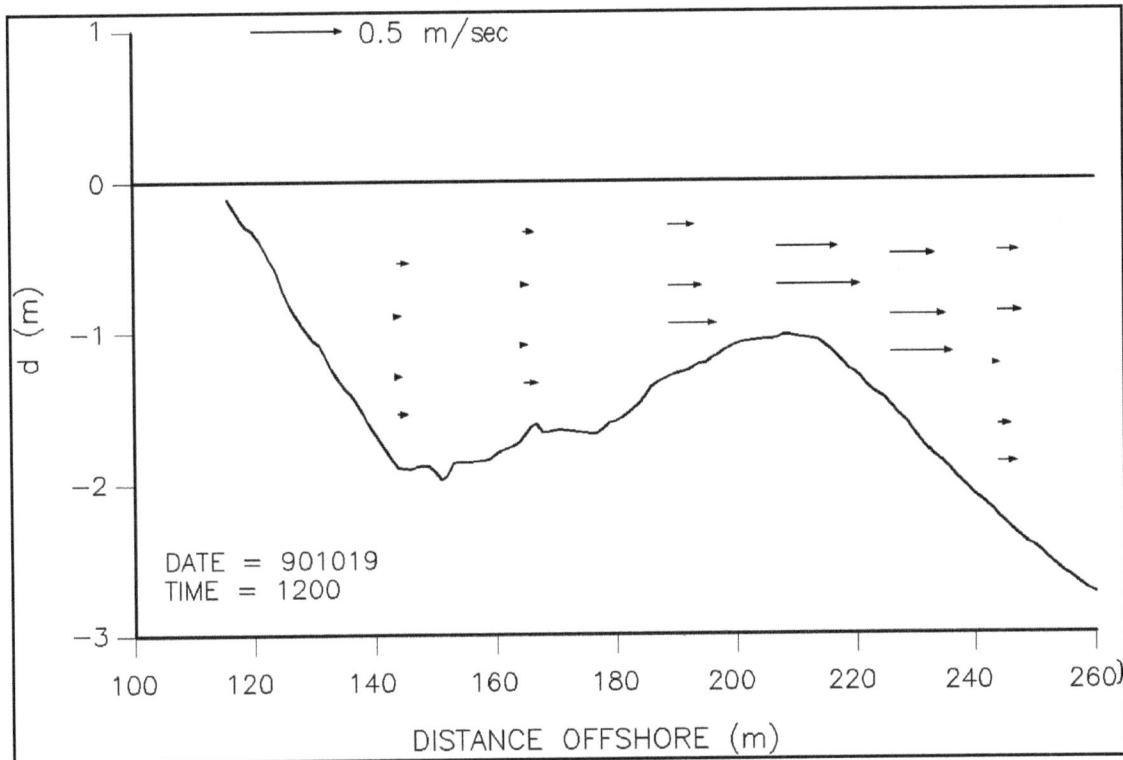

Figure II-4-18. **Field measurement of cross-shore flow on a barred profile (Duck, North Carolina, October 1990)**

(3) An alternate hypothesis for the generation of cell circulation is hydrodynamic instability (Hino 1974, LeBlond and Tang 1974, Miller and Barcilon 1978). Instability models are based on small, periodic perturbations in the setup and currents, with feedback between the currents and incident waves, to produce regular patterns of nearshore circulation.

(4) Several generation mechanisms for rip currents and cell circulation have been proposed. On a given beach, one or more of these mechanisms may drive the circulation pattern. Circulation patterns are dynamic, changing spatially and temporally. Presently, there is no proven method to predict rip current generation or the spacing between rips.

II-4-7. References

Baldock et al. 1998
Baldock, T. E., Holmes, P., Bunker, S., and Van Weert, P. 1998. "Cross-shore Hydrodynamics within an Unsaturated Surf Zone," *Coastal Engineering*, 34, pp 173-196.

Battjes 1972
Battjes, J. A. 1972. "Set-up Due to Irregular Waves," *Proceedings of the 13th Coastal Engineering Conference*, American Society of Civil Engineers, pp 1993-2004.

Battjes 1974
Battjes, J. A. 1974. "Surf Similarity," *Proceedings of the 14th Coastal Engineering Conference*, American Society of Civil Engineers, pp 466-480.

Battjes and Janssen 1978
Battjes, J. A., and Janssen, J. P. F. M. 1978. "Energy Loss and Setup due to Breaking of Random Waves," *Proceedings of the 16th Coastal Engineering Conference*, American Society of Civil Engineers, pp 569-587.

Birkemeier and Dalrymple 1975
Birkemeier, W. A., and Dalrymple, R. A. 1975. "Nearshore Water Circulation Induced by Wind and Waves," *Proceedings of Modeling '75 Conference*, American Society of Civil Engineers, pp 1062-1081.

Booij, Ris, and Holthuijsen 1999
Booij, N., Ris, R. C., and Holthuijsen, L. H. 1999. "A Third-Generation Wave Model for Coastal Regions 1. Model Description and Validation," *Journal of Geophysical Research*, 104(C4), pp 7649-7666.

Bowen 1969a
Bowen, A. J. 1969a. "The Generation of Longshore Currents on a Plane Beach," *Journal of Marine Research*, Vol 27, No. 1, pp 206-215.

Bowen 1969b
Bowen, A. J. 1969b. "Rip Currents, 1: Theoretical Investigations," *Journal of Geophysical Research*, Vol 74, No. 23, pp 5467-5478.

Bowen and Inman 1969
Bowen, A. J., and Inman, D. L. 1969. "Rip Currents, 2: Laboratory and Field Observation," *Journal of Geophysical Research*, Vol 74, No. 23, pp 5479-5490.

Camfield 1991
Camfield, F. E. 1991. "Wave Forces on Wall," *Journal of Waterway, Port, Coastal, and Ocean Engineering*, Vol 117, No. 1, pp 76-79.

Christensen and Deigaard 2001
Christensen, E. D., and Deigaard, R. 2001. "Large Eddy Simulation of Breaking Waves," *Coastal Engineering*, 42, pp 53-86.

Collins 1970
Collins, J. I. 1970. "Probabilities of Breaking Wave Characteristics," *Proceedings of the 12th Coastal Engineering Conference*, American Society of Civil Engineers, pp 1993-2004.

Dally 1990
Dally, W. R. 1990. "Random Breaking Waves: A Closed-Form Solution for Planar Beaches," *Coastal Engineering*, Vol 14, No. 3, pp 233-263.

Dally 1992
Dally, W. R. 1992. "Random Breaking Waves: Field Verification of a Wave-by-Wave Algorithm for Engineering Application," *Coastal Engineering*, Vol 16, pp 369-397.

Dally and Dean 1984
Dally, W. R., and Dean, R. G. 1984. "Suspended Sediment Transport and Beach Profile Evolution," *Journal of Waterway, Port, Coastal, and Ocean Engineering*, Vol 110, No. 1, pp 15-33.

Dally, Dean, and Dalrymple 1985
Dally, W. R., Dean, R. G., and Dalrymple, R. A. 1985. "Wave Height Variation Across Beaches of Arbitrary Profile," *Journal of Geophysical Research*, Vol 90, No. C6, pp 11917-11927.

Dalrymple 1975
Dalrymple, R. A. 1975. "A Mechanism for Rip Current Generation on an Open Coast," *Journal of Geophysical Research*, Vol 80, No. 24, pp 3485-3487.

Divoky, LeMéhauté, and Lin 1970
Divoky, D., LeMéhauté, B., and Lin, A. 1970. "Breaking Waves on Gentle Slopes," *Journal of Geophysical Research*, Vol 75, No. 9, pp 1681-1692.

Douglass 1990
Douglass, S. L. 1990. "Influence of Wind on Breaking Waves," *Journal of Waterway, Port, Coastal, and Ocean Engineering*, Vol 116, No. 6, pp 651-663.

Dyhr-Nielsen, and Sorensen 1970
Dyhr-Nielsen, M., and Sorensen, T. 1970. "Some Sand Transport Phenomena on Coasts with Bars," *Proceedings of the 12th Coastal Engineering Conference*, American Society of Civil Engineers, pp 1993-2004.

Ebersole 1987
Ebersole, B. A. 1987. "Measurement and Prediction of Wave Height Decay in the Surf Zone," *Proceedings of Coastal Hydrodynamics*, American Society of Civil Engineers, pp 1-16.

Ebersole and Dalrymple 1980
Ebersole, B. A., and Dalrymple, R. A. 1980. "Numerical Modelling of Nearshore Circulation," *Proceedings of the 17th Coastal Engineering Conference*, American Society of Civil Engineers, pp 2710-2725.

Freilich and Guza 1984
Freilich, M. H., and Guza, R. T. 1984. "Nonlinear Effects on Shoaling Surface Gravity Waves," *Philosophical Transactions*, Royal Society of London, A311, pp 1-41.

Freilich, Guza, and Elgar 1990
Freilich, M. H., Guza, R. T., and Elgar, S. L. 1990. "Observations of Nonlinear Effects in Directional Spectra of Shoaling Gravity Waves," *Journal of Geophysical Research*, Vol 95, No. C6, pp 9645-9656.

Gallagher 1971
Gallagher, B. 1971. "Generation of Surf Beat by Non-Linear Wave Interaction," *Journal of Fluid Mechanics*, Vol 49, Part 1, pp 1-20.

Galvin 1968
Galvin, C. J. 1968. "Breaker Type Classification on Three Laboratory Beaches," *Journal of Geophysical Research*, Vol 73, No. 12, pp 3651-3659.

Goda 1970
Goda, Y. 1970. "A Synthesis of Breaker Indices," *Transactions of the Japan Society of Civil Engineers*, Vol 2, Part 2, pp 227-230.

Goda 1975
Goda, Y. 1975. "Irregular Wave Deformation in the Surf Zone," *Coastal Engineering in Japan*, Vol 18, pp 13-26.

Goda 2002
Goda, Y. 2002. "A Fast Numerical Scheme for Unsaturated Random Breaking Waves in 3-D Bathymetry," *Proceedings*, 28th International Conference on Coastal Engineering. *World Scientific*, pp 508-520.

Gourley 1974
Gourley, M. R. 1974. "Wave Set-Up and Wave Generated Currents in the Lee of a Breakwater or Headland," *Proceedings of the 14th Coastal Engineering Conference*, American Society of Civil Engineers, pp 1976-1995.

Gourley 1976
Gourley, M. R. 1976. "Non-Uniform Alongshore Currents," *Proceedings of the 15th Coastal Engineering Conference*, American Society of Civil Engineers, pp 701-720.

Guza and Thornton 1982
Guza, R. T., and Thornton, E. G. 1982. "Swash Oscillations on a Natural Beach," *Journal of Geophysical Research*, Vol 87, No. C1, pp 483-491.

Hansen and Svendsen 1984
Hansen, J. B., and Svendsen, I. A. 1984. "A Theoretical and Experimental Study of Undertow," *Proceedings of the 19th Coastal Engineering Conference*, American Society of Civil Engineers, pp 2246-2262.

Hardy and Young 1991
Hardy, T. A., and Young, I. R. 1991. "Modelling Spectral Wave Transformation on a Coral Reef Flat," *Proceedings of the Australasian Conference on Coastal and Ocean Engineering*, pp 345-350.

Harris 1969
Harris, T. F. W. 1969. "Nearshore Circulations; Field Observation and Experimental Investigations of an Underlying Cause in Wave Tanks," *Proceedings of Symposium on Coastal Engineering*, Stellenbosch, South Africa, 11 pp.

Hino 1974
Hino, M. 1974. "Theory on Formation of Rip-Current and Cuspidal Coast," *Proceedings of the 14th Coastal Engineering Conference*, American Society of Civil Engineers, pp 901-919.

Holman 1986
Holman, R. A. 1986. "Extreme Value Statistics for Wave Run-up on a Natural Beach," *Coastal Engineering*, Vol 9, No. 6, pp 527-544.

Holman and Sallenger 1985
Holman, R. A., and Sallenger, A. H. 1985. "Setup and Swash on a Natural Beach," *Journal of Geophysical Research*, Vol 90, No. C1, pp 945-953.

Hubertz 1986
Hubertz, J. M. 1986. "Observations of Local Wind Effects on Longshore Currents," *Coastal Engineering*, Vol 10, No. 3, pp 275-288.

Hunt 1959
Hunt, I. A. 1959. "Design of Seawalls and Breakwaters," *Journal of the Waterways and Harbors Division*, Vol 85, No. WW3, pp 123-152.

Huntley 1976
Huntley, D. A. 1976. "Long-Period Waves on a Natural Beach," *Journal of Geophysical Research*, Vol 81, No. 36, pp 6441-6449.

Hwang and Divoky 1970
Hwang, L., and Divoky, D. 1970. "Breaking Wave Setup and Decay on Gentle Slope," *Proceedings of the 12th Coastal Engineering Conference*, American Society of Civil Engineers, pp 377-389.

Iversen 1952
Iversen, H. W. 1952. "Waves and Breakers in Shoaling Water," *Proceedings of the 3rd Coastal Engineering Conference*, American Society of Civil Engineers, pp 1-12.

Keller 1961
Keller, J. B. 1961. "Tsunamis -- Water Waves Produced by Earthquakes," *Proceedings of the Tsunami Meetings Associated with the 10th Pacific Science Congress*, International Union of Geodesy and Geophysics, pp 154-166.

Kennedy et al. 2000
Kennedy, A. B., Chen, Q., Kirby, J. T., and Dalrymple, R. A. 2000. "Boussinesq Modeling of Wave Transformation, Breaking and Runup. I: 1D," *Journal of Waterway, Port, Coastal and Ocean Engineering*, Vol 126, pp 39-47.

Kirby and Chen 2002
Kirby, J. T., and Chen, Q. 2002. "Examining the Low Frequency Predictions of a Boussinesq Wave Model," *Proceedings*, 28th International Conference on Coastal Engineering. *World Scientific,* pp 1075-1087.

Kobayashi and Wurjanto 1992
Kobayashi, N., and Wurjanto, A. 1992. "Irregular Wave Setup and Run-Up on Beaches," *Journal of Waterway, Port, Coastal, and Ocean Engineering*, Vol 118, No. 4, pp 368-386.

Komar 1979
Komar, P. D. 1979. "Beach-Slope Dependence of Longshore Currents," *Journal of Waterway, Port, Coastal, and Ocean Division*, Vol 105, No. WW4, pp 460-464.

Komar and Gaughan 1973
Komar, P. D., and Gaughan, M. K. 1973. "Airy Wave Theory and Breaker Height Prediction," *Proceedings of the 13th Coastal Engineering Conference*, American Society of Civil Engineers, pp 405-418.

Komar and Inman 1970
Komar, P. D., and Inman, D. L. 1970. "Longshore Sand Transport on Beaches," *Journal of Geophysical Research*, Vol 75, No. 30, pp 5914-5927.

Komar and Oltman-Shay 1990
Komar, P. D., and Oltman-Shay, J. 1990. "Nearshore Currents," *Handbook of Coastal and Ocean Engineering*, Vol. 2, J. Herbich, ed., Gulf Publishing Co., Houston, TX, pp 651-680.

Kuo and Kuo 1974

Kuo, C. T., and Kuo, S. T. 1974. "Effect of Wave Breaking on Statistical Distribution of Wave Heights," *Proceedings of Civil Engineering Oceans*, pp 1211-1231.

Larson and Kraus 1991

Larson, M., and Kraus, N. C. 1991. "Numerical Model of Longshore Current for Bar and Trough Beaches," *Journal of Waterway, Port, Coastal, and Ocean Engineering*, Vol 117, No. 4, pp 326-347.

Le Méhauté 1962

Le Méhauté, B. 1962. "On the Nonsaturated Breaker Theory and the Wave Run Up," *Proceedings of the 8th Coastal Engineering Conference*, American Society of Civil Engineers, pp 77-92.

LeBlond and Tang 1974

LeBlond, P. H., and Tang, C. L. 1974. "On Energy Coupling Between Waves and Rip Currents," *Journal of Geophysical Research*, Vol 79, No. 6, pp 811-816.

Lin and Liu 1998

Lin, P., and Liu, P. L.-F. 1998. "A Numerical Study of Breaking Waves in the Surf Zone," *Journal of Fluid Mechanics*, Vol 359, pp 239-264.

Longuet-Higgins 1970a

Longuet-Higgins, M. S. 1970a. "Longshore Currents Generated by Obliquely Incident Sea Waves, 1," *Journal of Geophysical Research*, Vol 75, No. 33, pp 6778-6789.

Longuet-Higgins 1970b

Longuet-Higgins, M. S. 1970b. "Longshore Currents Generated by Obliquely Incident Sea Waves, 2," *Journal of Geophysical Research*, Vol 75, No. 33, pp 6790-6801.

Longuet-Higgins and Stewart 1962

Longuet-Higgins, M. S., and Stewart, R. W. 1962. "Radiation Stress and Mass Transport in Gravity Waves, with Application to 'Surf Beats'," *Journal of Fluid Mechanics*, Vol 13, No. 4, pp 481-504.

Longuet-Higgins and Stewart 1963

Longuet-Higgins, M. S., and Stewart, R. W. 1963. "A Note on Wave Setup," *Journal of Marine Research*, Vol 21, No. 1, pp 4-10.

Longuet-Higgins and Stewart 1964

Longuet-Higgins, M. S., and Stewart, R. W. 1964. "Radiation Stresses in Water Waves: A Physical Discussion with Applications," *Deep Sea Research*, Vol 11, pp 529-562.

Madsen, Sørensen, and Schäffer 1997

Madsen, P. A., Sørensen, O. R., and Schäffer, H. A. 1997. "Surf Zone Dynamics Simulated by a Boussinesq Type Model; Part II: Surf Beat and Swash Oscillations for Wave Groups and Irregular Waves," *Coastal Engineering*, Vol 32, pp 289-319.

Mase 1988

Mase, H. 1988. "Spectral Characteristics of Random Wave Runup," *Coastal Engineering*, Vol 12, No. 2, pp 175-189.

Mase 1989
Mase, H. 1989. "Random Wave Runup Height on Gentle Slope," *Journal of Waterway, Port, Coastal, and Ocean Engineering*, Vol 115, No. 5, pp 649-661.

McCowan 1891
McCowan, J. 1891. "On the Solitary Wave," *Philosophical Magazine*, 5th Series, Vol 36, pp 430-437.

McDougal and Hudspeth 1983
McDougal, W. G., and Hudspeth, R. T. 1983. "Wave Setup/Setdown and Longshore Current on Non-Planar Beaches," *Coastal Engineering*, Vol 7, No. 2, pp 103-117.

Mei and Liu 1977
Mei, C. C., and Liu, P. L.-F. 1977. "Effects of Topography on the Circulation in and Near the Surf Zone - Linear Theory," *Journal of Estuary and Coastal Marine Science*, Vol 5, pp 25-37.

Miche 1951
Miche, M. 1951. "Le Pouvoir Réfléchissant des Ouvrages Maritimes Exposés à l'Action de la Houle," *Annals des Ponts et Chaussess*, 121e Annee, pp 285-319 (translated by Lincoln and Chevron, University of California, Berkeley, Wave Research Laboratory, Series 3, Issue 363, June 1954).

Miller and Barcilon 1978
Miller, C., and Barcilon, A. 1978. "Hydrodynamic Instability in the Surf Zone as a Mechanism for the Formation of Horizontal Gyres," *Journal of Geophysical Research*, Vol 83, No. C8, pp 4107-4116.

Munk 1949
Munk, W. H. 1949. "The Solitary Wave Theory and Its Applications to Surf Problems," *Annals of the New York Academy of Sciences*, Vol 51, pp 376-462.

Nadaoka, Hino, and Koyano 1989
Nadaoka, K., Hino, M., Koyano, Y. 1989. "Structure of the Turbulent Flow Field Under Breaking Waves in the Surf Zone," *Journal of Fluid Mechanics*, Vol 204, pp 359-387.

Nielsen and Hanslow 1991
Nielsen, P., and Hanslow, D. J. 1991. "Wave Runup Distributions on Natural Beaches," *Journal of Coastal Research*, Vol 7, No. 4, pp 1139-1152.

Noda 1974
Noda, E. K. 1974. "Wave-Induced Nearshore Circulation," *Journal of Geophysical Research*, Vol 79, No. 27, pp 4097-4106.

Nwogu 1996
Nwogu, O. G. 1996. "Numerical Prediction of Breaking Waves and Currents with a Boussinesq Model," *Proceedings*, 25th International Conference on Coastal Engineering, American Society of Civil Engineers, pp 4807-4820.

Nwogu and Demirbilek 2001
Nwogu, O.G., and Demirbilek, Z. 2001. "BOUSS-2D: A Boussinesq Wave Model for Coastal Regions and Harbors," ERDC/CHL TR-01-25. US Army Engineer Research and Development Center, Vicksburg, MS, 90 p.

Okayasu et al. 2002
Okayasu, A., Katayama, H., Tsuruga, H., and Iwasawa, H. 2002. "A Laboratory Experiment on Velocity Field Near Bottom Due to Obliquely Descending Eddies," *Proceedings*, International Conference on Coastal Engineering. *World Scientific,* pp 521-531.

Oltman-Shay and Hathaway 1989
Oltman-Shay, J., and Hathaway, K. K. 1989. "Infragravity Energy and its Implications in Nearshore Sediment Transport and Sandbar Dynamics," Technical Report CERC-89-8, U.S. Army Engineer Waterways Experiment Station, Coastal Engineering Research Center, Vicksburg, MS.

Raubenheimer et al. 1994
Raubenheimer, B., Guza, R.T., Elgar, S., and Kobayashi, N. 1994. "Swash on a Gently Sloping Beach," *Journal of Geophysical Research*, Vol 100, No. C5, pp 8751-8760.

Sasaki 1975
Sasaki, T. 1975. "Simulation on Shoreline and Nearshore Current," *Proceedings of Civil Engineering in the Oceans, III*, American Society of Civil Engineers, pp 179-196.

Savage 1958
Savage, R. P. 1958. "Wave Run-Up on Roughened and Permeable Slopes," *Journal of the Waterways and Harbors Division*, American Society of Civil Engineers, Vol 84, No. WW3, Paper 1640.

Saville 1956
Saville, T., Jr. 1956. "Wave Runup on Shore Structures," *Journal of the Waterways and Harbors Division*, American Society of Civil Engineers, Vol 82, No. WW2, Paper 925.

Schäffer, Madsen, and Deigaard 1993
Schäffer, H.A., Madsen, P.A., and Deigaard, R. 1993. "A Boussinesq Model for Waves Breaking in Shallow Water," *Coastal Engineering*, Vol 20, pp 185-202.

Singamsetti and Wind 1980
Singamsetti, S. R., and Wind, H. G. 1980. "Characteristics of Breaking and Shoaling Periodic Waves Normally Incident on to Plane Beaches of Constant Slope," Report M1371, Delft Hydraulic Laboratory, Delft, The Netherlands.

Smith and Kraus 1991
Smith, E. R., and Kraus, N. C. 1991. "Laboratory Study of Wave-Breaking Over Bars and Artificial Reefs," *Journal of Waterway, Port, Coastal, and Ocean Engineering*, Vol 117, No. 4, pp 307-325.

Smith, Resio, and Vincent 1997
Smith. J. M., Resio, D. T., and Vincent, C. L. 1997. "Current-Induced Breaking at an Idealized Inlet," *Proceedings*, Coastal Dynamics'97. American Society of Civil Engineers, pp 993-1002.

Smith 2001
Smith, J. M. 2001. "Breaking in a Spectral Wave Model," *Proceedings*, Waves 2001, American Society of Civil Engineers, pp 1022-1031.

Smith, Sherlock, and Resio 2001
Smith, J. M., Sherlock, A. R., and Resio, D. T. 2001. "STWAVE: Steady-State Spectral Wave Model User's Guide for STWAVE Version 3.0," ERDC/CHL SR-01-01, U.S. Army Engineer Research and Development Center, Vicksburg, MS, 80 pp.

Smith and Vincent 2002
Smith, J. M., and Vincent, C. L. 2002. "Application of Spectral Equilibrium Ranges in the Surf Zone," *Proceedings*, 28th International Conference on Coastal Engineering. *World Scientific*, pp 508-520.

Sonu 1972
Sonu, C. J. 1972. "Field Observation of Nearshore Circulation and Meandering Currents," *Journal of Geophysical Research*, Vol 77, No. 18, pp 3232-3247.

Stive and Wind 1986
Stive, M. J. F., and Wind, H. F. 1986. "Cross-shore Mean Flow in the Surf Zone," *Coastal Engineering*, Vol 10, No. 4, pp 325-340.

Sunamura 1980
Sunamura, T. 1980. "A Laboratory Study of Offshore Transport of Sediment and a Model for Eroding Beaches," *Proceedings of the 17th Coastal Engineering Conference*, American Society of Civil Engineers, pp 1051-1070.

Svendsen 1984
Svendsen, I. A. 1984. "Wave Heights and Setup in a Surf Zone," *Coastal Engineering*, Vol 8, No. 4, pp 303-329.

Svendsen, Madsen, and Hansen 1978
Svendsen, I. A., Madsen, P. A., and Hansen, J. B. 1978. "Wave Characteristics in the Surf Zone," *Proceedings of the 16th Coastal Engineering Conference*, American Society of Civil Engineers, pp 520-539.

Svendsen, Schäffer, and Hansen 1987
Svendsen, I. A., Schäffer, H. A., and Hansen, J. B. 1987. "The Interaction Between the Undertow and the Boundary Layer Flow on a Beach," *Journal of Geophysical Research*, Vol 92, No. C11, pp 11845-11856.

Symonds, Huntley, and Bowen 1982
Symonds, G., Huntley, D. A., and Bowen, A. J. 1982. "Two-Dimensional Surf Beat: Long Wave Generation by a Time-Varying Breakpoint," *Journal of Geophysical Research*, Vol 87, No. C1, pp 492-498.

Thornton 1970
Thornton, E. B. 1970. "Variation of Longshore Current Across the Surf Zone," *Proceedings of the 12th Coastal Engineering Conference*, American Society of Civil Engineers, pp 291-308.

Thornton and Guza 1983
Thornton, E. B., and Guza, R. T. 1983. "Transformation of Wave Height Distribution," *Journal of Geophysical Research*, Vol 88, No. C10, pp 5925-5938.

Thornton and Guza 1986
Thornton, E. B., and Guza, R. T. 1986. "Surf Zone Longshore Currents and Random Waves: Field Data and Models," *Journal of Physical Oceanography*, Vol 16, pp 1165-1178.

Vemulakonda 1984
Vemulakonda, S. R. 1984. "Erosion Control of Scour During Construction: Report 7, CURRENT -- a Wave-Induced Current Model," Technical Report HL-80-3, U.S. Army Engineer Waterways Experiment Station, Vicksburg, MS.

Wisser 1991
Visser, P. J. 1991. "Laboratory Measurements of Uniform Longshore Currents," *Coastal Engineering*, Vol 15, No. 5, pp 563-593.

Walton 1992
Walton, T. L., Jr. 1992. "Interim Guidance for Prediction of Wave Run-up on Beaches," *Ocean Engineering*, Vol 19, No. 2, pp 199-207.

Walton et al. 1989
Walton, T. L., Jr., Ahrens, J. P., Truitt, C. L., and Dean, R. G. 1989. "Criteria for Evaluating Coastal Flood-Protection Structures," Technical Report CERC-89-15, U.S. Army Engineer Waterways Experiment Station, Coastal Engineering Research Center, Vicksburg, MS.

Wang, Smith, and Ebersole 2002
Wang, P., Smith, E. R., and Ebersole. B. A. 2002. "Large-Scale Laboratory Measurements of Longshore Sediment Transport Under Spilling and Plunging Breakers," *Journal of Coastal Research*, Vol 18, No. 1, pp 118-135.

Watanabe and Saeki 1999
Watanabe, Y. and Saeki, H. 1999. "Three-Dimensional Large Eddy Simulation of Breaking Waves," *Coastal Engineering in Japan*, Vol 41, pp 281-301.

Weggel 1972
Weggel, J. R. 1972. "Maximum Breaker Height," *Journal of the Waterways, Harbors and Coastal Engineering Division*, Vol 98, No. WW4, pp 529-548.

Wind and Vreugdenhil 1986
Wind, H. G., and Vreugdenhil, C. B. 1986. "Rip-Current Generation Near Structures," *Journal of Fluid Mechanics*, Vol 171, pp 459-476.

Wright and Short 1984
Wright, L. D., and Short, A.D. 1984. "Morphodynamic Variability of Surf Zones and Beaches: A Synthesis," *Marine Geology*, Vol 56, pp 93-118.

Young 1989
Young, I. R. 1989. "Wave Transformation Over Coral Reefs," *Journal of Geophysical Research*, Vol 94, No. C7, pp 9779-9789.

II-4-8. Definitions of Symbols

α	Wave crest angle relative to bottom contours [deg]
β	Beach slope (tan β = length-rise/length-run)
β^*	Beach slope (tan β = length-rise/length-run) modified for wave setup
Γ	Empirical coefficient (= 0.4) (Equation II-4-14)
γ_b	Breaker depth index (Equation II-4-3) [dimensionless]
δ	Energy dissipation rate per unit surface area due to wave breaking
Δx	Shoreward displacement of the shoreline (Equation II-4-22) [length]
$\overline{\eta}$	Mean water surface elevation about the still-water level [length]
$\overline{\eta}_b$	Setdown at the breaker point [length]
$\overline{\eta}_{max}$	Setup at the mean shoreline (Equation II-4-22) [length]
$\overline{\eta}_s$	Setup at the still-water shoreline (Equation II-4-21) [length]
κ	Empirical decay coefficient (= 0.15) [dimensionless]
ξ	Surf similarity parameter (Equation II-4-1)
π	Constant (= 3.14159)
ρ	Mass density of water (salt water = 1,025 kg/m^3 or 2.0 slugs/ft^3; fresh water = 1,000kg/m^3 or 1.94 slugs/ft^3) [force-time^2/length^4]
Ω_b	Breaker height index (Equation II-4-4) [dimensionless]
a, b	Empirically determined dimensionless functions of beach slope (Equations II-4-6 and II-4-7)
C_f	Bottom friction coefficient with typical values in the range 0.005 to 0.01
C_g	Wave group velocity [length/time]
d	Water depth [length]
d_b	Water depth at breaking [length]
E	Wave energy per unit surface area [length-force/length^2]
F_{bx}, F_{by}	Cross-shore and longshore components of bottom friction [length/time^2]
f_m	Mean wave frequency (Equation II-4-17) [time^{-1}]
g	Gravitational acceleration [length/time^2]
h	Water depth [length]
H	Wave height [length]

$H_{1/10}$	Average of the highest 1/10 wave heights [length]
$H_{1/3}$	Significant wave height [length]
H_b	Wave height at incipient breaking [length]
$H_{m0,b}$	Zero-moment wave height at breaking (Equation II-4-10) [length]
H_{max}	Maximum wave height (Equation II-4-17) [length]
H_{rms}	Root-mean-square of all measured wave heights [length]
$H_{rms,b}$	Root-mean-square wave height at breaking (Equation II-4-9) [length]
H'_0	Equivalent unrefracted deepwater wave height [length]
K_r	Refraction coefficient [dimensionless]
L	Wave length [length]
L_x , L_y	Cross-shore and longshore components of lateral mixing [length/time^2]
n	Ratio of wave group speed and phase speed
$-_O$	The subscript 0 denotes deepwater conditions
Q_b	Percentage of waves breaking (Equation II-4-17)
R	Wave runup above the mean water level [length]
$\bar{R}$	Mean runup [length]
$R_{1/10}$	Average of the highest 1/10 of the runups [length]
$R_{1/3}$	Average of the highest 1/3 of the runups [length]
$R_{2\%}$	Runup exceeded by 2 percent of the runup crests [length]
R_{bx} , R_{by}	Cross-shore and longshore components of wave forcing [length/time^2]
R_{max}	Maximum wave runup [length]
R_{sx} , R_{sy}	Cross-shore and longshore components of wind forcing [length/time^2]
S_{xx}	Cross-shore component of the cross-shore directed radiation stress [force/length]
S_{xy}	Radiation stress component [force/length]
T	Wave period [time]
u	Total current in the surf zone (Equation II-4-30) [length/time]
U	Time- and depth-averaged cross-shore current [length/time]
u_a	Wind-driven current [length/time]
u_i	Oscillatory flow due to infragravity waves [length/time]
u_o	Oscillatory flow due to wind waves [length/time]

u_t Tidal current [length/time]

u_w Steady current driven by breaking waves [length/time]

V Longshore current speed (Equation II-4-36) [length/time]

V_0 Maximum current for the case without lateral mixing (Figure II-4-15) [length/time]

V_{mid} Longshore current at the mid-surf zone (Equation II-4-37) [length/time]

II-4-9. Acknowledgments

Author of Chapter II-4, "Surf Zone Hydrodynamics:"

Jane M. Smith, Ph.D., Coastal and Hydraulics Laboratory, Engineer Research and Development Center (CHL), Vicksburg, Mississippi.

Reviewers:

William R. Dally, Ph.D., Florida Institute of Technology, Melbourne, Florida, (currently private consultant).

Zeki Demirbilek, Ph.D., CHL.

Lee E. Harris, Ph.D., Department of Marine and Environmental Systems, Florida Institute of Technology, Melbourne, Florida.

Todd K. Holland, Ph.D., Naval Research Laboratory, Stennis Space Center, Mississippi.

Robert A. Holman, Ph.D., College of Oceanic & Atmospheric Sciences, Oregon State University, Corvallis, Oregon.

Bradley D. Johnson, CHL

Nicholas C. Kraus, Ph.D., CHL

Okey G. Nwogu, University of Michigan

Joan Pope, CHL.

Edward F. Thompson, Ph.D., CHL

Greg L. Williams, Ph.D., U.S. Army Engineer District, Wilmington, North Carolina.

C. Linwood Vincent, Ph.D., Office of Naval Research, Arlington, Virginia.

Table of Contents

Page

II-5-1. Introduction II-5-1
   a. Purpose II-5-1
   b. Applicability II-5-1
   c. Scope of manual II-5-1

II-5-2. Classification of Water Waves II-5-2
   a. Wave classification II-5-2
   b. Discussion II-5-5

II-5-3. Astronomical Tides II-5-5
   a. Description of tides II-5-5
      (1) Introduction II-5-5
      (2) Tide-producing forces II-5-6
      (3) Spring/neap cycle II-5-8
      (4) Diurnal inequality II-5-9
   b. Tidal time series analysis II-5-11
      (1) Introduction II-5-11
      (2) Harmonic constituents II-5-11
      (3) Harmonic reconstruction II-5-18
      (4) Tidal envelope classification II-5-23
   c. Glossary of tide elevation terms II-5-25

II-5-4. Water Surface Elevation Datums II-5-27
   a. Introduction II-5-27
   b. Tidal-observation-based datums II-5-27
   c. 1929 NGVD datum II-5-28
   d. Great Lakes datums II-5-32
   e. Long-term variations in datums II-5-32
   f. Tidal datums II-5-32
   g. Changes in lake level datums II-5-37
   h. Design considerations II-5-38

II-5-5. Storm Surge II-5-38
   a. Storm types II-5-38
      (1) Tropical storms II-5-39
      (2) Extratropical storms II-5-42
      (3) Surge interaction with tidal elevations II-5-43
   b. Storm event frequency-of-occurrence relationships II-5-43
      (1) Introduction II-5-43
      (2) Historical method II-5-46
      (3) Synthetic method II-5-47
      (4) Empirical simulation technique II-5-47

(a) Introduction II-5-47

(b) EST - tropical storm application II-5-47

(c) EST - extratropical storm application II-5-49

II-5-6. Seiches II-5-51

II-5-7. Numerical Modeling of Long-Wave Hydrodynamics II-5-54

a. Long-wave modeling II-5-54

b. Physical models II-5-54

c. Numerical models II-5-55

(1) Introduction II-5-55

(2) Example - tidal circulation modeling II-5-56

(3) Example - storm surge modeling II-5-57

II-5-8. References II-5-64

II-5-9. Definitions of Symbols II-5-69

II-5-10. Acknowledgments II-5-71

List of Tables

Page

Table II-5-1. Wave Classification (Ippen 1966) II-5-3

Table II-5-2. Hyperbolic Function Asymptotes II-5-4

Table II-5-3. NOS Tidal Constituents and Arguments II-5-16

Table II-5-4. Node Factors for 1970 through 1999 (Schureman 1924) II-5-17

Table II-5-5. Equilibrium Argument for Beginning of Years 2001 through 2010
(Schureman 1924) II-5-19

Table II-5-6. Summary of Harmonic Arguments for Sandy Hook, NJ (1 January 1984
at 0000 hr) II-5-23

Table II-5-7. Datums for Reference Tide Stations (Harris 1981) II-5-29

Table II-5-8. Low Water (chart) Datum for IGLD 1955 and IGLD 1985 . . . . II-5-35

List of Figures

Page

Figure II-5-1.   Long wave geometry (Milne-Thompson 1960) II-5-3

Figure II-5-2.   Variation of particle velocity with depth (Ippen 1966) II-5-4

Figure II-5-3.   Schematic representation of water particle trajectories (Ippen 1966) II-5-5

Figure II-5-4.   Schematic diagram of tidal potential (Dronkers 1964) II-5-7

Figure II-5-5.   Tide predictions for Boston, MA (Harris 1981) II-5-9

Figure II-5-6.   Spring and neap tides (Shalowitz 1964) II-5-10

Figure II-5-7.   The daily inequality (Dronkers 1964) II-5-11

Figure II-5-8.   Typical tide curves along the Atlantic and Gulf coasts (*Shore Protection Manual* 1984) II-5-12

Figure II-5-9.   Typical tide curves along Pacific coasts of the United States (*Shore Protection Manual* 1984) II-5-13

Figure II-5-10.   Tidal phase relationships II-5-18

Figure II-5-11.   Phase angle argument relationship II-5-20

Figure II-5-12.   NOS harmonic analysis for Sandy Hook, NJ II-5-21

Figure II-5-13.   Tide tables for Sandy Hook, NJ (NOAA 1984) II-5-22

Figure II-5-14.   Reconstructed tidal envelope for Sandy Hook, NJ II-5-24

Figure II-5-15.   Areal extent of tidal types (Harris 1981) II-5-25

Figure II-5-16.   Types of tides (*Shore Protection Manual* 1984) II-5-26

Figure II-5-17.   Reference and comparative tide stations, Atlantic, Gulf, and Pacific coasts (Harris 1981) II-5-31

Figure II-5-18.   Locations of tide stations used in establishing the National Geodetic Vertical Datum (NGVD) of 1929 (Harris 1981 (after Rappleye (1932)) II-5-33

Figure II-5-19.   Sample NOS description of tidal bench marks (Harris 1981) II-5-34

Figure II-5-20.   Vertical and horizontal relationships for the IGLD 1985 . . . . II-5-35

Figure II-5-21.   Sample NOS tabulation of tide parameters (Harris 1981) II-5-39

Figure II-5-22. Variations in annual MSL (Harris 1981) II-5-40

Figure II-5-23. Schematic diagram of storm parameters (U.S. Army Corps
of Engineers 1986) II-5-41

Figure II-5-24. Hurricane Gloria track from 17 September to 2 October 1985 (Jarvinen
and Gebert 1986) II-5-44

Figure II-5-25. Hurricane Gloria track offshore of Delaware and New Jersey (Jarvinen
and Gebert 1986) II-5-45

Figure II-5-26. Example phasing of storm surge and tide (Jarvinen and Gebert 1986) II-5-46

Figure II-5-27. Stage-frequency relationship - coast of Delaware II-5-50

Figure II-5-28. Sediment transport magnitude-frequency relationship - December 1992
Northeaster II-5-51

Figure II-5-29. Atlantic tropical storm tracks during the period 1886-1980 . . . . II-5-52

Figure II-5-30. Long wave surface profiles (*Shore Protection Manual* 1984) II-5-53

Figure II-5-31. First, second, and third normal modes of oscillation for Lake Ontario
(Rao and Schwab 1976) II-5-54

Figure II-5-32. Computational grid for the New York Bight II-5-58

Figure II-5-33. Model and prototype tidal elevation comparison at the Battery II-5-59

Figure II-5-34. Wind-induced circulation pattern II-5-60

Figure II-5-35. Global limits of ADCIRC computational grid II-5-61

Figure II-5-36. Blow-up of ADCIRC grid along Delaware coast II-5-62

Figure II-5-37. Model-to-prototype tidal comparison at Lewes, DE II-5-63

Figure II-5-38. Model-to-prototype surge comparison at Lewes, DE II-5-64

Chapter II-5
Water Levels and Long Waves

II-5-1. Introduction

a. Purpose.

(1) This chapter describes water levels and the various long wave components that contribute to a total water surface elevation. Vertical datums are also described to define some of the more commonly used reference datums.

(2) The following sections provide project engineers with sufficient guidance to develop a preliminary study approach and design procedure to analyze engineering projects that require consideration of water level elevations. References are provided from existing Engineer Manuals that describe generic design-criteria formulae for use in preliminary analyses. Additional references and approaches to problem solving are provided for complex projects that require detailed surface elevation and current input data for design. These data are generally provided by numerical models.

b. Applicability. Information contained in this chapter is directly applicable to any project requiring local water levels or currents as a primary design consideration. Applications include the design of coastal structures intended to provide protection against some pre-defined water surface elevation, specified according to an appropriate economic analysis and evaluation. Determining acceptable design elevations may require developing local stage relationships as opposed to frequency-of-occurrence relationships. This information can be generated through historical records or numerical modeling techniques to simulate the propagation of historical storm events. Additional examples of water surface and current variability include circumstances where tidal circulation patterns and surface elevations change as a result of structural or bathymetric modifications to existing coastal inlets or navigable waterways. These circulation-dominated problems can be addressed using either numerical or physical models.

c. Scope of manual.

(1) Water wave classification is used to describe the behavior of long waves and to distinguish between intermediate waves and short waves (described in Part II-2). This allows the reader to select which chapter of this manual is appropriate for the intended application. If long waves are appropriate, this chapter will provide a means of approximating basic wave characteristics such as celerity, current magnitudes, and surface elevation.

(2) The speed of propagation, surface profile, and vertical velocity distribution of long waves are different from those of short waves described in Part II-2. Because these properties of waves represent important design criteria, it is important to make a distinction between long and short waves. Therefore, Part II-5-2 reviews wave classification criteria and summarizes long wave properties.

(3) Tides are the most common and visible example of long wave propagation. Part II-5-3 summarizes tidal hydrodynamics and describes characteristic tidally induced long wave variability. This section includes a background description of the forces responsible for generating tides, gives examples of the variability of tides, and presents a methodology for harmonic reconstruction of tides.

(4) Many of the concepts described by tidal records are used as a basis for defining tidal datums. Part II-5-4 describes reference elevation datums commonly in use in the United States. Attention is also paid

to the change in coastal datums that may result from sea (or lake) level rise and/or land subsidence or rebound.

(5) Parts II-5-5 through II-5-7 describe nontidal variability in water surfaces. These fluctuations can be storm-generated, as in the case of tropical and extratropical storms; atmospheric- and geometry-related, as in the case of seiches or tidal bores; or be due to responses stemming from earthquake-generated tsunamis or other rapid changes in the environment.

(6) The primary goal of this chapter is to define tidal and storm-generated fluctuations in the water surface and describe the datums to which they are referenced. Seiches will only be briefly discussed and tsunamis are not addressed because a special report on tsumanis has been prepared by the Coastal and Hydraulics Laboratory (CHL) (Camfield 1980). In addition to Camfield, Engineer Manual 1110-2-1414, "Water Levels and Wave Heights for Coastal Engineering Design," addresses the propagation of tsunamis. However, because both seiches and tsunamis are classified as long waves, the numerical modeling techniques discussed in Part II-5-7 are an appropriate means of analysis.

II-5-2. Classification of Water Waves

a. Wave classification.

(1) The long wave descriptions that follow are based on small-amplitude wave theory solutions to the governing equations. This theory places certain criteria on the physical shape of the wave. For example, from Figure II-5-1, the amplitude is assumed small with respect to the depth (i.e., η/h ratio is small, and the surface slope $d\eta/dx$ is assumed small).

(2) Although wave amplitude is assumed small with respect to depth, the manner in which the wave propagates is a function of just how small this ratio is. The propagation of small-amplitude waves in water can now be described as a function of the wave length and the depth of water in which the wave is propagating. In fact, waves can be classified according to a parameter referred to as the "relative depth," defined as the ratio of water depth h to wave length L. When this ratio is less than approximately 1/20, waves can be classified as long waves or "shallow-water waves." Figure II-5-1 shows typical long wave geometry for a wave whose length L is large with respect to the depth of water h.

(3) Astronomical tides represent one important example of long waves. In Chesapeake Bay, for example, the M_2 primary lunar tidal constituent is contained completely within the Bay at a given instant in time, producing a wavelength of approximately 300 km. The mean depth of flow in the Bay is approximately 10 m; therefore, the relative depth is 3.3×10^{-5}. Long waves are not limited to what is normally considered shallow water because the relative depth is a function of wavelength. In fact, most tides are long waves over the entire ocean because their wavelengths are on the order of 1,000 km and depths are on the order of kilometers. Similarly, seismic-forced phenomena such as tsunamis propagate across the Pacific Ocean in depths of up to 20 km but have wavelengths on the order of hundreds of kilometers.

(4) Waves are classified as short waves, also referred to as "deepwater waves," when the relative depth is greater than approximately 1/2. Coastal waves described in Part II-2 are generally of this class. The geometry of short waves implies wave steepness great enough to cause them to break. The class of waves between short (deep) and long (shallow) are referred to as "intermediate waves." Table II-5-1 (Ippen

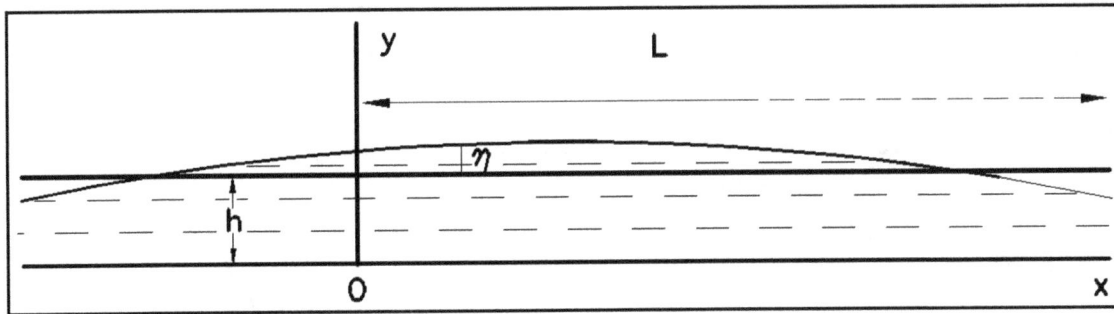

Figure II-5-1.   Long wave geometry (Milne-Thompson 1960)

Table II-5-1
Wave Classification (Ippen 1966)

Range of h/L	Range of kh=2πh/L	Types of waves
0 to 1/20	0 to π/10	Long waves (shallow-water wave)
1/20 to 1/2	π/10 to π	Intermediate waves
1/2 to ∞	π to ∞	Short waves (deepwater waves)

1966) summarizes wave classification criteria according to relative depth and the wave parameter kh defined below.

(5) Applying the relative depth and wave number parameter to the characteristics of long waves can be seen in the simplification to progressive small-amplitude wave theory solutions. For example, from Part II-1, the wave celerity, wave length, horizontal (x-direction) and vertical velocities can be written as

$$C^2 = \frac{g}{k} \tanh(kh) \qquad \text{(II-5-1)}$$

$$L = \frac{gT^2}{2\pi} \tanh(kh) \qquad \text{(II-5-2)}$$

$$u = \frac{agk}{\sigma} \frac{\cosh k(h+z)}{\cosh kh} \sin(kx - \sigma t) \qquad \text{(II-5-3)}$$

$$w = -\frac{agk}{\sigma} \frac{\sinh k(h+z)}{\cosh kh} \cos(kx - \sigma t) \qquad \text{(II-5-4)}$$

where k is the wave number ($2\pi/L$), σ is the angular frequency ($2\pi/T$ where T is the period of the wave), a is the amplitude of the wave, g the acceleration of gravity, h is the total depth, and z is the depth measured downward from the quiescent fluid surface. A schematic diagram of the variation of velocity as a function of depth is shown in Figure II-5-2.

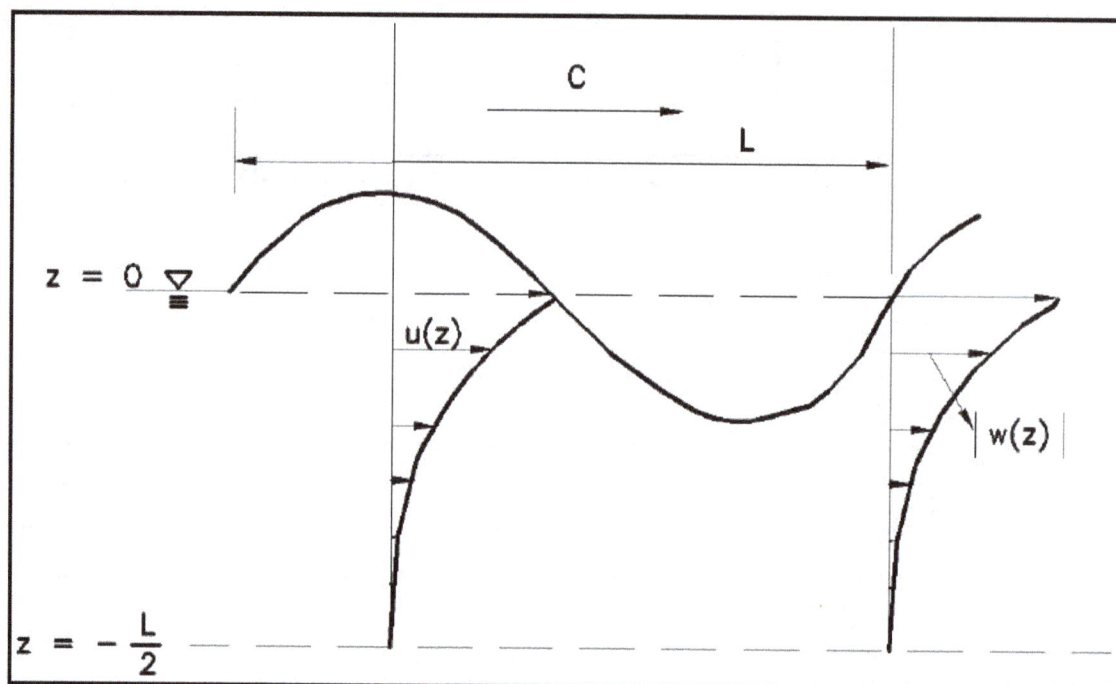

Figure II-5-2. Variation of particle velocity with depth (Ippen 1966)

(6) Certain simplifications to Equations II-5-1 through II-5-4 result from the asymptotic values of the hyperbolic functions. For example, Table II-5-2 (Ippen 1966) presents the hyperbolic functions contained in the long- and short-wave representations, as well as their asymptotes. No simplification results for intermediate waves.

Table II-5-2
Hyperbolic Function Asymptotes

Function	Asymptotes	
	Long waves	Short waves
$\sinh kh$	kh	$e^{kh}/2$
$\cosh kh$	1	$e^{kh}/2$
$\tanh kh$	kh	1

(7) The resulting long wave simplification for celerities and wave lengths is shown below.

$$C = \sqrt{gh} \tag{II-5-5}$$

$$L = T\sqrt{gh} \tag{II-5-6}$$

(8) Therefore, long waves propagate as the square root of gh. This relationship will be shown to be useful in analyzing and interpreting long-wave phase propagation data, because wave celerity is predictable for a given depth.

(9) Additionally, one important difference between long waves and short waves can be seen in the computed orbital velocities. Figure II-5-3 shows water particle trajectories for long, short, and intermediate waves as a function of depth. As can be seen, and computed from Equations II-5-3 and II-5-4, the horizontal velocity of a long wave is maintained throughout the water column, from the surface to the bottom. In the case of short waves, the strength of the horizontal and vertical component decreases with depth to the point that waves do not induce bottom currents. The fact that long waves affect the bottom is important in that bottom sediments can be eroded and transported by tidal and other long-wave currents. For example, tidal flood and ebb currents contribute to the transport of sediments to form ebb and flood shoals. Potential erosion and deposition considerations will be discussed in Part 3 of this manual.

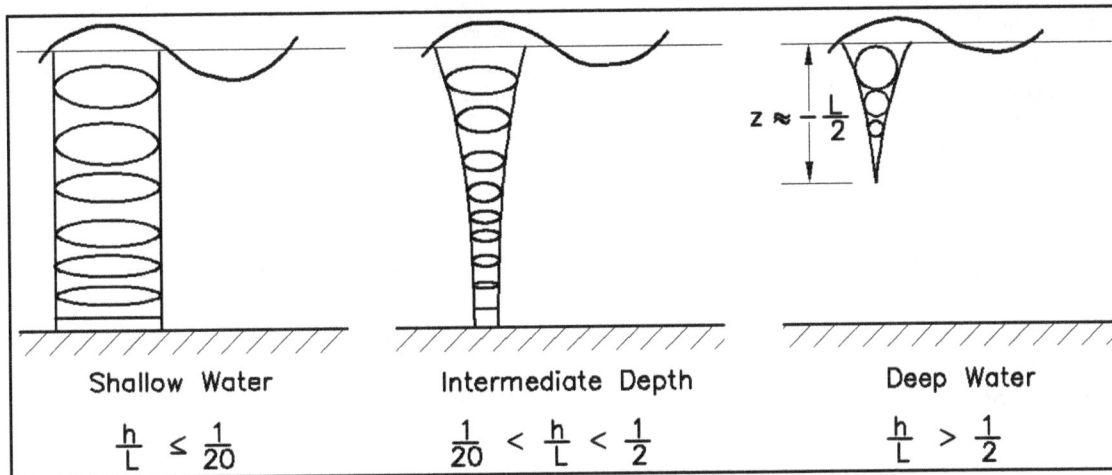

Figure II-5-3.   Schematic representation of water particle trajectories (Ippen 1966)

b. *Discussion.* Short waves impact nearshore water surface elevations by creating the wave setup condition. Details of wave setup and associated setdown are described in Part II-4. Including setup can be critical in developing water surface design criteria, as setup can create an elevated surface on which tide and storm surge propagate. Because the interactions of these components of the water surface are not linear, and because they are of different time scales (Part II-1), they are generally considered separately in the development of total water level design criteria. Methods for computing and combining these effects are given in Part II-3 and 4. The following sections concentrate on tidal and storm surge elevations.

II-5-3. Astronomical Tides

a. *Description of tides.*

(1) Introduction.

(a) Astronomical tides are observable as the periodic rising and falling of the surface of major water bodies on the earth. Tides are produced in response to the gravitational attraction of the moon, sun, and (to a considerably smaller extent) all other celestial bodies. Because of its relative closeness to the earth, the moon induces the strongest effect on the tides. Tidal currents are produced in response to differences in the water surface elevation.

(b) Tidal height, or the vertical distance between the maximum and successive minimum water surface elevation, is a function of the relative position of the moon and sun with respect to the earth, and varies from location to location. Typically, the dominant tidal cycle is related to the passage of the moon over a fixed meridian. This occurs an average of 50 min later each succeeding day. This passage of the moon produces approximately two tides per solar day (referred to as semidiurnal), with a maximum tide occurring approximately every 12 hr 25 min. However, differences in the relationship of the moon and sun in conjunction with local conditions can result in tides that exhibit only one tidal cycle per day. These are referred to as diurnal tides. Mixed tides exhibit characteristics of both semidiurnal and diurnal tides. At certain times in the lunar month, two peaks per day are produced, while at other times the tide is diurnal. The distinction is explained in the following paragraphs.

(c) The description of typical tidal variability begins with a brief background description of tide-producing forces, those gravitational forces responsible for tidal motion, and the descriptive tidal envelope that results from those forces. This sub-section will be followed by more qualitative descriptions of how the tidal envelope is influenced by the position of the moon and sun. Once this basic pattern is established, measured tidal elevations can, in part, be shown to be a function of the influence of the continental shelf and the coastal boundary on the propagating tide.

(2) Tide-producing forces.

(a) The law of universal gravitation was first published by Newton in 1686. Newton's law of gravitation states that every particle of matter in the universe attracts every other particle with a force that is directly proportional to the product of the masses of the particles and inversely proportional to the square of the distance between them (Sears and Zemansky 1963). Quantitative aspects of the law of gravitational attraction between two bodies of mass m_1 and m_2 can be written as follows:

$$F_g = f \frac{m_1 \, m_2}{r^2}$$

(II-5-7)

where F_g is the gravitational force on either particle, r is separation of distance between the centers of mass of the two bodies, and f is the universal constant with a value of 6.67×10^{-8} cm^3/gm sec^2. The gravitational force of the earth on particle m_1 can be determined from Equation II-5-7. Let $F_g = m_1 \, g$ where g is the acceleration of gravity (980.6 cm/sec^2) on the surface of the earth, and m_2 equal the mass of the earth E. By substitution, an expression for the gravitational constant can be written in terms of the radius of the earth a and the acceleration of gravity g.

$$f = g \frac{a^2}{E}$$

(II-5-8)

(b) Development of the tidal potential follows directly from the above relationship. The following variables are referenced to Figure II-5-4 (although Figure II-5-4 refers to the moon, an analogous figure can be drawn for the sun). Let M and S be the mass of the moon and sun, respectively. r_m and r_s are the distances from the center of the earth O to the center of the moon and sun. Let r_{mx} and r_{sx} be the distances of a point $X(x,y,z)$ located on the surface of the earth to the center of the moon and sun. The following relationships define the tidal potential at some arbitrary point X as a function of the relative position of the moon and sun.

(c) The attractive force potentials per unit mass for the moon and sun can be written as

$$V_M = \frac{fM}{r_{MX}} \qquad , \qquad V_S = \frac{fS}{r_{SX}}$$

(II-5-9)

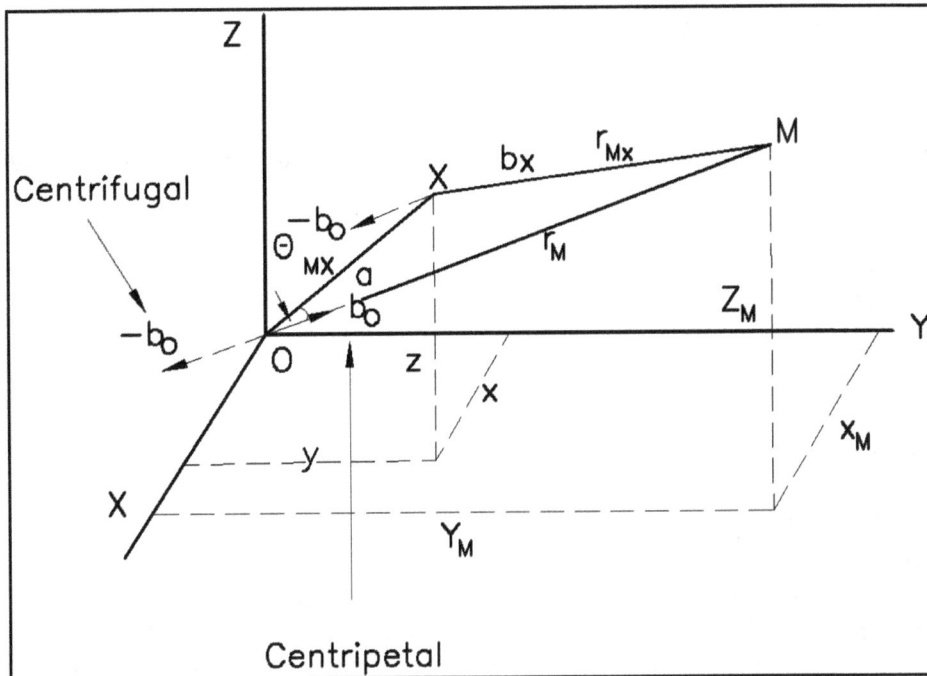

Figure II-5-4.   Schematic diagram of tidal potential (Dronkers 1964)

where the separation distance $r_{MX} = [(x_m - x)^2 + (y_m - y)^2 + (z_m - z)^2]^{1/2}$ with an equivalent expression for r_{SX}.

(d) The attractive force of the moon and sun at any point X is defined as

$$\vec{b}_X = \nabla [V_M(X) + V_S(X)] \tag{II-5-10}$$

where ∇ is the vector gradient operator defined as

$$\vec{\nabla} = \left(\partial \frac{}{\partial x} + \partial \frac{}{\partial y} + \partial \frac{}{\partial z}\right) \tag{II-5-11}$$

(e) From Figure II-5-4, the attractive force at the center of the earth (centripetal) b_O is balanced by the centrifugal force $-b_O$ (i.e., equal in magnitude but opposite in direction). Because any point on the earth experiences the same centrifugal force as that at O, the resultant force at any point X will be equal to $b_X - b_O$. This resultant force difference is the tide generating force, the force that causes the oceans to deform in order to balance the sum of external forces. Therefore, the difference between the tidal potential at point O and at point X becomes the tidal potential responsible for the tide-producing forces.

(f) The moon's tide-generating potential can be written as

$$V_M = fM\left[\frac{1}{r_{MX}} - \frac{1}{r_M} - \frac{a \cos \theta_{MX}}{r_M^2}\right] \tag{II-5-12}$$

with the tide potential for the sun written as

$$V_S = fS \left[\frac{1}{r_{SX}} - \frac{1}{r_S} - \frac{a \cos \theta_{SX}}{r_S^2} \right] \qquad \text{(II-5-13)}$$

where a is the mean radius of the earth. Various geometric relationships are used to write Equations II-5-12 and II-5-13 in the following forms:

$$V_M = fM \left(\frac{a^2}{r_M^3} \right) P_M \qquad \text{(II-5-14)}$$

$$V_S = fS \left(\frac{a^2}{r_S^3} \right) P_S \qquad \text{(II-5-15)}$$

where the terms P_M and P_S represent harmonic polynomial expansion terms that collectively describe the relative positions of the earth, moon, and sun. Note that in both cases, the tidal potential term is written as an inverse function of the distance between the earth and the moon or sun. Both Dronkers (1964) and Schureman (1924) present detailed derivations of the terms of Equations II-5-14 and II-5-15. For the purpose of this manual, however, the tidal potential terms shown here are adequate to describe the two most important features of a tidal record, the spring/neap cycle and the diurnal inequality.

(3) Spring/neap cycle.

(a) The semidiurnal rise and fall of tide can be described as nearly sinusoidal in shape, reaching a peak value every 12 hr and 25 min. This period represents one-half of the lunar day. Two tides are generally experienced per lunar day because tides represent a response to the increased gravitational attraction from the (primarily) moon on one side of the earth, balanced by a centrifugal force on the opposite side of the earth. These forces create a "bulge" or outward deflection in the water surface on the two opposing sides of the earth.

(b) The magnitude of tidal deflection is partially a function of the distance between the moon and earth. When the moon is in perigee, i.e., closest to the earth, the tide range is greater than when it is furthest from the earth, in apogee. For example, the potential terms in Equation II-5-14 contain the multiplier $1/r_M$, describing the distance of the moon from the earth. When the moon is closest to the earth, r_M is a minimum value and the tidal potential term is maximum. Conversely, when the moon is in apogee, the potential term is at a minimum value. This difference may be as large as 20 percent.

(c) The maximum water surface deflection of semidiurnal tides changes as the relative position of the moon and sun changes. The amplitude envelope connecting any two successive high tides (and low tides) gradually increases from some minimum height to a maximum value, and then decreases back to a minimum. Periods of maximum amplitude are referred to as spring tides, times of minimum amplitude are neap tides. This envelope of spring to neap occurs twice over a period of approximately 29 days. An example tidal signal for Boston, MA, is shown in Figure II-5-5 (Harris 1981) in which the normalized tidal signal exhibits two amplitude envelopes during the total time series.

(d) Spring tides occur when the sun and moon are in alignment. This occurs at either a new moon, when the sun and moon are on the same side of the earth, or at full moon, when they are on opposite sides of the earth. Neap tides occur at the intermediate points, the moon's first and third quarters. Figure II-5-6 is a schematic representation of these predominant tidal phases. Lunar quarters are indicated in the tidal time

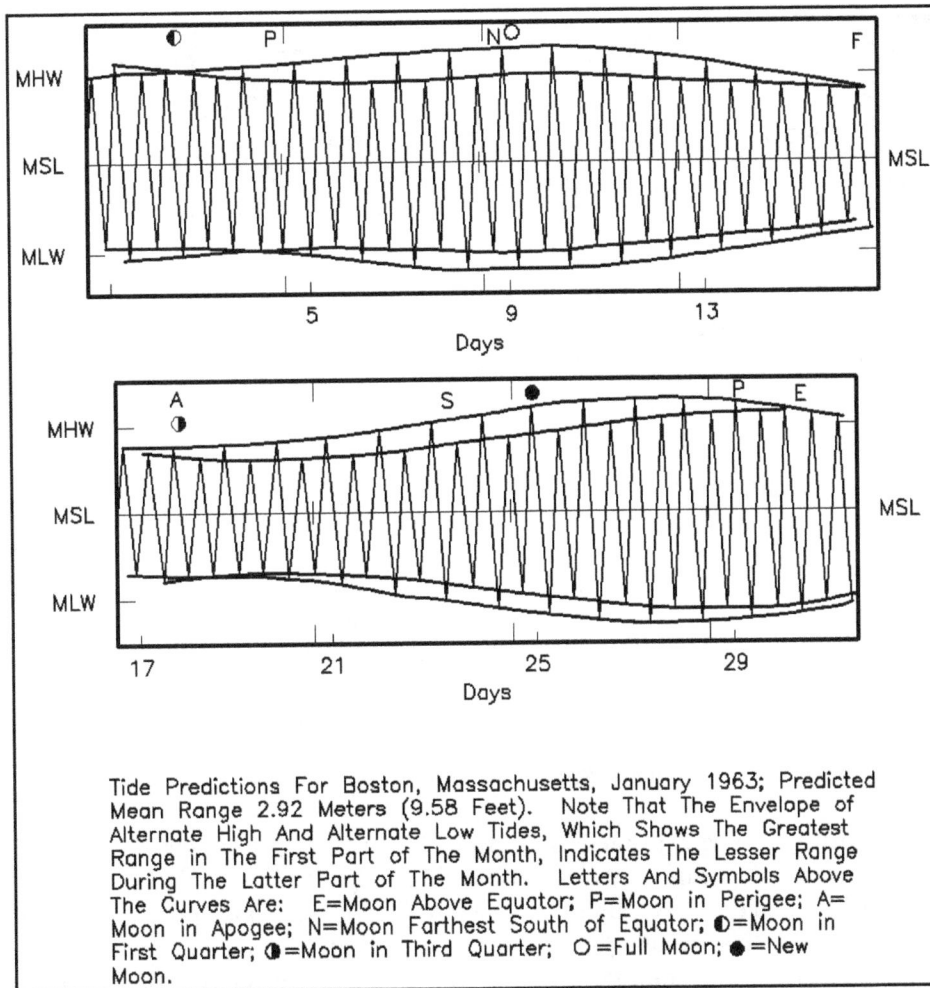

Tide Predictions For Boston, Massachusetts, January 1963; Predicted Mean Range 2.92 Meters (9.58 Feet). Note That The Envelope of Alternate High And Alternate Low Tides, Which Shows The Greatest Range in The First Part of The Month, Indicates The Lesser Range During The Latter Part of The Month. Letters And Symbols Above The Curves Are: E=Moon Above Equator; P=Moon in Perigee; A= Moon in Apogee; N=Moon Farthest South of Equator; ◐=Moon in First Quarter; ◑=Moon in Third Quarter; O =Full Moon; ● =New Moon.

Figure II-5-5.   Tide predictions for Boston, MA (Harris 1981)

series shown in Figure II-5-5. Note that in Figure II-5-5, the envelope that connects higher-high tide values for the first spring tide during the first 14.5 days becomes an envelope of the lower-high tide values during the second spring tide.

(4) Diurnal inequality.

(a) In the above example, the envelope of two successive high or low tides defines spring and neap conditions. Alternate tides were used because the ranges of two successive tides at a given location are generally not identical, but exhibit differences in height. Examples are evident in Figure II-5-5. These differences are referred to as the diurnal inequality of the tide and result from the relative position of the sun and moon as well as the specific location of an observer on the earth.

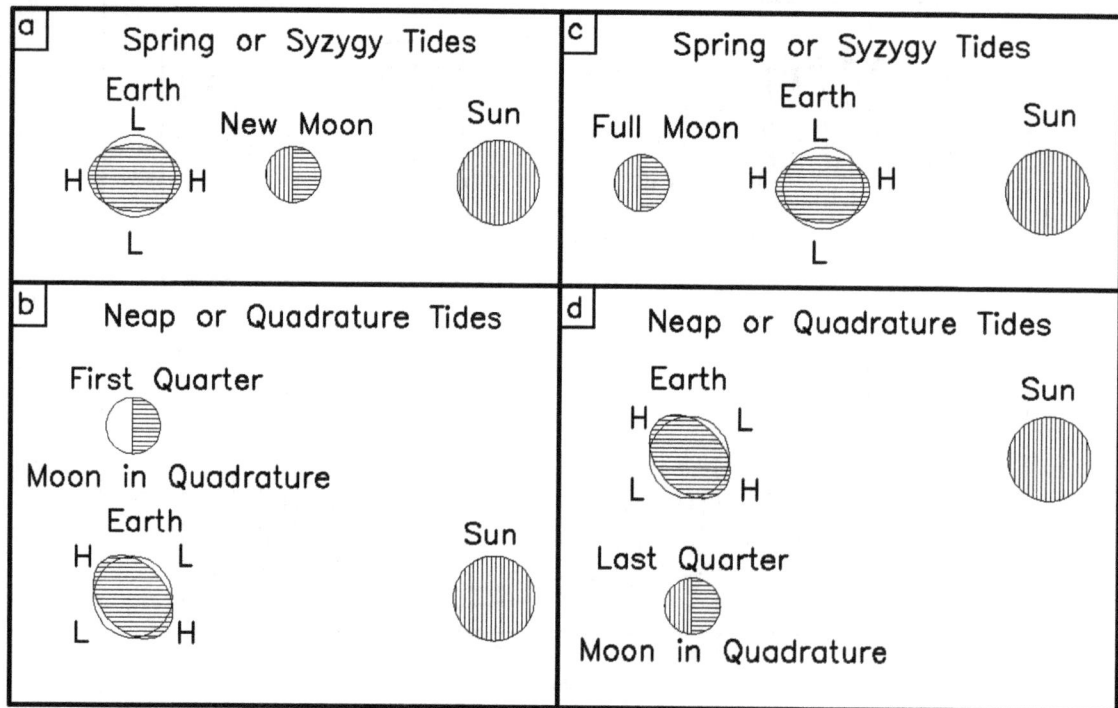

Figure II-5-6.   Spring and neap tides (Shalowitz 1964)

(b) Diurnal inequality can be explained as follows. The tidal bulge is centered along a line from the center of the moon or sun to the center of the earth. The tidal bulge at a given sublunar or subsolar (location on the earth nearest the moon or sun) location has an equivalent bulge on the opposite side of the earth, i.e., on a line drawn from the sublunar or subsolar point through the center of the earth on the opposite side of the equator. If the sublunar or subsolar point appears at a given north latitude, the peak of the corresponding tidal bulge on the opposite side of the equator will appear at a corresponding south latitude. Thus, a point on the same north latitude but 180 deg in longitude from the sublunar or subsolar point will show a reduced amplitude.

(c) A schematic example of the daily inequality is presented by Dronkers (1964) for the simple case of an earth-moon system. Referring to Figure II-5-7, the moon is located in the direction M and earth is rotating about the polar axis P. The deformed water surface resulting in response to the tide-producing forces is shown in the figure. Four locations (I - IV) are indicated to demonstrate the effect of location on the diurnal inequality. The fluctuating tide can be seen as the deviation in the deformed surface from a line at constant latitude on the undeformed spherical surface corresponding to each location. Location I corresponds to an observer on the equator. In this case, it can be seen that the tidal deformations from static conditions are equal; therefore, there is no diurnal inequality, each high tide is equal. However, at locations II and III, the inequality is evident with the second tide being substantially lower than the first. In the extreme case, location IV exhibits a diurnal tide only due to its location with respect to the deformed water surface.

(d) The combinations of astronomical forcing that define spring and neap cycles and diurnal inequalities is further modified by local bathymetry and shoreline boundary influences. All of these factors combine to produce tidal envelopes that vary from location to location. The result is site-specific tidal signatures, which can be classified as semidiurnal, diurnal, or mixed. Examples of these classes of tides are shown in Figures II-5-8 and II-5-9. Tides along the Atlantic coast are generally semidiurnal with a small diurnal inequality. Typical east coast envelopes for Boston, MA; New York, NY; Hampton Roads, (Hampton), VA;

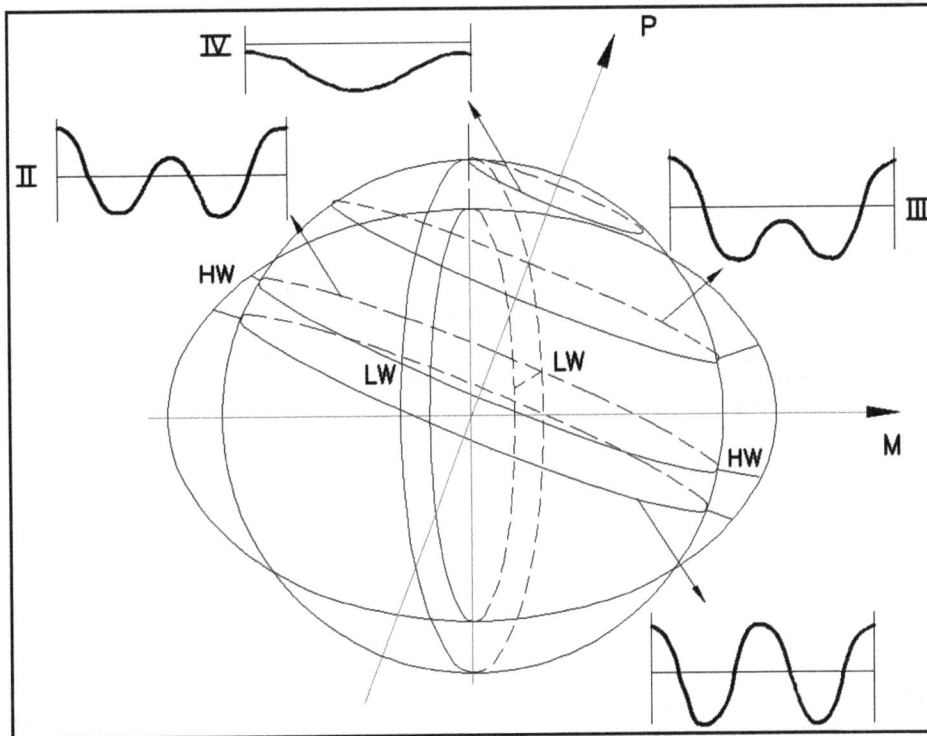

Figure II-5-7. The daily inequality (Dronkers 1964)

and the entrance to the Savannah River at Savannah, GA, are shown in the figure. Each time series exhibits two distinct and nearly equal tides per day. As one moves to Key West, FL, the character of the tide begins to change with a noticeable diurnal inequality. Tides inside the Gulf of Mexico range from semidiurnal at Key West, FL, to diurnal at Pensacola, FL, to mixed at Galveston, TX. Note that the Galveston data progresses from a diurnal tide during the first third of the record to a semidiurnal tide. Tides in the Gulf of Mexico are more complex than open ocean stations because astronomical forcing is modified by geometrically forced nodes and antinodes. These seiche-related phenomena are discussed in Part II-5-6. Pacific coast tides, shown in Figure II-5-9, are generally of larger amplitude than Atlantic and Gulf coast tides and often have a decided diurnal inequality.

   b. Tidal time series analysis.

   (1) Introduction. The equilibrium theory of tides is a hypothesis that the waters of the earth respond instantaneously to the tide-producing forces of the sun and moon. For example, high water occurs directly beneath the moon and sun, i.e., at the sublunar and subsolar points. This tide is referred to as an equilibrium tide. Part II-5-3 a (1), states that tide-producing forces are written in a polynomial expansion approximation for the exact solution of Equations II-5-12 and II-5-13. These expansion terms involve astronomical arguments describing the location of the sun and moon as well as the location of the observer on the earth. Although several variational forms of the series expansion have been published, the development presented in Schureman (1924) is given below. Alternate forms of expansion are discussed in Dronkers (1964).

   (2) Harmonic constituents.

   (a) According to equilibrium theory, the theoretical tide can be predicted at any location on the earth as a sum of a number of harmonic terms contained in the polynomial expansion representation of the

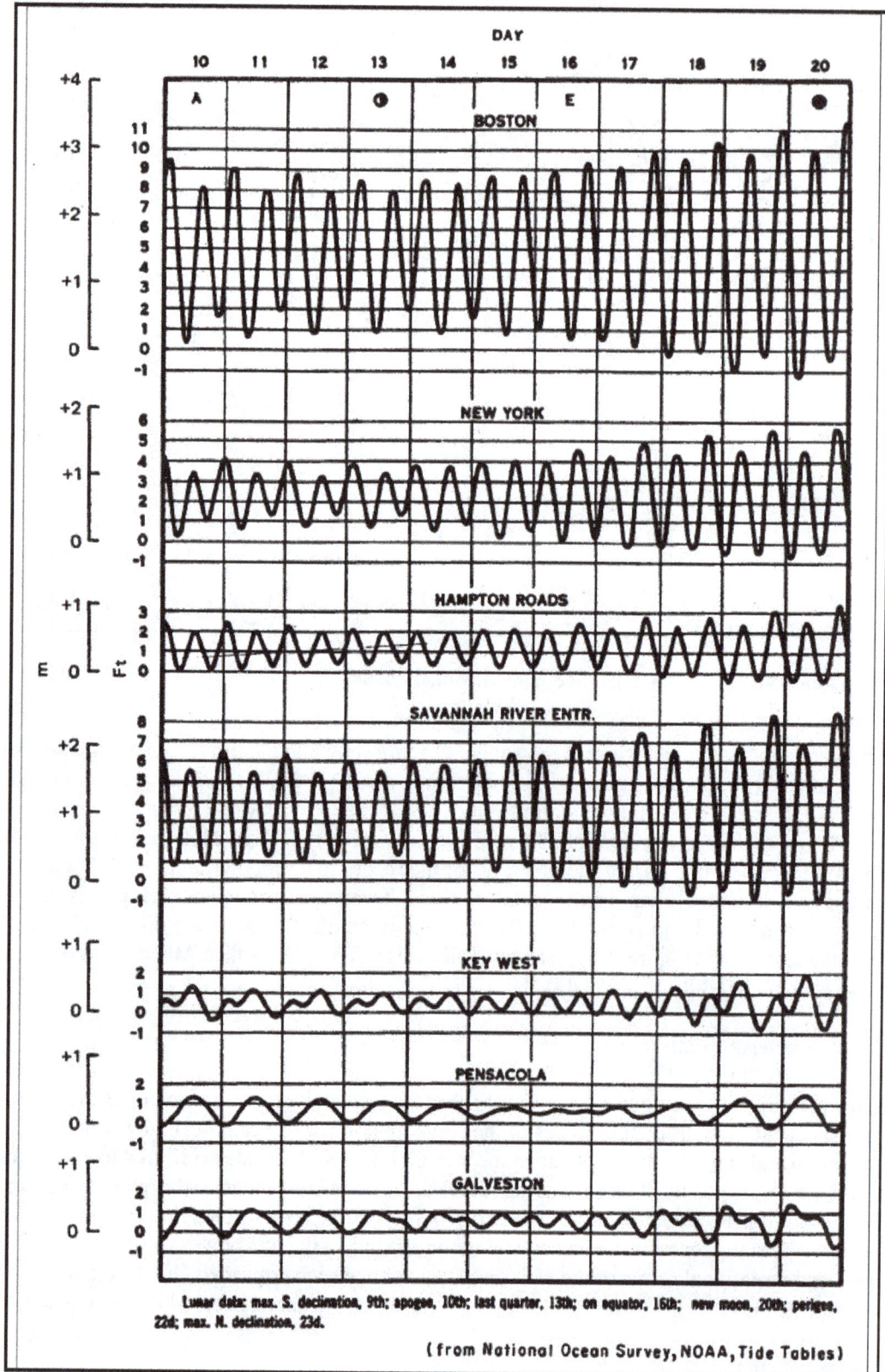

Figure II-5-8.   Typical tide curves along the Atlantic and Gulf coasts *(Shore Protection Manual* 1984)

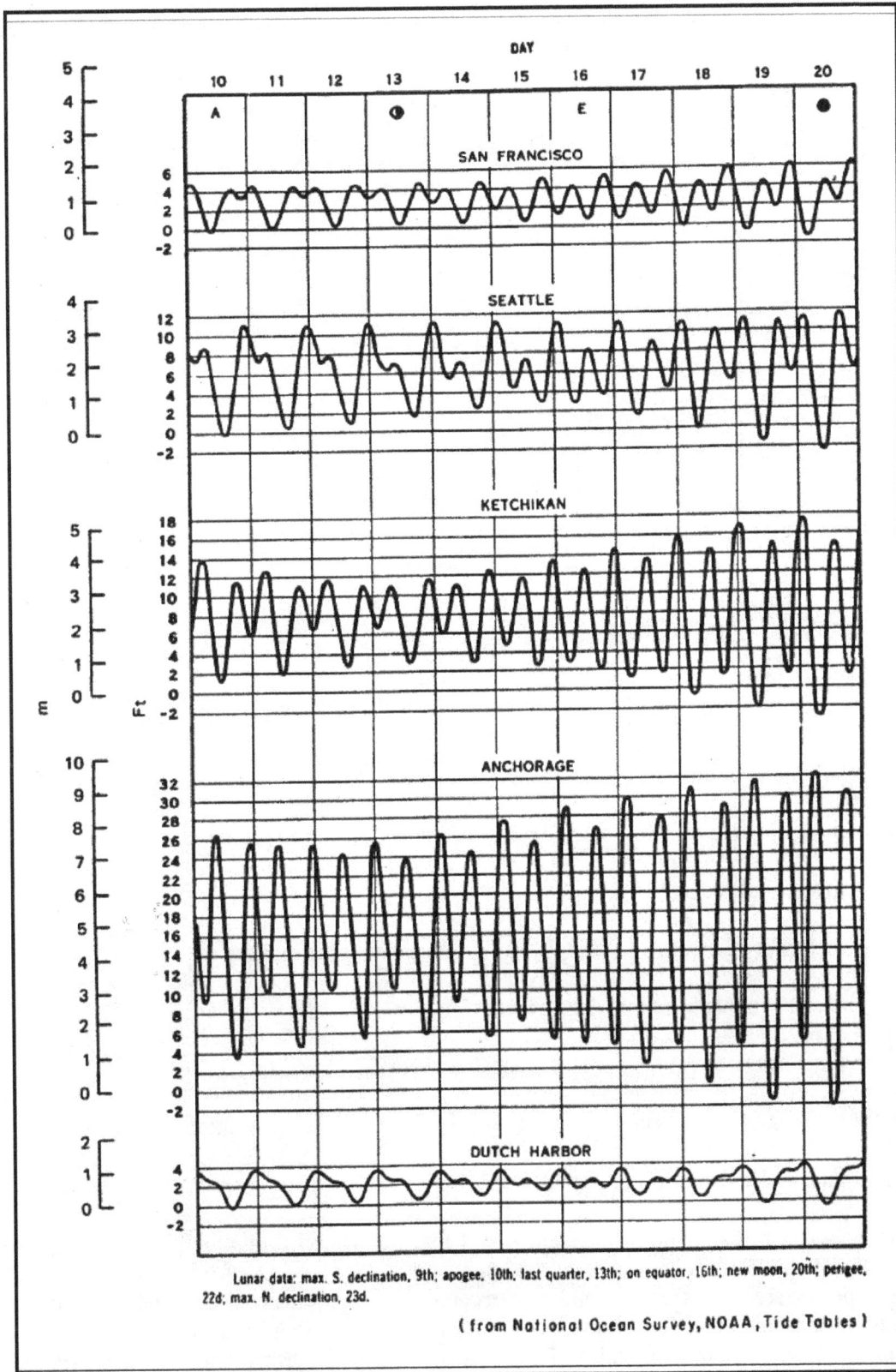

Figure II-5-9. Typical tide curves along Pacific coast of the United States (*Shore Protection Manual* 1984)

tide-producing forces. However, the actual tide does not conform to this theoretical value because of friction and inertia as well as differences in the depth and distribution of land masses of the earth.

(b) Because of the above complexities, it is impossible to exactly predict the tide at any place on the earth based on a purely theoretical approach. However, the tide-producing forces (and their expansion component terms) are harmonic; i.e., they can be expressed as a cosine function whose argument increases linearly with time according to known speed criteria. If the expansion terms of the tide-producing forces are combined according to terms of identical period (speed), then the tide can be represented as a sum of a relatively small number of harmonic constituents. Each set of constituents of common period are in the form of a product of an amplitude coefficient and the cosine of an argument of known period with phase adjustments based on time of observation and location. Observational data at a specific time and location are then used to determine the coefficient multipliers and phase arguments for each constituent, the sum of which are used to reconstruct the tide at that location for any time. This concept represents the basis of the harmonic analysis, i.e., to use observational data to develop site-specific coefficients that can be used to reconstruct a tidal series as a linear sum of individual terms of known speed. The following presentation briefly describes the use of harmonic constants to predict tides.

(c) Tidal height at any location and time can be written as a function of harmonic constituents according the following general relationship

$$H(t) = H_0 + \sum_{n=1}^{N} f_n H_n \cos [a_n t + (V_0 + u)_n - \kappa_n] \qquad \text{(II-5-16)}$$

where

$H(t)$ = height of the tide at any time t

H_0 = mean water level above some defined datum such as mean sea level

H_n = mean amplitude of tidal constituent n

f_n = factor for adjusting mean amplitude H_n values for specific times

a_n = speed of constituent n in degrees/unit time

t = time measured from some initial epoch or time, i.e., $t = 0$ at t_0

(V_0+u) = value of the equilibrium argument for constituent n at some location and when $t = 0$

κ_n = epoch of constituent n, i.e., phase shift from tide-producing force to high tide from t_0

(d) In the above formula, tide is represented as the sum of a coefficient multiplied by the cosine of its respective arguments. A finite number of constituents are used in the reconstruction of a tidal signal. Values for the site-specific arguments (H_0, H_n, and κ_n) are computed from observed tidal time series data, usually from a least squares analysis. The National Oceanic and Atmospheric Administration's (NOAA) National Ocean Survey (NOS) generally provides 37 constituents in their published harmonic analyses (generally based on an analysis of a minimum of 1 year of prototype data). The NOS constituents, along with the corresponding period and speed of each, are listed in Table II-5-3. The time-specific arguments (f_n and $V_0 + u$) are determined from formulas or tables as will be discussed below or through application of programs available through the Automated Coastal Engineering System (ACES) (Leenknecht, Szuwalski, and Sherlock 1992).

(e) Most of the constituents listed in Table II-5-3 are associated with a subscript indicating the approximate number of cycles per solar day (24 hr). Constituents with subscripts of 2 are semidiurnal constituents and produce a tidal contribution of approximately two high tides per day. Diurnal constituents occur approximately once a day and have a subscript of 1. Symbols with no subscript are termed long-period constituents and have periods greater than a day; for example, the Solar Annual constituent Sa has a period of approximately 1 year.

(f) In most harmonic analyses of tidal data in the continental United States, the majority of constituents shown above have amplitude contributions that are negligible with respect to the magnitude of the full tide. For example, in the Gulf of Mexico and east coast of the United States, well over 90 percent of the tidal energy can be represented by the amplitudes of the M_2, S_2, N_2, and K_2 semidiurnal and K_1, O_1, P_1, and Q_1 diurnal constituents. In other locations, many more tidal constituents are needed to adequately represent the tide. For example, over 100 constituents are needed for Anchorage, AK.

(g) Two categories of tidal constituents are necessary to reconstruct a tidal signal:

● Those that represent the elevation of the water surface.

● Those that specify a time and the phase shift associated with that time.

For example, the value for H_n in Equation II-5-16 is the mean constituent amplitude and is a function of both location and variations arising from changes in the latitude of the moon's node. The nodal effect of the moon is reflected by the introduction of the node factor f_n, which modifies each constituent amplitude to correspond to a specific time period. Mid-year values are usually specified for reconstructed time series because node factors vary slowly in time. Mid-year values for each constituent listed in Table II-5-3 are presented in Shureman (1924) for the years 1850 through 1999. An example is shown in Table II-5-4 for the years 1970 through 1999. Equations for computing f_n are given by Schureman.

(h) The second category of arguments specifies the phasing of high water for each constituent with respect to both time and location. These arguments are based on the fact that phases of the constituents of the observed tide do not coincide with the phases of the corresponding constituents of the equilibrium tide. For example, a high tide does not occur directly beneath the moon. There is a lag between the location of the tide-producing force (i.e., location of the moon) and the observed time of high water. This lag, due to frictional and inertial forces acting on the propagating tide, is referred to as the epoch of the constituent and is denoted by κ_n in Equation II-5-16.

(i) The relationship between the constituent arguments and high tide is shown in the schematic Figure II-5-10. In this figure, the cosine curve represents the surface elevation in the y-direction as a function of time or degrees of phase (maximum at 0 and 360 deg). For the M_2 tidal constituent, the cosine curve has a period of 12.42 hr (other constituent periods are indicated in Table II-5-3). Therefore, in Figure II-5-10, the horizontal axis represents either time or phase, both increasing to the right. The value of κ represents the actual phase lag required for the water surface to reach high water (HW) following the passing of the tide-producing force. In the case of the semidiurnal constituents, this force is the crossing of the moon.

Table II-5-3
NOS Tidal Constituents and Arguments

Symbol	Speed, deg/hr	Period, hr	Symbol	Speed, deg/hr	Period, hr
M_2	28.984	12.421	Mm	0.544	661.765
S_2	30.000	12.000	Ssa	0.082	4390.244
N_2	28.439	12.659	Sa	0.041	8780.488
K_1	15.041	23.935	Msf	1.015	354.680
M_4	57.968	6.2103	Mf	1.098	327.869
O_1	13.943	25.819	ρ_1	13.471	26.724
M_6	86.952	4.140	Q_1	13.398	26.870
$(MK)_3$	44.025	8.177	T_2	29.958	12.017
S_4	60.000	6.000	R_2	30.041	11.984
$(MN)_4$	57.423	6.269	$(2Q)_1$	12.854	28.007
ν_2	28.512	12.626	P_1	14.958	24.067
S_6	90.000	4.000	$(2SM)_2$	31.015	11.607
μ_2	27.968	12.872	M_3	43.476	8.280
$(2N)_2$	27.895	12.906	L_2	29.528	12.192
$(OO)_1$	16.139	22.306	$(2MK)_3$	42.927	8.386
λ_2	29.455	12.222	K_2	30.082	11.967
S_1	15.000	24.000	M_8	115.936	3.105
M_1	14.496	24.834	$(MS)_4$	58.984	6.103
J_1	15.585	23.099			

(j) The value κ is approximately constant at every location in the world because it represents the actual lag between the passing of the tide-producing force (i.e., moon) at a specific location and the following high-tide contribution of that force at that same location. It is computed as the sum of the theoretical phase or time lead of the tide-producing force relative to the observer at some fixed time and the measured phase lag ζ from the observer at that fixed time to the following high water. The theoretical location of the tide-producing force is referred to as the equilibrium argument $(V_0 + u)$. In Figure II-5-10, the tide-producing force and corresponding equilibrium tide at location M are located $(V_0 + u)$ degrees ahead of point T. Conversely, the equilibrium tide will be located at point T if shifted $(V_0 + u)$ degrees. The value of ζ represents the phase lag from point T to HW.

(k) The equilibrium argument $(V_0 + u)$ is computed from equations defining the time-varying relationship between the earth, moon, and sun. The value of ζ is computed from observed tidal time series data. As stated, the sum of these two values is approximately constant for any fixed location at any time.

(l) Values of the equilibrium argument for the constituents of Table II-5-3 relative to the passing of the tidal potential force at the Greenwich meridian for each calendar year from 1850 through 2000 are tabulated in Schureman (1924). An example is shown in Table II-5-5 for the years 1990 to 2000. Monthly and daily adjustment tables are also presented. Each of the values is computed according to the respective constituent speeds shown in Table II-5-3. The equilibrium arguments tabulated in Schureman are referenced to the meridian of Greenwich; therefore, the argument $(V_0 + u)$ represents the phase difference in degrees between the location of the tidal potential term (moon or sun) and Greenwich relative to some specific time.

Table II-5-4
Node Factors for 1970 through 1999 (Schureman 1924)

Constituent	1970	1971	1972	1973	1974	1975	1976	1977	1978	1979
J_1	1.155	1.132	1.097	1.051	0.995	0.936	0.881	0.842	0.827	0.839
K_1	1.105	1.088	1.063	1.029	0.991	0.951	0.916	0.891	0.882	0.890
K_2	1.289	1.232	1.150	1.055	0.957	0.871	0.804	0.763	0.748	0.760
L_2	0.882	0.668	1.118	1.270	1.014	0.808	0.988	1.179	1.169	0.994
M_1	1.987	2.176	1.503	1.012	1.535	1.777	1.428	0.870	0.874	1.361
M_2^1, N_2, $2N$, λ_2, μ_2, ν_2	0.966	0.973	0.983	0.995	1.008	1.020	1.029	1.035	1.038	1.036
M_3	0.950	0.960	0.975	0.993	1.012	1.029	1.044	1.054	1.057	1.054
M_4, MN	0.934	0.948	0.967	0.991	1.016	1.039	1.059	1.072	1.077	1.073
M_6	0.903	0.922	0.951	0.986	1.024	1.060	1.090	1.110	1.118	1.112
M_8	0.873	0.898	0.935	0.981	1.032	1.081	1.122	1.149	1.160	1.151
O_1, Q_1, $2Q$, ρ_1	1.170	1.143	1.101	1.047	0.984	0.920	0.863	0.822	0.806	0.819
OO	1.716	1.575	1.380	1.159	0.940	0.750	0.607	0.517	0.485	0.512
MK	1.068	1.059	1.045	1.024	0.998	0.970	0.943	0.923	0.915	0.922
2MK	1.032	1.031	1.028	1.020	1.006	0.989	0.970	0.956	0.950	0.955
Mf	1.417	1.341	1.233	1.102	0.962	0.831	0.723	0.652	0.625	0.647
Mm	0.882	0.906	0.940	0.982	1.025	1.067	1.100	1.123	1.131	1.121

Constituent	1980	1981	1982	1983	1984	1985	1986	1987	1988	1989
J_1	0.877	0.930	0.989	1.045	1.093	1.130	1.153	1.164	1.163	1.148
K_1	0.913	0.948	0.987	1.026	1.060	1.086	1.104	1.112	1.111	1.100
K_2	0.799	0.864	0.949	1.045	1.142	1.226	1.285	1.315	1.310	1.270
L_2	0.848	1.001	1.238	1.157	0.745	0.811	1.263	1.244	0.749	0.746
M_1	1.656	1.468	0.974	1.323	2.050	2.032	1.292	1.367	2.142	2.122
M_2^1, N_2, $2N$, λ_2, μ_2, ν_2	1.030	1.021	1.009	0.997	0.984	0.974	0.967	0.964	0.964	0.969
M_3	1.045	1.031	1.013	0.994	0.977	0.962	0.951	0.946	0.947	0.954
M_4, MN	1.061	1.042	1.018	0.993	0.969	0.949	0.935	0.928	0.930	0.939
M_6	1.092	1.063	1.027	0.989	0.954	0.924	0.904	0.894	0.896	0.910
M_8	1.125	1.085	1.036	0.986	0.939	0.901	0.874	0.862	0.864	0.881
O_1, Q_1, $2Q$, ρ_1	0.858	0.915	0.979	1.041	1.096	1.140	1.168	1.182	1.180	1.161
OO	0.596	0.735	0.921	1.137	1.361	1.560	1.706	1.778	1.766	1.668
MK	0.941	0.967	0.996	1.022	1.043	1.058	1.068	1.072	1.071	1.065
2MK	0.969	0.987	1.005	1.019	1.027	1.031	1.032	1.032	1.032	1.032
Mf	0.715	0.820	0.949	1.088	1.221	1.333	1.412	1.450	1.443	1.392
Mm	1.103	1.070	1.029	0.986	0.944	0.909	0.884	0.872	0.874	0.891

Constituent	1990	1991	1992	1993	1994	1995	1996	1997	1998	1999
J_1	1.120	1.080	1.030	0.972	0.914	0.864	0.833	0.829	0.852	0.896
K_1	1.079	1.051	1.015	0.976	0.937	0.905	0.886	0.883	0.897	0.926
K_2	1.203	1.115	1.016	0.922	0.842	0.785	0.754	0.750	0.772	0.821
L_2	1.216	1.248	0.898	0.801	1.077	1.208	1.107	0.921	0.893	1.096
M_1	1.334	1.156	1.778	1.829	1.282	0.800	1.083	1.487	1.560	1.214
M_2^1, N_2, $2N$, λ_2, μ_2, ν_2	0.977	0.988	1.000	1.013	1.024	1.032	1.037	1.038	1.034	1.027
M_3	0.966	0.982	1.000	1.019	1.036	1.048	1.056	1.057	1.051	1.040
M_4, MN	0.955	0.976	1.000	1.025	1.048	1.065	1.075	1.076	1.069	1.054
M_6	0.932	0.964	1.000	1.038	1.072	1.099	1.115	1.117	1.105	1.082
M_8	0.911	0.952	1.000	1.051	1.098	1.134	1.156	1.159	1.143	1.111
O_1, Q_1, $2Q$, ρ_1	1.128	1.081	1.024	0.960	0.897	0.844	0.812	0.808	0.832	0.879
OO	1.505	1.296	1.072	0.863	0.688	0.565	0.498	0.489	0.538	0.643
MK	1.054	1.038	1.015	0.988	0.959	0.934	0.918	0.916	0.928	0.950
2MK	1.030	1.025	1.015	1.000	0.982	0.964	0.952	0.951	0.959	0.976
Mf	1.303	1.184	1.048	0.910	0.786	0.691	0.636	0.629	0.669	0.752
Mm	0.918	0.956	0.998	1.042	1.081	1.110	1.128	1.130	1.117	1.091

[1] Factor f of MS, 28M, and M8f are each equal to factor f of M2.

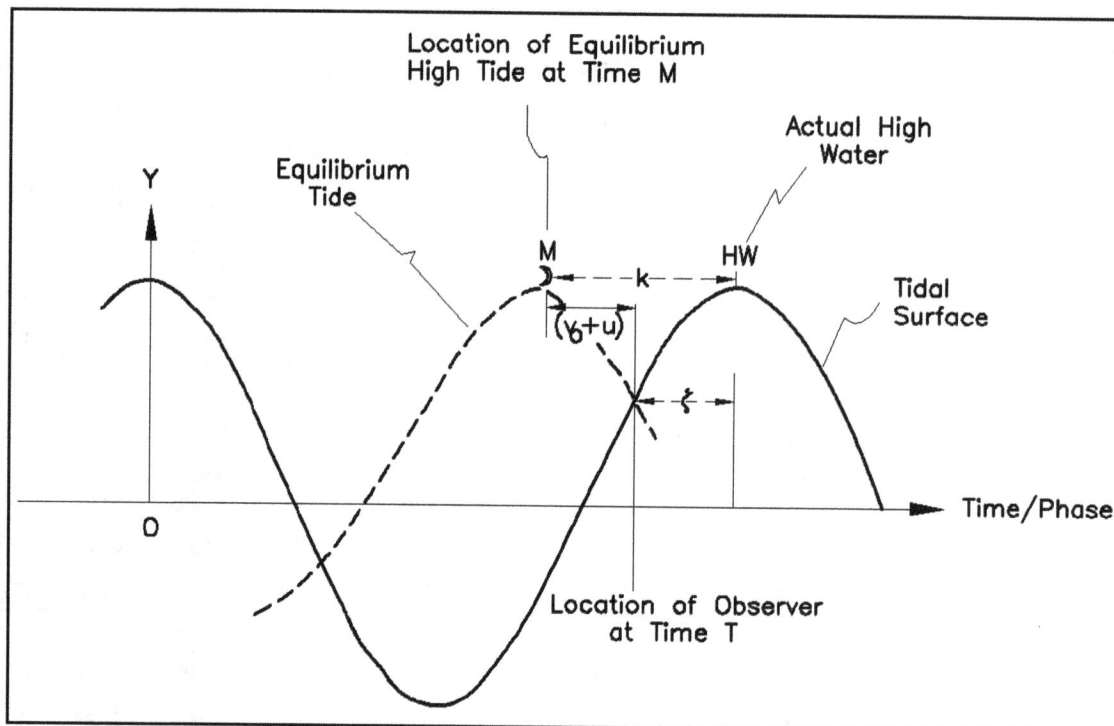

Figure II-5-10.   Tidal phase relationships

(m)   A tidal simulation computer program is available in ACES (Leenknecht, Szuwalski, and Sherlock 1992) to compute nodal factors and local (or Greenwich) equilibrium argument values for any time period and generate the corresponding water surface time series as a function of input constituent amplitudes. Formulas for computing the equilibrium arguments are found in Shureman (1924), but are too lengthy for this manual.

(3)   Harmonic reconstruction.

(a)   In order to reconstruct a tidal series for a specific time and location, the various phase arguments of Equation II-5-16 must be defined according to local conditions. Generally, local values of κ_n, H_0, and H_n are available from NOS harmonic analyses. Because the values of the nodal factors f_n are slowly varying, the yearly values determined according to Schureman are sufficiently accurate for the particular time of interest throughout the world. However, local values of $(V_0 + u)_n$ vary with the speed of the constituent and have to be determined for the location and time of interest. This information can be computed from tabulated equilibrium arguments relative to Greenwich such as those presented in Schureman or computed with programs developed for that purpose such as those contained in ACES.

(b)   Values of the local equilibrium arguments, i.e., local $(V_0 + u)_n$, represent the instantaneous value of each of the equilibrium tide-producing force constituents (in degrees) with respect to some specific point on the earth; for example, the time-varying location of the moon and sun with respect to some location on the earth. Referring to Figure II-5-11, the horizontal axis represents distance with the point G representing Greenwich, England, and O representing an observer located at some point west of Greenwich. The location of the moon with respect to Greenwich at longitude 0° at Greenwich time t_0 is indicated by the Greenwich equilibrium argument presented in Shureman, denoted as Greenwich $(V_0 + u)$, for time Greenwich t_0.

Table II-5-5
Equilibrium Argument for Beginning of Years 2001 through 2010

Constituent	2001	2002	2003	2004	2005	2006	2007	2008	2009	2010
J_1	227.9	316.9	47.0	138.0	243.7	335.7	67.8	159.8	265.5	356.5
K_1	1.9	1.8	2.6	4.1	7.1	9.3	11.7	13.9	16.8	18.3
K_2	184.0	183.3	184.6	187.6	193.7	198.5	203.5	208.2	214.3	217.1
L_2	269.1	105.4	267.4	94.8	295.2	141.6	297.3	121.4	321.6	165.8
M_1	145.4	52.1	303.7	204.7	139.6	58.2	311.4	210.3	141.0	64.7
M_2	210.8	311.4	52.4	153.5	230.4	331.9	73.4	174.8	251.7	352.9
M_3	316.1	287.1	258.5	230.2	165.7	137.8	110.0	82.2	17.6	349.3
M_4	61.5	262.9	104.7	307.0	100.9	303.8	146.7	349.6	143.5	345.7
M_6	272.3	214.3	157.1	100.5	331.3	275.6	220.1	164.4	35.2	338.6
M_8	123.1	165.7	209.4	254.0	201.7	247.5	293.4	339.2	286.9	331.4
N_2	340.5	353.5	4.7	17.1	352.3	5.0	17.7	30.4	5.6	18.0
$2N$	110.3	33.5	317.0	240.7	114.1	38.1	322.1	246.1	119.5	43.1
O_1	213.0	313.5	53.0	151.8	224.9	323.0	61.2	159.3	232.4	331.3
OO	322.4	222.3	125.6	31.5	326.3	234.6	143.3	51.6	346.3	251.9
P_1	349.3	349.5	349.8	350.0	349.3	349.5	349.7	350.0	349.2	349.5
Q_1	342.8	354.6	5.3	15.5	346.7	356.1	5.5	15.0	346.3	356.4
$2Q$	112.6	35.6	317.7	239.0	108.5	29.2	309.9	230.6	100.1	21.5
R_2	177.8	177.5	177.3	177.0	177.7	177.5	177.2	177.0	177.7	177.4
S_1	180.0	180.0	180.0	180.0	180.0	180.0	180.0	180.0	180.0	180.0
$S_{2,4,6}$	0.0	0.0	0.0	0.0	0.0	0.0	0.0	0.0	0.0	0.0
T_2	2.2	2.5	2.7	3.0	2.3	2.5	2.8	3.0	2.3	2.6
λ_2	307.7	218.9	130.3	42.0	300.8	212.8	124.8	36.7	295.5	207.1
μ_2	63.6	265.0	106.7	308.6	101.9	304.1	146.3	348.5	141.8	343.7
ν_2	293.8	224.0	154.4	85.0	340.1	271.0	202.0	132.9	28.0	318.6
ρ_1	296.1	226.1	155.0	83.3	334.5	262.2	189.7	117.4	8.6	297.0
MK	212.6	313.2	54.9	157.6	237.5	341.2	85.0	188.7	268.6	11.1
2MK	59.6	261.1	102.1	302.9	93.8	294.4	135.1	335.7	126.6	327.4
MN	191.3	303.9	57.0	170.6	222.7	336.9	91.1	205.2	257.3	10.9
MS	210.8	311.4	52.4	153.5	230.4	331.9	73.4	174.8	251.7	352.9
2SM	149.2	48.6	307.6	206.5	129.6	28.1	286.6	185.2	108.3	7.1
Mf	324.7	224.4	126.3	29.8	320.7	225.8	131.0	36.1	326.9	230.3
MSf	147.2	46.4	305.7	204.9	128.5	27.8	287.0	186.3	109.9	9.2
Mm	230.2	318.9	47.7	136.4	238.2	326.9	55.6	144.3	246.1	334.9
Sa	280.7	280.5	280.2	280.0	280.8	280.5	280.3	280.0	280.8	280.5
Ssa	201.4	201.0	200.5	200.0	201.5	201.0	200.5	200.1	201.6	201.1

Methodology based on Schureman 1924. Values computed May 2001.

(c)   The location of the moon at Greenwich time t_0 is different for an observer at some point O located at longitude L^o than it is for the observer located at Greenwich. For each constituent, the observer is located pL deg from Greenwich; therefore, the local equilibrium argument must be adjusted by $-pL$ to account for the difference in location between the point of interest (i.e., point O) and Greenwich.

(d)   The $-pL$ adjustment provides the necessary equilibrium argument correction for differences in location between some point and Greenwich, i.e., a local equilibrium argument corresponding to an observer at location O. However, the value of the equilibrium argument Greenwich $(V_0 + u)$ was specified with respect to Greenwich t_0. Therefore, Greenwich time must be written with respect to local time. Because local time for the observer located west of Greenwich is earlier than local time in Greenwich (t_0), the following adjustment is made to convert local time to Greenwich time.

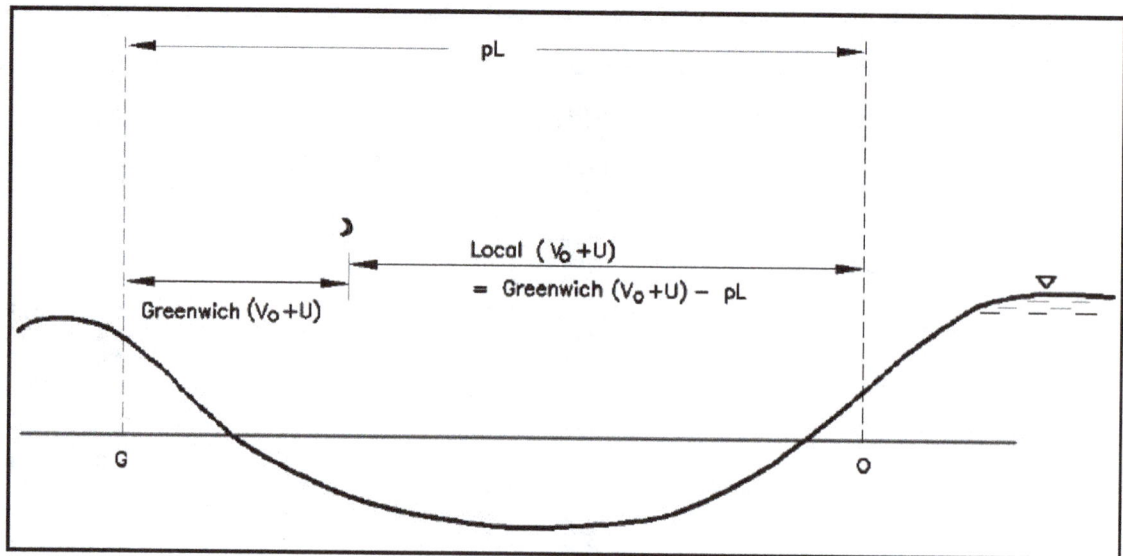

Figure II-5-11.   Phase angle argument relationship

$$\text{Greenwich } (t_0) = \text{local } (t_0) + \frac{S}{15} \qquad\qquad (\text{II-5-17})$$

where S is the local time meridian (shown on NOS analyses) and the number 15 indicates a time change of 1 hr per 15 deg longitude.

(e)   The speed of the time argument ($a_n t$ in Equation II-5-16) in degrees with respect to time is equal to the speed of the constituent a_n multiplied by Equation II-5-17. Therefore, the final relationship between local and Greenwich phase arguments that account for both differences in location ($-pL$) and local time ($aS/15$) can be written as:

$$\text{local } (V_0 + u) = \text{Greenwich } (V_0 + u) - pL \qquad\qquad (\text{II-5-18})$$

Therefore, a tide at any arbitrary location is computed as

$$H(t)_{local} = H_0 + \sum_{n=1}^{N} f_n H_n \cos [a_n t + \text{Greenwich } V_0 + u) - pL + \frac{a_n S}{15} - \kappa_n] \qquad\qquad (\text{II-5-19})$$

(f)   A reconstructed tidal time series of a published NOS harmonic analysis is presented for Sandy Hook, NJ. The NOS analysis is shown in Figure II-5-12. As can be seen by the reported amplitudes, the M_2, S_2, N_2, K_1, Sa, O_1, v_2, and K_2 constituents contain the majority of the tidal energy. These constituents are used to generate a 15-day tidal signal beginning on 1 January 1984 at 0000 hr Eastern Standard Time. Computed values are compared to the high and low tide predictions published in the Tide Tables for 1984 (NOAA 1984) shown in Figure II-5-13. Because all 37 constituents are not used in the reconstruction, the match is not perfect; however, it demonstrates the degree of accuracy that can be achieved by using only major constituents.

FORM C&GS-444
(11-67)

U.S. DEPARTMENT OF COMMERCE
ESSA- COAST AND GEODETIC SURVEY

~~TIDES~~
~~CURRENTS~~ STANDARD HARMONIC CONSTANTS FOR PREDICTION

STATION _Sandy Hook, N. J._

Lat. 40° 28.'1 N.
Long. 74° 0.'7 W.
Long. 74° 01 W.

COMPONENT	H AMPLITUDE	ε EPOCH	A ε'−ε	B 10⁴×H	C 360°−ε'	D −ε'	REMARKS
	ft. M.	°	°	*ft. M.*	°	°	
M₂	2.151	219.1	+ 3.1	2.237	+	−222.2	Time meridian 75° W. = +5 h.
S₂	0.448	246.0	− 2.0	0.466	+	−244.0	Extreme range .... { ft. kn.
N₂	0.473	204.1	+ 5.8	0.492	+	−209.9	Dial
K₁	0.319	102.2	− 1.2	0.332	+	−101.0	Marigram gear
M₄	0.036	347.4	+ 6.2	0.037	+	−353.6	Marigram scale
O₁	0.172	98.4	+ 4.3	0.179	+	−102.7	Z₀ 2.56 2.3 2.36 2.29 ft.
							Permanent current .... kn.
M₆	0.048	356.4	+ 9.3	0.050	+	− 5.7	The DATUM is a plane MLLW ft.
(MK)₃					+	−	below mean { low water springs / lower low water,
S₄	0.032	61.4	− 4.0	0.033	+	−52.4	
(MN)₄					+	−	
μ₂	0.109	191.1	+ 5.5	0.113	+	−196.6	First used for 1947 Tables
S₆					+	−	(modified 1966)
ν₂	0.073	243.7	+ 8.2	0.076	+	−251.9	
(2N)₂	0.063	189.1	+ 8.5	0.066	+	−197.6	
(OO)₁					+	−	
λ₂	0.015	231.6	+ 0.7	0.016	+	−232.3	Amplitudes of short period
S₁	0.037	56.2	− 1.0	0.038	+	−55.2	constituents increased 4%
M₁	0.012	100.3	+ 1.5	0.012	+	−101.8	
J₁	0.014	104.1	− 3.9	0.015	+	−100.2	
Mm					+	−	New 1960-78 Z₀ first used
Ssa	0.067	38.3	− 0.4	0.067	+	− 37.9	in 1985 T.T. (...G)
Sa	0.254	128.7	− 0.2	0.254	+	−128.5	
MSf					+	−	Z₀ = 2.56
Mf					+	−	1989 TT ✱
ρ₁					+	−	DATUM = MLLW
Q₁	0.037	104.5	+ 7.0	0.038	+	−111.5	
T₂	0.026	246.0	− 1.8	0.027	+	−244.2	
R₂					+	−	
(2Q)₁					+	−	
P₁	0.096	105.7	− 0.8	0.100	+	−104.7	
(2SM)₂					+	−	
M₃	0.032	198.6	+ 4.6	0.033	+	−203.2	
L₂	0.089	205.5	+ 0.4	0.093	+	−205.9	
(2MK)₃					+	−	
K₂	0.121	251.9	− 2.4	0.126	+	−249.5	
M₈					+	−	
(MS)₄					+	−	

Source of constants _3 - 369 day series (1933, 1937, 1940)_
Sa & Ssa from 10 years monthly MSL (1933-1942)
Compiled by _Rac_ _2-5-69_ (Date) Verified by .... (Date)

Figure II-5-12. NOS harmonic analysis for Sandy Hook, NJ

SANDY HOOK, N.J., 1984
Times and Heights of High and Low Waters

JANUARY

Day	Time (h m)	Height (ft)	Height (m)
1 Su	0619	5.1	1.6
	1242	-0.5	-0.2
	1840	3.9	1.2
2 M	0042	-0.4	-0.1
	0701	5.1	1.6
	1328	-0.5	-0.2
	1924	3.9	1.2
3 Tu	0128	-0.3	-0.1
	0741	5.0	1.5
	1411	-0.5	-0.2
	2001	3.8	1.2
4 W	0213	-0.2	-0.1
	0819	4.9	1.5
	1452	-0.5	-0.2
	2042	3.7	1.1
5 Th	0254	-0.1	0.0
	0859	4.7	1.4
	1531	-0.4	-0.1
	2124	3.7	1.1
6 F	0331	0.1	0.0
	0939	4.5	1.4
	1607	-0.3	-0.1
	2205	3.6	1.1
7 Sa	0408	0.2	0.1
	1019	4.3	1.3
	1645	-0.1	0.0
	2248	3.5	1.1
8 Su	0445	0.4	0.1
	1100	4.1	1.2
	1721	0.0	0.0
	2330	3.5	1.1
9 M	0530	0.6	0.2
	1141	3.9	1.2
	1802	0.2	0.1
10 Tu	0012	3.6	1.1
	0520	0.7	0.2
	1227	3.7	1.1
	1849	0.2	0.1
11 W	0100	3.6	1.1
	0726	0.7	0.2
	1316	3.5	1.1
	1943	0.2	0.1
12 Th	0150	3.8	1.2
	0831	0.6	0.2
	1412	3.4	1.0
	2039	0.2	0.1
13 F	0248	4.0	1.2
	0930	0.4	0.1
	1514	3.4	1.0
	2130	0.0	0.0
14 Sa	0350	4.3	1.3
	1026	0.1	0.0
	1621	3.5	1.1
	2222	-0.2	-0.1
15 Su	0449	4.6	1.4
	1120	-0.2	-0.1
	1720	3.7	1.1
	2316	-0.4	-0.1
16 M	0542	5.0	1.5
	1213	-0.5	-0.2
	1813	4.0	1.2
17 Tu	0009	-0.6	-0.2
	0632	5.4	1.6
	1306	-0.8	-0.2
	1903	4.3	1.3
18 W	0105	-0.8	-0.2
	0721	5.6	1.7
	1356	-1.1	-0.3
	1951	4.5	1.4
19 Th	0158	-0.9	-0.3
	0808	5.6	1.7
	1446	-1.3	-0.4
	2043	4.6	1.4
20 F	0250	-1.0	-0.3
	0859	5.6	1.7
	1533	-1.3	-0.4
	2134	4.7	1.4
21 Sa	0341	-1.0	-0.3
	0951	5.4	1.6
	1619	-1.2	-0.4
	2229	4.7	1.4
22 Su	0432	-0.8	-0.2
	1045	5.1	1.6
	1709	-1.0	-0.3
	2322	4.7	1.4
23 M	0529	-0.6	-0.2
	1139	4.7	1.4
	1802	-0.8	-0.2
24 Tu	0018	4.6	1.4
	0630	-0.3	-0.1
	1234	4.3	1.3
	1858	-0.5	-0.2
25 W	0114	4.5	1.4
	0735	-0.1	0.0
	1330	4.0	1.2
	1959	-0.3	-0.1
26 Th	0212	4.5	1.4
	0840	0.0	0.0
	1432	3.7	1.1
	2057	-0.2	-0.1
27 F	0311	4.4	1.3
	0942	0.0	0.0
	1535	3.5	1.1
	2153	-0.1	0.0
28 Sa	0414	4.5	1.4
	1038	-0.1	0.0
	1639	3.5	1.1
	2245	-0.1	0.0
29 Su	0510	4.6	1.4
	1131	-0.2	-0.1
	1736	3.6	1.1
	2336	-0.1	0.0
30 M	0558	4.7	1.4
	1221	-0.3	-0.1
	1822	3.7	1.1
31 Tu	0027	-0.2	-0.1
	0642	4.8	1.5
	1307	-0.4	-0.1
	1904	3.8	1.2

FEBRUARY

Day	Time (h m)	Height (ft)	Height (m)
1 W	0112	-0.2	-0.1
	0722	4.8	1.5
	1349	-0.4	-0.1
	1944	3.9	1.2
2 Th	0154	-0.2	-0.1
	0759	4.8	1.5
	1429	-0.5	-0.2
	2022	3.9	1.2
3 F	0235	-0.2	-0.1
	0838	4.7	1.4
	1504	-0.4	-0.1
	2058	3.9	1.2
4 Sa	0312	-0.1	0.0
	0914	4.5	1.4
	1539	-0.4	-0.1
	2135	3.9	1.2
5 Su	0347	0.0	0.0
	0949	4.3	1.3
	1611	-0.2	-0.1
	2211	3.9	1.2
6 M	0419	0.1	0.0
	1027	4.1	1.2
	1642	-0.1	0.0
	2250	3.9	1.2
7 Tu	0456	0.3	0.1
	1104	3.9	1.2
	1713	0.1	0.0
	2327	3.9	1.2
8 W	0535	0.4	0.1
	1146	3.7	1.1
	1750	0.2	0.1
9 Th	0010	3.9	1.2
	0630	0.6	0.2
	1232	3.5	1.1
	1839	0.3	0.1
10 F	0100	4.0	1.2
	0742	0.6	0.2
	1327	3.4	1.0
	1941	0.3	0.1
11 Sa	0200	4.1	1.2
	0852	0.4	0.1
	1432	3.3	1.0
	2051	0.2	0.1
12 Su	0306	4.3	1.3
	0956	0.2	0.1
	1548	3.4	1.0
	2153	0.0	0.0
13 M	0417	4.6	1.4
	1053	-0.2	-0.1
	1657	3.7	1.1
	2252	-0.3	-0.1
14 Tu	0520	5.0	1.5
	1150	-0.5	-0.2
	1753	4.1	1.2
	2352	-0.6	-0.2
15 W	0613	5.3	1.6
	1245	-0.9	-0.3
	1845	4.5	1.4
16 Th	0050	-0.9	-0.3
	0705	5.6	1.7
	1335	-1.2	-0.4
	1935	4.9	1.5
17 F	0145	-1.1	-0.3
	0754	5.7	1.7
	1424	-1.4	-0.4
	2024	5.1	1.6
18 Sa	0235	-1.2	-0.4
	0843	5.6	1.7
	1510	-1.4	-0.4
	2113	5.2	1.6
19 Su	0326	-1.2	-0.4
	0932	5.4	1.6
	1555	-1.3	-0.4
	2205	5.2	1.6
20 M	0416	-1.1	-0.3
	1024	5.1	1.6
	1642	-1.0	-0.3
	2256	5.1	1.6
21 Tu	0508	-0.7	-0.2
	1116	4.7	1.4
	1730	-0.7	-0.2
	2349	4.9	1.5
22 W	0604	-0.4	-0.1
	1209	4.2	1.3
	1824	-0.3	-0.1
23 Th	0041	4.7	1.4
	0705	0.0	0.0
	1303	3.8	1.2
	1923	0.0	0.0
24 F	0138	4.5	1.4
	0810	0.2	0.1
	1401	3.5	1.1
	2027	0.2	0.1
25 Sa	0240	4.3	1.3
	0915	0.2	0.1
	1509	3.3	1.0
	2127	0.3	0.1
26 Su	0343	4.2	1.3
	1014	0.2	0.1
	1617	3.3	1.0
	2223	0.3	0.1
27 M	0446	4.3	1.3
	1106	0.1	0.0
	1715	3.5	1.1
	2315	0.2	0.1
28 Tu	0536	4.5	1.4
	1156	0.0	0.0
	1805	3.8	1.2
29 W	0005	0.1	0.0
	0621	4.6	1.4
	1239	-0.2	-0.1
	1845	4.0	1.2

MARCH

Day	Time (h m)	Height (ft)	Height (m)
1 Th	0052	0.0	0.0
	0701	4.7	1.4
	1322	-0.3	-0.1
	1922	4.2	1.3
2 F	0134	-0.1	0.0
	0738	4.7	1.4
	1400	-0.4	-0.1
	1957	4.3	1.3
3 Sa	0214	-0.2	-0.1
	0812	4.7	1.4
	1436	-0.4	-0.1
	2031	4.3	1.3
4 Su	0250	-0.2	-0.1
	0848	4.6	1.4
	1507	-0.3	-0.1
	2104	4.3	1.3
5 M	0325	-0.1	0.0
	0923	4.4	1.3
	1539	-0.2	-0.1
	2136	4.3	1.3
6 Tu	0355	0.0	0.0
	0955	4.2	1.3
	1605	0.0	0.0
	2211	4.3	1.3
7 W	0429	0.1	0.0
	1032	4.0	1.2
	1632	0.1	0.0
	2248	4.3	1.3
8 Th	0504	0.3	0.1
	1112	3.8	1.2
	1704	0.1	0.0
	2330	4.3	1.3
9 F	0554	0.5	0.2
	1159	3.6	1.1
	1749	0.4	0.1
10 Sa	0023	4.3	1.3
	0703	0.6	0.2
	1257	3.4	1.0
	1857	0.5	0.2
11 Su	0125	4.3	1.3
	0821	0.5	0.2
	1407	3.4	1.0
	2021	0.5	0.2
12 M	0235	4.4	1.3
	0929	0.2	0.1
	1524	3.6	1.1
	2135	0.2	0.1
13 Tu	0350	4.6	1.4
	1031	-0.1	0.0
	1636	4.0	1.2
	2238	-0.1	0.0
14 W	0459	5.0	1.5
	1127	-0.5	-0.2
	1736	4.5	1.4
	2336	-0.5	-0.2
15 Th	0555	5.3	1.6
	1219	-0.8	-0.2
	1827	5.0	1.5
16 F	0035	-0.9	-0.3
	0646	5.6	1.7
	1310	-1.1	-0.3
	1915	5.4	1.6
17 Sa	0129	-1.1	-0.3
	0735	5.7	1.7
	1359	-1.3	-0.4
	2002	5.7	1.7
18 Su	0220	-1.3	-0.4
	0822	5.6	1.7
	1445	-1.3	-0.4
	2049	5.7	1.7
19 M	0309	-1.2	-0.4
	0910	5.3	1.6
	1529	-1.1	-0.3
	2138	5.6	1.7
20 Tu	0357	-1.0	-0.3
	0958	5.0	1.5
	1612	-0.8	-0.2
	2229	5.4	1.6
21 W	0445	-0.7	-0.2
	1049	4.5	1.4
	1658	-0.4	-0.1
	2317	5.1	1.6
22 Th	0538	-0.3	-0.1
	1143	4.1	1.2
	1749	0.1	0.0
23 F	0010	4.8	1.5
	0636	0.1	0.0
	1236	3.8	1.2
	1847	0.4	0.1
24 Sa	0105	4.5	1.4
	0740	0.4	0.1
	1336	3.5	1.1
	1955	0.7	0.2
25 Su	0204	4.2	1.3
	0845	0.5	0.2
	1440	3.4	1.0
	2059	0.8	0.2
26 M	0308	4.1	1.2
	0942	0.4	0.1
	1548	3.4	1.0
	2158	0.7	0.2
27 Tu	0409	4.1	1.2
	1035	0.3	0.1
	1646	3.7	1.1
	2251	0.5	0.2
28 W	0507	4.3	1.3
	1120	0.1	0.0
	1736	4.0	1.2
	2343	0.3	0.1
29 Th	0552	4.5	1.4
	1204	0.0	0.0
	1816	4.3	1.3
30 F	0026	0.1	0.0
	0632	4.6	1.4
	1246	-0.1	0.0
	1853	4.5	1.4
31 Sa	0109	0.0	0.0
	0712	4.7	1.4
	1325	-0.2	-0.1
	1927	4.7	1.4

Time meridian 75° W. 0000 is midnight. 1200 is noon.
Heights are referred to mean low water which is the chart datum of soundings.

Figure II-5-13.   Tide tables for Sandy Hook, NJ (NOAA 1984)

(g)   Reconstruction of the tide involves determining the equilibrium arguments, node factors, longitude and time adjustment for each constituent, and using the published values for κ, H_n, and H_0 in the NOS harmonic analysis. Table II-5-6 summarizes all necessary quantities.

Table II-5-6
Harmonic Arguments for Sandy Hook NJ (1 January 1984 at 0000 hr)

Symbol	G(V_0+ u)	pL	aS/15	κ	f	H
M_2	60.0	148.0	144.92	219.1	0.99	2.151
S_2	0.0	148.0	150.00	246.0	1.00	0.448
N_2	323.4	148.0	142.20	204.1	0.99	0.473
K_1	2.4	74.0	75.20	102.2	1.04	0.319
Sa	279.8	0.0	0.20	128.7	1.00	0.254
O_1	60.8	74.0	69.71	98.4	1.07	0.172
v_2	148.0	148.0	142.56	191.1	0.99	0.109
K_2	184.2	148.0	150.41	251.9	1.09	0.121

(h)   The NOS harmonic analysis indicates that a multiplier of 1.04 should be used for all short-period constituents and that the value of H_0 is 2.36 MSL. Also indicated on the analysis is the time meridian of 75° west longitude for use in computing the time zone compensation term $aS/15$. The 15-day tidal envelope is shown in Figure II-5-14. The open circles shown in the figure represent high- and low-water level predictions extracted from the tide tables in Figure II-5-13. As stated, the comparison is not exact because only eight constituents were used in the reconstruction. The match is, however, adequate for the majority of design applications.

(i)   The phase lag κ in Equation II-5-19 is called the local epoch in order to distinguish it from other forms of epochs (see Schureman (1924)). Some harmonic analyses use a modified form of the epoch that automatically accounts for the longitude and time meridian corrections. This modification is designated as κ' and is defined as shown below

$$\kappa' = \kappa + pL - \frac{aS}{15} \tag{II-5-20}$$

(j)   This modified form is usually included on NOS harmonic analyses as indicated on Figure II-5-12. Use of this form of epoch in the reconstruction of tides is as shown below

$$H(t)_{local} = H_0 + \sum_{n=1}^{N} f_n H_n \cos\left[a_n t + \text{Greenwich}\,(V_0 + u) - \kappa'_n\right] \tag{II-5-21}$$

(4)   Tidal envelope classification.

(a)   Semidiurnal, diurnal, and mixed tidal classifications were described in Part II-5-3a. Equation II-5-22 is a more quantitative delineation of tide types.

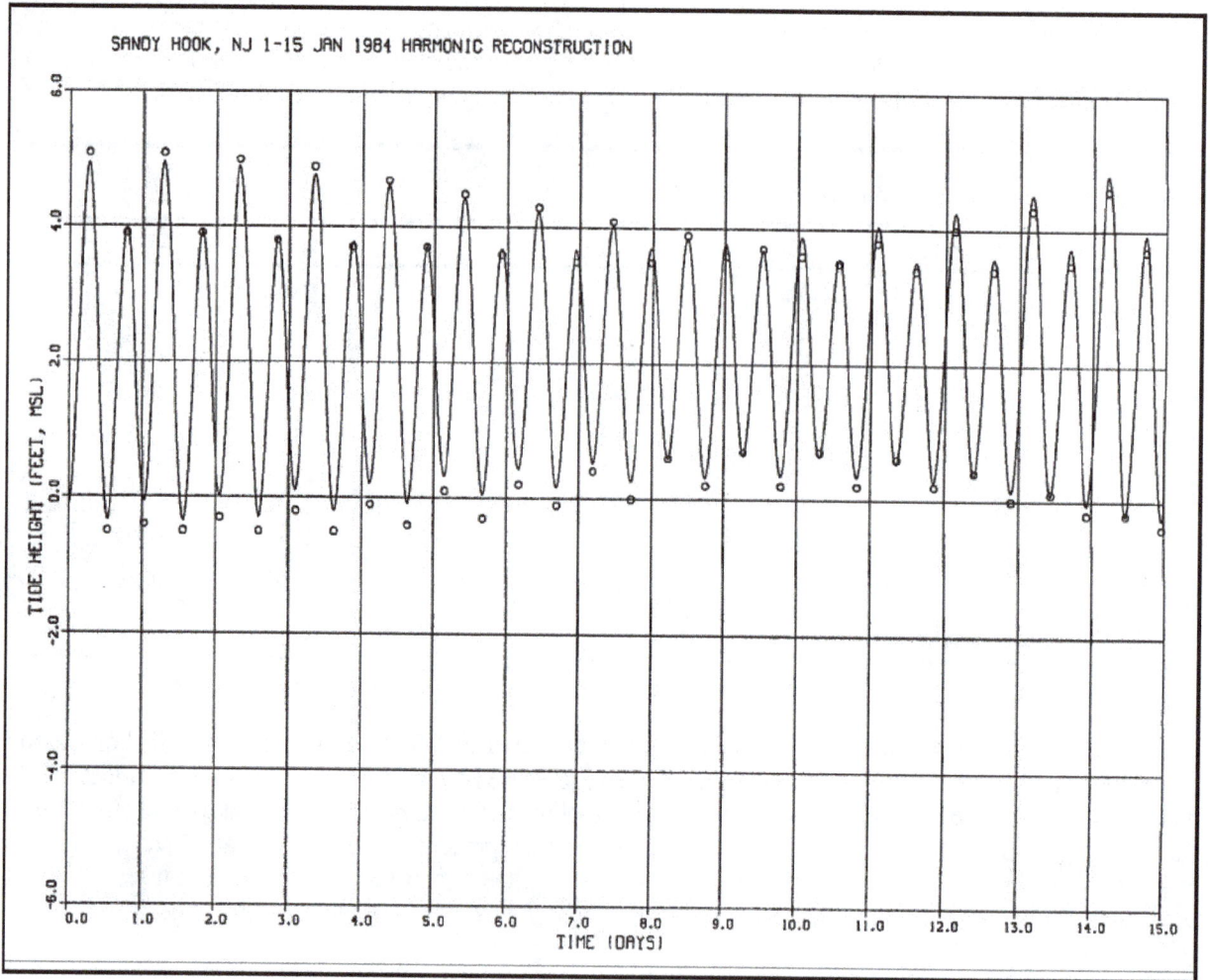

Figure II-5-14.   Reconstructed tidal envelope for Sandy Hook, NJ

$$R = \frac{A(K_1) + A(O_1)}{A(M_2) + A(s_2)}$$

(II-5-22)

where $A(K_1)$, $A(O_1)$, $A(M_2)$, and $A(S_2)$ represent the amplitudes of the corresponding constituents. A general classification of tides can be separated according to the following criteria:

$R \leq 0.25$ Semidiurnal
$0.25 < R \leq 1.50$ Mixed
$1.50 < R$ Diurnal

(b)   The tidal classification for the Sandy Hook, NJ, example can be computed as shown below:

$$R = \frac{A(K_1) + A(O_1)}{A(M_2) + A(s_2)} = \frac{0.319 + 0.172}{2.151 + 0.448} = 0.189$$

(II-5-23)

(c)   According to the classification criteria, the tides at Sandy Hook are semidiurnal. In fact, most tides along the northern east coast of the United States are semidiurnal. Tides in the lower east coast and Gulf of Mexico begin to change from semidiurnal, to mixed, to diurnal as shown in Figure II-5-15.

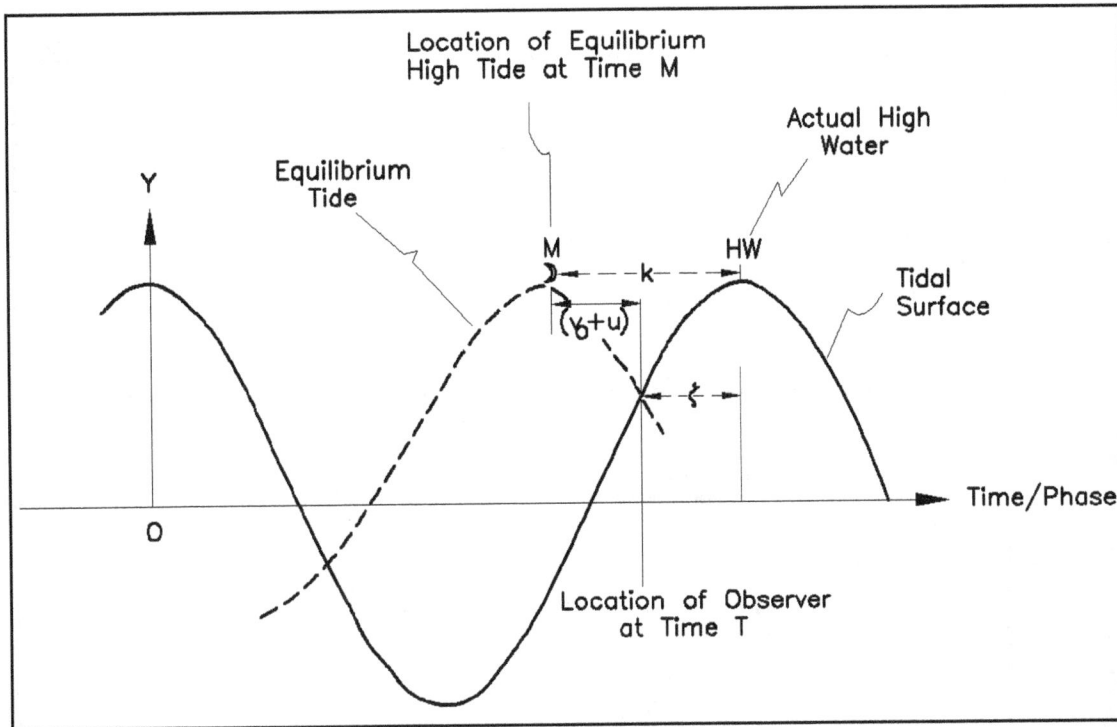

Figure II-5-15.   Areal extent of tidal types (Harris 1981)

c.   Glossary of tide elevation terms.

(1)   The differences in tidal envelopes have given rise to certain terminologies regarding the high and low crests of the tide. The following list of terms, reprinted from the "Glossary of Terms" of the *Shore Protection Manual* (1984), describes specific aspects of typical tidal records. Each term is referenced to Figure II-5-16.

(2)   DATUM. A base elevation from which vertical heights or depths are referenced. The reference elevation is locally defined; therefore, a list of commonly used datums is presented in Part II-5-4.

(3)   DIURNAL TIDE. A tide with one high water and one low water in a tidal day.

(4)   EBB CURRENT. The tidal current away from shore or down a tidal stream; usually associated with the decrease in height of a tide, associated with a falling tide.

(5)   EBB TIDE. The period of tide between high water and the succeeding low water: a falling tide.

(6)   FLOOD CURRENT. The tidal current toward shore or up a tidal stream, usually associated with the increase in height of the tide associated with a rising tide.

(7)   FLOOD TIDE. The period of tide between low water and the succeeding high water; a rising tide.

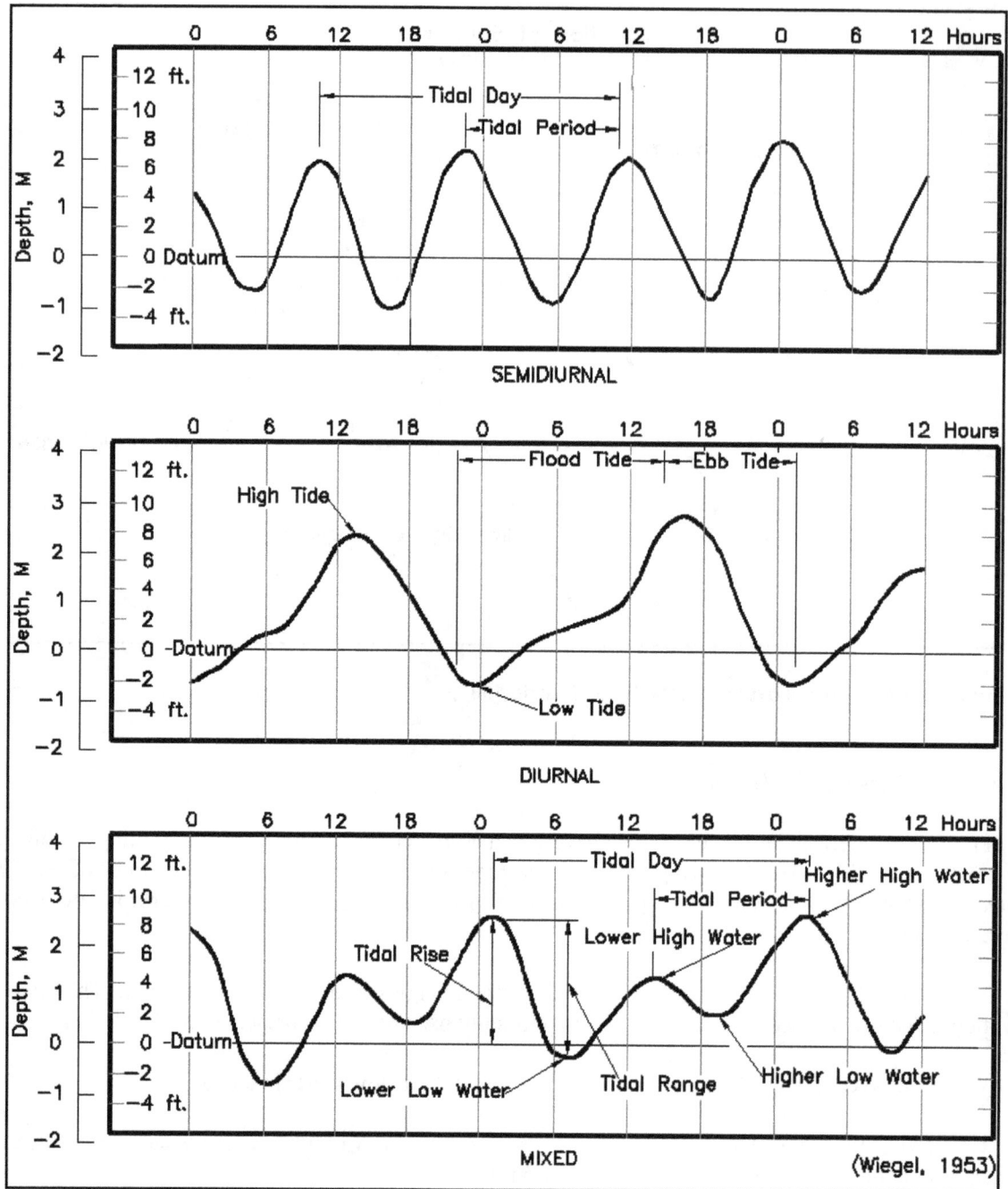

Figure II-5-16. Types of tides (*Shore Protection Manual* 1984)

(8) HIGHER HIGH WATER (HHW). The higher of the two high waters of any tidal day. The single high water occurring daily during periods when the tide is diurnal is considered to be a higher high water.

(9) HIGHER LOW WATER (HLW). The higher of two low waters of any tidal day.

(10) HIGH TIDE, HIGH WATER (HW). The maximum elevation reached by each rising tide.

(11) LOW TIDE, LOW WATER (LW). The minimum elevation reached by each falling tide.

(12) LOWER HIGH WATER (LHW). The lower of the two high waters of any tidal day.

(13) LOWER LOW WATER (LLW). The lower of the two low waters of any tidal day. The single low water occurring daily during periods when the tide is diurnal is considered to be a lower low water.

(14) MIXED TIDE. A tide in which the presence of a diurnal wave is conspicuous due to a large inequality in either the high or low water heights, with two high waters and two low waters usually occurring each tidal day. In strictness, all tides are mixed, but the name is usually applied without definite limits to the tide intermediate to those predominantly semidiurnal and those diurnal.

(15) SEMIDIURNAL TIDE. A tide with two high and two low waters in a tidal day with comparatively little diurnal inequality.

(16) TIDAL DAY. The time of the rotation of the earth with respect to the moon, or the interval between two successive upper transits of the moon over the meridian of a place, approximately 24.84 solar hours (24 hr and 50 min) or 1.035 times per solar day. Also called a lunar day.

(17) TIDAL PERIOD. The interval of time between two consecutive, like phases of the tide.

(18) TIDAL RANGE. The difference in height between consecutive high and low (or higher high and lower low) waters.

(19) TIDAL RISE. The height of the tide as referenced to the datum of a chart.

II-5-4. Water Surface Elevation Datums

  a. Introduction. Water level and its change with respect to time have to be measured relative to some specified elevation or datum in order to have a physical significance. In the fields of coastal engineering and oceanography this datum represents a critical design parameter because reported water levels provide an indication of minimum navigational depths or maximum surface elevations at which protective levees or berms are overtopped. It is therefore necessary that coastal datums represent some reference point which is universally understood and meaningful, both onshore and offshore. Ideally, two criteria should be expected of a datum: 1) that it provides local depth of water information, and 2) that it is fixed regardless of location such that elevations at different locations can be compared. These two criteria are not necessarily compatible. The following list of datums represents those commonly in use in the United States.

  b. Tidal-observation-based datums.

  (1) Tidal-observation-based datums are computed from measured time series of tidal elevations. As such, they vary with geographical location and exposure. Geographic variability of tide records can be seen in Figures II-5-8 and II-5-9. Although datums based on tidal series do indicate site-specific conditions, they

may not be comparable with similar measurements at other geographic locations because of the difficulty of assigning a comparable gauge zero. The most widely accepted of the datums computed from time series are described below.

(2)   Mean sea level (MSL) was widely adopted as a primary datum on the assumption that it could be accurately computed from tidal elevation records measured at any well-exposed tide gauge.   MSL determinations are based on the arithmetic average of hourly water surface elevations observed over a long period of time. The ideal length of record is approximately 19 years, a period that accounts for the 18- to 19-year long-term cycle in tides and is sufficient to remove most meteorological effects.   In order to fix the datum in time for a specific location, a common 19-year period is selected for computing MSL. The specific 19-year cycle was adopted by the National Ocean Survey as the official time segment for use in computing mean values for tidal datums.   The 19-year period, called the National Tidal Datum Epoch, is updated approximately every 25 years.

(3)   When estimates of MSL are required, but less than 19 years of data are available, computations should be based on an integral number of tidal cycles, for example, an integral number of years or 29-day spring/neap cycles. For gauges where hourly data are not available, or their use is impractical, MSL can be approximated as the tidal datum midway between MHW and MLW. This datum, referred to as Mean Tide Level (MTL), may differ from MSL depending on the local relative importance of the diurnal components of the tide.

(4)   Alternate tidal datums are based on low water tidal elevations.   These datums provide minimum depth information for navigational needs.   Two commonly used low-water datums in the United States are the Mean Low Water (MLW) for the Atlantic Coast and the Mean Lower Low Water (MLLW) for the Pacific coast.   The MLLW datum is currently being adopted to several locations along the Atlantic coast.   Both datums are defined as the average tidal height at low water or lower low water during the 19-year period. Additional datums, applicable to specific locations or purposes, include Mean High Water (MHW) and Mean Higher High Water (MHHW).   These are derived in a manner similar to MLW and MLLW.   For areas of primarily semidiurnal tides, the difference between MHW and MLW is called the mean tidal range.   The difference between MHHW and MLLW gives a corresponding diurnal range or the great diurnal range of the tide.   Both numbers provide an estimate of the magnitude of the local tidal range.   An example of the variability of the above datums is given by Harris (1981) and reproduced in Table II-5-7.   Reference tide stations used in the preparation of Table II-5-7 are shown in Figure II-5-17.

c.   *1929 NGVD datum.*

(1)   One difficulty with using any of the observation-based datums described above is that they vary considerably with location, as evidenced in Table II-5-7.   Also, because each datum can be computed independently, there is little or no connectivity between datum locations. This lack of a reference elevation for areas near the coast where no tide observations are available and at interior locations where tide observations are difficult to obtain led to the establishment of a national fixed datum.   This datum does not account for spatial variability in sea level. The following paragraphs describe the development of the 1929 National Geodetic Vertical Datum (NGVD).

(2)   First-order leveling lines established in the mid-1920's provided survey connections between the Atlantic and Pacific coasts. These surveys indicated that sea levels were higher on the Pacific coast than on the Atlantic coast and were also higher in the north than in the south on both coasts. The goal of developing a fixed reference datum was accomplished by defining a geodetic leveling-based datum whose "zero" coincided with local MSL at locations at which both MSL and geodetic leveling elevations were known.

Table II-5-7
Datums for Reference Tide Stations (Harris 1981)

Station	Normalizing Factor[2]	MSL	MTL	NGVD	MLLW	MLW	MHW	MHHW	Extremes of Record Highest	Extremes of Record Lowest	Interval for Establishing of Datum
						Atlantic and Gulf Coasts					
Eastport, Maine	M	9.2	9.10	9.00	--[3]	0.00	18.20	--[3]	23.1	-4.4	1941-61
Portland, Maine	M	4.5	4.50	4.28		0.00	9.00		13.9	-3.7	1941-59
Boston, Mass.	M	5.2	5.05	4.89		0.30[4]	9.80		14.2	-3.5	1941-59
Newport, R.I.	M	1.6	1.75	1.37		0.00	3.50		10.7[5]	-3.4	1941-59
New London, Conn.	M	1.4	1.30	0.97		0.00	2.60		10.7[5]	-3.4	1941-59
Bridgeport, Conn.	M	3.4	3.35	2.86		0.00	6.70		12.4	-3.5[5]	1967 (1 yr)
Willets Point, N.Y.	M	3.6	3.55	3.02		0.00	7.10		16.7	-4.1	1941-59
New York, N.Y. (The Battery)	M	2.3	2.25	1.81		0.00	4.50		10.2	-4.2	1941-59
A bany, N.Y.	M	2.5	2.5	--[6]		0.00	4.60		--[6]	--[6]	1941-59
Sandy Hook, N.J.	M	2.3	2.30	1.79		0.00	4.60		10.3	-4.4	1941-59
Breakwater Harbor, Del.	M	2.1	2.05	1.69		0.00	4.10		9.5	-3.9	1953-61
Reedy Point, Del.	M	2.8	2.75	2.45		0.00	5.50		10.0[5]	-6.3	1957-61
Philadelphia, Pa.	M	3.2	3.10	2.14		0.00	6.19[7]		10.7	-6.6	1941-59
Baltimore, Md.	M	1.0	0.97	0.57		0.42[8]	1.52		8.3	-4.5	1941-59
Washington, D.C.	M	1.97	1.97	1.43		0.52[9]	3.42		11.9	-4.2	1941-59
Hampton Roads, Va. (Sewells Point)	M	1.3	1.25	1.28		0.00	2.50		8.5	-3.1	1941-59
Wilmington, N.C.	M	1.9	2.10	1.52		0.00	4.20		8.2	-1.7	Jan. 1969 to Nov. 1973
Charleston, S.C.	M	2.7	2.91	2.65		0.31[10]	5.51		10.7	-3.3	1941-59
Savannah River Entrance, Ga. (Ft. Pulaski)	M	3.6	3.45	3.32		0.00	6.90		11.1	-4.4	1941-59
Savannah, Ga.	M	4.0	3.7	--[6]		0.00	7.40		--[6]	--[6]	1941-59
Mayport, Fla.	M	2.3	2.25	2.00		0.00	4.50		7.4	-3.2	1941-59
Miami Harbor Entrance, Fla.	M	1.3	1.25	0.96		0.00	2.50		6.4	-1.6	1941-59
Key West, Fla.	M	0.6	0.65	0.42		0.00	1.30		3.8	-1.6	1941-59
St. Petersburg, Fla.	D	1.2	1.15	0.83		0.00	2.30		5.3	-2.5[5]	1941-59

[1] Except as footnoted, chart datum is MLW for the Atlantic and gulf coasts and MLLW for the Pacific coast.

[2] D = diurnal; M = mean.

[3] MLLW and MHHW were not routinely derived for Atlantic and gulf coast stations for the 1941-59 epoch used for establishing most of the datums.

[4] Boston Low Water Datum (adopted about 1927).

[5] Estimated value.

[6] Data unavailable at the time of compilation.

[7] Datums and the mean range for Philadelphia were revised in July 1979 based on observations for the period 1969-77.

[8] Baltimore Low Water Datum (adopted 1922 by NOS and COE based on observations 1903-1921).

[9] Mean River Level.

[10] Charleston Low Water Datum (used since 1905 by NOS, and by COE in May 1929).

[11] Low Water Datum at the Presidio, Golden Gate, San Francisco, is based on miscellaneous observations before 1907.

(Continued)

Table II-5-7 (Concluded)

Station	Normalizing Factor[2]	MSL	MTL	NGVD	MLLW	MLW	MHW	MHHW	Extremes of Record		Interval for Establishing of Datum
									Highest	Lowest	
Atlantic and Gulf Coasts (Continued)											
St. Marks River Entrance, Fla.	D	1.8	1.8	--[6]		0.00	2.40		8.0	-3.5	1941-59
Pensacola, Fla.	D	0.6	0.65	0.33		0.00	1.30		8.9	-2.2	1941-59
Mobile, Ala.	D	0.8	0.75	--[6]		0.00	1.50		9.0[5]	-3.0[5]	1941-59
Galveston, Tex. (Ship channel)	D	0.8	0.70	0.70		0.00	1.40		11.4	-5.3	1941-59
San Juan, P.R.	D	0.6	0.55	--[6]		0.00	1.10		2.4	-1.1	Apr. 1962 to Dec. 1963
Pacific Coast											
San Diego, Calif.	D	3.0	2.95	2.79	0.00	0.90	5.00	5.70	8.3	-2.8	1941-59
Los Angeles, Calif. (Outer Harbor)	D	2.8	2.80	2.72	0.00	0.90	4.70	5.40	7.8	-2.6	1941-59
San Francisco, Calif. (Golden Gate)	D	3.0	3.30	3.06	0.20[11]	1.30	5.30	5.90	8.6	-2.5	1941-59
Humboldt Bay, Calif	D	3.4	3.75	--	0.00	1.20	5.70	6.40	9.5[5]	-3.0[5]	1962
Astoria, Oreg. (Tongue Point)[9]	D	4.3	4.35	3.05	0.00	1.10	7.60	8.30	12.1	-2.8	1941-59
Aberdeen, Wash.	D	5.5	5.45	--	0.00	1.50	9.40	10.10	14.9	-2.9	1955, 1956
Port Townsend, Wash.	D	4.8	5.10	--	0.00	2.50	7.70	8.40	12.0[5]	-4.5[5]	1972-74
Seattle, Wash.	D	6.6	6.60	6.25	0.00	2.80	10.40	11.30	14.8	-4.7	1941-59
Ketch kan, Alaska	D	8.0	7.95	--	0.00	1.50	14.40	15.30	21.2	-5.2	1941-59
Juneau, Alaska	D	8.6	8.50	--	0.00	1.60	15.40	16.40	23.2	-5.2	1966-72
Sitka, Alaska	D	5.2	5.30	--	0.00	1.40	9.10	9.90	14.6	-3.8	1941-59
Cordova, Alaska	D	6.6	6.45	--	0.00	1.40	11.50	12.40	16.8	-4.9	1965-74
Seldovia, Alaska	D	9.4	9.35	--	0.00	1.60	17.00	17.80	24.3	-6.2	1971-74
Anchorage, Alaska	D	16.8	15.25	--	0.00	2.20	28.30	29.00	35.5[5]	-6.5[5]	1964-68
Kodiak, Alaska	D	4.3	4.30	--	0.00	1.00	7.60	8.50	13.0[5]	-4.0[5]	1935-36
Dutch Harbor, Alaska	D	2.2	1.30	--	0.00	1.20	3.40	3.70	6.6	-2.7	1935-38
Sweeper Cove, Alaska (Adak Island)	D	2.1	1.85	--	0.00	--	--	3.70	7.0[5]	-3.3	1958-60; 1944, 1949
Massacre Bay, Alaska (Attu Island)	D	1.9	1.65	--	0.00	--	--	3.30	7.0[5]	-3.0[5]	1950, 1952, 1960
Nashagak Bay, Alaska (Clarks Pt)	D	10.3	10.15	--	0.00	2.50	17.80	19.50	24.5	-5.0	1958
St. Michael, Alaska	D	2.0	--	--	--	--	--	--	--	--	--
Honolulu, Hawaii	D	0.8	0.80	--	0.00	0.20	1.40	1.90	3.5	-1.3	1941-59

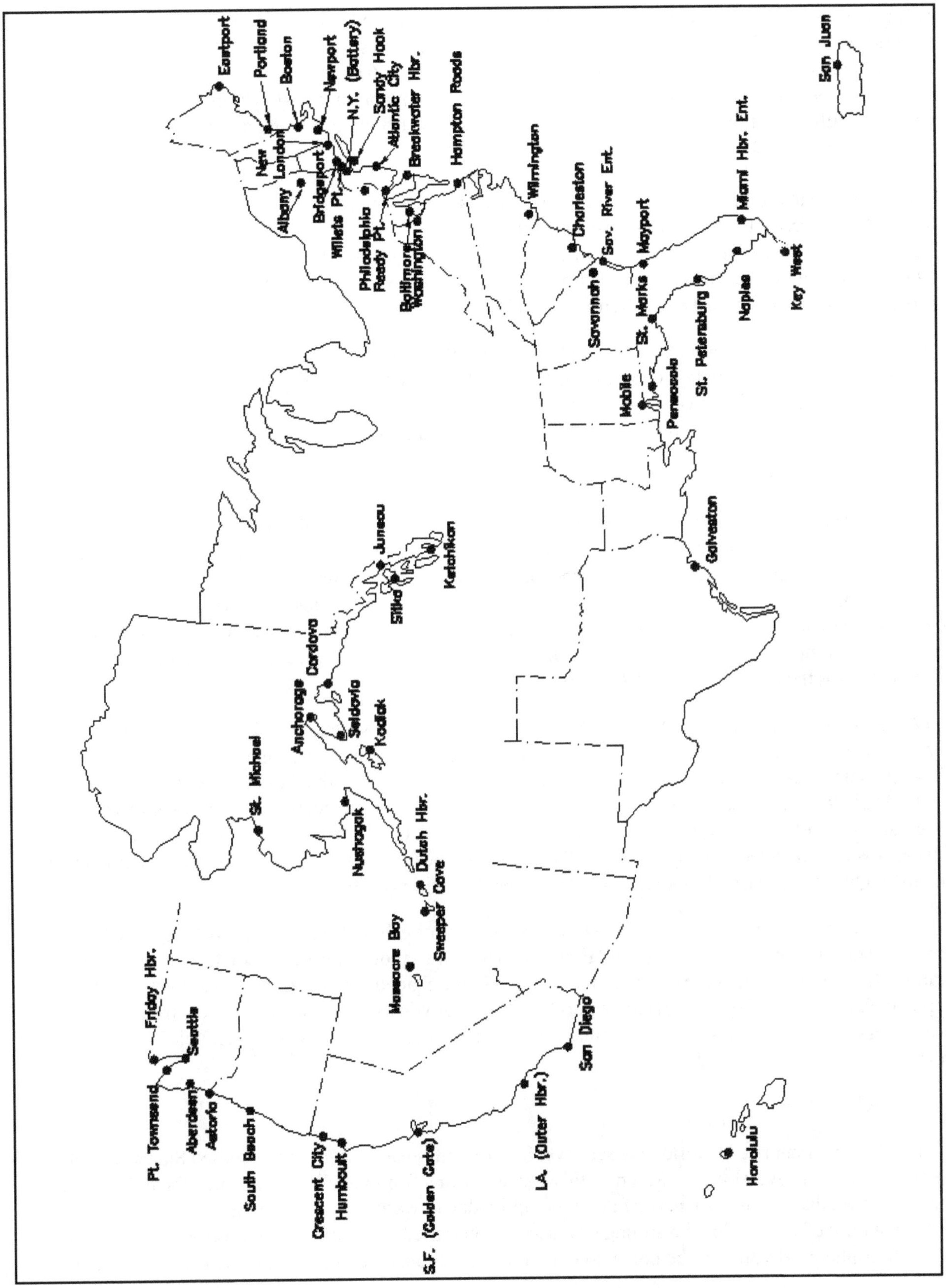

Figure II-5-17.   Reference and comparative tide stations, Atlantic, Gulf, and Pacific coasts (Harris 1981)

Thus, a general adjustment was made in 1929 in which it was assumed that the geodetic and local sea levels were equal to zero at 26 selected tide gauges in the United States and Canada (Rappleye 1932). The differences previously computed were treated as errors and were distributed over the network of observation points (Harris 1981). The tide gauge locations used in this computation are shown in Figure II-5-18. The datum, originally called the "Sea Level Datum of 1929," was officially renamed the "National Geodetic Vertical Datum (NGVD) of 1929" in 1963.

(3) Index maps of tidal benchmarks and lists of established references between the 1929 NGVD and the local MSL are available for each coastal state through the NOS. An NOS tidal benchmark sheet is shown in Figure II-5-19. These sheets often describe several benchmarks established in the vicinity of each tidal observation point and describe the relationship between the various datums common to the area. NGVD and tidal benchmarks are discussed more thoroughly in Harris (1981).

(4) The primary distinction between the NGVD and all other tidal datums is that the NGVD is defined as a fixed surface whose elevation does not change with time. Therefore, the procedure adopted to account for recognized changes in relative mean sea level is to compute updated MSL (or other) datum relationships. Relative changes in sea level will be discussed in the following section.

   d.   Great Lakes datums.

(1) A separate water surface elevation datum was established for the Great Lakes basin and St. Lawrence River. The datum was originally established by an international coordination committee composed of representatives of the U.S. Army Corps of Engineers, the National Oceanic and Atmospheric Administration, and the Department of the Environment, Canada. This first datum was called the International Great Lakes Datum (IGLD) of 1955. The "zero" of the datum was established as the average of all hourly water surface water level readings at Pointe-au-Pere, Quebec, located on the Gulf of St. Lawrence for the period between 1941 and 1956.

(2) First-order leveling lines from Pointe-au-Pere were used to systematically define datums for Lake Ontario, Lake Erie, Lakes Michigan and Huron, and Lake Superior. Lakes Michigan and Huron are assumed to have the same elevation because of the deep and wide connection of both lakes at the Straits of Mackinac. The IGLD of 1955 was replaced by the IGLD of 1985 to reflect certain corrections to the elevations assigned to the various lakes. Elevations of the IGLD of 1955 and IGLD of 1985 are given in Table II-5-8. This revision was implemented in January 1992. The zero reference of the IGLD 1985 has been specified to be Rimouski, Quebec. Revised lake elevations are shown in Figure II-5-20.

   e.   Long-term variations in datums. Water level observation based datums such as MSL or IGLD vary over time periods much longer than a tidal cycle. These variations can be seasonal or of much longer duration. Long-term changes are often described as a relative change in sea level and necessitate the 25-year interval updating of the 19-year tidal datum period. In the following section, some of the contributing factors to sea level change are discussed. In the final section, factors contributing to long-term elevation changes in the Great Lake are presented.

   f.   Tidal datums.

(1) The apparent rise in worldwide sea level has been of great concern to the United States, as well as other countries, for several years. Much of this concern stems from the claims of some climatologists and oceanographers that the rise will accelerate in the future due to warming of the atmosphere associated with the "greenhouse effect," a global warming produced by increased levels of carbon dioxide and other gasses in the atmosphere. Because of the potential consequences associated with sea level rise, a Committee on

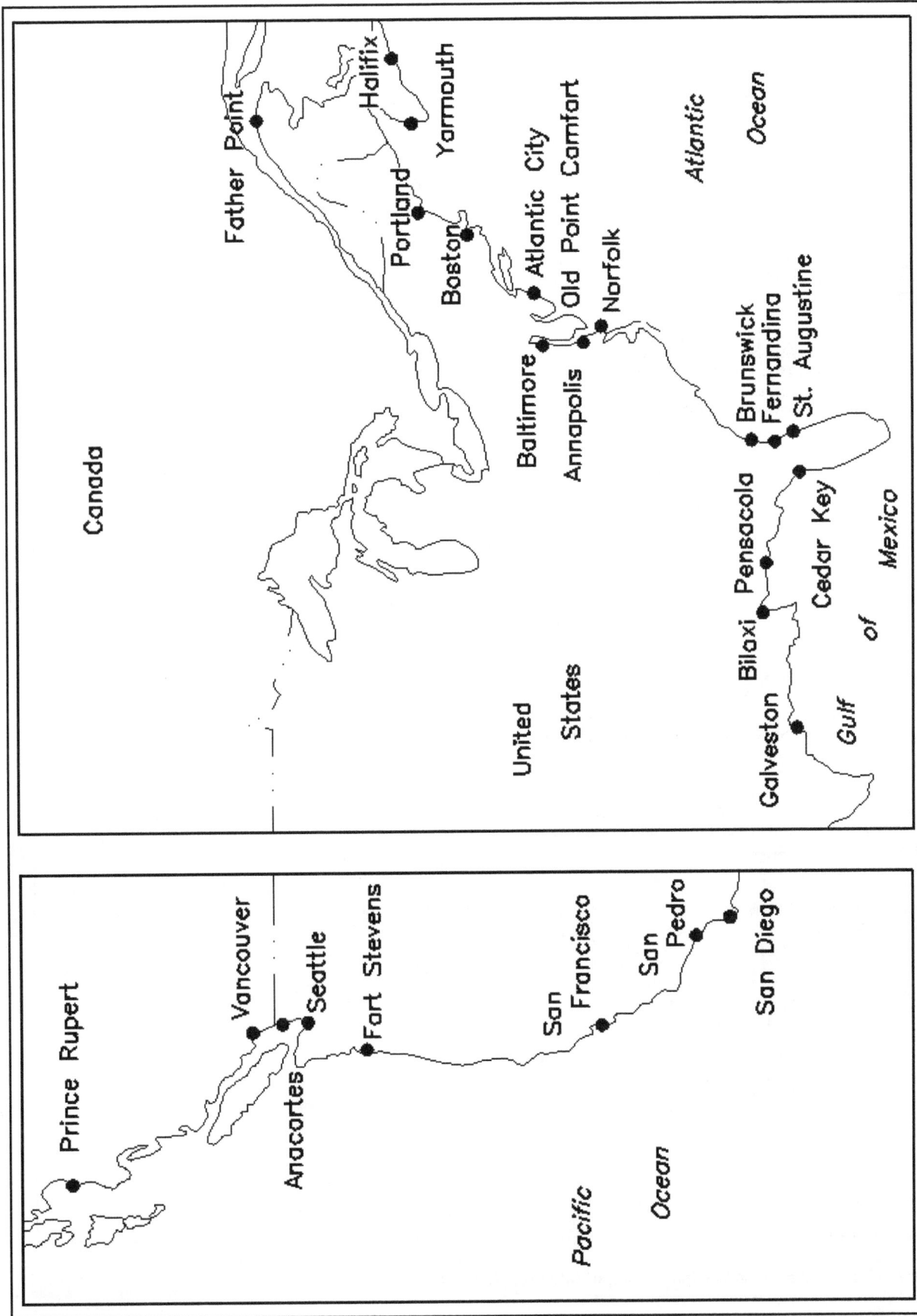

Figure II-5-18. Tide stations used in establishing the National Geodetic Vertical Datum (NGVD) of 1929 (Harris 1981 (after Rappleye (1932))

FLORIDA - I - 74

U.S. DEPARTMENT OF COMMERCE
NATIONAL OCEANIC AND ATMOSPHERIC ADMINISTRATION
NATIONAL OCEAN SURVEY

TIDAL BENCH MARKS

Miami Beach (City Pier)
Lat. 25 46′.1; Long 80 07′.9

BENCH MARK 4 (1928) is a standard disk, stamped "NO 4 1928", set vertically in the south face of the south post in the north-south fence line around a large city water tank. It is about 66 feet north of the extended centerline of Commerce Street, 36 feet west of the centerline of Jefferson Avenue, and 1/2 foot above ground level. Elevation: 5.62 feet above mean low water.

BENCH MARK 6 (1931) is a standard Corps of Engineers disk, stamped "BM NO 6," set in top of a 2-inch pipe surrounded at top with a 12-inch by 18-inch manhole frame with a removable cast iron cover, directly in centerline of a blacktop driveway which parallels Government Cut. It is at the U.S. Government Reservation on the north side of Government Cut, about 186 feet east of U.S. Engineers flagpole and 13 1/2 feet west of the center of a road junction. Elevation: 7.13 feet above mean low water.

BENCH MARK 7 (1937) is a standard disk, stamped "7 1937," set in the top of the northwest side of the concrete base to the east post of entrance gate to drive to the Corps of Engineers Office Building. It is about 100 yards south of intersection of Washington Avenue and Biscayne Street, 8 feet east of the extended centerline of the Avenue and 1/2 foot above ground level. Elevation: 5.03 feet above mean low water.

BENCH MARK 9 (1955) is a standard disk, stamped "9 1955," set in top of concrete deck along north edge of City Pier near the east end of Biscayne Street. It is about 122 yards east of the west end of pier, 39 feet northwest of the northeast corner of the ladies rest room and 1/2 foot south of south face of north guardrail. Elevation: 11.29 feet above mean low water.

BENCH MARK 10 (1956) is a standard disk, stamped "NO 10 1956," set on top of the northwest corner of the concrete base of light pole No. 166D6 about 68 yards west of the junction of Biscayne Street and Alton Road. It is near the northwest corner of the South Shore Recreation Park about 62 feet east of the west edge of the bulkhead on the water front and 9 1/2 feet northeast of the east edge of the north entrance to the Recreation Building. Elevation: 5.23 feet above mean low water.

BENCH MARK 11 (1956) is a standard disk, stamped "NO 11 1956," set in top of north corner of a concrete base which supports a 6 inch metal post near the City of Miami Beach Warehouse. It is near the intersection of Alton Road and First Street about 21 1/2 feet southwest of the southwest curb of Alton Road and 9 feet northwest of the northwest corner of the warehouse building. Elevation: 4.92 feet above mean low water.

Mean low water at Miami Beach is based on 19 years of records, 1941-1959. Elevations of other tide planes referred to this datum are as follows:

	Feet
Highest tide (observed)	
September 8, 1965	6.4
Mean high water	2.50
Mean tide level	1.25
NGVD, 1929	0.96
Mean low water	0.00
Lowest tide observed	
(March 24, 1936)	-1.6

Figure II-5-19.   Sample NOS description of tidal benchmarks (Harris 1981)

Water Levels and Long Waves

Table II-5-8
Low Water (chart) Datum for IGLD 1955 and IGLD 1985

Location	Low Water Datum in Meters	
	IGLD 55	IGLD 85
Lake Superior	182.9	183.2
Lake Michigan	175.8	176.0
Lake Huron	175.8	176.0
Lake St. Clair	174.2	174.4
Lake Erie	173.3	173.5
Lake Ontario	74.0	74.2
Lake St. Lawrence at Long Sault Dam, Ontario	72.4	72.5
Lake St. Francis at Summerstown, Ontario	46.1	46.2
Lake St. Louis at Pointe Claire, Quebec	20.3	20.4
Montreal Harbour at Jetty Number 1	5.5	5.6

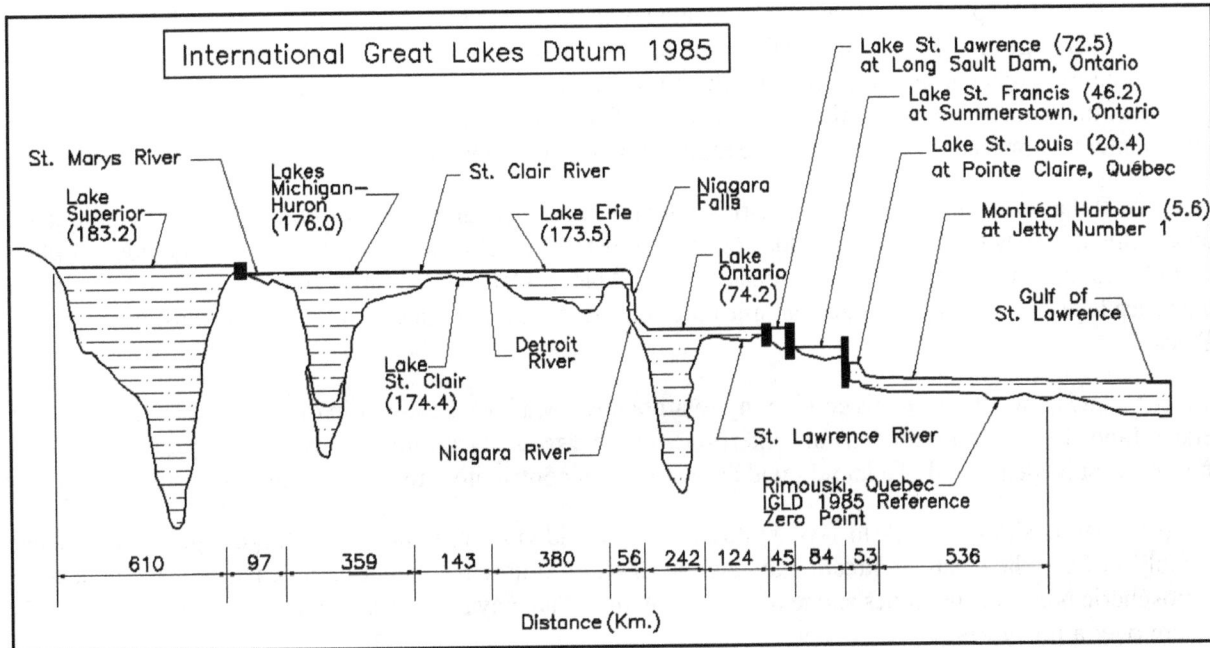

Figure II-5-20. Vertical and horizontal relationships for the IGLD of 1985 (Coordinating Committee on Great Lakes Basic Hydraulic and Hydrologic Data 1992)

Engineering Implications of Changes in Relative Mean Sea Level (CCMSL) was formed to examine existing knowledge concerning sea level change, to document existing relative rise rates, and to provide recommendations concerning their conclusions.

(2)   Relative mean sea level change can be defined as the difference between local changes in land elevation and global sea level changes. These changes result from a variety of processes, several of which can occur simultaneously. The following six processes can contribute to long-term relative mean sea level change; however, all processes do not necessarily apply to all geographic locations:

(a)   Eustatic rise. Refers to a global change in the oceanic water level. Examples of eustatic rise include melting of land-based glaciers and the expansion of near-surface ocean water due to global ocean warming.

(b)   Crustal subsidence or uplift from tectonic uplifting or downwarping of the earth's crust. These changes can result from uplifting or cooling of coastal belts, sediment loading and consolidation, or subsidence due to volcanic eruption loading.

(c)   Seismic subsidence. Caused by sudden and irregular incidence of earthquakes.

(d)   Auto-subsidence. Due to compaction or consolidation of soft underlying sediments such as mud or peat.

(e)   Climatic fluctuations. May also create changes in sea level; for example, surface changes produced by El Niño due to changes in the size and location of high pressure cells.

(3)   The above processes have been evaluated with respect to their historical and potential contribution to sea level change on U.S. coasts. The Committee report assesses changes in sea level as well as the affected hydrodynamic processes and the effect on the coastal zone. The report also investigates feasible response strategies that could be used to mitigate the effects of sea level change. Although it is beyond the scope of this chapter to reproduce the contents of the report, conclusions relevant to this chapter are reproduced below.

(a)   Relative mean sea level, on statistical average, is rising at the majority of tide gauge stations situated on continental coasts around the world. Relative mean sea level is generally falling near geological plate boundaries and in formerly glaciated areas such as Alaska, Canada, Scandinavia, and Scotland. Relative mean sea level is not rising in limited areas of the continental United States, including portions of the Pacific Coast.

(b)   The contrasting signals concerning relative mean sea level behavior in different parts of the United States (and the world in general) are interpreted as due to differing rates of vertical motion of the land surfaces. Subsidence and glacial rebound are significant contributors to vertical land displacements.

(c)   Large, short-term (2- to 7-year) fluctuations worldwide are related to meteorological phenomena, notably shifts in the mean jet-stream path and the El Niño-Southern Oscillation mechanisms, which lead to atmospheric pressure anomalies and temperature changes that may cause rise or fall of mean sea level by 15-30 cm over a few years.

(d)   Studies of a very small number of tide gauge records dating more than 100 years (the oldest being Amsterdam, started in 1682) show that after removal of the subsidence factor where known, mean sea level has been fluctuating through a range of not more than 40-150 cm (in long-term fluctuations) for at least 300 years.

(e)   The geological record over the last 6,000 years or so indicates that there has been a general, long-term rise (with short-term fluctuations) probably not exceeding 200 cm during the last 1,500 years.

(f)   Monitoring of relative mean sea level behavior is at present inadequate for measuring the possible global result of future climate warming due to rising greenhouse gases.

(g)   The risk of accelerated mean sea level rise is sufficiently established to warrant consideration in the planning and design of coastal facilities. Although there is substantial local variability and statistical uncertainty, average relative sea level over the past century appears to have risen about 30 cm relative to the east coast of the United States and 11 cm along the west coast, excluding Alaska, where glacial rebound has resulted in a lowering of relative sea level. Rates of relative sea level rise along the Gulf Coast are highly variable, ranging from a high of more than 100 cm/century in parts of the Mississippi delta plain to a low of less than 20 cm/century along Florida's west coast.

(h)   Accelerated sea level rise would clearly contribute toward a tendency for exacerbated beach erosion. However, in some areas, poor sand management practices or navigational modification at channel entrances has resulted in augmented erosion rates that are clearly much greater than would naturally occur. Thus, for some years into the future, sea level rise may play a secondary role in these areas.

(i)   As noted previously, the two response options to sea level rise are stabilization and retreat. Retreat is most appropriate in areas with a low degree of development. Given that a "proper" choice exists for each location, selecting an incorrect response alternative could be unduly expensive.

(j)   There does not now appear to be reason for emergency action regarding engineering structures to mitigate the effects of anticipated increases in future eustatic sea level rise. Sea level change during the design service life should be considered along with other factors, but it does not present such essentially new problems as to require new techniques of analysis. The effects of sea level rise can be accommodated during maintenance periods or upon redesign and replacement of most existing structures and facilities. There are very limited geographic areas where current subsidence rates may require near-term action as has been the case in Japan and Terminal Island, California.

(4)   The above conclusions represent the state of knowledge on the subject of relative sea level change. For additional information, the reader is referred to the Committee report. It presents a complete and comprehensive investigation of the subject based on known facts and engineering and scientific principles.

(5)   For the purposes of this report, the primary conclusion is that, with some regional exceptions, sea level is not rising at a rate to cause undue concern. Results of the report indicate an average sea level rise over the past century of approximately 30 cm/century on the east coast, and 11 cm/century on the west coast, and a range along the Gulf of Mexico coast of less than 20 cm/century along the west coast of Florida to more than 100 cm/century in parts of the Mississippi delta plain. The above summary remarks lead to the conclusion that normal design criteria should be followed in which the design life of a project should consider the possible local relative sea level rise rates shown above.

g.   Changes in lake level datums.   As with long-term change in the relative mean sea level of the oceans, lake levels also change in time. Although lake levels are subject to the same tectonic types of relative movement, the primary form of change is due to shorter-term phenomena. For example, lake levels are subject to seasonal and annual variations in precipitation and freshwater inflow resulting from regulated reservoirs. These hydrologically related effects produce water surface elevations in the Great Lakes that vary irregularly from one year to the next. The annual cycle consists of water surfaces that consistently fall in elevation to their lowest stage during winter. Falling stages are due to the fact that the majority of precipitation is in the form of snow or rainfall that freezes before becoming spring runoff into streams that empty into the lakes. Lake levels then begin to rise in the spring as the snow and ice melt into runoff. Lake levels generally reach their maximum during the summer. These long-term changes were the basis of the revised IGLD of 1985 described in the previous section.

h. Design considerations.

(1)   The datums described above and the reported variability of those datums represent design criteria considerations that directly impact the expected lifetime of a project. If, for example, a coastal project is to be situated in an area of known subsidence, then design elevations need to reflect additional freeboard as a factor-of-safety consideration. For example, if levee systems are to be situated in an area of known subsidence (i.e., the Gulf coast) and a crest elevation level of 20 ft NGVD is considered a minimum level of protection, the subsidence rate should be included in the design calculations to provide adequate protection for the life of the project. For example, if the relative sea level change is 1 cm/year, then the levee will subside by approximately 1 ft in 30 years. This rate of change should be accounted for in the design of projects expected to provide some predetermined level of protection. In this case, the levee should be designed with a 21-ft NGVD crest elevation if it is intended to provide a minimum of 20 ft MSL (present time) protection for 30 years.

(2)   If a design project needs to consider the possibility of long-term water level change, then accurate sources of data must be obtained for use in the design evaluation. The NOS continuously summarizes monthly MSL values, the mean and extreme high and low waters of the month, and many other tidal statistics. These records can be obtained from NOS in the form of either photocopies or magnetic media. Figure II-5-21 is a sample of NOS tide and sea level data. If data are available for a specific project location for a long period of time (i.e., 30-50 years), then relative sea level change rates for project design can be estimated. For example, plotted variations in annual sea level for several locations are shown in Figure II-5-22. These data can be used to indicate relative sea level change.

(3)   Although relative water surface elevations and datum changes are important over the life of a project, it is usually the short-term changes in water surface elevations that are responsible for project failure (although sudden elevation changes due to tectonic influence can be catastrophic). These short-term changes to elevation are the subject of the following sections.

II-5-5. Storm Surge

a.   Storm types. Storms are atmospheric disturbances characterized by low pressures and high winds. A storm surge represents the water surface response to wind-induced surface shear stress and pressure fields. Storm-induced surges can produce short-term increases in water level that rise to an elevation considerably above mean water levels. This is especially true when the storm front coincides with a local high spring tide. There are two types of storms that impact coastal regions. Storms that originate in the tropics are referred to as "tropical storms." These events primarily impact the east coast and gulf coast of the United States, the Caribbean Sea, and islands in the Pacific Ocean. Although infrequent, tropical events can also impact the western coast of Mexico and southern coast of California.

			Form 1794 DEPARTMENT OF COMMERCE COAST AND GEODETIC SURVEY Ed. August 1923	TIDES: MONTHLY MEANS OF _____ TL + SL													

Station: _Alamitos Bay Entrance, Calif._   Latitude _____ Longitude _____
Observations begin ____July 4, 1953____   Observations end _____
Datum is _____   which is ____14.28____ feet below B. M. __SL 25"__
Linear quantities in feet _____   Time in hours _____

	YEAR 1953	JAN.	FEB.	MAR.	APR.	MAY	JUNE	JULY	AUG. 4.49	SEPT. 4.59	OCT. 4.53	NOV. 4.50	DEC. 4.38	FOR YEAR SUM	MEAN	TOTAL SUM	MEAN
TL (1)	1954	4.36	4.34	4.19	4.12	4.48	4.68	4.66	4.62	4.64	4.40	4.48	4.40	53.37	4.45	4.45	4.46
(2)	1955	4.29	4.24	4.32	4.14	4.32	4.42	4.45	4.60	4.60	4.30	4.33	4.18	52.19	4.35	8.80	4.40
(3)	1956	4.18	4.24	4.14	4.24	4.28	4.54	4.58	4.71	4.63	4.50	4.52	4.50	53.06	4.42	13.22	4.41
(4)	1957	4.38	4.36	4.24	4.27	4.45	4.66	4.86	4.87	4.76	4.81	4.76	4.86	55.28	4.61	17.83	4.46
(5)	1958	4.70	4.62	4.51	4.52	4.50	4.55	4.68	4.88	5.00	5.08	4.88	4.76	56.68	4.72	22.55	4.51
(6)	1959	4.75	4.68	4.50	4.50	4.48	4.66	4.78	4.42	5.01	5.07	4.72	4.44	57.01	4.75	27.30	4.55
(7)	1960	4.57	4.28	4.44	4.38	4.44	4.56	4.72	4.76	4.86	4.69	4.61	4.66	54.97	4.58	31.88	4.55
(8)	1961	4.66	4.56	4.37	4.38	4.36	4.56	4.74	4.78	4.86	4.68	4.58	4.66	55.17	4.60	36.48	4.56
(9)	1962	4.61	4.63	4.23	4.36	4.43	4.51	4.61	4.78	4.89	4.89	4.61	4.62	55.17	4.60	41.08	4.56
(10)	1963	4.56	4.55	4.37	4.30	4.47	4.75	4.89	4.99	5.15	4.91	4.84	4.83	56.61	4.72	45.80	4.58
(11)																	
(12)	1964	4.74	4.42	4.40	4.28	4.28	4.44	4.64	4.86	4.85	(4.74)	4.58	4.58	54.81	4.57	50.37	4.58
(13)	1965	4.52	4.68	4.64	(4.66)	4.47	4.66	4.79	5.07	5.16	4.97	4.96	(5.08)	57.66	4.80	55.17	4.60
(14)	1966	4.89															
(15)	1953								4.49	4.57	4.49	4.45	4.33				
SL (16)	1954	4.36	4.29	4.19	4.09	4.43	4.62	4.62	4.61	4.60	4.34	4.44	4.37	52.96	4.41	4.41	4.44
(17)	1955	4.27	4.23	4.28	4.12	4.28	4.40	4.44	4.58	4.57	4.28	4.29	4.14	51.88	4.32	8.73	4.36
(18)	1956	4.17	4.24	4.13	4.25	4.52	4.59	4.69	4.61	4.49	(4.49)	4.47	52.86	4.40	13.13	4.38
(19)	1957	4.36	4.35	4.21	4.25	4.41	4.63	4.83	4.87	4.74	4.78	4.75	4.85	55.03	4.59	17.72	4.43
(20)	1958	4.71	4.60	4.49	4.51	4.46	4.53	4.65	4.85	4.99	5.04	4.84	4.72	56.39	4.70	22.42	4.48
(21)	1959	4.74	4.68	4.49	4.48	4.48	4.64	4.64	4.78	4.94	5.00	5.04	4.71	56.90	4.74	27.16	4.53
(22)	1960	4.52	4.28	4.37	4.37	4.43	4.58	4.71	4.69	4.84	4.65	4.58	4.64	54.66	4.55	31.71	4.53
(23)	1961	4.66	4.57	4.38	4.42	4.38	4.52	4.76	4.79	4.86	4.69	4.57	4.69	55.22	4.61	36.32	4.54
(24)	1962	4.60	4.62	4.22	4.34	4.40	4.47	4.60	4.78	4.93	4.86	4.57	4.59	54.78	4.56	40.90	4.54
(25)	1963	4.56	4.56	4.37	4.30	4.48	4.74	4.88	4.97	5.09	4.89	4.85	4.81	56.50	4.71	45.61	4.56
(26)																	
(27)	1964	4.72	4.40	4.41	4.24	4.21	4.41	4.63	4.83	4.83	(4.70)	4.51	4.54	54.44	4.54	50.15	4.56
(28)	1965	4.50	4.66	4.59	(4.64)	4.41	4.63	4.78	5.06	5.14	4.94	4.92	(5.04)	57.26	4.77	54.92	4.58
(29)	1966	4.86															
(30)																	

Figure II-5-21.   Sample NOS tabulation of tide parameters (Harris 1981)

Storms that result from the interaction of a warm front and a cold front are called "extratropical storms" and are often referred to as Northeasters. Extratropical events impact the east and west coasts of the United States as well as Alaska and the Great Lakes, but storm-generated surges are most significant along the upper east coast. Characteristics of both storm types and sources of detailed description follow.

(1)   Tropical storms.

(a)   Tropical storms are classified according to their intensity, the most severe of which is referred to as a hurricane. These storms have maximum sustained winds in excess of 74 mph. Similar storms in the Pacific ocean, west of the International Date Line, are referred to as typhoons. Storms in the Southern Hemisphere are called tropical cyclones. In order to indicate hurricane intensity, and associated potential damage, the NOAA National Hurricane Center has adopted the Saffir/Simpson (Saffir 1977, NOAA 1977) Hurricane Scale. This scale, based partially on maximum wind speeds shown below, ranges from category 1 through the most severe category 5 event. Associated damage, storm surge levels, and evacuation limits are described in the NOAA publication on tropical cyclones (NOAA 1981).

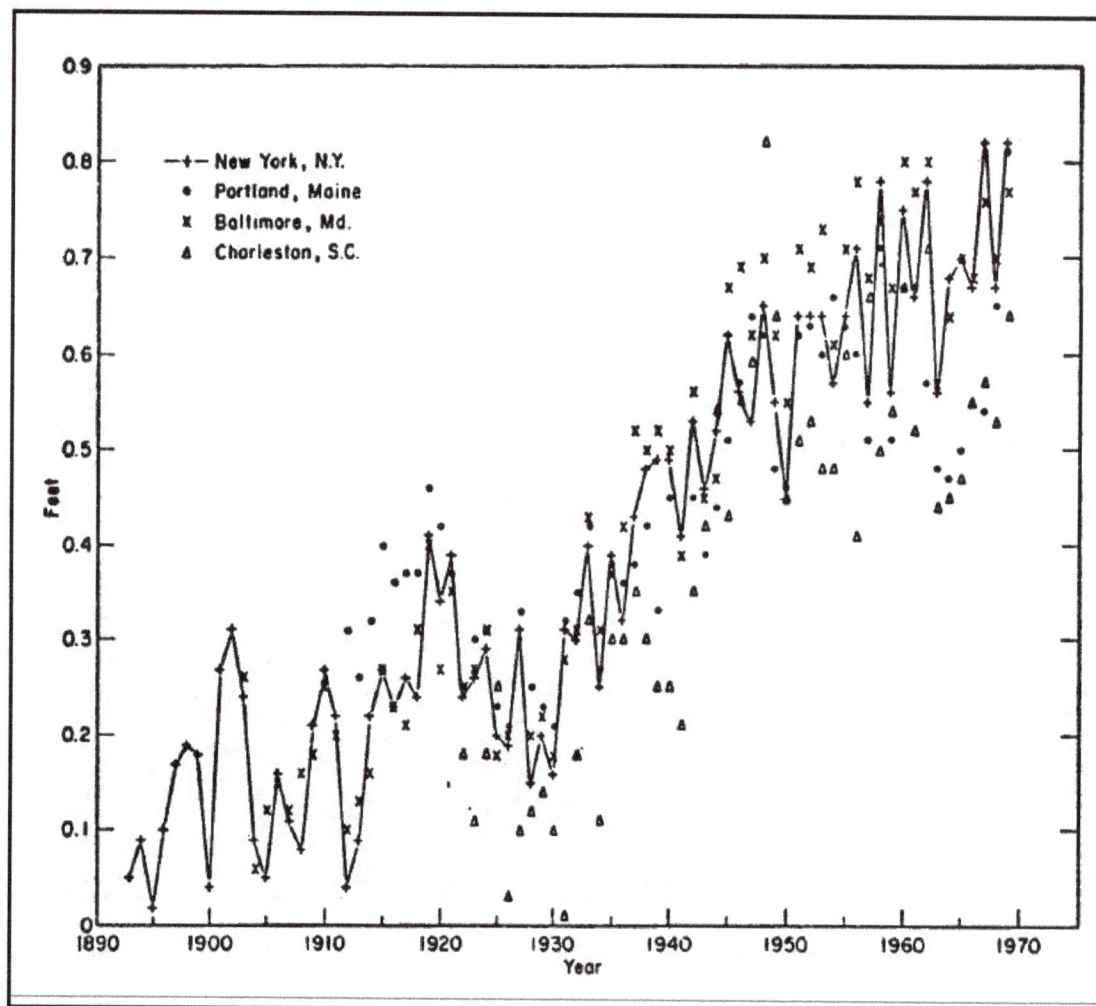

Figure II-5-22.   Variations in annual MSL (Harris 1981)

Category 1 - Winds of 74 - 95 mph
Category 2 - Winds of 96 - 110 mph
Category 3 - Winds of 111 - 130 mph
Category 4 - Winds of 131 - 155 mph
Category 5 - Winds over 155 mph

(b)   Hurricanes are characterized by circular wind patterns that rotate in a counterclockwise direction in the Northern Hemisphere and clockwise in the Southern Hemisphere. The direction of rotation is a consequence of the Coriolis effect induced by the rotation of the earth (see Part II-1).

(c)   Unlike extratropical storm events such as northeasters, tropical storms are well-organized with respect to their wind and pressure patterns. As a result, they can be reasonably well-described by several descriptive parameters listed below and shown on Figure II-5-23. Note that these parameters have been identified as descriptive of average events; precise modeling of some tropical events may require more detailed information.

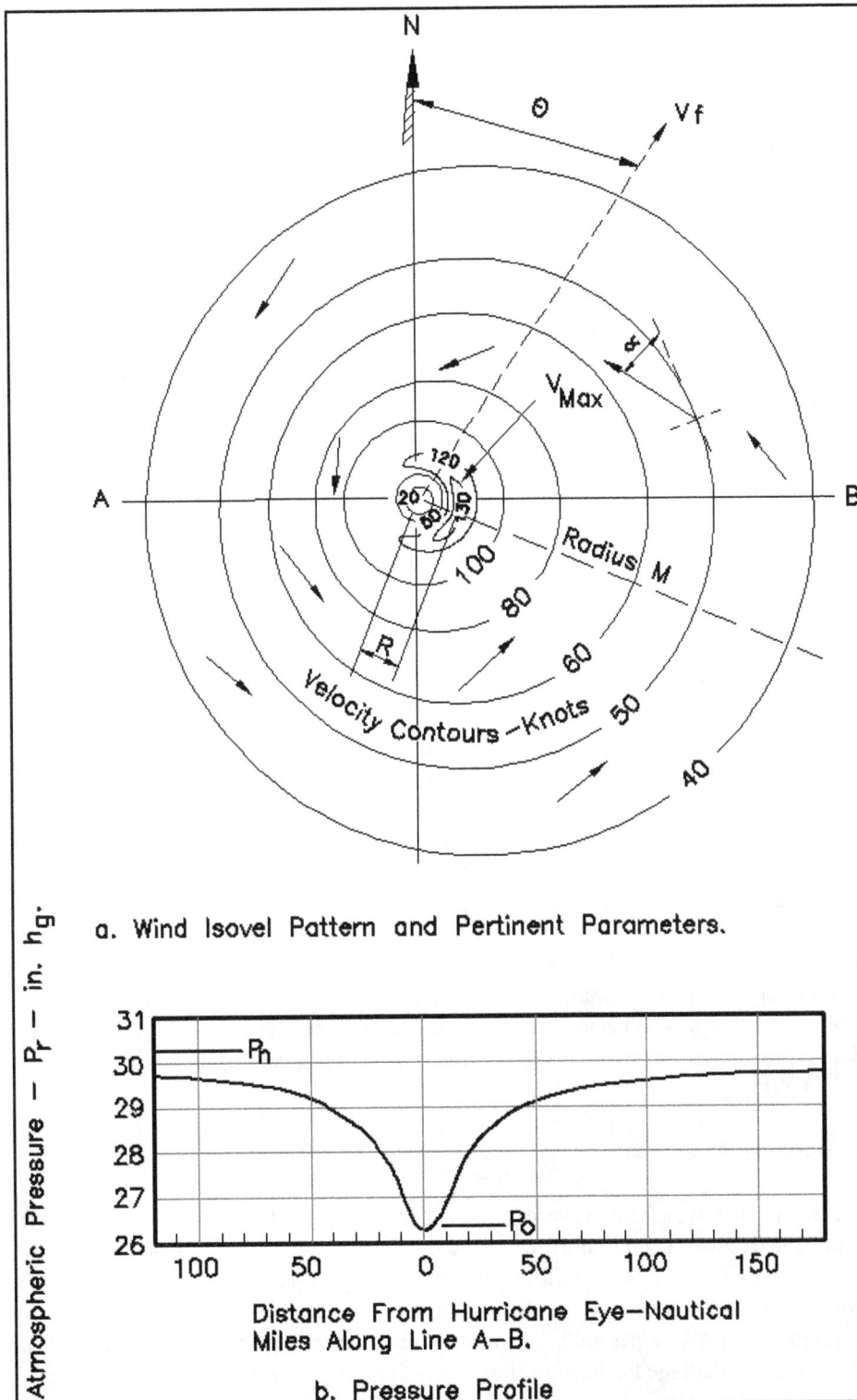

a. Wind Isovel Pattern and Pertinent Parameters.

b. Pressure Profile

Figure II-5-23.   Schematic diagram of storm parameters (U.S. Army Corps of Engineers 1986)

R = radius to maximum wind

V = maximum wind

V_f = forward speed of the eye of the storm

p_0 = central pressure of the eye of the storm

p_n = peripheral or far-field pressure

θ = angle of storm propagation

α = inflow angle

(d)   The Corps presently uses two approaches to model tropical storm wind fields. The first is the Standard Project Hurricane (SPH) model. The SPH is defined as a hurricane having a severe combination of storm parameters that will produce a storm with high sustained wind speeds that are reasonably characteristic of the particular location. Guidance on the selection of site-specific storm parameters is given in National Hurricane Research Project Report No. 33 (Graham and Nunn 1959) and NOAA Technical Report NWS 38 (Ho et al. 1987). The SPH model is an empirical model that produces steady-state hypothetical wind and atmospheric pressure fields as a function of the above parameters. The SPH model and its use are thoroughly described in the Coastal Modeling System (CMS) user's manual (Cialone et al. 1991).

(e)   The second approach to hurricane modeling is the Planetary Boundary Layer (PBL) model. The PBL model is based on the equations of motion for a vertically integrated boundary layer. The PBL model is currently supported by CHL for all tropical storm surge studies. This selection is due to the fact that the PBL model is based on the equations of motion and boundary layer physics. As such, the model is more flexible in its ability to incorporate a variety of land/sea boundary conditions to represent spatially variable wind and pressure fields that cannot be properly represented by the empirical SPH model.

(f)   Input to the PBL model is the time-varying location of the storm and all of the parameters given above, except maximum wind. Maximum winds are computed by the PBL model as a function of the above parameters. The model has provisions to account for spatial asymmetry of parameters as mentioned above. The PBL model is also supported by the CMS and the model, input, and application are described in detail by Cialone et al. (1991).

(2)   Extratropical storms.

(a)   Unlike hurricanes, which can severely impact local regions (typically less than 50 miles) for less than a day, extratropical storms such as northeasters can impose high winds with accompanying surges over large geographical areas (hundreds of miles) for extended periods of time, i.e., several days or more. Generally, extratropical events have lower wind magnitudes and generate smaller maximum surge elevations than hurricanes. Although lower storm surge elevations are associated with northeasters than with hurricanes, they can cause substantial damage because of their large area of influence and extended period of duration.

(b)   An additional design consideration for extratropical events is that they generally occur with a much greater frequency than hurricanes. For example, a hurricane with a peak storm surge elevation of 6 ft at Sandy Hook, NJ, has a return period of 10 years. A northeaster with a 6-ft surge has a return period of approximately 2 years (Gravens, Scheffner, and Hubertz 1989).

(c) Extratropical events cannot be parameterized in the same manner as hurricanes. Therefore, the development of stage-frequency relationships based on the Joint Probability Method (to be discussed in the following section) is not a viable approach for assigning frequency-of-occurrence relationships to extratropical events. As an alternative approach, frequency indexing is now established through application of a new statistical procedure called the empirical simulation technique. This approach will be discussed in the following section.

(d) Because a large number of extratropical storm events have impacted coastal areas of the United States, the 20- to 30-year database of hindcast storms of the Wave Information Study (WIS) is adequate for representing the historical population of storms along the coasts of the United States. Wind fields produced by the storms of this database are input to numerical hydrodynamic long-wave models to produce storm surges as a function of historically hindcast wind fields contained in the WIS database. Because of the frequency of events and large area of impact, this database is considered adequate for determining stage-frequency relationships for design use.

(3) Surge interaction with tidal elevations.

(a) A final consideration on the specification of tropical and extratropical storm surge elevations relates to the time of occurrence. The timing of storm events with respect to the phase of the astronomical tide cannot be overemphasized. When a storm surge coincides with a spring high tide, the resulting total surge can be many times more devastating than the surge alone. For example, a moderate event at low tide can become the storm surge of record at high tide. An example situation is shown for Hurricane Gloria, which moved along the Delaware and New Jersey coasts, making landfall on Long Island, NY, at 1100 EST on 27 September 1985. The total storm track is shown on Figure II-5-24, and a more detailed plot of the track in the vicinity of Long Island is shown on Figure II-5-25 (Jarvinen and Gebert 1986).

(b) The significance of timing between the surge peak and the tide phase can be seen in Figure II-5-26. The observed storm surge at Sandy Hook, NJ, had a peak value of 7.5 ft and occurred near low tide (1.0 ft). As a result, the total measured water level was 8.5 ft, corresponding to just 3.5 ft above normal high tide. If the storm surge had occurred 4.5 hr earlier to coincide with high tide (5.0 ft), the peak surge value would have been 12.5 ft. The differences in levels of damage resulting from an 8.5-ft versus a 12.5-ft surge level can be considerable, and the only difference between the two scenarios is a 4.5-hr difference in time of occurrence. As shown in this example, the phasing of the storm and tide impacts both the design of the structure and the probability analysis of the design.

b. Storm event frequency-of-occurrence relationships.

(1) Introduction.

(a) The majority of coastal structures are designed to provide some specific level of protection to the beach and its surrounding population and supporting structures. This level of protection is generally based on the frequency of occurrence of a storm surge of some specified maximum elevation selected by assessing the risks of structural failure or consequences of overtopping versus the economics of the cost of the design project. Therefore, one important aspect of coastal design criteria is the development of stage-frequency or frequency-of-occurrence relationships for the design area.

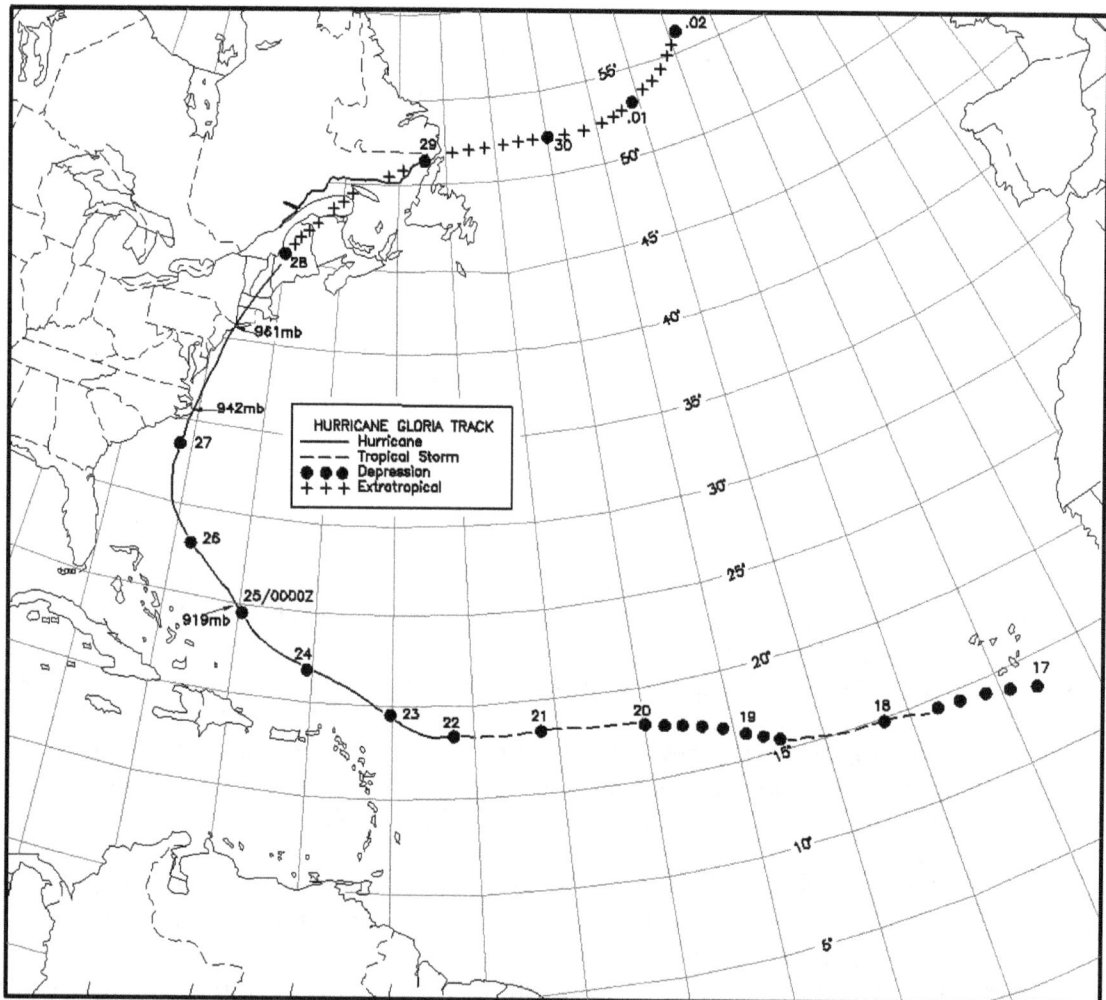

Figure II-5-24. Hurricane Gloria track from 17 September to 2 October 1985 (Jarvinen and Gebert 1986)

(b)   Three approaches are commonly used for developing site-specific frequency relationships. These are based on: 1) historical data, 2) synthetic data, and 3) empirical simulation technique (EST) approaches. The historical approach assumes that a database of historical storm events and their respective surge elevations is available for the study area. It is further assumed that the available database provides a representative sample of all possible events for that particular site. Many coastal locations do not have adequate historical data from which frequency-of-occurrence relationships can be developed. Even if a reasonable number of years of data are available, there is no assurance that the data represent a full population of possible storm events. Therefore, it is usually the case that historical data are insufficient for generating a reliable frequency-of-occurrence relationship. For this reason, synthetic or empirically based approaches are generally preferred.

(c)   The synthetic approach is based on the construction of a large number of hypothetical storm events based on assumed probability laws dictating the intensity of the storms and their associated frequency relationships. A more recent approach to the developing frequency-of-occurrence relationships is through the statistical approach known as the empirical simulation technique. This approach is a resampling or "bootstrap" scheme in which the probability relationships of a historical database are used to develop frequency relationships.

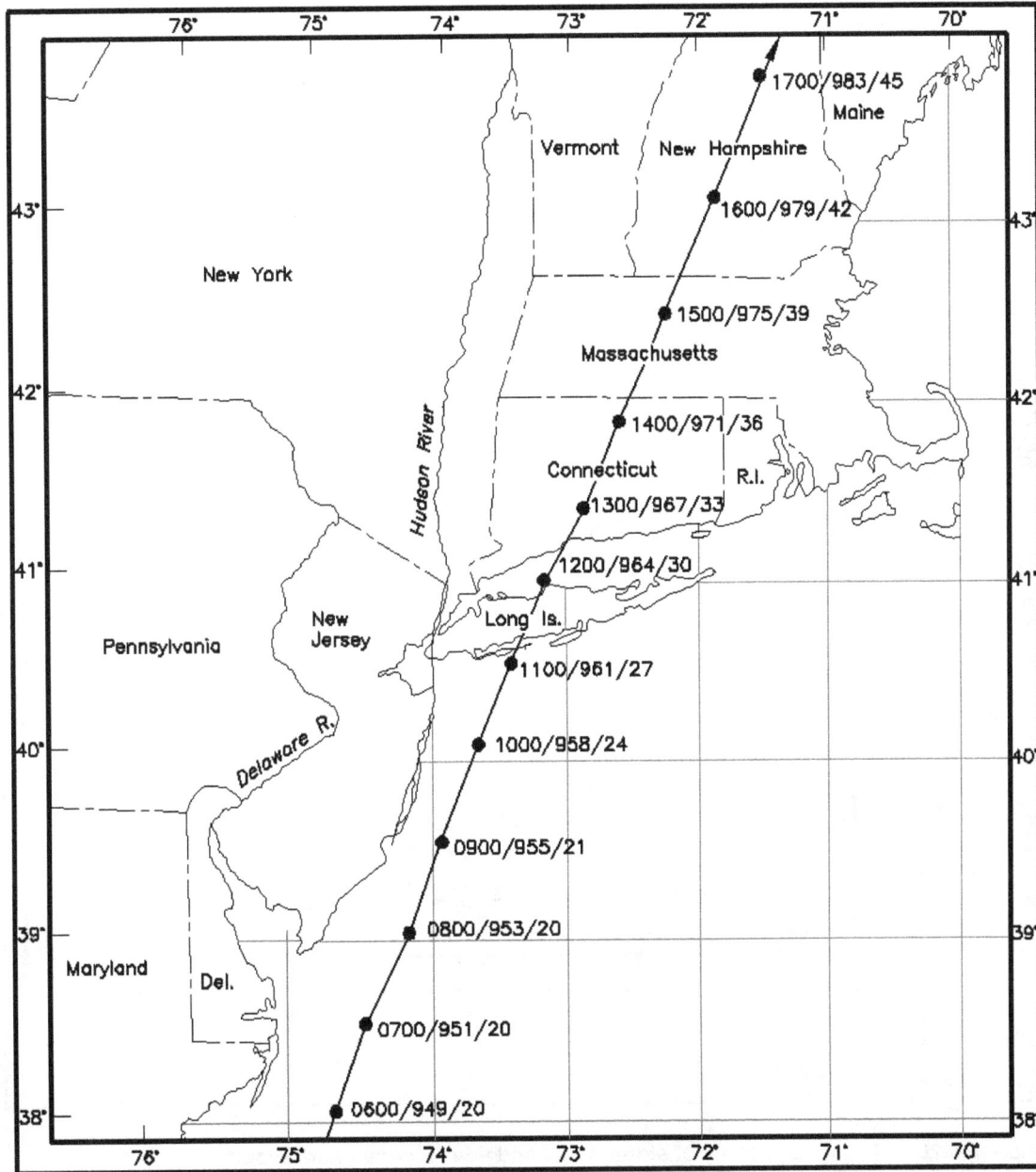

Figure II-5-25. Hurricane Gloria track offshore of Delaware and New Jersey (Jarvinen and Gebert 1986)

(d) CHL recommends the empirical simulation technique and uses this statistically based method for all frequency-related studies. This technique has been adopted because it generates multiple data sets of events from historical data and then generates not only frequency-of-occurrence relationships, but also error bands associated with those determinations. When computer facilities are not available for application of the EST, the other approaches can be successfully used; however, each has limitations which should be considered. All three approaches are briefly discussed in the following paragraphs.

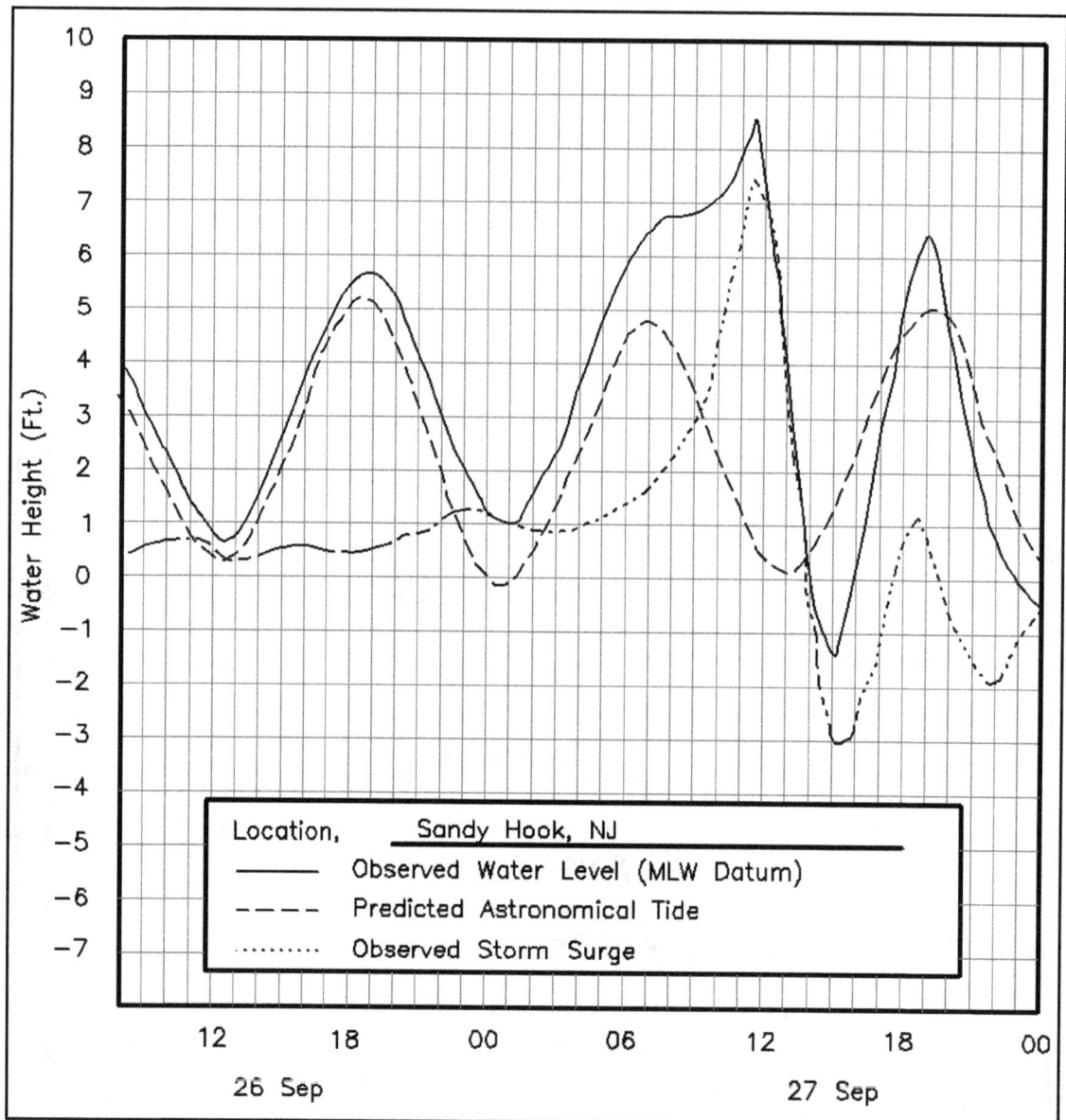

Figure II-5-26.   Example phasing of storm surge and tide (Jarvinen and Gebert 1986)

(2) Historical method.

(a) There are two basic approaches to the use of historical data for generating recurrence interval relationships. The first is the graphical method, the second makes use of analytical formulas to describe the behavior of the frequency relationship.

(b) The graphical method involves construction of a cumulative probability density function (or probability distribution function) in which the magnitude of the event is plotted against percent probability of exceedance.

(c) In addition to the graphical approach of developing a pdf, a variety of analytical methods exist for estimating these functions. Selection of a specific form depends on the behavior of the historical data and the purpose of the study. One function, the Pearson Type III distribution, is presented in detail in many texts on probability and statistics.

(3)   Synthetic method.

(a)   The synthetic method is based on the assumption that a particular event can be characterized by several descriptive (component) parameters, and that the probability associated with each of those parameters can be used to determine the joint probability of the total event. For example, tropical storm events were described by parameters such as radius to maximum wind, maximum wind, pressure deficit, etc. It is assumed that the frequency of the event can be described by the frequencies associated with each parameter.

(b)   The most commonly used form of the synthetic approach is the Joint Probability Method (JPM). The assumption basic to the JPM is that the probability of occurrence of a given storm event can be written as the product of the probabilities corresponding to each storm parameter and that the probabilities for each parameter are independent. For example, a common assumption is that the storm can be described by five storm parameters. If the parameters are statistically independent, then the joint probability density function for the five-dimensional set is the product of the individual probability density functions, i.e.,

$$f = f_D\, f_R\, f_\theta\, f_v\, f_L \tag{II-5-24}$$

where f_D, f_R, etc., represent the pdf for the central pressure deficit D (far-field pressure p_n less central pressure p_0), radius to maximum R, azimuth of the storm track θ, forward speed V, and closest distance to landfall (with respect to some specified location) of the eye of the storm L. The JPM has been widely used in the past for storm surge analyses. Details are well-documented in USACE (1986) and the reader is referred to that document for a more detailed description.

(4)   Empirical simulation technique.

(a)   Introduction. Few locations have an adequate historical database of storm events from which to develop reliable stage-frequency relationships. The JPM is subject to error because it is based on simplifying assumptions of parameter independence, and specifying the pdf of each parameter according to parametrically based relationships. The empirical simulation technique (EST) is a statistical resampling scheme that uses historical data to develop joint probability relationships among the various storm parameters. These relationships are based on data derived from actual storm events. There are no simplifying assumptions concerning the pdf's; the interdependence of parameters is computed directly from the respective parameter interdependencies contained in the historic data. In this manner, probabilities are site-specific, do not depend on fixed parametric relationships, and do not assume parameter independence. Thus, the EST is "distribution-free" and nonparametric. There are presently two approaches to applying the EST. The first approach is a multi-parameter simulation approach developed for application to tropical events. The second is a single-parameter approach developed for use in extratropical events. EST applications for tropical and extratropical events are discussed briefly below.

(b)   EST - tropical storm application.

(1)   The multi-parameter EST Program (Scheffner et al. 1999) utilizes historic data to generate a large number of multi-year simulations of possible future hurricane storm events for a specific location. The approach is based on resampling and interpolation of data contained in a database of events derived from historic events. The ensemble of simulations is consistent with the statistics and correlations of past storm activity at the site, but allows for random deviations in behavior that are likely to occur in the future.

(2) The simulation approach is based on the selection of a set of historic "training storms" extending over the range of storms that actually occurred at the site location. The training events are selected from the data set of historic storms as well as from storms that could occur, such as a historic event with a slightly shifted track. The descriptive storm parameters for each of these events (pressure deficit, radius to maximum wind, maximum wind, forward speed, angle of propagation, and tidal phase) are computed and referred to as input vectors describing each storm event.

(3) These input vectors define the multi-vector space of input parameters associated with each storm event. These parameters are input to the PBL storm model to produce a time-varying spatial distribution of wind and pressure fields subsequently used as input to a numerical hydrodynamic model for computing storm surge hydrographs over the computational domain. (Examples of numerical modeling of storm surge and tidal circulation are presented in Part II-5-7.) The maximum storm surge elevation reached at specified gauge locations is defined as the response vector of the storm at that location.

(4) Other output vectors such as maximum shoreline erosion, maximum dune recession, or maximum wave height can be described. The output vector(s) represents the environmental response to the storm. This response is defined at location X and is a direct consequence of the storm via the storm parameter values defined at the point of nearest proximity of the storm eye to point X. For the case of stage-frequency analyses, maximum surge is assumed to occur when the eye of the storm is nearest location X.

(5) Input to the EST model is the training set database of storm input and response vector(s) for a specific location. An example storm surge analysis for the coast of Delaware (Scheffner, Borgman, and Mark 1993) identified 33 tropical events that impacted the study region during the 104-year period of 1886 through 1989. Of these events, 15 were selected to be representative of the full set. All events were extracted from the NOAA (1981) database of historic storms. These data were used to develop input to the PBL model to generate input data to a hydrodynamic long-wave model. Storm surge hydrographs were computed for each of the 15 storm events. This preliminary analysis generated a set of input and response vectors for 15 storm events for input to the EST.

(6) Each storm surge is computed independent of the local tidal phase. Each event has an equal probability of occurring at high tide, MSL after flood, low tide, or MSL after ebb. Each of the computed surge elevations were linearly combined with the four phases of the tide to generate an expanded training set of 60 events, which included tidal elevation.

(7) A cumulative pdf for the 60 storm events is developed based on the assumption that each of the 60 events has an equal probability of occurrence, and that the number of hurricanes along the coast of Delaware is defined by the number of historical events, i.e., 33 events in 104 years. Each storm is considered a point event in time with each storm occurring independently. The empirical simulation technique utilizes a resampling scheme in which a random number seed from 0.0 to 1.0 is used to select a historic event from the training set. Each selected event is described according to its respective input vectors and the maximum surge elevation response vector.

(8) Input vectors corresponding to the randomly selected event are used to define a new set of input vectors, based on the initial selected storm values but weighted to reflect the value of the "nearest neighbor" parameter contained in a vector space representing parameters corresponding to all events in the 60-storm database. In this manner, a new storm is defined according to a new set of parameters that is similar to those defining the selected event but adjusted to reflect the parameter interrelationships of the full database. Multi-vector interpolation schemes use the new event input parameters to estimate a new response parameter - the peak storm surge at point X.

(9)   The EST simulation approach is to perform N-repetitions of M-years of simulation; for example, 100 simulations of a 200-year sequence of storms. Details of the EST are given in Borgman and Scheffner (1991). The number of storms simulated per year is specified according to a Poisson probability law of the form:

$$P[N=n] = \frac{e^{-\lambda T}(\lambda T)^n}{n!}$$ (II-5-25)

for $n = 0,1,2,3,....$ Equation II-5-25 defines the probability of having N storm events in T years. The variable λ defines the mean frequency of observed events per time period. For the example shown for the coast of Delaware, the value of λ was chosen to be 0.32 events per year, computed as the ratio of 33 observed events above some selected threshold intensity over a 104-year period.

(10) A 10,000-element array is initialized to the above Poisson distribution. The number corresponding to 0 storms per year from Equation II-5-25 is 0.7234; thus, if a random number selection is less than or equal to 0.7234 on an interval of 0.0 to 1.0, then no hurricanes would occur during that year of simulation.   If the random number is between 0.7234 and $0.7234 + P[N=1] = 0.7234 + 0.2343 = 0.9577$, one event is selected. Two events for $0.9577 + 0.0379 = 0.9956$, etc. When one or more storms are indicated for a given year, they are randomly selected from the expanded training set. In this manner, a randomly selected number of storms per year is computed for each year to generate a 200-year simulation.

(11) Each 200-year sequence is rank ordered, a cumulative pdf computed, and a frequency-of-occurrence relationship developed according to the approach described above. However, in the EST, tail functions (Borgman and Scheffner 1991) are used to define probabilities for events larger than the largest the 200-year simulation.

(12) This computation is repeated 100 times, resulting in 100 individual pdfs from which 100 stage-frequency relationships are computed. This family of curves is averaged and the standard deviation computed, resulting in the generation of a stage-frequency relationship containing a measure of variability of data spread about the mean. This variation is well-suited to the development of design criteria requiring the quantification of the element of risk associated with the frequency predictions. A computed stage-frequency relationship for a coastal station in the Delaware study is shown in Figure II-5-27.

(c)   EST - extratropical storm application.

(1)   Extratropical events cannot be easily parameterized; therefore, the multi-parameter application of the EST is not appropriate for events such as northeasters. An alternate small-parameter application of the EST (Palermo et al. 1998) was used for this class of events. Northeasters have a much greater frequency of occurrence than hurricanes. For example, several northeasters impact large regions of the east coast every year. The existing WIS database of northeasters represents an adequate population of storms from which frequency-of-occurrence relationships can be computed.

(2)   The extratropical EST approach is conceptually simpler than the multi-parameter version used for hurricanes because parameters such as tidal phase, wave height, and wave period are specified as input vectors. In this approach, a database of extratropical events is assembled and respective surge elevations computed for each storm in the training set using numerical modeling techniques. If 20-30 years of reliable storm surge elevation data are available for a particular location, the numerical modeling phase may not be necessary. The extratropical analogy to the tropical implementation is that there are fewer input vectors. The database of response vectors is combined with tide, rank ordered, and a pdf is computed.

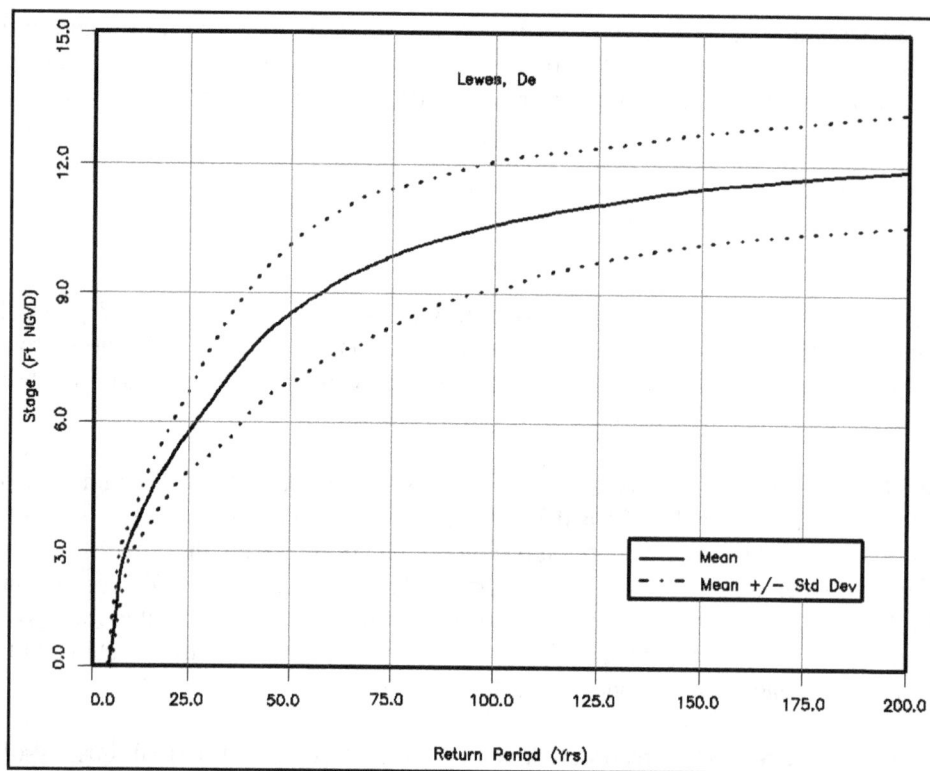

Figure II-5-27.   Stage-frequency relationship - coast of Delaware

(3)   The extratropical application of the EST can be used to develop recurrence relationships for any parameter if the database of parameters is sufficiently large to define the full population of possible events. A sample application was made to a dredged disposal site located in the New York Bight, on the east coast of the United States. The goal was to determine the recurrence interval of bottom erosion resulting from the 11 December 1992 extratropical storm event. An existing database of severe extratropical storms that impacted the Bight was available with corresponding wave conditions, maximum storm surge elevation, and tidal phase for each storm of the set. These data were used to define a database of peak sediment transport magnitudes, computed as a function of wave height, current (computed from surge elevations), depth, and grain size (Scheffner 1992). These data were input to the EST specifying only one input vector, the peak sediment transport magnitude (QX), to generate 1,000 repetitions of a 500-year simulation. For each 500-year simulation, a transport versus frequency-of-occurrence relationship was computed. As described above, an average curve was computed, which included variability. Results of the analysis are shown in the transport-frequency curve of Figure II-5-28 in which transport is given in $ft^3/sec/ft$-width $\times 10^{-4}$.

(4)   Historical data on storms in and around the coasts of the United States have been recorded and published in detail since the late 1800's. For example, NOAA publishes an atlas of tropical storms (NOAA 1981) that is continuously updated to reflect recent events. The distribution of east coast tropical events over the period of 1886-1980 is shown in Figure II-5-29, in which the tracks of the 793 Atlantic tropical cyclones reaching at least tropical storm intensity are shown. The Wave Information Study (WIS) of CHL has hindcast data for extratropical storms covering over 30 years.

(5)   Recent advances in numerical modeling techniques make it possible to use these very large databases of storm event track, wind, and pressure to simulate storm surges associated with each event. These sources provide data used to develop storm surge versus frequency-of-occurrence relationships.

Water Levels and Long Waves

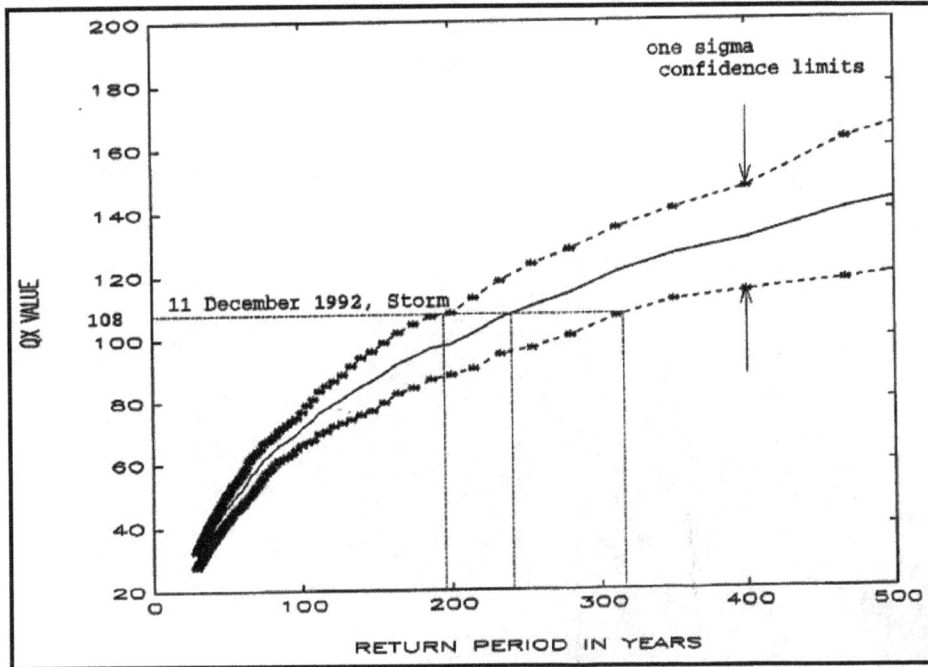

Figure II-5-28. Sediment transport magnitude-frequency relationship - December 1992 northeaster

II-5-6. Seiches

a. Seiches are standing waves or oscillations of the free surface of a body of water in a closed or semiclosed basin. These oscillations are of relatively long period, extending from minutes in harbors and bays to over 10 hr in the Great Lakes. Any external perturbation to the lake or embayment can force an oscillation. In harbors, the forcing can be the result of short waves and wave groups at the harbor entrance. Examples include 30- to 400-sec wave-forced oscillations in the Los Angeles-Long Beach harbor (Seabergh 1985). Dominant long-period seiche conditions on the Great Lakes have resonant modes with periods varying from 2 to 12 hr. These oscillations result primarily from changes in atmospheric pressure and the resultant wind conditions and occur over the entire basin. The frequency of oscillation is a function of the forcing, together with geometry and bathymetry, of the system.

b. In areas of simple geometry, modes of oscillation can be predicted from the shape of the basin. For example, Figure II-5-30 shows surface profiles for various simple geometric configurations.

c. For the one-dimensional case of a closed rectangular basin with vertical walls and uniform depth (Figure II-5-30b), the natural free oscillating period T_n can be written as

$$T_n = \frac{2l_B}{n\sqrt{gh}}$$

(II-5-26)

Water Levels and Long Waves

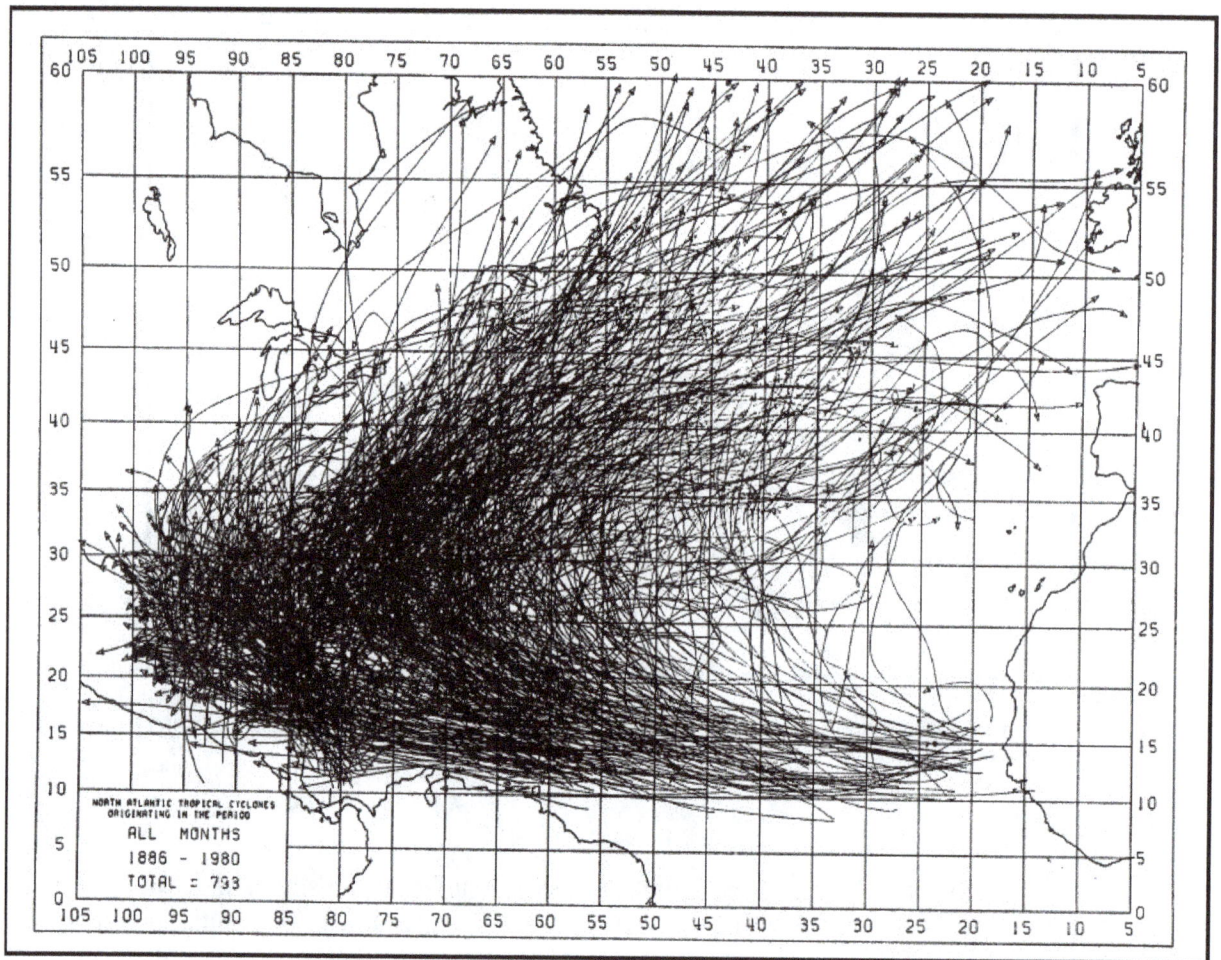

Figure II-5-29.   Atlantic tropical storm tracks during the period 1886-1980

where n is the number of nodes along the axis of the basin, h is the water depth, and l_B is the length of the basin. The fundamental and maximum period corresponds to T_n for $n = 1$. As shown in Figure II-5-30b, nodes (locations of no vertical deflection) are located at interior points and antinodes (locations of maximum deflection) are located on the boundaries of the basin. Longitudinal seiches oscillate from end to end, while transverse seiches oscillate from side to side in the basin.

  d.   For an open rectangular basin (Figure II-5-30c), the free oscillation period is written as follows:

$$T_n = \frac{4l_B}{(1 + 2n)\sqrt{gh}}$$

(II-5-27)

  e.   The fundamental mode corresponds to $n=0$. As shown in Figure II-5-30c, the node is located on the opening with an antinode on the opposite end.

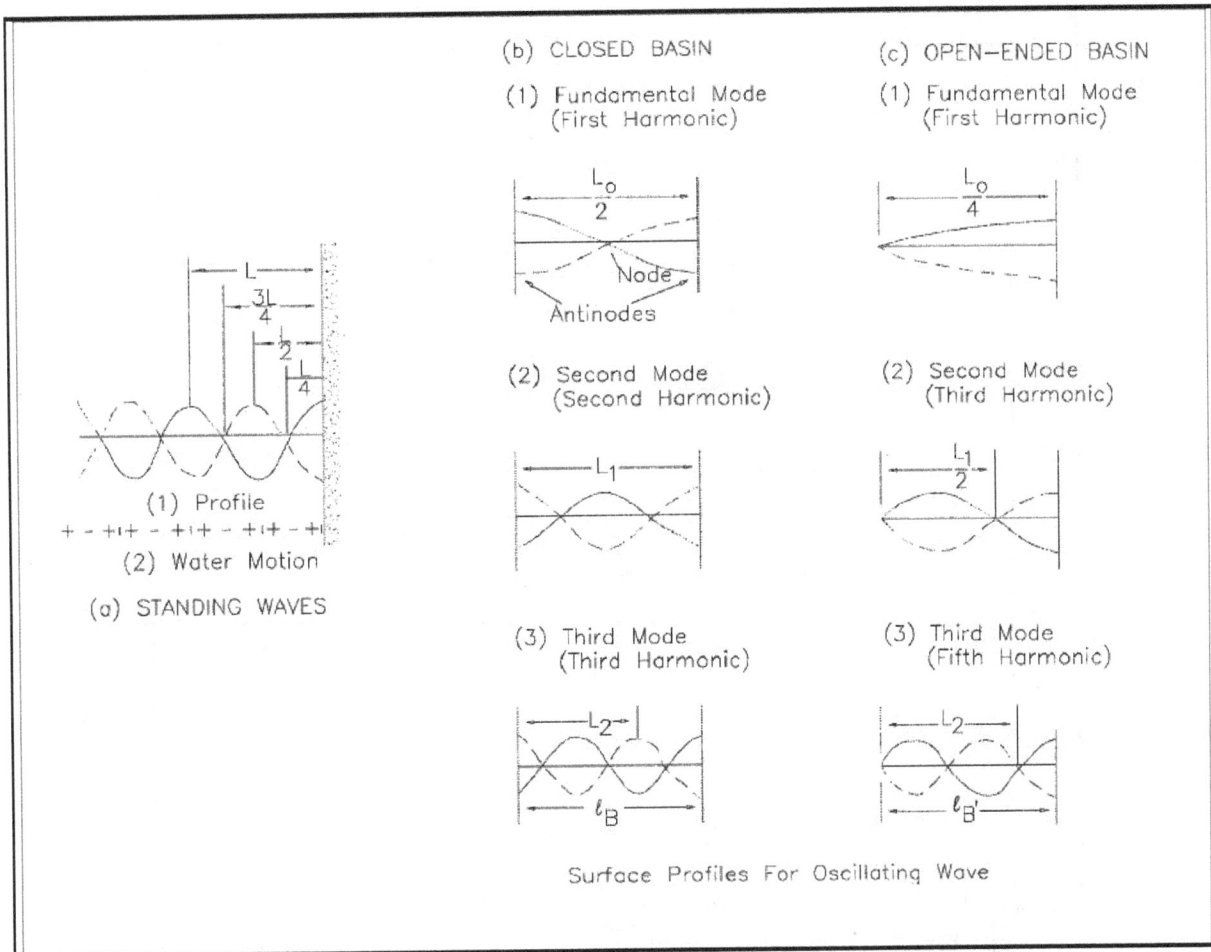

Figure II-5-30. Long wave surface profiles (*Shore Protection Manual* 1984)

f. Methods are available to estimate free oscillations in long narrow lakes of variable width and depth (Defant 1961); however, seiche periods and node/antinode locations are determined most accurately through long-wave modeling. For this class of problem, numerical solutions of the governing equations are necessary. An example is presented in Rao and Schwab (1976), in which a numerical model is used to investigate the normal modes of oscillation in an arbitrary enclosed basin on a rotating earth. Applications are made to Lakes Ontario and Superior. Rao and Schwab describe two distinct types of free oscillations, gravitational modes and rotational modes. Gravitational modes result from undulations of the free surface and are independent of the rotation of the earth. Rotational modes are a function of gravity via the Coriolis parameter. Figure II-5-31 represents the first, second, and third normal modes of oscillation for Lake Ontario. The periods of the computed six lowest modes in Lake Ontario are 5.11, 3.11, 2.13, 1.87, 1.78, and 1.46 hr for modes 1 through 6. The complexity of the nodal points and lines of phase demonstrate that idealized solutions such as those shown in Figure II-5-30 are often not adequate to explain natural phenomena. For applications requiring detailed information on maximum seiche elevations and nodal distributions, numerical models such as those described in the following section are the only viable approach to developing reliable solutions.

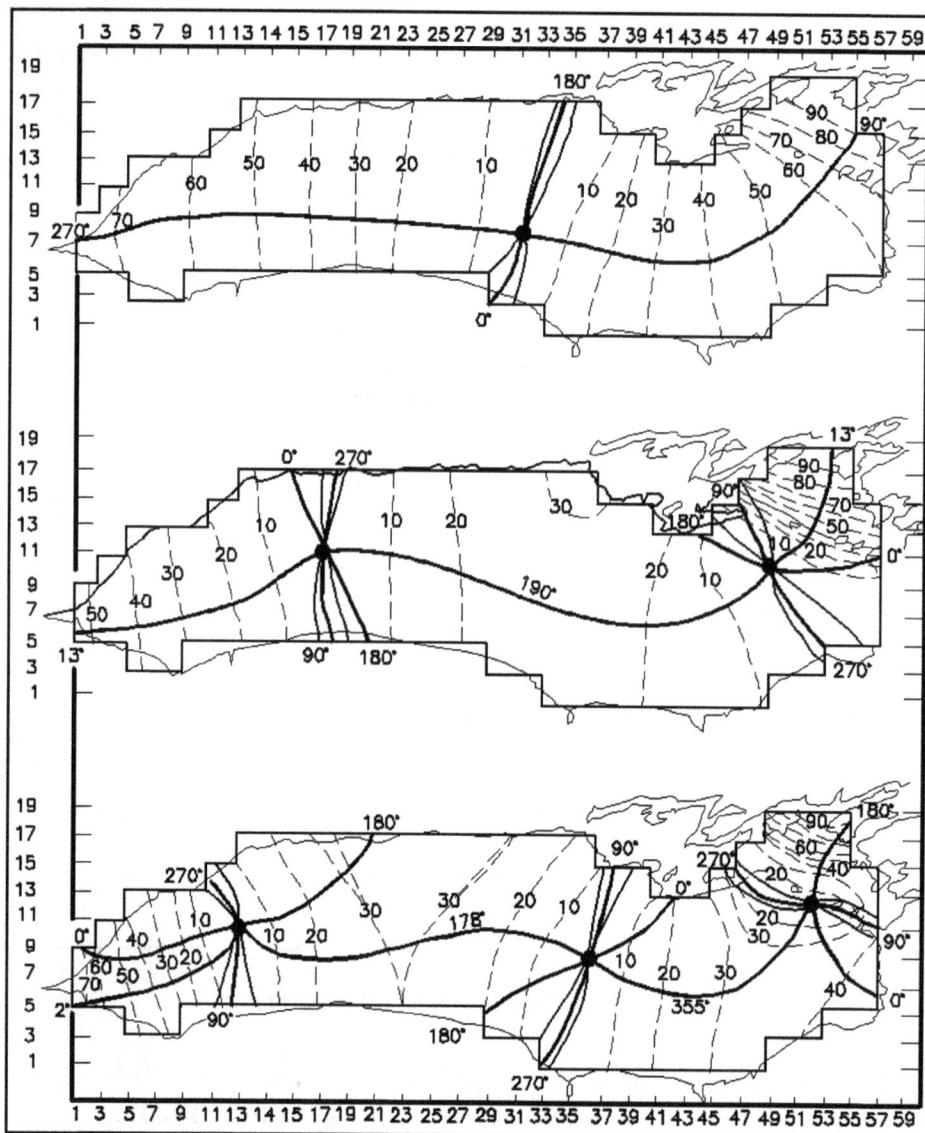

Figure II-5-31.   First, second, and third normal modes of oscillation for Lake Ontario (Rao and Schwab 1976)

II-5-7. Numerical Modeling of Long-Wave Hydrodynamics

a.   Long wave modeling. Most natural flow systems are extremely complex. They have variable bathymetry and irregular shorelines and are driven by a combination of tidal, wind, and pressure forcing, as well as temperature- and salinity-driven density gradient forcing. If a realistic approximation of flow hydrodynamics is necessary for the formulation of design criteria or testing of design alternatives (for example), numerical long-wave hydrodynamic models should be used. Although these models are generally not easy to apply and require mainframe computers, they provide the most accurate flow field approximation.

b.   Physical models. Prior to describing example applications of numerical models, physical models may provide a viable alternative to numerical approaches if the problem of concern is not wind-dominated. For example, if a physical model of a particular estuary exists, and the problem of interest is of tidal origin, then a physical model may be cost-effective. A substantial amount of literature on physical models and modeling is available. An extremely comprehensive source of information is Hudson et al. (1979).

c. Numerical models.

(1) Introduction.

(a) Numerical hydrodynamic models generally fall into two solution scheme categories, finite difference and finite element. Finite difference models use a rectangular orthogonal grid, although grid transformation schemes permit the mapping of the grid to conform to curvilinear boundaries. Finite element models use a variable-size, either triangular or rectangular, unstructured computational grid. The ability to define elements of variable size gives the finite element method an advantage over finite difference schemes for accurately representing areas of very complex geometry. However, both modeling techniques produce accurate results when applied correctly. Both models are in common use, as is demonstrated in the following two examples of numerical model applications to specific coastal and estuarine applications. These examples are intended to demonstrate the capability and accuracy of numerical models to reproduce natural hydrodynamic systems and generate an appreciation for the difficulty involved with applying them to specific flow-field situations.

(b) In addition to differences in computational schemes, hydrodynamic models are also categorized as 1-, 2-, and 3-dimensional. One-dimensional models provide a cross-sectional average solution to the governing equations. This class of solution is well-suited to pipeline or river flow problems, but not coastal/estuarine problems. Therefore, 1-D models are not addressed in this chapter because their solutions provide little detailed flow-field information that cannot be computed from the formulas given in Part II-1, "Water Wave Mechanics."

(c) Two-dimensional models are generally depth-integrated (depth-averaged); they provide a single velocity vector corresponding to each horizontal cell of a computational domain representing large areas of surface flow. Currents are defined at nodes or on cell faces, depending on the computational scheme used. Two-dimensional models are generally used in situations where the modeled system is well-mixed, i.e., currents are approximately uniform throughout the water column. These models are appropriate for studies in which changes in surface elevation are the primary concern. Typical examples include storm surge or lake seiche studies.

(d) An additional class of two-dimensional models is the laterally averaged model. In this case, instead of depth-averaged governing equations, a width-averaged set of equations is defined. In this form of solution, currents are defined at multiple locations through the water column; however, no horizontal distribution information is given. Laterally averaged models are generally used in conjunction with river flow, reservoir operation, and/or salinity intrusion applications and these models are not described in this manual.

(e) Three-dimensional applications are used when the vertical structure of currents is not uniform and the vertical distribution of currents is an important aspect of the study. Example applications include areas exhibiting flow reversal situations where surface currents and bottom currents flow in opposite directions. Included in this class of problems are cases in which vertical temperature and salinity gradients create density-driven flows. Although these non-tidal flows can be small in comparison to the tidal ebb and flood currents, they can contribute to residual circulation patterns that may affect the transport and dispersion of certain water quality parameters such as dissolved oxygen. An additional example may include channel-deepening effects on salinity intrusion.

(f)   Before presenting example applications of models, it is stressed that all models are not equal with respect to computational speed and accuracy. Depending on how the governing equations are written and solved and how and what boundary conditions are defined and used to drive the models, any two finite-element or finite-difference models can produce different results. Therefore, for some applications, one model may produce accurate results while another formulation will be unstable and produce totally unrealistic simulations. For this reason, any application of a numerical model should begin with a calibration and verification procedure in which the model is run to reproduce the hydrodynamic flow field corresponding to a specific time period during which prototype data (i.e., surface elevations and currents) have been collected. Additional data comparisons may include hydrodynamically driven parameters such as temperatures and salinities.

(g)   In the calibration phase, model parameters such as the friction factor distribution or depth resolution are adjusted to optimize the comparison of model-generated data to measured prototype data. Comparisons are generally made to surface elevations and velocities, but may include reproduction of temperature and salinity. In the verification procedure, the model is used to simulate an alternate time period containing additional prototype data. Model coefficients are not adjusted in the verification phase. An acceptable calibration and verification demonstrates that the model is capable of simulating the total flow regime over the entire computational domain. This ability is basic and necessary to the use of numerical models to quantify the flow-field response to proposed or existing changes in flow-field boundaries.

(h)   If data are available or can be collected, the calibration and verification procedures represent a minimum criterion required of a model before it can be applied to any specific problem. If these two steps are not performed in a satisfactory manner, model results can only be considered qualitative. Situations do exist where prototype data are not available and cannot be obtained within the time and cost limits of the budget. In these cases, models can be used to generate qualitative trends; however, unverified model results should not be the basis for quantitative decision making unless some data are available, even at a minimum level, to demonstrate that the model is producing realistic results. This aspect of modeling may be vital to a particular project because decisions based on unverified model results can be legally challenged and shown to be invalid.

(2)   Example - tidal circulation modeling.

(a)   The New York Bight project (Scheffner et al. 1993) is an excellent example of tidal circulation modeling. The goal of the study was to develop a hydrodynamic simulation tool that could be used to address questions concerning how certain modifications to the New York Bight may affect the local or global hydrodynamics of the system and how the computed flow fields affected the transport of certain water quality parameters. The model was therefore required to be capable of simulating the flow-field hydrodynamics, temperature, and salinity distribution over a large computational domain.

(b)   The model selected for this study was the CH3D (Curvilinear Hydrodynamics in Three (3) Dimensions) (Johnson et al. 1991) model, a three-dimensional, finite difference formulation model with boundary-fitted coordinates. The modeled area represents the region extending offshore from Cape May, NJ, and Nantucket Shoals to beyond the continental shelf, including Long Island Sound, the Hudson and East Rivers, and New York Harbor. Depths vary from less than 10 m to more than 2,000 m. The geographical boundaries are shown in Figure II-5-32a. The computational grid used to represent this area of interest is shown in Figure II-5-32b. The grid contains 76 cells in the alongshore direction and 45 cells in the cross-shore direction. There are 2,641 active computational cells in the horizontal and 10 in the vertical.

(c)   The model was preliminarily tested to demonstrate steady-state response to long-term circulation patterns, wind-induced circulation, and Hudson River inflow-induced circulation. Non-steady testing included modeling of the dynamic response to tidal constituents, both to the M_2 semidiurnal tide and to mixed tides. Calibration and verification were performed for the periods of April 1976 and May 1976, respectively.

(d)   Model results were compared to MESA project prototype data collected during the period from mid-September 1975 to mid-August 1976. Figure II-5-33 compares model-to-prototype tidal elevations at the Battery. A global circulation pattern of wind-field-induced circulation is shown in Figure II-5-34.

(3)   Example - storm surge modeling.

(a)   The coast of Delaware study (Mark, Scheffner, and Borgman 1993) provides an excellent example of storm surge modeling over very large computational domains. Goals of the effort were to generate stage-frequency relationships at a variety of locations along the open coast of Delaware and inside Delaware Bay. The study required tidal calibration and verification to demonstrate that the model could successfully reproduce tidal circulation. In addition to tide, the model was verified for storm surge by reproducing the storm surge hydrograph produced by Hurricane Gloria.

(b)   Storm track and pressure information for Hurricane Gloria was extracted from the National Hurricane Center's (NOAA 1981) database and used to generate input to the PBL model. The model selected for the hydrodynamic effort was the two-dimensional depth-integrated version of the Three-Dimensional **AD**vanced **CIRC**ulation (ADCIRC) finite-element model (Luettich, Westerink, and Scheffner 1993). This application demonstrates a very large-domain modeling capability (Luettich, Westerink, and Scheffner 1993), developed for the Dredging Research Program (DRP) to compute tidal and storm surge simulations along the open coast.

(c)   The DRP computational grid covers the east coast of the United States, Gulf of Mexico, and Caribbean Sea on the landward boundary. The offshore boundary is located in the mid-Atlantic Ocean, extending from Novia Scotia to Venezuela along the 60-deg west longitude line. "Continental-scale" computational domains are used in the model to minimize difficulties in specifying offshore and lateral boundary conditions.

(d)   The DRP computational grid was modified to provide increased resolution in Delaware Bay and within several small bays between the entrances of the Chesapeake and Delaware Bays. Figure II-5-35 shows the global limits of the 12,000-node grid. A blow-up of the coast of Delaware and Delaware Bay areas is shown in Figure II-5-36.

(e)   The ADCIRC model was verified to several locations at which NOAA tidal constituents were available. An example model-to-prototype (constituent) comparison to Lewes, DE, is shown in Figure II-5-38. Model-to-prototype data for the storm surge produced by Hurricane Gloria are also compared at Lewes, as shown in Figure II-5-39.

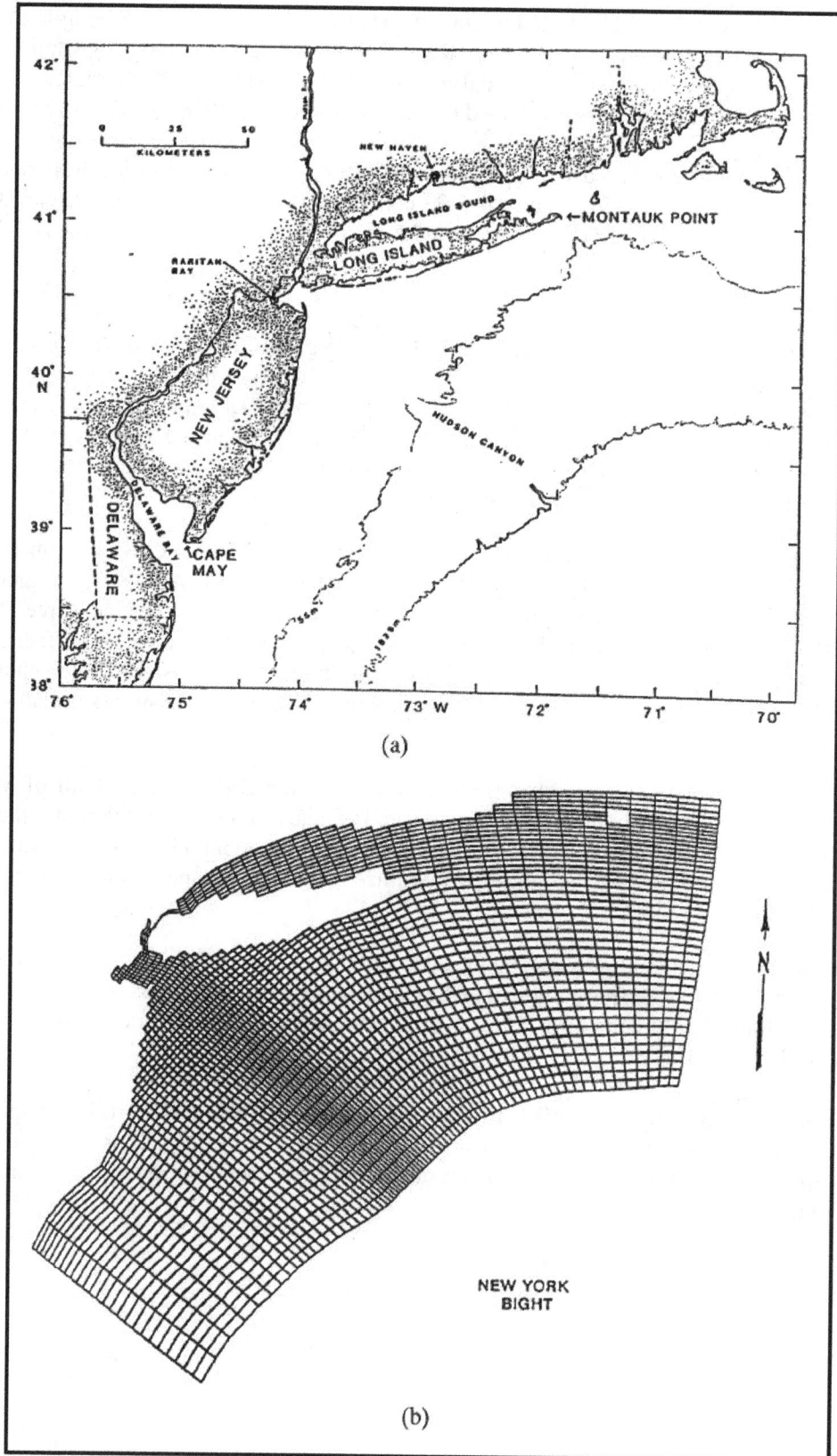

Figure II-5-32.   Computational grid for the New York Bight

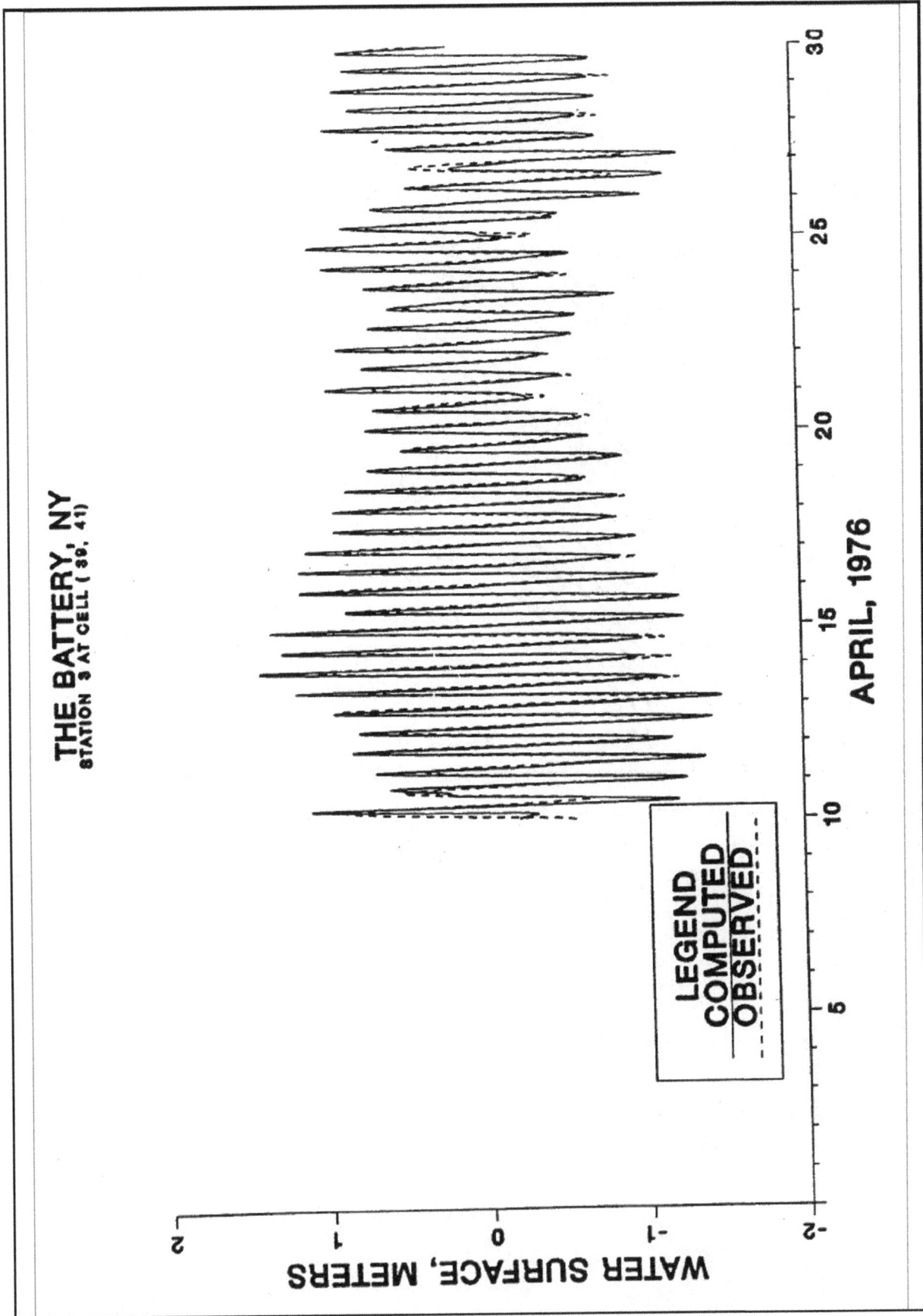

Figure II-5-33. Model and prototype tidal elevation comparison at the Battery

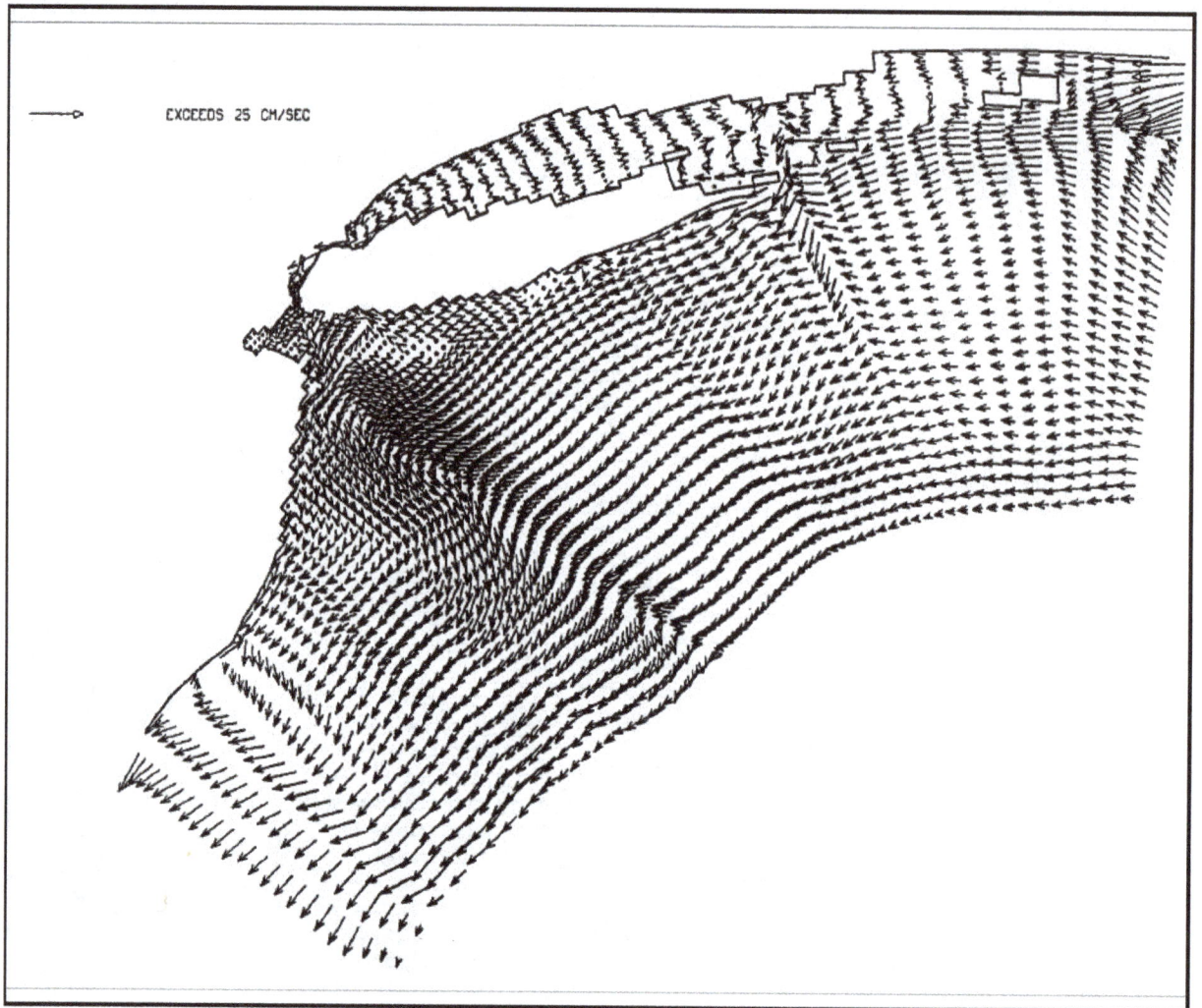

Figure II-5-34.   Wind-induced circulation pattern

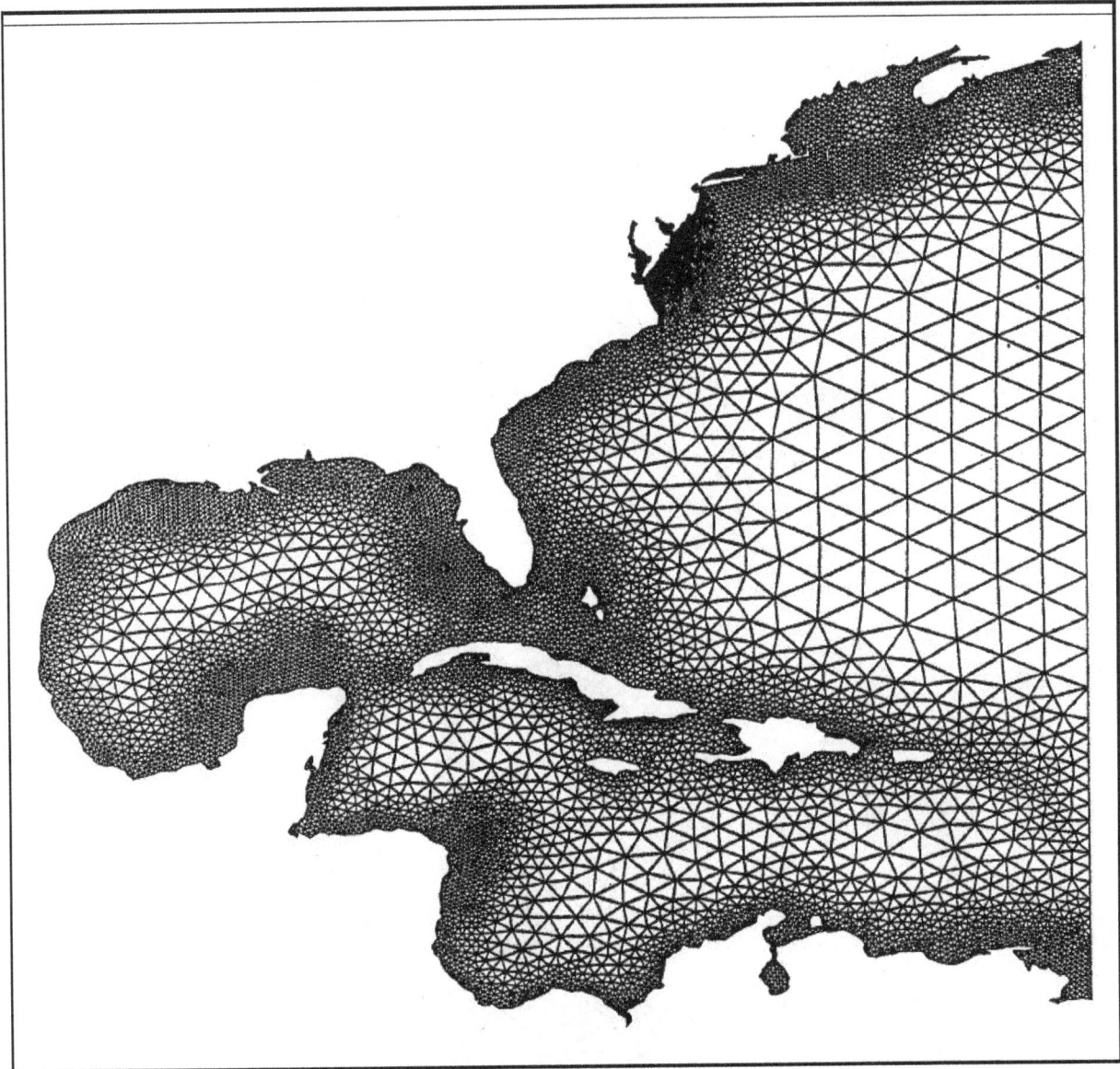

Figure II-5-35.   Global limits of ADCIRC computational grid

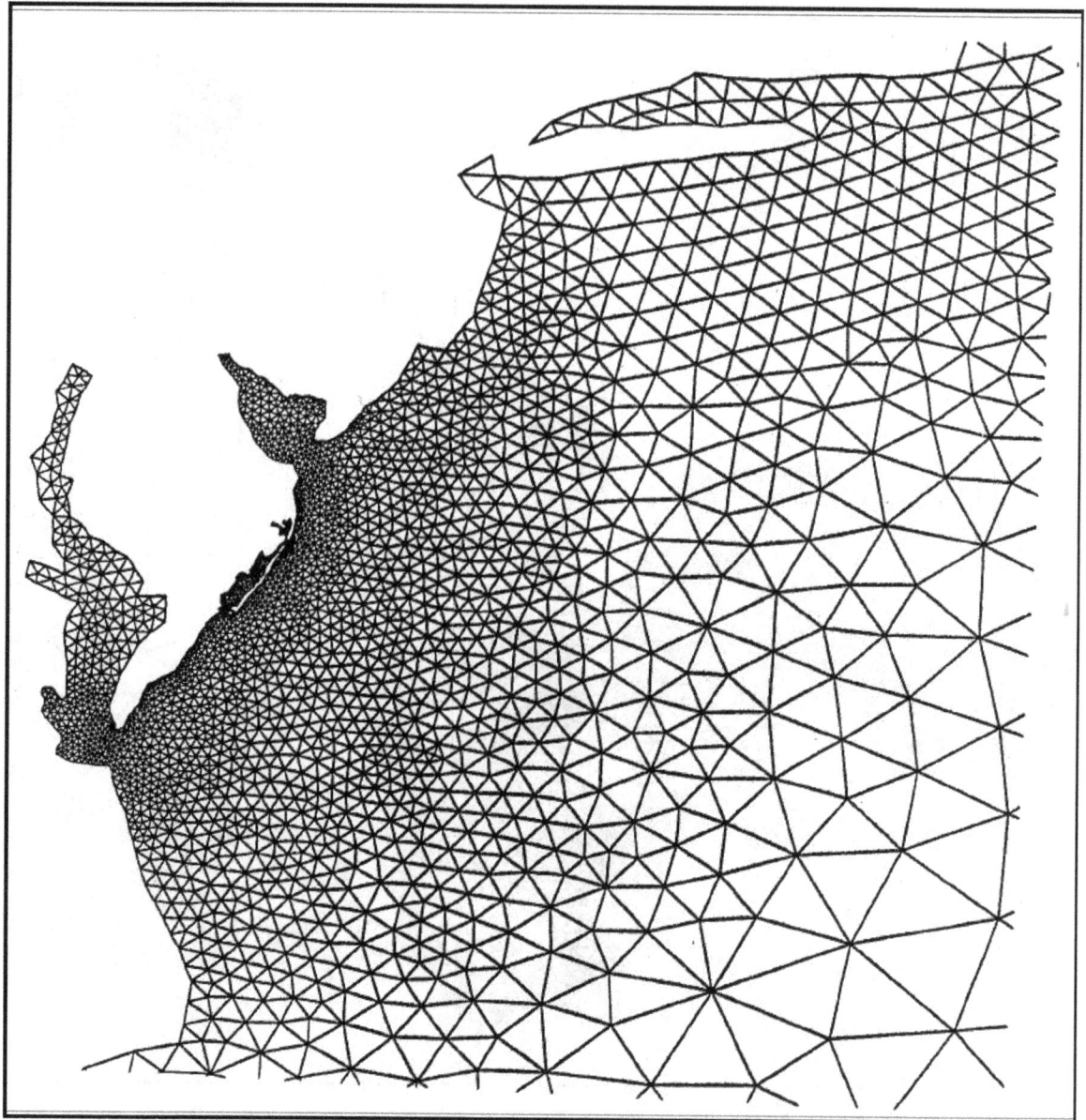

Figure II-5-36.   Blow-up of ADCIRC grid along Delaware coast

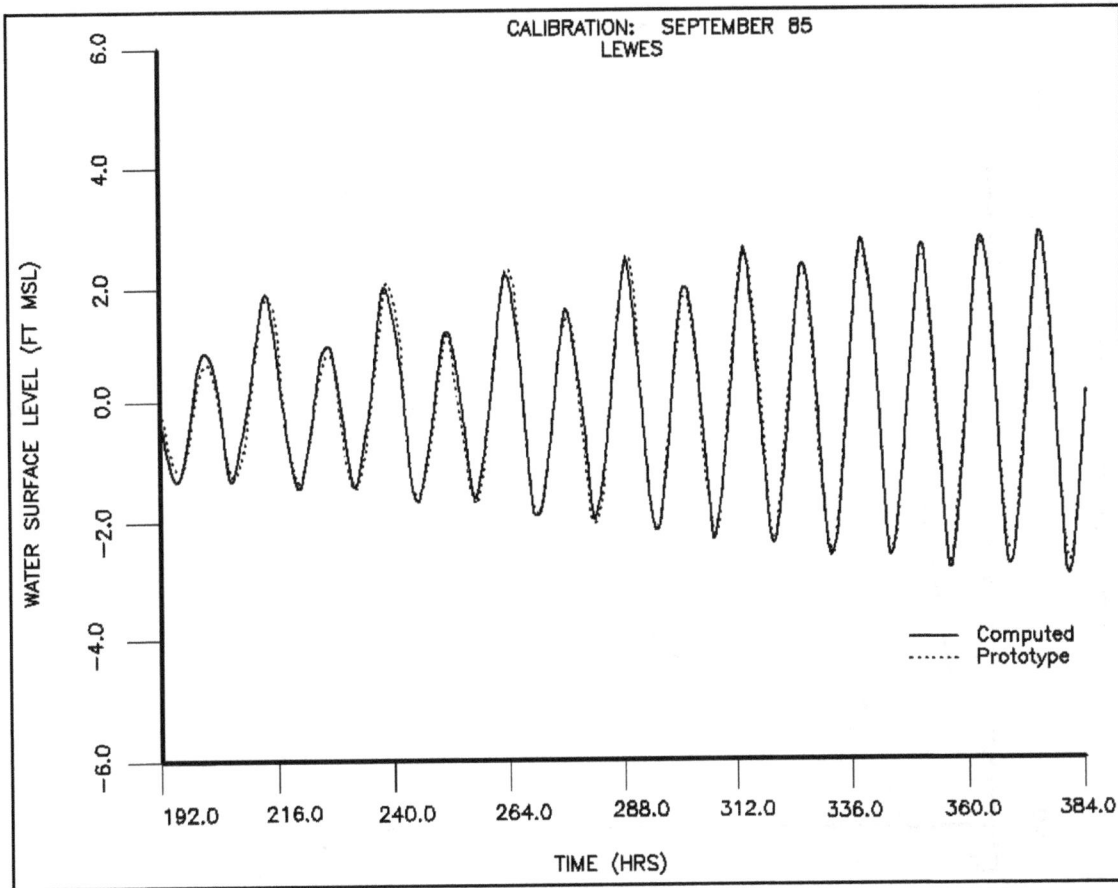

Figure II-5-37.   Model-to-prototype tidal comparison at Lewes, DE

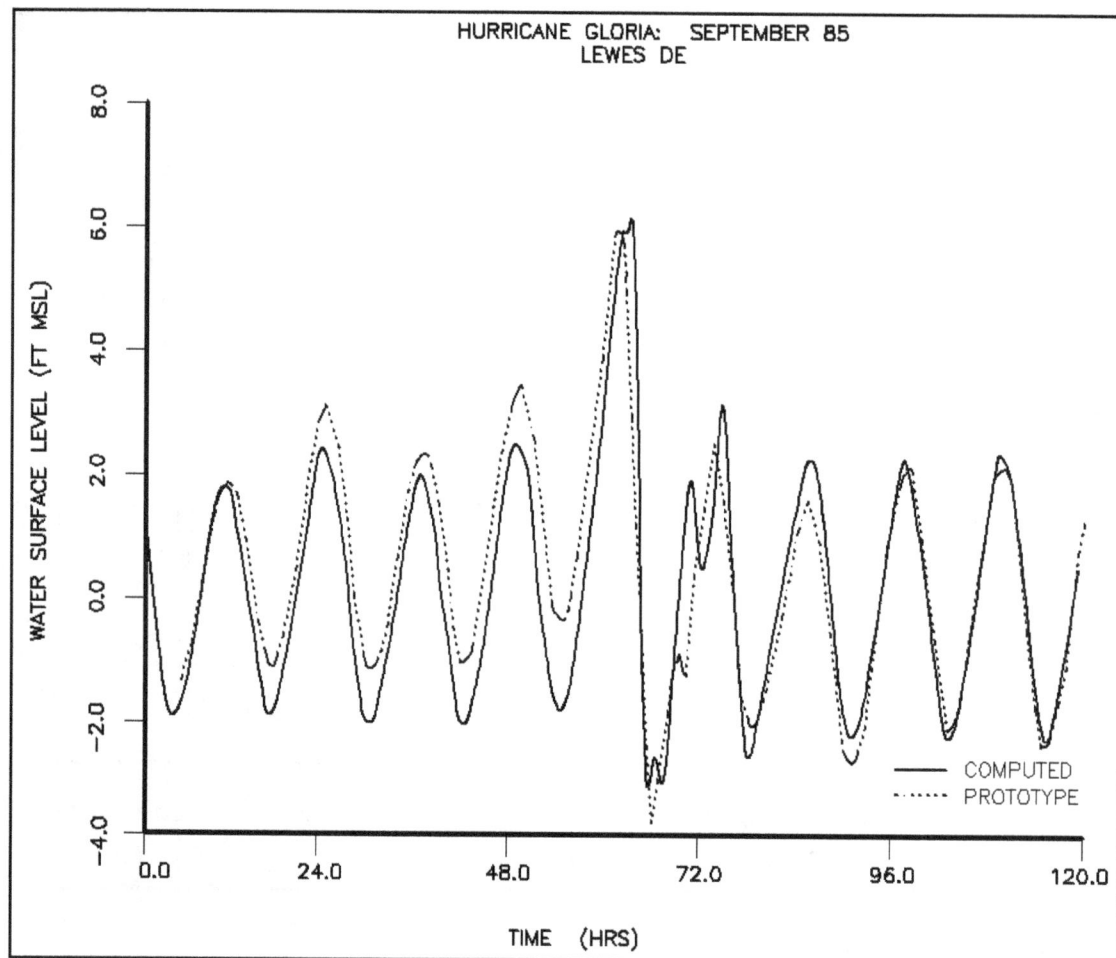

Figure II-5-38.   Model-to-prototype surge comparison at Lewes, DE

II-5-8. References

EM 1110-2-1415
Hydrologic Frequency Analysis

Borgman and Resio 1982
Borgman, L. E., and Resio, D. T. 1982. "Extremal Statistics in Wave Climatology," *Topics in Ocean Physics*, A. Osborne and P. M. Rizzoli, ed., Italian Physical Society, Bologna, Italy, pp 439-471.

Borgman and Scheffner 1991
Borgman, L. E., and Scheffner, N. W. 1991. "The Simulation of Time Sequences of Wave Height, Period, and Direction," Technical Report TR-DRP-91-2, U.S. Army Engineer Waterways Experiment Station, Vicksburg, MS.

Camfield 1980
Camfield, F. E. 1980. "Tsunami Engineering," Special Report No. 6, U.S. Army Engineer Waterways Experiment Station, Vicksburg, MS.

CCCMSL 1987
Committee on Engineering Implications of Changes in Relative Mean Sea Level. 1987. "Responding to Changes in Sea Level Engineering Implications," Marine Board, Commission on Engineering and Technical Systems, National Research Council. National Academy Press, Washington, DC, 148 p.

Cialone et al. 1991
Cialone, M. A., Mark, D. J., Chou, L. W., Leenknecht, D. A., Davis, J. A., Lillycrop. L. S., Jensen, R. E., Thompson, E. F., Gravens, M. B., Rosati, J. D., Wise, R. A., Kraus, N. C., and Larson, P. M. 1991. "Coastal Modeling System (CMS) User's Manual," Instruction Report CERC-91-1, U.S. Army Engineer Waterways Experiment Station, Vicksburg, MS.

Coordinating Committee on Great Lakes Basin Hydraulic and Hydrologic Data 1992
Coordinating Committee on Great Lakes Basin Hydraulic and Hydrologic Data. 1992. Brochure on the International Great Lakes Datum 1985, U.S. Government Printing Office, 1992-644-640.

Defant 1961
Defant, A. 1961. *Physical Oceanography*, Vol II, Pergamon Press, New York.

Dronkers 1964
Dronkers, J. J. 1964. "Tidal Computations in Rivers and Coastal Waters," North-Holland Publishing Company - Amsterdam, John Wiley & Sons, Inc., New York.

Graham and Nunn 1959
Graham, H. E., and Nunn, D. E. 1959. "Meteorological Considerations Pertinent to Standard Project Hurricane, Atlantic and Gulf Coasts of the United States," National Hurricane Research Project, Report No. 33, Department of Commerce, Washington, DC.

Gravens, Scheffner, and Hubertz 1989
Gravens, M. B., Scheffner, N. W., and Hubertz, J. M. 1989. "Coastal Processes from Asbury Park to Manasquan, New Jersey," Miscellaneous Paper CERC-89-11, U.S. Army Engineer Waterways Experiment Station, Vicksburg, MS.

Gumbel 1954
Gumbel, E. J. 1954. "Statistical Theory of Extreme Value and Some Practical Application," National Bureau of Standards Applied Math. Series 33, U.S. Gov. Publication, Washington, DC.

Harris 1981
Harris, D. L. 1981. "Tides and Tidal Datums in the United States," Special Report No. 7, U.S. Army Corps of Engineers, Fort Belvoir, VA., U.S. Government Printing Office (GPO Stock No. 008-022-00161-1).

Ho et al. 1987
Ho, F. P., Su, J. C., Hanevich, K. L., Smith, R. J., and Richards, F. P. 1987. "Hurricane Climatology for the Atlantic and Gulf Coasts of the United States," NOAA Technical Report NWS 38.

Hubertz et al. 1993
Hubertz, J. M., Brooks, R. M., Brandon, W. A., and Tracy, B. A. 1993. "Hindcast Wave Information for the U.S. Atlantic Coast," WIS Report 30, U.S. Army Engineer Waterways Experiment Station, Vicksburg, MS.

Hudson et al. 1979
Hudson, R. Y., Herrmann, F. A., Sager, R. A., Whalin, R. W., Keulegan, G. H., Chatham, C. E., and Hales, L. Z. 1979. "Coastal Hydraulic Models," Special Report No. 5, Coastal Engineering Research Center, U.S. Army Engineer Waterways Experiment Station, Vicksburg, MS.

Ippen 1966
Ippen, A. T. 1966. *Estuary and Coastline Hydrodynamics*. McGraw-Hill Book Company, Inc., New York.

Jarvinen and Gebert 1986
Jarvinen, B., and Gebert, J. 1986. "Comparison of Observed Versus Slosh Model Computed Storm Surge Hydrographs along the Delaware and New Jersey Shorelines for Hurricane Gloria, September 1985," NOAA Technical Memorandum NWS NHC 32, National Hurricane Center, Coral Gables, FL.

Johnson et al. 1991
Johnson, B. H., Heath, R. E., Hsieh, H. H., Kim, K. W., and Butler, H. L. 1991. "Development and Verification of a Three-Dimensional Numerical Hydrodynamic, Salinity, and Temperature Model of Chesapeake Bay; Volume I, Main Text and Appendix D," Technical Report HL-91-7, U.S. Army Engineer Waterways Experiment Station, Vicksburg, MS.

Leenknecht, Szuwalski, and Sherlock 1992
Leenknecht, D. A., Szuwalski, A., and Sherlock, A. R. 1992. "Automated Coastal Engineering System User Guide and Technical Reference, Version 1.07," U.S. Army Engineer Waterways Experiment Station, Vicksburg, MS.

Luettich, Westerink, and Scheffner 1993
Luettich, R. A., Westerink, J. J., and Scheffner, N. W. 1993. "ADCIRC: An Advanced Three-Dimensional Circulation Model for Shelves, Coasts and Estuaries; Report 1: Theory and Methodology of ADCIRC-2DDI and ADCIRC-3DL," Technical Report DRP-92-6, U.S. Army Engineer Waterways Experiment Station, Vicksburg, MS.

Mark and Scheffner 1997
Mark, D. J., and Scheffner, N. W. 1997. "Coast of Delaware Hurricane Stage-Frequency Analysis," Miscellaneous Paper CHL-97-1, U.S. Army Engineer Waterways Experiment Station, Vicksburg, MS.

Milne-Thompson 1960
Milne-Thompson, L. M. 1960. *Theoretical Hydrodynamics*. The MacMillan Company, New York.

Morang, Mossa, and Larson 1993
Morang, A., Mossa, J., and Larson, R. J. 1993. "Technologies for Assessing the Geologic and Geomorphic History of Coasts," Technical Report CERC-93-5, U.S. Army Engineer Waterways Experiment Station, Vicksburg, MS.

National Oceanic and Atmospheric Administration 1977
National Oceanic and Atmospheric Administration. 1977. "NOAA Federal Coordinator for Meteorological Services and Supporting Research, 1977: National Hurricane Operations Plan, ΦΧ 77-2."

National Oceanic and Atmospheric Administration 1981
National Oceanic and Atmospheric Administration. 1981. "Tropical Cyclones of the North Atlantic Ocean 1971-1980," U.S. Government Printing Office, Washington, DC.

National Oceanic and Atmospheric Administration 1984
National Oceanic and Atmospheric Administration. 1984. "Tide Tables 1984 High and Low Water Predictions, East Coast of North and South America, Including Greenland," National Ocean Service, Rockville, MD.

Palermo et al. 1998
Palermo, M. R., Clausner, J. E., Rollings, M. P., Williams, G. L., Myers, T. E., Fredette, T. J., and Randall, R. E. 1998. "Guidance for Subaqueous Dredged Material Capping," Technical Report DOER-1, U.S. Army Engineer Waterways Experiment Station, Vicksburg, MS.

Rao and Schwab 1976
Rao, D. B., and Schwab, D. J. 1976. "Two Dimensional Normal Modes in Arbitrary Enclosed Basins on a Rotating Earth: Application to Lakes Ontario and Superior," *Philosophical Transactions of the Royal Society of London*, Vol 281, pp 63-96.

Rappleye 1932
Rappleye, H. S. 1932. "The 1929 Adjustment of the Level Net," *The Military Engineer*, Vol 24, No. 138, pp 576-578.

Saffir 1977
Saffir, H. S. 1977. "Design and Construction Requirements for Hurricane Resistant Construction," American Society of Civil Engineers, Reprint Number 2830, New York.

Scheffner 1992
Scheffner, N. W. 1992. "A Numerical Simulation Approach to Estimating Disposal Site Stability," 1992 National Conference on Hydraulic Engineering, Baltimore, MD.

Scheffner, Borgman, and Mark 1993
Scheffner, N. W., Borgman, L. E., and Mark, D.J. 1993. "Applications of Large Domain Hydrodynamic Models to Generate Frequency-of-Occurrence Relationships," *3rd International Conference on Estuarine and Coastal Modeling*, Chicago, IL.

Scheffner et al. 1994
Scheffner, N. W., Vemulakonda, S. R., Kim, K. W., Mark, D. J., and Butler, H. L. 1994. "New York Bight Study; Report 1: Hydrodynamic Modeling," Technical Report CERC-94-4, U.S. Army Engineer Waterways Experiment Station, Vicksburg, MS.

Scheffner et al. 1999
Scheffner, N. W., Clausner, J. E., Militello, A., Borgman, L. E., Edge, B. L., and Grace, P. E. 1999. Use and Application of the Empirical Simulation Technique: Users Guide, Technical Report CHL-99-21, U.S. Army Engineer Waterways Experiment Station, Vicksburg, MS.

Seabergh 1985
Seabergh, W. C. 1985. "Los Angeles and Long Beach Harbors Model Study, Deep-Draft Dry Bulk Export Terminal, Alternative No. 6: Resonant Response and Tidal Circulation Studies," Miscellaneous Paper CERC-85-8, U.S. Army Engineer Waterways Experiment Station, Vicksburg, MS.

Sears and Zemansky 1963
Sears, F. W., and Zemansky, M. W. 1963. *University Physics*. Addison-Wesley Publishing Company, Inc., Reading, MA.

Shalowitz 1964

Shalowitz, A. L. 1964. "Shore and Sea Boundaries," Vol 2, Publication 10-1, U.S. Coast and Geodetic Survey, Washington, DC.

Shore Protection Manual 1984

Shore Protection Manual. 1984. 4th ed., 2 Vol, U. S. Army Engineer Waterways Experiment Station, U.S. Government Printing Office, Washington, DC.

Scheffner et al. 1999

Scheffner, N. W., Clausner, J. E., Militello, A., Borgman, L. E., Edge, B. L., and Grace, P. E. 1999. Use and Application of the Empirical Simulation Technique: Users Guide, Technical Report CHL-99-21, U.S. Army Engineer Waterways Experiment Station, Vicksburg, MS.

Schureman 1924

Schureman, P. 1924. "Manual of Harmonic Analysis and Prediction of Tides," Special Publication No. 98, U.S. Department of Commerce, Coast and Geodetic Survey, U.S. Government Printing Office, Washington, DC.

II-5-9. Definitions of Symbols

$\vec{\nabla}$	Vector gradient operator (Equation II-5-11)
α	Inflow angle (tropical storm parameter) [deg]
ζ	Tidal phase lag [deg]
θ	Angle of storm propagation (tropical storm parameter) [deg]
κ_n	Epoch of constituent n, i.e., phase shift from tide-producing force to high tide from t_0 [deg]
κ'	Modified form of the epoch that automatically accounts for the longitude and time meridian corrections (Equation II-5-20) [deg]
λ	Mean frequency of observed events per time period used in the Poisson probability law (Equation II-5-25)
σ	Angular frequency (= $2\pi/T$) [time^{-1}]
a	Wave amplitude [length]
a_n	Speed of tide constituent n [degrees/time]
$\vec{b}_X$	Attractive force of the moon and sun at any point X (Equation II-5-10)
C	Wave celerity [length/time]
f	Universal constant (= 6.67×10^{-8} cm^3/ gm sec^2)
F_g	Gravitational force (Equation II-5-7) [force-length/time^2]
f_n	Factor for adjusting mean tidal amplitude H_n values for specific times
g	Gravitational acceleration [length/time^2]
h	Water depth [length]
$H(t)$	Height of the tide at any time t (Equation II-5-16) [length]
H_0	Mean water level above some defined datum [length]
H_n	Mean amplitude of tidal constituent n [length]
k	Wave number (= $2\pi/L = 2\pi/CT$) [length^{-1}]
L	Wave length [length]
l_B	Length of a basin [length]
M	Mass of the moon
n	Number of nodes along the axis of a basin
N	Number of storm events used in the Poisson probability law (Equation II-5-25)
$-_0$	The subscript 0 denotes deepwater conditions

p_0	Central pressure of the eye of the storm (tropical storm parameter) [force/length^2]
P_M, P_S	Harmonic polynomial expansion terms that collectively describe the relative positions of the earth, moon, and sun (Equations II-5-14 and II-5-15)
p_n	Peripheral or far-field pressure (tropical storm parameter) [force/length^2]
$P[N=n]$	Poisson probability law (Equation II-5-25)
r	Distance between the centers of mass of two bodies (Equation II-5-7) [length]
R	Radius to maximum wind (tropical storm parameter) [length]
R	Tide classification parameter (Equation II-5-22) [dimensionless]
r_m	Distance from the center of the earth to the center of the moon [length]
r_{mx}	Distance of a point located on the surface of the earth to the center of the moon [length]
r_s	Distance from the center of the earth to the center of the sun [length]
r_{sx}	Distance of a point located on the surface of the earth to the center of the sun [length]
S	Local time meridian (Equation II-5-17)
S	Mass of the sun
t	Time measured from some initial epoch or time
T	Number of years used in the Poisson probability law (Equation II-5-25)
T	Wave period [time]
T_n	Natural free oscillating period of a basin (Equation II-5-26) [time]
u	Fluid velocity (water particle velocity) in the x-direction [length/time]
V	Maximum wind (tropical storm parameter) [length/time]
V_f	Forward speed of the eye of the storm (tropical storm parameter) [length/time]
V_M, V_S	Attractive force potentials per unit mass for the moon and sun (Equation II-5-9)
w	Fluid velocity (water particle velocity) in the z-direction [length/time]
z	Water depth below the SWL [length]

II-5-10. Acknowledgments

Author of Chapter II-5, "Water Levels and Long Waves:"

Norman W. Scheffner, Ph.D., Coastal and Hydraulics Laboratory, Engineer Research and Development Center (CHL), Vicksburg, Mississippi (retired).

Reviewers:

Zeki Demirbilek, Ph.D., CHL.
Lee E. Harris, Ph.D., Department of Marine and Environmental Systems, Florida Institute of Technology, Melbourne, Florida.
Edward F. Thompson, Ph.D., CHL.

Table of Contents

Page

II-6-1. Introduction to Inlet Hydrodynamic Processes II-6-1
   a. Inlet functions II-6-1
   b. Inlet characteristics II-6-3
   c. Inlet variables II-6-3
   d. Inlet flow patterns II-6-4

II-6-2. Inlet Hydrodynamics II-6-5
   a. Introduction II-6-5
   b. Inlet currents and tidal elevations II-6-7
   c. Tidal prism II-6-13
   d. Determining important inlet parameters II-6-19
      (1) Cross-sectional area II-6-19
      (2) Bay area II-6-21
      (3) Friction factor II-6-21
      (4) Inlet length II-6-22
      (5) Hydraulic radius II-6-23
   e. Evaluating inlet hydraulics with Keulegan K II-6-26
      (1) Introduction II-6-26
      (2) Jarrett's classification II-6-26
         (a) Class I: Keulegan $K < 0.3$ II-6-26
         (b) Class II: Keulegan $K > 0.80$ II-6-26
         (c) Class III: $0.3 < K < 0.8$ II-6-26
   f. Effect of freshwater inflow II-6-26
   g. Bay superelevation II-6-29
   h. The inlet as a filter, flow dominance, and net effect on sedimentation processes II-6-32
      (1) Introduction II-6-32
      (2) Tidal constituents II-6-32
      (3) Flow dominance II-6-36
   i. Multiple inlets II-6-37
   j. Tidal jets II-6-37
   k. Tidal dispersion and mixing II-6-39
   l. Wave-current interaction II-6-39
      (1) Wave-current interaction II-6-40
      (2) Current-channel interaction II-6-43
   m. Other methods of inlet analysis II-6-43
      (1) Automated Coastal Engineering System (ACES) II-6-43
      (2) DYNLET1 .... II-6-44
      (3) Coastal Modeling System (CMS) II-6-44
         (a) WIFM II-6-44
         (b) CLHYD II-6-46

(4) Other models II-6-47
(5) Physical models II-6-47

II-6-3. Hydrodynamic and Sediment Interaction at Tidal Inlets II-6-47
a. Introduction II-6-47
b. Tidal prism - channel area relationship II-6-47
c. Inlet stability analysis II-6-49
d. Scour hole problems II-6-53
e. Methods to predict channel shoaling II-6-55
(1) Introduction II-6-55
(2) Procedure II-6-56
(3) Phase I: Daily littoral materials transport volumes II-6-56
(4) Phase II: Channel sedimentation II-6-57
(5) Phase III: Regression analysis II-6-58
(6) Normalized, independent filling index II-6-58
(7) Ebb tidal energy flux II-6-58
(8) Hoerls special function distribution II-6-60
f. Inlet weir jetty hydraulic and sediment interaction II-6-60
(1) Weir location II-6-60
(2) Weir length II-6-60

II-6-4. References II-6-61

II-6-5. Definitions of Symbols II-6-70

II-6-6. Acknowledgments II-6-73

List of Tables

Page

Table II-6-1. Hydraulic Characteristics of Tidal Inlets by Cubature Method II-6-28

Table II-6-2. Semidiurnal and Shallow-Water Tidal Constituents and Harmonic Frequencies II-6-36

Table II-6-3. Tidal Prism-Minimum Channel Cross-sectional Area Relationships II-6-48

Table II-6-4. Inlet Stability Ratings II-6-52

List of Figures

Page

Figure II-6-1.   Typical structured and unstructured inlet II-6-2

Figure II-6-2.   Typical ebb-tidal delta morphology (Hayes 1980) II-6-4

Figure II-6-3.   Channel parameter measurement (Vincent and Corson 1980) II-6-5

Figure II-6-4.   Ebb delta area measurement (Vincent and Corson 1980) II-6-6

Figure II-6-5.   Minimum width cross-sectional area of channel A_{mw} versus channel
length L_{mw} (Vincent and Corson 1980) II-6-7

Figure II-6-6. A_{mw} versus ebb-tidal delta area (AED) (Vincent and Corson 1980) II-6-8

Figure II-6-7. A_{mw} versus maximum channel depth at minimum width section DMX
(Vincent and Corson 1980) II-6-10

Figure II-6-8. A_{mw} versus minimum controlling channel depth DCC (Vincent and
Corson 1980) II-6-11

Figure II-6-9.   Tidal prism-inlet area relationship II-6-12

Figure II-6-10.   Schematic diagram of flood and ebb currents outside an inlet
(O'Brien 1969) II-6-13

Figure II-6-11.   Wave refraction pattern in the vicinity of Merrimack River Estuary entrance
just south of the Merrimack Inlet (from Hayes (1971)) II-6-14

Figure II-6-12.   Ebb and flood flow patterns from a model study of Masonboro Inlet,
North Carolina II-6-15

Figure II-6-13.   Tidal current plus wave-generated currents approaching jettied inlet,
measured in physical model study II-6-16

Figure II-6-14.   Sediment transport gyres associated with both updrift and downdrift
portions of the delta at Essex River Inlet, Massachusetts (Note different
arrow types to denote wave or tide-generated sediment transportation
(Smith 1991)) II-6-17

Figure II-6-15.   Inlet bay system II-6-18

Figure II-6-16.   Variation of dimensionless parameters with Keulegan's repletion
coefficient K II-6-18

Figure II-6-17.   Sample tides and currents for an ocean-inlet-bay system that
satisfy the Keulegan assumptions (Keulegan 1967) II-6-19

Figure II-6-18.   Ratio of bay to sea tidal amplitude versus K_1 and K_2 . . . . II-6-20

Figure II-6-19.   Dimensionless maximum velocity versus K_1 and K_2 . . . . II-6-20

Figure II-6-20.   Bay tidal phase lag versus K_1 and K_2 . . . . II-6-21

Figure II-6-21.   Inlet flow net II-6-22

Figure II-6-22.   Inlet impedance (F) versus the ratio of inlet length to inlet hydraulic
radius to the 4/3 power ($l/R^{4/3}$) for various inlet width to inlet hydraulic
radius ratios, W/R II-6-23

Figure II-6-23.   Results from Example Problem II-6-1 . . . . II-6-25

Figure II-6-24.   Examples of hydraulic response of inlet and bay tide phasing and bay tide
amplitude for various Keulegan K values II-6-27

Figure II-6-25. $a_{b\,max}/a_o$ versus K for values of $Q_r{}'$ (river discharge model) II-6-29

Figure II-6-26. $a_{b\,min}/a_o$ versus K for values of $Q_r{}'$ (river discharge model) II-6-30

Figure II-6-27. $u'_{max\,e}$ versus K for values of $Q_r{}'$ (river discharge model) II-6-31

Figure II-6-28. $u'_{max\,f}$ versus K for values of $Q_r{}'$ (river discharge model) II-6-32

Figure II-6-29.   ε versus K for values of $Q_r{}'$ (river discharge model) II-6-33

Figure II-6-30.   Ratio of bay superelevation to ocean tide range as a function of the
coefficient of repletion K II-6-33

Figure II-6-31.   Relation between duration of outflow and bay superelevation II-6-36

Figure II-6-32.   Plume expansion II-6-38

Figure II-6-33.   Center-line velocity II-6-38

Figure II-6-34.   Contours of dimensionless wave height factor R_H given by Equation II-6-29.
Waves cannot propagate for values of F and Ω which lie in the forbidden
region (boundary line F = FM) II-6-40

Figure II-6-35.   Refraction of currents by channel II-6-44

Figure II-6-36.   Typical inlet flow nets II-6-45

Figure II-6-37.   Typical inlet cross section II-6-45

Figure II-6-38.   Sea and bay water elevations at inlet 1 . . . . II-6-46

Figure II-6-39.   Average velocity at minimum cross section II-6-46

Figure II-6-40.   Variations in cross-sectional area for Wachapreaque Inlet (Byrne,
De Alteris, and Bullock 1974) II-6-47

Figure II-6-41.   Tidal prism (P) versus jetty spacing (W) II-6-49

Figure II-6-42.   Escoffier (1940) diagram, maximum velocity and equilibrium velocity versus
inlet cross-sectional area II-6-51

Figure II-6-43.   Another version of Escoffier (1977) diagram, maximum velocity and
equilibrium velocity versus Keulegan repletion coefficient K II-6-51

Figure II-6-44.   Examples of channel geographic instability II-6-53

Figure II-6-45.   Little River Inlet geographical variability II-6-54

Figure II-6-46.   Example Problem II-6-4. Plot of P versus A_C from the hydraulic response
calculations and the stability equation II-6-55

Figure II-6-47.   1988 and 1989 bathymetry of Moriches Inlet II-6-56

Figure II-6-48.   Weir jetty flow patterns II-6-57

Chapter II-6
Hydrodynamics of Tidal Inlets

II-6-1. Introduction to Inlet Hydrodynamic Processes

a. Inlet functions.

(1) Inlets provide both man and nature with a means of access between the ocean and a bay. Commercial and recreational vessels need a navigable channel for safe transit to interior harbors. The flow of currents into and out of a bay through an inlet provides natural flushing to maintain good water quality and reasonable salinity levels. The migration of fish, fish larvae, and other sea life through the inlet conduit is also an important function of an inlet. Successful engineering of inlets requires knowledge of water and sediment movement in and adjacent to the inlet.

(2) Hydrodynamic conditions at tidal inlets can vary from a relatively simple ebb-and-flood tidal system to a very complex one in which tide, wind stress, freshwater influx, and wind waves (4- to 25-s periods) have significant forcing effects on the system. Figure II-6-1 shows a structured and an unstructured inlet with waves breaking on a shallow ebb shoal nearly surrounding the inlet itself. Flow enters the bay (or lagoon) through a constricted entrance, which is a relatively deep notch (usually 4 to 20 m at the deepest point). Entrance occurs after flow has traversed over a shallow shoal region where the flow pattern may be very complex due to the combined interaction of the tidal-generated current, currents due to waves breaking on the shallow shoal areas, wind-stress currents, and currents approaching the inlet due to wave breaking on adjacent beaches. The inlet acts to interrupt longshore current, which either reinforces or interferes with tidal currents, dependent on the time in the tidal cycle. The longshore current also is variable, dependent on wave conditions. Particularly during stormy conditions with strong winds, flow patterns may be highly complex. Also, the complicated two-dimensional flow pattern is further confounded because currents transverse to the coast tend to influence the propagation of waves, in some cases blocking them and causing them to break. The inlet has a complex shoal pattern that would cause odd refraction patterns and breaking regions, even if tidal flows were weak. Final complications are structures such as jetties, which cause wave diffraction patterns and reflections.

(3) In inlets with large open bays and small tidal amplitudes, flows can be dominated by wind stress (Smith 1977). In such cases, ebb conditions can last for days when winds pile up water near the bay side of the inlet, or long floods can occur when winds force bay water away from the inlet. Most inlet bays, however, are small and some are highly vegetated, so wind stress is not a dominant feature, except under storm conditions (this chapter will emphasize tidal components of the system; Part II-5 provides approaches for considering storm surges and seiche). Though not forced by tides, inlets on large lakes may have significant currents due to seiching (Seelig and Sorensen 1977), and may be studied by the methods in this chapter. Although many bays do not receive much fresh water relative to the volume of tidal flow, substantial freshwater input due to river flow can sometimes create vertically stratified flows through a tidal inlet. Typically, however, well-mixed conditions exist for most inlets. Neither the effects of wind on the bay nor stratified flows will be examined in this chapter; however, the gross effect of freshwater flow on inlet hydraulics will be discussed.

This chapter also provides simplified methods for estimating tidal hydraulics. In general, the inlet problem requires complex, complicated analyses, and reference will be made to such techniques.

Ocean City Inlet, Maryland

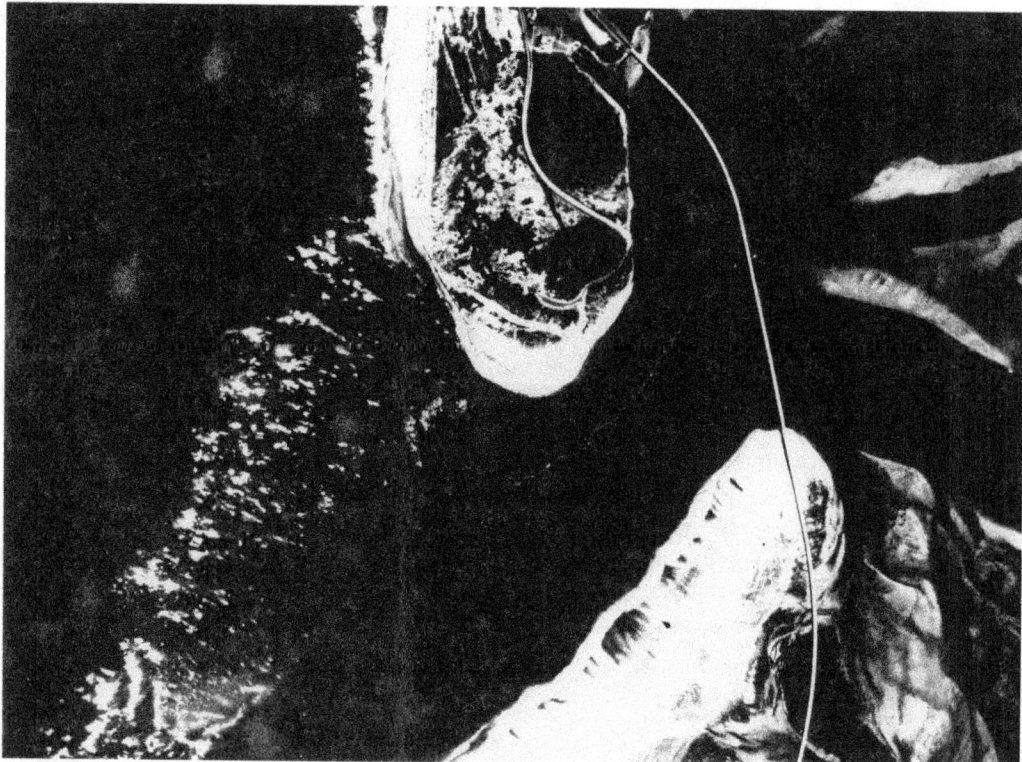

Oregon Inlet, North Carolina

Figure II-6-1. Typical structured and unstructured inlet

b. Inlet characteristics.

(1) Tidal inlets generally have a short, narrow channel passing between two sandy barrier islands (Figure II-6-1) and connect the ocean (or sea) to a bay. Some bays are small enough (on the order of tens of kilometers or less) for the water surface to rise and fall uniformly (co-oscillate) in response to the forcing ocean tide. Larger estuaries sometimes have broader junctions with the sea and may be long enough (hundreds of kilometers) to contain nearly an entire tidal wave length, thus having a variable water level at a given instant of time throughout the bay. Most methods discussed in this chapter apply to inlets that are closer to having a co-oscillating tide, but can be applied to most inlet systems as long as the tidal period is long compared to the time required for a shallow-water wave to propagate from the inlet to the farthest point in the bay, i.e.,

$$T >> \frac{L_b}{\sqrt{gd_b}}$$

(II-6-1)

where L_b is the distance to the farthest point, d_b is average bay depth, and T is typically taken as 12.42 hrs × 60 min/hr × 60 sec/min or 44,712 sec (for locations with semidiurnal or twice daily tides) or 89,424 sec for once-daily tides.

(2) The configuration of an individual inlet can vary significantly over time. Often the configuration is highly influenced by geology or peculiarities of the site, rather than a simple equilibrium of sediment and hydrodynamics. Convergence of flows from several directions at either side of the inlet can create strong turbulence that scours the channel deeply through the narrowest part of the inlet, called the inlet gorge, and silts in the channel on the bay and ocean sides. Maximum depths generally in the range of 4-15 m may occur in such channels, whereas seaward channel depths may diminish to 1.5-3 m where the flow has diffused and wave-driven sediment transport is important. Inside the inlet, water may diverge into one or more channels among shoal areas created by the deposition of sand from the ocean beaches. The resulting bathymetry can be a complex pattern of bars, shoals, and channels. Hayes (1980) shows the ocean-side ebb-tidal delta morphology for an unstructured inlet (Figure II-6-2).

c. Inlet variables. Although inlet systems can be quite complex, for the purpose of simple hydraulic analysis, the immediate inlet region can be approximated by key parameters which, although simplified, permit an analytical treatment of its hydraulics and a useful analysis of inlet systems. Analysis by Vincent and Corson (1980, 1981) shows the range of size of a number of inlet parameters based on 67 inlets that had been subjected to little or no human intervention. Figure II-6-3 defines an oceanside channel length L_{mw} (used for their particular study, and not to be confused with channel length defined later), depth at the crest of the outer bar in the channel (DCC), and location of inlet minimum width A_{mw}. Figure II-6-4 defines the area of the ebb tidal delta, AED, bounded by the depth contour of DCC (until it parallels the shoreline), the line joining this location and the shoreline and the line across the inlet minimum width. Cross-sectional area at the minimum inlet width (A_{mw}) is plotted against channel length (L_{mw}) in Figure II-6-5, against ebb tidal delta area (AED) in Figure II-6-6, against maximum channel depth (DMX) measured at minimum width section in Figure II-6-7, and against channel-controlling depth (DCC), the minimum depth across the outer bar, in Figure II-6-8. All parameters vary in log-linear relation to A_{mw} over several orders of magnitude. The 95-percent confidence bands are also plotted. O'Brien's (1931) observed that there is a direct relationship between the inlet's minimum cross-sectional flow area A (this is the minimum area A_c, not necessarily the area at the location of the minimum width), and the tidal prism (P) filling the bay (Figure II-6-9). This relationship will aid in defining the stability of inlet channels. The tidal prism in this case is defined as the volume of water entering through the inlet on a spring tide. Detailed methods to define inlet parameters will be discussed later.

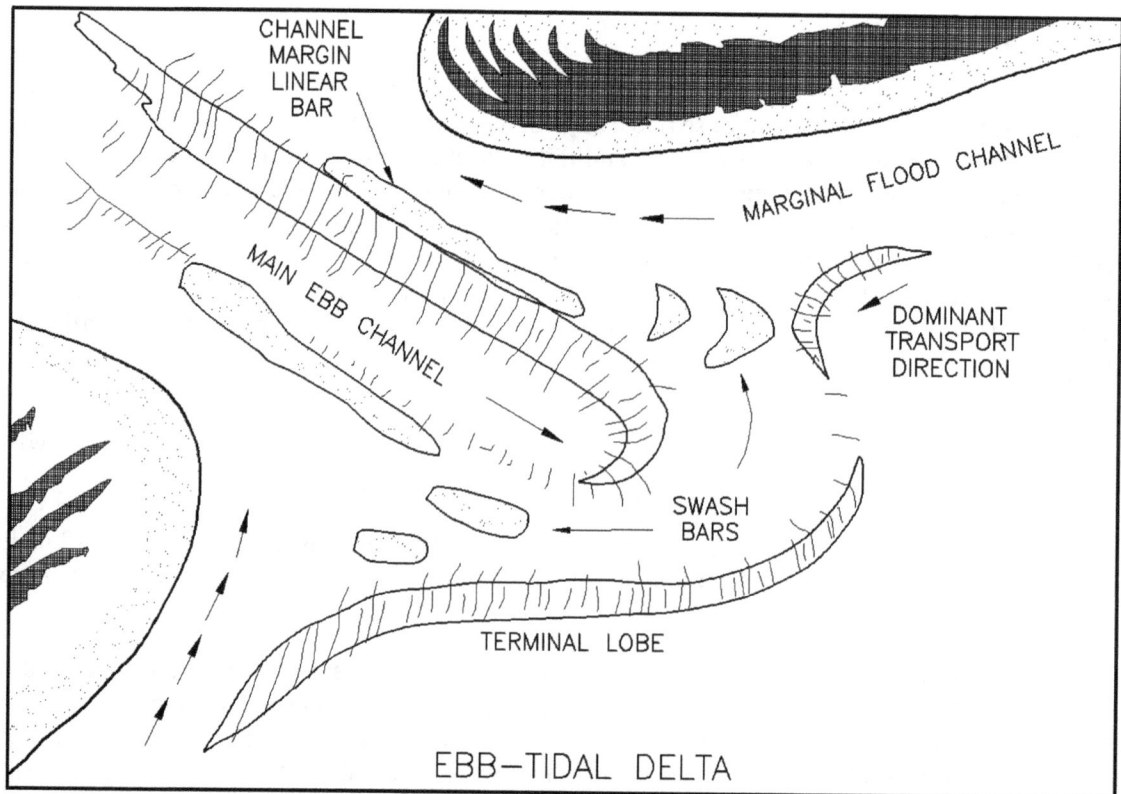

Figure II-6-2. Typical ebb-tidal delta morphology (Hayes 1980)

d. Inlet flow patterns.

(1) An inlet has a "gorge" where flows converge before they expand again on the opposite side. Shoal (shallow) areas that extend bayward and oceanward from the gorge depend on inlet hydraulics, wave conditions, and general geomorphology. All these interact to determine flow patterns in and around the inlet and locations where flow channels occur.

(2) Typical flood and ebb current patterns on the ocean side of a tidal inlet are shown in Figure II-6-10. The important aspect of this general circulation pattern is that currents usually flow toward the inlet near the shoreline (in the flood channels), even on ebb tide. The reason for this seeming paradox is the effect of wave-driven currents, (on the downdrift side of the inlet, breaking waves are turned toward the inlet due to refraction over the outer bar and on breaking, create currents toward the inlet). Further downdrift, currents are directed away from the inlet (an example of this is given in Figure II-6-11) and the effect of the ebb jet convecting or entraining ocean water as it exits the inlet creates an alongshore current at the base of the ebb jet. The "sink" flood flow pattern of the previous flood tide flow's momentum also helps sustain this flow pattern. Figure II-6-12 shows ebb and flood flow patterns from a model study of Masonboro Inlet, North Carolina, for both pre- and post-jetty conditions. Figure II-6-13 shows strength and direction of wave-generated currents plus tidal currents approaching a jetty at an inlet as measured in a physical model study. Figure II-6-14 shows the complexity of sediment flow patterns at an inlet as resolved by a field study at Essex River Inlet, Massachusetts (Smith 1991). Ebb and flood flow patterns are discussed in detail in Part II-6-2, paragraph *j*, *Tidal jets.*

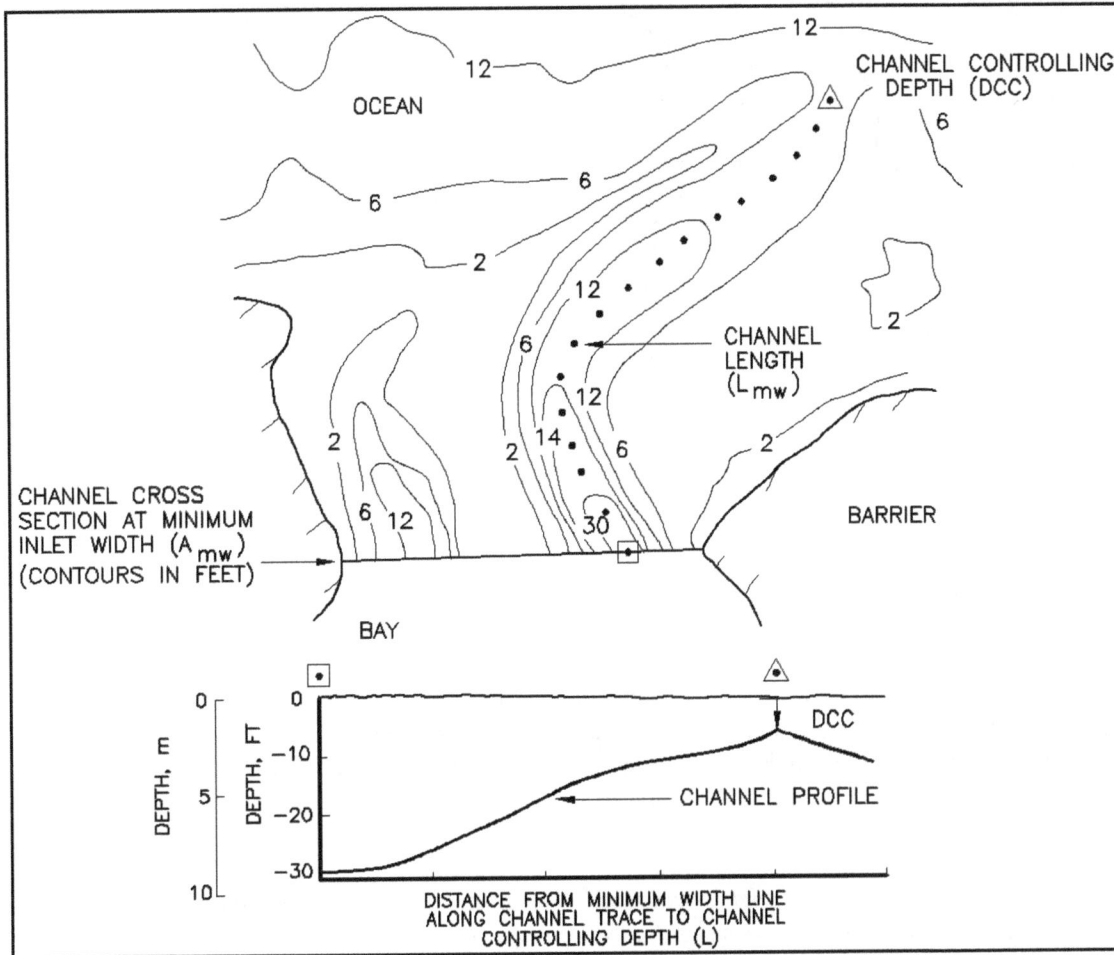

Figure II-6-3. Channel parameter measurement (Vincent and Corson 1980)

II-6-2. Inlet Hydrodynamics

a. Introduction.

(1) Many approaches are available to evaluate inlet hydrodynamics. Analytic expressions, numerical models, and physical models can be used. This section will present some simple analytic techniques to determine average velocities in a channel cross section due to the ocean tide and tidal elevation change in the bay. This section also will aid in understanding inlet response with regard to important parameters such as size of bay or channel area and length. Tidal inlet hydrodynamics are summarized by van de Kreeke (1988) and by Mehta and Joshi (1984).

(2) To characterize the development of inlet currents, COL E. I. Brown (1928) wrote the following:

To trace the characteristics of the flow of the tides through an inlet, assume the case of an inland bay being formed by the prolongation of a sandspit. It is quite evident that in the earlier stages of the growth of the spit the tide will rise and fall in the bay equally and simultaneously with its rise and fall in the surrounding ocean, and that a tidal current will be practically negligible. This state of affairs will continue for a long time, as long as the full and free propagation of the tidal wave is

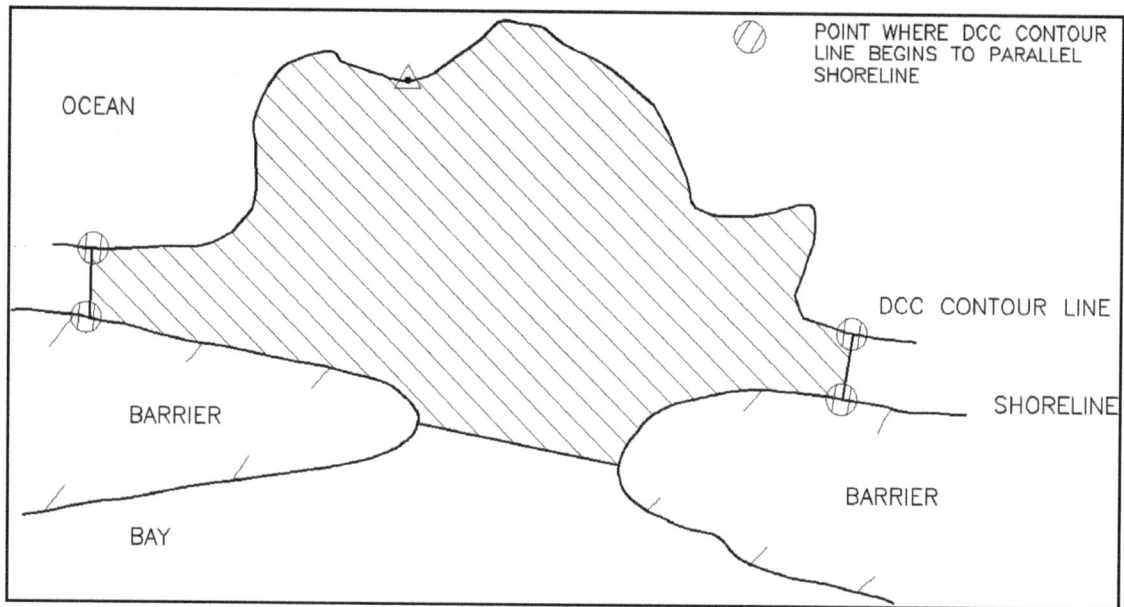

Figure II-6-4. Ebb delta area measurement (Vincent and Corson 1980)

unimpeded by the continually narrowing opening. When it does begin to be so impeded and the entrance acts as a barrier to the incoming tide, a delay in the advance of the tide will be caused, and a difference of head will occur between the water outside the inlet and that inside. This will create a hydraulic current into the bay, in addition to such movement as may be due solely to tidal wave propagation.

It is clear that as long as the inlet is wide and deep in proportion to the area of the bay, hydraulic currents will be small and tidal wave propagation will predominate. Tidal waves in the shallow waters near the shore will be essentially waves of translation, that is, the whole body of water moves with practically the same velocity horizontally. Now, if the size of the inlet becomes very small with respect to the bay area, tidal wave propagation will be negligible, the flow through the inlet will be hydraulic, that is, the water surface through the inlet will have a slope, causing a flow, and it can be considered mathematically in accordance with known hydraulic laws.

(3) With this in mind, early development of inlet hydraulics achieved reasonable results by using simplified approaches of steady-flow hydraulics to understand inlet currents and response of the bay (or lagoon) tide (Brown 1928). Keulegan (1951, 1967) solved the one-dimensional, depth-averaged shallow water wave equation for flow analytically. Others since have formulated a variety of analytical solutions for inlets (including van deKreeke (1967), Mota Oliveira (1970), Shemdin and Forney (1970), King (1974), Mehta and Özsöy (1978), Escoffier and Walton (1979), and DiLorenzo (1988)). Paralleling analytical development, physical models were used for detailed studies of inlet design (see below). More recently, numerical models have provided greater refinements and details using one-, two- and three-dimensional longwave equations of motion (including developments by Harris and Bodine (1977), Butler (1980), and Amein and Kraus (1991)). Some analytical models for inlets will be examined here to provide understanding of the inlet system and because they actually produce usable information with minimal effort. Application of the techniques of this chapter would include use of numerical models (e.g., Automated Coastal Engineering System (ACES); Leenknecht et al. 1992). ACES includes a spatially integrated one-dimensional numerical model. Additional information about ACES and other available models is provided in Part II-6-2, paragraph *m, "Other methods for inlet analysis."*

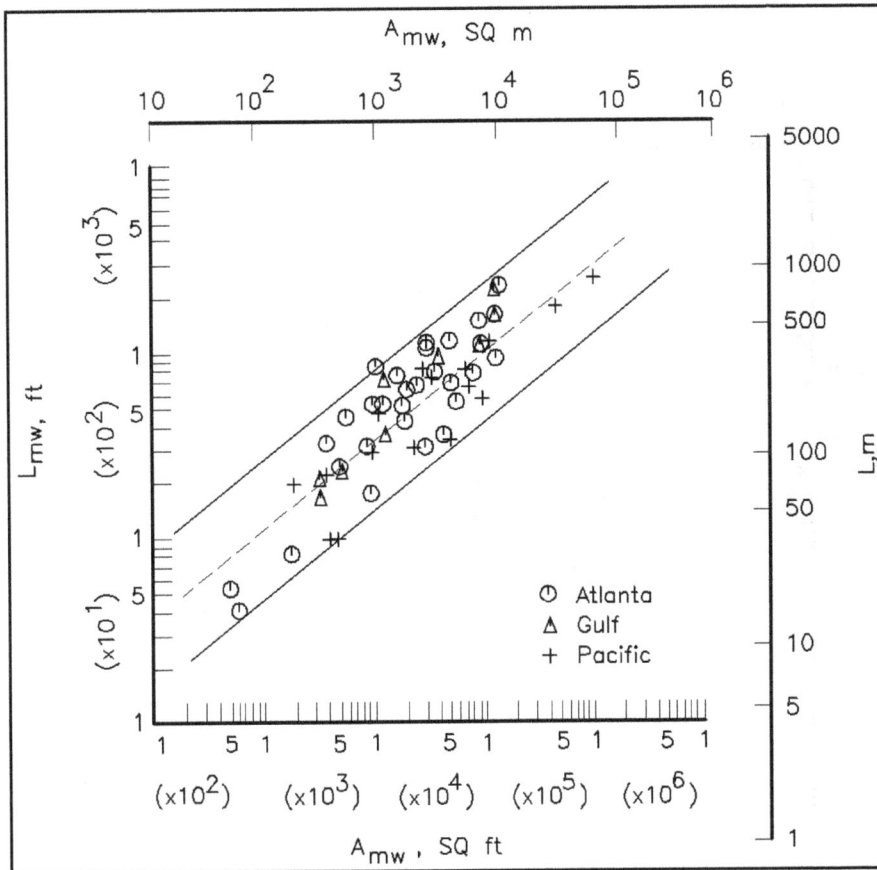

Figure II-6-5.   Minimum width cross-sectional area of channel A_{mw} versus
channel length L_{mw} (Vincent and Corson 1980)

b.   *Inlet currents and tidal elevations.*

(1) A simple introduction to inlet hydraulics will consider applying the one-dimensional equation of motion and the continuity equation. We want to find the maximum inlet current, the tide range of the bay and the phase lag of the bay tide relative to the tide in the ocean in terms of parameters which can be easily measured or determined, including inlet cross-sectional area, bay surface area, ocean tide amplitude and period, length of channel, and head loss coefficients. The simplified inlet system is shown in Figure II-6-15. Keulegan's assumptions (1967) were as follows:

(a) Walls of the bay are vertical.

(b) There is no inflow from streams.

(c) No density currents are present.

(d) Tidal fluctuations are sinusoidal.

(e) Bay water level rises uniformly.

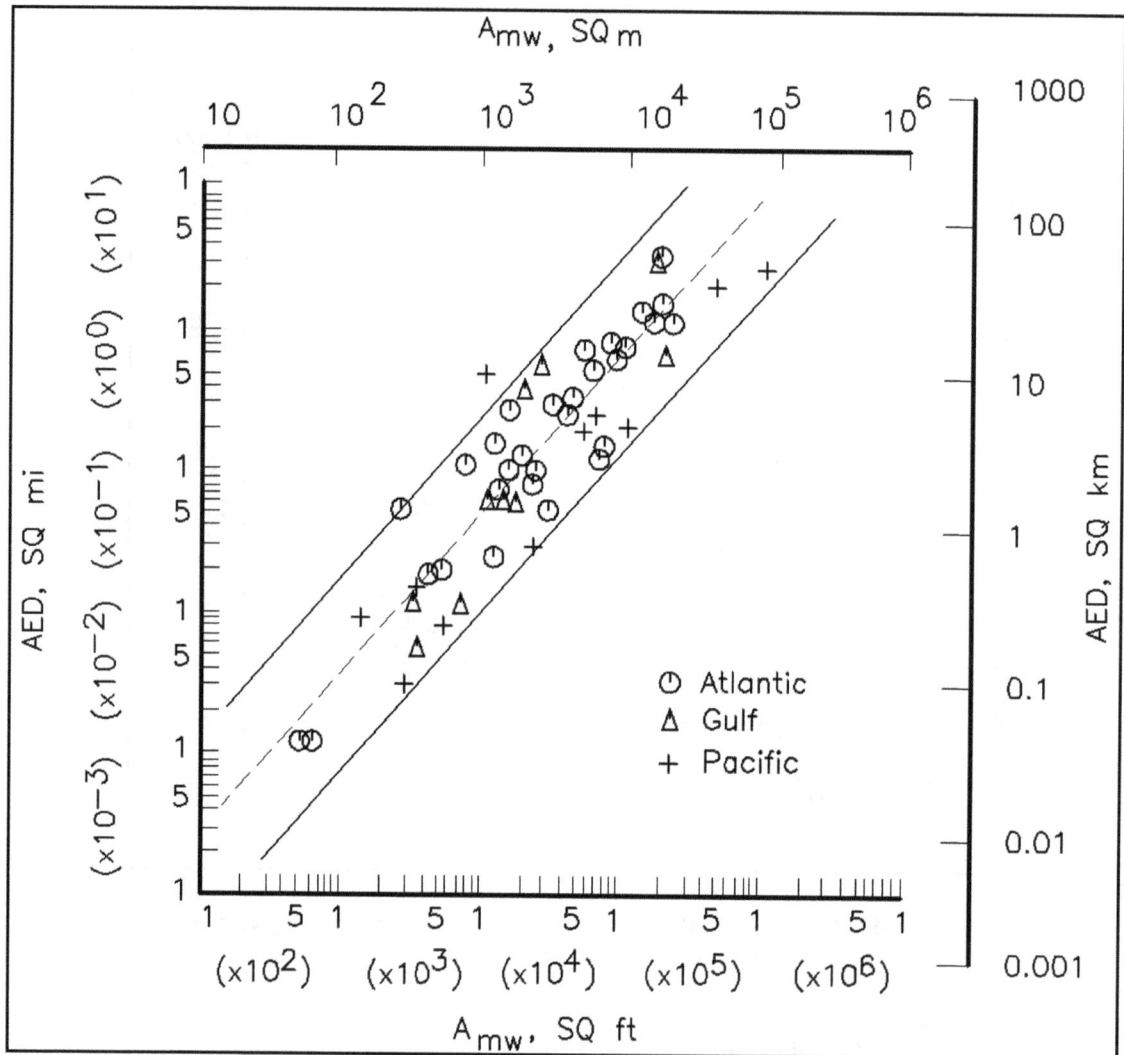

Figure II-6-6. A_{mw} versus ebb-tidal delta area (AED) (Vincent and Corson 1980)

(f) Inlet channel flow area is constant.

(g) Inertia of the mass of the water in the channel is negligible.

(2) Some assumptions may seem stringent (e.g., assumption (e), which essentially assumes a frictionless bay). This could be a problem if the entire bay is composed of very shallow tidal flats with no significant channelization. Also, in assumption (f), a relatively large tidal range compared to channel depth might occur. However, reasonable results can be determined for most cases. For more complex inlet systems, more sophisticated modeling should be performed, as discussed earlier.

(3) The one-dimensional equation of motion for flow in the channel is:

$$\frac{\partial V}{\partial t} + V\frac{\partial V}{\partial x} = -g\frac{\partial h}{\partial x} - \frac{f}{8R}V|V| \tag{II-6-2}$$

where

V = average velocity in channel

h = channel water surface elevation

f = Darcy - Weisbach friction term

R = hydraulic radius

g = acceleration due to gravity

(4) Keulegan neglected local acceleration (the first term in Equation II-6-2) and integrated the equation over the length of the inlet. Using the equation of continuity for flow through the inlet into the bay:

$$VA_{avg} = A_b \frac{dh_b}{dt} \tag{II-6-3}$$

where

A_{avg} = average area over the channel length

A_b = surface area of bay

dh_b/dt = change of bay elevation with time

(5) Combining Equations II-6-2 and II-6-3, Keulegan developed a solution for velocity and resulting bay tide which contained the dimensionless parameter K, known as the coefficient of repletion, or filling, which is defined as

$$K = \frac{TA_{avg}}{2\pi A_b} \sqrt{\frac{2g}{a_o \left[k_{en} + k_{ex} + \frac{fL}{4R} \right]}} \tag{II-6-4}$$

where A_{avg}, A_B, g, f, and R are as defined above, and

T = tidal period

a_O = ocean tide amplitude (one-half the ocean tide range)

k_{en} = entrance energy loss coefficient

k_{ex} = exit energy loss coefficient

L = inlet length

R = inlet hydraulic radius (see Equation II-6-18)

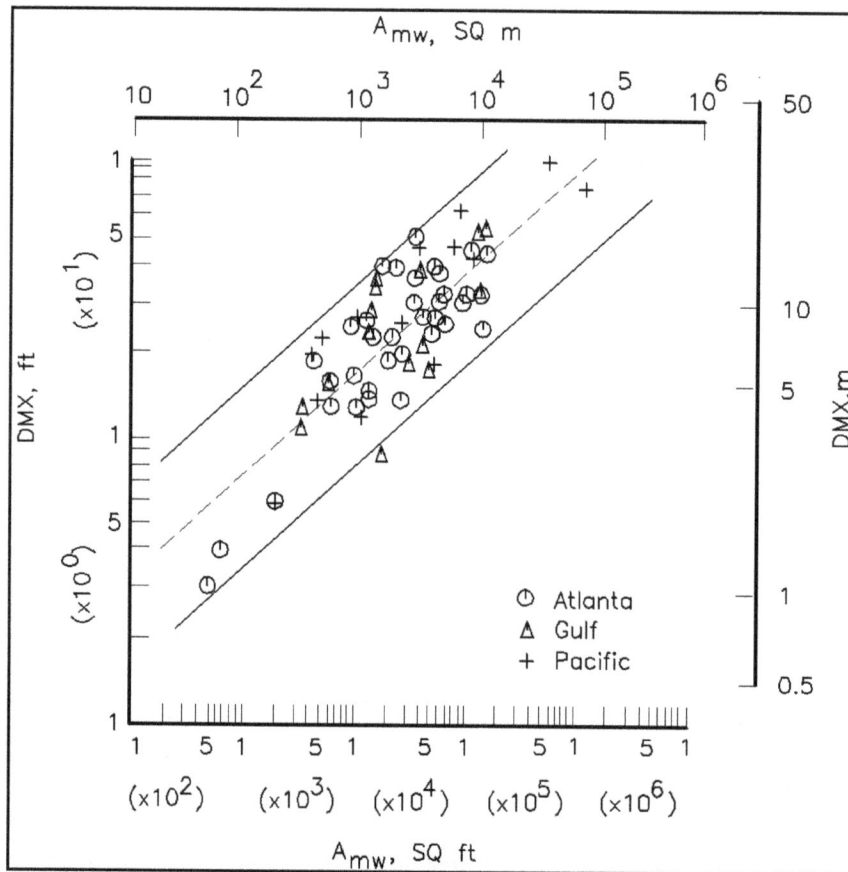

Figure II-6-7. A_{mw} versus maximum channel depth at minimum width section DMX (Vincent and Corson 1980)

(6) Keulegan's assumptions of prismatic channel cross section and vertical bay walls greatly simplify prototype conditions, because natural inlets generally have a complex morphology, making accurate determination of effective hydraulic radius, channel length, cross-sectional area, and bay area difficult. Considerable subjectivity is required to determine these values from bathymetric charts. Some aid is provided later in determining this information. Figure II-6-16 shows the variation of bay tide amplitude (a_b) to that of the ocean (a_o) and the phase lag (ε) for various values of the coefficient of filling K. Figure II-6-17 defines phase lag and provides a sample of output that can be determined from simple analytical models. Approximate values for k_{en}, k_{ex}, and f are given in Example Problem II-6-1. For flow entering an inlet channel, K_{en} is usually taken between 0.005 and 0.25. For natural inlets, which are rounded at the entrance, $K_{en} \simeq 0.05$ or less. For inlets with jetties and flow bending sharply as it enters the inlet channel, $K_{en} = 0.25$ may be appropriate. The exit flow is usually taken near unity, meaning kinetic head is fully lost. If there is significant flow inertia, as in a very channelized bag, K_{ex} may be less than 1.0.

(7) King (1974) solved the same equations but included the effect of inertia (first term of Equation II-6-2). If inertia effect is important, then at times when the tide curves of ocean and bay intersect, there still would be a flow into the bay, e.g., on flood flow there would still be movement of the water mass into the bay, even as the bay elevation dropped below that of the ocean. This would be likely to occur when L is large (channel is long, and therefore a large mass of water moving through the inlet has significant inertia to move against an opposing head difference). Also the possibility exists that the inlet system could have a Helmholtz frequency (or pumping mode, where the basin oscillates uniformly) that is tuned to the forcing

Hydrodynamics of Tidal Inlets

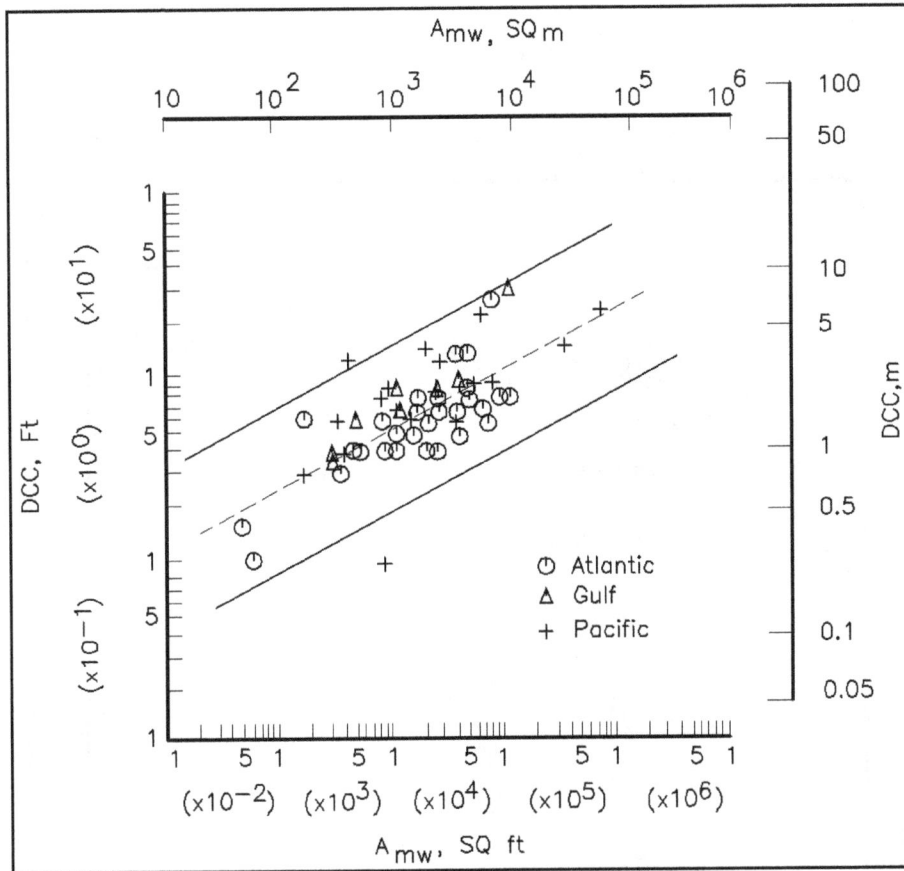

Figure II-6-8. A_{mw} versus minimum controlling channel depth DCC (Vincent and Corson 1980)

ocean tide and amplification of the bay tide could occur. This will be seen in King's solution also (and can be noted in Figure II-6-18 where the bay-ocean tide ratio is greater than one). King defines the dimensionless velocity as:

$$V'_m = \frac{A_{avg} T V_m}{2 \pi a_o A_b}$$

(II-6-5)

where V_m is the maximum cross-sectionally averaged velocity during a tidal cycle. (To determine V_m, use Equation II-6-5 and refer to Figure II-6-19.) Two parameters are calculated with the variables defined previously.

$$K_1 = \frac{a_o A_b F}{2 L A_{avg}}$$

(II-6-6)

$$K_2 = \frac{2\pi}{T} \sqrt{\frac{L A_b}{g A_{avg}}}$$

(II-6-7)

where F is defined as

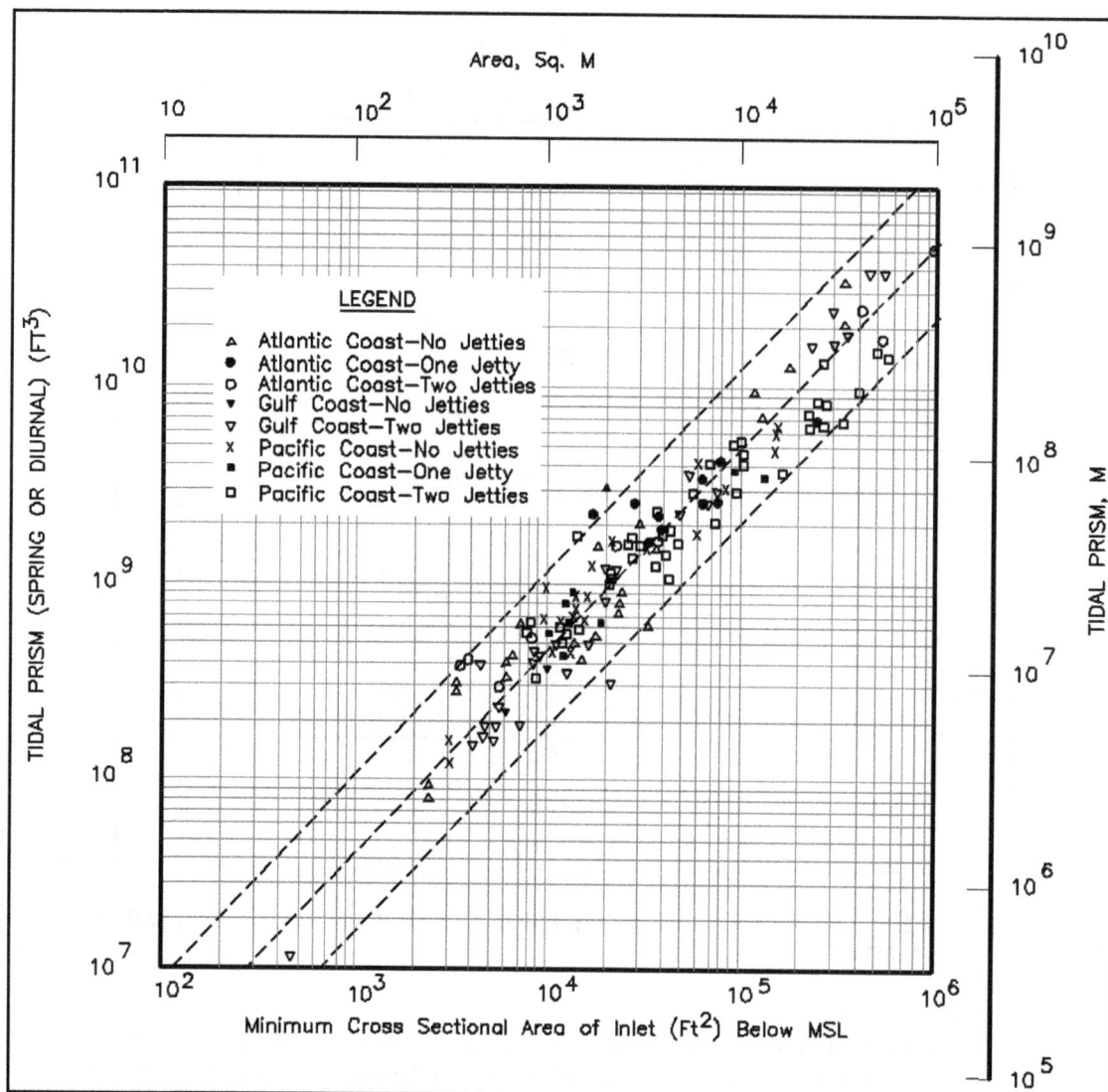

Figure II-6-9. Tidal prism-inlet area relationship

$$F = k_{en} + k_{ex} + \frac{fL}{4R} \tag{II-6-8}$$

(8) Figures II-6-18 through II-6-20 provide a means to determine the bay tide range and phase and velocity in the channel. A sample problem is presented in Part II-6-2, paragraph *d*, where determining important parameters is discussed.

(9) Since channel resistance is nonlinear, channel velocity and bay tide will not be sinusoidal (Keulegan 1967). Other effects such as a relatively large tidal amplitude-to-depth ratio through the channel and nonvertical walls in the channel and bay may be important in causing nonsinusoidal bay tides (Boon and Byrne 1981, Speer and Aubrey 1985, Friedrichs and Aubrey 1988). However, for a first approximation, channel velocity over time is represented as

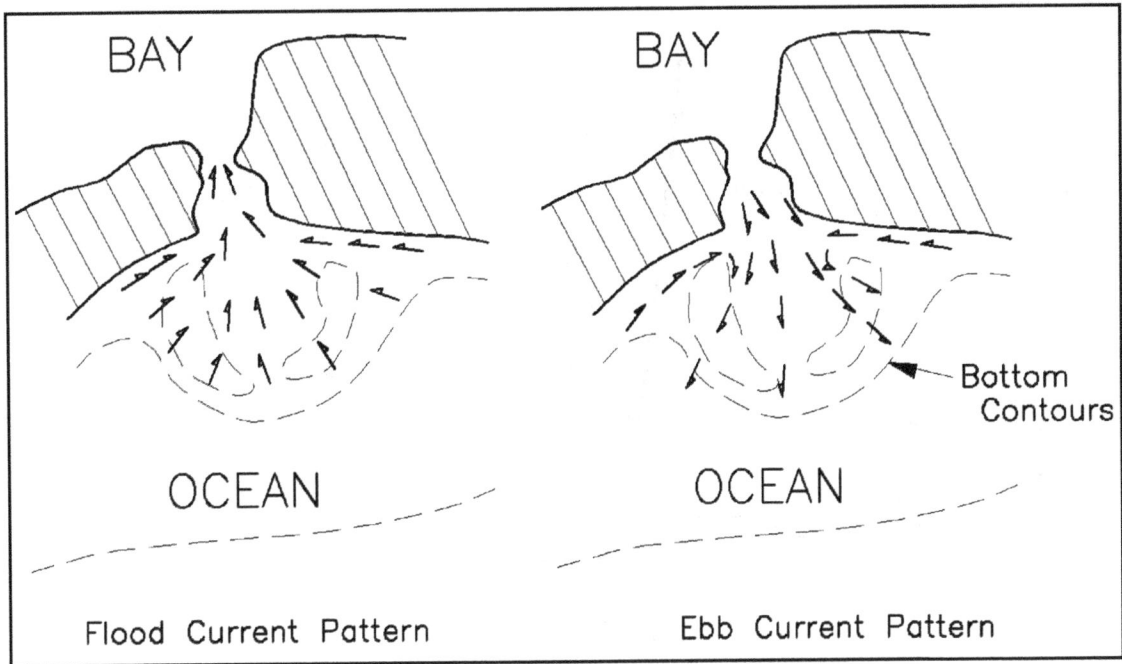

Figure II-6-10.   Schematic diagram of flood and ebb currents outside an inlet (O'Brien 1969)

$$V \approx V_m \sin \frac{2\pi t}{T} \tag{II-6-9}$$

and bay tide can be represented as (see Figure II-6-17)

$$h_b \approx a_b \cos\left(\frac{2\pi t}{T} - \varepsilon\right) \tag{II-6-10}$$

where ε is phase lag, a_b is bay tide amplitude (one-half the bay tide range), T is tide period, and t is time of interest during the tide cycle. Also King's K_1 and K_2 are related to the Keulegan repletion coefficient K by

$$K = \frac{1}{K_2}\sqrt{\frac{1}{K_1}} \tag{II-6-11}$$

   c.   *Tidal prism.*

   (1) The volume of water that enters through the inlet channel during flood flow and then exits during ebb flow is known as the tidal prism. If the hydraulic analysis as described above is used, then tidal prism P can be calculated as

$$P = 2\, a_b\, A_b \tag{II-6-12}$$

   (2) Another technique for prism estimation can use observed velocity (or discharge Q) data. Assuming a sinusoidal discharge in the channel and integrating over the flood or ebb portion of the tidal cycle

$$P = \frac{T Q_{max}}{\pi} \quad \equiv \quad P = \frac{T V_m A_{avg}}{\pi} \tag{II-6-13}$$

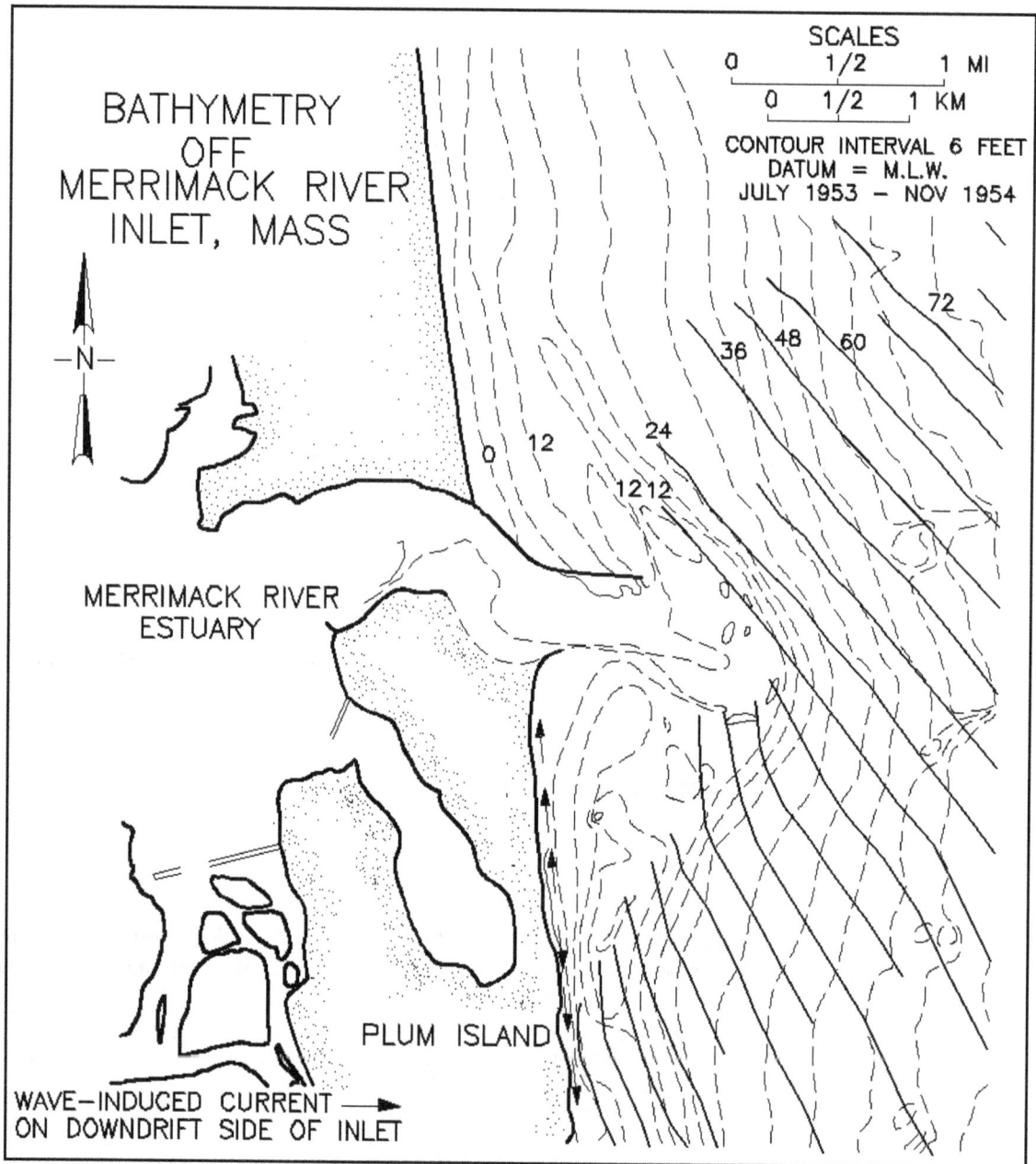

Figure II-6-11. Wave refraction pattern in the vicinity of the Merrimack River Estuary entrance, just south of the Merrimack Inlet (from Hayes (1971))

(3) To account for a non-sinusoidal character of the prototype flow, Keulegan determined

$$P = \frac{T Q_{max}}{\pi C} \equiv P = \frac{T V_m A_{avg}}{\pi C} \tag{II-6-14}$$

and found C varying between 0.81 and 1.0 for K, the filling coefficient, varying between 0.1 and 100. For practical application, if $0.1 < K < 1.8$, use $C = 0.86$, and for $K > 1.8$, use $C = 1.0$.

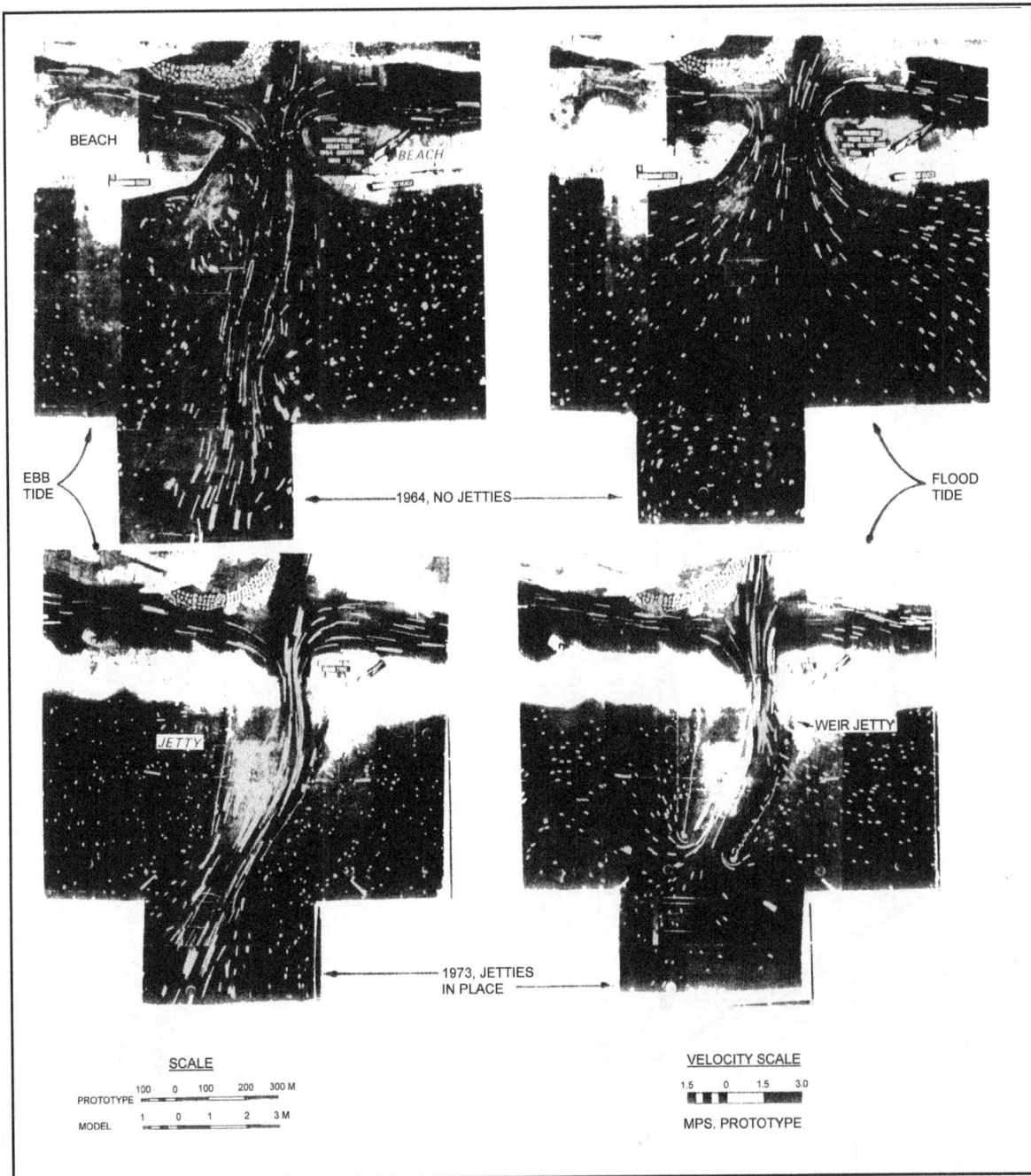

Figure II-6-12.   Ebb and flood flow patterns from a model study of Masonboro Inlet, North Carolina (Seabergh 1975)

(4) A more precise method of prism determination includes the cubature method (Jarrett 1976), which takes into account the time required for a tide to propagate through a bay and segments the bay into subareas rather than assuming a uniform rise and fall of the bay tide. Jarrett (1976) also recommends ways to relate point measurements of maximum velocity in the center of the channel (as can be obtained from NOS current tables) to velocities representative of the entire inlet, using

$$\frac{V_{avg}}{V_{meas}} = \left(\frac{R}{D} \right)^{\frac{2}{3}}$$

(II-6-15)

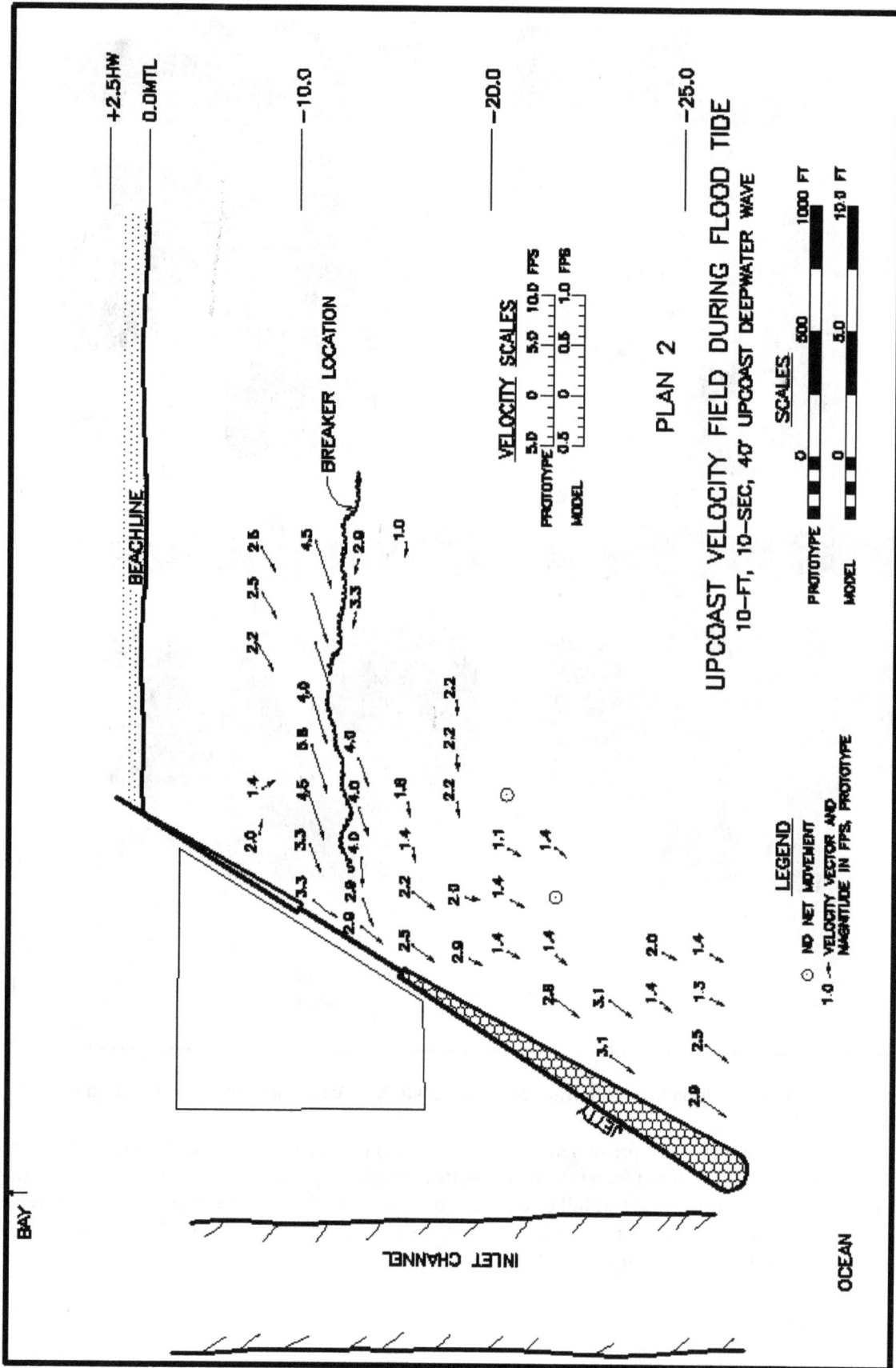

Figure II-6-13. Tidal current plus wave-generated currents approaching jettied inlet, measured in physical model study

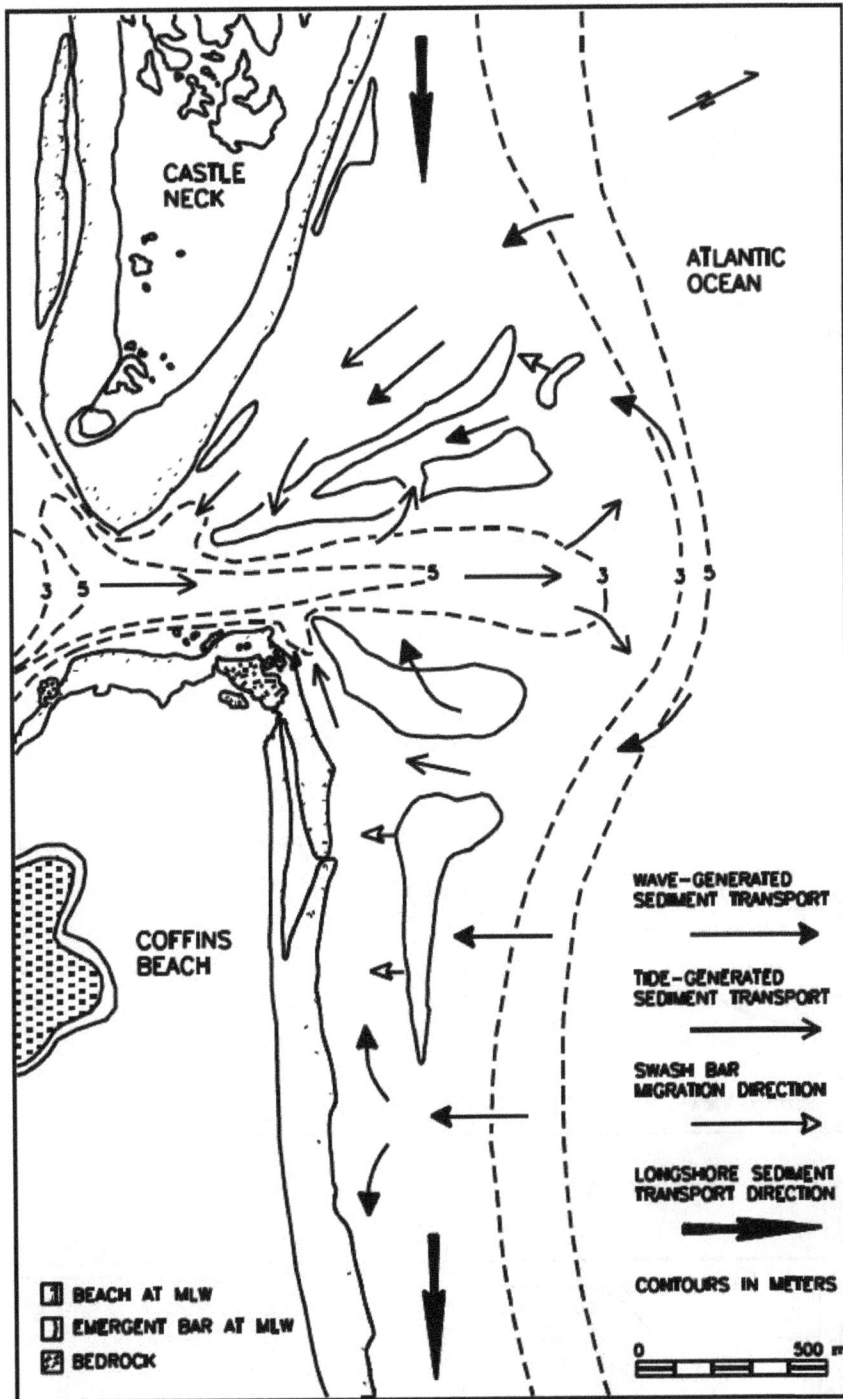

Figure II-6-14. Sediment transport gyres associated with both updrift and downdrift portions of the delta at Essex River Inlet, Massachusetts. (Note different arrow types to denote wave or tide-generated sediment transportation (Smith 1991)

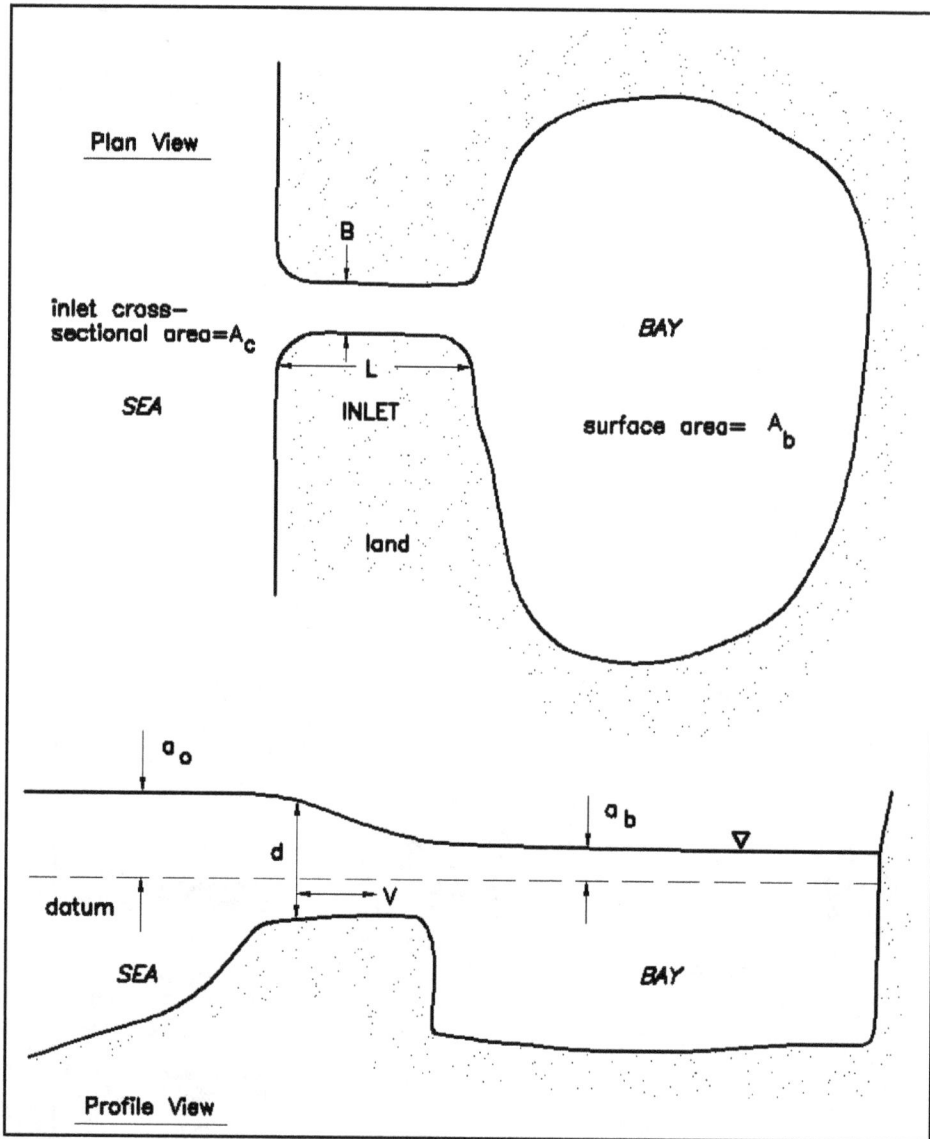

Figure II-6-15.   Inlet bay system

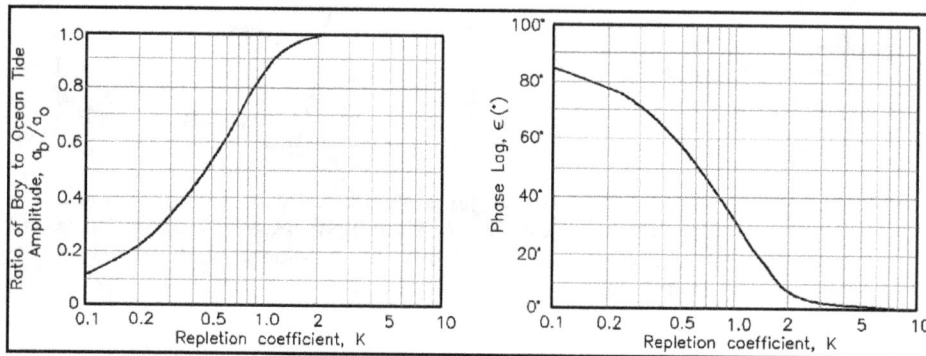

Figure II-6-16.   Variation of dimensionless parameters with Keulegan's repletion
coefficient, K

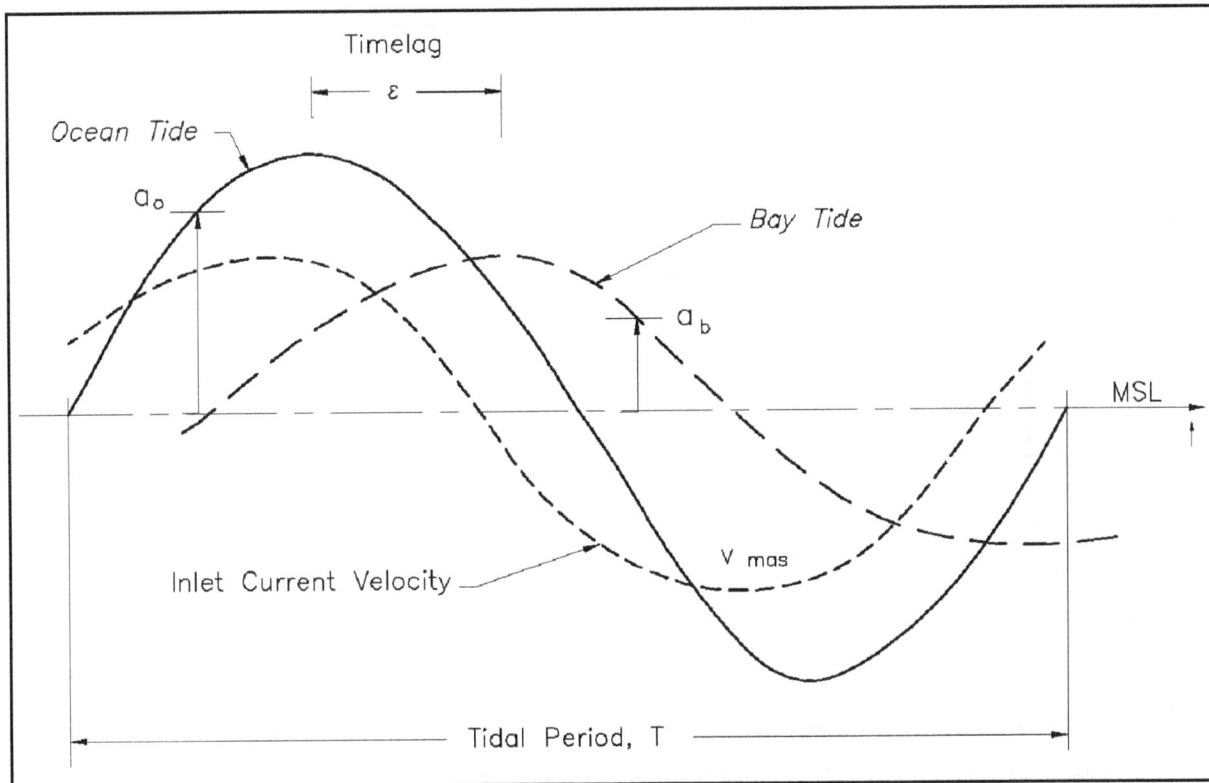

Figure II-6-17. Sample tides and currents for an ocean-inlet-bay system that satisfy the Keulegan assumptions (Keulegan 1967)

where

V_{avg} = maximum velocity averaged over entire cross section

V_{meas} = point measurement of maximum velocity

R = hydraulic radius of entire cross section

D = depth of water at current meter location

The V_{avg} value then can be used in Equation II-6-13.

d. Determining important inlet parameters. Graphs can easily be used to determine inlet hydraulics; however, evaluation of important parameters requires some effort. Mason (1975) determined appropriate techniques for defining cross-sectional area, bay area, inlet (channel) length, hydraulic radius, friction factor and exit and entrance loss coefficients. The following offers guidance in the evaluation of parameters.

(1) Cross-sectional area. Problems in defining location and magnitude of the cross-sectional area result in greatest variability in computed values of the repletion coefficient. To reduce such variability, the cross section should be determined in the following manner:

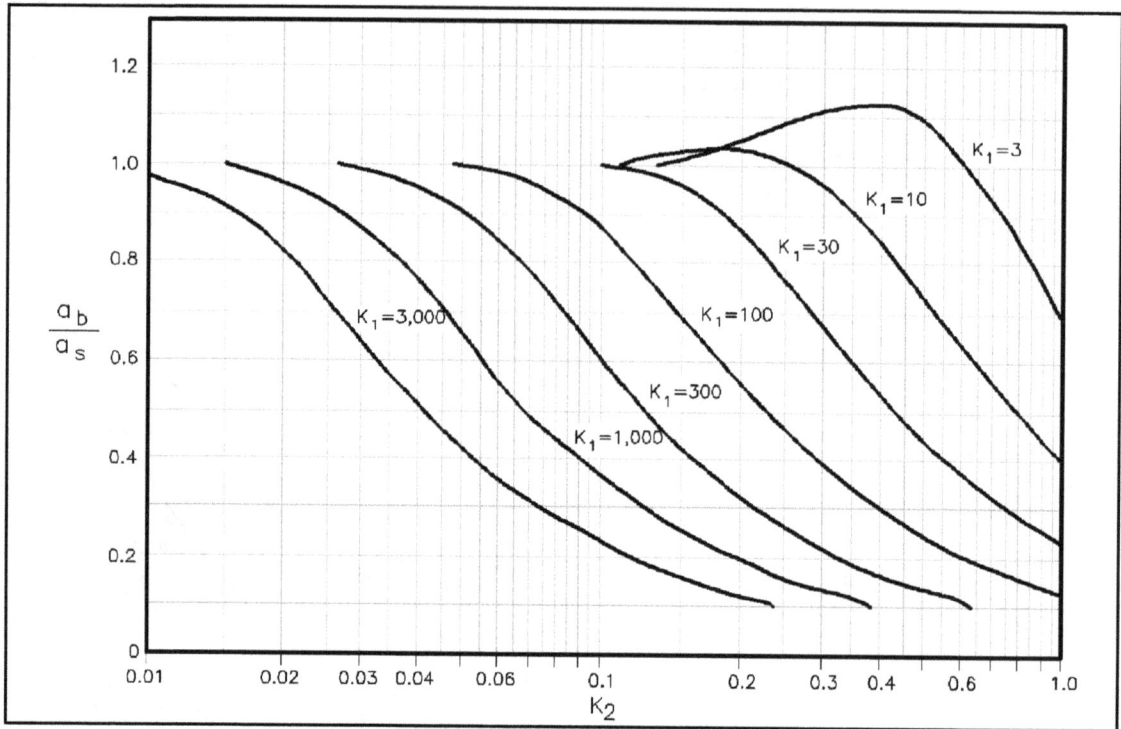

Figure II-6-18. Ratio of bay to sea tidal amplitude versus K_1 and K_2

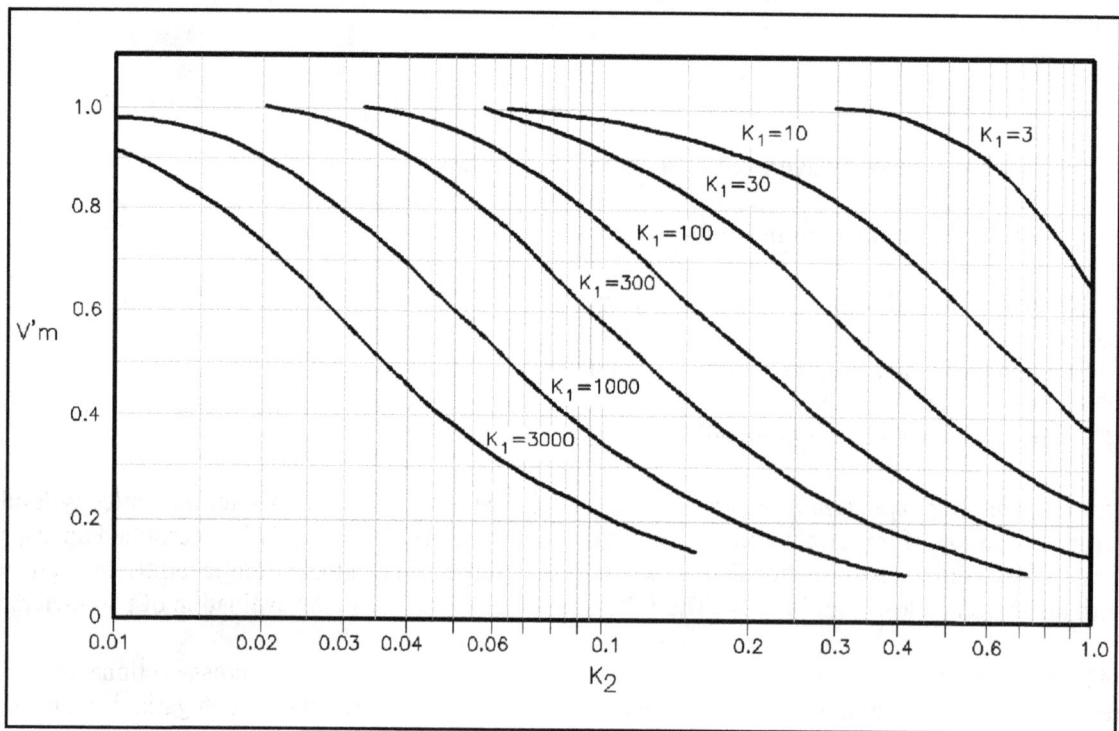

Figure II-6-19. Dimensionless maximum velocity versus K_1 and K_2

Hydrodynamics of Tidal Inlets

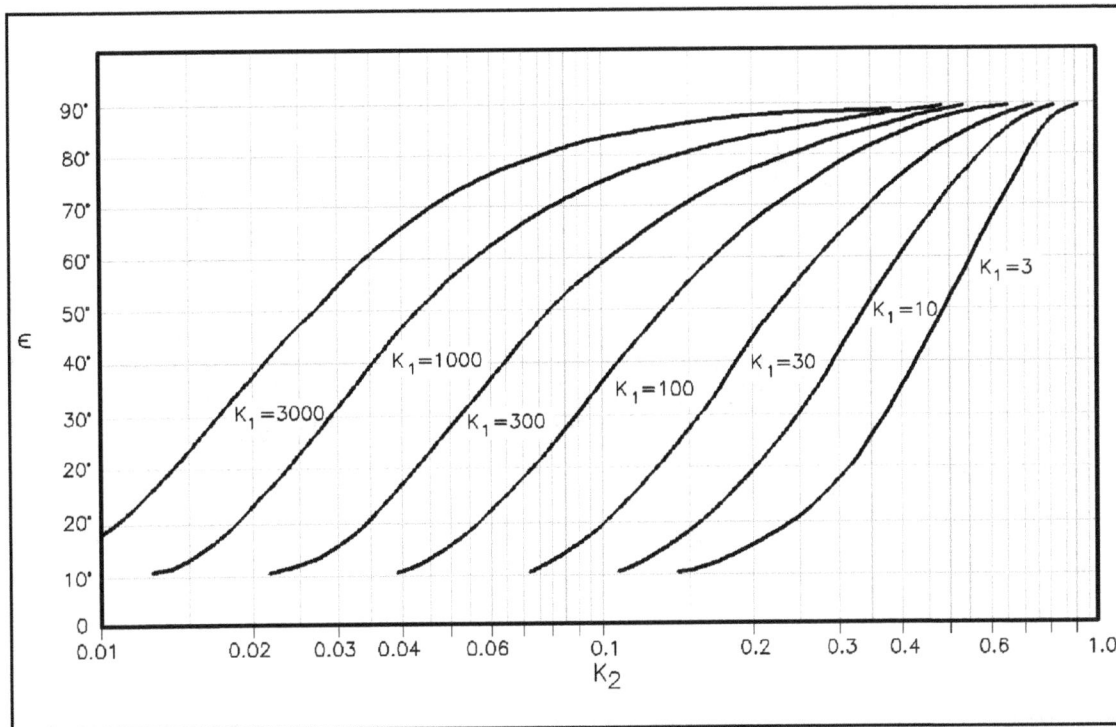

Figure II-6-20. Bay tidal phase lag versus K_1 and K_2

From a detailed bathymetric chart of the inlet, sketch lines that best approximate flood flow lines through the inlet as shown in Figure II-6-21. Establish 10 equally spaced cross-section ranges perpendicular to these flow lines, extending from MSL (mean sea level) on one side of the channel to MSL on the other (i.e. from one side of the inlet to the other). Measure cross-sectional area below MSL of each range and determine average area. Depths on most navigation charts are referenced to mean low or mean lower low water, and a correction factor should be added to these depths to determine the area below MSL.

(2) Bay area. Typically the problem is to define the extent of the inlet's influence throughout the bay. To do this, the following are recommended:

(a) Examine any existing tide records from stations located throughout the bay to define limit of the inlet's influence.

(b) Examine bathymetry of the bay to determine presence of natural sills, narrows, or shoal areas which can preclude significant tidal exchange.

(c) Locate any areas of heavy shoaling at points distant from the inlet, particularly in waterways connecting the bay of interest with more remote bodies of water, and define these sites as the effective bay limits.

(d) The surface area of the bay is usually varying to some degree, but for an initial iteration, the mean water line could be chosen (as well as it can be delineated).

(3) Friction factor.

(a) "f" is a dimensionless parameter that is a function of hydraulic radius R and Manning's roughness coefficient n:

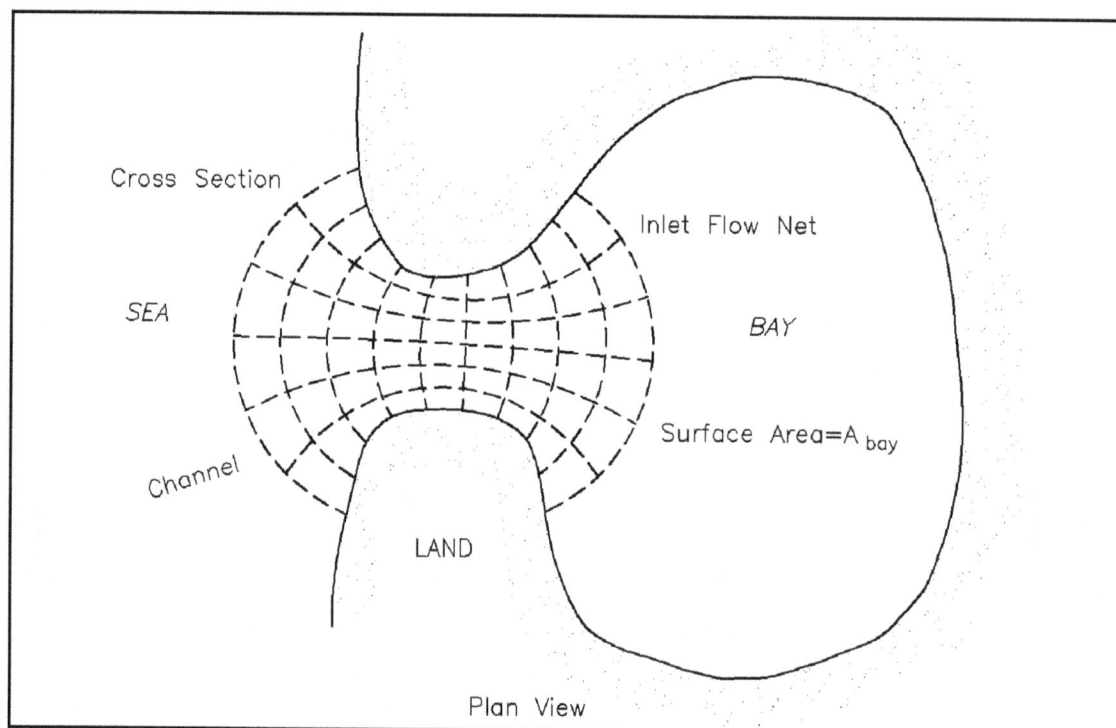

Figure II-6-21.   Inlet flow net

$$f = \frac{116n^2}{R^{1/3}}$$
(II-6-16)

(b) It seems reasonable to assume a constant n for an inlet in sand rather than a uniform f value. For open channels in hard-packed smooth sand, Chow (1959) recommends an n value of 0.020 (in English units, or 0.016 in SI). Since inlets often have rippled or duned bottoms, which generally increase frictional resistance, a higher n value might be expected, for instance between 0.025 and 0.030 (English units or 0.021 and 0.025, SI). Using an estimated value of n = 0.0275 (English units or 0.0225 SI), a friction factor defined by $f = 0.088/R^{1/3}$ (English units, R in feet, or $0.059/R^{1/3}$, SI, R in meters) is recommended for use in the equation for K.

Jarrett (1975) examined the friction aspect of Keulegan K (Equation II-6-4) and defined F, inlet impedance, as

$$F = k_{en} + k_{ex} + \frac{fL}{4R}$$
(II-6-17)

with k_{en}, k_{ex}, f, and L as defined earlier, and hydraulic radius R measured at the minimum cross section of the inlet. Figure II-6-22 plots inlet impedance against the ratio of inlet length to the 4/3 power of hydraulic radius for four ratios of minimum cross-section inlet width to hydraulic radius. These curves can be used for determining F.

(4) Inlet length. Inlet length is one of the more difficult parameters to define. The following standardized method is recommended to determine inlet length. On a detailed bathymetric chart, beginning in deep water just offshore from the inlet, mark the points of maximum depth moving across the bar, through the inlet, and into the bay. The bayward end point should be seaward of any <u>major</u> division of the inlet

Inlet Impedance, F vs L/R**(4/3)
For Ranges of Width to Hyd Radius Ratio

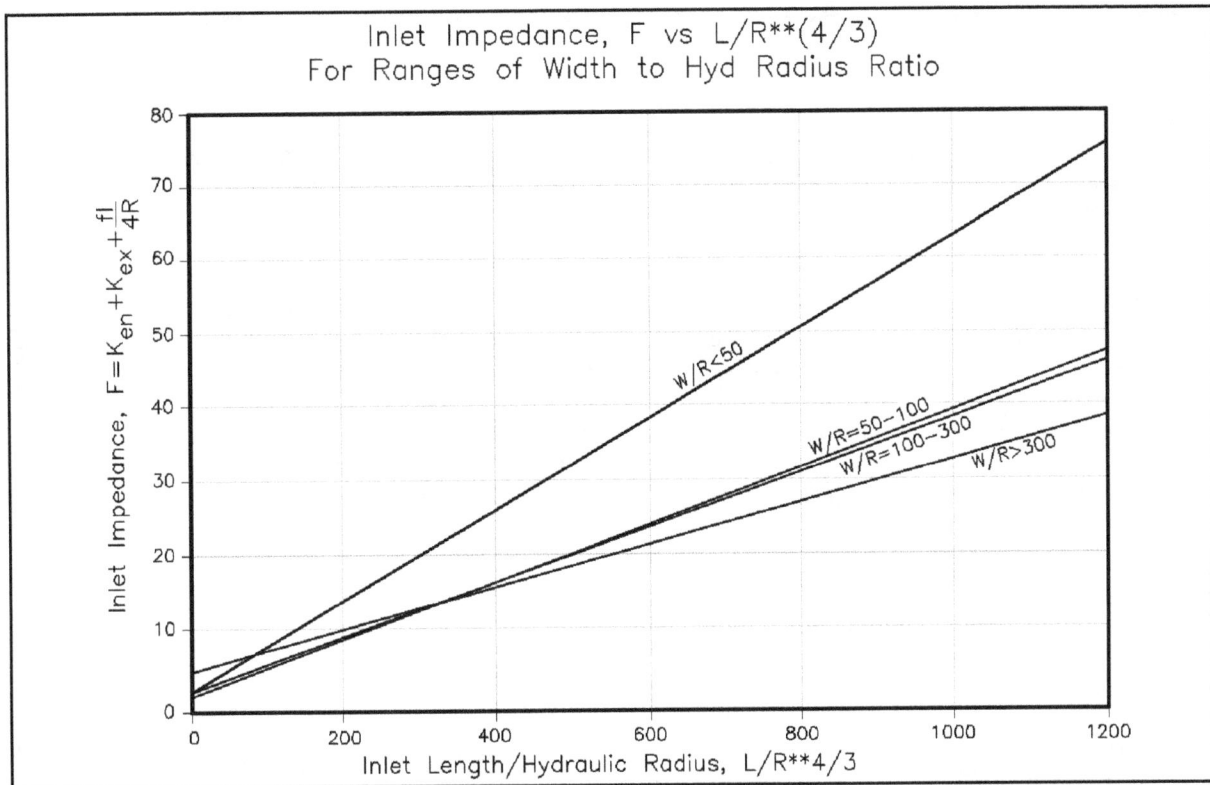

Figure II-6-22. Inlet impedance (*F*) versus the ratio of inlet length to inlet hydraulic radius to the 4/3 power (*l*/*R*$^{4/3}$) for various inlet width to inlet hydraulic radius ratios, *W/R*

channel. Connect these points of maximum depth with a series of straight lines to define the thalweg of the channel. Construct a plot of depth versus distance from the oceanward point along the thalweg. Define inlet length as the distance between the depth <u>minimum</u> oceanward of the deepest part of the inlet and the depth minimum bayward of the deepest part. Where the channel does not shoal near either end, the limit of the inlet must be estimated by considering the relationship of the inlet to adjacent beaches, i.e., draw a line perpendicular to the thalweg from MSL on one shoulder of the inlet to MSL on the other, and assume its intersection with the thalweg defines the channel end.

(5) Hydraulic radius. For a given cross-section area, the corresponding value of R should be computed from

$$R = \frac{A_{avg}}{Average\ Wetted\ Perimeter} \approx \frac{A_{avg}}{Average\ Width} \qquad (\text{II-6-18})$$

These methods are intended to provide a simplified approach to understanding inlet hydraulics in a rapid manner. If possible, field data should be examined to correlate with existing condition calculations, adjustments made to various controlling parameters, and calculations reiterated until reasonable results are achieved. Then calculations may proceed to evaluate design changes. It must be remembered that for extreme events, when wind and wave breaking may be significant factors influencing the hydrodynamics of the inlet system, more sophisticated approaches will be needed. Even a relatively well-behaved inlet is only approximately represented by these approaches, but much insight can be gained by these techniques.

EXAMPLE PROBLEM II-6-1

Find:

If a channel has a depth below MSL of 3.7 m (12.1 ft) and a width of 180 m (590 ft), what are the maximum flow velocity, maximum discharge bay tidal range, and the volume of water flowing into and out of the bay on a tidal cycle (tidal prism) for a tide having the spring range?

Given:

A bay with a surface area of 1.9×10^7 m^2 (2.0×10^8 ft^2) and an average depth of 6.0 m (19.7 ft) is located on the Atlantic coast. The tide is semidiurnal ($T = 12.42$ hr), with a spring range of 1.30 m (4.3 ft), as given by the National Ocean Survey Tide Tables (U.S. Department of Commerce 1989). An inlet channel, that will be the only entrance to the bay, is to be constructed across the barrier beach that separates the bay from the ocean. The inlet is to provide a navigation passage for small vessels, dilution water to control both salinity and pollution levels, and a channel for fish migration. The channel is to have a design length of 1,100.0 m (3,609 ft) with a pair of vertical sheet-pile jetties that will extend the full length of the channel.

Solution:

Using King's method, assume $k_{en} = 0.1$, $k_{ex} = 1.0$, and $f = 0.03$: $B = 180$ m and $d = 3.7$ m.

$$A_c = Bd = 180 \, (3.7) = 666 \text{ m}^2 \, (7166 \text{ ft}^2)$$

$$R = \frac{A_c}{(B + 2d)} = \frac{666.}{(180 + 2(3.7))} = 3.55 \text{ m } (11.54 \text{ ft})$$

$$F = k_{en} + k_{ex} + \frac{fL}{4R} = 1.0 + 0.1 + \frac{0.03 \, (1100)}{4 \, (3.55)} = 3.42$$

$$K_1 = \frac{a_o \, A_b \, F}{2 \, L \, A_c} = \frac{(1.30/2 \, (1.9) \, (10^7) \, 3.42}{2 \, (1100) \, (666)} = 28.8$$

$$K_2 = \frac{2\pi}{T} \sqrt{\frac{L \, A_b}{g \, A_c}} = \frac{2\pi}{12.4 \, (60 \, (60)} \sqrt{\frac{1100 \, (1.9) \, 10^7}{9.8 \, (666)}} = 0.25$$

From Figures II-6-18, II-6-19, and II-6-20 with the above values of K_1 and K_2

$$\frac{a_b}{a_s} = 0.78$$

$$V_m' = 0.66$$

$$\varepsilon = 53^o$$

(Continued)

Example Problem II-6-1 (Concluded)

Therefore, from Equation II-6-5

$$V_m = \frac{V_m' \, 2\pi \, a_s \, A_b}{A_c \, T}$$

$$V_m = \frac{0.66 \ (2) \ (3.14) \ (0.65) \ (1.9) \ 10^7}{666 \ (12.42) \ (3,600)} = 1.72 \ \text{m/s} \quad (5.64 \ \text{ft/s})$$

$$Q_m = V_m \, A_c = (1.72) \ (666) = 1145 \ \text{m}^3\text{/s} \ (40,430 \ \text{ft}^3\text{/s})$$

Since $a_b/a_s = 0.78$, $a_b = 0.78 \ (0.65) = 0.51$ m (1.67 ft) and the bay tidal range is 0.51 (2) or 1.02 m (3.35 ft).

The tidal prism is

$$2 \ a_b \ A_b = 2 \ (0.51) \ (1.90) \ (10^7) = 1.94 \times (10^7) \ \text{m}^3 \quad (6.27 \times 10^8 \ \text{ft}^3)$$

If the average depth of the bay is 6.0 m and the distance to the farthest point in the bay is 6.0 km, the time t_* it will take for the tide wave to propagate to that point is

$$t_* = \frac{L_b}{\sqrt{gd_b}} = \frac{6000}{\sqrt{9.8 \ (6.0)}} = 782 \ \text{sec or } 0.22 \ \text{hr}$$

Since this time is significantly less than 12.42 hr, the assumption that the bay surface remains horizontal is quite satisfactory.

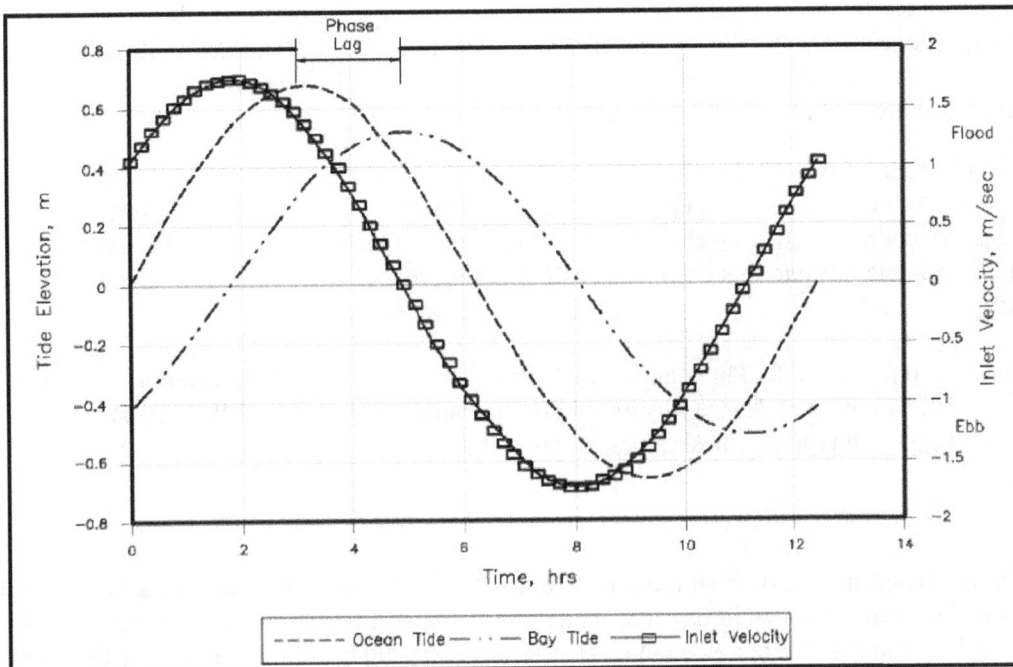

Figure II-6-23.   Results from Example Problem II-6-1

e. Evaluating inlet hydraulics with Keulegan K.

(1) Introduction. The Keulegan K value can be used to understand effects of gross geometric variations on inlet hydraulics. Figure II-6-24 shows ocean and bay tide curves and velocity in the channel during a tidal cycle. Three K values are represented. The lower the K or filling coefficient, the lower the tide range in the basin and the greater the phase lag (or time difference between high water (low water) in the ocean and high water (low water) in the bay. Also noted from the velocity curve, maximum current occurs during maximum head difference between ocean and bay tides and as K decreases, maximum ebb and flood currents shift from occurrence at a mid-tide level to tidal elevation extremes. For $K = 0.2$, maximum flood currents occur near high water and maximum ebb currents occur during low water. Therefore for low K inlets, flood flow occurs when depths in the channel and over tidal deltas are greatest (near high water), and flow is more broadly distributed. During ebb flow, depths are shallower (approaching low water) and flow is likely to be more channelized. As K increases, and the bay fills more completely, peak ebb and flood flows tend to occur near the same tide level. These variations can have implications with regard to sediment transport. Mota Oliveira (1970) found that, using an analysis similar to Keulegan but with variable entrance cross-sectional area, when $0.6 < K < 0.8$, the inlet had maximum sediment flushing ability. When $K > 0.8$, there was flood dominance in sediment transport capability, meaning net bayward transport and when $K < 0.6$, there was ebb dominance of sediment transport capability. Also when a variable bay surface area was considered, ebb sediment transport was enhanced to the detriment of flood transport. Part II-6-2, paragraph h contains more information on flow dominance.

(2) Jarrett's classification. Jarrett (1975) performed a "hydraulic classification" of inlets based on Keulegan's K after determining tidal characteristics of a large number of U.S. inlets (see Table II-6-1). The classification is as follows:

(a) Class I: Keulegan $K < 0.3$. Phase angle (ε) equal to or greater than 70 deg. For a semidiurnal tide, the phase lag between the ocean tidal extreme and slack water is equal to or greater than 2 hr 25 min. For the most part, bays associated with inlets in this class are open and relatively shallow with only one relatively small inlet (i.e., small ratio A_c/A_b connecting the ocean with the bay. This class also could include relatively long estuaries that are actually long-wave embayments.

(b) Class II: Keulegan $K > 0.80$. Phase angle (ε) equal to or less than 40 deg. For a semidiurnal tide, the phase lag would be equal to or less than 1 hr 25 min. Bays associated with this class are either short or consist of a system of relatively deep channels (Note: by virtue of the long tidal period in the Gulf of Mexico most Gulf Coast inlets fall into this class irrespective of bay shape). The ratio A_c/A_b is relatively large for these inlets.

(c) Class III: $0.3 < K < 0.8$. Phase angle lies between 40 and 70 deg. Characteristics are intermediate to classes I and II. Analysis by Mota Oliveira (1970) indicated that the natural flushing ability of a vertical bank lagoon reaches a maximum for K values of 0.6 to 0.8.

f. Effect of freshwater inflow.

(1) If a significant amount of fresh water is introduced into the bay relative to the tidal prism of the inlet, there will be significant changes in bay tide range and velocity through the inlet. King (1974) included freshwater inflow into a Keulegan type model (which did not include the inertial term in the 1-D equation of motion). He defined a dimensionless variable

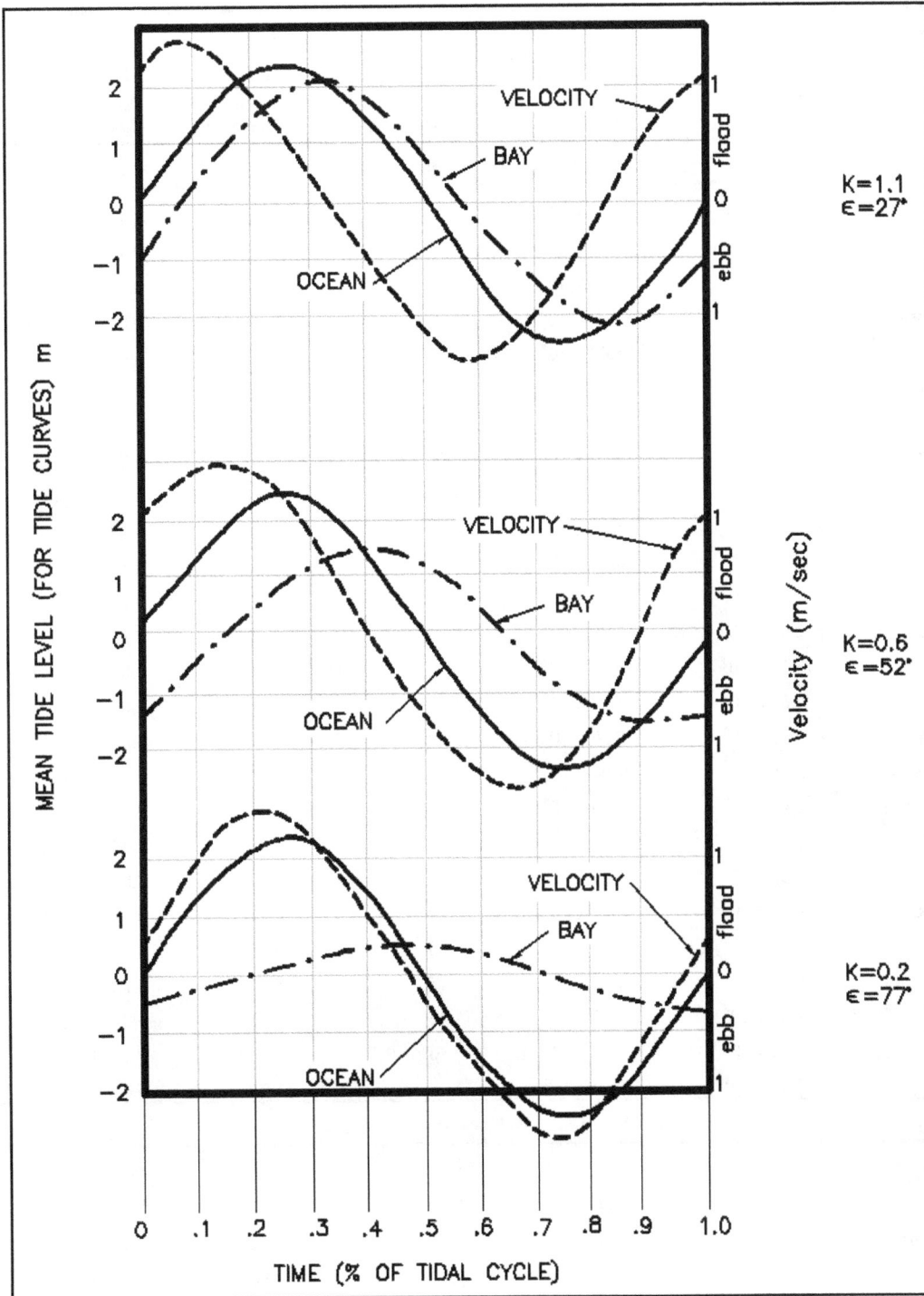

Figure II-6-24.   Hydraulic response of inlet and bay tide phasing and bay tide amplitude for various Keulegan *K* values

Table II-6-1
Hydraulic Characteristics of Tidal Inlets by Cubature Method

Inlet	Mean Tidal Prism P, ft^3	Bay Surface Area A_b, ft^2	Avg Phase Range in Bay, $2a_b$	Ocean Tidal Range $2a_o$	a_b/a_o	K	ε deg
Atlantic Coast							
Shinnecock	1.82×10^8	3.65×10^8	0.50	2.90	0.17	0.15	80
Fire Island	1.59×10^9	2.69×10^9	0.59	4.10	0.14	0.14	83
Jones	7.89×10^8	4.48×10^8	1.76	4.50	0.39	0.35	67
East Rockaway	4.03×10^8	1.06×10^8	3.80	4.50	0.84	0.95	33
Rockaway	2.24×10^9	4.64×10^8	4.83	4.70	1.03	--	--
Manasquan	1.40×10^8	4.91×10^7	2.85	4.30	0.66	0.65	49
Barnegat	4.91×10^8	1.34×10^9	0.37	4.20	0.09	0.09	85
Indian River	4.00×10^8	4.20×10^8	1.02	4.10	0.25	0.22	75
Beaufort	4.20×10^9	2.56×10^9	1.64	3.60	0.46	0.41	63
New River	1.59×10^8	5.31×10^8	0.30	3.60	0.08	0.08	86
Winyah Bay	2.47×10^9	9.14×10^8	2.70	4.60	0.59	0.56	54
Port Royal Sd	1.25×10^{10}	2.40×10^9	5.22	6.60	0.78	0.82	40
Calibogue Sd	3.05×10^9	5.28×10^8	5.78	6.60	0.88	1.07	28
Wassaw Sd	3.34×10^9	5.84×10^8	5.72	6.90	0.83	0.95	33
Ossabaw Sd	5.82×10^9	1.17×10^9	4.98	7.20	0.69	0.70	46
Sapelo Sd	6.36×10^9	9.66×10^8	6.59	6.90	0.94	1.30	20
St. Catherines Sd	5.94×10^9	1.04×10^9	5.73	7.10	0.81	0.89	36
Doboy Sd	3.43×10^9	5.19×10^9	6.62	6.80	0.97	1.50	15
Altamaha Sd	2.45×10^9	5.01×10^8	4.90	6.60	0.74	0.75	43
St. Simon Sd	5.52×10^9	8.51×10^8	6.50	6.60	0.98	1.90	0
St. Andrew Sd	8.34×10^9	1.41×10^9	5.92	6.60	0.90	1.12	26
St. Marys	4.11×10^9	7.93×10^8	5.19	5.80	0.89	1.10	27
Nassau Sd	1.87×10^9	4.40×10^8	4.25	5.70	0.75	0.76	43
St. Johns	1.50×10^9	1.22×10^9	1.23	5.20	0.24	0.21	76
Ft. Pierce	5.10×10^8	1.10×10^9	0.46	2.60	0.18	0.16	80
Lake Worth	7.00×10^8	4.00×10^8	1.75	2.60	0.67	0.66	48
Gulf Coast							
Venice Inlet	8.50×10^7	4.43×10^7	1.92	2.60	0.74	0.76	42
Midnight Pass	2.61×10^8	1.29×10^8	2.12	2.60	0.82	0.90	35
Sarasota Bay	2.46×10^9	1.16×10^9	2.12	2.60	0.82	0.90	35
Tampa Bay	1.95×10^{10}	1.01×10^{10}	1.95	2.60	0.75	0.76	43
Pensacola Bay	5.87×10^9	4.65×10^9	1.26	1.30	0.97	1.55	14
Mobile Bay	1.56×10^{10}	1.20×10^{10}	1.30	1.30	1.00	--	0
Galveston	5.94×10^9	8.36×10^9	0.71	2.10	0.33	0.30	70

$$Q_r^{\prime} = \frac{Q_r T}{2 \pi a_o A_b}$$

(II-6-19)

with Q_r being the river or freshwater inflow to the bay, and a_0, A_b, and T as defined previously. Figures II-6-25 and II-6-26 give high (or maximum) and low (minimum) dimensionless ratio water levels for values of $Q_r^{\prime}$ and K (Equation II-6-4). Using Figures II-6-27 and II-6-28, maximum average ebb and flood currents can be found and Figure II-6-29 is used to determine ocean high water to bay high water phase lag.

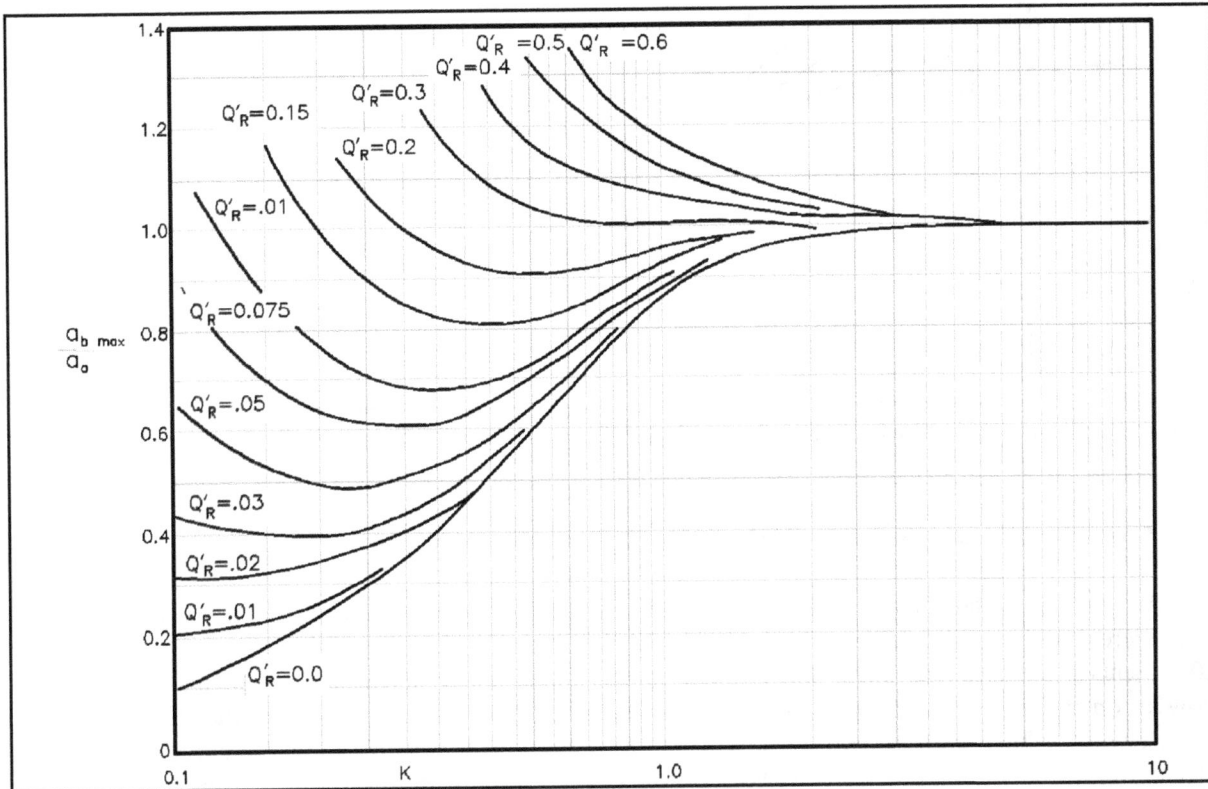

Figure II-6-25. $a_{b\,max}/a_o$ versus K for values of $Q_r^{\prime}$ (river discharge model)

(2) The above analysis assumes that waters in the inlet are well-mixed at the inlet minimum cross-section, where the salinity variation over a vertical section varies only a small fraction from the local mean salinity. As freshwater influence increases, it approaches a completely stratified flow, i.e., a flow that has a well-defined interface between fresh and saline waters, and in which flow may be simultaneously bidirectional (Wicker 1965). More sophisticated techniques should be used for such cases.

g. *Bay superelevation.*

(1) A superelevation of the bay means that the average level of the bay is greater than the average elevation of the ocean over a given time period. An obvious cause of this effect is the introduction of fresh water into the bay from local rivers and streams. The effect of long-term (months or years) rise in sea level causing a long-term rise in bay level is not included in this definition of bay superelevation, since both sea and bay are responding simultaneously. The bay may not rise at the same rate as the sea due to possible circulation effects if multiple inlets connect the bay to the ocean.

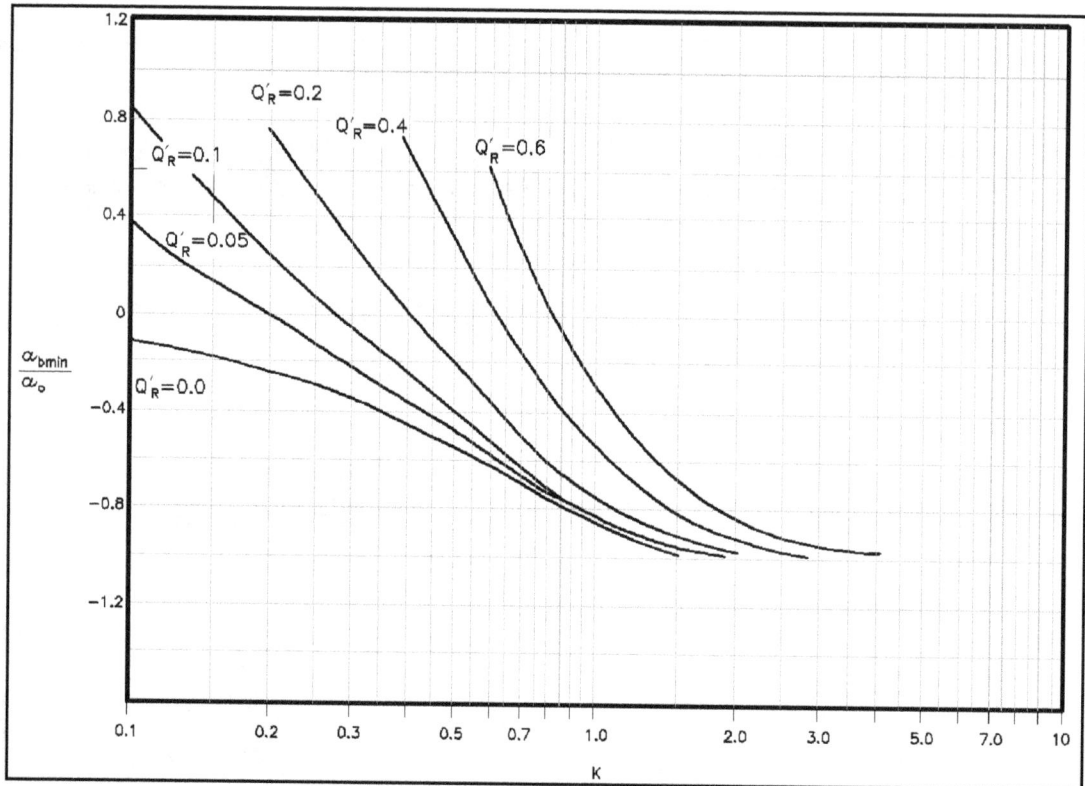

Figure II-6-26. $a_{b\,min}/a_o$ versus K for values of Q_r' (river discharge model)

(2) Laboratory and numerical studies (Mayor-Mora 1973; Mota Oliveira 1970) have indicated that there is an increasing bay superelevation as the coefficient K decreases (Figure II-6-30) and approaches nearly 20 percent of the ocean tide range. This basic cause of setup is due to increased frictional dissipation of ebb flow in comparison to flood flow as K decreases, with peak ebb flows in the channel occurring during lower water levels for inlets with low K values. This increased tractive stress for ebb flow relative to flood flow creates the setup, or increased head, necessary to preserve continuity and drive out the same tidal prism that entered the inlet.

(3) Model studies (Mayor-Mora 1973) show that as K decreases (and superelevation increases) the duration of ebb flow increases relative to flood flow in the inlet channel. Keulegan (1967) determined a relation between inflow duration and bay superelevation (due to any reason) given by

$$\frac{\Delta}{a_o} = \frac{\sin\left(\dfrac{2\pi t_i}{T}\right)}{1 - \cos\left(\dfrac{2\pi t_i}{T}\right)} \qquad\qquad \text{(II-6-20)}$$

where

Δ = superelevation

a_o = tidal amplitude in ocean

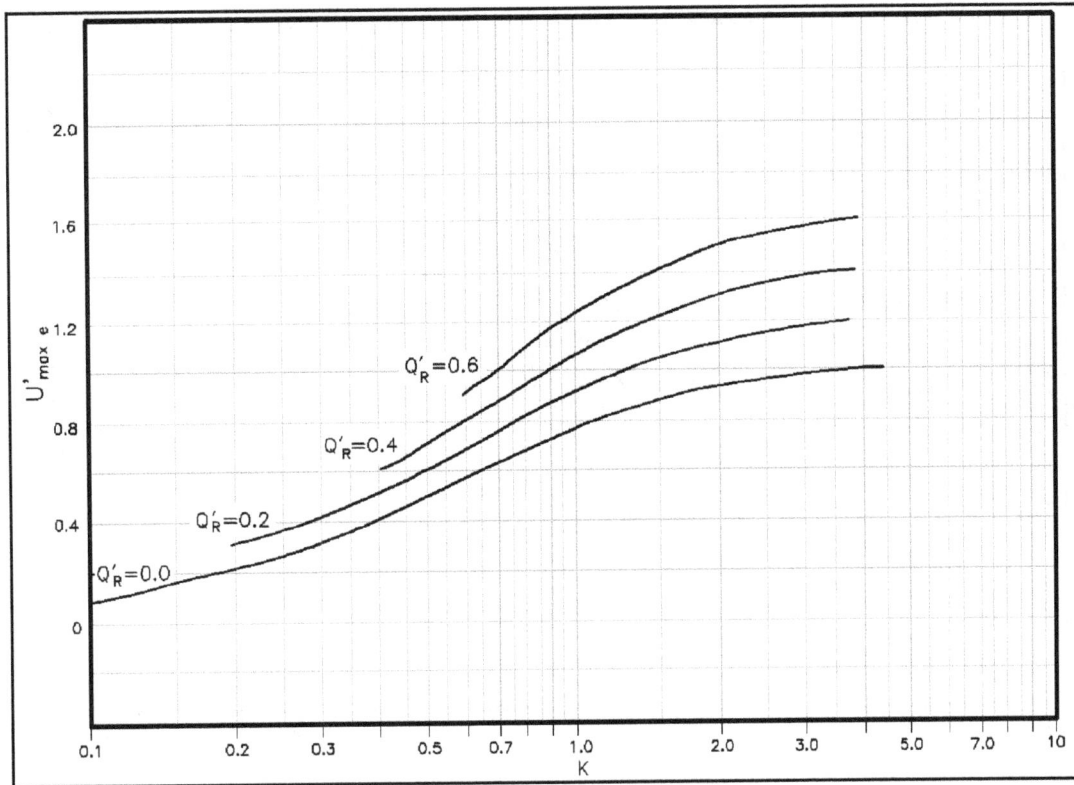

Figure II-6-27. $u'_{max\,e}$ versus *K* for values of Q_r' (river discharge model)

T = tidal period

t_i = duration of inflow

(4) Figure II-6-31 illustrates this. Mehta (1990) determined that superelevation due to varying entrance channel cross section (by itself only) is

$$\frac{\Delta}{a_o} = 1 - \frac{\left(\dfrac{a_b}{a_o}\right)^2}{4\left(\dfrac{d}{a_o}\right)} - \frac{a_o}{m\,W}\left[\frac{1}{2} - \left(\frac{a_b}{a_o}\right)\cos\varepsilon - \frac{3}{2}\left(\frac{a_b}{a_o}\right)^2 + 4\left(\frac{d}{a_o}\right)^2\right] \qquad \text{(II-6-21)}$$

where

Δ = superelevation

a_o = sea tide amplitude

a_b = bay tide amplitude

d = mean water depth in channel

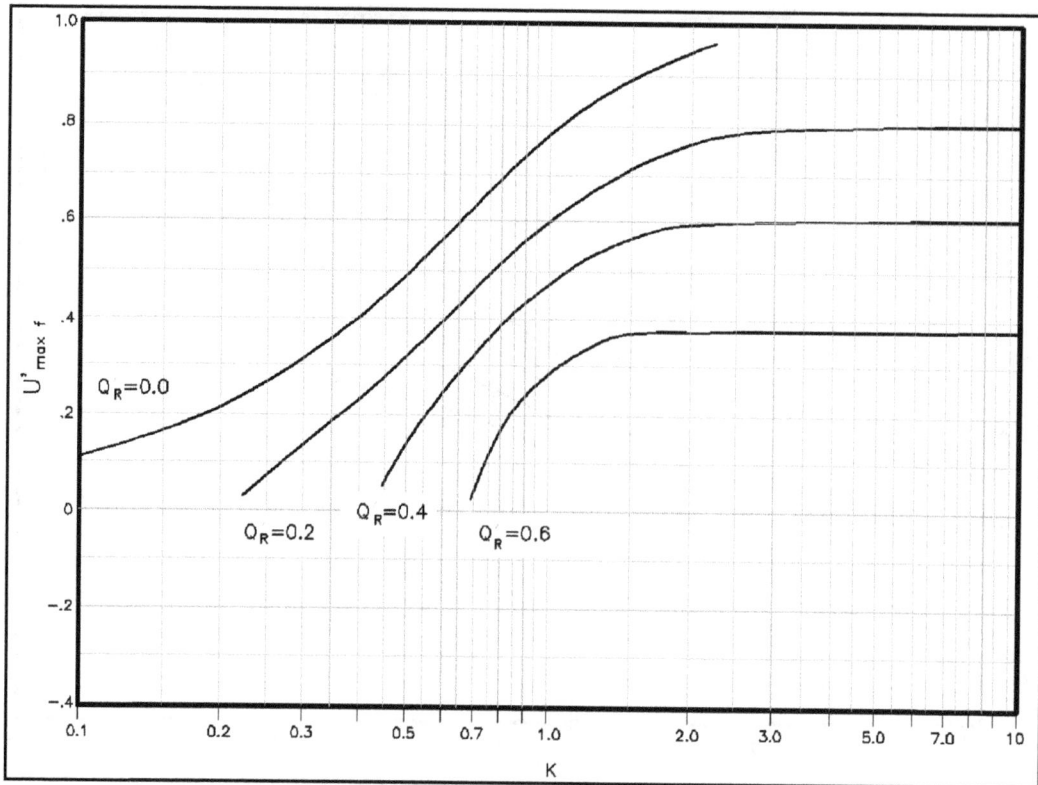

Figure II-6-28. $u'_{max\,f}$ versus K for values of Q_r' (river discharge model)

m = bank slope

W = channel width corresponding to mean water depth in channel

ε = phase lag between sea tide and bay tide

h. The inlet as a filter, flow dominance, and net effect on sedimentation processes.

(1) Introduction. As shown previously, an inlet filters some part of the ocean tide as it translates into the bay, depending on the characteristic inlet parameters defined earlier. Keulegan and King's analytical models, discussed earlier, assumed a sinusoidal ocean tide, and due to the nonlinear friction term (containing u^2), solutions of bay tide response implicitly contained higher-order harmonics (i.e., frequencies higher than that of the basic ocean tide). Other studies have shown that if a variable inlet cross section or variable bay surface area is considered, this will introduce higher-order harmonics. The forcing ocean tide, of course, is composed of many different frequency constituents (Part II-5), so there is interest in using this information to provide more accurate results. Also, investigators have shown that higher harmonics of current velocity (resulting from the forcing tides) are important in relation to sediment transport and its net movement through the inlet (King 1974; Aubrey 1986; DiLorenzo 1988). These effects are important due to the relationship of sediment transport to velocity raised to some power, e.g., V^5 (Costa and Isaacs 1977).

(2) Tidal constituents. Boon (1988) suggests that the seven tidal constituents in Table II-6-2 adequately represent shallow-water tide distortion in inlet-basin systems. Aubrey and Friedrichs (1988) and DiLorenzo (1988) used the amplitude of the ratio of M4 to M2 constituent amplitudes as a measure of nonlinear distortion. Variability in the spring-neap cycle, as well as seasonal mean water level changes, cause characteristic fluctuations in M4/M2.

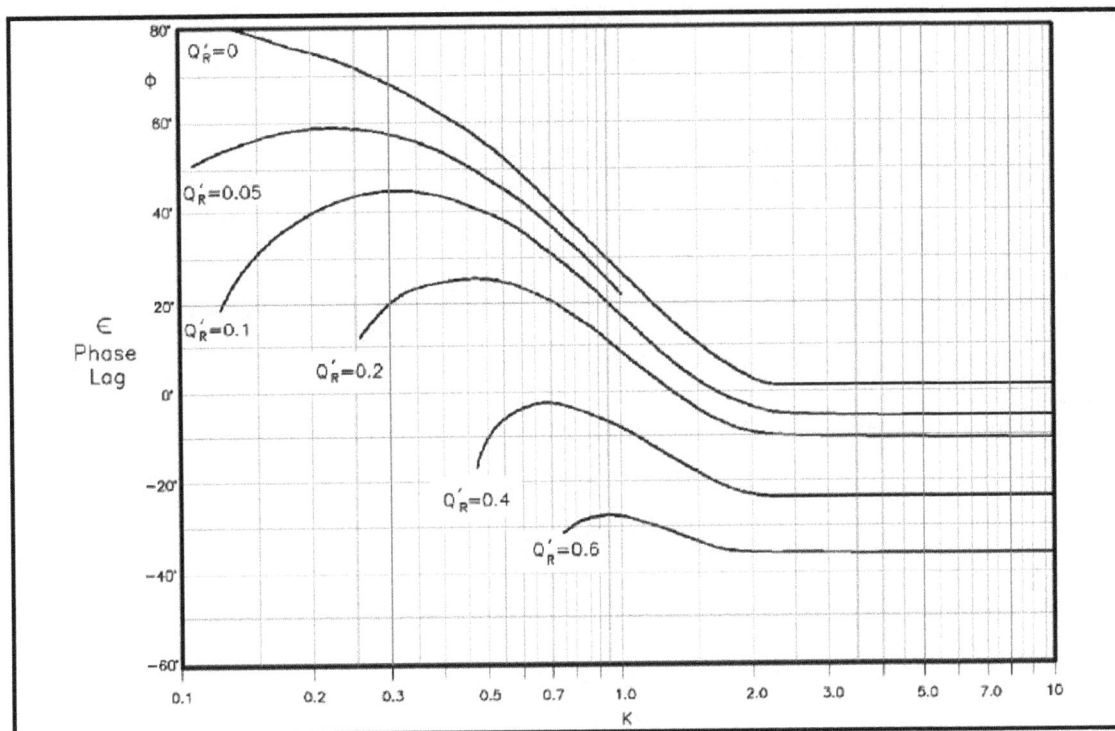

Figure II-6-29.   ε versus K for values of $Q_r{'}$ (river discharge model)

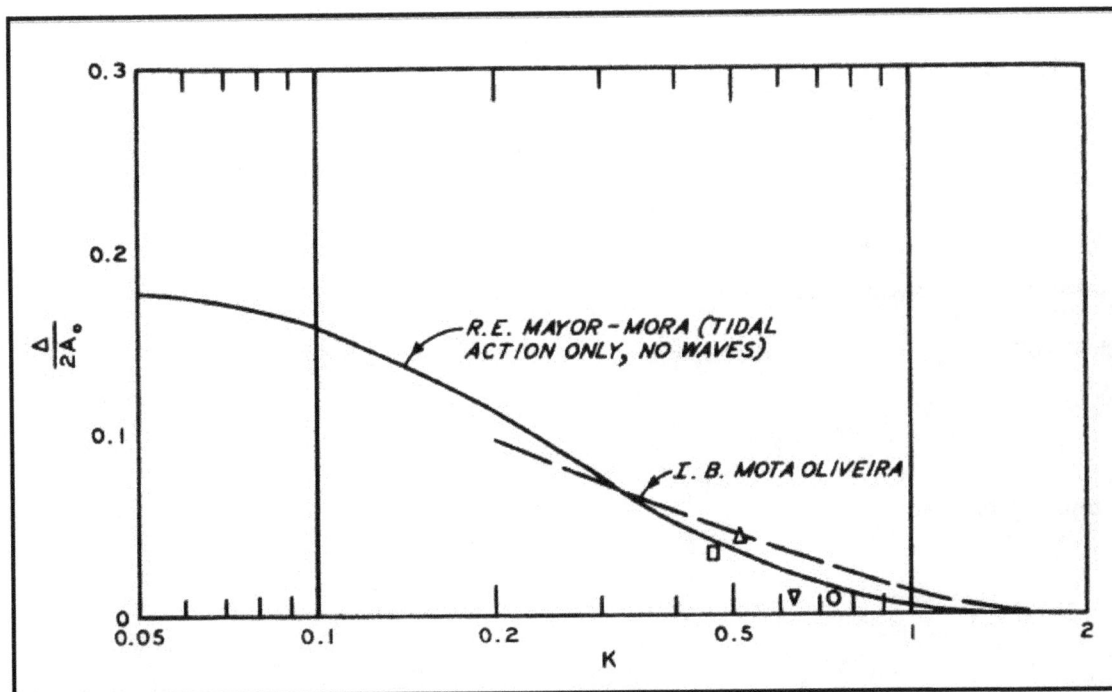

Figure II-6-30.   Ratio of bay superelevation to ocean tide range as a function of the coefficient of repletion K

EXAMPLE PROBLEM II-6-2

Given:

Using the same inlet system discussed in Example Problem II-6-1, consider the effect of a constant river discharge of 350 m⁻/s (12,360 ft³/s).

Find:

What are the effects of river discharge on high-water and low-water elevations in the bay, bay tide range, and maximum flood and ebb velocities, the maximum flood and ebb discharges, and the bay tide phase lag.

Solution:

Using Equation II-6-19 for dimensionless variable Q_r' with a_o, A_b, and T from Example Problem 1 and Q_r given above

$$Q_r' = \frac{Q_r T}{2\pi a_o A_b}$$

$$Q_r' = \frac{350\ (44712)}{2(3.14)\ (1.30/2)\ (1.90 \times 10^7)} = 0.20$$

Using Equation II-6-11, find the Keulegan K, knowing K_1 and K_2, as calculated in Example Problem 1.

$$K = \frac{1}{k_2}\sqrt{\frac{1}{K_1}}$$

$$K = \frac{1}{0.25}\sqrt{\frac{1}{28.8}} = 0.75$$

(or calculate Keulegan K from Equation II-6-4).

With $K = 0.75$ and $Q_r' = 0.20$, bay tidal response is determined from Figure II-6-25

$$\frac{a_{b\ max}}{a_o} = 0.92 \qquad\qquad a_{b\ max} = 0.92\ (0.65) = 0.60\ m$$

and from Figure II-6-26

$$\frac{a_{b\ min}}{a_o} = -0.56 \qquad\qquad a_{b\ min} = -0.56\ (0.65) = -0.36\ m$$

(Continued)

Example Problem II-6-2 (Concluded)

Bay tide range is 0.96 m and is not symmetric about mean sea level, due to river flow.

Maximum channel velocities are determined from Figure II-6-27 (maximum ebb velocity)

$$U'_{maxe} = 0.80$$

Using Equation II-6-5, and substituting U'_{maxe} for V'_m and U_{maxe} for V_m, and solving for U_{maxe}

$$U_{maxe} = \frac{U'_{maxe} \, 2\pi \, a_o A_b}{A_{avg} \, T} = \frac{0.80 \ (2) \ (3.14) \ (0.65) \ (1.90) \ 10^7}{666 \ (12.42) \ (3600)} = 2.08 \ \text{m/s}$$

From Figure II-6-28 (maximum flood velocity)

$$U'_{maxf} = 0.50$$

and in a manner similar to solving for ebb velocity, flood velocity is determined by

$$U_{maxf} = \frac{U'_{maxf} \, 2\pi \, a_o A_b}{A_{avg} \, T} = \frac{0.50 \ (2) \ (3.14) \ (0.65) \ (1.90) \ 10^7}{666 \ (12.42) \ (3600)} = 1.30 \ \text{m/s}$$

Maximum discharges are determined from

$$Q_{maxe} = U_{maxf} \, A_c = 2.08 \ (666) = 1385 \ \text{m}^3/\text{s}$$

$$Q_{maxf} = U_{maxf} \, A_c = 1.30 \ (666) = 866 \ \text{m}^3/\text{s}$$

River flow increases maximum ebb discharge and decreases maximum flood discharge, as one would expect. The values are nearly balanced around the no-river-discharge value of 1,145 m³/s of Example 1.

Using Figure II-6-29, the value of ε (phase lag of bay tide relative to ocean tide) is determined to be 19 deg, indicating high water occurs earlier than for the non-river-inflow condition of Example 1.

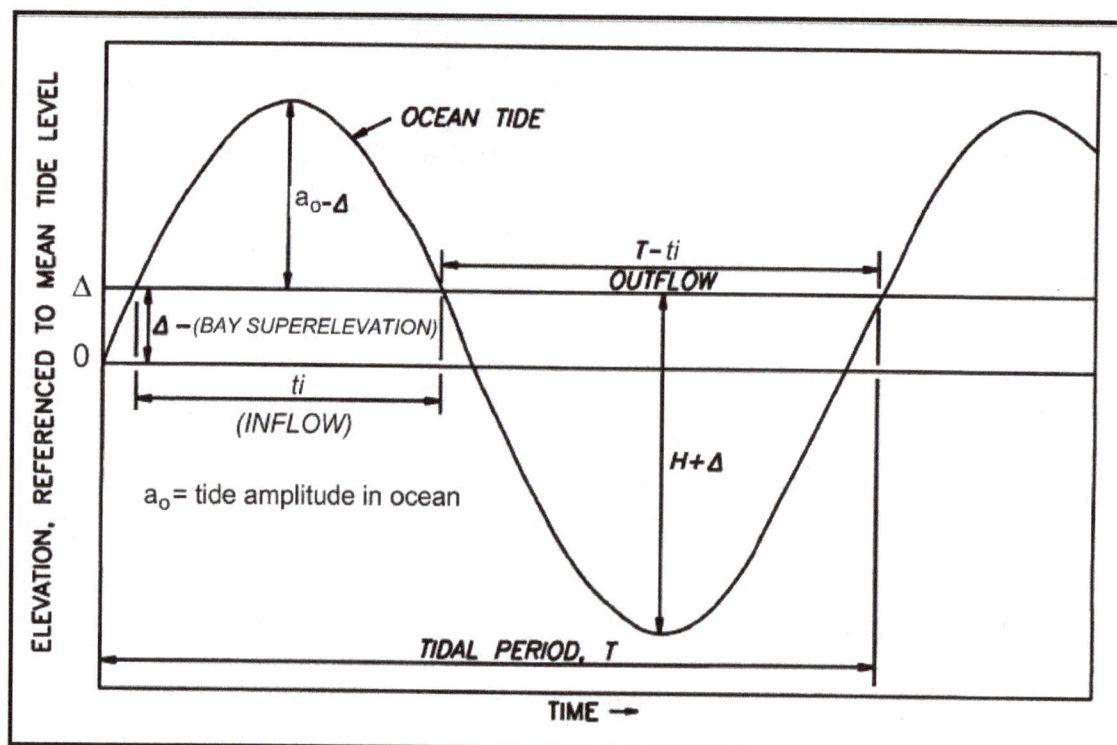

Figure II-6-31. Duration of outflow relative to bay superelevation

Table II-6-2
Semidiurnal and Shallow-Water Tidal Constituents and Harmonic Frequencies

Constituent	Frequency	Degrees/hr	Designation
M_2	a_1	28.9841	Lunar semidiurnal
S_2	a_2	30.0000	Solar semidiurnal
N_2	a_3	28.4397	Lunar elliptic
M_4	$2a_1$	57.9682	Lunar quarter-diurnal
S_4	$2a_2$	60.0000	Solar quarter-diurnal
MS_4	a_1+a_2	58.9841	Compound harmonic
MN_4	a_1+a_3	57.4238	Compound harmonic
MSf	a_2-a_1	1.0159	Lunar long-period

(3) Flow dominance. Entrance area and bay surface area variation impacts flow conditions at the inlet in terms of the continuity equation (Equation II-6-3), which indicates that

$$\frac{dh_b}{dt} \propto \frac{A_{avg}}{A_b} \qquad\qquad\qquad \text{(II-6-22)}$$

The rate of change of bay elevation is directly proportional to channel area and inversely proportional to surface area of the bay. If A_{avg} is held constant while A_b increases with the rising tide, then the rate of rise decreases, while the rate of fall increases over time. This means there is a longer rise and faster fall, indicating greater ebb velocity peaks (assuming the same amount of water enters and leaves the bay), which means currents are ebb-dominant when considering only the effect of variable bay area. Similar reasoning shows that variable entrance area increases the rate of rise of flood tide and decreases the rate fall, causing higher maximum flood velocities than ebb, creating a flood-dominant inlet. These conditions are important to help interpret the movement of sediment through inlets and the locations where shoals form.

    i.   Multiple inlets.

    (1) Often a lagoon is connected to the ocean by more than one inlet. The basic simplified hydraulics can be determined by methods introduced previously or may be numerically modeled for more detailed effects such as residual currents and net circulation (van de Kreeke 1976, 1978). Multiple inlets with a common bay can impact inlet stability if an inlet changes or if an inlet is added (Ward 1982). Changes to a multiple inlet system may be gradual (for example, one inlet gradually lengthens as littoral transport creates shoal regions in the inlet), or change may be rapid (e.g., a hurricane cuts a channel through a barrier beach and begins to rapidly expand in size, capturing a significant portion of the tidal prism from a previously existing inlet connected to the same bay).

    (2) For an existing system of multiple inlets, the K values may be added together

$$K_{inlet\ 1} + K_{inlet\ 2} + \ldots\ldots\ldots K_{inlet\ n} = K_{all\ inlets} \tag{II-6-23}$$

to determine the total K for the system, from which the bay tide amplitude and phase lag may be determined (Figures II-6-18 and II-6-20). A total maximum discharge (for the total of all inlets) from the V_{max} determined from Figure II-6-19 is

$$Q_{max} = \frac{2\pi}{T} a_o A_b V_{max} \tag{II-6-24}$$

    (3) This total maximum discharge then may be apportioned between all inlets

$$Q_{max} = Q_{1max} + Q_{2max} + Q_{3max} + \ldots \tag{II-6-25}$$

with the above equation knowing that, for example, for three inlets

$$Q_{1max} = \frac{Q_{max}}{1 + \dfrac{K_2}{K_1} + \dfrac{K_3}{K_1}} \tag{II-6-26}$$

Q_{2max} and Q_{3max} may be determined by placement of the appropriate K values in the denominator in the same relative positions as in the above equation.

    (4) For the case of an inlet connected to a bay that is segmented into separate internal bays connected by narrow openings, reference is made to Dean (1971) and Mota Oliveira (1970).

    j.   Tidal jets. Some analytical approaches to flow patterns at inlets treat parameters affecting the "plume" of water exiting the inlet (i.e., ebb flow). Joshi (1982) examined the effect of bottom friction, bathymetric changes, and lateral entrainment of water due to turbulent mixing. Generally, as bottom friction increases, the plume becomes wider and water velocity along the center line decreases. As bottom slope increases, the plume contracts. Özsöy (1977) and Özsöy and Ünlüata (1982) developed relationships to determine plume expansion and center-line velocity, as shown in Figures II-6-32 and II-6-33. Example

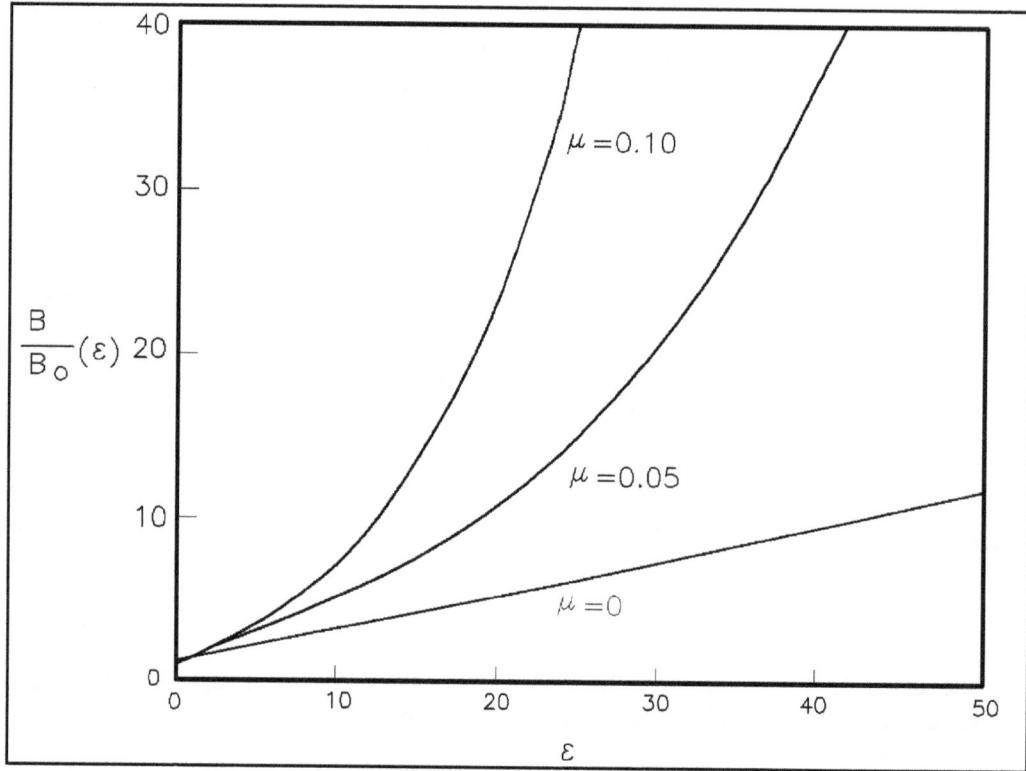

Figure II-6-32.   Plume expansion

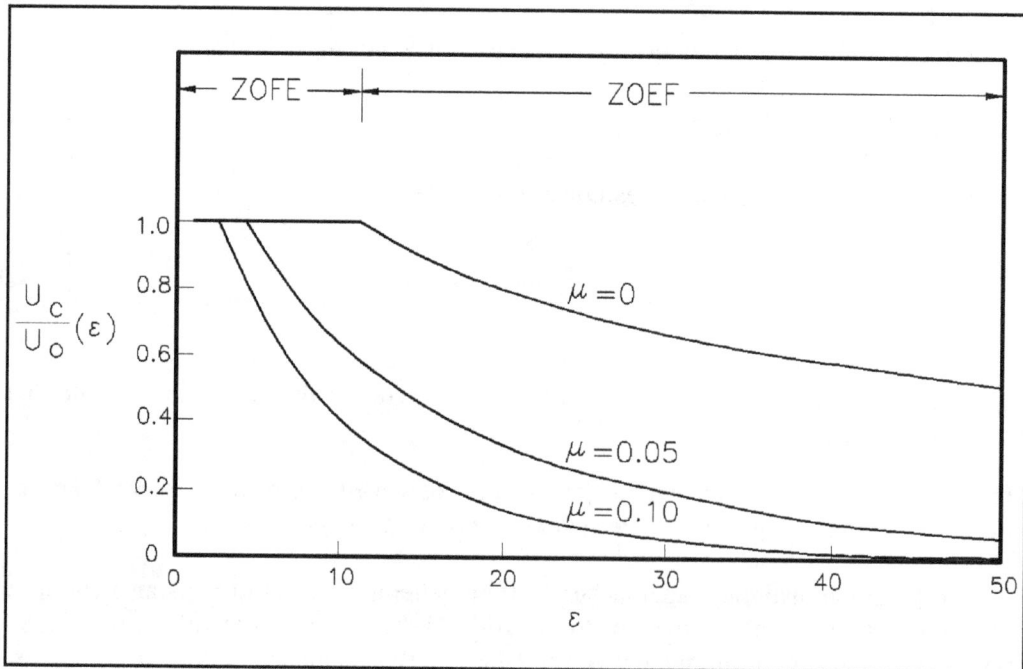

Figure II-6-33.   Center-line velocity

Problem II-6-2 shows the effects of a flat bottom with friction. Özsöy and Ünlüata provide additional information for sloping bottoms and the effect of a longshore current approaching perpendicular to the ebb jet, deflecting it downcoast. Joshi and Taylor (1983) examined the alongshore circulation created by tidal jets and predicted that currents toward the inlet along the shoreline (induced by the ebb tidal jet or plume) can range from 0.01 to 0.1 m/s within 10 half-widths (of the minimum inlet width) of the natural inlet entrance. With a jetty in place, the range of induced currents along the shoreline is reduced, finally becoming zero at the junction of the shoreline and jetty. The above work is based on a simplified bathymetry and does not include wave-induced longshore currents, which can be an order of magnitude larger than jet-induced longshore currents. Ismail and Wiegel (1983) found that waves increased the rate of ebb jet spread.

k. Tidal dispersion and mixing.

(1) Flow through inlets can play an important role in the exchange processes and flushing of the bay, both of which impact water quality and bay ecology. Tidal currents, along with wind-driven currents and in some cases density-driven circulation, can drive dispersion processes in the bay where particles of water (or other substances, e.g., pollutants) are scattered or diluted. Tidal dispersion and its role in flushing of bays has been studied by many investigators including Geyer and Signell (1992), Zimmerman (1988), van der Kreeke (1983), Dyer and Taylor (1973), and Sanford, Boicourt, and Rives (1992). There is typically a broad variance in the dispersion coefficient values that can be determined for the processes causing dispersion over a large area. Therefore, analytic techniques to determine flushing of a bay are limited to small basins, and assume complete internal mixing, with detailed numerical simulations required for larger bays. Tidal dispersion is important in regions where flow separations occur; e.g., in regions of abrupt geometric change, with the inlet entrance channel a prime example. The extent of dispersive influence is limited to the tidal excursion in the bay (the distance covered by a particle entering the bay at the start of flood tide until the end of flood tide). For tidal dispersion to influence large spatial scales, spacing between major topographic features must be less than the tidal excursion distance (Geyer and Signell 1992).

(2) For small basins, where it is assumed that waters in the basin are completely mixed at all times, van de Kreeke (1983) defined residence time as the average time for a particle to enter, then leave the basin, with this average based on many particles. For small basins, and with the following relationship met

$$10 > \frac{\langle V \rangle}{\varepsilon P} > 1.5 \qquad\qquad\qquad \text{(II-6-27)}$$

where <V> is the average volume of the bay over the tidal cycle, ε is the fraction of new water entering the bay from the sea each tidal cycle, and P is the tidal prism. A value of $\varepsilon = 0.5$ is often used when there is no other basis for estimation of this factor. Then residence time can be calculated by

$$\frac{\tau_r}{T} = \frac{\langle V \rangle}{\varepsilon P} \qquad\qquad\qquad \text{(II-6-28)}$$

with τ_r residence time and T tidal period. The time to reduce the mass of a substance's concentration by a factor of "e" (2.3) is equal to τ_r and to reduce it a factor of 10 is $2.3\tau_r$.

l. Wave-current interaction. The region where the navigation channel intersects and/or passes through the outer ebb shoal can be a region of intense wave-current interaction. Wave-current interaction can produce difficult navigation and dredge operation conditions. Also, interaction of waves, currents, and local bathymetry/channel typically produces the most severe channel shoaling problems at inlets. These interactions will be discussed in various combinations.

(1) Wave-current interaction.

(a) Horizontal currents can modify surface gravity waves by stretching or shrinking features of a wave train. This distortion can produce a "Doppler Shift" in wave period. Wave orthogonal, crest, and ray directions are modified (see Part II-1 for definitions). Also, a current modifies surface waves by causing an exchange of energy between wave and current. The pressure field accompanying the wave also is modified. Predicting current-modified wave energy, heights, directions, and pressures is typically a complex procedure still under research; however, a graphical method presented by Herchenroder (1981) based on work by Jonsson et al. (1970), can be used to estimate the effect of current on wave height H_A, where this is wave height without current.

Using the equation:

$$H = (R_H)H_A$$

(II-6-29)

(b) Figure II-6-34 may be used to determine R_H, the wave height factor, which depends on:

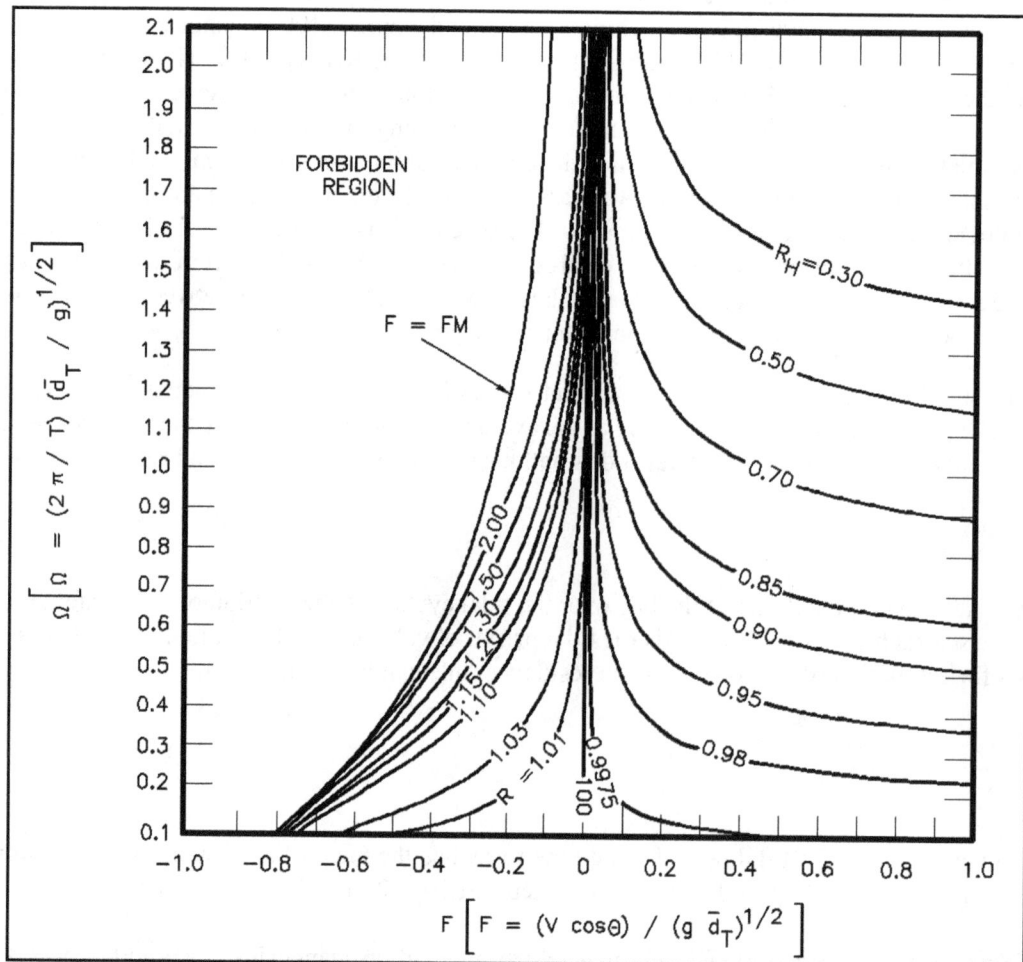

Figure II-6-34. Contours of dimensionless wave height factor R_H given by Equation II-6-29. Waves cannot propagate for values of F and Ω lying in the forbidden region (boundary line $F = FM$)

EXAMPLE PROBLEM II-6-3

FIND:
The width of the ebb jet and its maximum velocity along the centerline 2,000 m seaward from shore.

GIVEN:
The inlet channel width where the ebb jet exits the inlet is 200 m and the channel depth is 8 m. A flat bottom and Mannings $n = 0.02$ are assumed. The maximum average velocity in the channel is 1.5 m/sec.

SOLUTION:
The friction factor μ (from Özsöy and Ünlüata (1982)) is calculated as

$$\mu = \frac{g b_o n^2}{h_o^{0.75}} = \frac{(9.8)\ (100)\ (0.02)^2}{(8)^{0.75}} = 0.08$$

where b_o = half-width of channel = 200 m / 2 = 100 m and the dimensionless distance offshore parameter is

$$\xi = \frac{x}{b_o}$$

with x, the distance offshore, equal to 2000 m. Calculating these parameters, $\mu = 0.08$ and $\xi = 20$. Using Figures II-6-32 and II-6-33, and interpolating for μ, the results are

$$\frac{b}{b_o}(\xi) = 18.5 \quad \Rightarrow \quad b = 100\ (18.5) = 1850 \text{ m}$$

and

$$\frac{u_c}{u_o}(\xi) = 0.2 \quad \Rightarrow \quad u_c = 0.2\ (1.5) = 0.3 \text{ m/s}$$

Therefore, for a flat bottom, the half-width of the jet expands from 200 m to 1,850 m and the center-line velocity of the jet decreases from 1.5 m/s at the inlet to 0.3 m/s 2,000 m from shore.

EXAMPLE PROBLEM II-6-4

Find:
Determine the wave height in an inlet channel that is 5.0 m (16.4 ft) deep.

Given:
A 5.0-sec, 1.0-m (3.28-ft) wave is entering an inlet having a 1.0 m/s (3.3 ft/s) ebb current. The angle between current and wave orthogonal is 180 deg.

Solution:

$\Omega = (2\pi/T) \, (d_T/g)^{1/2}$

$F = (V \cos \theta)/(gd_T)^{1/2}$

T = wave period = 5 s

d_T = time-averaged water depth = 5.0 m

g = 9.8 m/s^2

V = horizontal current speed = 1.0 m/s

θ = angle between horizontal velocity vector and horizontal wave vector = 180°

Therefore:

$$\Omega = \left(\frac{2 \, (3.14)}{5.0} \right) \left(\frac{5}{9.8} \right)^{1/2} = 0.90$$

$$F = \frac{1.0 \, (\cos 180)}{(9.8 \cdot (5))^{1/2}} = -0.14$$

From Figure II-6-34, R_H = 1.35

Wave height H_A, originally 1 m, is modified by the current to a wave height of:

$$H = (R_H) \, H_A = 1.35 \, (1.0) = 1.35 \text{ m}$$

$$F = \frac{V \cos \theta}{\left(g\,d_T\right)^{\frac{1}{2}}}$$

(II-6-30)

and

$$\Omega = \left(\frac{2\pi}{T}\right) \left(\frac{d_T}{g}\right)^{\frac{1}{2}}$$

(II-6-31)

where

T = wave period

d_T = channel depth

θ = angle wave orthogonal makes with current

V = velocity in channel

(2) Current-channel interaction. As flow converges on an inlet entrance, the angle at which flow approaches a dredged channel can be important with regard to change in current direction and can ultimately relate to channel shoaling. The direction of current approach will depend on bottom configuration and structure(s) location. Boer (1985) developed a mathematical model to study currents in a dredged channel. He found that a current approaching obliquely to a channel is refracted within the channel and the streamlines contract within the channel causing a velocity increase (Figure II-6-35). This effect becomes relatively small for angles larger than 60° (angle between channel axis and current direction). This effect is largest near the bed and smallest near the surface. Due to continuity, depending on the relative depth of the channel to the surrounding depths, there is a decrease factor because of increased depth in the channel.

m. *Other methods of inlet analysis.* As mentioned in Part II-6-2, paragraph *a,* follow-up studies using techniques presented in previous sections and numerical and/or physical models, can be conducted after an initial examination of the inlet. Some available tools are discussed below.

(1) Automated Coastal Engineering System (ACES). ACES (Leenknecht et al. 1992) contains an inlet model that operates in the PC environment and estimates inlet velocities, discharges, and bay levels as a function of time. This model is designed for cases where the bay water level fluctuates uniformly. Seaward boundary conditions are specified as water level fluctuations associated with astronomical tides, storm surges, seiches, and tsunamis. Figure II-6-36 shows a flow net and Figure II-6-37 is a typical cross section indicating that the model can handle up to two inlet entrances to a single bay and that the entrance channel(s) can be divided into sub-channels to accommodate change in depth across the inlet. The model will permit up to seven sub-channels and up to sixteen cross sections of the flownet. Figures II-6-38 and II-6-39 show sample output from the program.

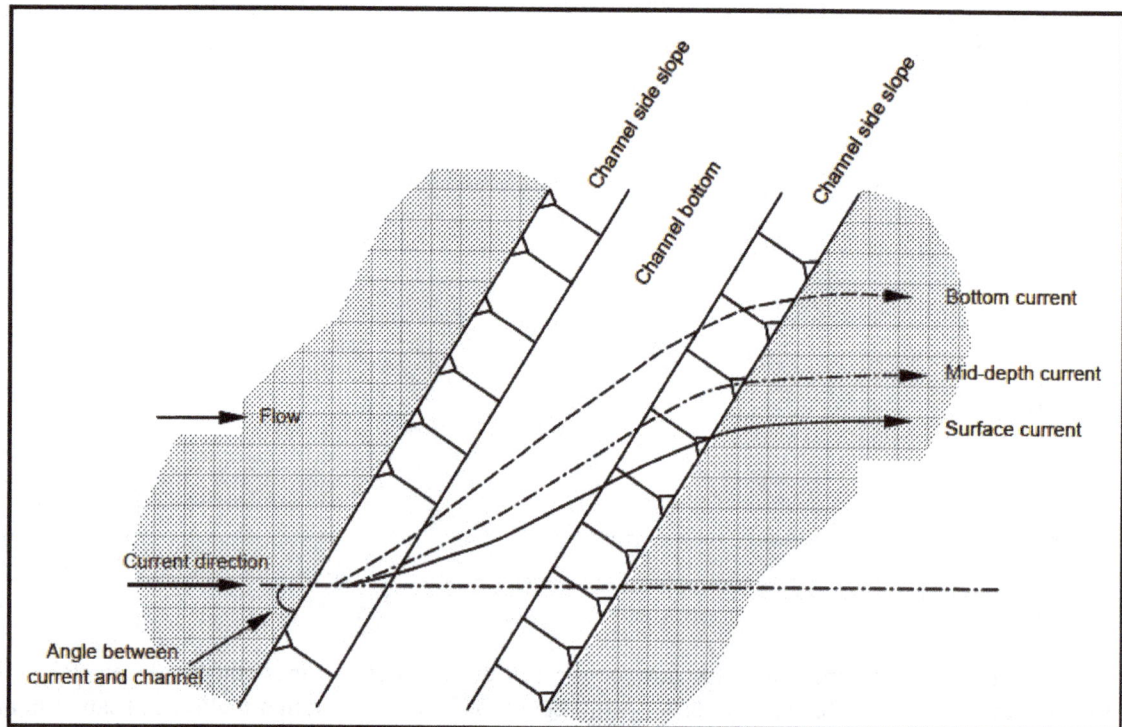

Figure II-6-35. Refraction of currents by channel

(2) DYNLET1. This model (Amein and Kraus 1991) is a one-dimensional PC model that can predict flow conditions in channels with varied geometry, and accepts varying friction factors across an inlet channel. Values of water surface elevation and average velocity are computed at locations across and along inlet channels. The inlet to be modeled may consist of a single channel connecting the sea to the bay, or it may be a system of interconnected channels, with or without bays.

(3) Coastal Modeling System (CMS). The CMS (Cialone et al. 1991) contains two hydrodynamic programs applicable to inlets. This modeling system is a supercomputer-based system of models and supporting software packages. Two models applicable to inlet studies are:

(a) WIFM. This model is a two-dimensional, time-dependent, long-wave model for solving the vertically integrated Navier-Stokes equations in a stretched Cartesian coordinate system. The model simulates shallow-water long-wave hydrodynamics such as tidal circulation, storm surges, and tsunami propagation. WIFM contains many useful features for studying these phenomena such as moving boundaries to simulate flooding/drying of low-lying areas and subgrid flow boundaries to simulate small barrier islands, jetties, dunes, or other structural features. Model output includes vertically integrated water velocities and water surface elevations.

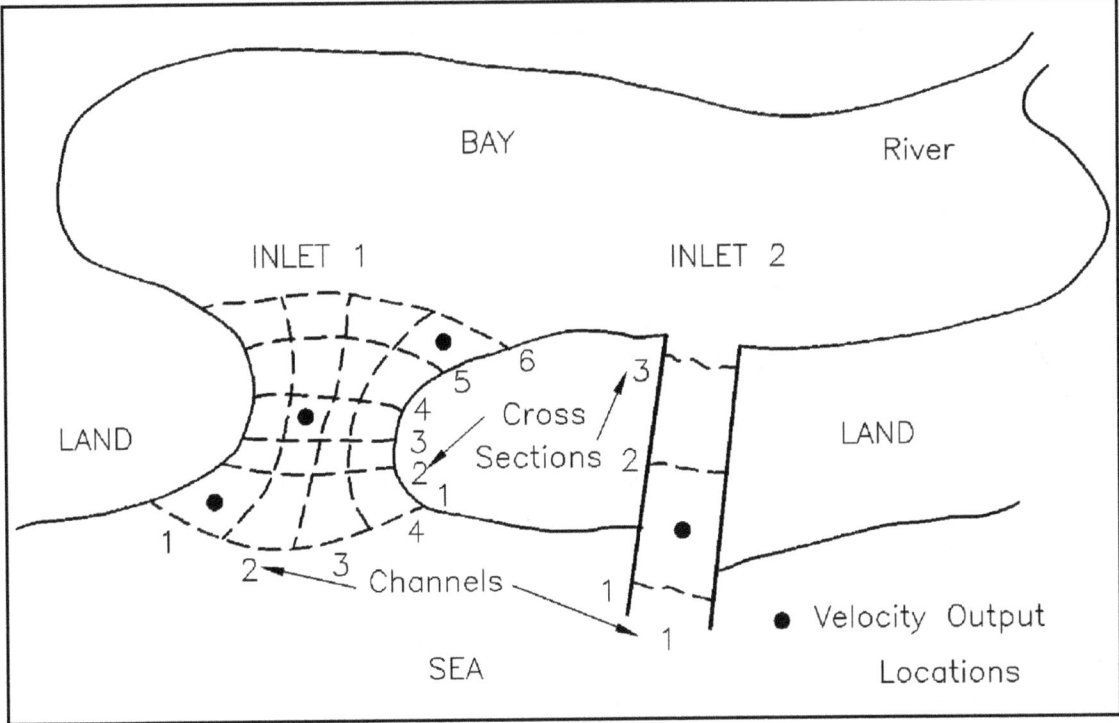

Figure II-6-36.   Typical inlet flow nets

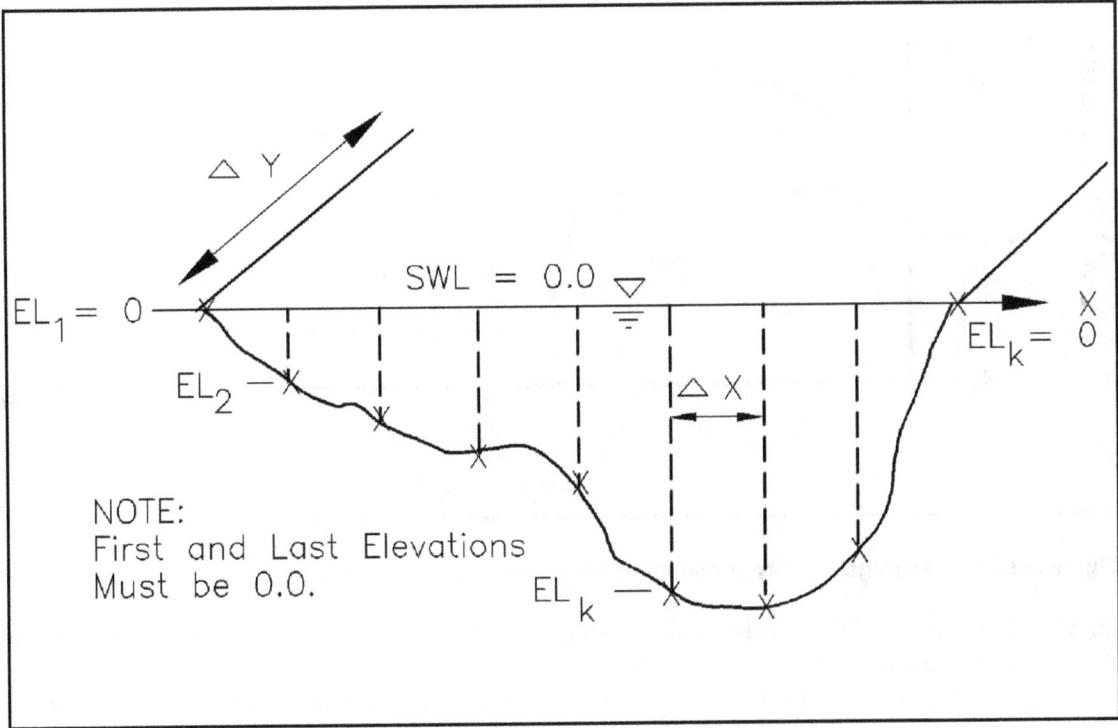

Figure II-6-37.   Typical inlet cross section

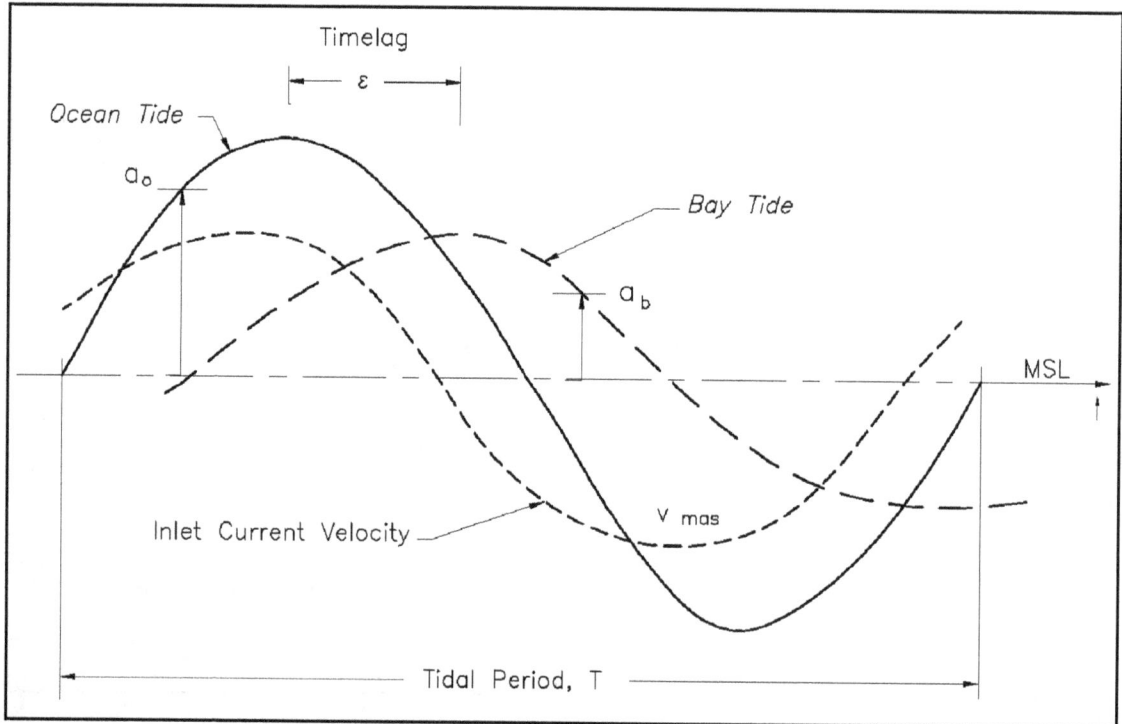

Figure II-6-38.   Sea and bay water elevations at inlet 1

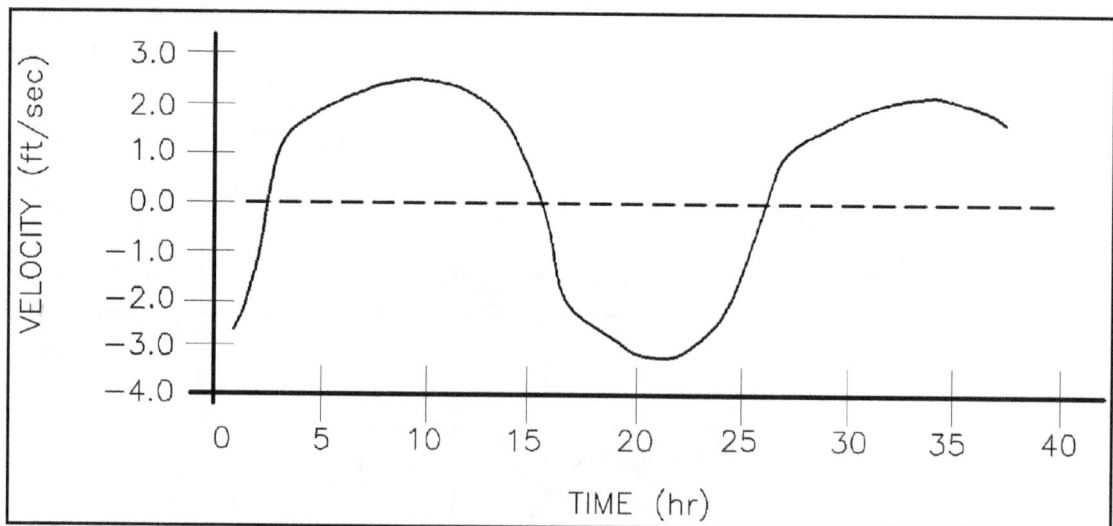

Figure II-6-39.   Average velocity at minimum cross section

(b) CLHYD. This model simulates shallow-water, long-wave hydrodynamics such as tidal circulation and storm surge propagation. It can simulate flow fields induced by wind fields, river inflows/outflows, and tidal forcing. CLHYD is similar to WIFM, with the added feature of operating on a boundary-fitted (curvilinear) grid system. However, CLHYD cannot simulate flooding/drying of low-lying areas, as WIFM can. Model output includes vertically integrated water velocities and water surface elevations.

(4) Other models. Other more sophisticated models that require considerable expertise to use already exist or are being developed. The reader is advised to consult current technical literature for details. Part II-5-7, "Numerical Modeling of Long Wave Hydrodynamics," discusses two such sophisticated models, CH3D and ADCIRC.

(5) Physical models. Physical models have also been used for both the hydraulic and sedimentation aspects of inlet studies, but numerical models usually have been used in recent years due to the expense of constructing the large bay system of an inlet and the need to examine processes difficult to reproduce in a physical model, such as storm surge, which would need to include wind effects. However, physical models are appropriate and cost-effective for examining structures and detailed flow patterns in the immediate vicinity of the inlet entrance (sample references include Bottin (1978, 1982), Seabergh (1975), Seabergh and Lane (1977), and Seabergh and McCoy (1982)).

II-6-3. Hydrodynamic and Sediment Interaction at Tidal Inlets

a. Introduction. Waves, currents, and sediment interact at tidal inlets over a large physical area, with varying effects over a tidal cycle (Smith 1991). At the throat or minimum cross-sectional area of the inlet and bayward, tidal currents are the predominant forcing agent interacting with sediments. However, moving seaward and alongshore away from the inlet, the effect of waves increases. Wave energy on the edges of the inlet channel can contribute sediment from adjoining beaches, with the flood tidal currents tending to move bayward to a flood tidal shoal (or middle-ground shoal), or the ebb currents jetting sediment seaward to settle on the ebb shoal. Sediment may again be moved to adjoining beaches by the combination of waves and currents, thus "bypassing" the inlet. The interaction of tidal currents, waves, and wave-generated currents is a complex process.

b. Tidal prism - channel area relationship.

(1) Initial research focused on the relationship to flow through the inlet and the minimum cross-sectional area that could be sustained by that flow. O'Brien (1931, 1969) originally determined a relationship between minimum throat cross-sectional area of an inlet below mean tide level and the tidal prism (i.e., the volume of water entering or exiting the inlet on ebb and flood tide) at spring tide (Figure II-6-40). This relationship was predominantly for Pacific coast inlets, where a mixed tidal pattern is observed and there typically is a strong ebb flow between higher-high water and lower-low water. Jarrett (1976) continued the analysis and performed regression analysis for various coastal areas with different tidal characteristics. A regression equation is given for each of the Atlantic, Pacific, and Gulf coasts. Jarrett also recommended using O'Brien's equation for dual-jettied inlets. The equations, in metric and English units, are given in Table II-6-3.

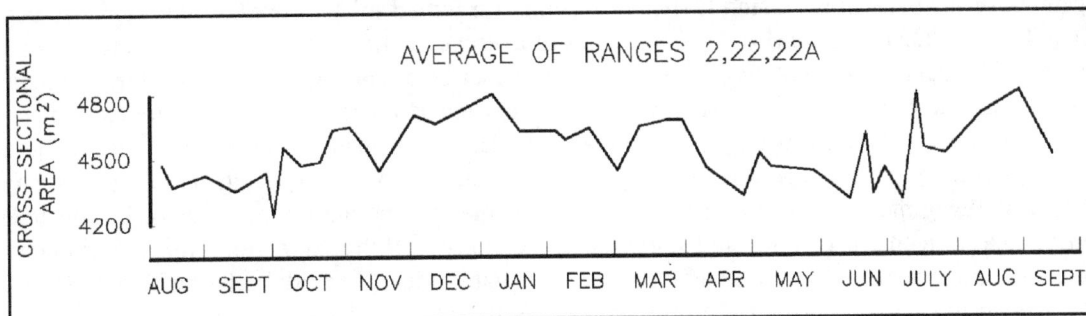

Figure II-6-40. Variations in cross-sectional area for Wachapreaque Inlet (Byrne, De Alteris, and Bullock 1974)

Table II-6-3
Tidal Prism-Minimum Channel Cross-sectional Area Relationships

Location	Metric Units	English Units
Atlantic Coast	$A_c = 3.039 \times 10^{-5} \ P^{1.05}$	$A_c = 7.75 \times 10^{-6} \ P^{1.05}$
Gulf Coast	$A_c = 9.311 \times 10^{-4} \ P^{0.84}$	$A_c = 5.02 \times 10^{-4} \ P^{0.84}$
Pacific Coast	$A_c = 2.833 \times 10^{-4} \ P^{0.91}$	$A_c = 1.19 \times 10^{-4} \ P^{0.91}$
Dual-Jettied Inlets (O'Brien)	$A_c = 7.489 \times 10^{-4} \ P^{0.86}$	$A_c = 3.76 \times 10^{-4} \ P^{0.86}$

A_c is the minimum cross-sectional area in square meters (square feet) and P is the tidal prism in cubic meters (cubic feet).

(2) Inlet tidal prisms versus minimum cross-sectional area for Jarrett's data is plotted in Figure II-6-9.

(3) Work by Byrne et al. (1980) indicated that for inlets with minimum cross sections less than 100 m² (1,076 ft²), there was a departure from the relationships developed above. They studied small inlets on the Atlantic Coast in Chesapeake Bay. Their relationship for area and tidal prism is

$$A_c = 9.902 \times 10^{-3} \ p^{061} \ \text{(metric units)} \qquad \text{(II-6-32)}$$

(4) Jarrett's (1976) work pertains to equilibrium minimum cross-sectional areas at tidal channels from one survey at a given date. Byrne, DeAlteres, and Bullock (1974) have shown that inlet cross section can change on the order of ± 10 percent in very short time periods (see Figure II-6-40). This, of course, is due to the variability in tidal currents and wave energy. A storm may bring a large amount of sediment to an inlet region, since the inlet tends to act as a sink for sediment and the tidal current may require a certain time period to return the minimum cross-sectional area to some quasi-equilibrium state. Brown (1928) notes that for Absecon Inlet, New Jersey, "a single northeaster has been observed to push as much as 100,000 cubic yards of sand in a single day into the channel on the outer bar, by the elongation of the northeast shoal, resulting in a decrease in depth on the centerline of the channel by 6 to 7 feet." Sediment changes such as this will affect the hydraulics of the inlet system, which in turn will remodify channel depth and location of sediment shoaling.

(5) Adding jetty structures to natural inlets modifies the inlet's morphology. Careful engineering design can reduce the amount of induced change. For example, jetties are placed so that the minimum cross-sectional area of the inlet is maintained, leaving the volume of water exchanged over a tidal cycle (i.e., the tidal prism) unchanged from the natural state. During design of inlet training structures, the question of appropriate spacing must be answered so that excessive scour does not occur and cause settling or displacement of the structures. Using O'Brien's formula for jettied inlets, the minimum distance between jetties can be calculated. Figure II-6-41 shows lines of average depth expected for given jetty spacing and tidal prism. The data points plotted on Figure II-6-41 describe actual field conditions for 44 inlets. No attempt was made to analyze or judge whether problems existed at a particular project. For example, a very large tidal prism with very narrow jetty spacing may develop problems of erosion along one of the jetties or very high velocities might exist, producing navigational difficulties. Also, if spacing of the jetties is too wide, the channel thalweg may meander in a sinuous manner, making navigation more difficult and reducing efficiency, so as to increase shoaling and lead to possible closure. If the same minimum area is maintained between the entrance channel's jetties that existed for the natural inlet, the tidal prism will be the same, and tides will flush out the bay behind the inlet as well as they did in the natural state. Actually, a more hydraulically efficient channel usually will exist at a jettied inlet because sediment influx is reduced and there are fewer shoals.

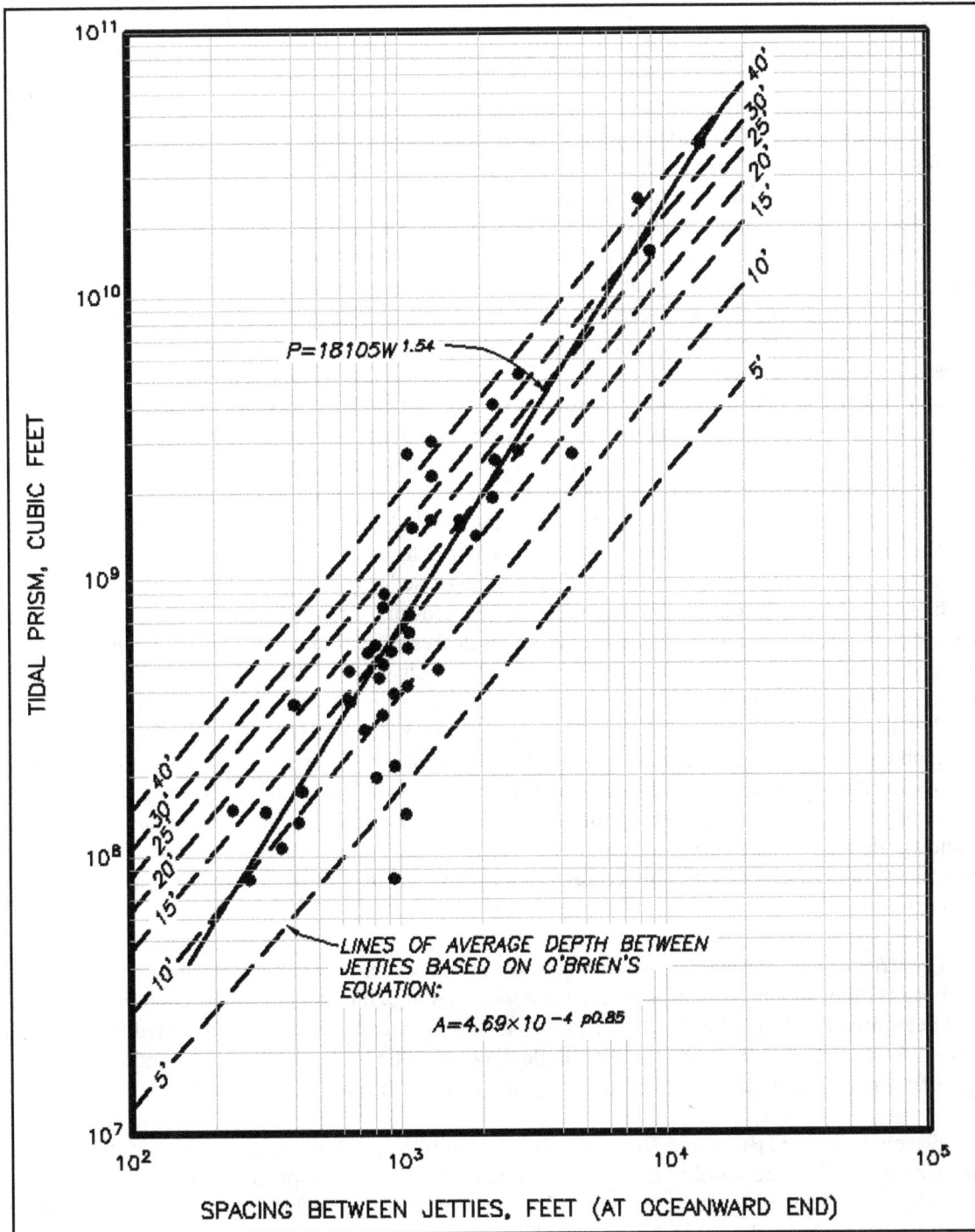

Figure II-6-41.   Tidal prism (*P*) versus jetty spacing (*W*)

c.   Inlet stability analysis.

(1) Dean (1971) equated the expression for tidal prism, $P = TV_m A_c / \pi$ to O'Brien's (1931) original tidal prism, inlet area relationship ($P = 5 \times 10^4 A_c$) and determined that for a tidal period of 44,640 sec, V_m for the inlet is about 1 m/sec. In other words, 1 m/sec might be interpreted as a level of velocity necessary to maintain an equilibrium flow area. Therefore, as wave action supplies sand to the inlet channel and tends to

reduce the cross-sectional area, the inlet flow will scour out any depositions that reduce the channel cross section below its equilibrium value. This concept was first developed analytically by Escoffier (1940, 1977). He proposed a diagram for inlet stability analysis in which two curves are initially plotted. The first is the velocity versus the inlet's cross-sectional flow area A_c. A single hydraulic stability curve represents changing inlet conditions when ocean tide parameters and bay and inlet plan geometry remain relatively fixed. As area approaches zero, velocity approaches zero due to increasing frictional forces, which are inversely proportional to channel area. As channel area increases, friction forces are reduced but, on the far right side of the curve, velocities decrease as the tidal prism has reached a maximum, and any area increases just decrease velocity as determined by the continuity equation. This curve can be constructed by calculating velocity V_m by varying channel area A_c. V_m can be determined by an analytical or numerical model, remembering that, if using Keulegan or King models, an average maximum velocity is determined. The continuity equation $V_{avg} A = V_m A_c$ must be used to determine maximum velocity at the minimum cross section. The other curve plotted as V_E is a stability criterion curve such as O'Brien and Jarrett's tidal prism versus cross-section area relationship. Escoffier (1940) originally proposed a constant critical velocity (e.g., 1 m/sec, which would plot as a straight horizontal line. If a P versus A_c relationship is used, the appropriate equation (Table II-6-3) can be used to relate V_m to tidal prism. The two curves are shown in Figure II-6-42. The possibilities of intersection of the two curves could possibly intersect at two locations or one location (a tangent), or there could be no intersection at all. In the first case, point b (see Figure II-6-42) is a stable root in that any deviation in area returns movement along the stability curve to its starting point. If channel area increases (move right on curve from point b) velocity will fall and more sediment can fill in the channel to bring it back to "equilibrium." If area decreases, velocity will increase scouring back to the equilibrium point. Point c is an unstable root, where if the area decreases, velocities decrease until the inlet closes. Moving to the right of point c, as velocity increases, area increases until the velocity starts falling and the stable root at point b is reached. If the stability curve falls tangent to or below the stability criterion curve, the inlet will close. Thus, if the inlet area is to the right of the unstable equilibrium point, and a storm occurs that provides a large sediment input to the inlet region, the inlet area could shift to the left of that point, and the inlet would close. van de Kreeke (1992) presents useful commentary on application of Escoffier's analysis where he notes that separation of stable and unstable inlets is not determined by the maximum in the maximum velocity curve (sometimes called the closure curve) of Figure II-6-42, but point c of that curve. Van de Kreeke emphasizes the integral use of O'Brien-type stability correlations with Esscofier's curve, rather than the use of stability equations alone.

(2) Escoffier (1977) presented work similar to the above, except the Keulegan K value was determined for the abscissa of the plot in place of the area. For a given set of parameters for a particular inlet, A_c is varied and K is determined to define the V_m curve. Note: If King's curves are used, K is related to K_1 and K_2 by Equation II-6-11. Figure II-6-43 shows an example diagram. The V_E curve then is determined by selecting an appropriate tidal-prism versus-area relationship and by determining V_{max} for various A values .

(3) A technique by O'Brien and Dean (1972) has also been used to calculate the stability of an inlet affected by deposition. The stability index ß is defined to represent the capacity of an inlet to resist closure under conditions of deposition. It incorporates a buffer storage area available in the inlet cross section, prior to deposition and includes the capability of the inlet to transport excess sand from its throat.

$$\beta = \int_{a_C}^{a_E} (V_{max} - V_T)^3 \, dA_c \tag{II-6-33}$$

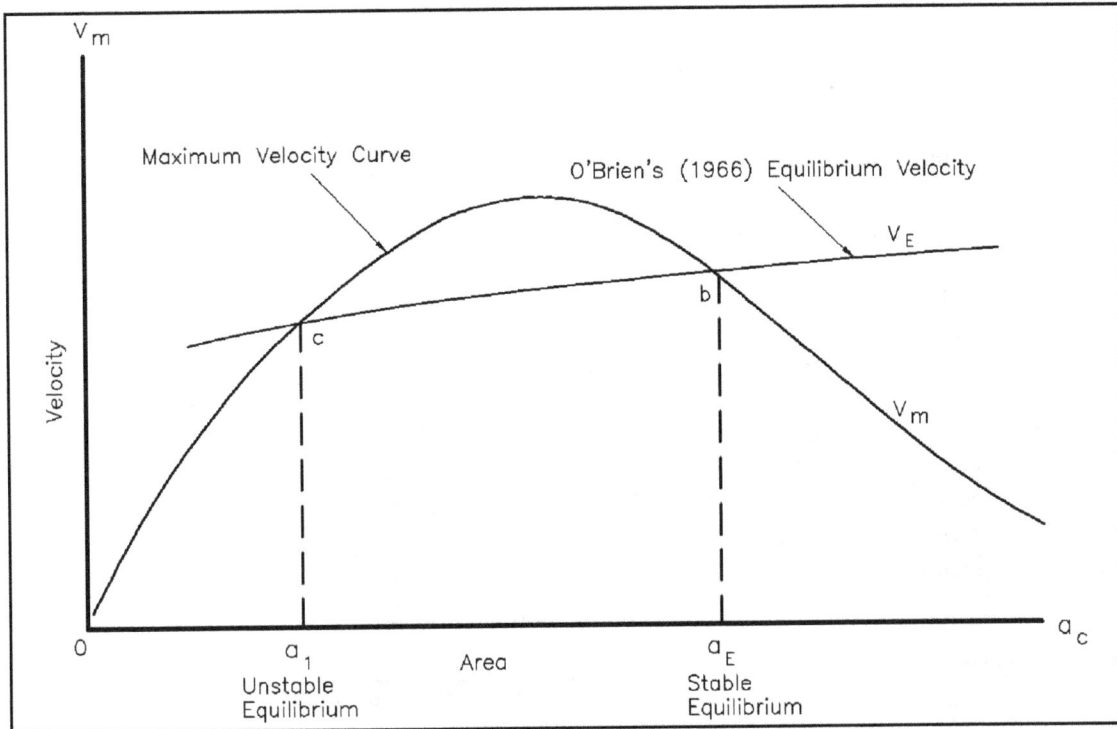

Figure II-6-42. Escoffier (1940) diagram, maximum velocity and equilibrium velocity versus inlet cross-sectional area

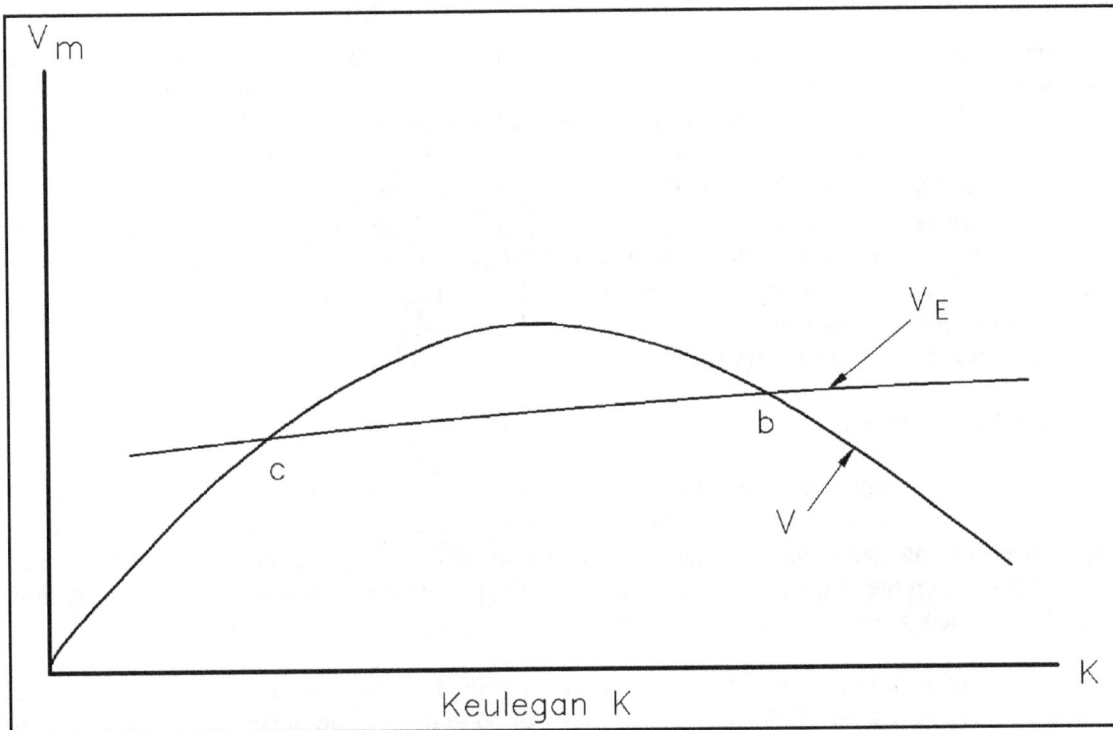

Figure II-6-43. Another version of Escoffier's (1977) diagram, maximum velocity and equilibrium velocity versus Keulegan repletion coefficient K

where

β = stability index (units of length^5/time^3)

a_E = equilibrium cross-sectional area of throat

V_{max} = maximum velocity in the throat

V_T = threshold velocity for sand transport

a_c = critical cross-sectional area (as from Figure II-6-42)

(4) Czerniak (1977) applied this technique successfully at Moriches Inlet, New York. Bruun and Gerritsen (1960) and Bruun, Mehta, and Jonsson (1978) developed an overall stability criteria for tidal inlets. They based their relationships on a ratio P/M, where P is the tidal prism and M is the total annual littoral drift (typical values for M and their determination are discussed in Part III). The stability of an inlet is rated good, fair, or poor as given in Table II-6-4.

Table II-6-4	
Inlet Stability Ratings	
$P/M_{tot} \geq 150$	Conditions are relatively good, little bar and good flushing
$100 \leq P/M_{tot} \leq 150$	Conditions become less satisfactory, and offshore bar formation becomes more pronounced
$50 \leq P/M_{tot} \leq 100$	Entrance bar may be rather large, but there is usually a channel through the bar
$20 \leq P/M_{tot} < 50$	All inlets are typical "bar-bypassers"
$P/M_{tot} \leq 20$	Descriptive of cases where the entrances become unstable "overflow channels" rather than permanent inlets

(5) Another aspect of inlet stability was discussed by Vincent, Corson, and Gingench (1991) in which inlet channel geographical location and horizontal topology were studied from aerial photography. They defined four stability indices, including minimum inlet width W, channel length L, change in geographical position N, and orientation variability of inlet channel E. Figure II-6-44 shows types of geographic channel instability. Figure II-6-45 shows the Little River Inlet, South Carolina, geographical variability. A stability limit was chosen so as to classify an inlet as stable or unstable. These limits were based on rate of change of width, length, position, and orientation. Records of 51 inlets over various time periods were examined and a change of 100 ft/month was selected as a reasonable arbitrary limit for stability in length and width of W and L. The 100-ft/month value was also selected as a divider between stable and unstable channel position. Most values examined were below the 100 ft/month rate.

d. Scour hole problems.

(1) Scour hole problems can occur at a variety of inlet locations and much remains to be understood about the mechanisms causing scour. Scour holes form as a result of the interaction of tidal currents, waves, wave-generated currents, sediment and adjacent structures (usually jetties). Scour holes have been observed at the tips of jetties, on the outside of the jetty trunk, and along the inner section of a jetty. Lillycrop and Hughes (1993) review scour problems at Corps projects.

(2) Scour at jetty tips (Figure II-6-47) can be caused by flow separation during flood flow, when turbulent eddies are generated. Ebb currents traversing through this region of scour also might contribute

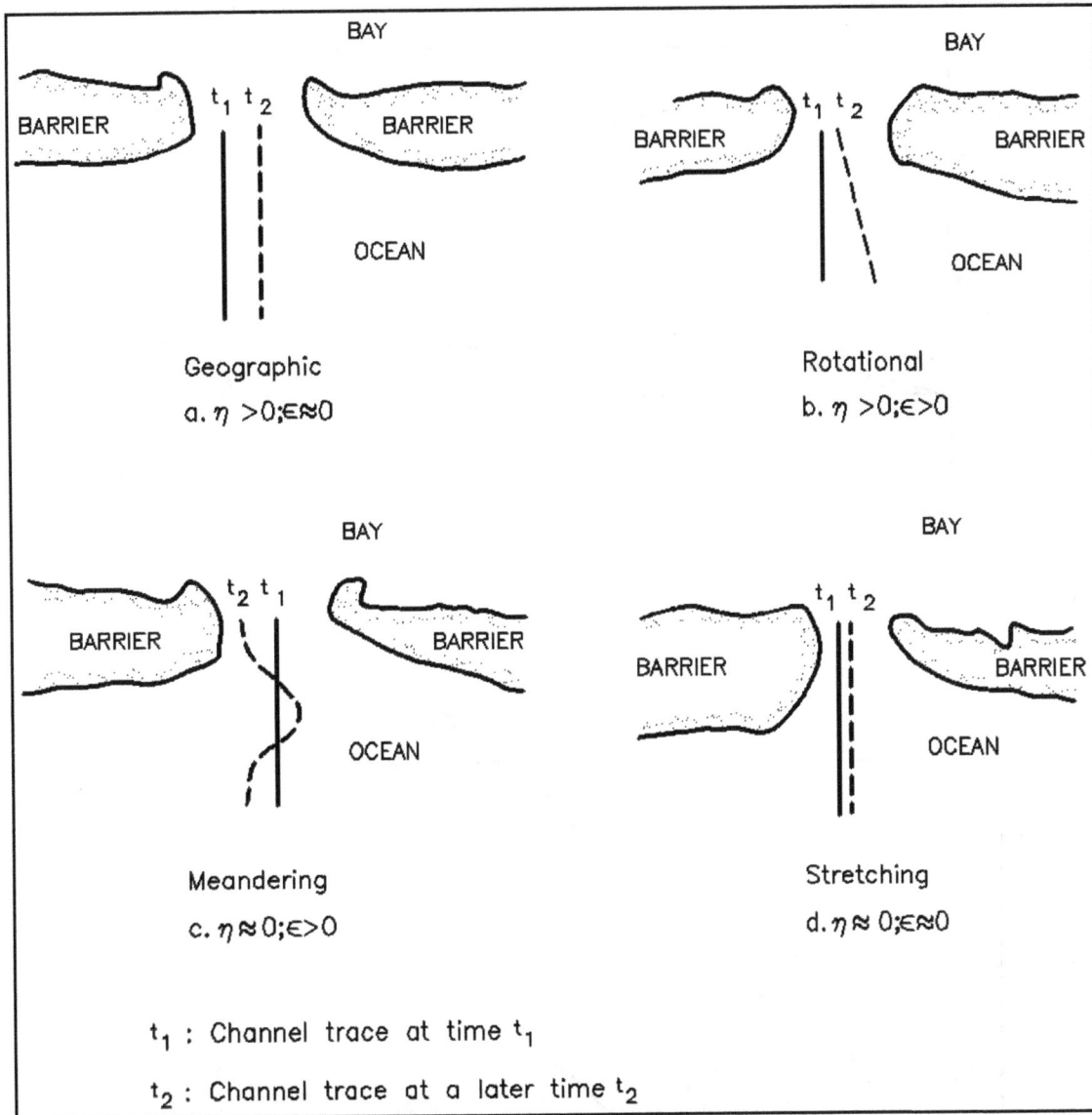

Figure II-6-44. Examples of channel geographic instability

to maintaining a scour hole if channelization in the shoreward portion of the inlet throat directs ebb currents toward the scour hole.

(3) Scour along the inside of jetties can occur as the bathymetry of a newly constructed jetty system adjusts. Little River Inlet (Chasten and Seabergh 1992) had an initial scour hole that was a remnant of the construction process. After the conclusion of construction, scour increased in this area, but as the inlet adjusted, velocities in that scour region decreased.

e. Methods to predict channel shoaling.

(1) Introduction. Entrance channel infill rates can be calculated by a variety of methods (U.S. Army Corps of Engineers 1984). One method is presented to show the parameters involved. A method to determine entrance channel infill rates is called the "Oregon Inlet Method." It is an analytic/empirical approach documented in a U.S. Army Engineer District, Wilmington (1980) report concerning the dredging

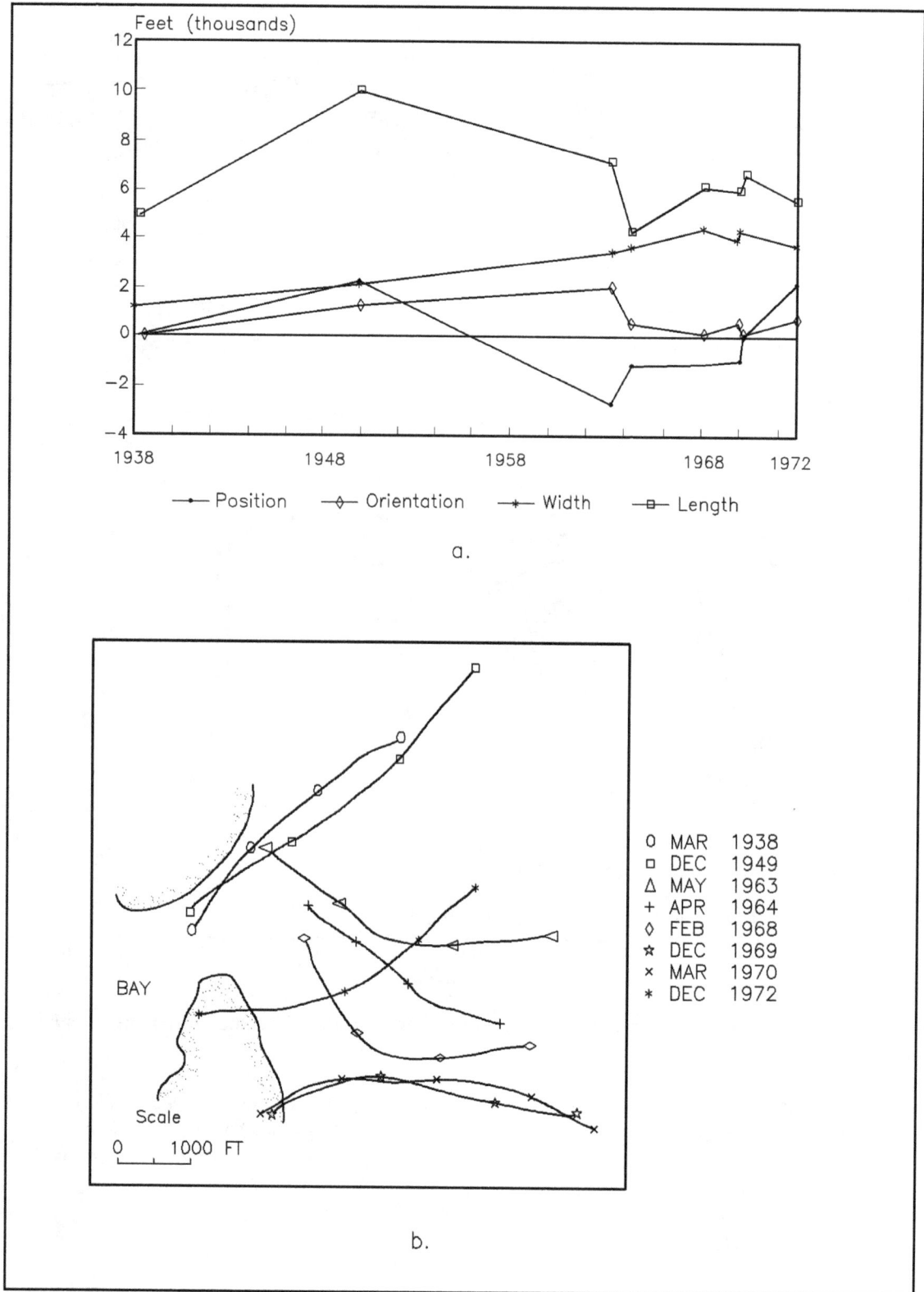

Figure II-6-45.   Little River Inlet geographical variability

EXAMPLE PROBLEM II-6-5

FIND:

Using the information provided in Example Problem 1, find the potential stability of the proposed channel cross section. Remember the channel has vertical sheet-pile walls, so its cross section can only change in the vertical.

SOLUTION:

By varying the cross-sectional area of the channel A_C, assuming that the channel width B remains constant and varying the channel depth d and recalculating the tidal prism as described above, the effect of channel area on the bay tidal prism can be evaluated and compared with the appropriate equation from Table II-6-3 (dual jettied inlets, $A_C = 7.489 \times 10^{-4} \, p^{0.86}$). This is done graphically in Figure II-6-46, which shows a plot of P versus A_C from the hydraulic response calculations and from the stability equation. The intersection of the two curves on the right side is the stable solution. It yields a channel cross-sectional area of 1,440 m² or a depth of 8 m. This shows that the 180-m by 3.7-m design channel would be unstable, with a strong tendency to erode.

Where the hydraulic response curve lies above the stability curve (as in the example) the tidal prism is too large for the inlet channel area and erosion will likely occur until a stable channel develops. If the hydraulic response curve crosses the stability curve twice, the lower point is an unstable equilibrium point from which the channel can either close or scour to the upper stability point. If the hydraulic response curve is substantially below the stability curve at all points, a stable inlet channel is unlikely to develop and the channel should eventually close.

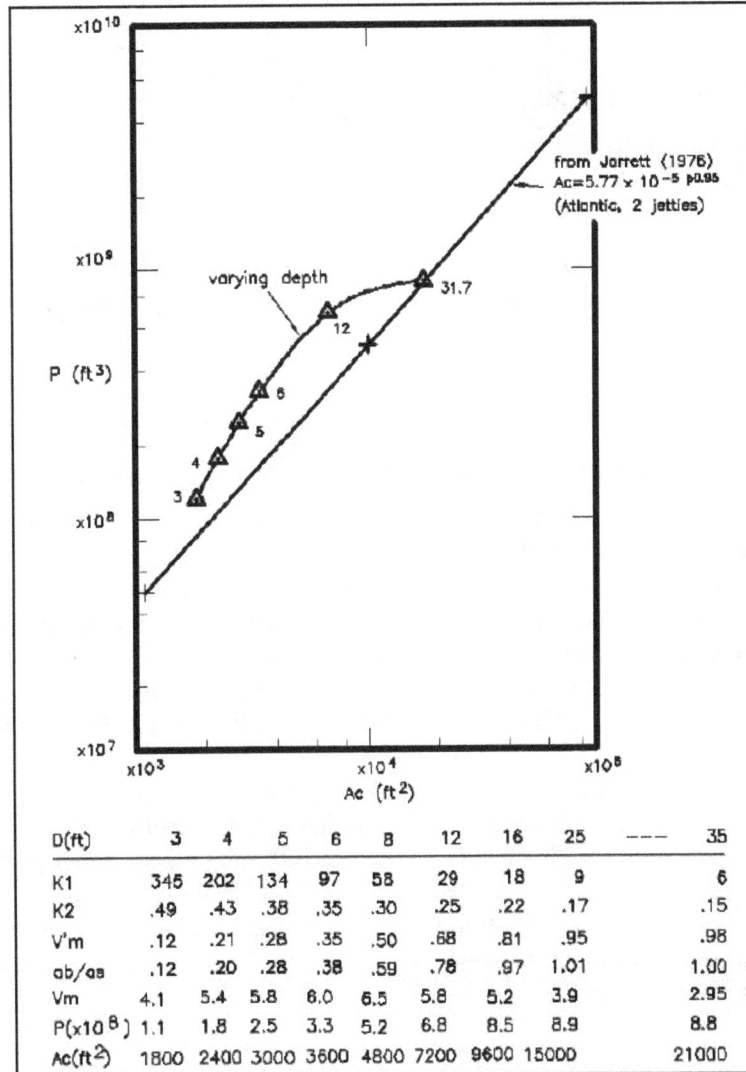

D(ft)	3	4	5	6	8	12	16	25	---	35
K1	345	202	134	97	58	29	18	9		6
K2	.49	.43	.38	.35	.30	.25	.22	.17		.15
V'm	.12	.21	.28	.35	.50	.68	.81	.95		.98
ab/as	.12	.20	.28	.38	.59	.78	.97	1.01		1.00
Vm	4.1	5.4	5.8	6.0	6.5	5.8	5.2	3.9		2.95
P(x10⁸)	1.1	1.8	2.5	3.3	5.2	6.8	8.5	8.9		8.8
Ac(ft²)	1800	2400	3000	3600	4800	7200	9600	15000		21000

Figure II-6-46. Example Problem II-6-4. Plot of P versus A_C from the hydraulic response calculations and from the stability equation

The stable inlet cross-sectional area depends on other factors (e.g., wave climate, monthly tidal range variations, surface runoff) besides the spring or diurnal tidal prism. As a result, the tidal prism-inlet area equations given in Table II-6-3 only serve as an indication of the approximate stable cross-sectional area. The analysis performed in the example demonstrates that the design channel is very likely to erode to a greater depth; however, that depth, which will fluctuate with time, can vary substantially from the indicated depth of 8 m.

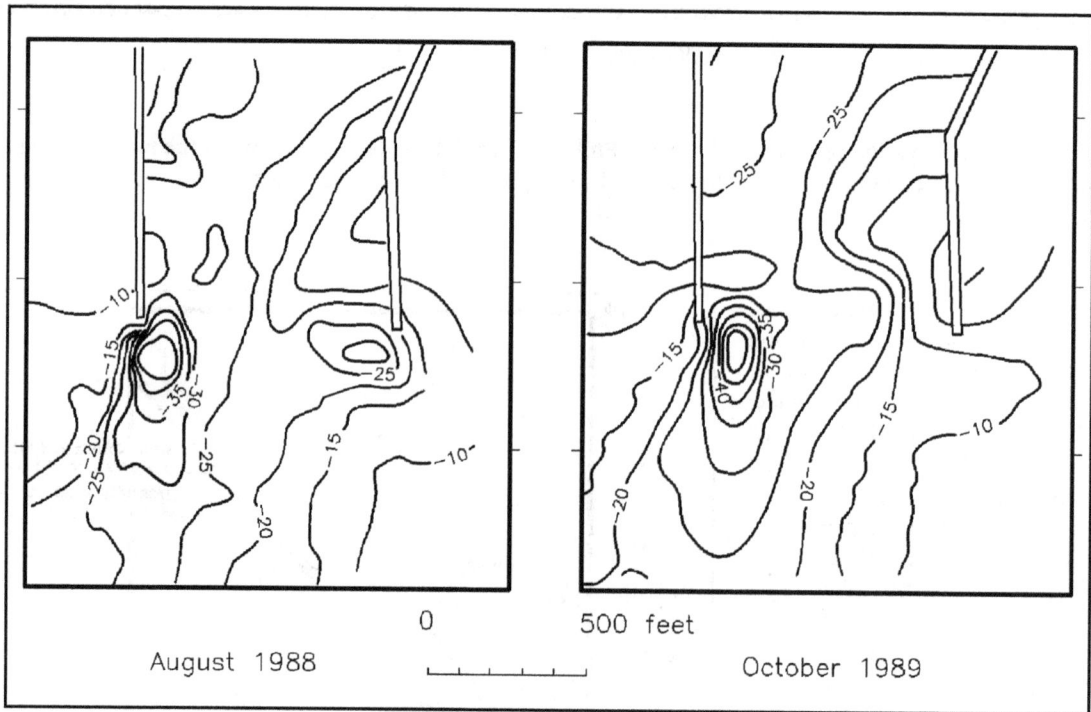

Figure II-6-47.   1988 and 1989 bathymetry of Moriches Inlet

of Oregon Inlet, North Carolina. It is based on the premise that the natural geometry of an inlet's ebb tide delta on the ocean bar results from the integrated effects of tidal currents, wave action, and associated sediment transport and deposition. Of particular importance is the natural elevation of the ocean bar, which represents both the limiting elevation of sediment accumulation resulting from the influx of littoral materials from adjacent shores and the level below which littoral sediments will collect if the ocean bar is dredged to provide a navigation channel. Sediment accumulation should increase with increasing channel depth until it becomes so deep that the channel traps all sediment crossing the inlet.

(2) Procedure. In order to develop a day-by-day simulation of siltation processes and dredging effects, a procedure consisting of three main phases was developed. The phases are (a) daily littoral materials transport volumes, (b) channel sedimentation, and (c) regression analysis.

(3) Phase I: Daily littoral materials transport volumes. It is assumed that the relative magnitude of alongshore sediment transport can be estimated simply on the basis of wave height squared in accordance with the relationship

$$Q_i = Q_g \left(\frac{H_i^2}{\sum_{i=1}^{n} H_i^2} \right)$$

(II-6-34)

where

i = number of the day during the year

Q_i = littoral transport occurring on day i [L^3]

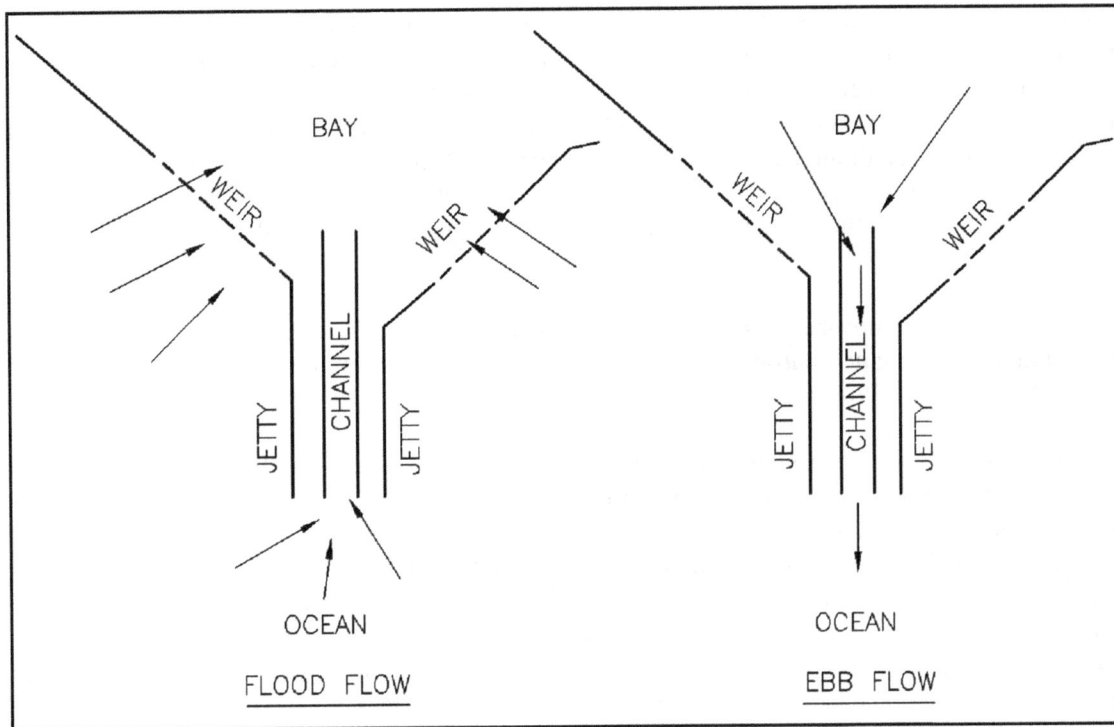

Figure II-6-48.   Weir jetty flow patterns

Q_g = total volume of alongshore sediment transport to the inlet each year [L³]

H_i = average wave height for the ith day of the year [L]

n = total number of days during the year [T]

(4) Phase II: Channel sedimentation. The first portion of the channel sedimentation procedure is

$$Transport\ ratio\ = \left(\frac{d_1}{d_2}\right)^{\frac{5}{2}} \tag{II-6-35}$$

where the transport ratio equals the ratio of the sediment transport potential in the dredged cut to the sediment transport on the bar before dredging and

d_1 = depth before dredging [L]

d_2 = depth after dredging [L]

On a daily basis, the amount of material that would be retained in the ocean bar channel of the inlet is given by:

$$Daily\ shoal\ volume\ =\ CQ_g \left(\frac{H_i^2}{\sum\limits_{i=1}^{n} H_i^2}\right) \left[1\ -\ \left(\frac{d_1}{d_2}\right)^{\frac{5}{2}}\right]\ \ [L^3] \tag{II-6-36}$$

in which all terms have been previously defined except for "C," which is termed the "potential shoaling factor." As stated previously, the quantity of material entering and depositing in the channel increases with the depth of the cut. However, until sufficient channel depth is attained to intercept all materials moving into the inlet, some material continues to bypass the inlet via the sloping face of the ocean bar below the bottom of the channel cut. Accordingly, C is defined as the portion of the total alongshore transport to the inlet entering the domain of the channel. In order to determine C, a regression analysis was conducted to correlate the inlet bar siltation with those factors judged to be most influential in the filling and flushing of such channels.

(5) Phase III: Regression analysis. The following three factors were selected as dominant in influencing the magnitude of shoaling at a particular inlet site having a dredge-maintained ocean bar channel. These factors are described as follows:

(a) Ebb tide flow is the primary factor acting to flush intrusive littoral materials from the inlet environment. Its influence in the analysis is represented by the symbol $E_{\Delta T}$, which is the difference between the mean ebb tide flow energy flux across the ocean bar at its natural elevation and the mean ebb tide flow energy flux through the cross section of the excavated ocean bar navigation channel. It is assumed that the tidal discharge is not significantly altered from one condition to the other. A basic concept in the tidal energy flux difference is that the tidal flow velocities directed seaward over the ocean bar at its natural elevation, in combination with wave agitation, are rapid enough to prevent accumulation of sediments above the natural bar depth. If a section of the ocean bar is deepened by a navigation channel, the related average local flow velocity is diminished and sediment deposition is initiated.

(b) Wave energy reaching the littoral zones adjacent to the inlet is the primary factor controlling the quantity of littoral material moving toward the inlet and, as such, determines the shoaling characteristics of an ocean bar navigation channel. In the analysis, the unrefracted wave energy flux per unit width of wave crest offshore of the area of interest, designated E_w, is the basic measure of sediment transport toward the inlet.

(c) The depth of an ocean bar channel determines the degree to which a channel will trap the littoral sediments entering the inlet environment. In the analysis, the amount of channel entrenchment, and hence the measure of sediment entrapment potential, is taken as the ratio D_R of the depth of the channel to the depth at which the seaward slope of the ocean bar meets the sea bottom. Each of these depths is measured from the natural elevation of the ocean bar; therefore, the ratio D_R represents the extent to which the ocean bar's seaward slope has been incised by the channel.

(6) Normalized, independent filling index. The channel sedimentation potential increases or decreases as each of the three factors $E_{\Delta T}$, E_w, and D_R increases or decreases, respectively. The factors were combined as follows to establish a normalized, independent variable F_I (filling index) for the regression analysis.

$$F_I = \frac{(E_{\Delta T} \cdot E_W \cdot D_R)}{10^{14}} \quad \left[\frac{ML^2}{T^2} \right] \tag{II-6-37}$$

The normalized, dependent variable selected for the analysis was the ratio of the volume of channel infill to the computed volume of the total alongshore sediment influx to the inlet multiplied by 100. This percentage value V_R is referred to as the "volume ratio." The regression analysis was based on data from four dredged-maintained inlets within the boundaries of the Wilmington District: Oregon, Beaufort, Masonboro, and Lockwoods Folly Inlets. Information available consisted of: (a) measured tidal discharges or inlet throat cross-sectional areas, which permitted tidal discharge computations by means of tidal prism-inlet area relations, (b) site wave statistics representing one or more years of wave gauge records, (c) detailed alongshore sediment transport analyses for the shorelines adjacent to each of the inlets, and (d) one or more sets of inlet hydrographic surveys, each set consisting of two surveys taken at different times, thereby allowing measurement of the volume of sediment filling within the navigation channels in the time periods between surveys.

(7) Ebb tidal energy flux. For each inlet selected, the parameter $E_{\Delta T}$, representing the change of ebb tidal energy flux across the ocean bar between natural and channel conditions, was computed and fixed as constant for the given inlet using the following equation:

$$E_{\Delta T} = \frac{4T}{3\pi} \, q_{max}^3 \, \frac{d_2^2 - d_1^2}{d_1^2 \, d_2^2} \qquad (II\text{-}6\text{-}38)$$

where

T = tidal wave period [T]

q_{max} = maximum instantaneous ebb tide discharge per unit width [L³/T/L]

d_2 = depth of navigation channel [L]

d_1 = natural depth of ocean bar [L]

The next parameter is E_w, the average daily wave energy flux per unit width of wave crest, offshore of the site of interest. This parameter represents the intensity of sediment transport from the adjacent beaches to the inlet environment. Daily wave energy flux can be expressed as

$$E_W = K \times 10^6 \, (H_0)^2 \, T \quad \left[\frac{ML^2}{T^2}\right] \qquad (II\text{-}6\text{-}39)$$

where

K = constant (K = 1.769 slugs/sec in English engineering units or 0.121 Kg/sec) [M/T]

H_o = deepwater wave height equivalent to observed shallow-water wave, if unaffected by refraction and friction [L]

T = wave period [T]

The last parameter is a depth ratio D_R, which is given as

$$D_R = \frac{d_2 - d_1}{d_3 - d_1} \qquad (II\text{-}6\text{-}40)$$

where

d_3 = water depth where seaward tip of ocean bar meets offshore sea bottom [L]

(8) Hoerls special function distribution. Tests of several regression equations and related curves revealed that the best fit of the data from the four inlets was attained by a "Hoerls" special function distribution, given in general form by:

$$V_R = a \ F_I^b \ e^{cF} \ I \tag{II-6-41}$$

where a, b, and c are the best-fit coefficients. The regression equation was solved for filling index F_I values corresponding to various channel depths, which allows for the proper "C" value to be used in the "daily shoal volume" equation developed in Phase II of this method.

    f.   Inlet weir jetty hydraulic and sediment interaction. The subject of weir jetty design is dealt with in detail by Seabergh (1983) and Weggel (1981), who discuss several hydraulic and sedimentary aspects including flow over the weir and its elevation, location, length, jetty alignment and orientation, as well as the shoreward end of the weir. The function of the weir jetty is to ideally collect the <u>net</u> sediment movement behind the weir section, where it can then be handled in protected waters and bypassed. A sedimentation benefit may occur in that sediment-laden longshore currents are kept away from the channel. Also, the ebb currents may be hydraulically enhanced, as seen in Figure II-6-48, creating ebb dominance of channel currents, which can aid in flushing sediment seaward.

    (1) Weir location. With respect to tidal currents within the jetty system, the farther the weir is from the navigation channel, the less likely it is to capture channel ebb currents that are directed seaward. This can depend on the location of the predominant ebb channels and their orientation. Care must be taken in evaluation of ebb flow direction because once the jetty system is constructed, the channel orientation may change due to removal of wave effects and consequent sediment movement, which, for example, may have deflected the ebb channel downcoast.

    (2) Weir length. Primary transport over the weir exists at its intersection with the shoreline. If the wave climate is mild, the weir should only be as long as necessary to prevent a chance of closure, something that has not yet been noted to occur in existing weir jetty systems. The length to prevent closure would need to be evaluated based on wave conditions, beach slope, orientation of the structure, etc. (Weggel 1981). If the wave climate is highly variable, the weir should probably extend further oceanward so as to include a large percentage of the breaker zone since heavy transport over the weir will occur at the breaker location for larger wave conditions; otherwise, the sediment will move offshore along the jetty. Another factor influencing weir length will be consideration of the amount of flow that is desired in the system. If a design objective is to obtain high ebb dominance of flow in the navigation channel, then the weir should be longer, if this would not interfere with other constraints, such as placement of a portion of the weir too close to the channel. The complete hydraulic flow situation must be considered to determine whether the additional flow provided by a wider weir will substantially augment ebb flow predominance and provide additional scouring ability in the entrance channel (see Seabergh (1983)).

II-6-4. References

Amein and Kraus 1991
Amein, A., and Kraus, N. C. 1991. "DYNLET1: Dynamic Implicit Numerical Model of One-Dimensional Tidal Flow Thought Inlets," Technical Report CERC-91-10, U.S. Army Engineer Waterways Experiment Station, Vicksburg, MS.

Aubrey 1986
Aubrey, D. G. 1986. Hydrodynamic Controls on Sediment Transport in Well-mixed Bays and Estuaries," J. van de Kreeke, ed., *Physics of Shallow Estuaries and Bays*, Springer-Verlag, New York.

Aubrey and Friedrichs 1988
Aubrey, D. G., and Friedrichs, C. T. 1988. "Seasonal Climatology of Tidal Non-Linearities on a Shallow Estuary," Lecture Notes on Coastal and Estuarine Studies, Vol 29, D. G. Aubrey and L. Weishar, eds., Springer-Verlag, New York.

Boer 1985
Boer, S. 1985. "The Flow Across Trenches at Oblique Angles to the Flow," Report 5490 Delft Hydraulics Laboratory, Delft, The Netherlands.

Boon 1988
Boon, J. D. 1988. "Temporal Varication of Shallow Water Tides in Basin-Inlet Systems," Lecture Notes on Coastal and Estuarine Studies, Vol 29, D. G. Aubrey and L. Weishar, eds., *Hydrodynamics and Sediment Dynamics of Tidal Inlets*, Springer-Verlag, New York.

Boon and Byrne 1981
Boon, J. D., and Byrne, R. J. 1981. "On Basin Hypsometry and the Morphodynamic Response of Coastal Inlet Systems," *Marine Geology*, Vol 40, pp 27-48.

Bottin 1978
Bottin, R. R. 1978. "Design for Harbor Entrance Improvements, Wells Harbor, Maine, Hydraulic Model Investigation," Technical Report H-78-18, U.S. Army Engineer Waterways Experiment Station, Vicksburg, MS.

Bottin 1982
Bottin, R. R. 1982. "Design for the Prevention of Shoaling, Flood Control, and Wave Protection, Rogue River, Oregon, Hydraulic Model Investigation," Technical Report HL-82-18, U.S. Army Engineer Waterways Experiment Station, Vicksburg, MS.

Brown 1928
Brown, E. I. 1928. "Inlets on Sandy Coasts," *Proceedings of the American Society of Civil Engineers*, Vol 54, Part I, pp 505-523.

Bruun 1967
Bruun, P. 1967. "Tidal Inlets Housekeeping," *Journal of the Hydraulic Division*, American Society of Civil Engineers, Vol 93, pp 167-184.

Bruun and Gerritsen 1960

Bruun, P., and Gerritsen, F. 1960. *Stability of Coastal Inlets*, North Holland Publishing Co., Amsterdam, The Netherlands.

Bruun, Mehta, and Jonsson 1978

Bruun, P., Mehta, A. J., and Jonsson, I. G. 1978. *Stability of Tidal Inlets: Theory and Engineering*, Elsevier Scientific Publishing Co., Amsterdam, The Netherlands.

Butler 1980

Butler, H. L. 1980. "Evolution of a Numerical Model for Simulating Long-Period Wave Behavior in Ocean Estuarine Systems," *Estuarine and Wetland Processes: With Emphasis on Modeling; Marine Science,* Vol 11, Plenum Press, New York.

Byrne, DeAlteris, and Bullock 1974

Byrne, R. J., DeAlteris, J. R., and Bullock, P. A. 1974. "Channel Stability in Tidal Inlets: A Case Study," *Proceedings, 14th Coastal Engineering Conference*, American Society of Civil Engineers, New York, pp 1585-1604.

Byrne, Gammisch and Thomas 1980

Byrne, R. J., Gammisch, R. A., and Thomas, G. R. 1980. "Tidal Prism-Inlet Area Relations for Small Tidal Inlets," *Proceedings of the Seventeenth Coastal Engineering Conference*, American Society of Civil Engineers, Sydney, Australia, Vol III, Ch 151, pp 23-28.

Chasten and Seabergh 1992

Chasten, M. A., and Seabergh, W. C. 1992. "Engineering Assessment of Hydrodynamics and Jetty Scour at Little River Inlet North and South Carolina," Miscellaneous Paper CERC-92-10, U.S. Army Engineer Waterways Experiment Station, Vicksburg, MS.

Chow 1959

Chow, V. T. 1959. *The Open-Channel Hydraulics*, McGraw-Hill Book Company, New York.

Cialone et al. 1991

Cialone, M. A., Mark, D. J., Chou, L. W., Leenknecht, D. A., Davis, J. A., Lillycrop. L. S., Jensen, R. E., Thompson, E., Gravens, M. B., Rosati, J. D., Wise, R. A., Kraus, N. C., and Larson, P. M. 1991. "Coastal Modeling System (CMS) User's Manual," Instruction Report CERC-91-1, U.S. Army Engineer Waterways Experiment Station, Vicksburg, MS.

Costa and Isaacs 1977

Costa, S. L., and Isaacs, J. D. 1977. "The Modification of Sand Transport in Tidal Inlet Inlets," *Coastal Sediments 77, Fifth Symposium of the Waterway, Port Coastal and Ocean Division,* American Society of Civil Engineers, Nov 2-4, Charleston, SC.

Czerniak 1977

Czerniak, M. T. 1977. "Inlet Interaction and Stability Theory Verification," *Coastal Sediments 77*, American Society of Civil Engineers, Charleston, SC.

Dean 1971
Dean, R. G. 1971. "Hydraulics of Inlets," COEL/UFL-71/019, Department of Coastal and Oceanographic Engineering, University of Florida, Gainsville.

DiLorenzo 1988
DiLorenzo, J. L. 1988. "The Overtide and Filtering Response of Small Inlet/Bay Systems," Lecture Notes on Coastal and Estuarine Studies, Vol 29, D. G. Aubrey and L. Weishar, eds., Springer-Verlag, New York.

Dyer and Taylor 1973
Dyer, K. R., and Taylor, P. A. 1973. "A Simple Segmented Prism Model of Tidal Mixing in Well-Mixed Estuaries," *Estuarine and Coastal Marine Science*, Vol 1, pp 411-418.

Escoffier 1940
Escoffier, F. F. 1940. "The Stability of Tidal Inlets," *Shore and Beach*, Vol 8, No. 4, pp 114-115.

Escoffier 1977
Escoffier, F. F. 1977. "Hydraulics and Stability of Tidal Inlets," GITI Report 13, U.S. Army Engineer Waterways Experiment Station, Vicksburg, MS.

Escoffier and Walton 1979
Escoffier, F. F., and Walton, T. L. 1979. "Inlet Stability Solutions for Tributary Inflow," *Journal of the Waterway, Port, Coastal and Ocean Division*, American Society of Civil Engineers, Vol 105, No. WW4, Proc. Paper 14964, pp 341-355.

Fischer, List, Koh, Imberger and Brooks 1979
Fischer, H. B., List, E. J., Koh, R. C. Y., Imberger, J., and Brooks, N. H. 1979. *Mixing in Inland and Coastal Waters*, Academic Press, New York.

French 1960
French, J. L. 1960. "Tidal Flow in Entrances," Technical Bulletin No. 3, U.S. Army Engineer Waterways Experiment Station, Committee on Tidal Hydraulics, Vicksburg, MS.

Friedrichs and Aubrey 1988
Friedrichs, C. T., and Aubrey, D. G. 1988. "Non-Linear Tidal Distortion in Shallow Well-Mixed Estuaries: A Synthesis," *Estuarine, Coastal and Shelf Science*, Vol 27, pp 521-545.

Friedrichs, Lynch, and Aubrey 1992
Friedrichs, C. T., Lynch, D. R., and Aubrey, D. G. 1992. "Velocity Asymmetries in Frictionally Dominated Tidal Embayments: Longitudinal and Lateral Variability," *Dynamics and Exchanges in Estuaries and the Coastal Zone, Coastal and Estuarine Studies*, D. Prandle, ed., American Geophysical Union, Washington, DC, Vol 40, pp 277-312.

Galvin 1971
Galvin, C. J., Jr. 1971. "Wave Climate and Coastal Processes," *Water Environments and Human Needs*, A. T. Ippen, ed., M.I.T. Parsons Laboratory for Water Resources and Hydrodynamics, Cambridge, MA, pp 48-78.

Geyer and Signell 1992
Geyer, W. R., and Signell, R. P. 1992. "A Reassessment of the Role of Tidal Dispersion in Estuaries and Bays," *Estuaries*, Vol 15, pp 97-108.

Harris and Bodine 1977

Harris, D. L., and Bodine, B. R. 1977. "Comparison of Numerical and Physical Hydraulic Models, Masonboro Inlet, North Carolina," GITI Report 6, U.S. Army Engineer Waterways Experiment Station, Vicksburg, MS.

Hayes 1971

Hayes, M. O. 1971. "Lecture Notes for Course on Inlet Mechanics and Design (unpublished)," 10-20 May 1971, Ft. Belvoir, VA, Coastal Engineering Research Center.

Hayes 1980

Hayes, M. O. 1980. "General Morphology and Sediment Patterns in Tidal Inlets," *Sedimentary Geology*, Vol 26, pp 139-156.

Herchenroder 1981

Herchenroder, B. E. 1981. "Effects of Currents on Waves," Coastal Engineering Technical Aid No. 81-14, October 1981, U.S. Army Engineer Waterways Experiment Station, Vicksburg, MS.

Ismail and Wiegel 1983

Ismail, N. M., and Wiegel, R. L. 1983. "Opposing Wave Effect on Momentum Jet Spreading Rate," *Journal of the Waterway Port, Coastal and Ocean Division*, American Society of Civil Engineers, Vol 109, No. WW4, pp 465-487.

Jarrett 1975

Jarrett, J. T. 1975. "Analyses of the Hydraulic Characteristics of Tidal Inlets," Unpublished memorandum for record, U.S. Army Engineer Waterways Experiment Station, Vicksburg, MS.

Jarrett 1976

Jarrett, J. T. 1976. "Tidal Prism-Inlet Area Relationships, GITI Report 3, U.S. Army Engineer Waterways Experiment Station, Vicksburg, MS.

Jonsson, Skovgaard, and Wang 1970

Jonsson, I. G., Skovgaard, C., and Wang, J. D. 1970. "Interaction Between Waves and Currents," *Proceedings of the 12th Conference on Coastal Engineering*, American Society of Civil Engineers, Vol 1, pp 489-509.

Joshi 1982

Joshi, P. B. 1982. "Hydromechanics of Tidal Jets," *Journal of the Waterway, Port, Coastal and Ocean Division*, American Society of Civil Engineers, Vol 108, No. WW3, Proc. Paper 17294, pp 239-253.

Joshi and Taylor 1983

Joshi, P. B., and Taylor, R. B. 1983. "Circulation Induced by Tidal Jets," *Journal of the Waterway, Port, Coastal and Ocean Division*, American Society of Civil Engineers, Vol 109, No. WW4.

Keulegan 1951

Keulegan, G. H. 1951. "Third Progress Report on Tidal Flow in Entrances, Water Level Fluctuations of Basins in Communication with Seas," Report No. 1146, National Bureau of Standards, Washington, DC.

Keulegan 1967
Keulegan, G. H. 1967. "Tidal Flow in Entrances Water-Level Fluctuations of Basins in Communications with Seas," Technical Bulletin No. 14, Committee on Tidal Hydraulics, U.S. Army Engineer Waterways Experiment Station, Vicksburg, MS.

King 1974
King, D. B. 1974. "The Dynamics of Inlets and Bays," Technical Report No. 2, Coastal and Oceanographic Engineering Laboratory, University of Florida, Gainesville.

Leenknecht, Szuwalski, and Sherlock 1992
Leenknecht, D. A., Szuwalski, A., and Sherlock, A. R. 1992. "Automated Coastal Engineering System User Guide and Technical Reference, Version 1.07," U.S. Army Engineer Waterways Experiment Station, Vicksburg, MS.

Lillycrop and Hughes 1993
Lillycrop, W. J., and Hughes, S. A. 1993. "Scour Hole Problems Experienced by the Corps of Engineers; Data Presentation and Summary," Miscellaneous Paper CERC-93-2, U.S. Army Engineer Waterways Experiment Station, Vicksburg, MS.

Mason 1975
Mason, C. 1975. "Tidal Inlet Repletion Coefficient Variability," unpublished report, copy available at Coastal and Hydraulics Laboratory, U.S. Army Engineer Waterways Experiment Station, Vicksburg, MS.

Mayor-Mora 1973
Mayor-Mora, R. E. 1973. "Hydraulics of Tidal Inlets on Sandy Coasts," HEL-24-16, Hydraulic Engineering Laboratory, University of California, Berkeley.

Mehta 1990
Mehta, A. J. 1990. "Significance of Bay Superelevation in Measurement of Sea Level Change," *Journal of Coastal Research*, Vol 6, No. 4, pp 801-813.

Mehta and Joshi 1984
Mehta, A. J., and Joshi, P. B. 1984. "Review of Tidal Inlet Hydraulics UFL/COBL - TR/054," Coastal and Oceanographic Engineering Department, University of Florida, Gainsville.

Mehta and Özsöy 1978
Mehta, A. J., and Özsöy, E. 1978. "Inlet Hydraulics," *Stability of Tidal Inlets: Theory and Engineering*, P. Bruun, Elsevier Scientific Publishing Co., Amsterdam, The Netherlands, pp 83-161.

Mota Oliveira 1970
Mota Oliveira, I. B. 1970. "Natural Flushing Ability in Tidal Inlets," *Proceedings of the Twelfth Coastal Engineering Conference*, American Society of Civil Engineers, Washington, DC, Vol 3, Ch. 111, pp 1827-1845.

O'Brien 1931
O'Brien, M. P. 1931. "Estuary Tidal Prisms Related to Entrance Areas," *Civil Engineering*, pp 738-739.

O'Brien 1969
O'Brien, M. P. 1969. "Equilibrium Flow Areas of Inlets on Sandy Coasts," *Journal of the Waterways and Harbors Division*, American Society of Civil Engineers, No. WWI, pp 43-52.

O'Brien and Clark 1974
O'Brien, M. P., and Clark, R. R. 1974. "Hydraulic Constants of Tidal Entrances," *Proceedings of the Fourteenth Coastal Engineering Conference*, American Society of Civil Engineers, Copenhagen, Denmark, Vol 2, Ch. 90, pp 1546-1565.

O'Brien and Dean 1972
O'Brien, M. L., and Dean, R. G. 1972. "Hydraulics and Sedimentary Stability of Coastal Inlets," *Proceedings of the Thirteenth Coastal Engineering Conference,* American Society of Civil Engineers, Vancouver, Canada.

Özsöy 1977
Özsöy, E. 1977. "Flow and Mass Transport in the Vicinity of Tidal Inlets," Technical Report No. TR-036, Coastal and Oceanographic Engineering Laboratory, University of Florida, Gainesville.

Özsöy and Ünlüata 1982
Özsöy, E., and Ünlüata, E. 1982. "Ebb-Tidal Flow Characteristics Near Inlets," *Estuarine, Coastal and Shelf Science*, Vol 14, No. 3, pp 251-263.

Seabergh 1975
Seabergh, W. C. 1975. "Improvement for Masonboro Inlet, North Carolina, Hydraulic Model Investigation, Vols I and II," Technical Report H-76-4, U.S. Army Engineer Waterways Experiment Station, Vicksburg, MS.

Seabergh 1983
Seabergh, W. C. 1983. "Weir Jetty Performance: Hydraulic and Sedimentary Considerations, Hydraulic Model Investigation," Technical Report HL-83-5, U.S. Army Engineer Waterways Experiment Station, Vicksburg, MS.

Seabergh and Lane 1977
Seabergh, W. C., and Lane, E. F. 1977. "Improvements for Little River Inlet, South Carolina, Hydraulic Model Investigation," Technical Report H-77-21, U.S. Army Engineer Waterways Experiment Station, Vicksburg, MS.

Seabergh and McCoy 1982
Seabergh, W. C., and McCoy, J. W. 1982. "Design for Prevention of Shoaling at Little Lake Harbor, Michigan, Hydraulic Model Investigation," Technical Report HL-82-16, U.S. Army Engineer Waterways Experiment Station, Vicksburg, MS.

Seelig and Sorensen 1977
Seelig, W. N., and Sorensen, R. M. 1977. "Hydraulics of Great Lakes Inlets," Technical Paper No. 77-8, U.S. Army Engineer Waterways Experiment Station, Vicksburg, MS.

Shemdin and Forney 1970
Shemdin, O. H., and Forney, R. M. 1970. "Tidal Motion in Bays," *Proceedings of the Twelfth Coastal Engineering Conference*, American Society of Civil Engineers, Washington, DC, Vol 3, Ch. 134, pp 2225-2242.

Smith 1977
Smith, N. P. 1977. "Meteorological and Tidal Exchanges Between Corpus Christi Bay, Texas, and Northwestern Gulf of Mexico," *Estuarine and Coastal Marine Science*, Vol 5, pp 511-520.

Smith 1991
Smith, J. B. 1991. "Morphodynamics and Stratigraphy of Essex River Ebb-Tidal Delta: Massachusetts," Technical Report CERC-91-11, U.S. Army Engineer Waterways Experiment Station, Vicksburg, MS.

Sorensen 1977
Sorensen, R. M. 1977. "Procedures for Preliminary Analysis of Tidal Inlet Hydraulics and Stability," Coastal Engineering Technical Aid 77-8, U.S. Army Engineer Waterways Experiment Station, Vicksburg, MS.

Sorensen 1980
Sorensen, R. M. 1980. "The Corps of Engineer's General Investigation of Tidal Inlets," *Proceedings of the Seventeenth Coastal Engineering Conference*, American Society of Civil Engineers, Sydney, Australia, Vol III, Ch. 154, pp 2565-2580.

Sorensen and Seelig 1976
Sorensen, R. M., and Seelig, W. N. 1976. "Hydraulics of Great Lakes Inlet-Harbor Systems," *Proceedings of the Fifteenth Coastal Engineering Conference*, American Society of Civil Engineers, Honolulu, HI, Vol 2, Ch. 96, pp 1646-1665.

Speer and Aubrey 1985
Speer, P. E., and Aubrey, D. G. 1985. "A Study of Non-Linear Tidal Propagation in Shallow Inlet/Estuarine Systems; Part II: Theory," *Estuarine, Coastal and Shelf Science*, Vol 21, pp 207-224.

Taylor and Dean 1974
Taylor, R. B., and Dean, R. G. 1974. "Exchange Characteristics of Tidal Inlets," *Proceedings of the Fourteenth Coastal Engineering Conference*, American Society of Civil Engineers, Copenhagen, Denmark, Vol 3, Ch. 132, pp 2268-2289.

U. S. Army Corps of Engineers 1984
U.S. Army Corps of Engineers. 1984. "Entrance Channel Infill Rates," Engineer Technical Letter No. 1110-2-293, March 1984, Washington, DC.

U. S. Army Engineer District, Wilmington 1980
U.S. Army Engineer District, Wilmington. 1980. "Manteo (Shallowbag) Bay, North Carolina," General Design Memorandum, Phase II, Appendix 5, Wilmington, NC.

Ünlüata and Özsöy 1977
Ünlüata, U. A., and Özsöy, E. 1977. "Tidal Jet Flows Near Inlets," *Proceedings of the Hydraulics in the Coastal Zone Conference*, American Society of Civil Engineers, Texas A&M University, College Station, TX, pp 90-98.

van de Kreeke 1967
van de Kreeke, J. 1967. "Water Level Fluctuations and Flows in Tidal Inlets," *Journal of the Waterways, Harbors and Coastal Engineering Division*, American Society of Civil Engineers, Vol 93, No. WW4, Proc. Paper 5575, pp 97-106.

van de Kreeke 1976
van de Kreeke, J. 1976. "Increasing the Mean Current in Coastal Channels." *Journal of the Waterways Harbors and Coastal Engineering Division*, American Society of Civil Engineers, Vol 102, No. WW2, pp 222-234.

van de Kreeke 1978
van de Kreeke, J. 1978. "Mass Transport in a Coastal Channel, Marco River, Florida," *Estuarine and Coastal Marine Science,* No. 7, pp 203-214.

van de Kreeke 1983
van de Kreeke, J. 1983. "Residence Time: Application to Small Boat Basins," *Journal of Waterway, Port, Coastal and Ocean Engineering,* American Society of Civil Engineers, Vol 109, pp 416-428.

van de Kreeke 1988
van de Kreeke, J. 1988. "Hydrodynamics of Tidal Inlets," Lecture Notes on Coastal and Estuarine Studies, Vol 29, *Hydrodynamic and Sediment Dynamics of Tidal Inlets,* Springer-Verlag, New York, pp 1-23.

Vincent and Corson 1980
Vincent, C. L., and Corson, W. D. 1980. "The Geometry of Selected US Tidal Inlets," GITI Report 20, U.S. Army Corps of Engineers, Washington, DC.

Vincent and Corson 1981
Vincent, C. L., and Corson, W. D. 1981. "Geometry of Tidal Inlets: Empirical Equations," *Journal of the Waterway, Port, Coastal and Ocean Division,* American Society of Civil Engineers, Vol 107, No. WW1, Proc. Paper 16032, pp 1-9.

Vincent, Corson, and Gingerich 1991
Vincent, C. L., Corson, W. D., and Gingerich, K. J. 1991. "Stability of Selected United States Tidal Inlets," General Investigation of Tidal Inlets (GITI) Report 21, U.S. Army Engineer Waterways Experiment Station, Vicksburg, MS.

Walton and Escoffier 1981
Walton, T. L., and Escoffier, F. F. 1981. "Linearized Solution to Inlet Equation with Inertia," *Journal of the Waterway, Port, Coastal and Ocean Division,* Paper 16414, Vol 105, No. WW4, pp 191-195.

Ward 1982
Ward, G. H., Jr. 1982. "Pass Cavallo, Texas: Case Study of Tidal-Prism Capture, *Journal of the Waterway Port Coastal and Ocean Division,* American Society of Civil Engineers, Vol 108, No. WW4, pp 513-525.

Watson and Behrens 1976
Watson, R. L., and Behrens, E. W. 1976. "Hydraulics and Dynamics of New Corpus Christi Pass, Texas: A Case History, 1973-1975," GITI Report 9, U.S. Army Engineer Waterways Experiment Station, Vicksburg, MS.

Weggel 1981
Weggel, J. R. 1981. "Weir Sand-Bypassing Systems," *Special Report No. 8,* U.S. Army Engineer Waterways Experiment Station, Vicksburg, MS.

Wicker 1965
Wicker, C. F., ed. 1965. "Evaluation of Present State of Knowledge of Factors Affecting Tidal Hydraulics and Related Phenomena," Committee on Tidal Hydraulics, Report No. 3, U.S. Army Corps of Engineers, Washington, DC.

Zimmerman 1988
Zimmerman, J. T. F. 1988. "Estuarine Residence Times," *Hydrodynamics of Estuaries; Vol 1: Estuarine Physics,* B. Kjerfve, ed., CRC Press, Boca Raton, FL, pp 75-84.

II-6-5. Definitions of Symbols

β	Stability index of an inlet (Equation II-6-33) [length^5/time^3]
Δ	Bay superelevation (Equations II-6-20 and II-6-21) [length]
ε	Fraction of new water entering the bay from the sea each tidal cycle [dimensionless]
ε	Phase lag between sea tide and bay tide [deg]
θ	Angle wave orthogonal makes with current entering an inlet [deg]
μ	Friction factor
ξ	Dimensionless distance offshore parameter
τ_r	Residence time (Equation II-6-28)
a_0	Ocean tide amplitude (one-half the ocean tide range) [length]
A_{avg}	Average area over a channel length [length^2]
a_b	Bay tide amplitude (one-half the bay tide range) [length]
A_b	Surface area of a bay or basin [length^2]
a_c	Critical cross-sectional area [length^2]
A_c	Channel cross-sectional flow area of an inlet [length^2]
a_E	Equilibrium cross-sectional area of inlet throat [length^2]
A_{mw}	Cross-sectional area of an inlet at the minimum inlet width [length^2]
B	Channel width between jetties [length]
C	Potential shoaling factor (Equation II-6-36)
d	Mean water depth in channel [length]
D	Depth of water at current meter location [length]
d_1, d_2, d_3	Natural depth of ocean bar, depth of navigation channel, and water depth where seaward tip of ocean bar meets offshore sea bottom [length]
d_b	Average bay depth [length]
D_R	Ratio of depth of the channel to the depth at which the seaward slope of the ocean bar meets the sea bottom (Equation II-6-40)
d_T	Channel depth [length]
$E_{\Delta T}$	Difference between the mean ebb tide flow energy flux across the ocean bar at its natural elevation and the mean ebb tide flow energy flux through the cross section of the excavated ocean bar navigation channel (Equation II-6-38)

E_W	Average daily wave energy flux per unit width of wave crest, offshore of the site of interest (Equation II-6-39)
f	Darcy - Weisbach friction term
f	Dimensionless parameter that is a function of the hydraulic radius R and Manning's roughness coefficient n (Equation II-6-16)
F	Inlet impedance (Equation II-6-17) [dimensionless]
F_I	Filling index (Equation II-6-37)
g	Gravitational acceleration [length/time^2]
h	Inlet channel water surface elevation [length]
H_A	Wave height entering inlet [length]
H_i	Average wave height for the ith day of the year [length]
K	Keulegan repletion coefficient [dimensionless]
K	Keulegan's dimensionless coefficient of repletion or filling (Equation II-6-4)
K_1 , K_2	King's inlet friction coefficient (Equation II-6-6) and King's inlet frequency coefficient (Equation II-6-7) [dimensionless]
k_{en} , k_{ex}	Inlet entrance and exit energy loss coefficient [dimensionless]
L	Inlet length [length]
L_b	Distance from a inlet to the farthest pint in the bay [length]
L_{mw}	Oceanside inlet channel length [length]
m	Bank slope [length-rise/length-run]
M	Total annual littoral drift [length^3/year]
n	Manning's roughness coeeficient
n	Total number of days during the year
$-_O$	The subscript 0 denotes deepwater conditions
P	Tidal prism filling the bay (Equation II-6-12) [length^3]
Q	Inlet discharge [length^3/time]
Q_g	Total volume of alongshore sediment transport to the inlet each year [length^3]
Q_i	Littoral transport occurring on day i (Equation II-6-34) [length^3]
q_{max}	Maximum instantaneous ebb tide discharge per unit width [length^3/time-length]
Q_r	River or freshwater inflow to a bay [length^3/time]
$Q_r{'}$	King's dimensionless variable (Equation II-6-19)

R	Inlet hydraulic radius [length]
R_H	Inlet current wave height factor (Figure II-6-34) [dimensionless]
T	Tidal period [time]
t_i	Duration of inflow [time]
$U_{max\,f}, U_{max\,e}$	Maximum channel flood and ebb velocity [length/time]
$U'_{max\,f}, U'_{max\,e}$	Dimensionless maximum channel flood and ebb velocity
V	Inlet channel velocity [length/time]
$\langle V \rangle$	Average volume of the bay over the tidal cycle [length^3/time]
V_{avg}	Maximum velocity averaged over entire inlet cross section [length/time]
V_m	Maximum cross-sectionally averaged velocity during a tidal cycle [length/time]
V_{max}	Maximum velocity in the inlet throat [length/time]
V_{meas}	Point measurement of maximum velocity [length/time]
V_R	Hoerls special function distribution (Equation II-6-41)
V_T	Threshold velocity for sand transport [length/time]
V'_m	King's dimensionless velocity (Equation II-6-5)
W	Channel width corresponding to mean water depth in channel [length]

II-6-6. Acknowledgments

Author of Chapter II-6, "Hydrodynamics of Tidal Inlets:"

William C. Seabergh, Coastal and Hydraulics Laboratory, Engineer Research and Development Center, Vicksburg, Mississippi.

Reviewers:

Lee E. Harris, Ph.D., Department of Marine and Environmental Systems, Florida Institute of Technology, Melbourne, Florida.

Tom Jarrett, U.S. Army Engineer District, Wilmington, Wilmington, South Carolina (retired).

Ashish J. Metha, Ph.D., Department of Civil and Coastal Engineering, University of Florida, Gainesville, Florida.

John Oliver, U.S. Army Engineer District, Portland, Portland, Oregon, (retired).

C. Linwood Vincent, Ph.D., Office of Naval Research, Arlington, Virginia.

Table of Contents

Page

II-7-1. **Introduction** II-7-1

II-7-2. **Wave Diffraction** II-7-3
    a. Definition of diffraction II-7-3
    b. Diffraction analysis II-7-3
    c. Diffraction at a harbor entrance II-7-3
        (1) Waves passing a single structure II-7-3
        (2) Waves passing through a structure gap II-7-6
    d. Irregular wave diffraction II-7-9
    e. Combined refraction-diffraction in harbors II-7-12
    f. Combined diffraction - reflection in harbors II-7-13

II-7-3. **Wave Transmission** II-7-19
    a. Definition of transmission II-7-19
    b. Transmission over/through structures II-7-19
        (1) Rubble-mound structures-subaerial II-7-19
        (2) Rubble-mound structures-submerged II-7-20
        (3) Permeable rubble-mound structures II-7-21
        (4) Floating breakwaters II-7-23
        (5) Wave barriers II-7-24

II-7-4. **Wave Reflection** II-7-26
    a. Definition of reflection II-7-26
    b. Reflection from structures II-7-28
    c. Reflection from beaches II-7-29
    d. Reflection patterns in harbors II-7-29
    e. Reflection problems at harbor entrances II-7-30

II-7-5. **Harbor Oscillations** II-7-31
    a. Introduction II-7-31
    b. Mechanical analogy II-7-33
    c. Closed basins II-7-34
    d. Open basins - general II-7-36
    e. Open basins - simple shapes II-7-36
    f. Open basins - complex shapes II-7-40
    g. Open basins - Helmholtz resonance II-7-44

II-7-6. **Flushing/Circulation** II-7-48
    a. Statement of importance II-7-48
    b. Flushing/circulation processes II-7-48
        (1) Tidal action II-7-48
        (2) Wind effects II-7-49
        (3) River discharge II-7-51
    c. Predicting of flushing/circulation II-7-51
        (1) Numerical models II-7-51
        (2) Physical model studies II-7-52
        (3) Field studies II-7-53

II-7-7. **Vessel Interactions** II-7-54
    a. Vessel-generated waves II-7-54
    b. Vessel motions II-7-57
        (1) Response to waves II-7-57
        (2) Response to currents II-7-58
        (3) Wave-current interaction II-7-58
        (4) Vessel sinkage and trim II-7-59
        (5) Ship maneuverability in restricted waterways II-7-62
    c. Mooring II-7-64
        (1) Wave forcing mechanism II-7-64
        (2) Mooring configurations II-7-64
        (3) Mooring lines II-7-65
        (4) Fenders II-7-65
        (5) Surge natural period II-7-65
        (6) Mooring forces II-7-67

II-7-8. **References** II-7-73

II-7-9. **Definitions of Symbols** II-7-87

II-7-10. Acknowledgments II-7-91

II-7-11. Note to Users, Vessel Buoyancy II-7-91

List of Tables

Page

Table II-7-1.   Wave Reflection Equation Coefficient Values Structure II-7-28

Table II-7-2.   Harbor Oscillation Characteristics II-7-32

Table II-7-3.   Flushing Characteristics of Small-Boat Harbors II-7-50

Table II-7-4.   Advantages of Physical and Numerical Models II-7-53

Table II-7-5.   Selected Vessel-Generated Wave Heights II-7-57

Table II-7-6.   Drag Coefficients for Wind Force II-7-70

List of Figures

Page

Figure II-7-1.   Harbor siting classifications II-7-2

Figure II-7-2.   Wave diffraction, definition of terms II-7-4

Figure II-7-3.   Wave diffraction diagram - 60^0 wave angle II-7-5

Figure II-7-4.   Wave diffraction through a gap II-7-6

Figure II-7-5.   Contours of equal diffraction coefficient gap width = 0.5 wavelength
(B/L = 0.5) II-7-7

Figure II-7-6.   Wave incidence oblique to breakwater gap II-7-8

Figure II-7-7.   Diffraction for a breakwater gap of one wavelength width where $\phi = 0$ deg . . . II-7-8

Figure II-7-8.   Diffraction for a breakwater gap of one wavelength width where
$\phi = 15$ deg II-7-9

Figure II-7-9.   Diffraction for a breakwater gap of one wavelength width where
$\phi = 30$ deg II-7-10

Figure II-7-10.   Diffraction for a breakwater gap of one wavelength width where
$\phi = 45$ deg II-7-11

Figure II-7-11.   Diffraction for a breakwater gap of one wavelength width where
$\phi = 60$ deg II-7-12

Figure II-7-12.   Diffraction for a breakwater gap of one wavelength width where
$\phi = 75$ deg II-7-13

Figure II-7-13.   Diffraction diagram of a semi-infinite breakwater for directional
random waves of normal incidence II-7-14

Figure II-7-14.   Diffraction diagrams of a breakwater gap with B/L = 1.0 for directional
random waves of normal incidence II-7-15

Figure II-7-15.   Diffraction diagrams of a breakwater gap with B/L = 2.0 for directional
random waves of normal incidence II-7-16

Figure II-7-16.   Diffraction diagrams of a breakwater gap with B/L = 4.0 for directional
random waves of normal incidence II-7-17

Figure II-7-17.   Diffraction diagrams of a breakwater gap with B/L = 8.0 for directional
random waves of normal incidence II-7-18

Figure II-7-18.   Schematic breakwater profile and definition of terms II-7-20

Figure II-7-19. Wave transmission for a low-crested breakwater (modified from
Van der Meer and Angremond (1992)) II-7-22

Figure II-7-20. Common types of floating breakwaters II-7-24

Figure II-7-21. Wave transmission coefficient for selected floating breakwaters
(Giles and Sorensen 1979; Hales 1981) II-7-25

Figure II-7-22. Wave transmission coefficient for vertical wall and vertical thin-wall
breakwaters where $0.0157 \le d_s \; lgt^2 \le 0.0793$ II-7-25

Figure II-7-23. Complete and partial reflection II-7-27

Figure II-7-24. Reflected wave crest pattern II-7-30

Figure II-7-25. Reflection of a diffracted wave II-7-31

Figure II-7-26. Surface profiles for oscillating waves (Carr 1953) II-7-33

Figure II-7-27. Behavior of an oscillating system with one degree of freedom II-7-34

Figure II-7-28. Behavior of an oscillating system with one degree of freedom II-7-36

Figure II-7-29. Motions in a standing wave II-7-37

Figure II-7-30. Theoretical response curves of symmetrical, narrow, rectangular harbor
(Raichlen 1968) II-7-38

Figure II-7-31. Resonant length and amplification factor of symmetrical rectangular
harbor (from Raichlen and Lee (1992); after Ippen and Goda (1963)) II-7-39

Figure II-7-32. Node locations for a dominant mode of oscillation in a square harbor:
a) fully open; b) asymmetric, constricted entrance II-7-41

Figure II-7-33. Response curves for rectangular harbor with flat and sloping bottom
(Zelt 1986) II-7-42

Figure II-7-34. Resonant response of idealized harbors with different geometry (Zelt
1986) II-7-43

Figure II-7-35. Photograph of physical model, Barbers Point Harbor, HI
(Briggs et al. 1994) II-7-44

Figure II-7-36. Numerical model grid for Barbers Point Harbor, HI (Briggs et al. 1994) II-7-45

Figure II-7-37. Amplification factors for five resonant periods, Barbers Point Harbor, HI
(Briggs et al. 1994) II-7-46

Figure II-7-38. Phases for five resonant periods, Barbers Point Harbor, HI
(Briggs et al. 1994) II-7-47

Figure II-7-39.   Exchange coefficients - rectangular harbor, TRP = 0.4 (modified from
Falconer (1980)) II-7-50

Figure II-7-40.   Wave crest pattern generated at a vessel bow moving over deep water II-7-54

Figure II-7-41.   Typical vessel-generated wave record II-7-56

Figure II-7-42.   Definition of terms, vessel drawdown II-7-60

Figure II-7-43.   Vessel sinkage prediction II-7-61

Figure II-7-44.   Pressure fields for moving vessels (vessels moving left to right) II-7-63

Figure II-7-45.   Mooring fiber rope elongation curves II-7-67

Chapter II-7
Harbor Hydrodynamics

II-7-1. Introduction

a. A harbor is a sheltered part of a body of water deep enough to provide anchorage for ships or a place of shelter; refuge. The purpose of a harbor is to provide safety for boats and ships at mooring or anchor and to provide a place where upland activities can interface with waterborne activities. Harbors range in complexity from the basic harbor of refuge, consisting of minimal or no upland support and only moderate protective anchorage from storm waves to the most complex, consisting of commercial port facilities, recreational marinas, and fuel docks linked to the sea through extensive navigation channels and protective navigation structures. Key features of all harbors include shelter from both long- and short-period open ocean waves, easy and safe access to the ocean in all types of weather, adequate depth and maneuvering room within the harbor, shelter from storm winds, and minimal navigation channel dredging.

b. Harbors can be classified according to location relative to the shoreline or coast. Figure II-7-1 illustrates six harbor classifications. The inland basin and offshore basin require considerable construction, including protective navigation structures and harbor and channel dredging to provide adequate protection. They are usually constructed where no natural features exist but where a facility is required. Examples of such harbors are Port Canaveral, Florida, Marina del Rey, California (inland basin), and Gulfport, Mississippi.

c. The natural geography can provide partial protection or headlands that can be augmented to construct a protective harbor. This approach may reduce the initial costs of construction. Examples of such harbors are Half Moon Bay Harbor, California; and Barcelona Harbor, New York (bay indentation); Crescent City Harbor, California; and Palm Beach Harbor, Florida (offshore island).

d. In some locations, the land can provide protective harbors requiring minimal modification. Examples include inside estuaries and up rivers such as at Panama City Harbor, Florida; Kings Bay, Georgia, (natural harbor); Port of Portland, Oregon; and New Orleans, Louisiana (river harbor).

e. The U.S. Army Corps of Engineers has constructed hundreds of harbor projects that include protective structures such as breakwaters, jetties, and navigation channels. Projects are classified by depth and range from deep-draft projects with navigation channel depths greater than -45 ft, to intermediate-depth projects with depths between -20 ft and -45 ft, to shallow-draft projects with depths less than -20 ft. Currently, USACE operates and maintains over 25,000 miles of navigation channels in association with hundreds of harbor projects.

f. This chapter covers basic harbor hydrodynamics. Harbor design is covered in Part V. The chapter covers wave diffraction, wave transmission and reflection, harbor oscillations, flushing and circulation, and vessel interactions. These are important elements that must be understood in order to design a safe harbor that is operationally efficient.

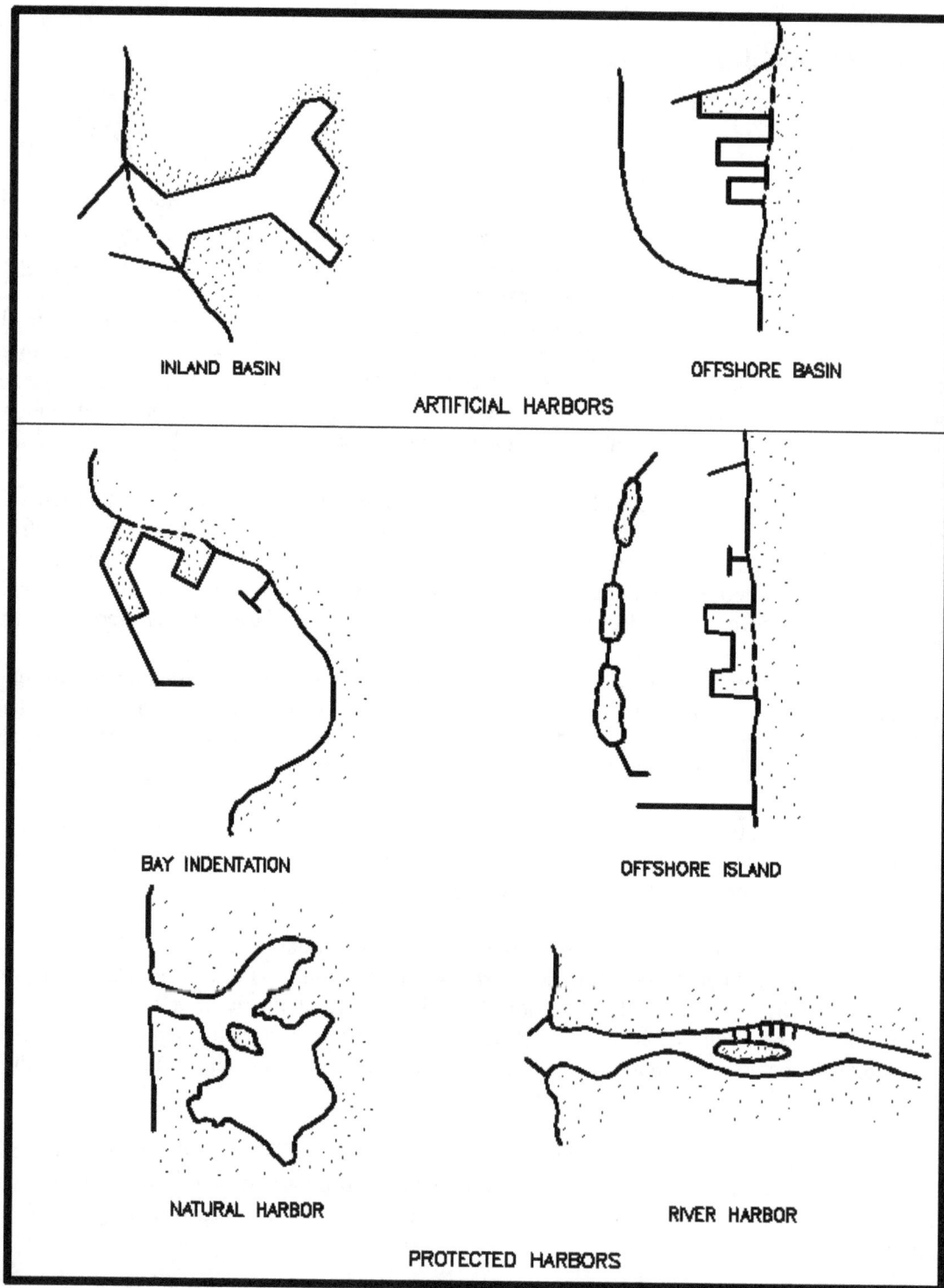

Figure II-7-1.   Harbor siting classifications

II-7-2. Wave Diffraction

a. Definition of diffraction.

(1) Consider a long-crested wave that has a variable height along its crest. As this wave propagates forward, there will be a lateral transfer of wave energy along the crest (perpendicular to the direction of wave propagation). The energy transfer will be from points of greater to lesser wave height. This process is known as wave diffraction.

(2) Nearshore wave refraction will cause concentrations of wave energy at points where wave orthogonals converge. Diffraction will lessen this refraction-induced energy concentration by causing wave energy to transfer across the orthogonals away from the region of concentration. Consequently, wave diffraction can have a small effect on the resulting heights of waves that approach harbor entrances.

(3) Diffraction has a particularly significant effect on wave conditions inside a harbor. When waves propagate past the end of a breakwater, diffraction causes the wave crests to spread into the shadow zone in the lee of the breakwater. The wave crest orientations and wave heights in the shadow zone are significantly altered.

b. Diffraction analysis.

(1) Much of the material developed for wave diffraction analysis employs monochromatic waves. Ideally, an analysis should employ the directional spectral conditions. But, for a preliminary design analysis, one or a set of monochromatic wave diffraction analyses is often used to represent the more complex result that occurs when a directional spectrum of waves diffracts at a harbor.

(2) In the material presented below, monochromatic results are presented first. Then, some of the available results for diffraction of irregular waves are presented. These results are based on the superposition of several monochromatic waves having a range of representative frequencies and directions. This type of analysis requires significant effort, but it can be carried out where the situation so requires. Also, physical model tests employing a directional wave spectrum can be used for a harbor diffraction analysis.

c. Diffraction at a harbor entrance. A major concern in the planning and design of coastal harbors is the analysis of wave conditions (height and direction) that occur inside the harbor for selected incident design waves. These waves may shoal and refract after they pass through the harbor entrance; but, the dominant process affecting interior wave conditions is usually wave diffraction. Two generic types of conditions are most commonly encountered: wave diffraction past the tip of a single long breakwater and wave diffraction through a relatively small gap in a breakwater.

(1) Waves passing a single structure.

(a) Figure II-7-2 shows a long-crested monochromatic wave approaching a semi-infinite breakwater in a region where the water depth is constant (i.e. no wave refraction or shoaling). A portion of the wave will hit the breakwater where it will be partially dissipated and partially reflected. The portion of the wave that passes the breakwater tip will diffract into the breakwater lee. The diffracted wave crests will essentially form concentric circular arcs with the wave height decreasing along the crest of each wave. The region where wave heights are affected by diffraction will extend out to the dashed line in Figure II-7-2.

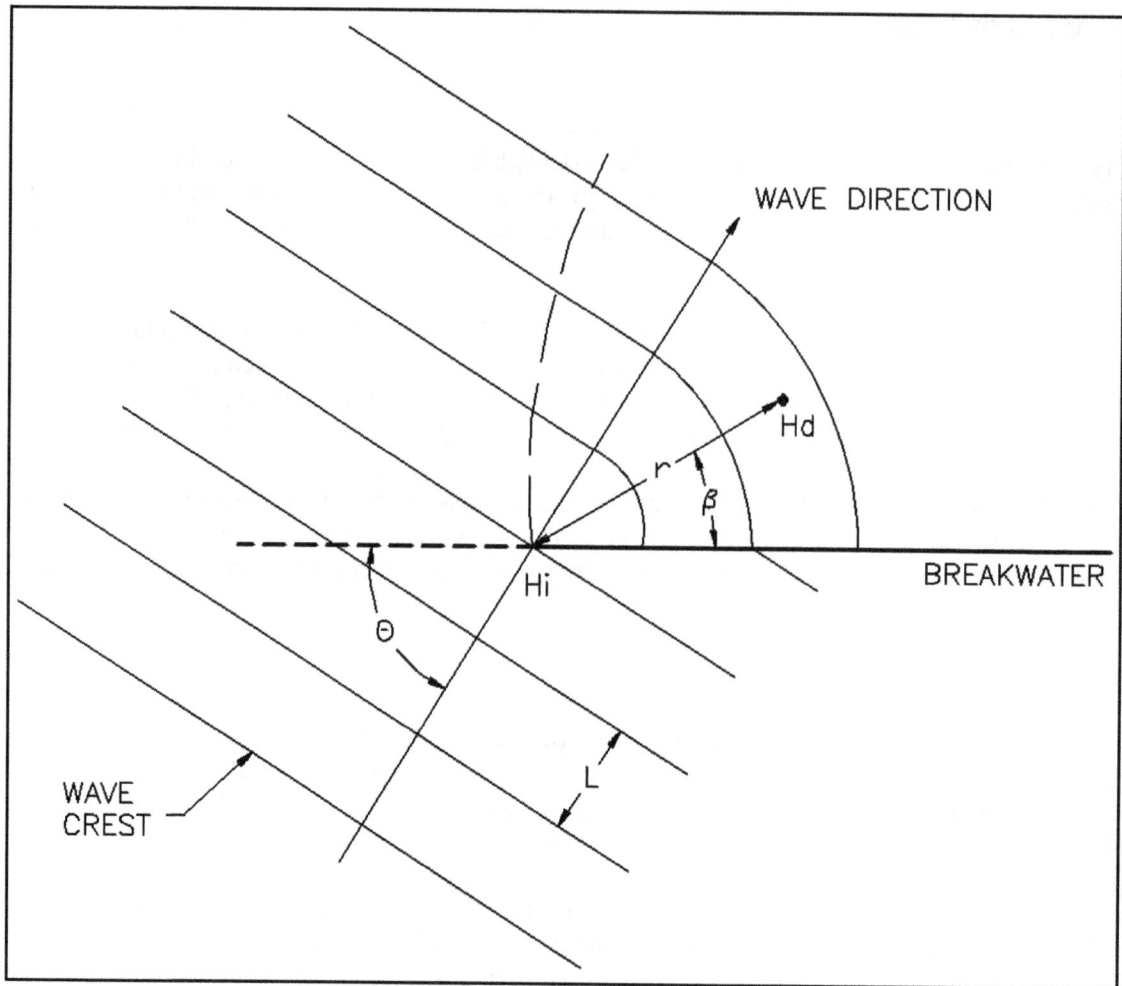

Figure II-7-2.   Wave diffraction, definition of terms

(b) The reflected wave crests (not shown in the figure) would also diffract to form concentric wave crests that curl around the breakwater tip into the lee. These waves are typically much lower than the incident waves and are more affected by diffraction when they reach the breakwater lee, so they typically have a very small height in the lee of the breakwater.

(c) A diffraction coefficient $K' = H_d/H_i$ can be defined where H_d is the diffracted wave height at a point in the lee of the breakwater and H_i is the incident wave height at the breakwater tip. If r is the radial distance from the breakwater tip to the point where K' is to be determined and β is the angle between the breakwater and this radial, then $K' = fcn(r/L, \beta, \theta)$ where θ defines the incident wave direction (see Figure II-7-2) and L is the wave length. Consequently, for a given location in the lee of the breakwater, the diffraction coefficient is a function of the incident wave period and direction of approach. So, for a spectrum of incident waves, each frequency component in the wave spectrum would have a different diffraction coefficient at a given location in the breakwater lee.

(d) The general problem depicted in Figure II-7-2 was originally solved by Sommerfeld (1896) for the diffraction of light passing the edge of a semi-infinite screen. Penny and Price (1952) showed that the same solution applies to the diffraction of linear surface waves on water of constant depth that propagate past the end of a semi-infinite thin, vertical-faced, rigid, impermeable barrier. Thus, the diffraction coefficients in the structure lee include the effects of the diffracted incident wave and the much smaller diffracted wave that reflects completely from the structure. Wiegel (1962) summarizes the Penny and Price (1952) solution and tabulates results of this solution ($K' = fct(r/L, \beta, \theta)$) for selected values of r/L, β and θ. Figure II-7-3 shows Wiegel's (1962) results for an approach angle θ of 60 deg. Plots of approach angles θ varying by 15-deg intervals from 15 to 180 deg can be found in Wiegel (1962) and the *Shore Protection Manual* (1984).

(e) An interesting feature demonstrated by Figure II-7-3 is that for this approach angle, the value of the diffraction coefficient along a line in the lee of the breakwater that extends from the breakwater tip in the direction of the approaching wave is approximately 0.5. This is true not only for the approach angle of 60 deg, but for any approach angle. Note also that for a given location in the lee of a breakwater, a one-dimensional spectrum of waves that comes from the same direction will undergo a greater decrease in height(energy density) for successively higher frequency waves in the spectrum. Increasing frequencies mean shorter wavelengths and consequently larger values of r/L (for given values of β and θ). Thus the diffracted spectrum will have a shift in energy density towards the lower frequency portion of the spectrum.

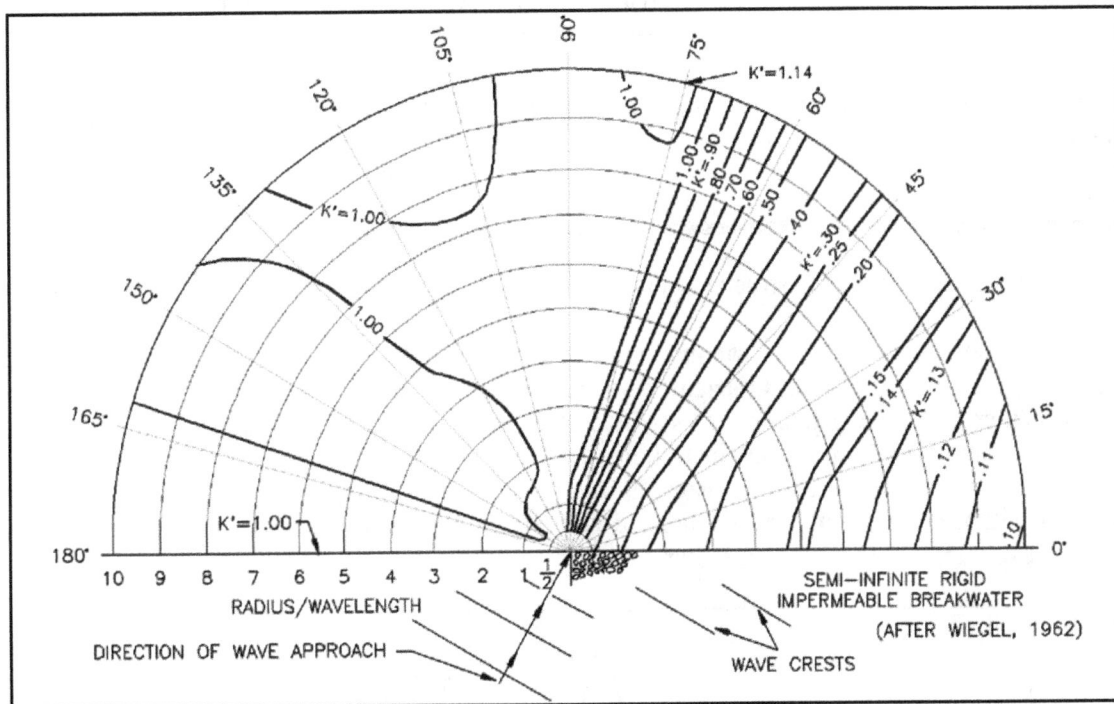

Figure II-7-3. Wave diffraction diagram - 60° wave angle

(f) Wave tank measurements of diffracted heights for waves passing a semi-infinite barrier were made by Putnam and Arthur (1948). They considered six approach directions for each of two incident wave periods. Their measurements generally confirm the diffraction theory. But, the diffraction theory assumes small-amplitude waves and Putnam and Arthur (1948) employed relatively small-amplitude waves in their experiments. For steeper waves, finite amplitude effects would cause the results to differ somewhat from the diffraction theory based on small-amplitude waves.

EXAMPLE PROBLEM II-7-1

FIND:
    The wave height in the lee of the breakwater at a point specified by $\beta = 30$ deg and r = 100 m.

GIVEN:
    A train of 6-sec-period 2-m-high waves is approaching a breakwater at an angle $\theta = 60$ deg. The water depth in the lee of the breakwater is 10 m.

SOLUTION:
    From the linear wave theory for a period of 6 sec and a water depth of 10 m, the wave length can be calculated to be 48.3 m. Thus, $r/L = 100/48.3 = 2.07$. From Figure II-7-3 at $\beta = 30$ deg and r/L = 2.07, $K' = 0.28$. This yields a diffracted wave height = (0.28)2 = 0.56 m.

    The diffracted wave would be propagating in the direction $\beta = 30$ deg and would have a continually diminishing wave height as indicated by Figure II-7-3. If the breakwater has a reflection coefficient that is less than one, the above result would still be reasonable, since the diffracted height of the reflected wave would be very small at the point of interest.

(2) Waves passing through a structure gap.

(a) When waves pass through a gap in a breakwater, diffraction occurs in the lee of the breakwater on both sides of the gap. As the waves propagate further into the harbor, the zone affected by diffraction grows toward the center line of the gap until the two diffraction zones interact (see Figure II-7-4). The wider the gap, the further into the harbor this interaction point occurs. For typical harbor conditions and gap widths greater than about five wavelengths, Johnson (1952) suggests that the diffraction patterns at each side of the gap opening will be independent of each other and Wiegel (1962) may be referenced for diffraction analysis. For smaller gap widths, an analysis employing the gap geometry must be used.

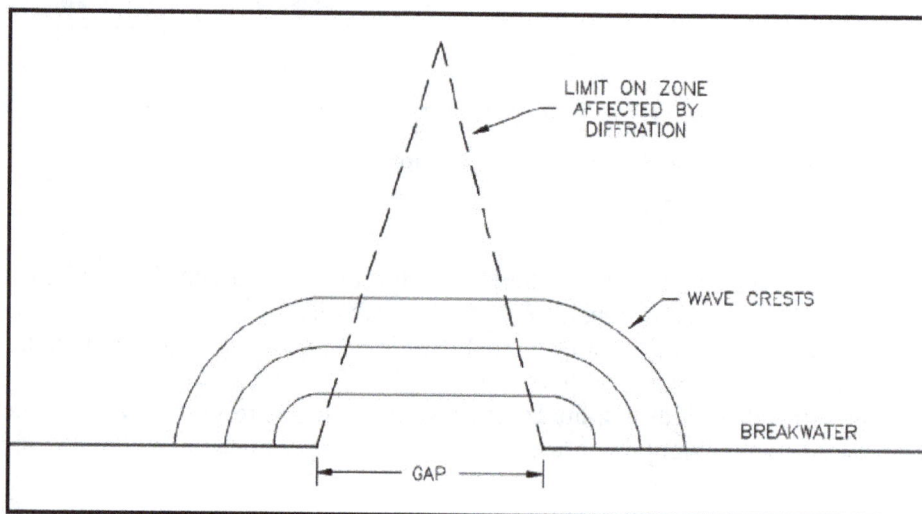

Figure II-7-4.   Wave diffraction through a gap

(b) Penny and Price (1952) developed a solution for the diffraction of normally incident waves passing through a structure gap by superimposing the solutions for two mirror image semi-infinite breakwaters. Johnson (1952) employed their solution to develop diagrams that give diffraction coefficients for gap widths (B) that are between one half and five times the incident wavelength. The lateral coordinates (x,y) are again nondimensionalized by dividing by the wavelength. Figure II-7-5 is an example of one of the diagrams Johnson (1952) developed. Only one half of the pattern is shown; the other half would be a mirror image. The reader is referred to Johnson (1952) and the *Shore Protection Manual* (1984) when dealing with gap widths other than 0.5 wavelength.

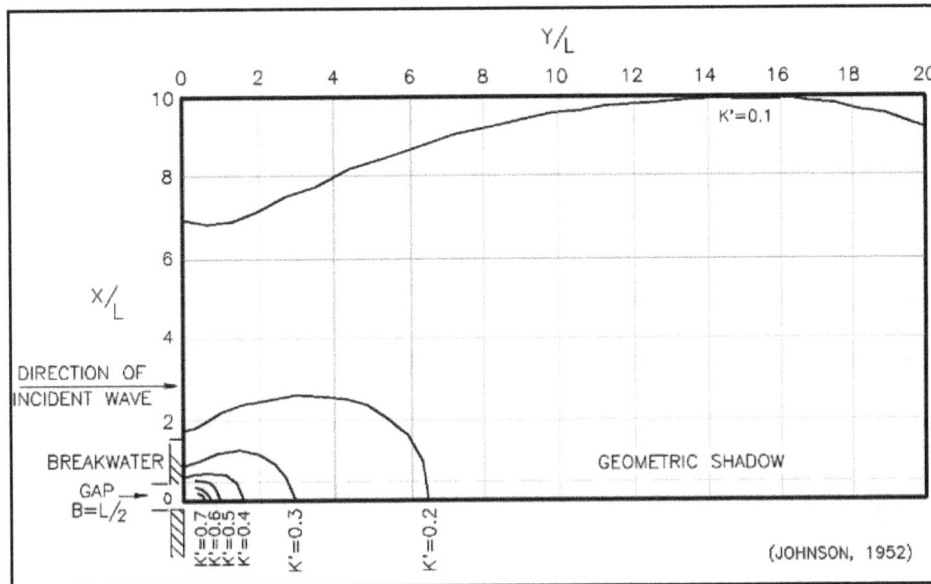

Figure II-7-5. Contours of equal diffraction coefficient gap width = 0.5 wavelength (B/L = 0.5)

(c) It may be necessary to know the wave crest orientation in the diffraction zone. Up to about six wavelengths beyond the gap, it is recommended (*Shore Protection Manual* 1984) that the wave crest position be approximated by two arcs that are centered on the two breakwater tips (as for a semi-infinite breakwater) and that the arcs be connected by a smooth curve that is approximately a circular arc centered on the midpoint of the gap. Beyond eight wavelengths, the crest position may be approximated by a single circular arc centered on the gap midpoint.

(d) The waves approaching a breakwater gap will usually not approach in a direction normal to the gap. Johnson (1952) found that the results presented above for normally incident waves can be used as an approximation for oblique waves by employing an imaginary equivalent gap having an orientation and width B' as defined in Figure II-7-6. Carr and Stelzriede (1952) employed a different analytical approach than that developed by Sommerfeld (1896) and also developed diffraction pattern solutions for barrier gaps that are small compared to the incident wavelength. Johnson (1952) used their approach to develop diffraction coefficient diagrams for a range of wave approach angles and a gap width equal to one wavelength. These are shown in Figures II-7-7 through II-7-12 for approach angles between 0 and 75 deg.

(e) Often, harbor entrance geometries will be different from the semi-infinite and gap geometries presented above. Still, approximate but useful results can be achieved by applying these solutions with some ingenuity in bracketing the encountered entrance geometry and interpolating results.

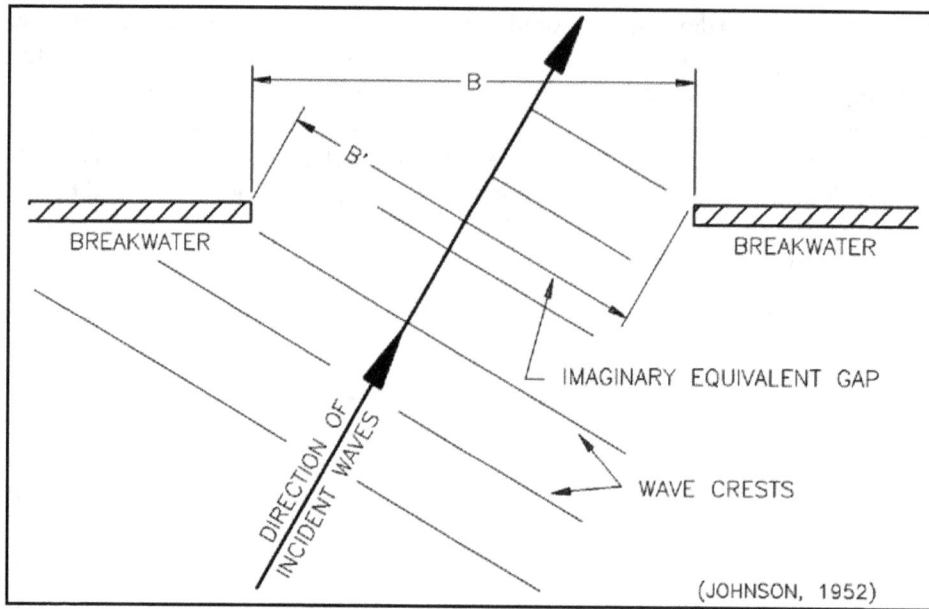

Figure II-7-6.   Wave incidence oblique to breakwater gap

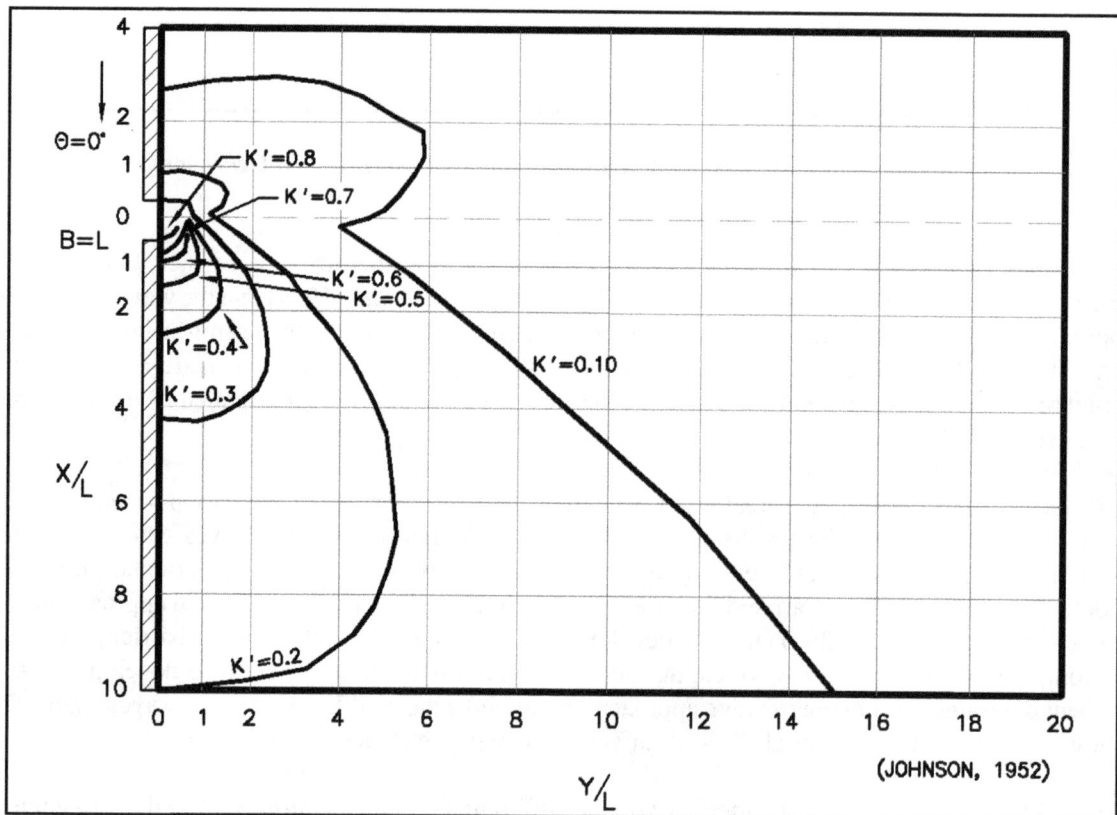

Figure II-7-7.   Diffraction for a breakwater gap of one wavelength width where φ = 0 deg

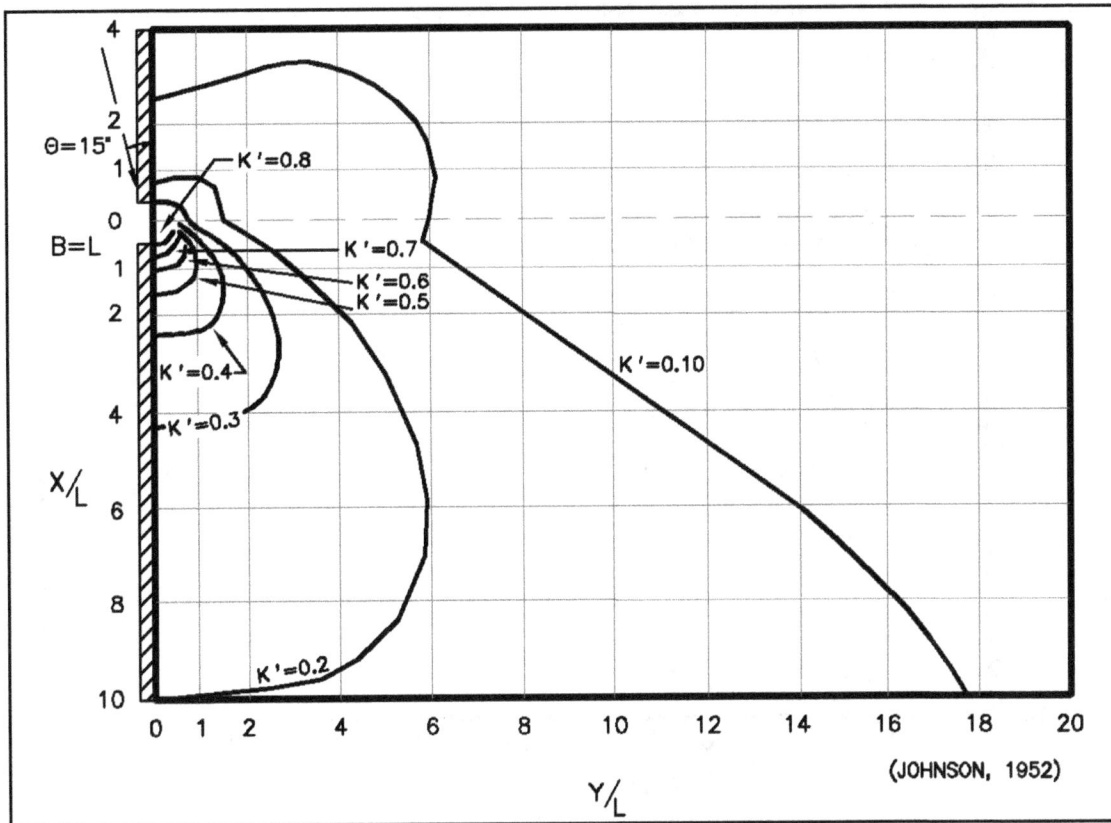

Figure II-7-8. Diffraction for a breakwater gap of one wavelength width where ϕ = 15 deg

(f) Bowers and Welsby (1982) conducted a physical model study of wave diffraction through a gap between two breakwaters whose axes are angled (rather than being collinear, i.e. 180 deg). Breakwater interior angles of 90 deg and 120 deg were employed. As would be expected, angling the breakwaters increased the heights behind the breakwater compared with the results for collinear breakwaters. But the increases were relatively small - up to 15 percent for 120-deg interior angles and up to 20 percent for 90-deg interior angles, when gap widths were in excess of half a wavelength.

(g) Memos (1976, 1980a, 1980b, 1980c) developed an approximate analytical solution for diffraction through a gap formed at the intersection of two breakwaters having axes that are not collinear but intersect at an angle. The point of intersection of the breakwater axes coincides with the tip of one of the breakwaters. Memos' solution can be developed for various angles of wave approach.

d. Irregular wave diffraction.

(1) The preceding discussion of wave diffraction was concerned with monochromatic waves. The effects of wave diffraction on an individual wave depend on the incident wave frequency and direction. Thus, each component of a directional wave spectrum will be affected differently by wave diffraction and have a different K' value at a particular point in the lee of a breakwater.

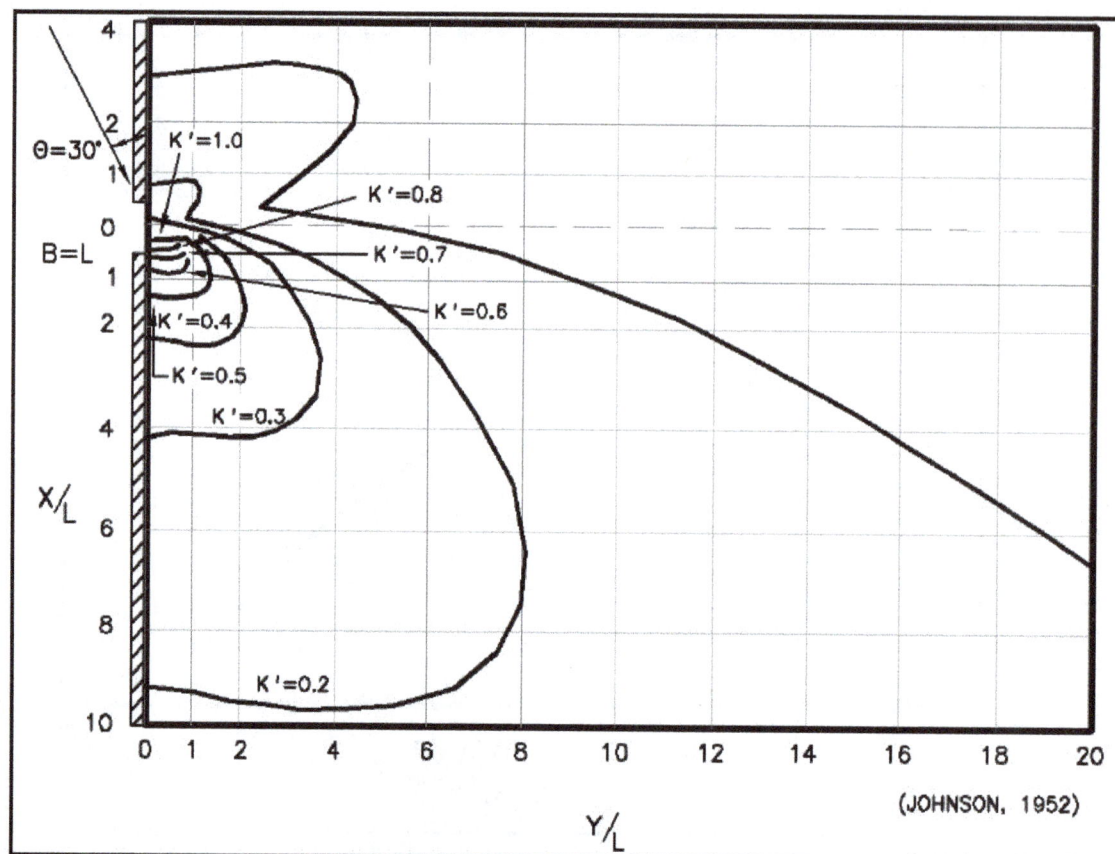

Figure II-7-9.   Diffraction for a breakwater gap of one wavelength width where φ = 30 deg

(2) To evaluate the effect of diffraction on a directional wave spectrum, Goda, Takayama, and Suzuki (1978) calculated diffraction coefficients for a semi-infinite breakwater and a breakwater gap by breaking the spectrum into number of frequency (10) and direction (20 to 36) components and combining the result at points in the breakwater lee. This produced an effective diffraction coefficient defined by

$$K_e' = \left[\frac{1}{M_o} \int_0^\infty \int_{\theta_{min}}^{\theta_{max}} S(f,\theta)\, (K')^2\, d\theta\, df \right]^{\frac{1}{2}} \tag{II-7-1}$$

where K' is the diffraction coefficient for each frequency/direction component when acting as a monochromatic wave, M_o is the zero moment of the spectrum, df and $d\theta$ are the frequency and direction ranges represented by each component of the spectrum, θ_{max} and θ_{min} are the limits of the spectral wave component directions, and $S(f, \theta)$ is the spectral energy density for the individual components. The spectral frequency distribution they employed was similar to most typical storm spectra such as the JONSWAP spectrum. The directional spread of the spectrum was characterized by a directional concentration parameter S_{max}, which equals 10 for widely spread wind waves and 75 for swell with a long decay distance, so the directional spread is quite limited.

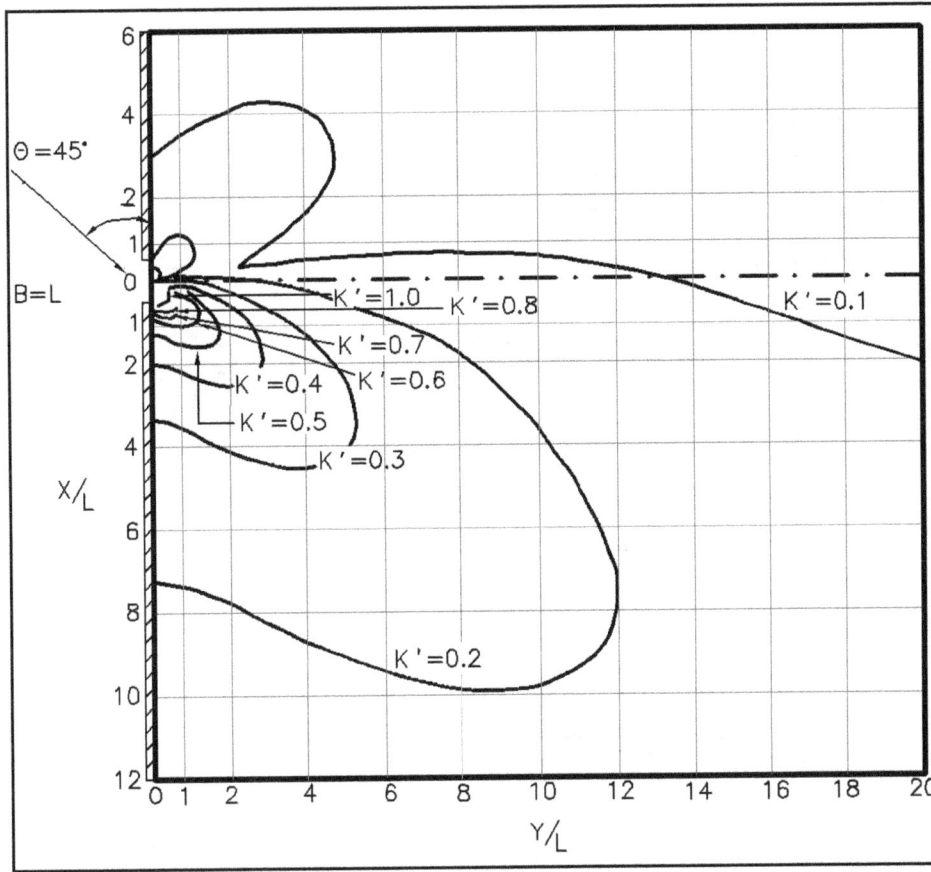

Figure II-7-10. Diffraction for a breakwater gap of one wavelength width where φ = 45 deg

(3) The results for waves approaching perpendicular to a semi-infinite breakwater are shown in Figure II-7-13 where the solid lines define values of K'_e and the dashed lines are the ratio of peak spectral period for the diffracted to incident waves. The results for waves approaching normal to a breakwater gap are shown in Figures II-7-14 to II-7-17, where the ratio of gap width to incident wavelength varies from 1 to 8. Note that these diagrams are normalized by dividing by both the gap width and the wavelength, so two different horizontal scales are given for each case. Also, the left-hand side gives the peak period ratio and the right-hand side gives values of K'_e.

(4) The spectral diffraction diagrams for the semi-infinite breakwater show a small change in the peak period ratio as the waves extend into the breakwater lee. (For monochromatic waves the wave period would not change.) The same holds for the breakwater gap. For the semi-infinite breakwater the values of K'_e are generally higher than the equivalent values of K' for monochromatic waves. For a breakwater gap, the spatial variation of K'_e values is smoothed out by the directional spread of the incident waves. That is, there is less variation in K'_e values for the spectral case than in K' for the monochromatic case. For waves approaching a breakwater gap at some oblique angle, the imaginary equivalent gap approach depicted in Figure II-7-6 can be used.

(5) Thus, if the one-dimensional or directional spectrum for the design waves is known at a harbor entrance, Equation II-7-1 can be used with the monochromatic wave diffraction diagrams to more effectively evaluate wave diffraction in the harbor. The spectrum can be broken into a number of direction and/or frequency components, each component can be analyzed as a diffracting monochromatic wave, and the results can be recombined using Equation II-7-1.

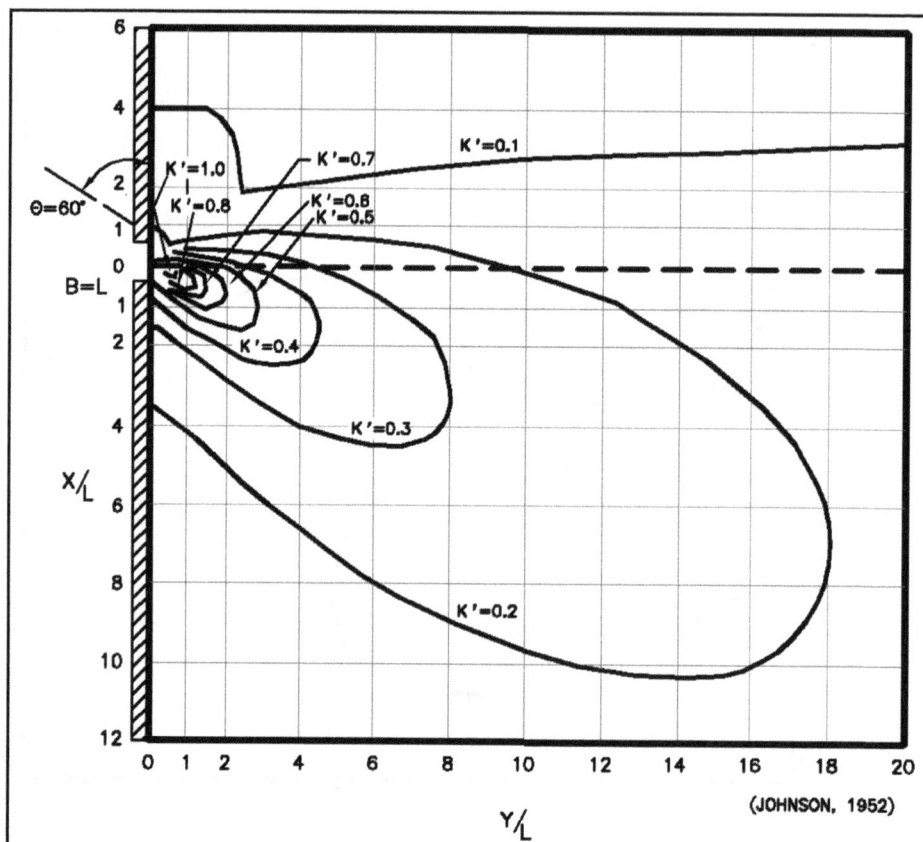

Figure II-7-11. Diffraction for a breakwater gap of one wavelength width where ɸ = 60 deg

e. Combined refraction-diffraction in harbors.

(1) In most harbors, the depth is relatively constant so the diffraction analyses discussed above are adequate to define the resulting wave conditions. However, if the depth changes significantly, then wave amplitudes will change because of shoaling effects. If the harbor bottom contours are not essentially parallel to the diffracting wave crests, then wave amplitudes and crest orientations will be affected by refraction.

(2) Where depth changes in a harbor are sufficient for combined refraction and diffraction effects to be significant, the resulting wave height and direction changes can be investigated by either a numerical or a physical model study. For examples of numerical model studies of combined refraction-diffraction in the lee of a structure, see Liu and Lozano (1979), Lozano and Liu (1980), and Liu (1982). Physical models that investigate the combined effects of refraction and diffraction are routinely conducted (see Hudson et al. 1979). The one major limitation on these models for wind wave conditions is that the model cannot have a distorted scale (i.e., horizontal and vertical scale ratios must be the same). Sometimes, lateral space limitations or the need to maintain an adequate model depth to avoid viscous and surface tension scale effects make a distorted scale model desirable. But such a model cannot effectively investigate combined refraction-diffraction problems.

(3) In many cases, the depth near the entrance to a harbor is relatively constant with the significant depth changes occurring further from the entrance (in the vicinity of the shoreline). Then an approximate (but often adequate) analysis can be carried out using the techniques discussed herein. The diffraction analysis would be carried out from near the harbor entrance to the point inside the harbor where significant

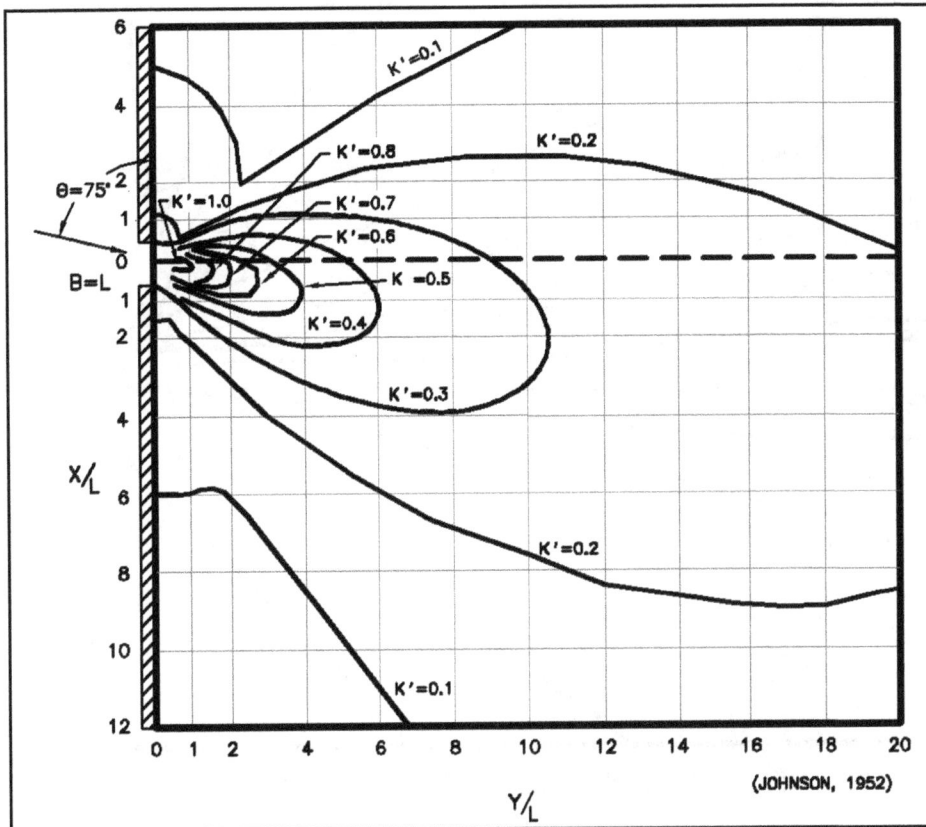

Figure II-7-12. Diffraction for a breakwater gap of one wavelength width where φ = 75 deg

depth changes commence. Hopefully, this will be a distance of at least three or four wavelengths from the entrance. This diffraction analysis will define the wave heights and crest orientation at the point where significant shoaling-refraction effects commence. From this point landward, a refraction-shoaling analysis using the procedures described in Part II-3 can be used to carry the wave to the point of breaking or interaction with a land boundary.

f. Combined diffraction - reflection in harbors.

(1) A computer program for dealing with combined diffraction and reflection by a vertical wedge has been developed by Seelig (1979, 1980) and is available in the Automated Coastal Engineering System (ACES) (Leenknecht et al. 1992). This package estimates wave height modifications due to combined diffraction and reflection caused by a structure. It has the ability to simulate a single straight, semi-infinite breakwater, corners of docks, and rocky headlands. Assumptions include monochromatic, linear waves, and constant water depth.

(2) The user has the ability to vary the wedge angle from 0 to 180 deg, where 0 deg would represent the case of a single straight, semi-infinite breakwater and 90 deg, the corner of a dock.

(3) The required input includes incident wave height, wave period, water depth, wave angle, wedge angle, and X and Y coordinates (location of desired calculation). The range of X and Y should be limited to plus or minus 10 wavelengths.

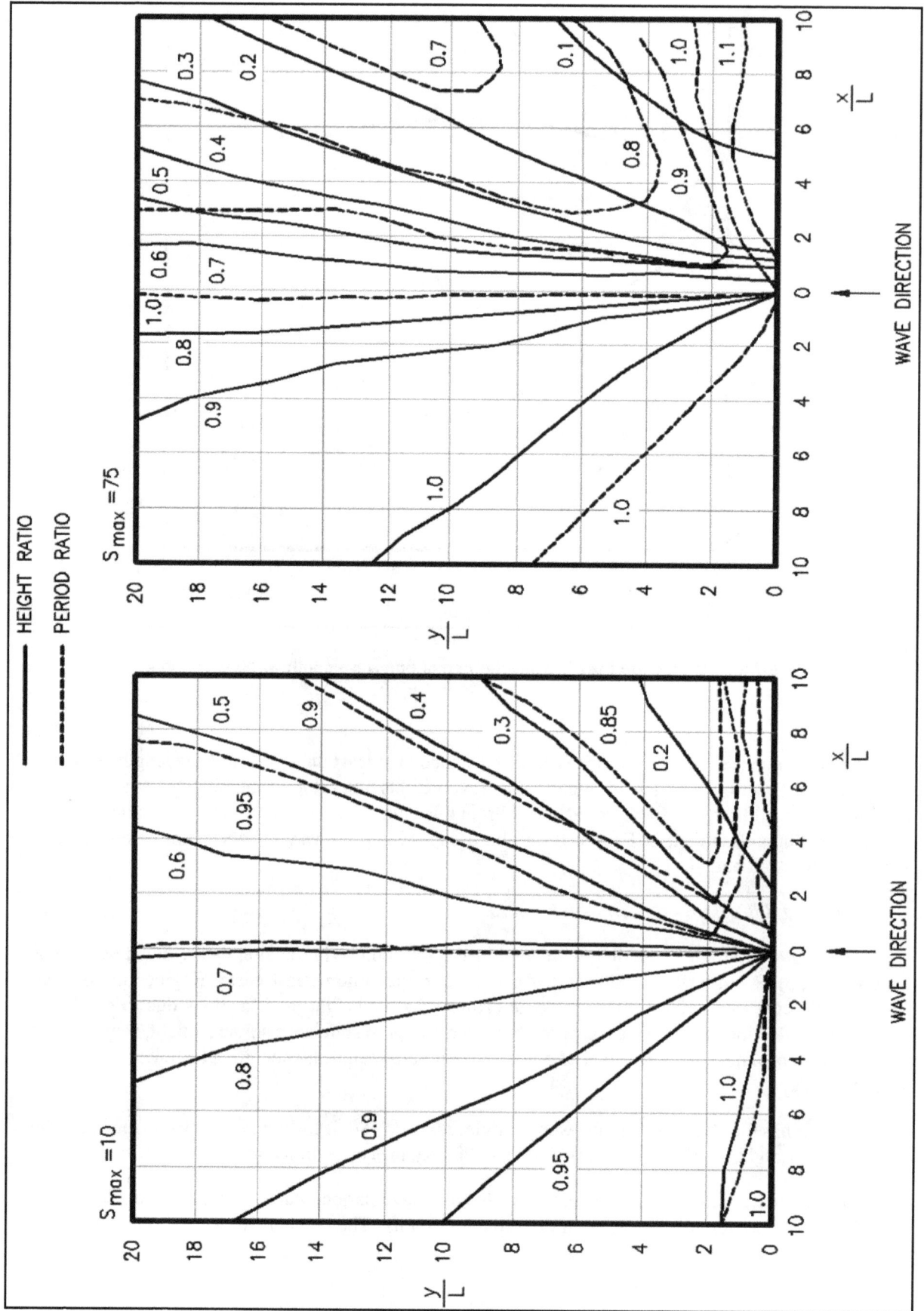

Figure II-7-13. Diffraction diagram of a semi-infinite breakwater for directional random waves of normal incidence (Goda 2000)

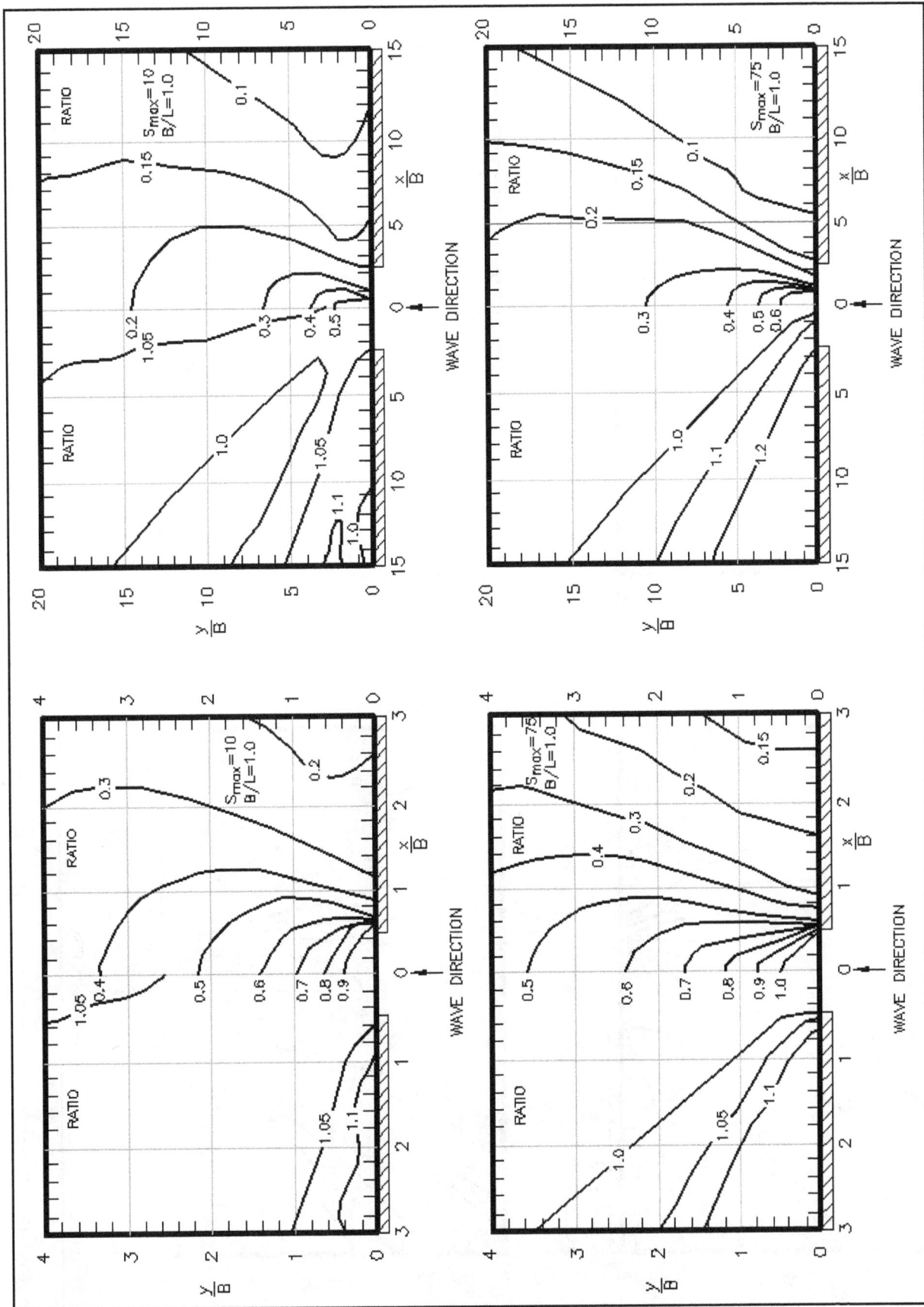

Figure II-7-14. Diffraction diagrams of a breakwater gap with $B/L = 1.0$ for directional random waves of normal incidence (Goda 2000)

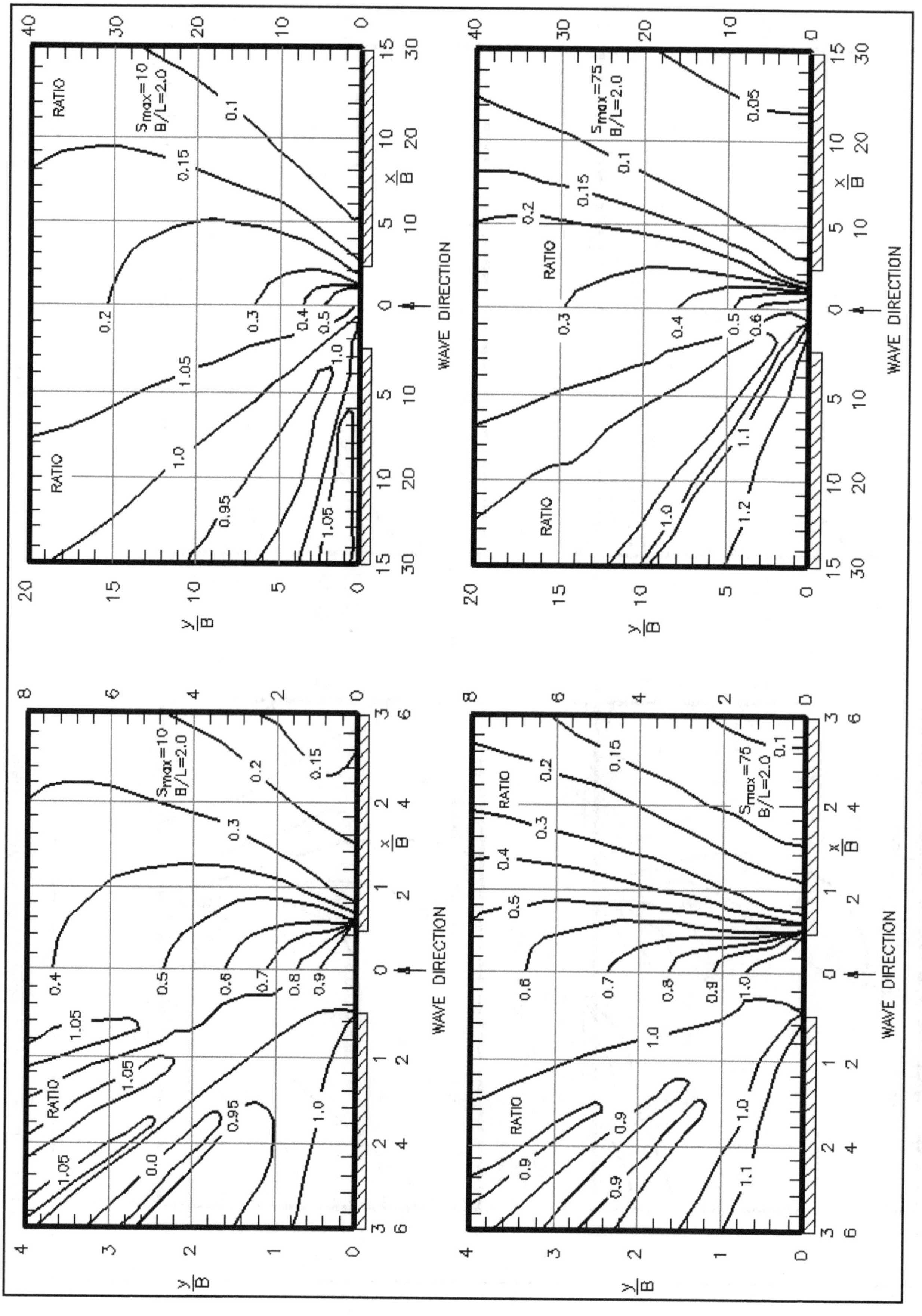

Figure II-7-15.   Diffraction diagrams of a breakwater gap with *B/L* = 2.0 for directional random waves of normal incidence (Goda 2000)

Harbor Hydrodynamics

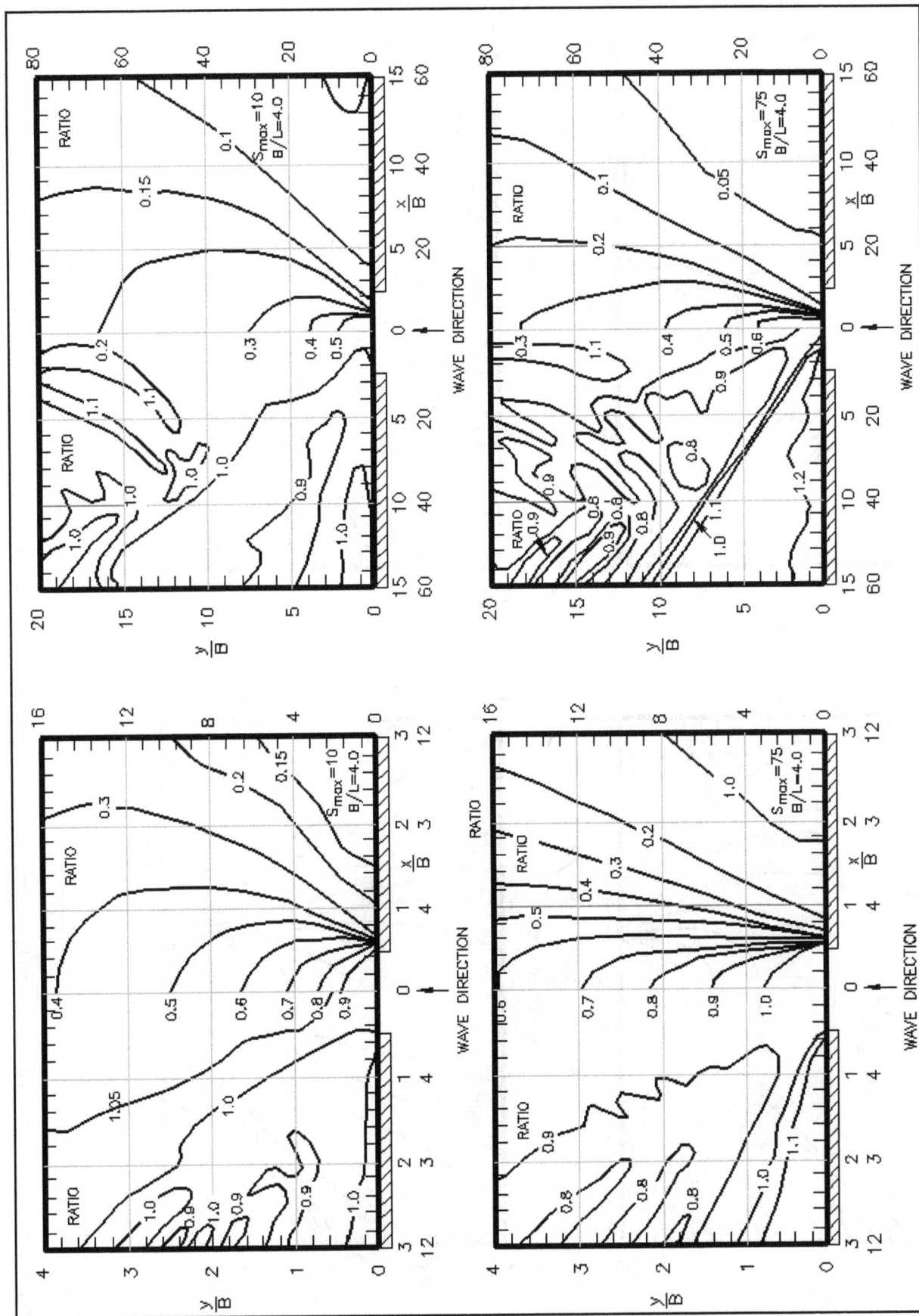

Figure II-7-16. Diffraction diagrams of a breakwater gap with *B/L* = 4.0 for directional random waves of normal incidence (Goda 2000)

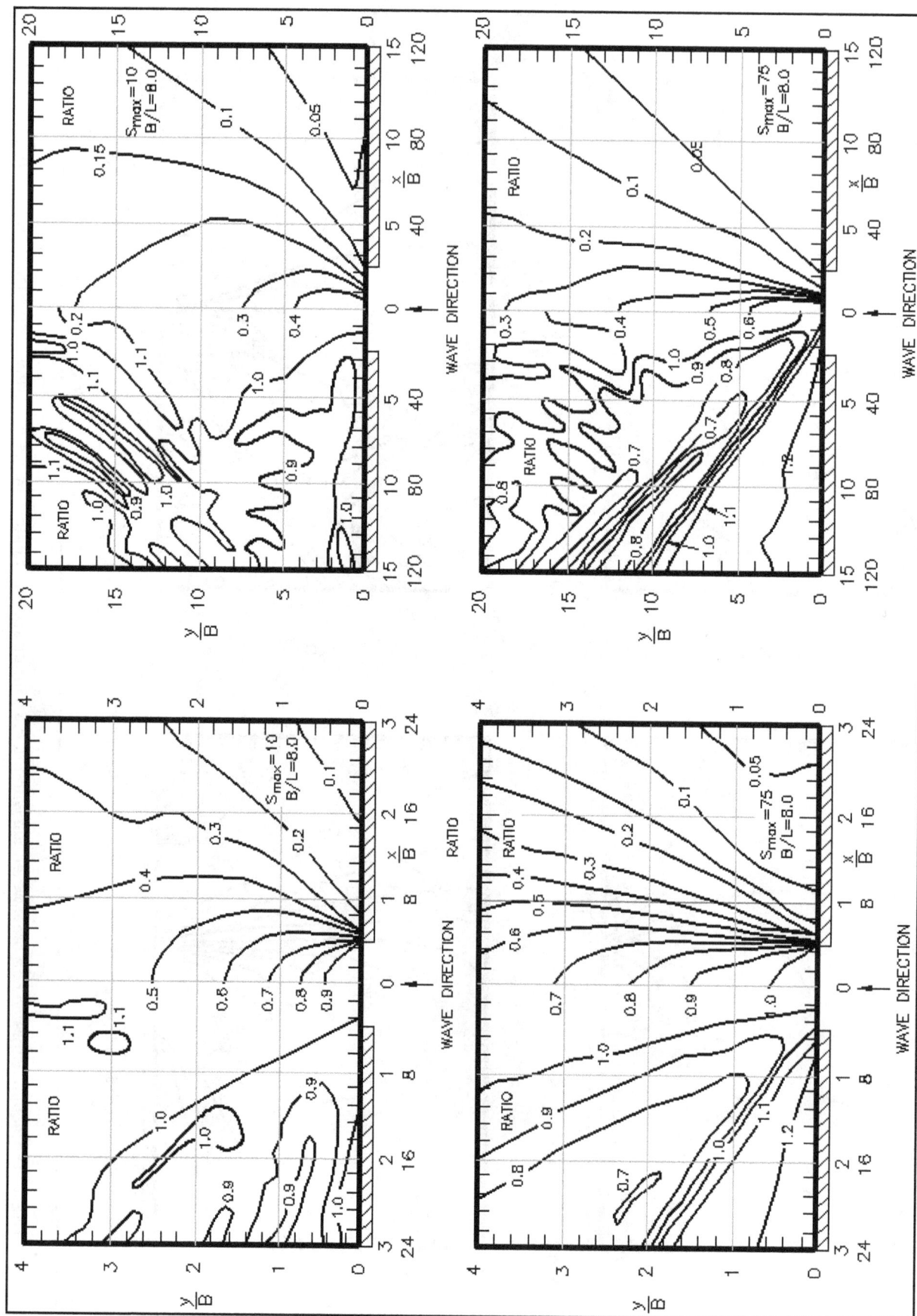

Figure II-7-17. Diffraction diagrams of a breakwater gap with B/L = 8.0 for directional random waves of normal incidence (Goda 2000)

(4) The wavelength, ratio of calculated wave height to incident wave height, wave phase, and modified wave height are given as output data. For further information on the ACES system, the reader is referred to Leenknecht et al. (1992).

II-7-3. Wave Transmission

a. Definition of transmission.

(1) When waves interact with a structure, a portion of their energy will be dissipated, a portion will be reflected and, depending on the geometry of the structure, a portion of the energy may be transmitted past the structure. If the crest of the structure is submerged, the wave will simply transmit over the structure. However, if the crest of the structure is above the waterline, the wave may generate a flow of water over the structure which, in turn, regenerates waves in the lee of the structure. Also, if the structure is sufficiently permeable, wave energy may transmit through the structure. When designing structures to protect the interior of a harbor from wave attack, as little wave transmission as possible should be allowed, while optimizing the cost versus performance of the structure.

(2) Transmitted wave height will be less than incident wave height, and wave period will usually not be identical for transmitted and incident waves. Laboratory experiments conducted with monochromatic waves typically show that the transmitted wave has much of its energy at the same frequency as the incident wave, but a portion of the transmitted energy has shifted to the higher harmonic frequencies of the incident wave. For a given incident wave spectrum, there would be a commensurate shift in the transmitted wave spectrum to higher frequencies.

(3) The degree of wave transmission that occurs is commonly defined by a wave transmission coefficient $C_t = H_t/H_i$ where H_t and H_i are the transmitted and incident wave heights, respectively. When employing irregular waves, the transmission coefficient might be defined as the ratio of the transmitted and incident significant wave heights or some other indication of the incident and transmitted wave energy levels.

(4) Most quantitative information on wave transmission past various structure types has necessarily been developed from laboratory wave flume studies. Historically, most of the early studies employed mono-chromatic waves; but, during the past two decades there has been a significant growth in information based on studies with irregular waves.

b. Transmission over/through structures.

(1) Rubble-mound structures-subaerial.

(a) Figure II-7-18 is a schematic cross section of a typical rubble-mound structure. The freeboard F is equal to the structure crest elevation h minus the water depth at the toe of the structure d_s (i.e., $F = h - d_s$). Also shown is the wave runup above the mean water level R that would occur if the structure crest elevation was sufficient to support the entire runup. When $F < R$, wave overtopping and transmission will occur. The parameter F/R is a strong indicator of the amount of wave transmission that will occur. Procedures for determining wave runup are presented elsewhere in the CEM.

(b) A number of laboratory studies of wave transmission by overtopping of subaerial structures have been conducted (see *Shore Protection Manual* (1984)). The most recent and comprehensive of these studies was conducted by Seelig (1980), who also studied submerged breakwaters.

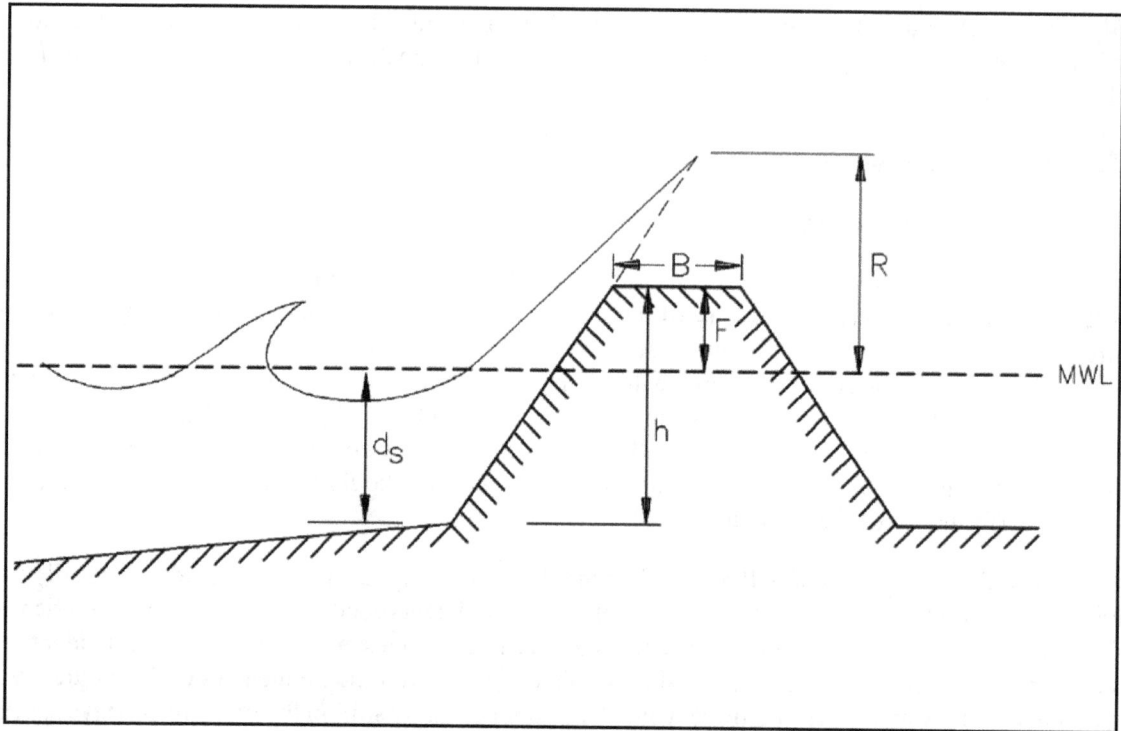

Figure II-7-18.   Schematic breakwater profile and definition of terms

(c) Seelig presented a simple formula for estimating the wave transmission coefficient for subaerial stone mound breakwaters, which is valid for both monochromatic and irregular waves:

$$C_t = C\left(1 - \frac{F}{R}\right)$$

(II-7-2)

(d) The coefficient C is given by

$$C = 0.51 - \frac{0.11B}{h}$$

(II-7-3)

where B is the structure crest width and h is the crest elevation above the bottom (see Figure II-7-18). It is recommended that Equation II-7-2 be applied to the relative depth (d_s/gT^2) range of 0.03 to 0.006 and Equation II-7-3 to the range of B/h between 0 and 3.2, because these are the ranges of experimental data employed to develop the equations.

(e) The report by Seelig (1980) contains laboratory results for 19 different breakwater cross-section geometries. Specific results for one or more of these geometries may closely relate to the prototype structure being analyzed. For irregular waves, the wave transmission coefficients were defined in terms of the incident and transmitted spectral energies.

(2) Rubble-mound structures-submerged.

(a) Rubble-mound structures having their crest at or below the mean water level have seen increasing use recently. Often, they simply consist of a homogeneous wide-graded mass of stone. A functional advantage for these lower-cost structures is that they may have a relatively high transmission coefficient for everyday lower waves, but as the height of the incident waves increases, the transmission coefficient will generally decrease. For harbor installations, they have been used in tandem with a conventional subareal breakwater placed in their lee - the combined cost of the two structures being less than a single structure with the same operational criteria (Cox and Clark 1992). A number of laboratory experiments on wave transmission past submerged rubble-mound structures have been conducted. Results are summarized in Seelig (1980) and Van der Meer and Angremond (1992).

(b) Van der Meer and Angremond (1992) summarized the available data plus some of their own data to present a comprehensive procedure for predicting wave transmission for low-crested breakwaters. For irregular waves they defined the transmission coefficient as the ratio of incident and transmitted significant wave heights. They correlated C_t with F/H_i where F would have positive values for low subaerial breakwaters and negative values for breakwaters with a submerged crest. Correlations with parameters that also included the incident wave period did not improve results. Figure II-7-19 gives C_t versus F/H_i.

(c) Van der Meer and Angremond (1992) were able to improve the correlations based on experimental data by introducing the median diameter D_{50} of the armor stone used to build the structure. (There is a relationship between D_{50} and the design wave height for a stable structure.) They then correlated C_t with F/D_{50} and secondary factors H_s/gT_p^2, H_i/D_{50}, and B/D_{50}. These relationships are presented in the form of somewhat complex formulas.

(3) Permeable rubble-mound structures.

(a) Wave energy may transmit through a rubble-mound structure, particularly if it is constructed solely of a homogeneous mass of large diameter stone. If the structure contains a number of stone layers including a core of fine stones, the wave transmission will be much less. Also, wave energy transmission through a stone mound structure would be significant for long-period, low waves, but much less for shorter-period waves (e.g. the tide would only be slightly reduced by a rubble-mound structure, but steep, wind-generated waves would have negligible transmission through the structure).

EXAMPLE PROBLEM II-7-2

FIND:
The transmission coefficients and the transmitted wave heights for incident wave heights of 0.5, 1.0, 2.0 and 4.0 m.

GIVEN:
A submerged offshore breakwater situated where the bottom is 4.0 m below the design still-water level. The structure crest elevation is 3.0 m above the bottom.

SOLUTION:
The breakwater freeboard F is 3.0 - 4.0 = -1.0 m. Employing Figure II-7-19 we have:

$H_{i(m)}$	F/H_i	C_t	$H_{t(m)}$
0.5	-2.0	0.82	0.41
1.0	-1.0	0.74	0.74
2.0	-0.5	0.60	1.20
4.0	-0.25	0.52	2.10

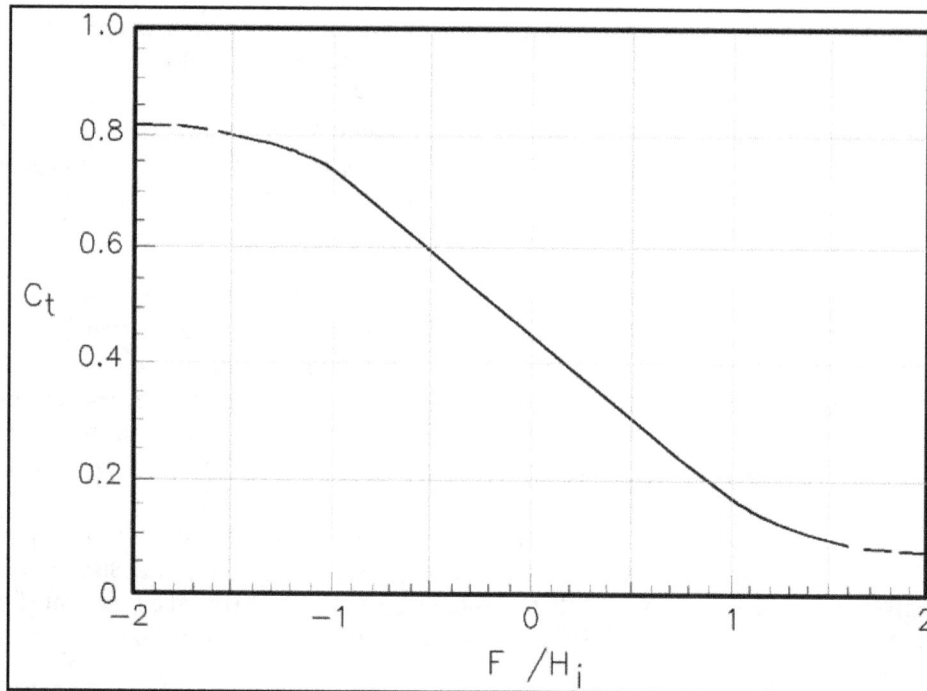

Figure II-7-19.   Wave transmission for a low-crested breakwater (modified from Van der Meer and Angremond (1992))

(b) When a porous rubble-mound structure suffers wave transmission caused by wave overtopping and by wave propagation through the structure, the resulting combined transmission coefficient would be

$$C_t = \sqrt{C_{tt}^2 + C_{t0}^2}$$

(II-7-4)

where C_{tt} is the coefficient for wave transmission through the structure and C_{t0} is the coefficient for wave transmission by flow over the structure.

(c) Potential scale effects make it difficult to conduct scaled laboratory experiments to measure wave transmission through rubble-mound structures. (Wave motion requires Froude similitude, while flow through porous media requires Reynolds similitude, but the two are incompatible.) Consequently, the best procedure for determining C_{tt} for a rubble-mound structure is a numerical procedure developed by Madsen and White (1976). A computer program for applying this procedure has been developed by Seelig (1979,1980) and is available in the ACES system (Leenknecht et al. 1992).

(d) The procedure developed by Madsen and White (1976) first calculates the amount of wave dissipation caused by wave runup/rundown on the seaward face of the structure. (It is assumed that the wave does not break - a good assumption for longer waves.) Wave reflection from the structure is also determined. The remaining energy propagates into the structure and is partially dissipated by turbulent action. The procedure then determines this rate of turbulent energy dissipation assuming a rectangular homogeneous breakwater cross section that is hydraulically equivalent to the actual layered breakwater. This leads to the transmitted wave height and C_{tt}. Application of this procedure requires a knowledge of the incident wave height and period, the water depth, the breakwater layer geometry, and the stone sizes and porosities for each layer. Seelig (1980) found that this procedure could be applied to irregular waves by using the mean wave height and spectral peak period of the incident waves in the calculation.

EXAMPLE PROBLEM II-7-3

FIND:
The transmission coefficient for incident waves having periods of 2 and 5 sec.

GIVEN:
The catamaran breakwater whose performance is depicted in Figure II-7-21.

SOLUTION:
For a water depth of 7.6 m, using the linear wave theory, wavelengths would be calculated to be 6.24 m (for T = 2 sec) and 34.4 m (for T = 5 sec). For a width W of 6.4 m, this yields W/L = 1.03 and 0.186, respectively. From Figure II-7-21, this yields C_t = 0.2 (extrapolating) for the 2-sec wave and C_t = 0.8 for the 5-sec wave. Thus, this catamaran is quite effective for the 2-sec wave but quite ineffective for the 5-sec wave.

(4) Floating breakwaters.

(a) Moored floating breakwaters have some distinct advantages for harbor installations. They are more adaptable to the water level changes that occur at harbors that are built on reservoirs and in coastal areas having a large tidal range. They are usually more economical than fixed breakwaters for deep-water sites, and they interfere less with water circulation and fish migration. But they also have some significant limitations. Since they are articulating structures, they are prone to damage at connecting points between individual breakwater units and between these units and mooring lines. And, their performance is very dependent on the period of the incident waves. This last factor will establish relatively severe limits on where floating breakwaters can effectively be deployed.

(b) Several different types of floating breakwaters have been proposed. Hales (1981) presents a survey of these various types and their performance. Most of the floating breakwaters in use are of three generic types - prism, catamaran, and scrap tire assembly (see Figure II-7-20). Figure II-7-21 plots the wave transmission coefficients for a typical representative of each of these three types. The transmission coefficient is plotted versus the breakwater's characteristic dimension in the direction of wave propagation W divided by the incident wave length L at the breakwater. The data plotted in Figure II-7-21 are all derived from laboratory experiments. The prism results are for a concrete box having a 4.88-m width (W), a draft of 1.07 m, and a water depth of 7.6 m (Hales 1981). The catamaran (Hales 1981) has two pontoons that are 1.07 m wide with a draft of 1.42 m and a total width (W) of 6.4 m. The water depth was 7.6 m. The tire assembly (Giles and Sorensen 1979) had a width of 12.8 m (W), a nominal draft of one tire diameter, and was tested in water 3.96 m deep. The three breakwaters were all moored fore and aft; other mooring arrangements would somewhat alter the transmission coefficient.

(c) Example Problem II-7-3 demonstrates a major limitation of floating breakwaters. For typical breakwater sizes (i.e., W equals 5 to 10 m) the incident wave period must not exceed 2-3 sec for the breakwater to be very effective. Thus, for typical design wind speeds, the fetch generating the waves to which the structure is exposed cannot be very large. Sorensen (1990) conducted an analysis to determine general wind speed, fetch, and duration guidelines for the three floating breakwaters from Figure II-7-20. He assumed an allowable transmitted wave height of 2 ft (0.61 m). For example, for a wind speed of 60 mph (26.8 m/sec) having a duration of 20-30 min, the fetch must not exceed 2-3 miles (3.2-4.8 km) if the transmitted wave height is to be less than 2 ft.

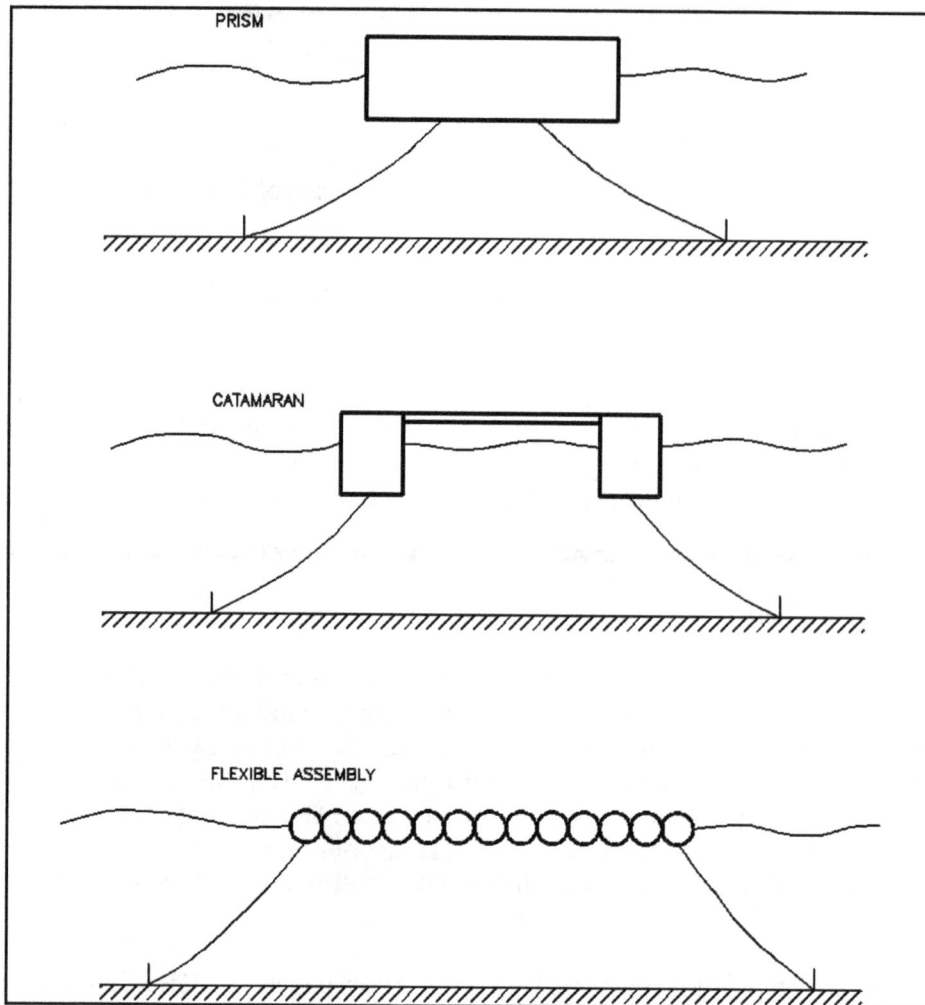

Figure II-7-20. Common types of floating breakwaters

(d) The wave-induced loading on the floating breakwater's mooring/anchor system is also an important design concern. Hales (1981) and Harms et al. (1982) present mooring load data for various types of floating breakwaters. The peak mooring load that develops for a given breakwater geometry and mooring arrangement depends primarily on the incident wave height and increases significantly as the wave height increases; the wave period or length is only of secondary importance in determining the mooring load. If the mooring load cannot be adequately estimated from published information on similar breakwaters, physical model tests may be required to determine these loads for the anticipated design wave conditions.

(5) Wave barriers.

(a) Vertical thin semirigid barriers are used as breakwaters at some harbors, particularly where wave loading is relatively small and wave reflection would not be a problem. For some small-craft harbor installations, the barrier is open at the bottom, which allows water to circulate in and out of the harbor and reduces structure costs (see Gilman and Nottingham (1992) and Lott and Hurtienne (1992). If the incident wave period is relatively short (d/L relatively large), wave action is focused near the surface and wave transmission would be relatively small.

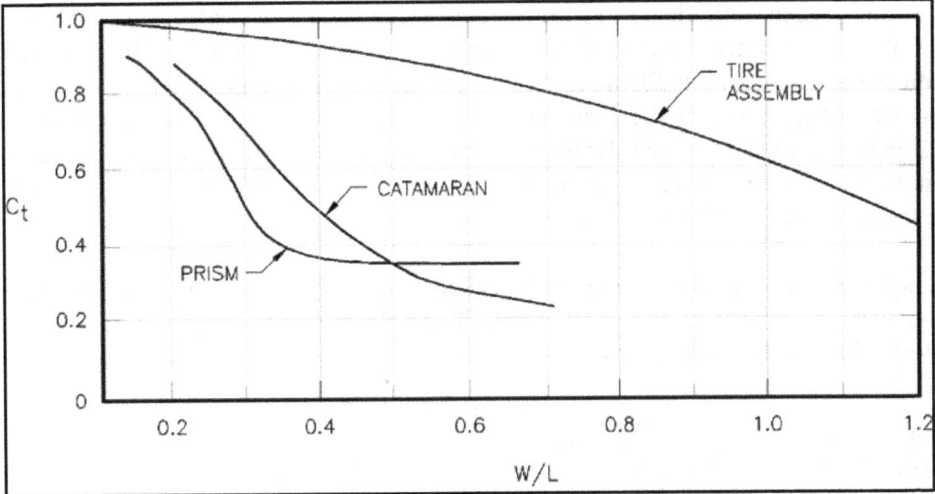

Figure II-7-21. Wave transmission coefficient for selected floating breakwaters (Giles and Sorensen 1979; Hales 1981)

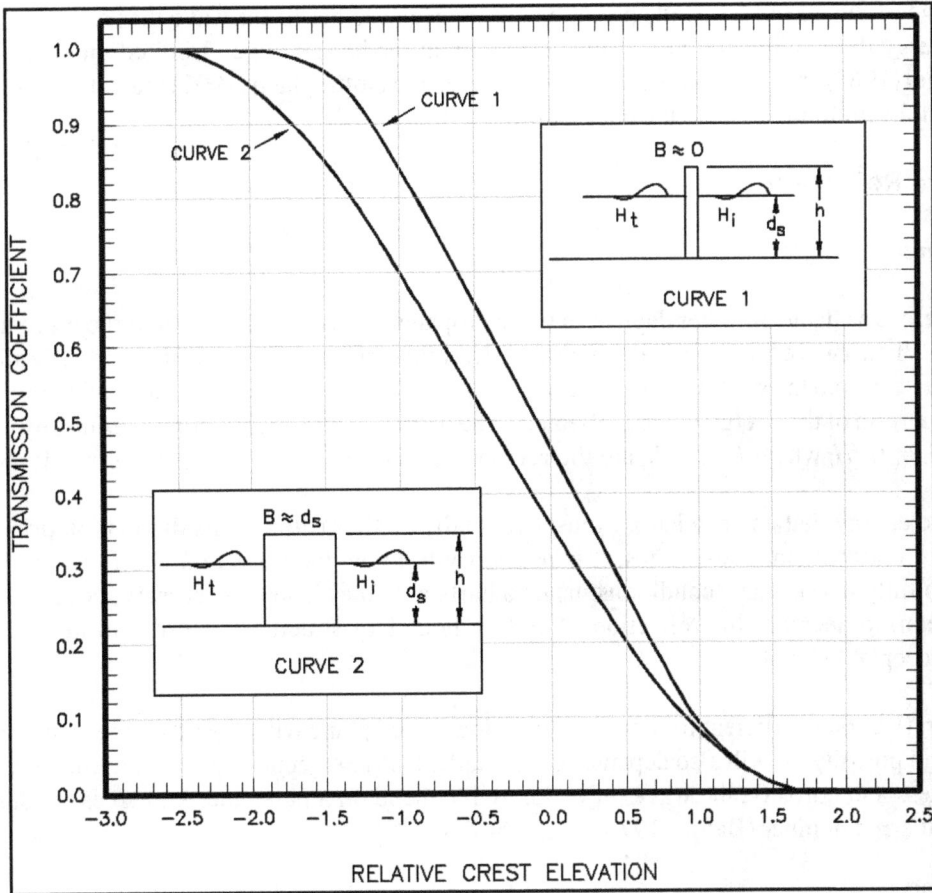

Figure II-7-22. Wave transmission coefficient for vertical wall and vertical thin-wall breakwaters where $0.0157 \leq d_s / gt^2 \leq 0.0793$ (based on Goda 2000)

(b) Wave transmission coefficients for a thin and a relatively wide vertical barrier that has its crest above or below the water surface were determined from experiments by Goda (1969). The results, which are for monochromatic waves, are shown in Figure II-7-22. Later tests with irregular waves, where the transmission coefficient is defined in terms of the incident and transmitted significant wave heights, showed that Figure II-7-22 can also be used for irregular wave conditions (Goda 2000). As would be expected, a portion of the transmitted wave energy shifted to higher frequencies so that the transmitted significant wave period was less than the incident significant period.

(c) Wiegel (1960) developed a simple analytical formulation for wave transmission past a thin rigid vertical barrier that does not extend to the bottom. For monochromatic waves and no overtopping, the resulting transmission coefficient is given by

$$C_t = \left[\frac{\dfrac{2k(d-y)}{\sinh 2kd} + \dfrac{\sinh 2k(d-y)}{\sinh 2kd}}{1 + \dfrac{2kd}{\sinh 2kd}} \right]^{\frac{1}{2}}$$

(II-7-5)

where d is the water depth, y is the vertical extent of the barrier below the still-water surface, and $k = 2\pi/L$. In developing Equation II-7-5, Wiegel assumed that the portion of the wave power in the water column below the lower edge of the barrier transmits past the barrier. He used the linear wave theory in this analysis and neglected energy dissipation caused by flow separation at the barrier edge. Limited monochromatic wave tests by Wiegel (1960) and irregular wave tests by Gilman and Nottingham (1992) indicate that Equation II-7-5 can be used for preliminary design calculations.

II-7-4. Wave Reflection

a. Definition of reflection.

(1) If there is a change in water depth as a wave propagates forward, a portion of the wave's energy will be reflected. When a wave hits a vertical, impermeable, rigid surface-piercing wall, essentially all of the wave energy will reflect from the wall. On the other hand, when a wave propagates over a small bottom slope, only a very small portion of the energy will be reflected. The degree of wave reflection is defined by the reflection coefficient $C_r = H_r/H_i$ where H_r and H_i are the reflected and incident wave heights, respectively.

(2) Wave energy that enters a harbor must eventually be dissipated. This dissipation primarily occurs at the harbor interior boundaries. Thus, it is necessary to know the reflection coefficients of the interior boundaries to fully define wave conditions inside a harbor. It may also be necessary, because of excessive wave reflection, to decrease the reflection of certain boundary structures in order to keep interior wave agitation at acceptable levels.

(3) The reflection coefficient for a surface-piercing sloped plane will depend on the slope angle, surface roughness, and porosity. It will also depend on the incident wave steepness H/L. Consequently, for a given slope roughness and porosity, the wave reflection will depend on a parameter known as the surf similarity number or Iribarren number (Battjes 1974)

$$I_r = \frac{\tan\alpha}{\sqrt{\dfrac{H_i}{L_o}}}$$

(II-7-6)

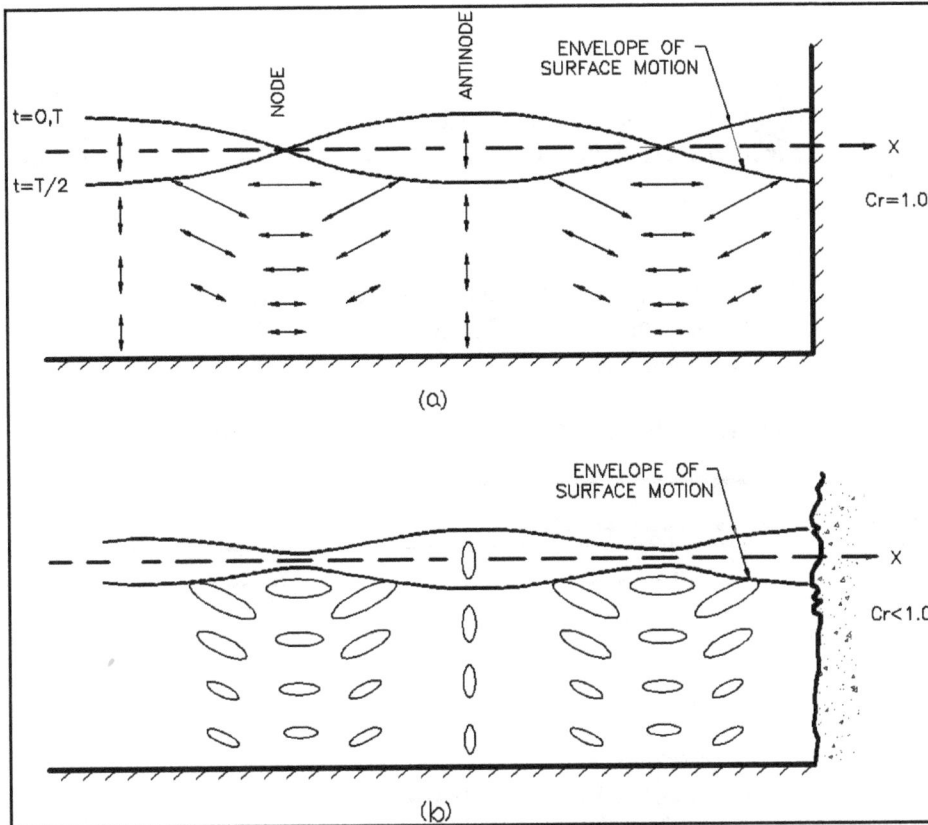

Figure II-7-23. Complete and partial reflection

where α is the angle the slope forms with the horizontal and L_0 is the length of the incident wave in deep water.

(4) Figure II-7-23a is a profile view of the water surface envelope positions for a wave reflecting from a wall that has a reflection coefficient equal to unity (i.e., $H_i = H_r$). The water surface amplitude is given by

$$N = H_i \cos\frac{2\pi x}{L} \cos\frac{2\pi t}{T} \qquad \text{(II-7-7)}$$

where L and T are the incident and reflected wave length and period respectively, x is the horizontal ordinate and t is the time elapsed. The figure also shows the water particle paths at key points. At nodal points, water particle motions are horizontal and at antinodes, water particle motions are vertical. At $t = 0$, $T/2$, and T, the water is instantaneously still and all of the wave energy is potential energy. At $t = T/4$ and $3T/4$, the water surface is horizontal and all of the energy is kinetic energy.

(5) When the wall reflection coefficient is less than unity, the water surface envelope positions and particle paths are as depicted in Figure II-7-23b. As the reflection coefficient decreases toward zero, the water surface profile and water particle path changes toward the form of a normal progressive wave.

b. *Reflection from structures.*

(1) Most of the interior boundaries of many harbors are lined with structures such as bulkheads or reveted slopes. Recent laboratory investigations (Seelig and Ahrens 1981; Seelig 1983; Allsop and Hettiarachchi 1988) indicate that the reflection coefficients for most structure forms can be given by the following

$$C_r = \frac{a I_r^2}{b + I_r^2}$$

(II-7-8)

where the values of coefficients a and b depend primarily on the structure geometry and to a smaller extent on whether waves are monochromatic or irregular. The Iribarren number employs the structure slope and the wave height at the toe of the structure.

(2) Table II-7-1 presents values for the coefficients a and b collected from the above references.

Table II-7-1
Wave Reflection Equation Coefficient Values Structure

Structure	a	b
Plane slope-monochromatic waves	1.0	5.5
Plane slope-irregular waves	1.1	5.7
Rubble-mound breakwaters[1]	0.6	6.6
Dolos-armored breakwaters - monochromatic waves	0.56	10.0
Tetrapod-armored breakwaters - irregular waves	0.48	9.6

[1]This is an average conservative value. Seelig and Ahrens (1981) recommend a range of values for a and b that depend on the number of stone layers, the relative water depth (d/L), and the ratio of incident wave height to breaker height.

EXAMPLE PROBLEM II-7-4

FIND:
The height of the reflected wave.

GIVEN:
A wave in deep water has a height of 1.8 m and a period of 6 sec. It propagates toward shore without refracting or diffracting to reflect from a rubble-mound breakwater located in water 5 m deep. The breakwater front slope is 1:1.75 (29.7 deg).

SOLUTION:
From linear wave theory shoaling calculations (Part II-1) the wave height at the structure would be 1.70 m (this is H_i). From the linear wave theory, the deepwater wave length is $L_0 = 56.2$ m. Then, from Equation II-7-6, the Iribarren number is

$$I_r = \frac{\tan 29.7^\circ}{\sqrt{1.70/56.2}} = 3.28$$

For the coefficient values $a = 0.6$ and $b = 6.6$ (from Table II-7-1), Equation II-7-8 yields

$$C_r = \frac{0.6 (3.28)^2}{6.6 + (3.28)^2} = 0.37$$

Thus, the reflected wave height $H_r = C_r H_i = 0.37(1.70) = 0.63$ m.

---

(3) Note that Equation II-7-8 indicates that C_r approaches the value of a at higher values of the Iribarren number. Thus, the highest reflection coefficients for stone and concrete unit armored structures are around 0.5. As expected, vertical plane slopes (infinite Iribarren number) would have a reflection coefficient near unity. For typical rigid vertical bulkheads, owing to the irregularity of the facing surface, a value of $C_r = 0.9$ might be used.

   c. Reflection from beaches.

   (1) When the reflecting slope becomes very flat, the incident wave will break on the slope, causing an increase in energy dissipation and commensurate decrease in the reflection coefficient. Thus, beaches are generally very efficient wave absorbers, particularly for shorter period wind-generated waves. An added complexity is that as the incident wave conditions change, the beach profile geometry will change, in turn changing the reflection coefficient somewhat. Laboratory measurements of wave reflection from beaches suffer from scale effects in trying to replicate the prototype beach profile, surface roughness, and porosity. Also, it is harder to select a beach slope angle for the Iribarren number owing to the complexity of beach profiles. Thus, data plots follow the form of Equation II-7-8, but with significant scatter in the data points. Seelig and Ahrens (1981) suggest that $a = 0.5$ and $b = 5.5$ be used for beaches. Since the slope angles are small, the Iribarren number will be relatively small, yielding relatively low reflection coefficients.

   (2) An interesting phenomenon (known as Bragg reflections after a similar phenomenon in optics) occurs when the shallow nearshore seabed has uniformly spaced bottom undulations. A resonance develops between the incident surface waves of certain periods and the bottom undulations, causing a reflection of a portion of the incident wave energy (see Davies and Heathershaw (1984), Mei (1985), Kirby (1987)). Resonance and reflection are maximum for that portion of the incident wave spectrum having a wavelength that is twice the length of the bottom undulations. There is an approximately linear increase in the reflection coefficient with the increase in the number of bottom undulations. Reflection also increases if the amplitude of the bottom undulations increases or the water depth over the undulations decreases. For appropriate undulation geometries and water depths, reflection coefficients in excess of 0.5 are possible.

   (3) It has been suggested (Mei 1985) that Bragg reflections that develop on a nearshore bar system will set up a standing wave pattern seaward of the bar system which, in turn, causes the bar system to extend in the seaward direction. Theoretical, laboratory, and field studies indicate that it is possible to build a series of shore-parallel submerged bars tuned to the dominant incoming wave periods, that will act as a shore protection device. The feasibility of building these bar systems and their economics need to be studied further.

   d. Reflection patterns in harbors.

   (1) Figure II-7-24 depicts the plan view of a wave crest approaching and reflecting from a barrier that has a reflection coefficient C_r. The incident wave crest is curved owing to refraction and possibly to diffraction. The actual bottom contours in front of the barrier are as shown. The reflected wave crest pattern can be constructed by:

   (a) Constructing imaginary mirror image hydrography on the other side of the barrier.

   (b) Constructing the wave crest pattern that would develop as the wave propagates over this imaginary hydrography (this would involve the use of refraction and diffraction analyses as described in previous sections).

   (c) Constructing the mirror image of the imaginary wave crest pattern to define the real reflected wave.

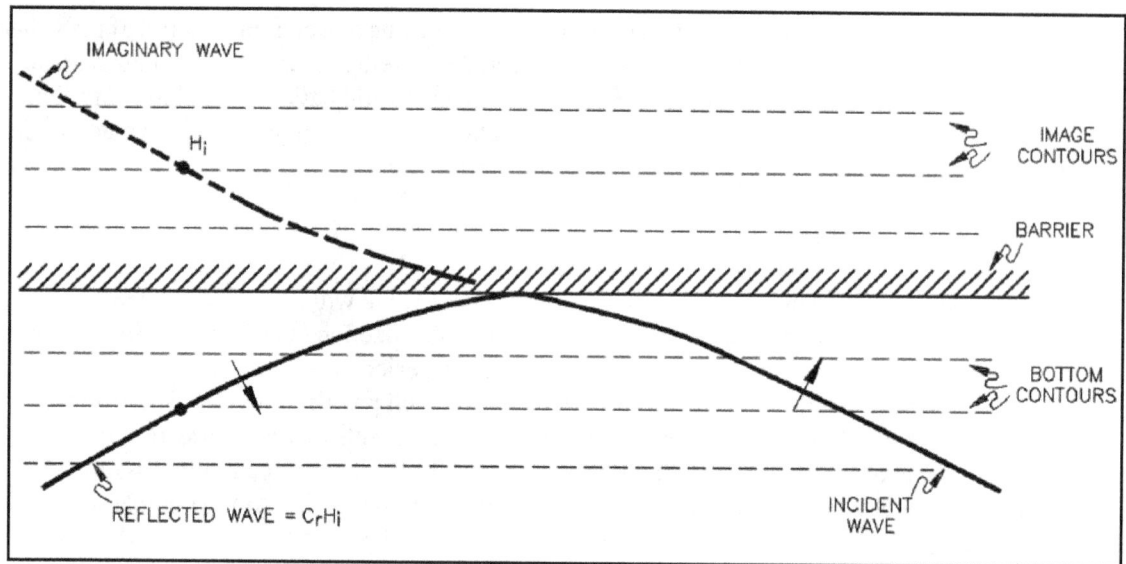

Figure II-7-24. Reflected wave crest pattern

(2) The reflected wave height at any point would be the height at the corresponding point in the imaginary area multiplied by the reflection coefficient at the point on the barrier where that segment of the wave reflected.

(3) Figure II-7-25 shows a wave diffracting in the lee of a breakwater and then reflecting off a wall $(C_r > 0)$. The inner end of the reflected wave then hits a beach where it is effectively dissipated; the outer end of the reflected wave diffracts around the breakwater tip and escapes the harbor. The height the reflected wave would have at point A would equal the diffracted height at A' times the reflection coefficient of the wall. By applying the concepts demonstrated in Figures II-7-24 and II-7-25, one can develop the reflection patterns and resulting wave heights for more complex harbor situations (see Carr (1952) and Ippen (1966)).

e. Reflection problems at harbor entrances.

(1) Generally speaking, wave energy that penetrates a harbor entrance should be dissipated as soon as possible, to prevent its subsequent reflection and propagation further into and about the harbor. A number of mechanisms for doing so are discussed in Bruun (1956, 1989). One mechanism is to construct sections of beach, stone mounds or other specially designed wave dissipating structures having low reflection coefficients at appropriate positions just inside the harbor entrance. Another mechanism is to construct resonance chambers at the entrance that are tuned to the dominant frequencies of the incident wave spectrum to set up oscillations that dissipate the penetrating wave energy. Often the general layout of the harbor can be designed to minimize the amount of wave energy that penetrates through the entrance and reflects and rereflects around the harbor.

(2) When significant wave reflection occurs in the vicinity of a harbor entrance (either inside or outside) the incident and reflected waves cross to form a complex "diamond-shaped" wave pattern. Hsu (1990) presents an extensive discussion of the kinematics of the resulting short-crested waves. This form of wave action can cause navigation difficulties and unusual sediment transport and scour patterns (see Silvester and Hsu (1993)).

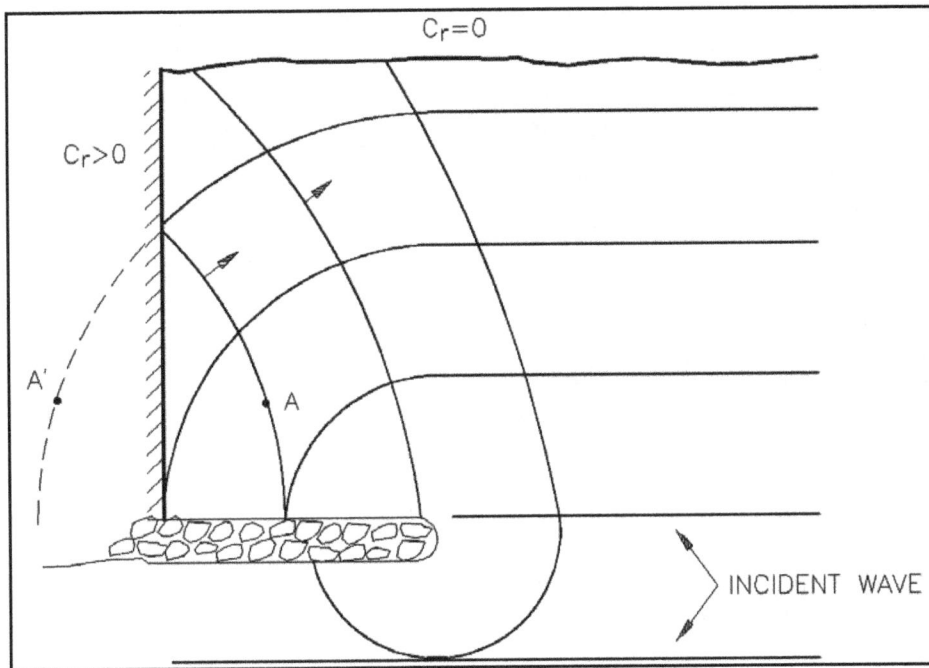

Figure II-7-25. Reflection of a diffracted wave

II-7-5. Harbor Oscillations

a. Introduction.

(1) Harbor oscillations are long-period wave motions that sometimes disrupt harbor activities. The oscillations are standing waves with typical periods between 30 sec and 10 min. Vertical motions are generally small, but horizontal motions can be large. Oscillation characteristics are generally controlled by basin size, shape, and water depth. Oscillations are most damaging when the period coincides with a natural resonant period of the harbor. The phenomenon is also referred to as harbor resonance, surging, seiching, and resonant oscillations.

(2) Harbor oscillations can be a significant problem for inner harbor components and moored vessels within a harbor basin. Resonant periods characteristic of moored vessels often fall into the same range of periods as harbor oscillations. Thus, harbor oscillations can create dangerous mooring conditions including breaking of mooring lines, damage to fender systems, vessel collisions, and delays of loading and unloading operations at port facilities.

(3) Processes and estimation procedures for harbor oscillations are discussed in this section. Discussion of harbor oscillations is generally presented in terms of the characteristics in Table II-7-2.

Table II-7-2
Harbor Oscillation Characteristics

Description	Alternatives	
Basin boundaries	Closed	Open
External forcing	Free	Forced
Dimensionality	2-dimensional	3-dimensional
Basin planform	Simple	Complex

(4) Characteristics are defined as follows:

(a) <u>Closed basin</u> - basin is completely enclosed.

(b) <u>Open basin</u> - basin is semi-enclosed, but open to a larger water body along at least part of one side.

(c) <u>Free oscillations</u> - oscillations that occur without external forcing (although some external forcing was applied earlier to initiate the oscillations).

(d) <u>Forced oscillations</u> - oscillations in response to external forcing.

(e) <u>2-dimensional</u> - oscillations are independent of one horizontal dimension.

(f) <u>3-dimensional</u> - oscillations vary in both horizontal dimensions.

(g) <u>Simple</u> - basin planform is a simple geometrical shape, such as a square, rectangle, or circle.

(h) <u>Complex</u> - basin planform is an irregular shape.

(5) A harbor basin generally has several modes of oscillation with corresponding natural resonant frequencies (or periods) and harmonics. Figure II-7-26 illustrates the fundamental, second, and third harmonic modes of oscillation in idealized, perfectly reflecting, closed and open two-dimensional basins.

(6) Following this introduction, the process of resonance is discussed in terms of a more intuitive, but analogous, mechanical system. Closed basins are covered next, mainly in terms of free oscillations and simple shapes. Although they are not closed basins, harbors or parts of harbors can behave much like closed basins under some conditions. The presentation is also applicable to enclosed water bodies such as lakes and reservoirs.

(7) The last parts of the section are devoted to open basins. Open basins are susceptible to oscillations forced across the open boundary. Because of the limited size of harbors, other types of forcing, such as meteorological forcing in the harbor, are generally not considered. Both simple and complex shapes are presented. The final part describes Helmholtz resonance, a very long-period, non-standing wave phenomenon that causes water levels over the entire harbor to oscillate up and down in unison. Practical consequences of harbor oscillation, such as vessel motions, mooring line forces, and fender forces, are not presented in this section. Motions of small boats moored in resonant conditions and possible mitigation measures have been investigated by Raichlen (1968). Comprehensive reviews of harbor oscillations are given by Raichlen and Lee (1992) and Wilson (1972).

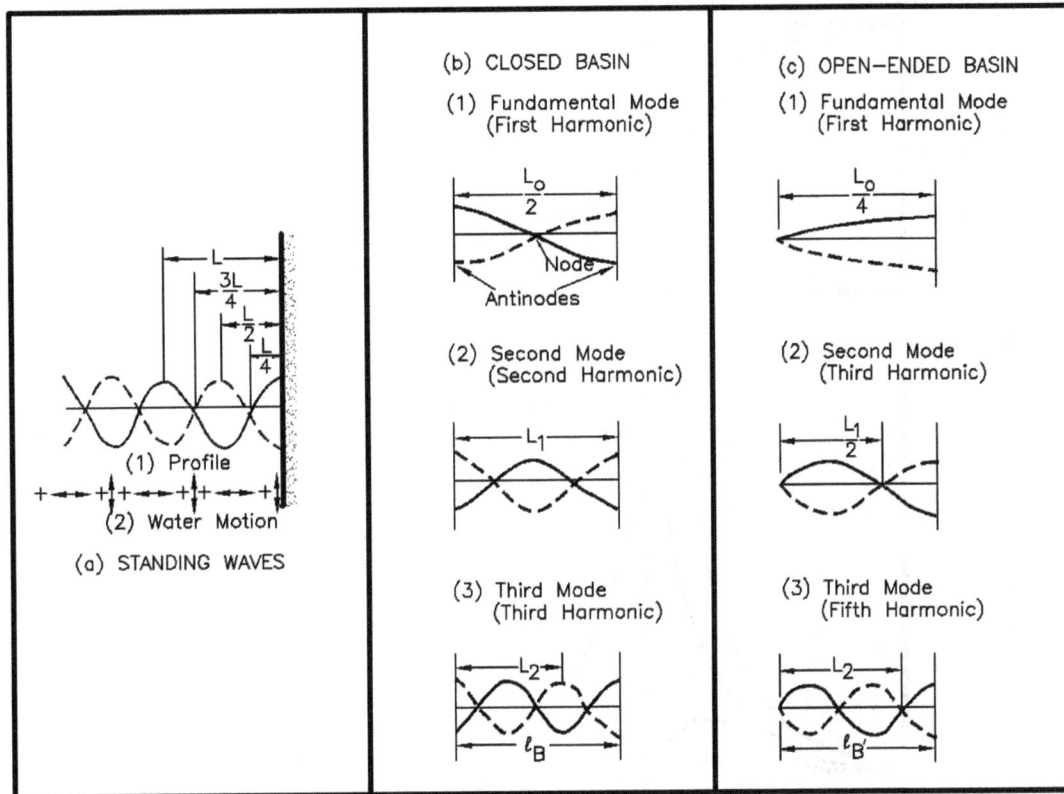

Figure II-7-26.   Surface profiles for oscillating waves (Carr 1953)

b.   Mechanical analogy.

(1) The basic theory for basin oscillations is similar to that of free and forced oscillations experienced by some mechanical, acoustical, and other fluid systems. Certain systems respond to a disturbance by developing a restoring force that reestablishes equilibrium in the system. A pendulum is a good example of such a system. A free oscillation at the system's natural period or frequency is initiated if the system is carried by inertia beyond the equilibrium condition. If the forces responsible for the initial disturbance are not sustained, free oscillations at the natural frequency will continue, but their amplitude decays exponentially due to friction. The system eventually comes to rest. Forced oscillations can occur at non-natural frequencies if cyclic energy is applied to a system at non-natural frequencies. Continuous excitation at frequencies at or near the natural frequency of a system generally causes an amplified response. The response magnitude depends on the proximity of the excitation to the natural frequency and the frictional characteristics of the system.

(2) The response of a linearly damped, vibrating spring-mass system with one degree of freedom provides a good illustration (Figure II-7-27). Terms are defined as follows: A = amplification factor (ratio of mass displacement to excitation displacement); T = excitation (and response) period; T_n = natural period of the mass-spring system; and ϕ = phase angle by which the mass displacement lags the excitation displacement. The mass responds directly to the excitation if the excitation period is much greater than the natural period of the system; that is, $A = 1$ and $\phi = 0°$ when $T_n/T \ll 1$. The harbor equivalence to this case is the harbor response to astronomical tides. The mass responds very little and out of phase with the excitation if the excitation period is much shorter than the natural period of the system; that is, $A \ll 1$ and $\phi \approx 180°$ when $T_n/T \gg 1$. The mass response is amplified and a phase lag develops as the excitation period approaches the natural period of the system. The ratio T_n/T determines the degree of amplification and phase

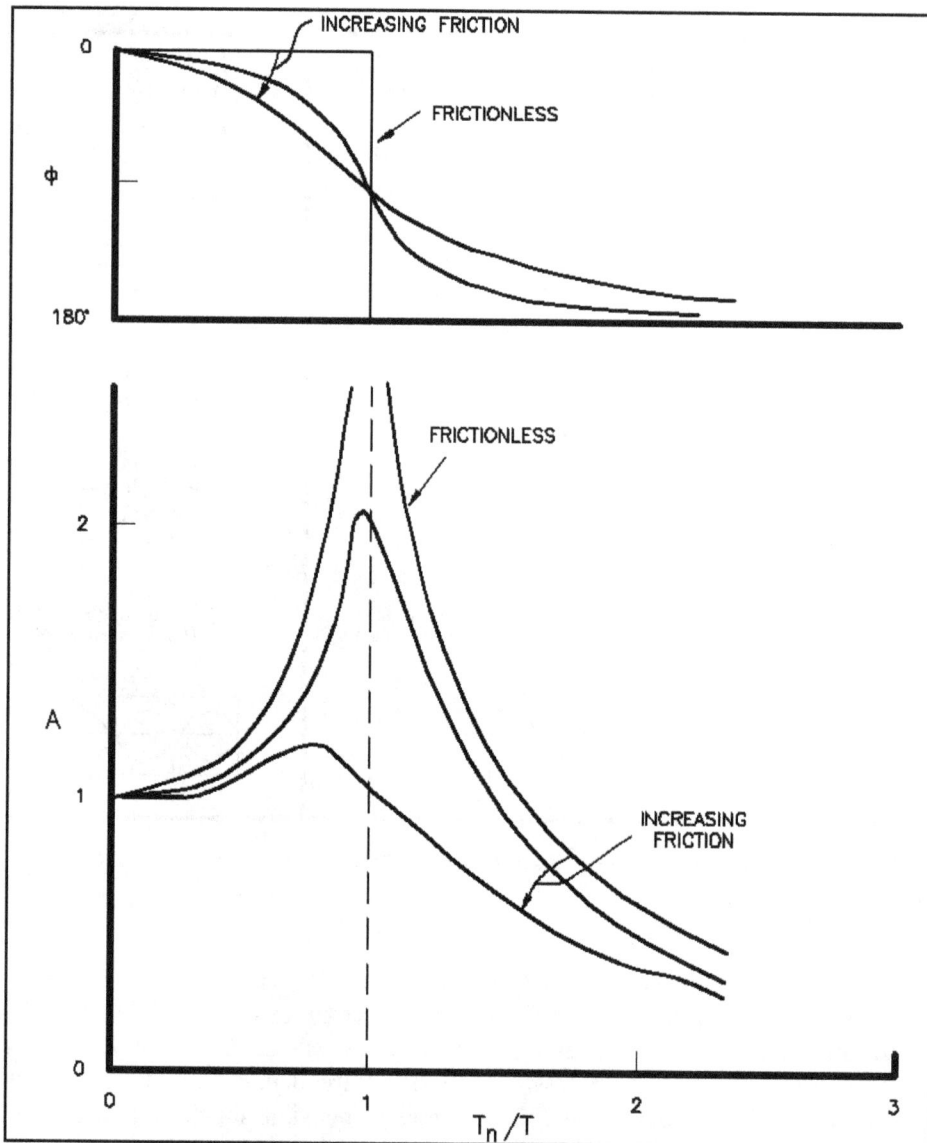

Figure II-7-27. Behavior of an oscillating system with one degree of freedom

lag. The greatest amplification occurs when $T_n/T = 1$ and $\phi = 90$ deg. With a cyclic excitation at period T_n, the amplification factor increases with time until the rate of energy input equals the rate of energy dissipation by friction. When energy input is stopped, the response amplitude decreases exponentially with time as a result of friction. Amplification due to resonance in a mass-spring system and the respective phase relationships are further developed and discussed in Raichlen (1968), Meirovitch (1975), Sorensen (1986), and Wilson (1972).

   c. *Closed basins.*

   (1) Enclosed basins can experience oscillations due to a variety of causes. Lake oscillations are usually the result of a sudden change, or a series of intermittent-periodic changes, in atmospheric pressure or wind velocity. Oscillations in canals can be initiated by suddenly adding or subtracting large quantities of water. Harbor oscillations are usually initiated by forcing through the entrance; hence, they deviate from a true closed basin. Local seismic activity can also create oscillations in an enclosed basin.

(2) Basins are generally shallow relative to their length. Hence, basin oscillations involve standing waves in shallow water. The simplest basin geometry is a narrow rectangular basin with vertical sides and uniform depth. The natural free oscillating period for this simple case, assuming water is inviscid and incompressible, is given by

$$T_n = \frac{2\,\ell_B}{n\,\sqrt{gd}} \qquad \text{Closed basin} \qquad (II\text{-}7\text{-}9)$$

where

T_n = natural free oscillation period

n = number of nodes along the long basin axis (Figure II-7-46)

ℓ_B = basin length along the axis

g = acceleration due to gravity

d = water depth

(3) This equation is often referred to as Merian's formula. The maximum oscillation period T_1 corresponding to the fundamental mode is given by setting $n = 1$ as

$$T_1 = \frac{2\,\ell_B}{\sqrt{gd}} \qquad (II\text{-}7\text{-}10)$$

(4) If the rectangular basin has significant width as well as length (Figure II-7-28), both horizontal dimensions affect the natural period, given by

$$T_{n,m} = \frac{2}{\sqrt{gd}}\left[\left(\frac{n}{\ell_1}\right)^2 + \left(\frac{m}{\ell_2}\right)^2\right]^{\left(-\frac{1}{2}\right)} \qquad \text{Closed basin} \qquad (II\text{-}7\text{-}11)$$

where

$T_{n,m}$ = natural free oscillation period

n,m = number of nodes along the x- and y-axes of basin

ℓ_1, ℓ_2 = basin dimensions along the x- and y-axes

(5) Equation II-7-11 reduces to Equation II-7-9 for the case of a long narrow basin, in which $m = 0$. Further discussion is provided in Raichlen and Lee (1992) and Sorensen (1993). Closed basins of more complex shape require other estimation procedures. Raichlen and Lee (1992) present procedures for a circular basin and approximate solution methods for more arbitrary basin shapes. Defant (1961) outlines a method to determine the possible periods for two-dimensional free oscillations in long narrow lakes of variable width and depth. Locations of nodes and antinodes can also be determined. Usually numerical models are used to properly estimate the response of complex basins.

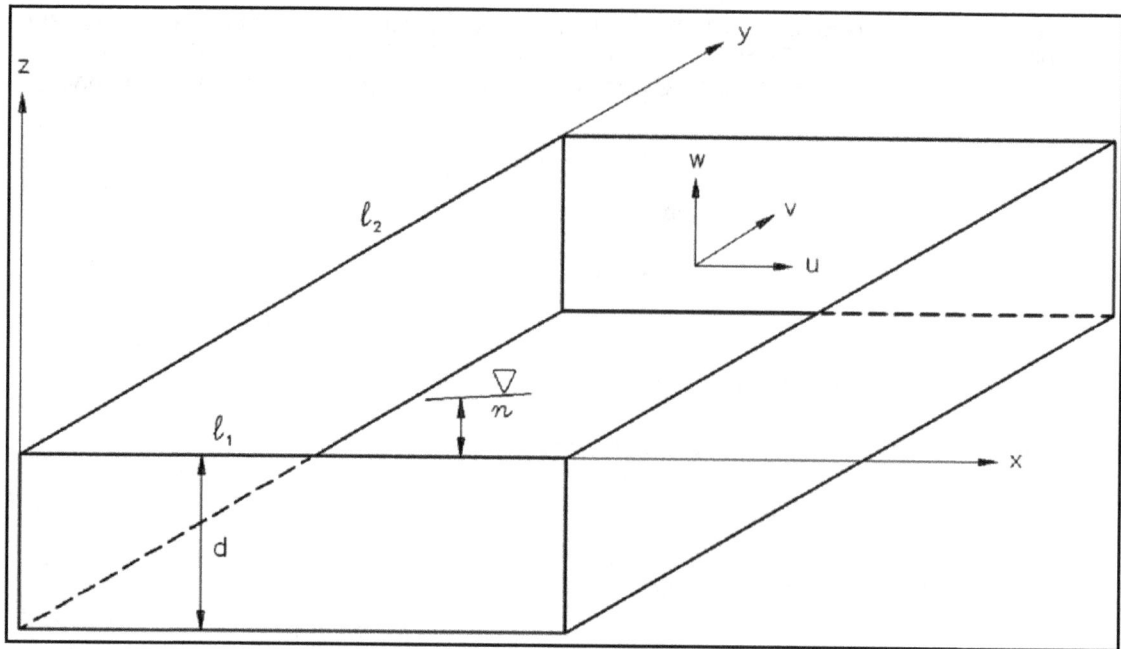

Figure II-7-28. Behavior of an oscillating system with one degree of freedom

(6) Forced oscillations of closed basins can be a concern when the basin is large enough to be affected by moving pressure gradients or strong surface winds, which is generally not true for harbors. Winds can be especially important in large shallow basins, such as Lake Erie. The subject of forced oscillations of closed basins is discussed by Raichlen and Lee (1992).

  d.   Open basins - general.

(1) Open basins, typically harbors or bays, are most susceptible to oscillations forced across the open boundary. Typical forcing mechanisms include infragravity waves, eddies generated by currents moving past a harbor entrance, and tsunamis (Sorensen 1986). Local seismic activity can also generate oscillations within the basin. Meteorological forces (changes in atmospheric pressure and wind) can initiate oscillations in large bays, but they are not usually a concern over areas the size of a harbor.

(2) The period of a true forced oscillation is the same as the period of the exciting force. However, forced oscillations are usually generated by intermittent external forces, and the period of oscillation is greatly influenced by the basin dimensions and mode of oscillation.

  e.   Open basins - simple shapes.

(1) In many cases, the geometry of a harbor can be approximated by an idealized, simple shape such as a rectangle or circle. Then the approximate response characteristics can often be determined from guidance based on analytic solutions. Simple harbors are also very helpful for developing an understanding of the general behavior of an open basin.

(2) As with closed basins, the simplest, classical case is a narrow, rectangular basin with uniform depth. The basin has vertical walls on three sides and is fully open at one end. The fundamental mode of resonant oscillation occurs when there is one-quarter of a wave in the basin (Figure II-7-26). The general expression for the free oscillation period in this case is

$$T_n = \frac{4\,\ell_B}{(1 + 2n)\,\sqrt{gd}} \qquad \text{Open basin} \qquad \text{(II-7-12)}$$

(3) The number of nodes in the basin n does not include the node at the entrance. The period for the fundamental mode ($n = 0$) is

$$T_0 = \frac{4\,\ell_B}{\sqrt{gd}} \qquad \text{(II-7-13)}$$

(4) Maximum horizontal velocities and particle excursions in a standing wave occur at the nodes. Thus there is potential for troublesome harbor conditions in the vicinity of nodes. Some additional useful relationships for standing waves are as follows (Figure II-7-29) (see Sorensen (1986) for details)

$$V_{max} = \frac{H}{2}\sqrt{\frac{g}{d}} \qquad \text{(II-7-14)}$$

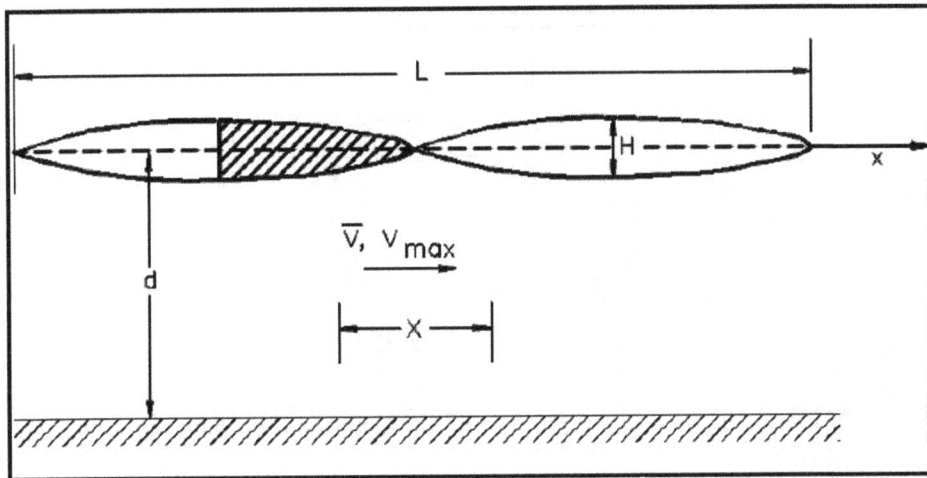

Figure II-7-29. Motions in a standing wave

where

V_{max} = maximum horizontal velocity at a node

H = standing wave height

$$X = \frac{H\,T_n}{2\pi}\sqrt{\frac{g}{d}} \qquad \text{(II-7-15)}$$

where X = maximum horizontal particle excursion at a node

$$\bar{V} = \frac{H\,L}{\pi\,d\,T_n} \qquad \text{(II-7-16)}$$

where $\bar{V}$ = average horizontal velocity at a node

(5) For example, when $H = 1$ ft, $T_I = 200$ sec, and $d = 30$ ft, the maximum horizontal velocity and particle excursion are $V_{max} = 0.5$ ft/sec and $X = 33$ ft.

(6) The resonant response of the simple rectangular harbor, presented by Ippen and Goda (1963) in terms of the amplification factor A (assuming no viscous dissipation) and the relative harbor length $k\ell$, where $k = 2\pi/L$, illustrates other important aspects of harbor oscillation. The amplification factor for harbor oscillations is traditionally defined as the ratio of wave height along the back wall of the harbor to standing wave height along a straight coastline (which is twice the incident wave height). The response curve for a long, narrow harbor is given in Figure II-7-30. The left portion of the curve resembles that for the mechanical analogy in Figure II-7-26. Resonant peaks for higher order modes are also shown. The three curves correspond to a fully open harbor and two partially open harbors with different degrees of closure. The $k\ell$ value at resonance decreases as the relative opening width decreases. It is bounded by the value for a closed basin, also shown in the figure. Amplification factors for the inviscid model are upper bounds on those experienced in a real harbor. Because amplification factor decreases at each successive higher order mode, simple analysis methods often focus only on the lowest order modes.

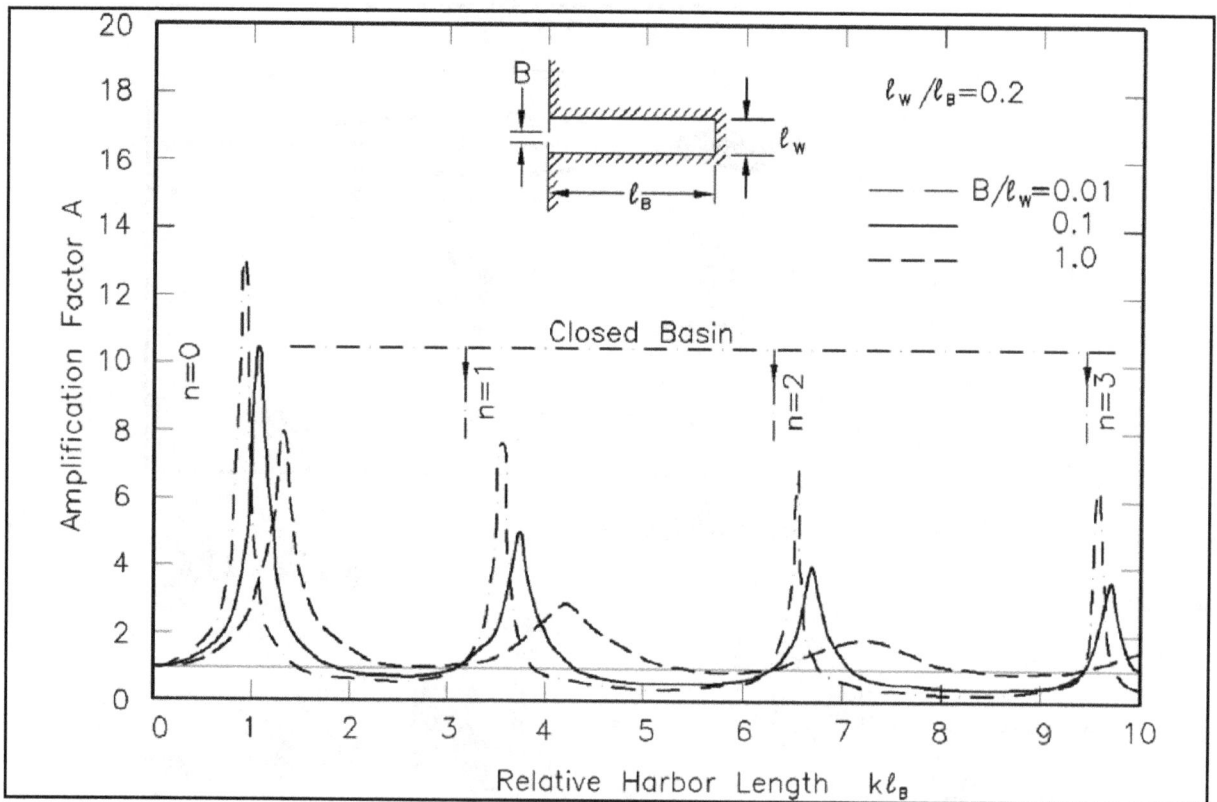

Figure II-7-30. Theoretical response curves of symmetrical, narrow, rectangular harbor (Raichlen (1968); after Ippen and Goda (1963))

(7) Practical guidance for assessing the strength and period of the first two resonant modes in a partially enclosed rectangular harbor with a symmetric entrance is given in Figure II-7-31. A wide range of relative harbor widths (aspect ratios, b/ℓ) and relative entrance widths (d/b) is represented. The $k\ell$ values can be converted to resonant periods by

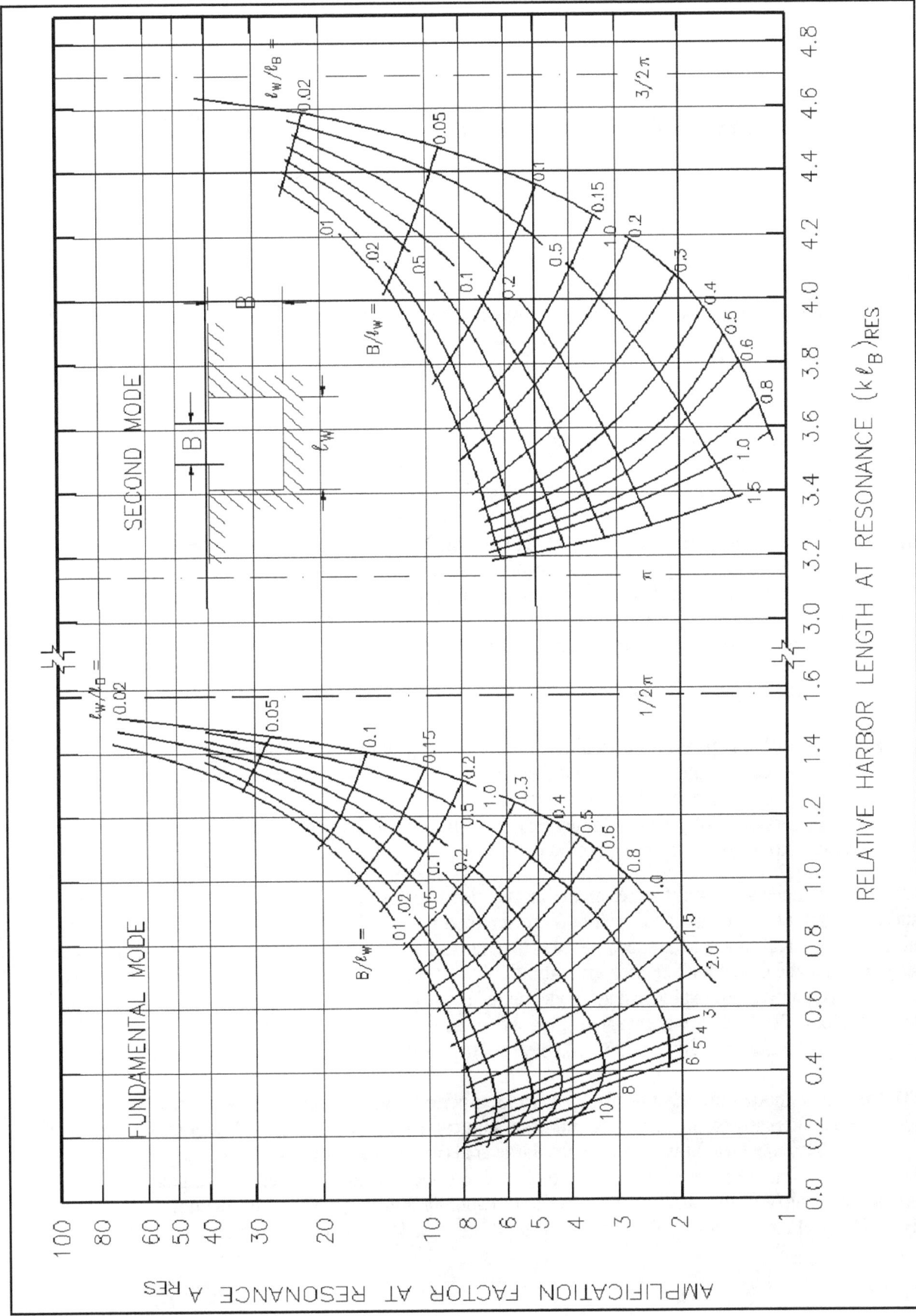

Figure II-7-31.   Resonant length and amplification factor of symmetrical rectangular harbor (from Raichlen and Lee (1992); after Ippen and Goda (1963))

$$T = \left(\frac{2\pi\, \ell}{\sqrt{gd}}\right)\left(\frac{1}{k\ell}\right) \qquad\qquad (\text{II-7-17})$$

(8) It is interesting to note that the resonant period increases as the aspect ratio increases, even for the fully open harbor. The change is due to the effect of the confluence of the entrance and the open sea. Resonant periods estimated from Figure II-7-30 generally differ from the simple approximation in Equation II-7-12.

(9) The simple guidance has some important limitations relative to real harbors. Amplification factors are upper bounds, since no frictional losses were modelled. Harbors that are not narrow experience transverse oscillation modes that are not represented in the simple guidance. Harbors with asymmetric entrances experience additional transverse modes of oscillation and possible increases in amplification factor (Ippen and Goda 1963). In addition to introducing new resonant frequencies and modified amplification, transverse oscillations change the locations of nodes, as illustrated in Figure II-7-32 based on numerical model calculations.

(10) Harbor oscillation information is available for some other simple harbor shapes. Circular harbors were investigated by Lee (1969) and reviewed by Raichlen and Lee (1992). Zelt (1986) and Zelt and Raichlen (1990) developed a theory to predict the response of arbitrary shaped harbors with interior sloping boundaries, including runup. The dramatic effect of a sloping rather than flat bottom on the behavior of a rectangular harbor is illustrated in Figure II-7-33. The sloping bottom greatly increases A and greatly reduces $k\ell$ values at resonance. For example, $k\ell$ for the second resonant peak drops from 4.2 for the flat bottom case to 2.6 for the sloping bottom case. The first two resonant modes for six symmetric, fully open configurations are given in Figure II-7-34. The A for the first mode is much more affected by the sloping bottom than by the harbor planform. Details are available in Zelt (1986).

f. Open basins - complex shapes.

(1) Real harbors never precisely match the simple shapes; usually they differ significantly. Complex harbors can be analyzed with physical and numerical models. These modeling tools should be applied even for relatively simple harbor shapes when the study has large economic consequences and accurate results are essential. A combination of physical and numerical modeling is usually preferred for investigating the full range of wave conditions in a harbor (Lillycrop et al. 1993).

(2) Physical models generally represent the shorter-period harbor oscillations more accurately. The harbor and immediately surrounding coastal areas are sculpted in cement in a three-dimensional model basin (Figure II-7-35). Breakwaters and other structures that transmit significant amounts of long-wave energy are properly scaled in the model for size and permeability. Currents can be introduced when needed to represent field conditions. Limitations to physical modeling include cost, no direct simulation of frictional dissipation, and the inherent difficulties in working with long waves in an enclosed basin. Special care is required to properly generate long waves. Once generated, they tend to produce undesirable reflections from basin walls.

(3) Numerical models are most useful for very long-period investigations, initial studies, comparative studies of harbor alternatives, and revisiting harbors documented previously with field and/or physical model data. They are also most useful when unusually large areas and/or very long waves are to be studied. For example, numerical models have been used effectively to select locations for field wave gauges (to avoid nodes) and to identify from many alternatives the few most promising harbor modification plans for fine tuning in physical model tests. Lillycrop et al. (1993) suggested that numerical modeling is preferable to

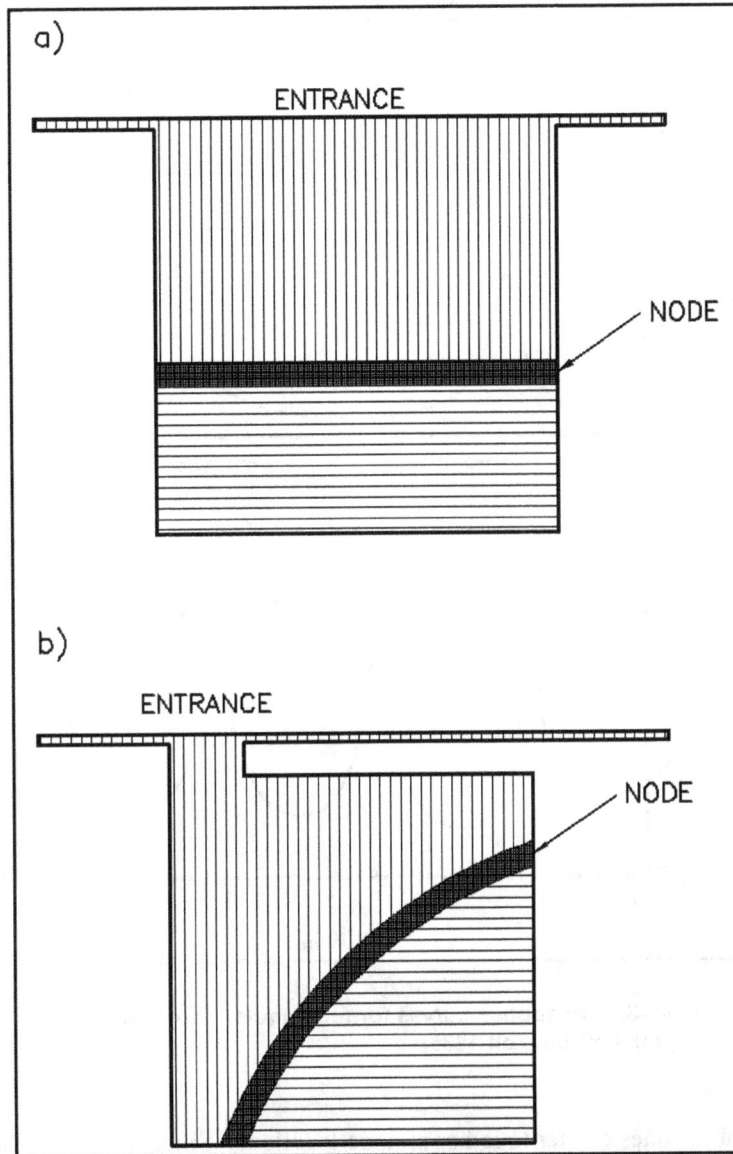

Figure II-7-32.   Node locations for a dominant mode of oscillation in a square harbor: a) fully open; b) asymmetric, constricted entrance

physical modeling for periods longer than 400 sec. Both modeling tools can be used effectively for the shorter-period oscillations.

(4) Numerical models can reproduce the geometry and bathymetry of a harbor area reasonably well and estimate harbor response to long waves. Figure II-7-36 is an example numerical model grid. This grid is finer than would normally be required for harbor oscillation studies because it was designed for both wind waves and long waves. Numerical models can be used to generate harbor response curves as in Figure II-7-33 at various points in the harbor. Results for resonance conditions of particular interest can be displayed over the whole harbor to show oscillation patterns. For example, amplification factors and phases calculated with the example grid are presented for five wave periods, corresponding to resonant peaks in the main harbor basin (Figures II-7-37 and II-7-38). Phases are relative to the incident wave. Phase plots are useful because phases in a pure standing wave are constant up to a node and then change 180 deg across the

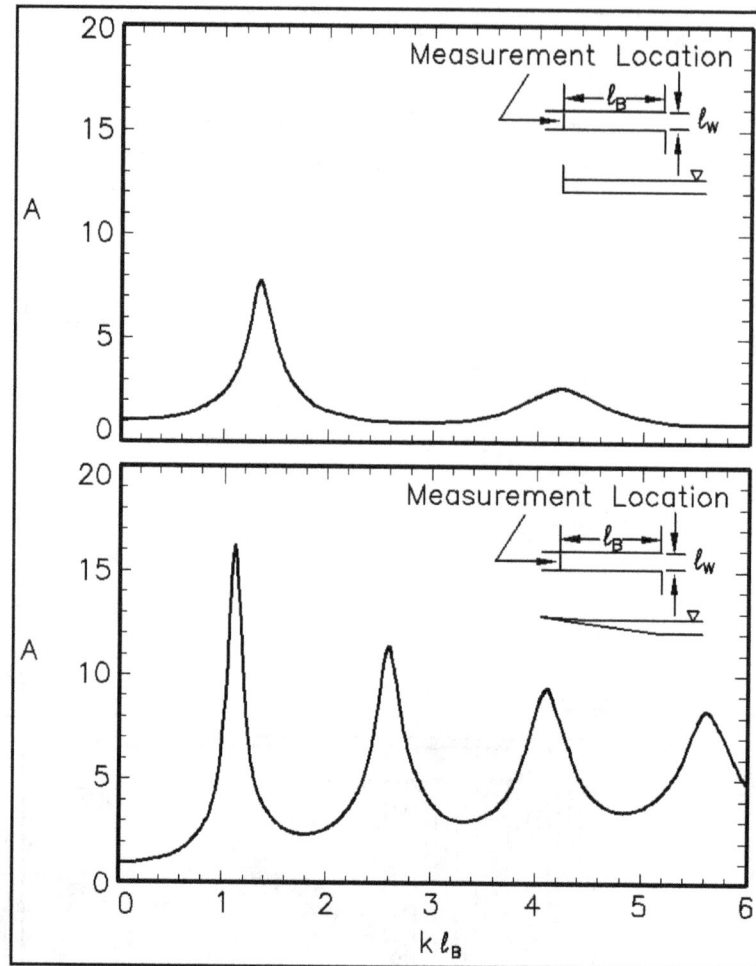

Figure II-7-33. Response curves for rectangular harbor with flat and sloping bottom (Zelt 1986)

node. Thus, phase contour lines cluster together at node locations. Since regions of similar phase, plotted with similar gray shades (or colors on a computer screen), move up and down together, the resonant modes of oscillation can be much more easily visualized. This information about node locations and behavior of resonant modes is quite useful in analyzing harbor oscillations.

(5) Limitations to numerical models relate to approximations in formulating the following items: incident wave conditions, wave evolution during propagation over irregular bathymetry, wave interaction with structures, and harbor boundary features. Numerical model technology is reviewed by Raichlen and Lee (1992) and Abbott and Madsen (1990).

(6) The Corps of Engineers has traditionally used a steady-state, hybrid-element model based on the mild slope equation (Chen 1986; Chen and Houston 1987; Cialone et al. 1991). Variable boundary reflection and bottom friction are included. The seaward boundary is a semicircle outside the harbor entrance. Monochromatic waves are incident along the semicircle boundary. Since the model is linear, results from multiple monochromatic wave runs can be recombined to simulate a spectral response. The model does not account for entrance losses (Thompson, Chen, and Hadley 1993).

Harbor Geometry	First Resonant Mode		Second Resonant Mode	
	$k\ell_B$	A_{RES}	$k\ell_B$	A_{RES}
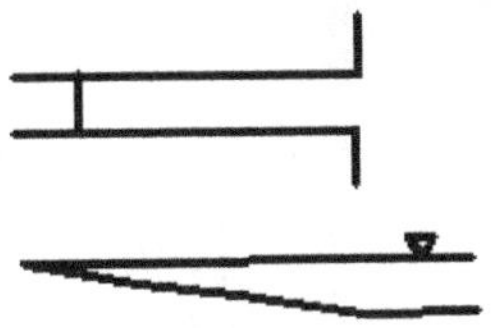	1.089	16.43	2.565	11.45
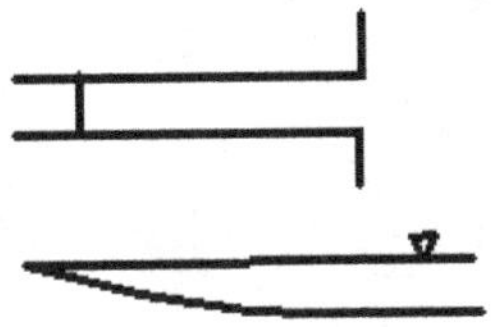	1.229	10.96	3.177	7.61
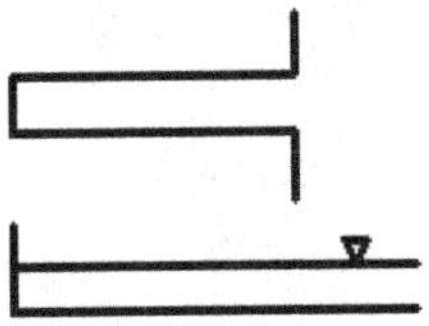	1.315	7.81	4.182	2.68
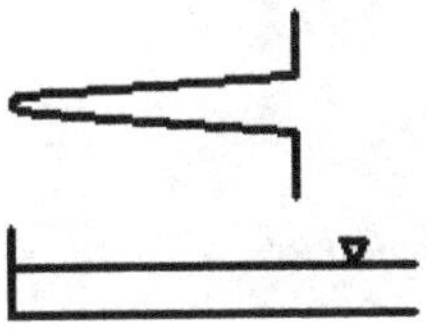	1.696	8.12	4.559	4.32
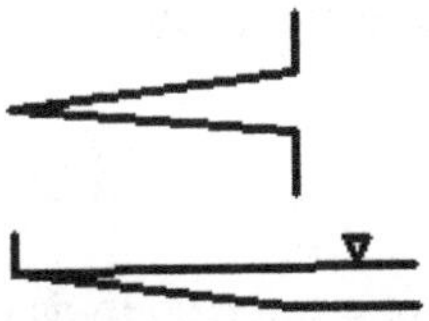	1.757	21.85	3.280	32.18
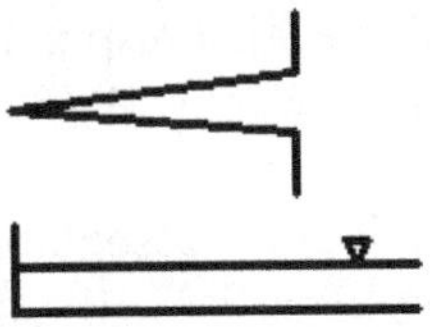	2.050	8.50	4.926	6.19

Figure II-7-34. Resonant response of idealized harbors with different geometry (Zelt 1986)

Figure II-7-35. Photograph of physical model, Barbers Point Harbor, HI (Briggs et al. 1994)

   g.   *Open basins - Helmholtz resonance.*

   (1) A harbor basin open to the sea through an inlet can resonate in a mode referred to as the Helmholtz or grave mode (Sorensen 1986b). This very long period mode appears to be particularly significant for harbors responding to tsunami energy and for several harbors on the Great Lakes that respond to long-wave energy spectra generated by storms (Miles 1974; Sorensen 1986; Sorensen and Seelig 1976).

   (2) Water motion characterizing the Helmholtz mode is like that of a Helmholtz resonator in acoustics. It is analogous to the spring-mass system with one degree of freedom, discussed earlier in this section, where the spring is similar to the basin water surface and the mass represents water in the inlet channel. During Helmholtz resonance, the basin water surface uniformly rises and falls while the inlet channel water oscillates in and out. The period of this mode is greater than the fundamental mode. The resonant period is given by (Carrier et al. 1971).

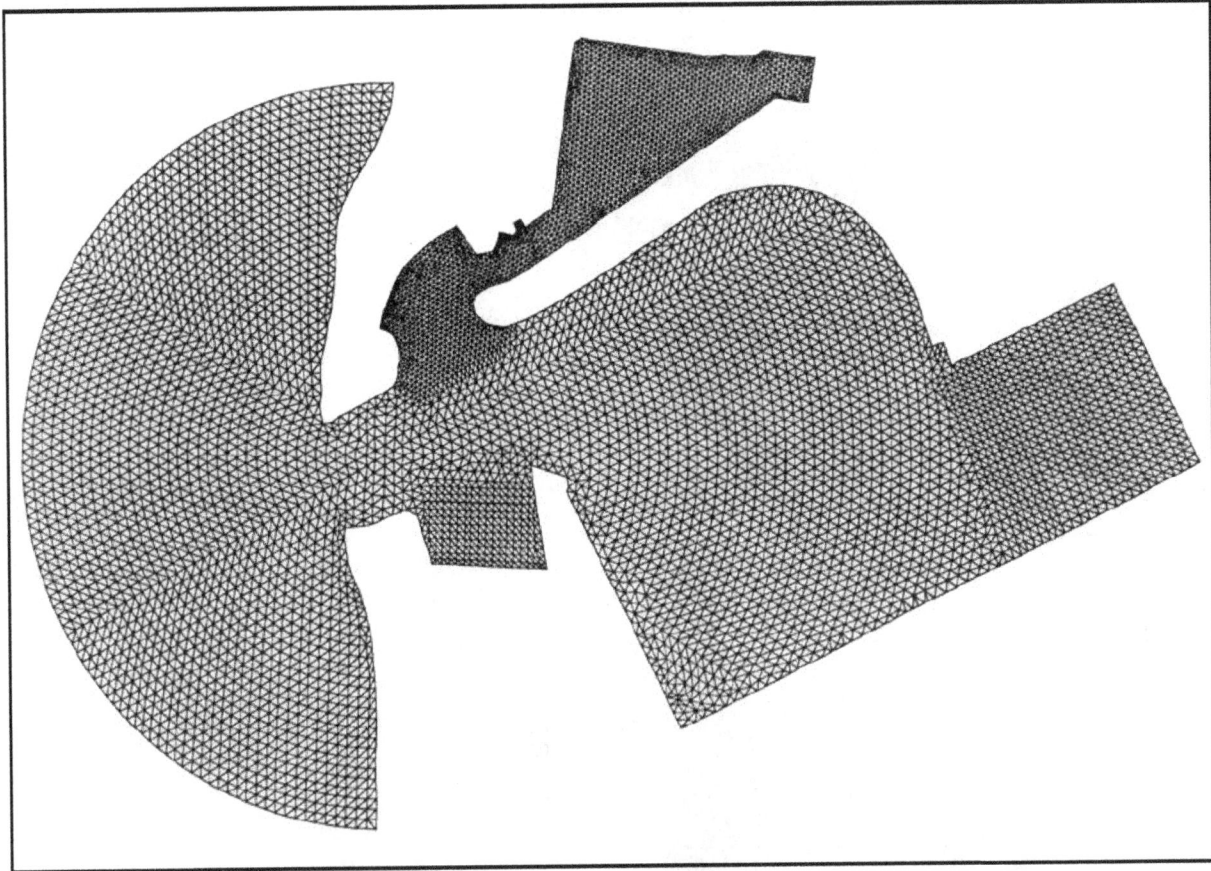

Figure II-7-36. Numerical model grid for Barbers Point Harbor, HI (Briggs et al. 1994)

$$T_H = 2\pi \sqrt{\frac{(\ell_c + \ell'_c)\, A_b}{g A_c}}$$

(II-7-18)

where

T_H = resonant period for Helmholtz mode

ℓ_c = channel length

ℓ'_c = additional length to account for mass outside each end of the channel

A_b = basin surface area

A_c = channel cross-sectional area

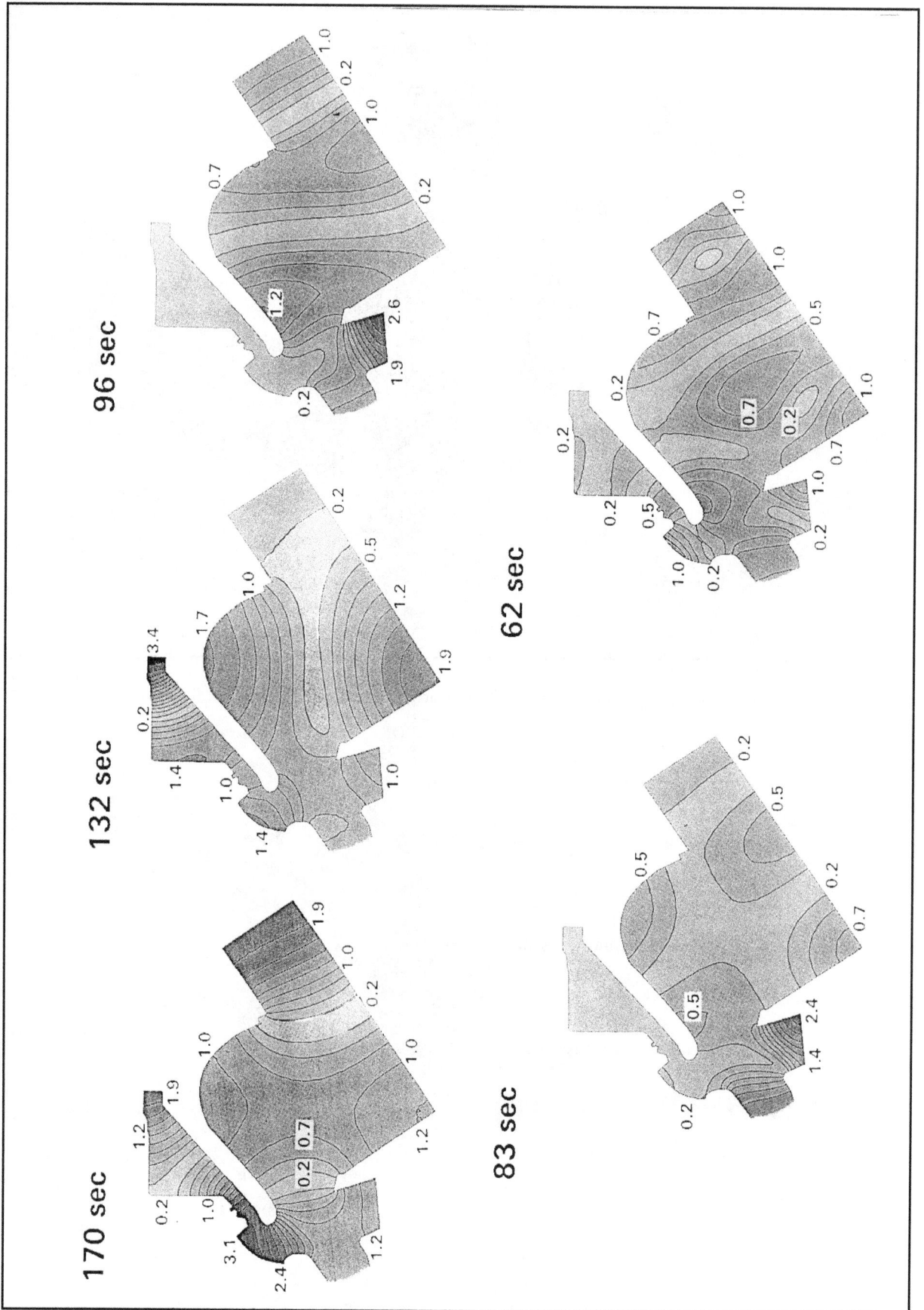

Figure II-7-37.   Amplification factors for five resonant periods,   Barbers Point Harbor,   HI   (Briggs et al. 1994)

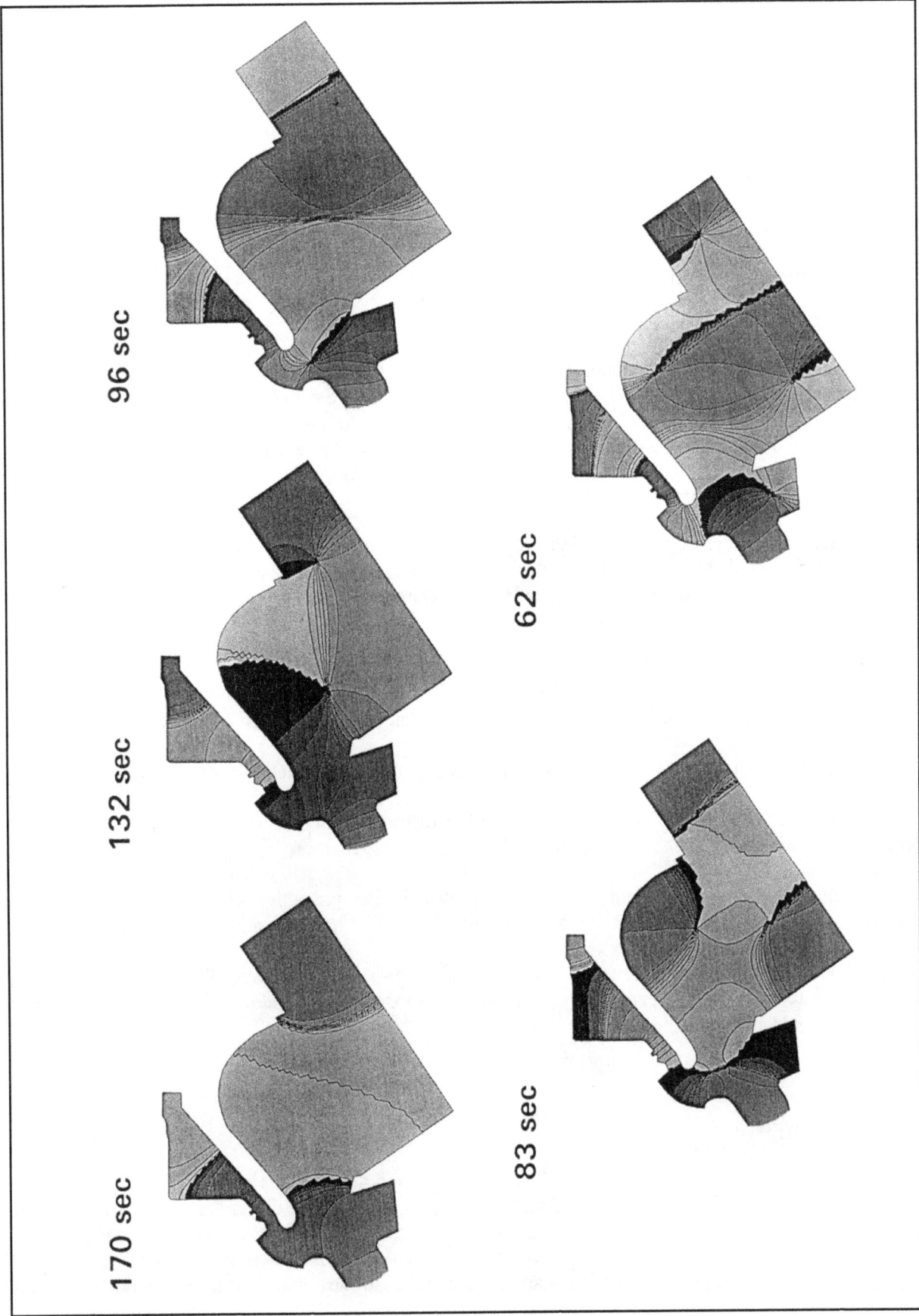

Figure II-7-38. Phases for five resonant periods, Barbers Point Harbor, HI (Briggs et al. 1994)

(3) The added length is given by (Sorensen 1986, adopted from Miles (1948))

$$\ell'_c = \frac{-W}{\pi} \ln\left(\frac{\pi W}{\sqrt{gd_c}\, T_H}\right) \tag{II-7-19}$$

where

W = channel width

d_c = channel depth

II-7-6. Flushing/Circulation

  a. Statement of importance.

(1) An important aspect of harbor design is the water circulation that occurs within the harbor and between the harbor and the surrounding water body. Water exchange with the surrounding water body will produce a flushing action in the harbor. Flushing is important to reduce the level of chemical, biological, and floating solids pollution in the harbor. Typically, a criterion such as the exchange of X percent of the water in the harbor within a certain time period or, alternatively, exchange of Y percent of the water in the harbor each tidal cycle might be set as a design goal.

(2) Circulation patterns within a harbor should eliminate areas of stagnant water where pollution levels will rise and fine sediment deposition may occur. This can occur, for example, in closed-end channels such as pier slips and residential canals. Strong circulation is desired as long as adverse high-velocity currents are not generated.

(3) Natural harbor circulation and flushing should be optimized. This can be accomplished by:

(a) Siting the harbor to make use of ambient currents that will pass the harbor entrance.

(b) Considering water circulation when determining the number, size, and placement of harbor entrances.

(c) Establishing the harbor planform and internal structure locations so that circulation is optimized and potential pollution sources are located in areas of strong water circulation.

(d) Employing (where incident wave action allows) harbor protective structures such as floating breakwaters and vertical barriers with bottom openings that permit flow into and out of the harbor.

(e) Installing culvert pipes through more massive harbor protective structures. This natural flushing can be supplemented by installing pumping systems that bring exterior water into dead areas of the harbor or remove polluted water from these areas.

  b. Flushing/circulation processes.

(1) Tidal action.

(a) As the tide rises, ambient water will enter a harbor and mix with the water in the harbor. On the subsequent falling tide, a portion of this ambient/harbor mixture will leave the harbor. The net result is the exchange of some harbor water with water from outside the harbor. The efficiency of this exchange depends primarily on two factors. One factor is the ratio of the volume of water that enters on one tidal cycle (the tidal prism) to the total volume of water in the harbor. This ratio, in turn, will depend on the tide range, the hydraulic efficiency of the harbor entrance, and the water depths in the harbor. Generally, the larger the value of this ratio, the more exchange that will occur. The other factor is the momentum of the incoming jet of water on the rising tide, and the consequent amount of penetration of this jet and its resulting angular momentum as it establishes a rotating gyre inside the harbor. The strength of this jet is related to the amount of flushing that will occur and the strength of the tidally induced circulation in the harbor.

(b) The most commonly used factor for defining the effectiveness of tidally induced harbor flushing is the average per cycle exchange coefficient E (Nece and Richey 1975) given by

$$E = 1 - \left(\frac{C_i}{C_o} \right)^{\frac{1}{i}} \qquad \text{(II-7-20)}$$

where C_o is the initial concentration of some substance in the harbor water and C_i is the concentration of this substance after i tidal cycles. E may be defined at a point in the harbor by using concentration values at that point or may be defined for the entire harbor by using spatially average concentration values for the entire harbor. E is the fraction of harbor water removed each tidal cycle. Equation II-7-20 assumes essentially repetitive identical tides and no further addition of the marker substance to the harbor.

(c) As an example of the use of the exchange coefficient, consider Figure II-7-39, which is modified from Falconer (1980) and based on physical and numerical model results. Figure II-7-39 shows the spatial average exchange coefficient for a rectangular harbor having the dimensions L and B shown in the insert. Note where the harbor entrance is located. E has a peak value of about 0.5 when L/B is near unity (i.e. a square harbor). For L/B values outside the range of 0.3 to 3, the flushing of the harbor is much less effective. The tidal prism ratio (TPR), defined as the tidal prism divided by the total harbor volume at high tide, was 0.4 for the results shown in Figure II-7-39.

(d) Another factor used to define the amount of flushing is the "flushing efficiency," defined as the exchange coefficient divided by the tidal prism ratio (i.e. E/TPR) expressed as a percentage (Nece, Falconer, and Tsutumi 1976). Flushing efficiency compares the actual water exchange in one cycle with the volume of exchange that would occur if the incoming tidal prism completely mixed with the harbor water on each cycle. Table II-7-3 shows some results for five small-boat harbors in the state of Washington. These results are based on model tests (Nece, Smith, and Richey 1980).

(e) Note that the exchange coefficients are all around 0.2, which is below the value for the more idealized rectangular harbors (Figure II-7-39). And, for relatively consistent exchange coefficients, the flushing efficiency showed a much greater range of values. For the rectangular harbor in Figure II-7-39 where the TPR is 0.4, the flushing efficiency would vary from 0 up to about 125 percent. The desirable exchange coefficient or flushing efficiency for a given harbor would depend on the level of pollution in the harbor versus the desired harbor water quality. Although higher exchange coefficients and flushing efficiencies generally indicate better harbor flushing by tidal action, there may still be small pockets of stagnant water in a generally well-flushed harbor.

(2) Wind effects.

(a) Wind acting on the water surface will generate a surface current that, in water depths typically found in harbors, will essentially be in the direction of the wind. Owing to Coriolis effects, the current will be a few degrees to the right of the wind in the Northern Hemisphere; e.g., see Neumann and Pierson (1966) or Bowden (1983). If the distance over which the wind blows and the wind duration are sufficient, the surface

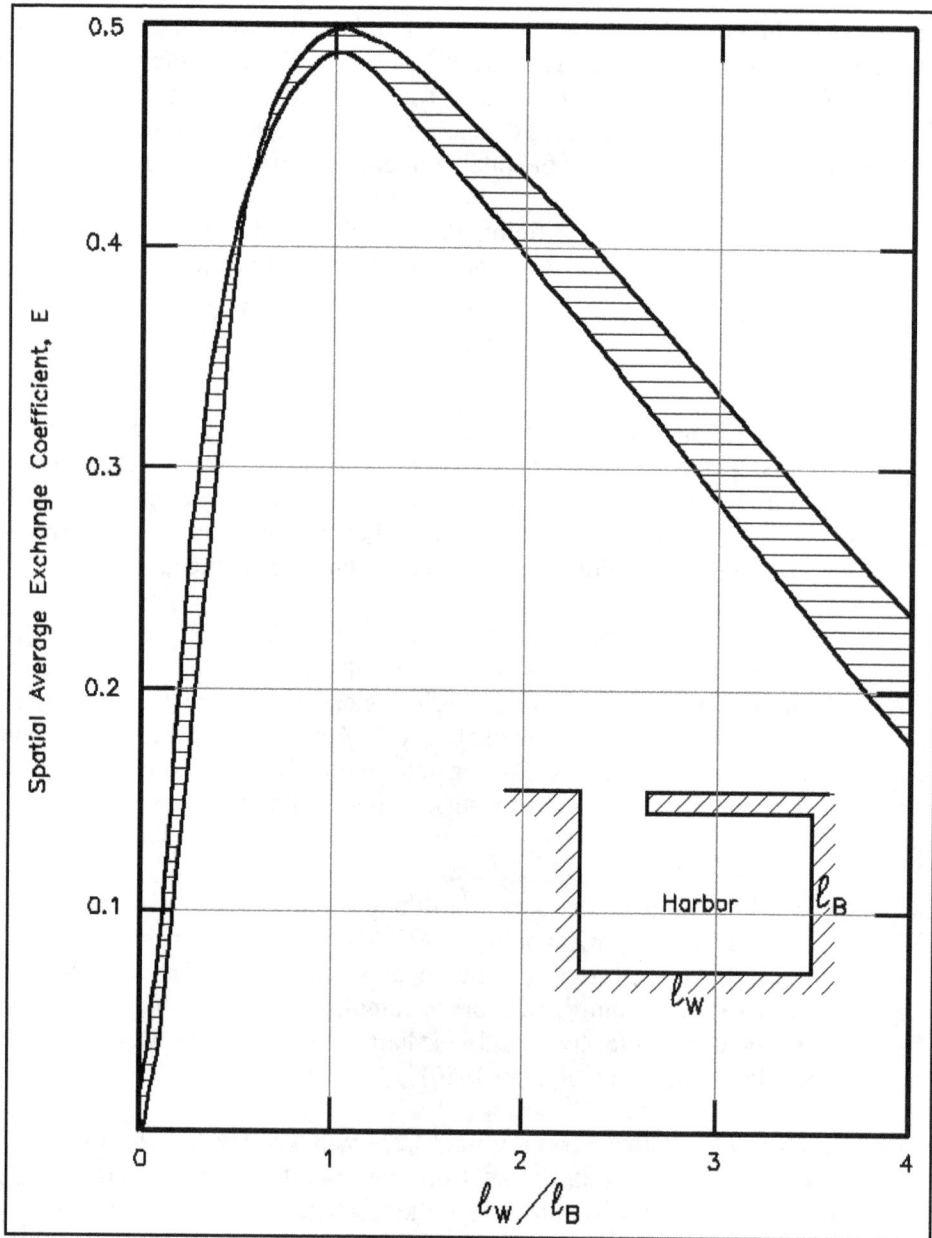

Figure II-7-39. Exchange coefficients - rectangular harbor, TRP = 0.4 (modified from Falconer (1980))

Table II-7-3
Flushing Characteristics of Small-Boat Harbors

Harbor	Tide Range (m)	Exchange Coefficient	Flushing Efficiency (%)
Des Moines	1.07	0.22	123
Edmonds	0.82	0.14	100
Lagoon Point	0.91	0.16	63
Penn Cove	1.22	0.19	75
Lake Crockett	1.83	0.22	78

Harbor Hydrodynamics

current will approach a velocity equal to 2 to 3 percent of the wind speed. But, given the lengths of open water found in most harbors, it is likely that the resulting current velocity will be less than this magnitude.

(b) The wind-induced surface current would cause a lateral or bottom return flow to develop, resulting in some circulation in a harbor under sustained winds. The resulting current pattern would be very dependent on the wind direction and the harbor geometry. Some flow in or out of the harbor entrance could develop. Also, the surface current will skim floating materials to the down-wind areas of the harbor. Generally, wind-induced circulation and flushing will be much less effective than circulation and flushing induced by tidal action.

(3) River discharge.

(a) Fresh water may enter a harbor from the land side as surface runoff or as concentrated flow in rivers that enter the harbor. Surface runoff will generally contribute more to the harbor pollution load because of the agricultural or urban pollutants in the water than it will to alleviating pollution by contributing to harbor flushing.

(b) Likewise, the efficacy of river discharge in easing harbor pollution will depend on the quality of the river water. If there are lower chemical and biological pollution loads in the river water than in the harbor, the flushing effect will be positive. But this could be counterbalanced if the river has a suspended silt load that will deposit in the slower moving harbor water. Silt suspension is a particular problem inside sections of the harbor such as pier slips. The silt-laden water may set up a turbidity current that will force bottom water having higher silt concentrations into a dead area, where it then deposits (Lin, Lott, and Mehta 1986). The silt load in a river may be relatively low during periods of normal river flow, but the concentration of suspended sediment may increase by orders of magnitude during storms.

(c) Some harbors are built inland from the coast along river/estuary systems. River flow past the harbor entrance can have conflicting effects. The flow past the entrance can generate a rotary flow system at the entrance that produces some flushing of the harbor. But, if the river silt load is significant, there can be a net deposition of sediment in the harbor near the entrance.

c. *Predicting of flushing/circulation.* Harbor flushing rates and circulation patterns can be predicted by numerical and physical models and by field studies. Often, some combination of these efforts is the most effective approach. Numerical and physical models benefit from the collection of field data that is usually required to calibrate and verify the models.

(1) Numerical models.

(a) A numerical model of the hydrodynamics of a harbor can be developed by employing finite difference solutions of the equation of mass balance and the two horizontal component momentum equations. With the appropriate boundary conditions at the fixed boundaries of the harbor and the tide as a forcing function at the harbor entrance, one can compute the time-dependent water flow velocities and water surface elevations at selected grid points in the harbor. If so desired, the surface wind stress on the water can also be employed as a forcing function. And river flow into the harbor can be specified at points along the harbor boundary. From the computed time-dependent grid of flow velocities, the resulting circulation patterns in the harbor can be defined.

(b) Commonly, the two-dimensional depth-integrated forms of the equations are used. That is, the horizontal flow velocities are averaged over the water depth and it is assumed that the water column is well mixed so there is no vertical density stratification. Also, it is assumed that vertical flow accelerations are small compared to the acceleration of gravity, so the pressure is hydrostatic and vertical components of flow velocity may be ignored. This latter limitation would require that there be no abrupt significant depth changes in the harbor. These equations are known as the long-wave equations (see Part II-5). Harris and Bodine (1977) present a derivation of these equations and discuss their formulation for numerical solution.

(c) If the solute advection-diffusion equation is added to the numerical hydrodynamic model, the movement and distribution of pollutants in the harbor can be computed. From this, the harbor exchange coefficients and flushing efficiency can be determined. An interesting application of numerical modeling to investigate harbor circulation, flushing, and variations in dissolved oxygen has been carried out by the Waterways Experiment Station for Los Angeles and Long Beach Harbor in California (Vemulakonda, Chou, and Hall 1991). Typical two-dimensional and a quasi-three-dimensional numerical model investigations studied the impact of deepening channels and constructing landfills in the harbor.

(d) Chiang and Lee (1982), Spaulding (1984), and Falconer (1980, 1984, 1985) provide other examples of applying the long-wave equations to calculate harbor hydrodynamics and adding the solute transport equation to determine the flushing characteristics of harbors.

(e) These numerical models have a number of advantages - they are flexible in that it is easy to adjust input tide and wind conditions as well as harbor bottom and lateral boundary conditions, and they do not have some of the scale/distortion problems found in physical models. But they also have disadvantages - they are a two-dimensional representation of the flow field in the harbor, and since calculations are done for a grid, flow details that are smaller than the grid dimension are not represented. This latter disadvantage makes it difficult to investigate, for example, the eddies generated by flow separation at the harbor entrance or at internal structures. It can be overcome to some extent by decreasing the grid point spacing in key segments of the harbor such as around the harbor entrance (Falconer and Mardapitta-Hadjipandeli 1986).

(f) Numerical models require that a number of empirical coefficients (e.g. surface wind stress and bottom stress, eddy viscosity, and component diffusion coefficients) be defined in order to run the model. Thus, confidence in the model can be significantly increased if field data are available to calibrate the model and verify subsequent model results.

(2) Physical model studies.

(a) Physical model studies have been conducted to investigate flushing and circulation patterns for existing and proposed harbors (Nece and Richey 1972, Schluchter and Slotta 1978, Nece 1985, Nece and Layton 1989) and for basic planform patterns of idealized harbors (Nece, Falconer, and Tsutumi 1976; Jiang and Falconer 1985).

(b) Physical models of harbors are designed to investigate tidal flushing, so they are based on Froude similitude (Hudson et al. 1979). They typically have a distorted scale, with the vertical scale being larger than the horizontal scale. Common model scale ratios that have been used are 1:30 to 1:50 for the vertical scale and 1:300 to 1:500 for the horizontal scale. It is assumed that wind effects are negligible and that the water column has no density stratification. Also, the effects of Coriolis acceleration are not modeled. Most harbors are sufficiently small that Coriolis effects can be neglected.

(c) For models using water and having the typical harbor model/prototype scale ratios, Froude and Reynolds similitude are incompatible. Consequently, model Reynolds numbers are underscaled. Thus, inertial effects are scaled but turbulent diffusion is not scaled. To minimize these effects, some experimenters have installed roughness strips at the model harbor entrance to generate turbulence. Thus, local diffusion-dispersion of solutes is not accurately replicated but advective transport of solutes is replicated. The latter typically dominates. Fine details of the internal flow circulation are not replicated, but gross circulation patterns are.

(d) Circulation patterns are typically measured by photographing the movement of floats. Model exchange coefficients and flushing efficiencies are measured by adding dye to the water at the start of tests and then measuring the decrease of the dye concentration during subsequent tidal cycles. While exchange coefficient values from the model and model circulation patterns do not precisely equate to prototype conditions, these harbor model studies can provide a basis for comparing alternative design features for a proposed harbor or guidance in modifying an existing harbor.

(e) Table II-7-4 lists, for ease of comparison, the relative advantages of physical and numerical models for harbor flushing and circulation studies. The table gives a general comparison, but for a specific application, modelling experts should be consulted before deciding whether to use a numerical or physical model or some combination of both.

Table II-7-4
Advantages of Physical and Numerical Models

Physical Models

Provide good visual demonstration of flow patterns
Some three-dimensional effects can be represented relatively easily
Intricate harbor boundaries can be easily simulated

Numerical Models

Wind stress and Coriolis effects can be simulated
Lower model development and maintenance costs
Easy to store model for future use
Easy to adjust or expand model boundary conditions
Extensive output data can easily be obtained

---

(3) Field studies.

(a) Limited field studies at an existing harbor may be conducted to obtain sufficient data to calibrate a numerical model of that harbor so the model can be run to investigate a range of other tide and wind conditions. This would provide a detailed look at the flushing and circulation characteristics of the harbor and some insight into possible remedial efforts that may be necessary. Or the model may then be run to evaluate proposed modifications of the harbor.

(b) An alternative is to run more extensive field studies as the sole effort to evaluate conditions at a harbor. This would generally be more costly than the hybrid field-model approach, but it may provide some detail that cannot be achieved from model studies alone.

(c) Also, field studies have been done to support the general development of physical and numerical modelling techniques for the study of harbor flushing and circulation.

(d) Field measurements include those that define the hydrodynamics of a harbor and supplementary measurements to quantify harbor flushing. The former include measurements of tide levels inside and outside the harbor, current velocity measurements at the entrance to quantify flow rates into and out of the harbor, and flow velocity measurements throughout the harbor and/or drogue studies to define circulation patterns in the harbor. If tidal flushing is the primary concern, these measurements would be conducted on days when the wind velocity is low. Otherwise, a directional anemometer would also be used to measure the wind speed and direction.

(e) To determine exchange coefficients throughout the harbor and the harbor's flushing efficiency, the harbor would be uniformly seeded with a harmless detectable solute such as a fluorescent dye and then sampled periodically at several points in the harbor for a period of several tidal cycles. Initial and subsequent dye concentrations (see Equation II-7-20) can be measured in situ by a standard fluorometer. The dye Rhodamine WT has been used in a number of harbor flushing studies (see Callaway (1981) and Schwartz and Imberger (1988)).

II-7-7. Vessel Interactions

a. Vessel-generated waves.

(1) As a vessel travels across the water surface, a variable pressure distribution develops along the vessel hull. The pressure rises at the bow and stern and drops along the midsection. These pressure gradients, in turn, generate a set of waves that propagate out from the vessel bow and another generally lower set of waves that propagate out from the vessel stern. The heights of the resulting waves depend on the vessel speed, the bow and stern geometry, and the amount of clearance between the vessel hull and channel bottom and sides. The period and direction of the resulting waves depend only on the vessel speed and the water depth. For a detailed discussion of the vessel wave-generating process and the resulting wave characteristics, see Robb (1952), Sorensen (1973a, 1973b), and Newman (1978).

(2) The pattern of wave crests generated at the bow of a vessel that is moving at a constant speed over deep water is depicted in Figure II-7-40. There are symmetrical sets of *diverging* waves that move obliquely out from the vessel's sailing line and a set of *transverse* waves that propagate along the sailing line. The *transverse* and *diverging* waves meet along the cusp locus lines that form an angle of 19°28' with the sailing line. The largest wave heights are found where the *transverse* and *diverging* waves meet. If the speed of the vessel is increased, this wave crest pattern retains the same geometric form, but expands in size as the individual wave lengths (and periods) increase.

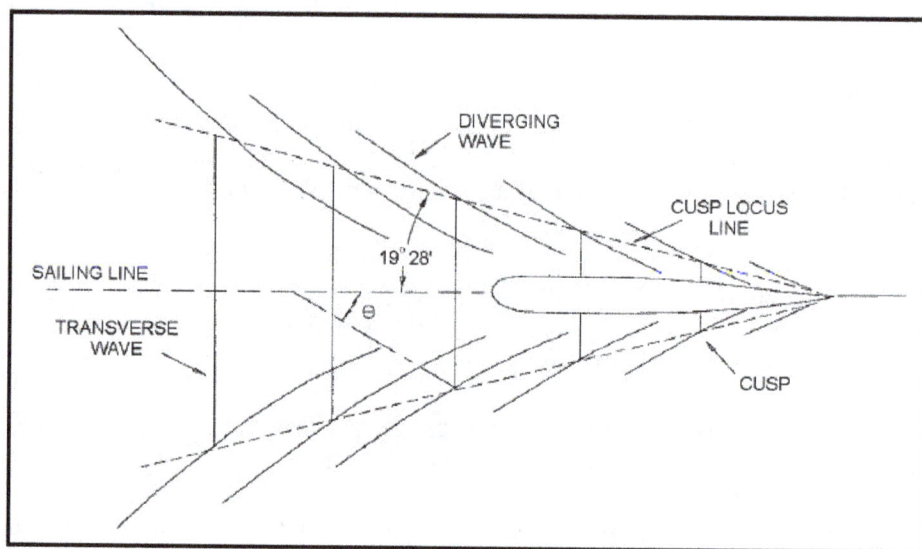

Figure II-7-40. Wave crest pattern generated at a vessel bow moving over deep water

(3) The fixed pattern of wave crests requires that individual wave celerities C be related to the vessel speed V_s by

$$C = V_s \cos\theta$$

(II-7-21)

where θ is the angle between the sailing line and the direction of wave propagation (Figure II-7-40). Thus, the *transverse* waves travel at the same speed as the vessel and, in deep water, θ has a value of 35°16' for the *diverging* waves.

(4) At increasing distances from the vessel, diffraction causes the wave crest lengths to continually increase and the resulting wave heights to continually decrease. It can be shown (Havelock 1908) that the wave heights at the cusp points decrease at a rate that is inversely proportional to the cube root of the distance from the vessel's bow (or stern). *Transverse* wave heights at the sailing line decrease at a rate proportional to the square root of the distance aft of the bow (or stern). Consequently, the *diverging* waves become more pronounced with distance from the vessel.

(5) The above discussion applies to deep water, i.e. water depths where the particle motion in the vessel-generated waves does not reach to the bottom. This condition holds for a Froude number less than approximately 0.7, where the Froude number F is defined by

$$F = \frac{V_s}{\sqrt{gd}} \tag{II-7-22}$$

(6) As the Froude number increases from 0.7 to 1.0, wave motion is affected by the water depth and the wave crest pattern changes. The cusp locus line angle increases from 19°28' to 90° at a Froude number of one. The *diverging* wave heights increase more slowly than do the *transverse* wave heights, so the latter become more prominent as the Froude number approaches unity. At a Froude number of one, the *transverse* and *diverging* waves have coalesced and are oriented with their crest perpendicular to the sailing line. Most of the wave energy is concentrated in a single large wave at the bow. Owing to propulsion limits (Schofield 1974), most self-propelled vessels can only operate at maximum Froude numbers of about 0.9. Also, as a vessel's speed increases, if the vessel is sufficiently light (i.e. has a shallow draft), hydrodynamic lift may cause the vessel to plane so that there is no significant increase in the height of generated waves for vessel speeds in excess of the speed when planing commences.

(7) For harbor design purposes, one would like to know the direction, period, and height of the waves generated by a design vessel moving at the design speed. For Froude numbers up to unity, Weggel and Sorensen (1986) show that the direction of wave propagation θ (in degrees) is given by

$$\theta = 35.27 \left(1 - e^{12(F-1)}\right) \tag{II-7-23}$$

(8) Then, from Equation II-7-21, the *diverging* wave celerity can be calculated, and the wave period can be determined from the linear wave theory dispersion equation.

(9) Figure II-7-41 is a typical wave record produced by a moving vessel. Most field and laboratory investigations of vessel-generated waves (Sorensen and Weggel 1984; Weggel and Sorensen 1986) report the maximum wave height (H_m, see Figure II-7-41) as a function of vessel speed and type, water depth, and distance from the sailing line to where the wave measurement was made. Table II-7-5 (from Sorensen (1973b)) tabulates selected H_m values for a range of vessel characteristics and speeds at different distances from the sailing line. These data indicate the range of typical wave heights that might occur for common vessels and show that vessel speed is more important than vessel dimensions in determining the height of the wave generated.

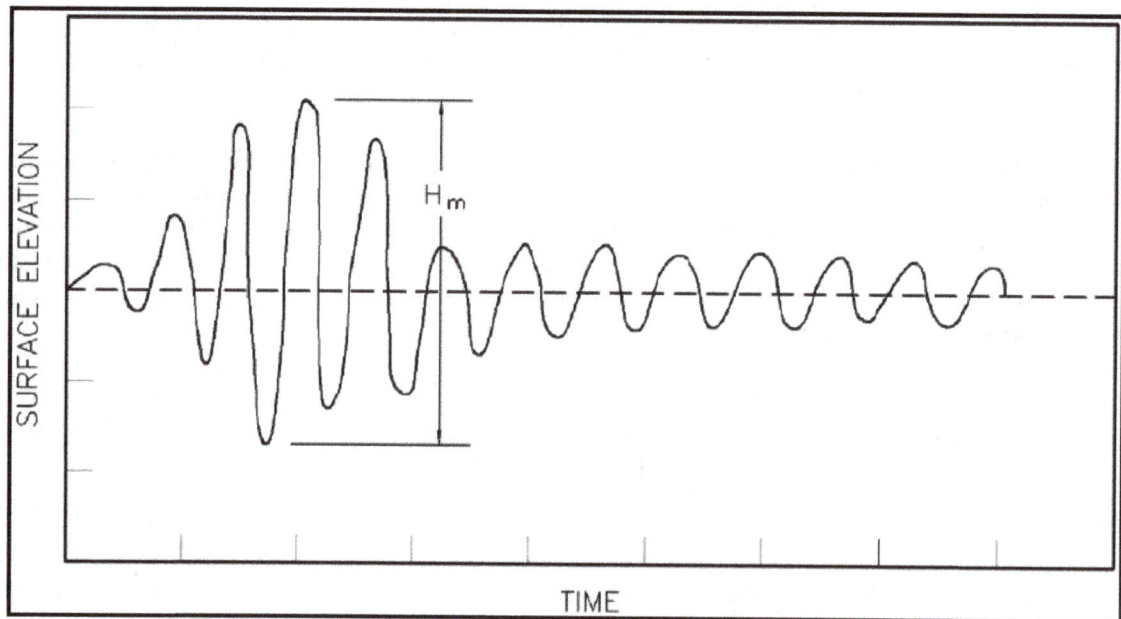

Figure II-7-41. Typical vessel-generated wave record

EXAMPLE PROBLEM II-7-5

FIND:
   The period of the *diverging* waves generated by the vessel.

GIVEN:
   A vessel is moving at a speed of 10 knots (5.157 m/sec) over water 5 m deep.

SOLUTION:
   The vessel Froude number is

$$F = \frac{5.157}{\sqrt{9.81\ (5)}} = 0.73$$

so Equation II-7-23 gives a direction of propagation

$$\theta = 35.27\ [1 - e^{12\,(0.73-1)}] = 33.88°$$

and Equation II-7-21 gives a wave celerity

$$C = 5.157\ \cos(33.88°) = 4.28\ m/s$$

The linear wave dispersion equation can be written

$$C = \frac{gT}{2\pi}\ \tanh \frac{2\pi d}{CT}$$

Inserting known values for C, g, and d into the dispersion equation leads to a trial solution for T of 2.8 sec. This is a typical period for vessel-generated waves and demonstrates why floating breakwaters are usually effective in protecting against vessel waves.

(10) A number of quasi-empirical procedures for predicting vessel-generated wave heights have been published (see Sorensen (1986) and Sorensen (1989) for a summary). Most procedures are restricted to a certain class or classes of vessels and specific channel conditions. A comparison (Sorensen 1989) of predicted H_m values for selected vessel speeds and water depths showed significant variation among the results predicted by the various procedures. The best approach for design analyses appears to be to review the published vessel wave measurement data to compare with the vessel, vessel speed, and channel conditions that most closely approach the design condition and select a conservative value of H_m from these data. If this is not possible, then the values in Table II-7-5 can be used as rough estimates for the different types of vessels.

Table II-7-5
Selected Vessel-Generated Wave Heights (Sorensen 1973b)

Vessel	Speed (m/s)	H_m(m) at 30 m	H_m(m) at 150 m
Cabin Cruiser length-7.0 m beam-2.5 m draft-0.5 m	3.1 5.1	0.2 0.4	0.1 0.2
Coast Guard Cutter length-12.2 m beam-3.0 m draft-1.1 m	3.1 5.1 7.2[1]	0.2 0.5 0.7	0.3
Tugboat length-13.7 m beam-4.0 m draft-1.8 m	3.1 5.1	0.2 0.5	0.1 0.3
Air-Sea Rescue Vessel length-19.5 m beam-3.9 m draft-0.9 m	3.1 5.1 7.2[1]	0.1 0.4 0.6	0.2 0.3
Fireboat length-30.5 m beam-8.5 m draft-3.4 m	3.1 5.1 7.2	0.1 0.5 0.9	0.1 0.3 0.8
Tanker length-153.6 m beam-20.1 m draft-8.5 m	7.2 9.3		0.5 1.6

Note: The above data are from tests conducted at water depths ranging from 11.9 to 12.8 m.
[1] Denotes that the vessel was starting to plane.

b. Vessel motions.

(1) Response to waves.

(a) Wave action will excite a floating vessel to oscillate in one or more of six components of motion or degrees of freedom. These are translated in the three coordinate directions (surge, sway, and heave) and rotation around the three principal axes (roll, pitch, and yaw). Which of these motion components is excited and to what extent depends primarily on the direction of wave incidence relative to the primary vessel axes and on the incident wave frequency spectrum compared to the resonant frequencies of the six motion components (Wehausen 1971). If the vessel is moored, the arrangement of the mooring lines and their taughtness will influence the resonant periods and the response amplitudes of the vessel motions. If the vessel is moving, the effective or encounter period of wave agitation is the wave period relative to the ship rather than to a fixed observation point. Wave mass transport will also cause a slow drift of the vessel in the direction of wave propagation.

(b) Small vessels, such as the recreational vessels found in marinas, will commonly respond to shorter wind-wave periods. An analytical study, coupled with some field measurements for seven small boats

(Raichlen 1968), indicated that the periods of free oscillation were less than 10 sec. Larger seagoing deep-draft vessels, depending on the oscillation mode being excited, will respond to the entire range of wind-wave periods. Field measurements by van Wyk (1982) on ships having lengths between 250 and 300 m and beams of about 40 m found maximum roll and pitch responses at encounter periods between 10 and 12 sec. By properly designing the mooring system, the periods and amplitudes of vessel motion can be significantly modified.

(c) The wave-induced lateral and vertical motions of the design vessel will affect the required channel horizontal and depth dimensions, respectively. The problem of wave-induced vessel oscillations has been addressed by analytical/numerical means (Andersen 1979; Madsen, Svendsen, and Michaelsen 1980; Isaacson and Mercer 1982). These efforts usually employ small-amplitude, monochromatic waves and some limitations on vessel geometry and the incident wave directions relative to the vessel.

(d) Some field measurement programs yield valuable design information. Wang and Noble (1982) describe an investigation of vessels entering the Columbia River channel. Pitch, roll, heave, yaw, and horizontal position were measured for selected vessels as they traversed the channel. The data were analyzed statistically to define extreme limits of vessel motion for various wave and other conditions (Noble 1982). van Wyk (1982) reports on a field study of vessel response to wave action at two South African ports. The data were analyzed statistically so extreme motion probabilities could be evaluated. Other field studies of moving large vessels have been reported by Greenstreet (1982) and Zwamborn and Cox (1982). Raichlen (1968) and Northwest Hydraulic Consultants (1980) discuss field measurements of small moored vessels in marinas.

(e) Most of the major coastal engineering labs have also conducted model studies of vessel response to wave motion. Some of these tests are discussed in Mansard and Pratte (1982), Isaacson and Mercer (1982), Zwamborn and Cox (1982), and Briggs et al. (1994).

(2) Response to currents.

(a) There are several possible causes for currents in harbors and in the vicinity of harbor entrances. Wind, wave-induced radiation stress, rivers, and tides can all generate currents in the vicinity of a coastal harbor entrance. Flow from a river that enters a harbor or flows past the entrance to a harbor located in an estuary can generate significant currents. Ebb and flood tide can generate strong reversing currents through a harbor entrance. Tidally induced longshore currents around islands can cause navigation problems at harbor entrances. Long-wave resonant oscillations in harbors can generate noticeable currents at nodal points, which can seriously affect moored vessels and create hazardous navigation conditions if this location is at a place where the flow is constricted.

(b) Currents will directly affect vessel operation, particularly when they act oblique to the sailing line of the vessel. These currents are particularly troublesome when the vessel speed is low and vessel maneuverability is commensurately reduced. The situation is made even more difficult when there are strong winds acting from a different direction than the currents. Physical model studies and numerical simulations of vessel motion have been used to predict vessel paths under various wind and current conditions, particularly as vessels enter a harbor (Bruun 1989, Briggs et al. 1994). Currents can increase vessel sinkage and trim in restricted channels (see next section).

(3) Wave-current interaction.

(a) At harbor entrances, currents can also exert an indirect effect on vessel navigation through their effect on waves. Ebb currents will steepen incoming waves, making the waves more hazardous to the stability of small vessels in particular. The ebb currents may be of sufficient strength to induce wave breaking and turbulence in the entrance, a condition that is particularly hazardous to vessel operation.

(b) Bruun (1978) summarizes earlier literature and discusses the problem of sediment transport. Mehta and Özsöy (1978) found that the effects of bottom friction are important in the hydrodynamics of two-dimensional turbulent jets because it increases the rate of spreading of the jet, which has a significant effect on incident wave characteristics. Sakai and Saeki (1984) measured the effect of opposing currents on wave height transformation over a 1:30 sloping beach for a range of wave periods and steepness. They found an increase in wave height and decay rate in the presence of the opposing current. Willis (1988) conducted monochromatic wave and ebb current tests in a 1-m-wide rectangular entrance channel cut in a 1:30 sloping beach. Current measurements were averaged over 3 min to obtain a quantitative picture of the mean currents. They experienced major problems with the stability of the current field due to large-scale meandering motions.

(c) Lai, Long, and Huang (1989) conducted flume tests of kinematics of wave-current interactions for strong interactions with waves propagating with and against the current. They found the influence of the waves on the mean current profiles was small, although opposing waves would give a slightly lower current. They observed a drastic change in the spectral shape, especially higher harmonics, following wave breaking in the presence of opposing currents. Their experiments confirmed blockage of waves by a current when the ratio of depth-averaged current velocity to wave celerity without currents approaches -0.25.

(d) Raichlen (1993) conducted a laboratory investigation of waves propagating on an adverse jet to simulate the effect of ebb currents on incoming waves at a tidal inlet. He tested regular waves (depth-to-wavelength ratios from 0.086 to 0.496) for a range of relative channel entrance velocities to wave celerity for locations both upstream (20 channel widths) and downstream (15 channel widths) of the channel entrance. Wave height increased by a factor of 2.5 near the channel entrance in the presence of a current that was only 7 percent of the phase speed, primarily caused by wave refraction. He also found that the wave height decreases significantly as waves propagate up the entrance channel. Refraction and entrainment of the still fluid by the ebb current jet produces a lateral variation in the wave height across the channel width downstream of the entrance.

(e) Briggs and Green (1992) and Briggs and Liu (1993) conducted three-dimensional laboratory experiments of the interaction of regular waves with ebb currents offshore of a tidal entrance channel (6 channel widths) on a 1:30 plane beach. Current velocity to wave celerity ratios ranged from 0.06 to 0.34. Under the influence of ebb currents, waves experienced increases in steepness and corresponding wave height up to a factor of nearly 2 for currents that were 20 to 30 percent of the phase speed. As waves shoal and break, higher harmonics are formed as the wave becomes more nonlinear. Energy is transferred from the fundamental mode due to nonlinear coupling between frequencies. Briggs and Liu found that ebb currents also promote the nonlinear growth of the fundamental frequency, higher harmonics, and subharmonics of the incident wave. This shift in energy can change the response characteristics of vessels in the entrance channel.

(4) Vessel sinkage and trim.

(a) The pressure distribution that develops around the hull of a moving vessel results in an above-average pressure at the bow and stern and a below-average pressure at the midsection of the vessel. The reduced pressure at the midsection dominates and causes a net sinkage of the vessel. The vessel sinkage is also referred to as squat. Since there is usually an imbalance of upward forces between the bow and stern, the vessel will often also trim by the bow (i.e. the vessel bow is lowered relative to the stern) or by the stern.

(b) Sinkage and trim of a vessel in a navigation channel depend on the factors that control the pressure distribution along the vessel hull. Primarily, these are the vessel speed, the ratio of the channel cross section area to the vessel wetted cross section area (known as the section coefficient), and the ratio of the water depth to the vessel draft. Other factors that might affect sinkage and trim include the vessel hull geometry, the at-rest trim of the vessel, the cross-section geometry of the channel, the sailing line of the vessel relative to the centerline of the channel, vessel acceleration or deceleration, the presence of currents in the channel, and the passing of other vessels in the channel.

(c) A wide variety of analytical, quasi-analytical, and empirical methods for predicting vessel sinkage and a few for predicting vessel trim have been published (Garthune et al. 1948, Constantine 1960, Tuck 1966, Tothill 1967, Sharp and Fenton 1968, Dand and Furgeson 1973, McNown 1976, Beck 1977, Gates and Herbich 1977, Eryuzlu and Hausser 1978, Blauuw and van der Knapp 1983, Ferguson and McGregor 1986). The earliest and most basic approach to predicting vessel sinkage employs the one-dimensional energy and continuity equations. This approach gives generally acceptable results for uniform channels and vessel hull geometries that are not too irregular. The continuity and energy equations are written between a point ahead of the vessel and a point at the vessel midsection. The water surface drawdown is thus calculated and assumed equal to the vessel sinkage. The results for a rectangular channel cross section are presented below; for a trapezoidal channel cross section see Tothill (1967) and for a parabolic channel cross section, see McNown (1976).

(d) Consider a vessel in a channel with the relevant dimensions as defined in Figure II-7-42 where b is the channel width, A_m is the vessel's midsection wetted cross-section area, the undisturbed water depth is d, and the vessel drawdown is Δd. V_s and g are the vessel speed and the acceleration of gravity as defined earlier. The continuity and energy equations in terms of the Froude number F, dimensionless drawdown D, and vessel blockage ratio S are solved by

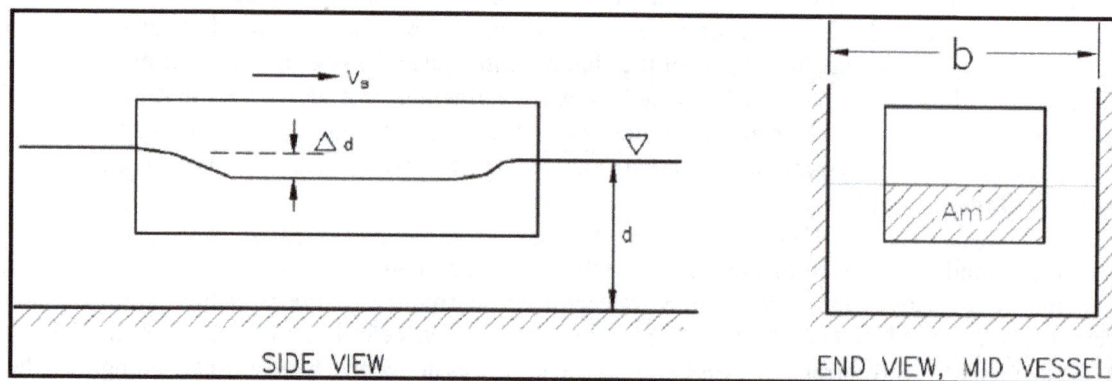

Figure II-7-42. Definition of terms, vessel drawdown

$$F = \sqrt{\frac{2D(1-D-S)^2}{1-(1-D-S)^2}} \qquad \text{(II-7-24)}$$

where

$$F = \frac{V_s}{\sqrt{gd}} \qquad\qquad D = \frac{\Delta d}{d} \qquad\qquad S = \frac{A_m}{bd}$$

(e) Figure II-7-43 plots F versus D for selected values of S. Given the channel cross-section dimensions, the water depth, the vessel speed, and the midsection wetted cross-section area of the vessel, D can be determined from the figure. This, in turn, yields the water surface drawdown Δd, which is taken as the vessel sinkage.

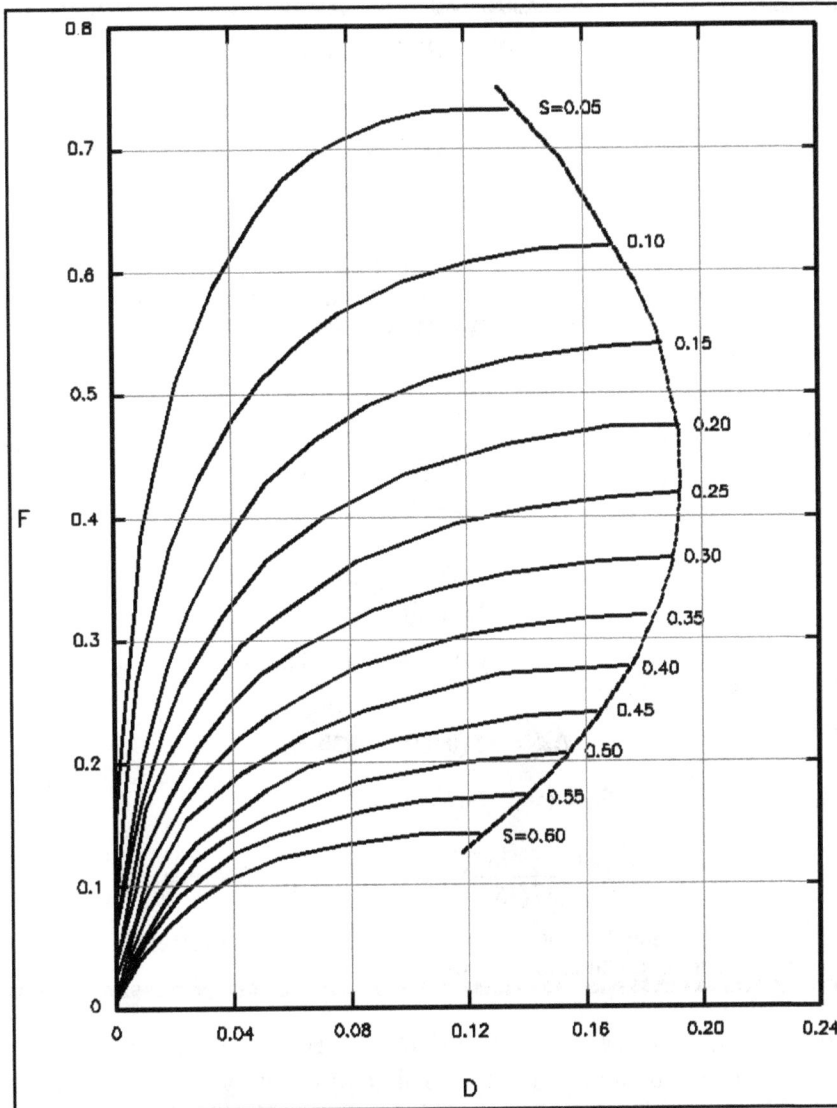

Figure II-7-43. Vessel sinkage prediction

(f) Design of safe channel depths is a function of the loaded draft of the ship, vessel sinkage, minimum underkeel clearance, and effect of pitch and roll in outer channels exposed to wave and current conditions (Herbich 1992).

(g) A self-propelled vessel cannot exceed the critical speed (see Constantine (1960)). However, a vessel can be towed faster than the critical speed, the result being the generation of a bore wave that propagates ahead of the vessel. Before reaching the critical speed, the vessel sinkage increases rapidly. If the initial vessel under-keel clearance is small, the vessel may hit bottom before coming close to the critical speed. Also, near the critical speed, there is a sharp increase in the power required to propel the vessel.

EXAMPLE PROBLEM II-7-6

FIND:
The vessel sinkage and return flow velocity between the vessel hull and the channel bottom and sides.

GIVEN:
A C9 containership has a length of 262 m, a beam of 32 m, and a fully loaded draft of 12 m. The C9 containership with a midsection wetted cross section area of 384 m^2 is moving at a speed of 6 knots (3.09 m/s) in a channel 137 m wide and 15 m deep. This is a typical scenario for deep-draft vessels entering a harbor through a one-way entrance channel.

SOLUTION:
The vessel Froude number is

$$F = \frac{3.09}{\sqrt{9.81\ (15)}} = 0.26$$

and the blockage ratio is

$$S = \frac{384}{137(15)} = 0.19$$

From Figure II-7-43 at the intersection of F = 0.26 and S = 0.19, the dimensionless drawdown is approximately D = 0.02. The vessel sinkage is then given by

$$\Delta d = 15(0.02) = 0.30\,m$$

The return velocity is calculated from Equation II-7-25 as

$$V_r = 3.09 \left[\frac{137(15)}{137(15-0.3)-384} - 1 \right] = 0.81\ m/s$$

Thus, channel depths must include additional clearance to accommodate the vessel drawdown and a substantial return velocity flow can lead to scour along the channel banks.

(h) As a vessel travels in a channel, continuity of flow causes a return flow velocity to develop between the vessel hull and the channel bottom and sides. This return flow velocity V_r can be calculated for a rectangular channel and vessel cross section (as depicted in Figure II-7-42) from

$$V_r = V_s \left[\frac{bd}{b(d - \Delta d) - A_m} - 1 \right] \tag{II-7-25}$$

(i) Particularly for vessel/channel conditions that produce a large blockage ratio (S) and for high vessel speeds, a substantial return flow velocity can develop. Channel bottom scour and damage to reveted channel sides may result.

(5) Ship maneuverability in restricted waterways.

(a) The pressure distribution along the hull of a vessel will also cause lateral forces and horizontal moments to act on the vessel when it passes another vessel or moves in close proximity to a channel bank or other large structure. These effects affect all sizes of ships, but are particularly dangerous for larger vessels. Figure II-7-44 shows a pressure field for moving vessels, where the plus and minus signs indicate the relative pressures along the side of the hull.

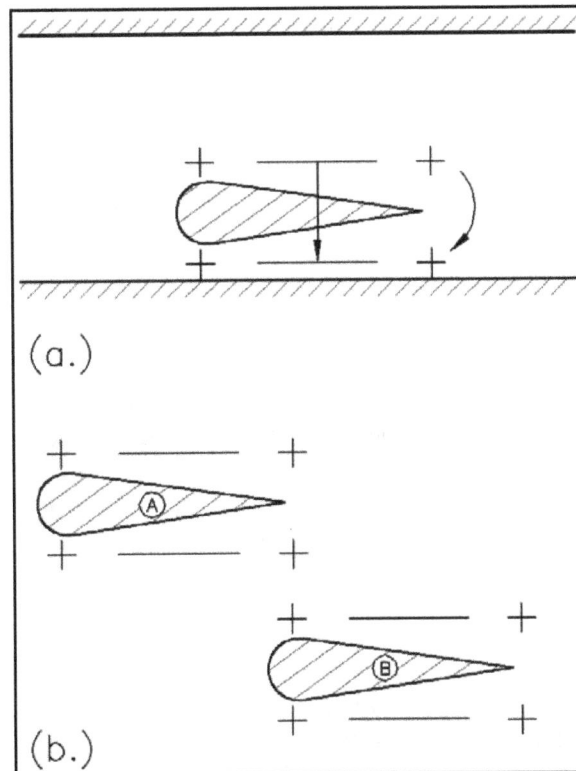

Figure II-7-44. Pressure fields for moving vessels (vessels moving left to right)

(b) Figure II-7-44a shows a vessel sailing close to a channel bank. Owing to the "Bernoulli effect," larger positive and negative pressures occur on the side closest to the channel bank, with the negative pressure dominating to draw the vessel toward the bank and the larger bow pressure than stern pressure causing the bow to rotate (yaw) away from the bank. Lateral force and yawing moment increase as the vessel sails closer to the bank and they are approximately a function of the vessel speed squared. Manipulation of the ship's rudder can correct for these forces, but it is difficult to travel close to a channel bank for any long period of time (Kray 1970).

(c) Figure II-7-44b shows two similar vessels, with vessel A passing vessel B in unrestricted water. For the positions shown, the bow of vessel A and the stern of vessel B would deflect outward because of the positive pressures at both bow and stern. When the ships are even, they would be drawn together because of the negative pressures midships, and their bows would be deflected slightly apart because of the positive pressures at the bows. And as vessel A completes passing vessel B, the stern of B and the bow of A would be deflected outward. Thus, the interacting forces and moments are complex, continually changing while the ships are in the proximity of each other, and dependent on hull sizes and geometries; vessel speeds, directions, and lateral separation; and water depth. Vessel interaction effects are more significant for vessels that have a large cross-section area relative to the channel cross-section area and for higher vessel speeds.

(d) The effects of vessel interaction have been studied for specific conditions by carrying out model studies (e.g. Taylor 1909, Robb 1949, Garthune et al. 1948, Delft Hydraulics Laboratory 1965, Moody 1970, Dand 1976). Some analytical studies that allow calculation of the resulting forces and moments, but not the vessel response, were conducted by Silverstein (1958), Hooft (1973), Tuck and Newman (1974), and Dand (1976).

(e) Vessel interactions with other vessels and with the channel bank must be accounted for in determining required navigation channel widths. For example, guidance given in Engineer Manual (EM) 1110-2-1613, which specifies channel width as a function of the design vessel beam dimension for various channel conditions, includes these effects. The total channel width consists of a maneuvering lane, bank clearance, and ship clearance, if two-way traffic is involved. This specification is based on model studies of large vessels in deep-draft navigation channels and discussions with ship pilots (Garthune et al. 1948).

c. Mooring.

(1) Wave forcing mechanism.

(a) Infragravity waves (wave periods typically between 25 and 300 sec) force long-period oscillations or seiche in harbors. If the natural period of the ship corresponds to a harbor resonance mode and they are moored in the vicinity of the node, excessive ship motion can prevent loading and unloading of the ship for a number of days. In some cases, extensive damage to the ship and pier can result if the mooring lines fail.

(b) Infragravity energy can be divided into bound and free wave energy. Bound or forced infragravity waves are nonlinearly coupled to wave groups, traveling at the group velocity of the wind waves, and phase locked to sea and swell waves. Free infragravity waves radiate to and from deep water after being reflected from the shoreline or are generated by nonlinear interactions and wave breaking of incident wind waves and are refractively trapped in shallow water, propagating in the longshore direction. According to numerous investigators (Herbers et al. 1992; Elgar et al. 1992; Okihiro, Guza, and Seymour 1992), bound and free wave energies increase with increasing swell energy and decreasing water depth. Bound wave contributions are usually more significant when energetic swell conditions exist, but free waves dominate when more moderate conditions prevail.

(2) Mooring configurations.

(a) A vessel moored in a harbor or at some point offshore commonly has one of three types of mooring arrangements:

- A single-point mooring where the vessel is tied to a buoy by a single line from the bow and is thus free to rotate around the buoy (i.e. weathervane) in response to environmental forces.

- A multiple- point mooring where the ship is tied by several fore and aft lines to anchors or buoys.

- A conventional pier anchorage, where the vessel is tied fore and aft to the pier and separated from the pier by a fendering system.

(b) For a deep-draft vessel at a pier, the mooring line system will typically consist of 8 to 12 lines in a symmetrical pattern, half from the bow and half from the stern of the ship. One to two *breast* lines are positioned on both bow and stern. The breast line(s) is perpendicular to the ship and dock and presses the ship against the dock and fenders. Two *head* lines make an angle of 60-70 deg to the *breast* line and go forward from the bow. Two *stern* lines are analogous to the *head* lines, but originate from the stern of the ship. These four lines are on the order of 100 ft long between ship and dock attachment points. Finally, two or four *spring* lines make an angle of 85 deg to the *breast* line and go toward midships. These lines can vary in length from 100 to 200 ft. The *spring* lines, in combination with the *breast* lines, provide the most efficient ship mooring. The deck of the vessel is typically 3 to 8 m above the pier, with the bow being higher than the stern.

(3) Mooring lines.

(a) Mooring lines are made from steel, natural fibers, and synthetic materials. Steel ropes may be made of different strength grades and galvanized for protection against corrosion. Natural fibers include manilla, sisal, and coir. Commonly used synthetic materials are nylon, dacron, polypropylene, Kevlar, and Karastan (Herbich 1992).

(b) Synthetic lines are easy to handle, do not corrode, and have excellent strength-to-weight ratios. Different construction types include stranded, plaited, braided, and parallel yarns. Stranded is the least satisfactory for mooring lines because it tends to unlay under free end conditions. Each has different mechanical properties that make it appropriate for different applications. These include strength, weight, stretch, endurance, and resistance to abrasion and cuts.

(c) Care should be taken not to mix different materials and lengths in mooring arrangements (Oil Companies International Marine Forum (OCIMF) 1978). Elasticity is a measure of a mooring line's ability to stretch under load. It is a function of material, diameter, and length. The ultimate breaking strength has been related to the square of the nominal rope diameter (Wilson 1967). If two lines of different elasticity but similar lengths and orientations are combined at the same point, the stiffer one will assume more of the load. Also, lines of different length will carry different amounts of the total load. Thus, one of the lines may be near breaking while the other one is carrying almost no load.

(4) Fenders. Fenders are like bumpers on cars, designed to protect vessels and piers during berthing and mooring against forces due to winds, waves, and currents. They are designed to absorb impact energy of the vessel through deflection and dissipation. Some respond very fast and violently and others more slowly. The latter are the more desirable because they produce smaller fender forces. Highly elastic, recoiling type fenders should be replaced with the non-recoiling type if possible. Fenders can be continuous or placed in certain areas where vessels land. The mooring system should be designed based on the combined response of mooring lines and fenders to ensure that resonance effects with the environmental forces are minimized. The interested reader should consult Bruun (1989) for additional information on types, materials, characteristics, selection, forces, deformation, and energy absorption of fenders.

(5) Surge natural period.

(a) For moored vessels at a dock or quay, surge is one of the most important parameters to consider. Ranges of allowable movements for different vessels have been given by Bruun (1989).

(b) The motion of a moored ship in surge can be described by the motion of a linear system with a single degree of freedom. Restoring or reaction forces due to the change in position, velocity, and acceleration of the ship from equilibrium are assumed linear. The exciting force is due to the drag force of the water flowing past the ship. The motion of the ship in surge is assumed to be independent of other directions of motion. Damping is assumed to be small for the low-frequency motions of a ship in surge. Solving for the undamped natural period in surge T_s

$$T_n = 2\pi \sqrt{\frac{m_v}{k_{tot}}} \qquad (II-7-26)$$

(c) The virtual mass of the ship m_v is the sum of the actual mass or displacement of the ship m and the added mass m_a due to inertial effects of the water entrained with the ship. For a ship in surge, m_a is approximately 15 percent of the actual mass m, which is based on the ship's displacement

$$m_v = m + m_a = 1.15\,m \tag{II-7-27}$$

(d) A mooring system is composed of many lines, but only those in tension contribute to the stiffness or effective spring constant k_{tot} given by

$$k_{tot} = \sum_n k_n \sin\theta_n \cos\phi_n \tag{II-7-28}$$

where the index n sums over all head, stern, and spring lines in tension during surge motion (breast lines are conservatively assumed to provide no restoring force in surge), θ_n is the angle the line makes in the horizontal plane with the perpendicular to the ship, and ϕ_n is the angle the line makes in the vertical plane between the ship and the dock.

(e) For a taut mooring line in which sag is negligible and deflections are small, the individual stiffness k_n is defined by

$$k_n = \frac{T_n}{\Delta l_n} \tag{II-7-29}$$

where T_n is the axial tension or load and Δl_n is the elongation in the mooring line. The elastic behavior of fiber ropes is difficult to ascertain since it is a function of material, construction, size, load and load history, time, and environmental conditions. Typically, manufacturers supply elongation curves based on experimental data, which show percent elongation ε_n as a function of load as a percent of the breaking strength of the mooring line (Figure II-7-45). A new rope undergoes construction stretch or permanent strain, which occurs when initial loading places the fibers in paths different from their initial construction. Elastic stretch occurs for subsequent loading and is repeatable each time the rope is loaded. Under high loads for a long time, the rope may undergo cold flow of the fibers and eventually break. Thus, previously elongated rope does not stretch as much as new rope and separate elongation curves may be provided. Percent elongation is related to Δl_n by

$$\varepsilon_n = 100\left(\frac{\Delta l_n}{l_n}\right) \tag{II-7-30}$$

where l_n is the length of the mooring line. This formulation assumes that cable dynamics can be neglected, and that the natural frequency of the mooring line in longitudinal and transverse vibration is much higher than the surge frequency of the ship.

(f) Therefore, the natural period of a moored ship in surge is a function of displacement, and number, type, length, size, and tension of the mooring lines. As the ship is off-loaded, displacement of the ship will decrease and this will change the ship response characteristics. Proper ballasting can be used to prevent surge conditions from developing. If this is not possible, other remedies can be sought. The natural period of the moored ship can be adjusted by changing the mooring line configuration or tension. Increased tension will make the moored ship stiffer and will reduce its resonant period of oscillation. A decrease in the mooring line tension will make the moored ship less prone to shorter-period resonant modes. If this is not practical, the number and type of mooring lines can be changed to affect the response of the moored ship.

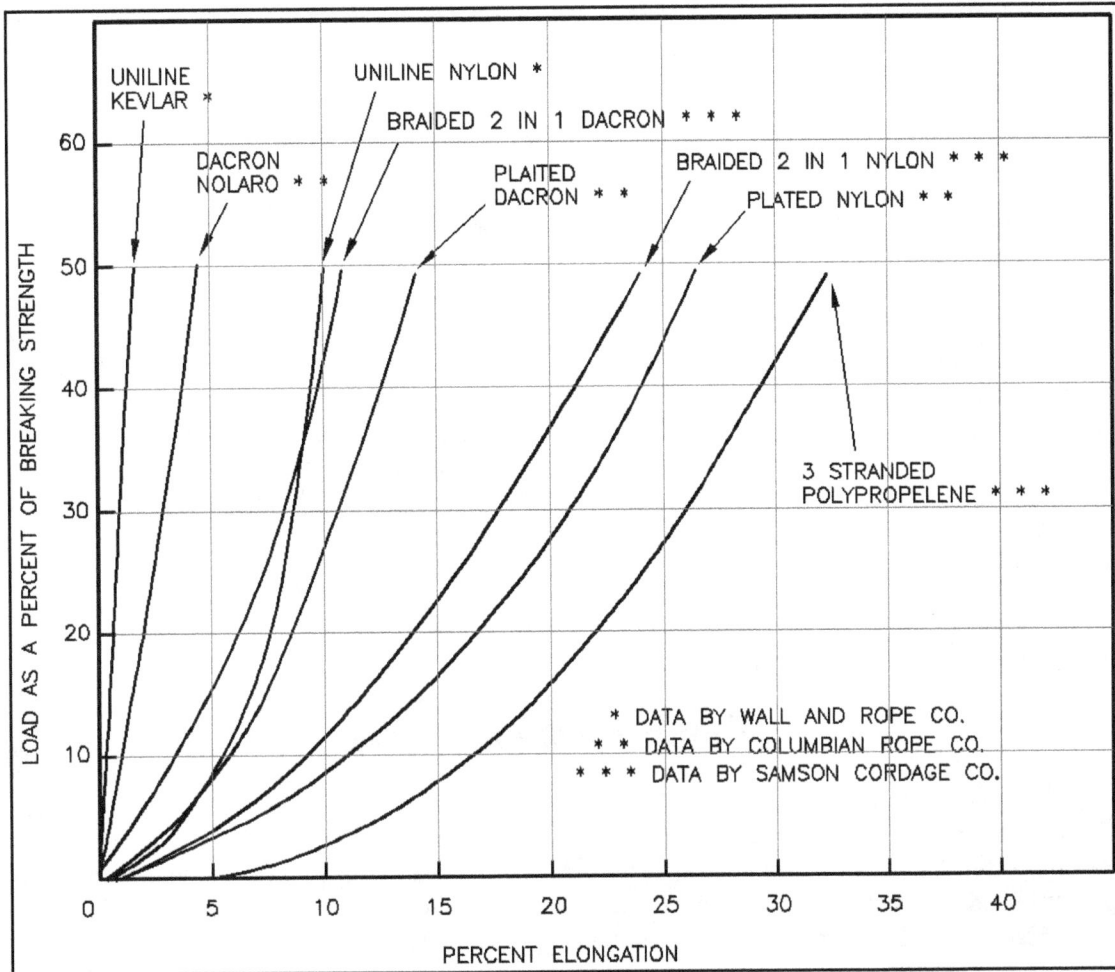

Figure II-7-45. Mooring fiber rope elongation curves

(6) Mooring forces.

(a) Design of any of these systems requires that the primary hydrodynamic loads on the vessel caused by the wind, currents, and waves be determined, usually on a probabilistic basis. Other loads that may be important in certain situations are ice forces and forces induced by passing vessels. The appropriate forces are employed in a dynamic analysis of the mooring system to determine expected loads in the mooring lines and on the fender systems. This analysis can be done by a physical model study or by a computer simulation of the system dynamics equations (see Bruun (1989) and Gaythwaite (1990)). Detailed discussions of the primary vessel mooring forces, those caused by the wind, currents, and waves, can be found in Tsinker (1986), Bruun (1989), and Gaythwaite (1990).

(b) The wind load is determined from the usual drag equation

$$F_w = \frac{1}{2}\rho_a C_D A V_{10}^2 \tag{II-7-31}$$

where F_w is the drag force due to the wind, ρ_a is the air density, C_D is a drag coefficient, A is the projected area of the vessel above the waterline, and V_{10} is the wind speed at the standard elevation of 10 m above the water's surface.

Harbor Hydrodynamics

EXAMPLE PROBLEM II-7-7

FIND:
    The natural period of the vessel in surge for used mooring lines in good condition.

GIVEN:
    A C9 containership is moored at a pier in Barbers Point Harbor, Oahu, Hawaii. Its fully-loaded displacement is 54,978 long tons (55,860,231 kg). Karat Estalon fiber ropes with a 7-1/2-in. circumference (6.3-cm-diam) are used for the two head, two stern, and two spring lines. Head and stern lines are 30.5 m long and make an angle of 70 deg with the perpendicular breast lines. The spring lines are 45.7 m long and form an angle of 85 deg with the breast lines. The tension in these lines is maintained at 20 tons (177,900 N). The deck of the C9 is assumed to be 7 m above the dock at the bow and 4.5 m at the stern.

SOLUTION:
    Virtual mass is calculated from Equation II-7-27

$$m_v = 1.15(55,860,200) = 64,239,200 \text{ kg}$$

The next step is to calculate the effective spring constant. Karat lines are manufactured by Columbian Rope Company, Guntown, MS, under license from Akzo, Holland. Estalon is a fiber that is a copolymer of polyester and polypropylene. According to manufacturer's literature, the breaking strength for Karat ropes is 526,000 N. The load as a percent of breaking strength is

$$Load = 100 \left[\frac{177,900}{526,000} \right] = 33.8\%$$

From the elongation chart for Karat lines (not shown), the percent elongation is 8.7 percent. The elongation for the head and stern lines is the same and is calculated from Equation II-7-30

$$\Delta l_{Hd} = \Delta l_{St} = \frac{8.7(30.5)}{100} = 2.7 \text{ m}$$

Likewise, the elongation for the spring lines is given by

$$\Delta l_{Sp} = \frac{8.7(45.7)}{100} = 4.0 \text{ m}$$

Example Problem II-7-7 (Continued)

Example Problem II-7-7 (Concluded)

Individual stiffness for the head and stern lines is calculated from Equation II-7-29

$$k_{Hd} = k_{St} = \frac{177,900}{2.7} = 65,900 \text{ N/m}$$

Similarly, the stiffness for individual spring lines is

$$k_{Sp} = \frac{177,900}{4.0} = 44,500 \text{ N/m}$$

The vertical angles from the bow, stern, and spring lines are

$$\phi_{Hd} = \arcsin\left[\frac{7.0}{30.5}\right] = 13°$$

$$\phi_{St} = \arcsin\left[\frac{4.5}{30.5}\right] = 8°$$

$$\phi_{Sp} = \arcsin\left[\frac{4.5}{45.7}\right] = 6°$$

When the ship surges forward, the two stern lines and a spring line provide restoring forces. Similarly, when the surge is in reverse, two head lines and a spring line provide the restoring forces. The effective spring constant is calculated both ways and averaged from Equation II-7-28

$$k_{tot} = 2(65,900) \sin 70° \cos 8° + 44,500 \sin 85° \cos 6°$$
$$= 122,600 + 44,100 = 166,700 \text{ N/m}$$

$$k_{tot} = 2(65,900) \sin 70° \cos 13° + 44,500 \sin 85° \cos 6°$$
$$= 120,700 + 44,100 = 164,800 \text{ N/m}$$

$$k_{tot} = \frac{166,700 + 164,800}{2} = 165,750 \text{ N/m}$$

Finally, the natural period in surge is calculated from Equation II-7-26

$$T_s = 2(3.14)\sqrt{\frac{64,239,200}{165,750}} = 124 \text{ sec}$$

If the rope had been new, the percent elongation would have been higher at 10.7 percent. This would have resulted in a slightly higher surge natural period of approximately 136 sec.

(c) During a storm, it is likely that some water will be entrained in the air. This would significantly increase the air density. Some investigators have argued that the entrained water will also slow down the air so there would be no net increase in wind drag. This issue is still unresolved. Current practice is to neglect entrained water when specifying the air density in the wind drag calculation.

(d) Values for the drag coefficient depend on the vessel geometry and orientation to the wind, and have been determined from wind tunnel tests. C_D values show significant variation, so reference should be made to Benham et al. (1977), Gaythwaite (1990), Isherwood (1973), Naval Facilities Engineering Command (1968), Owens and Palo (1981), and Palo (1983) for information on drag coefficients for a variety of vessels. These references also give some information on projected areas for common types of vessels. Table II-7-6 lists drag coefficients for wind (Bruun 1989).

Table II-7-6
Drag Coefficients for Wind Force

Wind Direction	C_D		
	Max.	Min.	Mean
Crosswise	1.40	0.80	1.11
Bow	1.04	0.62	0.82
Stern	1.02	0.64	0.77

(e) The wind speed at the standard elevation (10 m) is commonly used because this would be a good reference elevation for most larger vessels. If the center of pressure of the vessel is at a significantly different elevation than 10 m, the 1/7th power velocity profile law may be used to correct the wind velocity. It is defined by

$$V_z = V_{10} \left(\frac{z}{10} \right)^{0.11} \tag{II-7-32}$$

where V_z is the velocity at the desired elevation z. Typically, a 50-year return period wind speed is used or the limiting wind speed for which a vessel might remain moored would be used.

(f) The gustiness of the wind must also be considered; i.e., what duration wind gust is sufficient to envelop the vessel. To account for gustiness, the drag coefficient is often increased by a factor of about 1.4 (Tsinker 1986).

(g) Commonly, current drag forces are determined with less certainty than wind drag forces. Current speeds and directions at the mooring site may be difficult to predict, particularly in a harbor having a complex layout. Currents may be continually shifting if dominated by the tide or river flow. The Reynolds number for currents acting on a vessel is usually in the fully turbulent region, but often close to the transition point where drag coefficients show a wider range of scatter. The vessel draft and resulting bottom clearance can also have a significant impact on the resulting drag force. Gaythwaite (1990) presents several sources for current drag coefficient and vessel projected area information. Seelig, Kreibel, and Headland (1992) have published a recent analysis of new scale-model drag studies and other data collected over the past five decades.

EXAMPLE PROBLEM II-7-8

FIND:

The wind drag for a Beaufort 5 fresh breeze of 20 knots (10.31 m/s) at an angle of 30 deg to the wind. Assume gustiness can be accounted for by applying a factor of 1.20 to the wind speed. Beaufort 5 is usually the limiting wind for port operations of loading and unloading.

GIVEN:

A C9 containership is moored at a pier with an asymmetrical container distribution of 5 high in the bow half of the vessel and none in the aft portion. The useable length of the ship for cargo handling is 246 m. The height above the water is 10.9 m. The containers are 2.7 m tall.

SOLUTION:

The forward and aft areas exposed to the wind are

$$A_{aft} = \frac{246}{2}\, 10.9 = 1{,}340\,\text{m}^2$$

$$A_{forward} = 123\big[10.9 + 5(2.7)\big] = 3{,}000\,\text{m}^2$$

The net effective area is then

$$A_{eff} = \big[1{,}340 + 3{,}000\big]\sin 30° = 2{,}170\,\text{m}^2$$

The wind speed, including a gustiness factor, is

$$V_w = 1.20(10.31) = 12.4\,\text{m/s}$$

Substituting these values into Equation II-7-31, we get

$$F_w = \frac{1}{2}(1.2)(1.0)(2{,}170)(12.4)^2 = 200\,\text{kN}$$

where we have assumed $\rho_a = 1.2\ \text{kg/m}^3$ at 20 °C and $C_D = 1.0$ as a reasonable value since the mean drag coefficient for bow and transverse winds varies between 0.82 to 1.11.

(h) Longitudinal current loads on ships are taken from procedures by NAVFAC Design Manual DM26.5. The total longitudinal current load $F_{c,tot}$ is composed of form drag $F_{c,form}$, skin friction $F_{c,fric}$, and propeller $F_{c,prop}$ drag components. All three components have similar equations.

$$F_{c,tot} = F_{c,form} + F_{c,fric} + F_{c,prop} \tag{II-7-33}$$

(i) Form drag is due to the flow of water past the vessel's cross-sectional area and is defined as

$$F_{c,form} = -\frac{1}{2}\,\rho_w\,C_{c,form}\,B\,T\,V_c^2\,\cos\theta_c \tag{II-7-34}$$

where ρ_w is the water density, $C_{c,form}$ is the form drag coefficient = 0.1, B is the vessel beam, T is the vessel draft, V_c is the average current speed, and θ_c is the angle of the current relative to the longitudinal axis of the vessel. Skin friction drag is due to the flow of water over the wetted surface area of the vessel and is given by

$$F_{c,fric} = -\frac{1}{2} \rho_w \, C_{c,fric} \, S \, V_c^2 \, \cos\theta_c \qquad\qquad \text{(II-7-35)}$$

where $C_{c,fric}$ is the skin friction coefficient and S is the wetter surface area of the vessel. The skin friction coefficient is a function of the Reynolds number R_n

$$C_{c,friction} = \frac{0.075}{(\log_{10} R_n - 2)^2} \qquad\qquad \text{(II-7-36)}$$

$$R_n = \frac{V_c \, L_{wl} \, \cos\theta_c}{\nu} \qquad\qquad \text{(II-7-37)}$$

where L_{wl} is the waterline length of the vessel and ν is the kinematic viscosity of water. The wetted surface area is defined as

$$S = (1.7 T \, L_{wl}) + \frac{D}{T\gamma_w} \qquad\qquad \text{(II-7-38)}$$

where D is the displacement of the vessel in Long Tons (=2240 lb = 1016.0 kg), and γ_w = 1.01 Long Tons / m^3 (LT/m^3).

(j) Finally, the propeller drag is due to the form drag of the propeller with a locked shaft and is given by

$$F_{c,prop} = -\frac{1}{2} \rho_w \, C_{c,prop} \, A_p \, V_c^2 \, \cos\theta_c \qquad\qquad \text{(II-7-39)}$$

where $C_{c,prop}$ is the propeller drag coefficient = 1 and A_p is the expanded or developed blade area of the propeller defined as

$$A_p = \frac{L_{wl} B}{0.838 \, A_r} \qquad\qquad \text{(II-7-40)}$$

(k) The area ratio A_r is the ratio of the waterline length times the beam to the total projected propeller area. Typical values range between 100 for a destroyer and 270 for a tanker.

(l) Thus, to calculate current loads, the following input parameters are needed: the vessel's beam, draft, waterline length, displacement, and propeller area ratio; and average current speed and direction.

(m) It is difficult to evaluate wave loadings on a moored vessel. Both wind-wave and long-period waves are oscillatory; so there is a complex interaction with the mooring system dynamics. This interaction depends on the incident wave frequencies, the vessel added mass and drag characteristics, and the elastic characteristics of the mooring lines and fender system. The vessel can respond in one or more of the six modes of oscillation (see above). Rather than an analytical treatment of the problem in which actual wave loads are determined, scale model testing is often employed. In the latter, the system response in terms of mooring line loads, fender loads, and vessel response modes are directly determined for given incident wave spectra. Sarpkaya and Isaacson (1981) and Gaythwaite (1990) discuss this further.

II-7-8. References

EM 1110-2-1613
Hydraulic Design of Deep-Draft Navigation Projects

Abbott and Madsen 1990
Abbott, M. B., and Madsen, P. A. 1990. "Modelling of Wave Agitation in Harbors," *The Sea,* Volume 9, Part B, B. Le Méhauté and D. M. Hanes, ed., John Wiley & Sons, New York.

Allsop and Hettiarachchi 1988
Allsop, N. W. H., and Hettiarachchi, S. S. L. 1988. "Reflections from Coastal Structures," *Proceedings, 21st International Conference on Coastal Engineering*, American Society of Civil Engineers, pp 782-794.

Andersen 1979
Andersen, P. 1979. "Ship Motions and Sea Loads in Restricted Water Depth," *Ocean Engineering*, Vol 6, pp 557-569.

Bailard et al. 1990
Bailard, J. A., DeVries, J., Kirby, J. T., and Guza, R. T. 1990. "Bragg Reflection Breakwater: A New Shore Protection Method?" *Proceedings, 22nd International Conference on Coastal Engineering*, American Society of Civil Engineers, Delft, The Netherlands, pp 1702-1715.

Bailard et al. 1992
Bailard, J. A., DeVries, J. W., and Kirby, J. T. 1992. "Considerations in Using Bragg Reflection from Storm Erosion Protection," *Journal, Waterway, Port, Coastal and Ocean Eng. Division,* American Society of Civil Engineers, Jan/Feb, pp 62-74.

Battjes 1974
Battjes, J. A. 1974. "A Computation of Set-Up, Longshore Currents, Run-Up and Overtopping Due to Wind-Generated Waves," Ph.D. diss., Delft University of Technology, The Netherlands.

Beck 1977
Beck, R. F. 1977. "Forces and Moments on a Ship in a Shallow Channel," *Journal of Ship Research*, June.

Benham et al. 1977
Benham, F. A., et al. 1977. "Wind and Current Shape Coefficients for Very Large Crude Carriers," *Proceedings, Offshore Technology Conference,* Paper 2739.

Blaauw and van der Knapp 1983
Blaauw, H. G., and van der Knapp, F. M. C. 1983. "Prediction of Squat of Ships Sailing in Restricted Water," *Proceedings, 8th International Harbor Conference*, Antwerp, June.

Bowden 1983
Bowden, K. F. 1983. *Physical Oceanography of Coastal Waters*, John Wiley, New York.

Bowers and Welsby 1982
Bowers, E. C., and Welsby, J. 1982. "Experimental Study of Diffraction through a Breakwater Gap," Report IT 229, Hydraulics Research Station, Wallingford, U.K.

Briggs and Green 1992
Briggs, M. J., and Green, D. R. 1992. "Experimental Study of Monochromatic Wave-Ebb Current Interaction," Technical Report CERC-92-9, U.S. Army Engineer Waterways Experiment Station, Vicksburg, MS, p 91.

Briggs and Liu 1993
Briggs, M. J., and Liu, P. L-F. 1993. "Experimental Study of Monochromatic Wave-Ebb Current Interaction," *Proceedings, Waves '93 Conference,* American Society of Civil Engineers, New Orleans, LA.

Briggs et al. 1994
Briggs, M. J., Lillycrop, L. S., Harkins, G. S., Thompson, E. F., and Green, D. R. 1994. "Physical and Numerical Model Studies of Barbers Point Harbor, Oahu, Hawaii," Technical Report CERC-94-14, U.S. Army Engineer Waterways Experiment Station, Vicksburg, MS.

Bruun 1956
Bruun, P. 1956. "Destruction of Wave Energy by Vertical Walls," *Journal, Waterways and Harbors Division,* American Society of Civil Engineers, WW1, pp 1-13.

Bruun 1978
Bruun, P. 1978. "Stability of Tidal Inlets, Theory and Engineering," *Developments in Geotechnical Engineering 23*, Elsevier Scientific Publishing, Amsterdam, The Netherlands.

Bruun 1989
Bruun, P. 1989. *Port Engineering*, 4th ed., Vol 1, Gulf Publishing Company, Houston, TX.

Callaway 1981
Callaway, R. J. 1981. "Flushing Study of South Beach Marina, Oregon," *Journal, Waterway, Port, Coastal and Ocean Division*, American Society of Civil Engineers, WW2, pp 47-58.

Carr 1952
Carr, J. H. 1952. "Wave Protection Aspects of Harbor Design," Report E-11, Hydrodynamics Laboratory, California Institute of Technology.

Carr and Stelzriede 1952
Carr, J. H., and Stelzriede, M. E. 1952. "Diffraction of Water Waves by Breakwaters," *Gravity Waves*, Circular 521, National Bureau of Standards, Washington, DC, pp 109-125.

Carrier 1971
Carrier, G. F., Shaw, R. P., and Miyata, M. 1971. "The Response of Narrow-Mouthed Harbors in a Straight Coastline to Periodic Incident Waves," *J. Appl. Mech.,* Vol 38, Series E, Nov 2, 1971.

Chen 1986
Chen, H. S. 1986. "Effects of Bottom Friction and Boundary Absorption on Water Wave Scattering," *Applied Ocean Research*, Vol 8, No. 2, pp 99-104.

Chen and Houston 1987
Chen, H. S., and Houston, J. R. 1987. "Calculation of Water Oscillation in Coastal Harbors; HARBS and HARBD User's Manual," Instruction Report CERC-87-2, U.S. Army Engineer Waterways Experiment Station, Vicksburg, MS.

Chiang and Lee 1982
Chiang, W.-L., and Lee, J.-J. 1982. "Simulation of Large-scale Circulation in Harbors," *Journal, Waterway, Port, Coastal and Ocean Division*, American Society of Civil Engineers, WW1, pp 17-31.

Cialone et al. 1991
Cialone, M. A., Mark, D. J., Chou, L. W., Leenknecht, D. A., Davis, J. A., Lillycrop. L. S., Jensen, R. E., Thompson, Gravens, M. B., Rosati, J. D., Wise, R. A., Kraus, N. C., and Larson, P. M. 1991. "Coastal Modeling System (CMS) User's Manual." Instruction Report CERC-91-1, U.S. Army Engineer Waterways Experiment Station, Vicksburg, MS.

Constantine 1960
Constantine, T. 1960. "On the Movement of Ships in Restricted Waterways," *Journal of Fluid Mechanics*, Vol 9, pp 247-256.

Cox and Clark 1992
Cox, J. C., and Clark, G. R. 1992. "Design Development of a Tandem Breakwater System for Hammond Indiana," *Coastal Structures and Breakwaters*, Institution of Civil Engineers, Thomas Telford, London, pp 111-121.

Dand 1976
Dand, I. W. 1976. "Ship-Ship Interaction in Shallow Water," *11th Symposium on Naval Hydrodynamics*, University College, London, April.

Dand and Furgeson 1973
Dand, I. W., and Furgeson, A. M. 1973. "The Squat of Full Ships in Shallow Water," *Transactions, Royal Institute of Naval Architects*, April.

Davies and Heathershaw 1984
Davies, A. G., and Heathershaw, A. D. 1984. "Surface-Wave Propagation over Sinusoidally Varying Topography," *Journal of Fluid Mechanics*, Vol 144, pp 419-443.

Defant 1961
Defant, A. 1961. *Physical Oceanography*, Vol II, Pergamon Press, New York.

Delft Hydraulics Laboratory 1965
Delft Hydraulics Laboratory. 1965. "Report on Model Tests, North Sea Canal, Part III," Delft, The Netherlands.

Elgar et al. 1992
Elgar, S., Herbers, T. C., Okihiro, M., Oltman-Shay, J., and Guza, R. T. 1992. "Observations of Infragravity Waves," *Journal of Geophysical Research*, Vol 97, No. C10, pp 15573-15577.

Eryuzlu and Hausser 1978
Eryuzlu, N. E., and Hausser, R. 1978. "Experimental Investigation Into Some Aspects of Large Vessel Navigation in Restricted Waterways," *Proceedings, Symposium on Aspects of Navigability*, Delft, The Netherlands.

Falconer 1980
Falconer, R. A. 1980. "Modelling of Planform Influence on Circulation in Harbors," *Proceedings, 17th International Conference on Coastal Engineering*, American Society of Civil Engineers, pp 2726-2744.

Falconer 1984
Falconer, R. A. 1984. "A Mathematical Model Study of the Flushing Characteristics of a Shallow Tidal Bay," *Proceedings, Institution of Civil Engineers*, Vol 77, pp 311-332.

Falconer 1985
Falconer, R. A. 1985. "Application of Numerical Models in the Hydraulic Design of Four U.K. Harbors," *Proceedings, Conference on Numerical and Hydraulic Modelling of Ports and Harbors*, British Hydraulic Research Association, pp 1-11.

Falconer and Mardapitta-Hadjipandeli 1986
Falconer, R. A., and Mardapitta-Hadjipandeli, L. 1986. "Application of a Nested Numerical Model to Idealized Rectangular Harbors," *Proceedings, 20th International Conference on Coastal Engineering*, American Society of Civil Engineers, pp 176-192.

Ferguson and McGregor 1986
Ferguson, A. M., and McGregor, R. C. 1986. "On the Squatting of Ships in Shallow and Restricted Water," *Proceedings, 20th International Conference on Coastal Engineering*, American Society of Civil Engineers.

Garthune et al. 1948
Garthune, R. S., Rosenberg, B., Cafiero, D., and Olson, C.R. 1948. "The Performance of Model Ships in Restricted Channels in Relation to the Design of a Ship Canal," Report 601, U.S. Navy David Taylor Model Basin, Washington, DC.

Gates and Herbich 1977
Gates, E. T., and Herbich, J. B. 1977. "The Squat Phenomenon and Related Effects of Channel Geometry," *Proceedings, Hydraulics in the Coastal Zone Conference*, American Society of Civil Engineers.

Gaythwaite 1990
Gaythwaite, J. W. 1990. *Design of Marine Facilities-For the Berthing, Mooring and Repair of Vessels*, Von Nostrand Reinhold, New York.

Giles and Sorensen 1979
Giles, M. L., and Sorensen, R. M. 1979. "Determination of Mooring Loads and Wave Transmission for a Floating Tire Breakwater," *Proceedings, Coastal Structures '79 Conference*, American Society of Civil Engineers, pp 1069-1085.

Gilman and Nottingham 1992
Gilman, J. F., and Nottingham, D. 1992. "Wave Barriers: An Environmentally Benign Alternative," *Proceedings, Coastal Engineering Practice'92 Conference*, American Society of Civil Engineers, pp 479-486.

Goda 1969

Goda, Y. 1969. "Reanalysis of Laboratory Data on Wave Transmission Over Breakwaters," Port and Harbor Research Institute Report, Vol 8, No. 3, pp 3-18.

Goda 2000
Goda, Y. 2000. *Random Seas and Design of Maritime Structures*, 2nd ed., World Scientific Pub. Co., Teaneck, NJ.

Goda, Takayama, and Suzuki 1978
Goda, Y., Takayama, T., and Suzuki, Y. 1978. "Diffraction Diagrams for Directional Random Waves," *Proceedings, 16th International Conference on Coastal Engineering*, American Society of Civil Engineers, pp 628-650.

Greenstreet 1982
Greenstreet, G. 1982. "Study of Ships Track and Motions at Port Taranaki," *Proceedings, 18th International Conference on Coastal Engineering*, American Society of Civil Engineers, pp 2763-2772.

Hales 1981
Hales, L. Z. 1981. "Floating Breakwaters: State-of-the-Art Literature Review," Technical Report 81-1, U.S. Army Engineer Waterways Experiment Station, Vicksburg, MS.

Harms et al. 1982
Harms, V. W., Westerink, J. J., Sorensen, R. M., and McTamany, J. E. 1982. "Wave Transmission and Mooring Force Characteristics of Pipe-Tire Floating Breakwaters," Technical Paper 82-4, U.S. Army Engineer Waterways Experiment Station, Vicksburg, MS.

Harris and Bodine 1977
Harris, D. L., and Bodine, B. R. 1977. "Comparison of Numerical and Physical Hydraulic Models, Masonboro Inlet, North Carolina," GITI Report 6, General Investigation of Tidal Inlets, U.S. Army Engineer Waterways Experiment Station, Vicksburg, MS.

Havelock 1908
Havelock, T. H. 1908. "The Propagation of Groups of Waves in Dispersive Media, with Application to Waves on Water Produced by a Travelling Disturbance," *Proceedings, Royal Society of London*, Series A.

Herbers et al. 1992
Herbers, T. C., Elgar, S., Guza, R. T., and O'Reilly, W. C. 1992. "Infragravity-Frequency (0.005-0.05 Hz) Motions on the Shelf," *Proceedings of 23rd International Conference on Coastal Engineering (ICCE)*.

Herbich 1992
Herbich, J. B. 1992. "Harbors, Navigational Channels, Estuaries, Environmental Effects," *Handbook of Coastal and Ocean Engineering, Vol 3*, Gulf Publishing Company, Houston, TX, p 1340.

Hooft 1973
Hooft, J. P. 1973. "Maneuvering Large Ships in Shallow Water," *Journal of Navigation*, Vol 26.

Hsu 1990
Hsu, J. R. C. 1990. "Short-Crested Waves," *Handbook of Coastal and Ocean Engineering,Vol 1*, J. B. Herbich, ed., Gulf Publishing Company, Houston, TX.

Hudson et al. 1979

Hudson, R. Y., Herrmann, F. A., Sager, R. A., Whalin, R. W., Keulegan, G. H., Chatham, C. E., and Hales, L. Z. 1979. "Coastal Hydraulic Models," Special Report No. 5, U.S. Army Engineer Waterways Experiment Station, Vicksburg, MS.

Ippen 1966
Ippen, A. T. 1966. *"Estuary and Coastline Hydrodynamics*, McGraw-Hill Book Company, Inc., New York.

Ippen and Goda 1963
Ippen, A. T., and Goda, Y. 1963. "Wave Induced Oscillations in Harbors: The Solution for a Rectangular Harbor Connected to the Open-sea," Report 59, Hydrodynamics Lab., Mass. Inst. Technology, Cambridge, MA.

Isaacson and Mercer 1982
Isaacson, M., and Mercer, A. G. 1982. "The Response of Small Craft to Wave Action," *Proceedings, 18th International Conference on Coastal Engineering*, American Society of Civil Engineers, pp 2723-2742.

Isherwood 1973
Isherwood, R. M. 1973, "Wind Resistance of Merchant Ships," *Transactions, Royal Institute of Naval Architects*, pp 327-338.

Jiang and Falconer 1985
Jiang, J. X., and Falconer, R. A. 1985. "The Influence of Entrance Conditions and Longshore Currents on Tidal Flushing and Circulation in Model Rectangular Harbors," *Proceedings, International Conference on Numerical and Hydraulic Modelling of Ports and Harbors*, British Hydraulic Research Association, pp 65-74.

Johnson 1952
Johnson, J. W. 1952. "Generalized Wave Diffraction Diagrams," *Proceedings, Second Conference on Coastal Engineering*, The Council on Wave Research, Berkeley, CA, pp 6-23.

Kirby 1987
Kirby, J. T. 1987. "A Program for Calculating the Reflectivity of Beach Profiles," Report UFL/COEL-87/004, University of Florida, Gainesville.

Kray 1970
Kray, C. J. 1970. "Supership Effect on Waterway Depth and Alignment," *Journal, Waterways and Harbors Division*, American Society of Civil Engineers.

Lai, Long, and Huang 1989
Lai, R. J., Long, S. R., and Huang, N. E. 1989. "Laboratory Studies of Wave-Current Interaction: Kinematics of the Strong Interaction," *Journal of Geophysical Research*, Vol 94, No. C11, November, pp 16,201-16,214.

Lee 1969
Lee, J. J. 1969. "Wave Induced Oscillations in Harbors of Arbitrary Shape," Report No. KH-R-20, W. M. Keck Lab. of Hydraulics and Water Resources, Calif. Inst. of Technology, Pasadena, CA.

Leenknecht, Szuwalski, and Sherlock 1992
Leenknecht, D. A., Szuwalski, A., and Sherlock, A. R. 1992. "Automated Coastal Engineering System, User Guide and Technical Reference, Version 1.07," U.S. Army Engineer Waterways Experiment Station, Vicksburg, MS.

Lillycrop et al. 1993
Lillycrop, L. S., Briggs, M. J., Harkins, G. S., Boc, S. J., and Okihiro, M. S. 1993. "Barbers Point Harbor, Oahu, HI, Monitoring Study," Technical Report CERC-93-18, U.S. Army Engineer Waterways Experiment Station, Vicksburg, MS.

Lin, Lott, and Mehta 1986
Lin, C.-P., Lott, J. W., and Mehta, A. J. 1986. "Turbidity-Sedimentation in Closed-End Channels," *Proceedings, 20th International Conference on Coastal Engineering*, American Society of Civil Engineers, pp 1336-1350.

Liu 1982
Liu, P. L-F. 1982. "Combined Refraction and Diffraction: Comparison Between Theory and Experiments," *Journal of Geophysical Research*, Vol 87, No. C8, pp 5723-5730.

Liu and Lozano 1979
Liu, P. L-F., and Lozano, C . 1979. "Combined Wave Refraction and Diffraction," *Proceedings, Coastal Structures 79 Conference*, American Society of Civil Engineers, pp 978-997.

Lott and Hurtienne 1992
Lott, J. W., and Hurtienne, A. M. 1992. "Design, Construction and Performance of a Baffled Breakwater," *Proceedings, Coastal Engineering Practice '92 Conference*, American Society of Civil Engineers, pp 487-502.

Lozano and Liu 1980
Lozano, C., and Liu, P.L-F. 1980. "Refraction-Diffraction Model for Linear Surface Waves," *Journal of Fluid Mechanics*, Vol 101, Pt. 4.

Madsen and White 1976
Madsen, O. S., and White, S. M. 1976. "Reflection and Transmission Characteristics of Porous Rubble Mound Breakwaters," Miscellaneous Report 76-5, U.S. Army Engineer Waterways Experiment Station, Vicksburg, MS.

Madsen, Svendsen, and Michaelsen 1980
Madsen, P. A., Svendsen, I. A., and Michaelsen, C. 1980. "Some Recent Results for Wave-Induced Motions of a Ship in Shallow Water," *Proceedings, 17th International Conference on Coastal Engineering*, American Society of Civil Engineers, pp 3043-3062.

Mansard and Pratte 1982
Mansard, E. P. D., and Pratte, B. D. 1982. "Moored Ship Response in Irregular Waves," *Proceedings, 18th International Conference on Coastal Engineering*, American Society of Civil Engineers, pp 2621-2640.

McNown 1976
McNown, J. S. 1976. "Sinkage and Resistance for Ships in Channels," *Journal, Waterways, Harbors and Coastal Engineering Division*, American Society of Civil Engineers.

Mehta and Özsöy 1978
Mehta, A. J., and Özsöy, E. 1978. Chapter 3, "Inlet Hydraulics," *Stability of Tidal Inlets: Theory and Engineering*, P. Bruun, ed., Elseveir Scientific Publishing Co., Amsterdam, The Netherlands, pp 83-161.

Mei 1985
Mei, C. C. 1985. "Resonant Reflection of Surface Water Waves by Periodic Sand Bars," *Journal of Fluid Mechanics*, Vol 152, pp 315-335.

Memos 1976
Memos, C. D. 1976. "Diffraction of Waves Through a Gap Between Two Inclined Breakwaters," Ph.D. diss., University of London.

Memos 1980a
Memos, C. D. 1980a. "An Extended Approach to Wave Scattering Through a Harbor Entrance," *Bulletin, Permanent International Association of Navigation Congresses*, Vol 1, No. 35, pp 20-26.

Memos 1980b
Memos, C. D. 1980b. "Energy Transmission by Surface Waves Through an Opening," *Journal of Fluid Mechanics*, Vol 97, Pt. 3, pp 557-568.

Memos 1980c
Memos, C. D. 1980c. "Water Waves Diffracted by Two Breakwaters," *Journal of Hydraulic Research*, Vol 18, No. 4, pp 343-357.

Meirovitch 1975
Meirovitch, L. 1975. *Elements of Vibration Analysis*, McGraw-Hill, Inc.

Miles 1948
Miles, J. W. 1948. "Coupling of a Cylindrical Tube to a Half Space," *Journal, Acoustic Society of America*, pp 652-664.

Miles 1974
Miles, J. W. 1974. "Harbor Seiching," *Annual Review of Fluid Mechanics*, Vol 6, pp 17-35.

Moody 1970
Moody, C. G. 1970. "Study of the Performance of Large Bulk-Cargo Ships in a Proposed Interoceanic Canal," Report 374-H-01, U.S. Naval Ship Research and Development Center, Carderock, MD.

Naval Facilities Engineering Command 1968
Naval Facilities Engineering Command. 1968. *Harbors and Coastal Facilities*, Design Manual-26, Alexandria, VA.

Nece 1985
Nece, R. E. 1985. "Physical Modeling of Tidal Exchange in Small-Boat Harbors," *Proceedings, International Conference on Numerical and Hydraulic Modelling of Ports and Harbors*, British Hydraulic Research Association, pp 33-41.

Nece and Layton 1989
Nece, R. E., and Layton, J. A. 1989. "Mitigating Marina Environmental Impacts Through Hydraulic Design," *Proceedings, International Conference on Marinas*, Computational Mechanics Institute, pp 435-449.

Nece and Richey 1972
Nece, R.E., and Richey, E.P. 1972. "Flushing Characteristics of Small-Boat Marinas," *Proceedings, 13th International Conference on Coastal Engineering*, American Society of Civil Engineers, pp 2499-2512.

Nece and Richey 1975
Nece, R. E., and Richey, E. P. 1975. "Application of Physical Tidal Models in Harbor and Marina Design," *Proceedings, Symposium on Modelling Techniques*, American Society of Civil Engineers, pp 783-801.

Nece, Falconer, and Tsutumi 1976
Nece, R. E., Falconer, R. A., and Tsutumi, T. 1976. "Planform Influence on Flushing and Circulation in Small Harbors," *Proceedings, 15th International Conference on Coastal Engineering*, American Society of Civil Engineers, pp 3471-3486.

Nece, Smith, and Richey 1980
Nece, R. E., Smith, H. N. and Richey, E. P. 1980. "Tidal Circulation and Flushing in Five Western Washington Marinas," Technical Report 63, Harris Hydraulics Laboratory, University of Washington, Seattle.

Neumann and Pierson 1966
Neumann, G., and Pierson, W. J. 1966. *Principles of Physical Oceanography*, Prentice-Hall, Englewood Cliffs, NJ.

Newman 1978
Newman, J. N. 1978. *Marine Hydrodynamics*, MIT Press, Cambridge, MA.

Noble 1982
Noble, S. 1982. "Ship Motions Related to Deep Draft Channel Design," *Proceedings, 18th International Conference on Coastal Engineering*, American Society of Civil Engineers, pp 2662-2680.

Northwest Hydraulic Consultants 1980
Northwest Hydraulic Consultants. 1980. "Study to Determine Acceptable Wave Climate in Small Craft Harbors," Report to Department of Fisheries and Oceans Government of Canada, Ottawa.

Oil Companies International Marine Forum 1978
Oil Companies International Marine Forum. 1978. "Guidelines and Recommendations for the Safe Mooring of Large Ships at Pier and Sea Island," Witherby and Co. Ltd., London.

Okihiro, Guza, and Seymour 1992
Okihiro, M., Guza, R. T., and Seymour, R. J. 1992. "Bound Infragravity Waves," *Journal of Geophysical Research*, Vol 97, No. C7, pp 11453-11469.

Owens and Palo 1981
Owens, R., and Palo, P. A. 1981. "Wind-Induced Steady Loads on Moored Ships," Technical Memo M-44-81-7, Navy Civil Engineering Laboratory, Pt Hueneme, CA.

Palo 1983

Palo, P. A. 1983. Steady Wind and Current Induced Loads on Moored Vessels," *Proceedings, Offshore Technology Conference*, Paper 4530.

Penny and Price 1952
Penny, W. G., and Price, A. T. 1952. "The Diffraction Theory of Sea Waves by Breakwaters, and the Shelter Afforded by Breakwaters," *Philosophical Transactions*, Royal Society of London, Series A, Vol 244, pp 236-253.

Putnam and Arthur 1948
Putnam, J. A., and Arthur, R.S. 1948. "Diffraction of Water Waves by Breakwaters," *Transactions, American Geophysical Union*, Vol 29, No. 4, pp 481-490.

Raichlen 1968
Raichlen, F. 1968. "The Motions of Small Boats in Standing Waves," *Proceedings, 11th Conference on Coastal Engineering*, American Society of Civil Engineers, pp 1531 -1554.

Raichlen 1993
Raichlen, F. 1993. "Waves Propagating on an Adverse Jet," *Proceedings, Waves '93 Conference*, American Society of Civil Engineers, New Orleans, LA.

Raichlen and Lee 1992
Raichlen, F., and Lee, J.J. 1992. "Oscillation of Bays, Harbors and Lakes," *Handbook of Coastal and Ocean Engineering,* Vol 3, J. B. Herbich, ed., Gulf Publishing Company, Houston, TX.

Robb 1949
Robb, A. M. 1949. "Interaction Between Ships," *Transactions, Institute of Naval Architects*, Vol 91.

Robb 1952
Robb, A. M. 1952. *Theory of Naval Architecture*, Charles Griffen and Co., London.

Sakai and Saeki 1984
Sakai, S., and Saeki, H. 1984. "Effects of Opposing Current on Wave Transformation on a Sloping Bed," *Proc. 19th Conf. on Coastal Engineering*, American Society of Civil Engineers, Vol 1, pp 1132-1148.

Sarpkaya and Isaacson 1981
Sarpkaya, T., and Isaacson, M. 1981. *Mechanics of Wave Forces on Offshore Structures*, Van Nostrand Reinhold Co., New York.

Schluchter and Slotta 1978
Schluchter, S. S., and Slotta, L. 1978. "Flushing Studies of Marinas," *Proceedings, Coastal Zone '78 Conference*, American Society of Civil Engineers, pp 1878-1896.

Schofield 1974
Schofield, R. B. 1974. "Speed of Ships in Restricted Navigation Channels," *Journal, Waterways, Harbors and Coastal Engineering Division*, American Society of Civil Engineers, May.

Schwartz and Imberger 1988
Schwartz, R. A., and Imberger, J. 1988. "Flushing Behavior of a Coastal Marina," *Proceedings, 21st International Conference on Coastal Engineering*, American Society of Civil Engineers, pp 2626-2640.

Seelig 1979
Seelig, W. N. 1979. "Estimation of Wave Transmission Coefficients for Permeable Breakwaters," CETA 79-6, U.S. Army Engineer Waterways Experiment Station, Vicksburg, MS.

Seelig 1980
Seelig, W. N. 1980. "Two-dimensional Tests of Wave Transmission and Reflection Characteristics of Laboratory Breakwaters," Technical Report 80-1, U.S. Army Engineer Waterways Experiment Station, Vicksburg, MS.

Seelig 1983
Seelig, W. N. 1983. "Wave Reflection from Coastal Structures," *Proceedings, Coastal Structures '83*, American Society of Civil Engineers, pp 961-973.

Seelig and Ahrens 1981
Seelig, W. N., and Ahrens, J. P. 1981. "Estimation of Wave Reflection and Energy Dissipation Coefficients for Beaches, Revetments and Breakwaters," Technical Paper 81-1, U.S. Army Engineer Waterways Experiment Station, Vicksburg, MS.

Seelig, Kreibel, and Headland 1992
Seelig, W. N., Kreibel, D., and Headland, J. 1992. "Broadside Current Forces on Moored Ships," *Proceedings, Civil Engineering in the Oceans V Conference*, American Society of Civil Engineers, pp 326-340.

Sharp and Fenton 1968
Sharp, B. B., and Fenton, J. D. 1968. "A Model Investigation of Squat," *The Dock and Harbor Authority*, November.

Shore Protection Manual 1984
Shore Protection Manual. 1984. 4th ed., 2 Vol, U. S. Army Engineer Waterways Experiment Station, U.S. Government Printing Office, Washington, DC.

Silverstein 1958
Silverstein, B. L. 1958. "Linearized Theory of the Interaction of Ships," Institute of Engineering Research, University of California, Berkeley.

Silvester and Hsu 1993
Silvester, R., and Hsu, J.R.C. 1993. *Coastal Stabilization-Innovative Concepts*, Prentice-Hall, Englewood Cliffs, NJ.

Sommerfeld 1896
Sommerfeld, A. 1896. "Mathematische Theorie der Diffraction," *Mathematische Annalen*, Vol 47, pp 317-374.

Sorensen 1973a
Sorensen, R. M. 1973a. "Ship-Generated Waves," *Advances in Hydroscience*, Academic Press, New York, Vol 9, pp 49-83.

Sorensen 1973b
Sorensen, R. M. 1973b. "Water Waves Produced by Ships," *Journal, Waterways,Harbors and Coastal Engineering Division*, American Society of Civil Engineers.

Sorensen 1976
Sorensen, R. M., and Seelig, W. N. 1976. "Hydraulics of Great Lakes Inlet-Harbor Systems," *Proceedings, Fifteenth Conference on Coastal Engineering,* American Society Civil Engineers, Honolulu, p 20.

Sorensen 1986
Sorensen, R. M. 1986. "Bank Protection for Vessel-Generated Waves," Report IHL-117-86, Lehigh University, prepared for U.S. Army Engineer Waterways Experiment Station, Vicksburg, MS.

Sorensen 1989
Sorensen, R. M. 1989. "Port and Channel Bank Protection from Ship Waves," *Proceedings, Ports '89 Conference,* American Society of Civil Engineers.

Sorensen 1990
Sorensen, R. M. 1990. "The Deployment of Floating Breakwaters: Design Guidance," *Proceedings, 12th Coastal Society Conference,* San Antonio, TX.

Sorensen 1993
Sorensen, R. M. 1993. *Basic Wave Mechanics,* John Wiley & Sons, New York.

Sorensen and Weggel 1984
Sorensen, R. M., and Weggel, J. R. 1984. "Development of Ship Wave Design Information," *Proceedings, 19th International Conference on Coastal Engineering,* American Society of Civil Engineers.

Spaulding 1984
Spaulding, M. L. 1984. "A Vertically Averaged Circulation Model Using Boundary-Fitted Coordinates," *Journal of Physical Oceanography,* American Meteorological Society, Vol 14, pp 973-982.

Taylor 1909
Taylor, D. W. 1909. "Some Model Experiments on Suction of Vessels," *Society of Naval Architects and Marine Engineers,* Vol 17.

Thompson, Chen and Hadley 1993
Thompson, E. F., Chen, H. S., and Hadley, L. L. 1993. "Numerical Modeling of Waves in Harbors," *Proceedings, WAVES 93,* American Society of Civil Engineers, New Orleans, LA.

Tothill 1967
Tothill, J. T. 1967. "Ships in Restricted Channels- A Correlation of Model Tests, Field Measurements, and Theory," *Marine Technology,* April.

Tsinker 1986
Tsinker, G. P. 1986. *Floating Ports-Design and Construction Practices,* Gulf Publishing, Houston, TX.

Tuck 1966
Tuck, E. O. 1966. "Shallow-Water Flow Past Slender Bodies," *Journal of Fluid Mechanics,* Vol 26, pp 81-96.

Tuck and Newman 1974
Tuck, E. O., and Newman, J. N. 1974. "Hydrodynamic Interactions Between Ships," *10th Naval Hydrodynamics Symposium,* Massachusetts Institute of Technology, Cambridge.

van Wyk 1982
van Wyk, A. C. 1982. "Wave-induced Ship Motions in Harbor Entrances - A Field Study," *Proceedings, 18th International Conference on Coastal Engineering*, American Society of Civil Engineers, pp 2681-2699.

Van der Meer and Angremond 1992
Van der Meer, J. W., and Angremond, K. 1992. "Wave Transmission at Low-Crested Structures," *Coastal Structures and Breakwaters*, Institution of Civil Engineers, Thomas Telford, London, pp 25-41.

Vemulakonda, Chou, and Hall 1991
Vemulakonda, S. R., Chou, L. W., and Hall, R. W. 1991. "Los Angeles and Long Beach Harbors Additional Plan Testing - Numerical Modeling of Tidal Circulation and Water Quality," Technical Report CERC-91-2, U.S. Army Engineer Waterways Experiment Station, Vicksburg, MS.

Wang and Noble 1982
Wang, S., and Noble, S. 1982. "Columbia River Entrance Channel Ship Motion Study," *Journal, Waterways, Port, Coastal and Ocean Division*, American Society of Civil Engineers, pp 291-305.

Weggel and Sorensen 1986
Weggel, J. R., and Sorensen, R. M. 1986. "Ship Wave Prediction for Port and Channel Design," *Proceedings, Ports '86 Conference*, American Society of Civil Engineers.

Wehausen 1971
Wehausen, J. V. 1971. "The Motion of Floating Bodies," *Annual Review of Fluid Mechanics*, Annual Reviews, Palo Alto, Vol 3, pp 237-268.

Wiegel 1960
Wiegel, R. L. 1960. "Transmission of Waves Past a Rigid Vertical Thin Barrier," *Journal, Waterways and Harbors Division*, American Society of Civil Engineers, pp 1-12.

Wiegel 1962
Wiegel, R. L. 1962. "Diffraction of Waves by Semi-infinite Breakwater," *Journal of the Hydraulics Division*, American Society of Civil Engineers, Vol 88, No. HY1, pp 27-44.

Willis 1988
Willis, D. H. 1988. "Experimental and Theoretical Studies of Wave-Current Systems," Technical Report, Hydraulics Laboratory, National Research Council of Canada, Ottawa, Ontario, Canada.

Wilson 1967
Wilson, B. W. 1967. "Elastic Characteristics of Moorings," *ASCE Journal of the Waterways and Harbors Division*, Vol 93, No. WW4, pp 27-56.

Wilson 1972
Wilson, B. W. 1972. "Seiches," Advances in Hydroscience, Academic Press, New York, Vol 8, pp 1-94.

Zelt 1986
Zelt, J. A. 1986. "Tsunamis: The Response of Harbors with Sloping Boundaries to Long Wave Excitation," Report KH-R-47, W.M. Keck Lab. of Hydraulics and Water Resources, Calif. Inst. of Technology, Pasadena, CA.

Zelt and Raichlen 1990

Zelt, J. A., and Raichlen, F. 1990. "A Lagrangian Model for Wave-Induced Harbor Oscillations," *Journal of Fluid Mechanics*, Vol 213.

Zwamborn and Cox 1982

Zwamborn, J. A., and Cox, P. J. 1982. "Operational Procedures-Richards Bay Harbor," *Proceedings, 18th International Conference on Coastal Engineering*, American Society of Civil Engineers, pp 2700-2722.

II-7-9. Definitions of Symbols

α	Angle a surface-piercing sloped plane forms with the horizontal [deg]
β	Angle between the breakwater and the radial distance from the breakwater tip to the point where the diffraction coefficient (K') is to be determined (Figure II-7-2) [deg]
Δd	Vessel drawdown [length]
Δl_n	Elongation in the mooring line [length]
ϵ_n	Percent elongation in a mooring line
θ	Angle between the sailing line and the direction of wave propagation [deg]
θ_c	Angle of the current relative to the longitudinal axis of the vessel [deg]
θ_n	Angle a mooring line makes in the horizontal plane with the perpendicular to the vessel [deg]
ν	Kinematic viscosity of water [length^2/time]
ρ_a	Mass density of air [force-time^2/length^4]
ρ_w	Mass density of water (salt water = 1,025 kg/m^3 or 2.0 slugs/ft^3; fresh water = 1,000kg/m^3 or 1.94 slugs/ft^3) [force-time^2/length^4]
ϕ	Angle a mooring line makes in the vertical plane between the vessel and the dock [deg]
ϕ	Phase angle by which a mass displacement lags the excitation displacement [deg]
A	Amplification factor (ratio of mass displacement to excitation displacement) [dimensionless]
A	Projected area of a vessel above the waterline [length^2]
A_b	Surface area of a bay or basin [length^2]
A_c	Channel cross-sectional flow area of an inlet [length^2]
A_m	Vessel's midsection wetted cross-sectional area [length^2]
A_p	Expanded or developed blade area of a propeller (Equation II-7-40) [length^2]
A_r	Dimensionless ratio of the waterline length times the beam to the total projected propeller area
b	Channel width [length]
B	Breakwater gap (Figure II-7-6} [length]
B	Structure crest width [length]
B	Vessel beam [length]
B'	Imaginary breakwater gap [length] (Figure II-7-6)

C	Dimensionless coefficient in the Seelig equation (Equation II-7-2) for the wave transmission coefficient (Equation II-7-3)
C	Individual wave celerities created by a moving vessel (Equation II-7-21) [length/time]
$C_{c,form}$	Form drag coefficient [dimensionless]
$C_{c,prop}$	Propeller drag coefficient [dimensionless]
$C_{c,fric}$	Skin friction coefficient [dimensionless]
C_D	Coefficient of drag for winds measured at 10-m [dimensionless]
C_o, C_i	Initial concentration of some substance in the harbor water and concentration of the substance after i tidal cycles (Equation II-7-20)
C_r	Reflection coefficient [dimensionless]
C_t	Wave transmission coefficient [dimensionless]
C_{t0}	Coefficient for wave transmission by flow over the structure [dimensionless]
C_{tt}	Coefficient for wave transmission through the structure [dimensionless]
d	Water depth [length]
D	Dimensionless drawdown
D	Displacement of a vessel [length^3]
D_{50}	Median diameter of armor stone [length]
d_c	Channel depth [length]
d_s	Water depth at the toe of the structure [length]
E	Dimensionless average per cycle exchange coefficient, a factor for defining the effectiveness of tidally induced harbor flushing (Equation II-7-20)
F	Freeboard [length]
F	Froude number
$F_{c,form}$	Form drag of a vessel (Equation II-7-34) [force]
$F_{c,fric}$	Skin friction of a vessel (Equation II-7-35) [force]
$F_{c,prop}$	Vessel propeller drag (Equation II-7-39) [force]
$F_{c,tot}$	Total longitudinal current load on a vessel (Equation II-7--33) [force]
F_w	Drag force due to wind (Equation II-7-31) [force]
g	Gravitational acceleration [length/time^2]
h	Structure crest elevation [length]

H	Standing wave height [length]
H_d	Diffracted wave height at a point in the lee of a breakwater [length]
H_i	Incident wave height [length]
H_r	Reflected wave height [length]
H_t	Transmitted wave height [length]
I_r	Surf similarity number or Iribarren number (Equation II-7-6) [dimensionless]
k_n	Individual stiffness of a mooring line (Equation II-7-29) [force/length]
k_{tot}	Effective spring constant (Equation II-7-28) [force/length]
K'	Diffraction coefficient [dimensionless]
K'_e	Effective diffraction coefficient (Equation II-7-1) [dimensionless]
L	Wave length [length]
ℓ_B	Basin length along the axis B [length]
ℓ_c	Channel length [length]
ℓ'_c	Additional length to account for mass outside each end of the channel [length]
l_n	Length of a mooring line [length]
L_{wl}	Waterline length of a vessel [length]
m	Mass of a vessel [force]
M_0	Zero moment of the spectrum
m_a	Mass of a vessel due to inertial effects of the water entrained with the vessel [force]
m_v	Virtual mass of a vessel (Equation II-7-27) [force]
N	Water surface amplitude (Equation II-7-7) [length]
$-_0$	The subscript 0 denotes deepwater conditions
r	Radial distance from the breakwater tip to the point where the diffraction coefficient (K') is to be determined (Figure II-7-2) [length]
R	Wave runup above the mean water level [length]
R_n	Reynolds number
S	Vessel blockage ratio [dimensionless]
S	Wetted surface area of a vessel [length^2]
S_{max}	Directional concentration parameter characterizing the directional spread of the wave spectrum [dimensionless]

T	Excitation (and response) period [time]
T	Vessel draft [length]
T	Wave period [time]
T_H	Resonant period for Helmholtz mode (Equation II-7-18)[time]
T_n	Axial tension or load on a mooring line [force]
$T_{n,m}$	Natural free oscillating period (n, m is the number of nodes along the x- and y-axes of a basin [time]
T_n	Undamped natural period of a vessel (Equation II-7-26) [time]
$\overline{V}$	Average horizontal velocity at a node (Equation II-7-16) [length/time]
V_{10}	Wind speed at the standard elevation of 10 m above the water surface [length/time]
V_c	Average current speed [length/time]
V_{max}	Maximum horizontal velocity at a node [length/time]
V_r	Return flow velocity (Equation II-7-25), velocity that develops between the vessel hull and the channel bottom and sides [length/time]
V_s	Vessel speed [length/time]
V_z	Wind speed at elevation z [length/time]
W	Breakwater's characteristic dimension in the direction of wave propagation [length]
W	Channel width [length]
X	Maximum horizontal particle excursion at a node in a standing wave (Equation II-7-15) [length]
y	Vertical extent of the barrier below the still-water surface [length]

II-7-10. Acknowledgments

Author of Chapter II-7, "Harbor Hydrodynamics:"

Robert Sorensen, Ph.D., Fritz Engineering Laboratory, Lehigh University, Bethlehem, Pennsylvania.
Edward F. Thompson, Ph.D., Coastal and Hydraulics Laboratory (CHL), Engineer Research and Development Center, Vicksburg, Mississippi.

Reviewers:

Zeki Demirbilek, Ph.D., CHL.
Lee E. Harris, Ph.D., Department of Marine and Environmental Systems, Florida Institute of Technology, Melbourne, Florida.
Michael J. Briggs, Ph.D., CHL.
Monica A. Chasten, U.S. Army Engineer District, Philadelphia, Philadelphia, Pennsylvania.
Linda S. Lillycrop, U.S. Army Engineer District, Mobile, Mobile, Alabama.

II-7-11. Note to Users, Vessel Buoyancy

The determination of the buoyancy of and the weight of a floating vessel requires determining the immersed volume at any waterline and multiplying that volume by the specific weight of the water, generally assumed as 64 lb/ft³. This is nothing but an expression of Archimedes' principle that the weight of any floating body is equal to the weight of the fluid displaced by the body. In naval architecture, the buoyancy is most commonly called the displacement, but although the use of the word displacement suggests volume, the displacement itself is customarily given in tons. The displacement of a vessel is the ratio of the immersed volume divided by the number of cubic feet per ton of water. The immersed volume is generally in cubic feet (ft³). The vessel displacement is denoted by the symbol Δ, and is in tons. The vessel displacement to any waterline depends on the density or specific weight of water, γ_w. For salt water, $\gamma_w = 64$ and the divisor is $2240/64 = 35$. For fresh water $\gamma_w = 62.4$ and the divisor is $2240/62.4 = 35.85 \sim 36.0$. The volume of displacement is denoted by the symbol ∇, and is customarily in ft³. Therefore, we have

$$\Delta = \nabla * \frac{\gamma_w lb / ft^3}{2240 lb / ton} = \frac{\nabla}{r}$$

where $r = \dfrac{2240}{\gamma_w}$ and γ_w is in lb/ft³. For salt water r = 35 and for fresh water r = 36.

In general, the wetted surface area of a vessel whose geometry is approximated by a rectangular solid can be estimated as

$$\nabla = L * B * T * C_B$$

and

$$S = 2 * L * T + \frac{\nabla}{T} = 2 * L * T + L * B * C_B$$

In 1895, Mumford developed a similar expression from the analysis of calculations for a number of ships:

$$s = 1.7 * L * T + \frac{\nabla}{T} = 1.7 * L * T + L * B * C_B$$

Example 1: For a cargo vessel with the values of L = 401 ft, B = 54.83 ft (molded), T = 23.83 ft (molded), C_B = 0.7685 and C_P = 0.7763. We find S =33,150 ft^2.

Example 2: For fast passenger liner with values of L = 513.4 ft, B = 70.54 ft (molded), d = 24.61 (molded), C_B = 0.5751 and C_P = 0.6084. We find S = 42,300 ft^2.

Chapter 8
HYDRODYNAMIC ANALYSIS AND
DESIGN CONDITIONS

EM 1110-2-1100
(Part II)
30 April 2002

Table of Contents

Page

II-8-1. Overview of Chapter II-8-1
   a. Objectives II-8-1
   b. Contents II-8-1
   c. Relationship to other chapters and parts II-8-1

II-8-2. Identifying Meteorological and Hydrodynamic Processes
Impacting Design II-8-1
   a. Brief review of processes II-8-1
   b. Identifying relevant processes II-8-2
   c. Interaction between processes II-8-2

II-8-3. Acquiring Information II-8-2
   a. Identifying available information II-8-2
   b. Consideration of collecting field measurements II-8-2
   c. Numerical and physical modeling possibilities II-8-2

II-8-4. Statistical Methods - Short-term II-8-3
   a. Introduction II-8-3
   b. Probability distribution functions II-8-3
   c. Statistical parameters II-8-4

II-8-5. Statistical Methods - Long-term II-8-4
   a. Introduction II-8-4
   b. Stochastic time history II-8-6
   c. Extremal probability distribution functions II-8-6
   d. Empirical simulation technique II-8-6
   e. Methods for fitting distributions to data II-8-6
      (1) Data selection II-8-6
      (2) Estimating parameters in extremal distribution functions II-8-9
      (3) Approaches to estimating parameters II-8-10
      (4) Outliers II-8-11

      (5) Choosing an extremal distribution function II-8-12
      (6) Confidence intervals II-8-12
   f. Return period and encounter probability II-8-12
   g. Extrapolation of data II-8-13

II-8-6. Analysis of Key Meteorological and Hydrodynamic Processes in Design II-8-13
    a. Introduction II-8-13
    b. Wind II-8-14
        (1) Design importance II-8-14
        (2) General climate II-8-14
        (3) Storms II-8-14
    c. Extreme waves II-8-20
        (1) Design importance II-8-20
        (2) Deep water II-8-20
        (3) Intermediate-depth water II-8-25
        (4) Shallow water (depth-limited) II-8-25
        (5) Extreme individual wave characteristics II-8-26
    d. Wave climate II-8-28
        (1) Design importance II-8-28
        (2) General climate II-8-28
        (3) Storms II-8-30
        (4) Persistence of high and low wave conditions II-8-31
    e. Long waves II-8-31
        (1) Tsunamis II-8-31
        (2) Seiche II-8-32
        (3) Infragravity waves II-8-32
    f. Extreme water level II-8-32
        (1) Design importance II-8-32
        (2) Estimation procedures II-8-32
    g. Water level climate II-8-32
        (1) Design importance II-8-33
        (2) Estimation procedures II-8-33
        (3) Long-term changes II-8-33
    h. Currents II-8-33
        (1) Design importance II-8-33
            (a) Nearshore shelf II-8-33
            (b) Surf zone II-8-33
            (c) Inlets II-8-33
            (d) Harbors II-8-33
        (2) Estimation procedures II-8-33
    i. Design example II-8-34

II-8-7. Interdependence of Processes During Severe Events II-8-57
    a. Design importance II-8-57
    b. Procedures for estimating realistic design events, probabilities and return periods . . . II-8-57

II-8-8. References II-8-57

II-8-9. Definition of Symbols II-8-61

II-8-10. Acknowledgments II-8-62

List of Tables

Page

Table II-8-1.   Definition of Short-Term and Long-Term Statistics II-8-4

Table II-8-2.   Parameters in Extremal Distribution Functions II-8-9

Table II-8-3.   Percent Chance for H_s Equaling or Exceeding Return Period H_s II-8-13

Table II-8-4.   Probability Distribution Functions for Meteorological and Hydrodynamic
Processes II-8-14

Table II-8-5.   Standard Deviation and Confidence Interval Relationships II-8-16

Table II-8-6.   Return Period Adjustment Factor II-8-19

Table II-8-7.   Conversion of Extreme Wind Speeds from Fastest-Mile to 1-hr Average II-8-19

Table II-8-8.   Averaging Time Adjustments for Extreme Wind Speeds II-8-20

Table II-8-9.   Maximum Wind Speed by Month, in m/s II-8-21

Table II-8-10. Wind Speed Statistics II-8-23

Table II-8-11. Wind Speed at 25- and 50-year Return Period II-8-24

Table II-8-12. Tidal Datum Information II-8-35

Table II-8-13. Significant Storm Events II-8-40

Table II-8-14. Tidal Water Levels for Jetty Design II-8-43

Table II-8-15. Calculation of d_{ijetty} for Storm Event 1 . . . . II-8-44

Table II-8-16. Calculation of d_{jetty} for Storm Event 1 . . . . II-8-45

Table II-8-17. Calculation of H_{sjetty} for Storm Event 1 . . . . II-8-46

Table II-8-18. Calculation of Upper Bound H_{sjetty} Based on Maximum Tide Level II-8-49

Table II-8-19. Calculation of Upper Bound H_{sjetty} Based on Maximum During Shoaling II-8-50

Table II-8-20. Design Significant Wave Heights at Jetty Head II-8-51

Table II-8-21. Calculation of H_{stoe} for Storm Event 1 at Jetty Head II-8-52

Table II-8-22. Calculation of Wave Height Modification by Currents II-8-53

List of Figures

Page

Figure II-8-1.   Probability distribution functions for short-term statistics II-8-5

Figure II-8-2.   Probability distribution functions for long-term statistics II-8-7

Figure II-8-3.   Probability distribution functions for long-term statistics II-8-9

Figure II-8-4.   Selection of extreme values for a partial duration series II-8-10

Figure II-8-5.   Plotting position formulas II-8-11

Figure II-8-6.   Example wind rose (Leffler et al. 1990) II-8-15

Figure II-8-7.   Extreme fastest-mile wind speeds with 50-year return period across the
U.S. at 10-m elevation (ASCE 1993) II-8-17

Figure II-8-8.   Milepost map for use with Figure II-8-9; coastal distance intervals marked
in nautical miles (1 nautical mile = 1.9 km) II-8-18

Figure II-8-9.   Extreme fastest-mile hurricane wind speeds blowing from any direction
at 10 m above ground in open terrain near the coastline for various return
periods (after National Bureau of Standards (1980)) II-8-19

Figure II-8-10. Example Problem II-8-1 probability distribution of wind speeds II-8-22

Figure II-8-11. Probability distribution of H_s on shallow water II-8-26

Figure II-8-12. Maximum value of H_s in the surf zone (Goda 1985) II-8-27

Figure II-8-13. Water depth at which H_s is maximum in the surf zone (Goda 1985) II-8-28

Figure II-8-14. Crest elevation at the 2-percent probability level of exceedance
(Seelig, Ahrens, and Grosskopf 1983) II-8-29

Figure II-8-15. Wave climate summary (Hubertz et al. 1993) II-8-30

Figure II-8-16. Persistence of storm waves II-8-31

Figure II-8-17. Example of short and long-term surf zone current data (Leffler et al. 1990) II-8-34

Figure II-8-18. Jetty plan II-8-35

Figure II-8-19. Probability distribution of astronomical tide levels II-8-36

Figure II-8-20. Probability distribution of storm surge II-8-37

Figure II-8-21. Probability distribution of high tides and combined tide and storm surge II-8-38

Figure II-8-22. Estimation of shoaling coefficient (Goda 1985) II-8-41

Figure II-8-23. Astronomical tide probability referenced to MSL II-8-42

Figure II-8-24. Estimation of wave setup (Goda 1985) II-8-43

Figure II-8-25. Estimation of wave height at the jetty (Goda 1985) II-8-46

Figure II-8-26. Identification of surf zone conditions II-8-47

Figure II-8-27. Return period wave heights, 165 events, jetty head II-8-48

Figure II-8-28. Return period wave heights, H_0' II-8-49

Figure II-8-29. Return period wave heights, 165 events including upper bounds, jetty head II-8-50

Figure II-8-30. Return period wave heights, toe design, jetty head II-8-52

Figure II-8-31. Wave height modification by currents II-8-54

Figure II-8-32. Joint distribution of H_{sjetty} and T_p from 33 storm events, jetty head II-8-55

Figure II-8-33. Example storm event (Leffler et al. 1990) II-8-56

Chapter II-8
Hydrodynamic Analysis and Design Conditions

II-8-1. Overview of Chapter

a. Objectives. Previous chapters in Part II provide detailed descriptions of the various processes involved in coastal hydrodynamics. The purpose of this chapter is to draw together the key aspects of these processes for design.

b. Contents. The following section gives a brief review of hydrodynamic processes covered in earlier chapters and their relative importance to design. Part II-8-3 summarizes approaches to acquiring information needed for design. Statistical methods needed for design analysis of short-term (single sea state) and long-term information are discussed in Part II-8-4 and 8-5. Design aspects of key meteorological and hydrodynamic processes are discussed in detail in Part II-8-6. Two design-related example problems are included. Finally, the interdependence of processes during severe events, which often subjects a project to extreme conditions of multiple processes (for example, extreme winds, waves, and water levels may occur simultaneously in a very severe storm), is discussed in Part II-8-7. References are given in Part II-8-8.

c. Relationship to other chapters and parts.

(1) Earlier chapters in Part II generally present descriptions, statistics, and probability distribution functions for *short-term* hydrodynamic processes. For example, the statistics of individual wave heights in a sea state, a statistical definition of *significant wave height*, and the Rayleigh distribution are given in Part II-1. The short-term variability of hydrodynamic processes can be important in design and results of earlier chapters are briefly summarized in this chapter.

(2) The primary concern for design is long-term variation of processes, particularly extreme occurrences over long time periods. For example, the highest significant wave height to be expected over a 25-year time period (or other long-term time period) is often a critical design parameter. Long-term extremes are the main focus of this chapter. This subject is generally not addressed in earlier chapters, with the notable exception of Part II-5. Long-term water level analysis procedures presented in Part II-5 are referenced and briefly summarized in the general hydrodynamic design framework given in this chapter.

(3) This chapter provides information on the hydrodynamic aspects of design, which are part of the more comprehensive design process and procedures of Parts V and VI. The treatment of hydrodynamics in this chapter is fairly general, whereas Parts V and VI develop more specific hydrodynamic design applications for particular types of projects. This chapter provides necessary background for the material in Parts V and VI. Also, broad design considerations, which encompass more than just the hydrodynamics, are deferred to Part V.

II-8-2. Identifying Meteorological and Hydrodynamic Processes Impacting Design

a. Brief review of processes. The first step in hydrodynamic analysis for design is to identify the meteorological and hydrodynamic processes that are likely to be important for design. The candidate processes, discussed in previous chapters of Part II, are briefly reviewed here with emphasis on design applications. The introduction to Part II-3 also provides perspective on processes.

b. Identifying relevant processes. In any particular design application, some (but not all) of the meteorological and hydrodynamic processes will be relevant for analysis. These processes are identified by a combination of general understanding of coastal behavior and insight into the design needs of the specific project. Some processes are almost always a concern for certain project types. For example, circulation and flushing are generally evaluated in a harbor project.

c. Interaction between processes. It is important to remember that although meteorological and hydrodynamic processes are often discussed *individually*, they impact a project in *combination*. This combination of processes leads to two strong implications for design:

(1) Processes mutually interact and cannot always be treated as being independent of each other. For example, water level affects waves by influencing shallow-water transformation and breaking.

(2) Extreme occurrences of one process often coincide with strong or extreme occurrences of some other processes. For example, strong winds, large waves, and elevated water levels often occur together during a severe storm.

II-8-3. Acquiring Information

a. Identifying available information. The available meteorological and hydrodynamic information for coastal design is rarely adequate for direct use. Typically all reasonable information sources on relevant processes are identified. Then the information is modified and carefully interpreted in various ways to arrive at the required design conditions. Part II-8 provides a guide to available sources of wind, wave, water level, and current information.

b. Consideration of collecting field measurements. Field measurements can be very helpful at a project site, especially if the information already available is seriously limited in quality and representativeness. Often the needs and scope of coastal projects justify some level of field study. Field measurements must be carefully planned so that the project schedule can accommodate the time and cost required for collecting and analyzing data. Measuring any extreme storm events during the life of a project level study is a matter of chance, but at least some routine storm events are typically recorded. Field measurement options are discussed in Part VII-3. Field measurements have the following potential advantages:

(1) Direct documentation of processes at the project site.

(2) On-site data can be correlated with a better documented, related location (such as a point for which long-term measurements or hindcasts are available).

(3) Onsite data for calibration and validation in model studies.

c. Numerical and physical modeling possibilities. Numerical and physical models offer powerful tools to assist in design analysis. They are used for evaluating existing conditions at a site and various alternative modifications. Both modeling tools are discussed in Part VII. In terms of meteorology and hydrodynamics, numerical models are typically used for:

(1) Extending the length of record.

(2) Hindcasting extreme events not included in the available information.

(3) Synthesizing hypothetical, but possible, extreme events (such as historical hurricanes with modified tracks).

(4) Transferring information from a related, better documented location to the project site.

(5) Providing a more comprehensive assessment of processes over an area than is generally possible with point measurements.

Physical models, properly scaled, often provide a helpful representation of the hydrodynamics of a complex project site, including wave shoaling, wave breaking, wave reflection, and wave-current interaction. They are typically considered for:

(1) Transformation of waves, tides, and/or currents from offshore to a complex project site.

(2) Design forces and overtopping of breakwaters (Part VI).

II-8-4. Statistical Methods - Short-Term

a. Introduction.

(1) Hydrodynamic conditions in a coastal area at any instant in time may be viewed as a *sea state*, that is, a state in which conditions are relatively constant over some short time period. That time period is typically from 1 to 6 hr before the sea transitions to some significantly different state of waves and, in the nearshore area, water level and currents. Since waves are typically the most intensely varying factor outside the nearshore area, the term sea state is often intended to mean waves. Sea states change in response to changing local and offshore winds, tides, and other factors. Thus wave measurement and hindcasting programs often gather wave conditions at 1-hr to 6-hr intervals to adequately sample the range of sea states.

The constancy of a sea state is best defined in statistical terms because components of the sea state, particularly wind waves and swell, have strong variations over time periods of seconds. By contrast, wave theories and many physical model tests performed more than about 10 years ago represent a sea state as a regular wave (e.g., Part II-1). This deterministic representation masks some important aspects of wave behavior.

(2) Most design is based on long-term hydrodynamic statistics representing many years of record. However, statistical variations *within a sea state*, referred to as *short-term* variations, can also be critical for design applications in which damage is a highly sensitive response to individual extreme waves rather than an integrated response to the overall sea state. For example, damage to a pier or platform deck will occur only if a wave crest is high enough to hit it. Damage in this example results from a combination of extremes in *long-term* statistics (extreme combination of H_s and water level) and *short-term* statistics (extreme value of individual wave height). Another example in which both long and short-term statistics are important is the case of waves overtopping a seawall. Only the individual waves that run up over the seawall crest will cause damage behind the wall. Distinctions between short-term and long-term statistics are summarized in Table II-8-1.

b. Probability distribution functions. Short-term statistics are discussed in earlier chapters of Part II. Rayleigh, Gaussian, and normal distribution functions are useful tools. The normal distribution is a normalized version of the Gaussian distribution in which the mean is zero and the standard deviation is one. Characteristics of distribution functions for various short-term statistics are summarized in Figure II-8-1.

Table II-8-1
Definition of Short-Term and Long-Term Statistics

Statistics	Processes Represented	Typical Time Period
Short-Term	Variations within an individual sea state	1 hr
Long-Term	Variations over a long-term collection of sea states	20 yr

    c. Statistical parameters. For most design applications, it is sufficient to represent a sea state by a few parameters. A mean value usually suffices for most processes, such as water level, current, wind speed, and direction. Wave parameters require special consideration (see Part II-1). Typically the parameters H_{m0}, T_p, and θ_p are used to represent either the total sea state or each major wave component in the sea state, which may include a locally generated sea and several independent swell components. It is especially important to note in shallow water design applications that the energy-based parameter H_{m0} and the height-based parameter $H_{1/3}$ may differ substantially (see Part II-1). Wave height is often the most critical design parameter and the choice of parameter must be consistent with the particular design application. For example, H_{m0} is appropriate for sediment transport and beach erosion applications, but $H_{1/3}$, H_{10}, or H_1 may be preferred for estimating wave forces on a pier.

II-8-5. Statistical Methods - Long-term

    a. Introduction.

    (1) Statistical methods for analyzing *long-term* information covering many years are an integral part of most design applications. These methods deal with the various values assumed by selected statistical parameters representing short-term information. Typically the largest parameter values are the primary design concern. For example, the largest value of the statistical parameter H_{m0} to be expected over a 25-year time interval might be needed.

    (2) Extreme events are often highly variable in terms of intensity and sequencing. By definition, they are rare. Thus, long-term statistical methods must deal with the problem of using a small, variable sample to estimate parameters that often have a major impact on design. Engineers are continually reminded that events such as the 100-year extreme storm can by chance occur during a much shorter data collection effort. Conversely, a 10-year record may not contain any events that equal or exceed the long-term 10-year extreme. Long-term statistical methods typically address the following two, related problems:

    (a) How to extend available information to a longer time period; e.g., how to use 10 years of data to estimate the 25-year extreme.

    (b) How to use a small sample of extreme events to get unbiased extreme estimates and some measure of confidence or variability that can be expected in the estimates.

    (3) In some applications, the preferred approach is to extend the available information base to longer time periods by generating additional realizations of the process and ensuring that the realizations are statistically consistent with known information. One adaptation of this approach is described in the following section on stochastic time histories.

RAYLEIGH

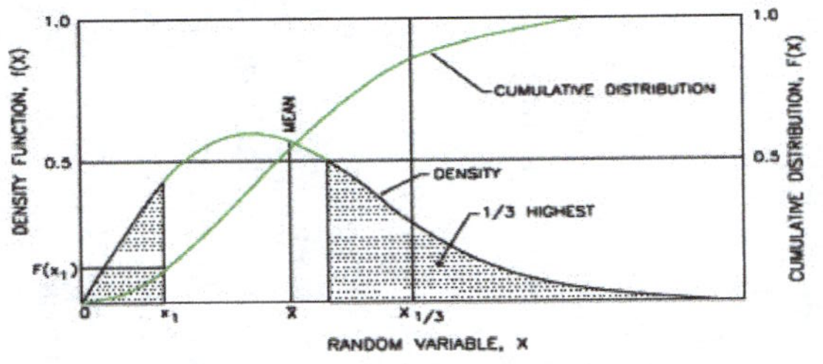

$$f(x) = 2 \alpha x e^{(-\alpha x^2)}$$

$$F(x) = 1 - e^{(-\alpha x^2)}$$

$$\bar{x} = \frac{0.886}{\sqrt{\alpha}}$$

$$\sigma_X = \frac{0.463}{\sqrt{\alpha}}$$

$$coeff.\ of\ var. = 0.52$$

GAUSSIAN

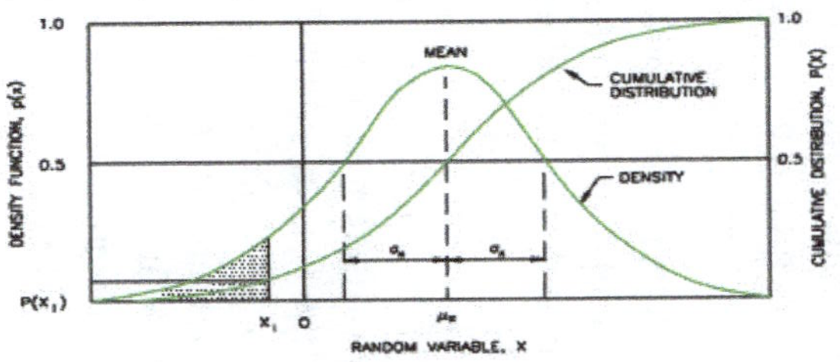

$$f(x) = \frac{1}{\sigma\sqrt{2\pi}} e^{-\left[\frac{(x-\mu)^2}{2\sigma^2}\right]}$$

$$\bar{x} = \mu$$

$$\sigma_X = \sigma$$

$$coeff.\ of\ var. = \frac{\sigma}{\mu}$$

NORMAL

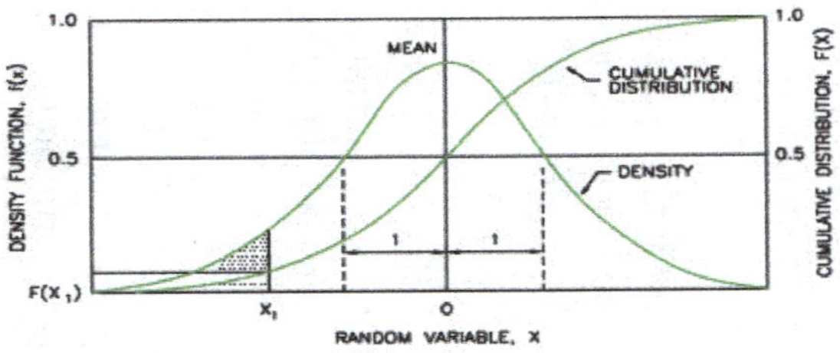

$$f(x) = \frac{1}{\sqrt{2\pi}} e^{\left(-\frac{x^2}{2}\right)}$$

$$\bar{x} = 0$$

$$\sigma_x = 1$$

Figure II-8-1. Probability distribution functions for short-term statistics

b. Stochastic time history.

(1) Applications such as long-term shoreline evolution depend partially on major storm events and partially on a wide variety of day-to-day conditions that also influence sediment movement. Further, the *sequencing* of wave and water level conditions is important in progressive beach erosion and recovery. Thus attention must be focussed on the long-term *time history* of conditions.

(2) A statistical framework is available for using a limited, but multi-year wave information base to define statistical characteristics at the location and then synthesize an unlimited number of additional years of information (Scheffner and Borgman 1992). The synthesized information matches the known information in a statistical sense but includes random variability also present in the process.

c. Extremal probability distribution functions. Many design applications focus only on extreme conditions. Because these conditions are typically difficult to estimate accurately and they often have large economic implications, a number of different probability distribution functions have been used to find a best fit to available data (Figures II-8-2 and II-8-3). The Fisher-Tippett Type I and II (FT-I and FT-II) distributions were derived from statistical theory of extremes, and hence are true extremal distributions. The Weibull distribution with $k=2$ is equivalent to the Rayleigh distribution. The parameters A, B, and k are known as the *scale, location,* and *shape* parameters, respectively (Table II-8-2). Typical values for the shape parameter in coastal engineering applications (e.g. Goda 1988) are given along with the general distribution functions. Expressions for the mean and standard deviation in terms of the distribution function parameters and vice-versa are also included if they can be written in compact form. Of the distributions shown in Figure II-8-3, choosing the Weibull distribution with $k=0.75$ clearly leads to the highest extremal estimates. Choosing the Weibull distribution with $k=2.0$ leads to the lowest estimates. The FT-I distribution gives estimates intermediate to the Weibull with k values of 1.0 and 1.4.

d. Empirical simulation technique. The *Empirical Simulation Technique (EST)* offers a powerful tool for estimating extreme responses, especially when multiple input parameters are important and the linkages between inputs and response are complex. This technique makes use of relationships embedded in the input information. There is no requirement for selecting distribution functions or assuming that input parameters are mutually independent. The EST is described in Part II-5-5-b-(3) in relation to storm surge estimation. In addition to providing the traditional stage-frequency relationship, the method gives valuable information on variability about the mean relationship. The information can be used to assess the level of risk associated with surge heights selected for design within the limits of the range of events simulated. The EST can be extended to design applications besides storm surge, such as beach erosion caused by tropical storms (Farrar et al. 1994).

e. Methods for fitting distributions to data. Selecting data for extremal analysis, estimating parameters in the distribution function, and choosing an extremal distribution function must be done carefully. Each of these choices can significantly influence the estimated extreme values, especially those for very rare events.

(1) Data selection.

(a) Data used for extreme analysis should be taken only from significant events in the recorded time history. Further, each data value should be from a different event to ensure statistical independence between values. The events should be representative of the *type* of events (though of lesser intensity) expected to cause the extremes of design concern. It is assumed that the statistics of extreme events are stationary over the period of record and in the future (e.g. no systematic increase in number and severity of extreme storms due to such possible effects as global warming). A full climatological data set (such as observations every 3 hr over 20 years) is not recommended for extreme analysis. Such data sets include multiple data values from each major storm, and one or several very severe storms can dominate the extremes.

Distribution Function	Mathematical Expression	Mean & Standard Deviation	Parameters
Fisher-Tippett I (FT-I) or Gumbel	$F(x) = e^{-e^{-\left(\frac{x-B}{A}\right)}}$	$\bar{x} = B + 0.5772A$ $\sigma = 1.283A$	$A = 0.779\sigma_x$ $B = \bar{x} - 0.4500\,\sigma_x$
Weibull	$F(x) = 1 - e^{-\left(\frac{x-B}{A}\right)^k}$ *general*	$\bar{x} = B + A\,\Gamma\left(1 + \frac{1}{k}\right)$ $\sigma_x = A\sqrt{\Gamma\left(1 + \frac{2}{k}\right) - \Gamma^2\left(1 + \frac{1}{k}\right)}$	
	$k = 0.75$	$\bar{x} = B + 1.191A$ $\sigma_x = 1.611A$	$A = 0.621\sigma_x$ $B = \bar{x} - 0.740\sigma_x$
	$k = 1.0$	$\bar{x} = B + A$ $\sigma_x = A$	$A = \sigma_x$ $B = \bar{x} - \sigma_x$
	$k = 1.4$	$\bar{x} = B + 0.911A$ $\sigma_x = 0.660A$	$A = 1.515\sigma_x$ $B = \bar{x} - 1.380\sigma_x$
	$k = 2.0$	$\bar{x} = B + 0.886A$ $\sigma_x = 0.463A$	$A = 2.160\sigma_x$ $B = \bar{x} - 1.914\sigma_x$
Fisher-Tippett II (FT-II) or Frechet	$F(x) = e^{-\left(\frac{x}{A}\right)^{-k}}$ *general*	$\bar{x} = A\,\Gamma\left(1 - \frac{1}{k}\right)$ $\sigma_x = A\sqrt{\Gamma\left(1 - \frac{2}{k}\right) - \Gamma^2\left(1 - \frac{1}{k}\right)}$	
	$k = 2.5$		
	$k = 3.3$		
	$k = 5.0$		
	$k = 10.0$		

Figure II-8-2.   Probability distribution functions for long-term statistics - Part 1 (Continued)

Distribution Function	Mathematical Expression	Mean & Standard Deviation	Parameters
Log-Normal	$f(x) = \dfrac{1}{Ax\sqrt{\pi}} \, e^{-\left(\frac{lnx-B}{A}\right)^2}$	$\bar{x} = e^{\left(B + \frac{1}{2}A^2\right)}$ $\sigma_x = \sqrt{e^{(2B+A^2)}(e^{A^2}-1)}$	
Log Pearson Type III (Kite 1978)	$f(x) = \dfrac{1}{Ax\Gamma(k)}\left(\dfrac{lnx-B}{A}\right)^{k-1} e^{-\left(\frac{lnx-B}{A}\right)}$	$ln\bar{x} = B + Ak$ $\sigma_{lnx} = A\sqrt{k}$	$k = skew\ of\ ln\ x\ corrected$ $for\ bias$ $A = \dfrac{\sigma_{lnx}}{\sqrt{k}}$ $B = lnx - \sigma_{lnx}\sqrt{k}$
Pearson Type III	$f(x) = \dfrac{1}{A\Gamma(k)}\left(\dfrac{x-B}{A}\right)^{k-1} e^{-\left(\frac{x-B}{A}\right)}$	$\bar{x} = B + Ak$ $\sigma_x = A\sqrt{k}$	$k = \left(\dfrac{2}{G}\right)^2$ $G = skew\ coefficient$ $A = \dfrac{\sigma_x}{\sqrt{k}}$ $B = \bar{x} - \sigma_x\sqrt{k}$
Binomial (discrete)	$f(x) = \dfrac{N!}{x!\,(N-x)!}\, p^x (1-p)^{N-x}$	$\bar{x} = Np$ $\sigma_x = \sqrt{Np(1-p)}$	
Poisson (discrete)	$f(x) = \dfrac{\lambda^x e^{-\lambda}}{x!}$	$\bar{x} = \lambda$ $\sigma_x = \sqrt{\lambda}$	

Figure II-8-2. Probability distribution functions for long-term statistics - Part 2 (concluded)

Hydrodynamic Analysis and Design Conditions

Table II-8-2
Parameters in Extremal Distribution Functions

Symbol	Name
A	Scale
B	Location
k	Shape

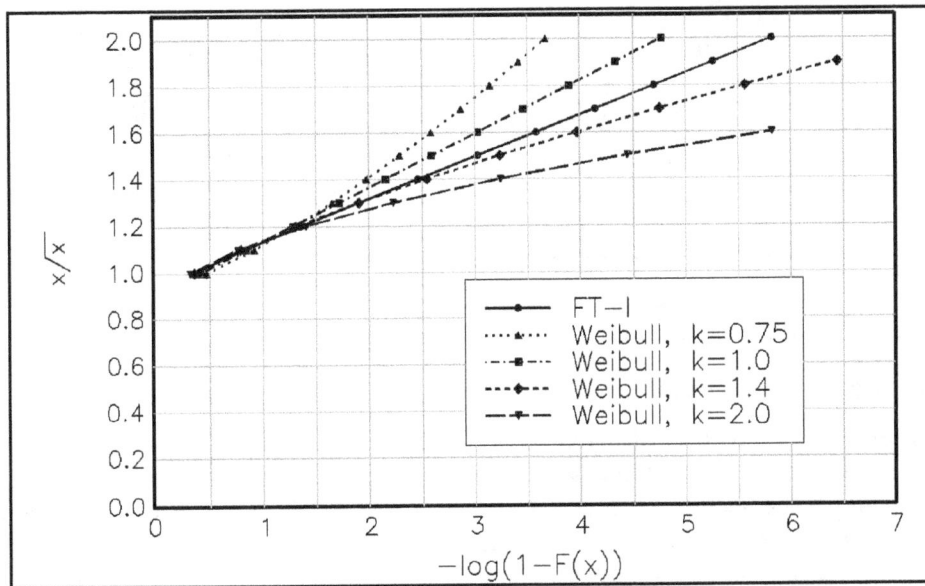

Figure II-8-3. Probability distribution functions for long-term statistics

(b) The preferred approach to data selection is to take the maximum value from each event to create a *partial duration series* of extreme values. Typically, the events are storms, ranging from small, weak events to the most severe storms of record. Small events can be difficult to identify. Further, they are of little interest if an adequate number of bigger storms is available in the record. Often the partial duration series will be *censored* to exclude data values less than some threshold value (Figure II-8-4). Thus the extremal analysis can focus on a smaller series representing truly significant events. The threshold is often chosen so that the number of data values in the series is greater than the number of years of record (generally 1-3 times the number of years of record).

(c) Another accepted approach is to take the maximum value from each year of record to form an *annual maximum series*. A record length on the order of 20 years or more (yielding at least 20 data values) is needed for this approach. In coastal engineering, record lengths are often shorter than this requirement. Another drawback to the annual maximum series arises from the possibility of multiple severe storms occurring in the same year, as in El Niño/Southern Oscillation events, which sporadically distort winter storm seasons along the U.S. Pacific Coast (see Part II-2). Only the maximum event in each year is considered. Other storms, which may be bigger than the maximum event in many other years, will be ignored.

(2) Estimating parameters in extremal distribution functions.

(a) Generally an extreme data value series (either partial duration series or annual maximum series) is treated as a sample from a process that follows one of the extremal distribution functions presented in

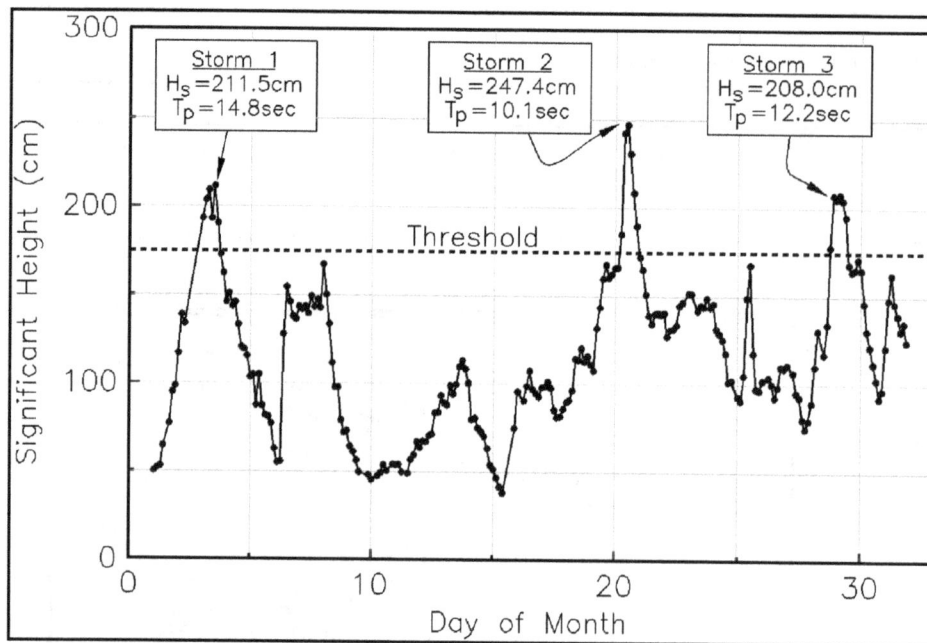

Figure II-8-4. Selection of extreme values for a partial duration series

Part II-8-5-c. There is no strong theoretical reason for preferring one distribution function over another. The single sample cannot be expected to fit the true distribution function exactly, especially for the few largest events. For some processes, such as water levels, one particular distribution function is generally accepted for all applications. For other processes, such as significant wave heights, a best-fitting distribution function is often chosen from among several candidates.

(b) Typically, extreme data values are sorted into descending order. A nonexceedance probability must be assigned to each extreme data value. These *plotting positions* should be chosen so that the distribution function can be accurately estimated. Figure II-8-5 gives the commonly used traditional plotting position formula. The figure also gives formulas developed to remove bias and minimize *rms* errors when fitting to specific distribution functions (Goda 1988, Goda and Kobune 1990).

(3) Approaches to estimating parameters. The following approaches can be used to determine parameters for each candidate distribution function:

(a) Graphical approach. Traditionally, the goodness of fit was determined visually by plotting the data along with candidate distribution functions. By scaling the plotting axes to make a candidate distribution appear as a straight line, the parameters of a visually optimum distribution function can more easily be determined.

(b) Computational approach. An automated computational approach is more objective (though not necessarily more accurate) and often easier to apply than the relatively tedious graphical approach. Three alternatives are the least squares method, the maximum likelihood method, and the method of moments. The least squares method is simplest and, with two-parameter distribution functions, it is often used. It is included in the ACES software package. One caution with this approach is that it is sensitive to even one or two extreme points that deviate greatly from the general trend of the data (*outliers*). The maximum likelihood method has the advantage of being less likely to produce erratic results when the data contain outliers or differ somewhat from the distribution function (Mathiesen et al. 1994). More information on computational approaches is also available from Goda (1988, 1990). Regardless of the method used, it is

Application of Formula	Plotting Position Formula
Traditional (Gumbel 1958)	$$\hat{F}_m = 1 - \frac{m}{N + 1}$$
Fisher-Tippett I (FT-I) (Gringorten 1963)	$$\hat{F}_m = 1 - \frac{m - 0.44}{N + 0.12}$$
Weibull Distribution Function (Goda 1988)	$$\hat{F}_m = 1 - \frac{m - 0.20 - \dfrac{0.27}{\sqrt{k}}}{N + 0.20 + \dfrac{0.23}{\sqrt{k}}}$$
Fisher Tippett II (FT-II) or Frechet Distribution Function (Goda and Kobune 1990)	$$\hat{F}_m = 1 - \frac{m - 0.11 - \dfrac{0.52}{k}}{N + 0.12 - \dfrac{0.11}{k}}$$
Log-Normal Distribution Function (Blom 1958)	$$\hat{F}_m = 1 - \frac{m - 0.375}{N + 0.25}$$

Parameter definitions:
F = probability that the mth highest data value will not be exceeded
m = rank of data value in descending order ($m = 1$ for largest, etc.)
N = number of events[1]
k = parameter in Weibull distribution function
[1] For censored data, N should represent the total number of events over the time interval considered (not just the number of censored events)

Figure II-8-5. Plotting position formulas

prudent to plot the computed distribution function and data together and ensure that the fit is consistent with good engineering judgement.

(4) Outliers. Outliers are retained in the data, but they should receive special scrutiny, as follows:

(a) Ensure accuracy. Each outlier should be checked to ensure that it is a valid data value, rather than a measurement or modeling error.

(b) Examine each event that produces a high outlier. Typical causes are very severe winter storms or direct impact of an intense hurricane. If extreme events at the site are produced by distinctly different natural processes (different statistical *populations*), it may be preferable to divide the data values into several series, one for each process, and analyze each series separately (e.g. Goda 1988). For example, winter storms and hurricanes should not necessarily be expected to produce extremes that follow the same extremal probability distribution function. Extreme data values can be analyzed as separate populations only if sufficient data values are available in *each* population.

(c) Consider whether an adjustment to the probability assigned to a high outlier data value can be justified. Often a high outlier is due to a storm event and extreme hydrodynamic response which are much more severe than would normally be expected over the length of hydrodynamic record. Meteorological records generally cover a much longer historical time period than hydrodynamic records. By carefully analyzing storm probabilities and longer-term records from nearby sites if available, it may be possible to assign a more realistic (lower) probability to the hydrodynamic outlier. Then a more valid distribution function fit can be obtained.

(5) Choosing an extremal distribution function. When several candidate distribution functions are under consideration, usually one is selected as a best fit to the data. The selection criteria can range from visual inspection of plotted results and simple statistics such as the correlation between data and model (e.g., Leenknecht, Szuwalski, and Sherlock 1992) to more elaborate statistical tests (Mathiesen et al. 1994). An objective approach to selecting a distribution function for significant wave heights is given by Goda and Kobune (1990).

(6) Confidence intervals. Confidence intervals associated with the chosen distribution function should also be estimated, preferably with a computer program (Leenknecht, Szuwalski, and Sherlock 1992; Goda 1988; Goda 1990; Mathiesen et al. 1994). They depend on the distribution function and number of data values. Confidence in computed values can also be influenced by random and systematic errors in the data and physical site characteristics such as long-term variability of water level and climate, possible extreme events not represented in the recorded population, and physical limits on extremes (such as the depth-imposed limit on wave height in shallow water).

f   Return period and encounter probability.

(1) Extreme conditions in coastal engineering are often described in terms of *return values* and *return periods*. The *return period* is the average time interval between successive events of the design wave being equalled or exceeded. For example, a 25-year significant wave height is that height that is equalled or exceeded an average of once during a 25-year time period. Return period is expressed as

$$T_r = \frac{t}{1 - P(\hat{H}_s)} \tag{II-8-1}$$

where

T_r = return period, in years

t = time interval associated with each data point, in years

$P(\hat{H}_s)$ = cumulative probability that $H_s \leq \hat{H}_s$

$\hat{H}_s$ = design significant wave height

(2) A related concept, *encounter probability*, gives the probability that waves with H_s equal to or greater than $\hat{H}_s$ will occur during the design life or other time period. It is given by

$$P_e = 1 - \left(1 - \frac{t}{T_r}\right)^{\frac{L}{t}} \tag{II-8-2}$$

where

P_e = encounter probability

L = desired time period, in years

(3) Values of P_e expressed in percent for some typical coastal engineering concerns are given in Table II-8-3.

Table II-8-3
Percent Chance for H_s Equaling or Exceeding Return Period H_s

Return Period	Desired Time Period (year)					
	2	5	10	25	50	100
2	75	97	100	100	100	100
5	36	67	89	100	100	100
10	19	41	65	93	99	100
25	8	18	34	64	87	98
50	4	10	18	40	64	87
100	2	5	10	22	39	63

g. *Extrapolation of data.* The main objective in determining an appropriate extremal distribution function is to get the best possible estimates of extreme conditions at desired return periods. Often the data must be extrapolated to probabilities beyond the record length to match design return periods. Extrapolation beyond 2-3 times the data record length should be avoided if possible. For example, 10 years of data should be used for estimating return values of 20-30 years or less.

II-8-6. Analysis of Key Meteorological and Hydrodynamic Processes in Design

a. *Introduction.*

(1) Key processes in design are reviewed in this chapter. Applicable chapters of Part II are cited. The design importance of each process is briefly stated. Critical design concerns are discussed, including two design-related examples.

(2) Statistical information about the processes is summarized in Table II-8-4. For each process (as applicable and available), the table includes representative statistical distribution functions and general expressions for parameters of the distribution function.

(3) It is important to bear in mind that the most extreme event of record may not merely be an intensified version of lesser extreme events. Most experienced coastal and ocean engineers and scientists can remember at least one catastrophic event that was distinctly different from typical storm events. Often the catastrophic event arises from an unusual interaction between several major weather features. The "Halloween Storm" that occurred in the northwestern Atlantic Ocean in October 1991 is a good example (U.S. Department of Commerce 1992). Three significant meteorological systems, including a hurricane and an intense winter storm, combined to create very strong winds over an extremely long fetch, which lasted for a period of days. This type of event is difficult to anticipate, but it should be recognized that such things can occur. They may appear as outliers in extreme data distributions.

Table II-8-4
Probability Distribution Functions for Meteorological and Hydrodynamic Processes

Parameter	Representative Distribution Functions
Short-term Statistics	
Surface elevations	Gaussian
Individual wave heights	Rayleigh: $x=H$; $\alpha=1/H_{rms}^2$
Individual wave periods	Rayleigh: $x=T^2$; $\alpha=1/\overline{T}^4$
Wave runup	Rayleigh
Long-term Statistics	
Extreme wind	FT-I
H_s	We bull: A= standard deviation of H_s; B= minimum value of H_s; k=1
T_p	We bull: B= minimum value of T_p; A, k estimated from data
Extreme H_s	FT-I; Weibull
Water level	Log-Pearson Type III

b. Wind. Wind is discussed in detail in Part II-2.

(1) Design importance. Wind at a design site may be important for local wave generation, nearshore current generation both inside and outside the surf zone, modification of nearshore breaking waves, nearshore water level, nearshore sediment transport, subaerial sediment transport, intensification of runup, overtopping, and flooding, sail forces on moored and moving boats, and harbor circulation and flushing.

(2) General climate. Winds at a coastal site are determined by some combination of large-scale weather systems, smaller-scale systems, land-sea breeze circulations, land/water roughness differences, and orographic effects. Winds can vary greatly over short distances along a coast. Thus, local measurements *at the project site* are very helpful in establishing the climate. As with waves, 2-3 years of data are generally sufficient for climatological purposes (i.e. annual or monthly mean and standard deviation of wind speed). Even one month of data can be useful, though certainly not ideal, for estimating relationships between the project site and winds at a long-term measurement station within the same region. Often wind climate information is summarized in the form of *wind roses*, which can represent months, seasons, or years (Figure II-8-6).

(3) Storms. Storms are a natural part of the wind climate at a site. They can vary greatly in size and intensity. Frequency of occurrence and intensity of storms are important concerns for functional design. The occurrence of extreme storms is a necessary concern for structural design. The distribution of extreme wind speeds has been modeled with FT-I, FT-II, and Weibull distribution functions (Figure II-8-2). The FT-I distribution function seems to be the preferred choice, especially when the annual extremes do not include rare, unusually powerful events arising from distinctly different meteorology (such as hurricanes) (Simiu and Scanlan 1986).

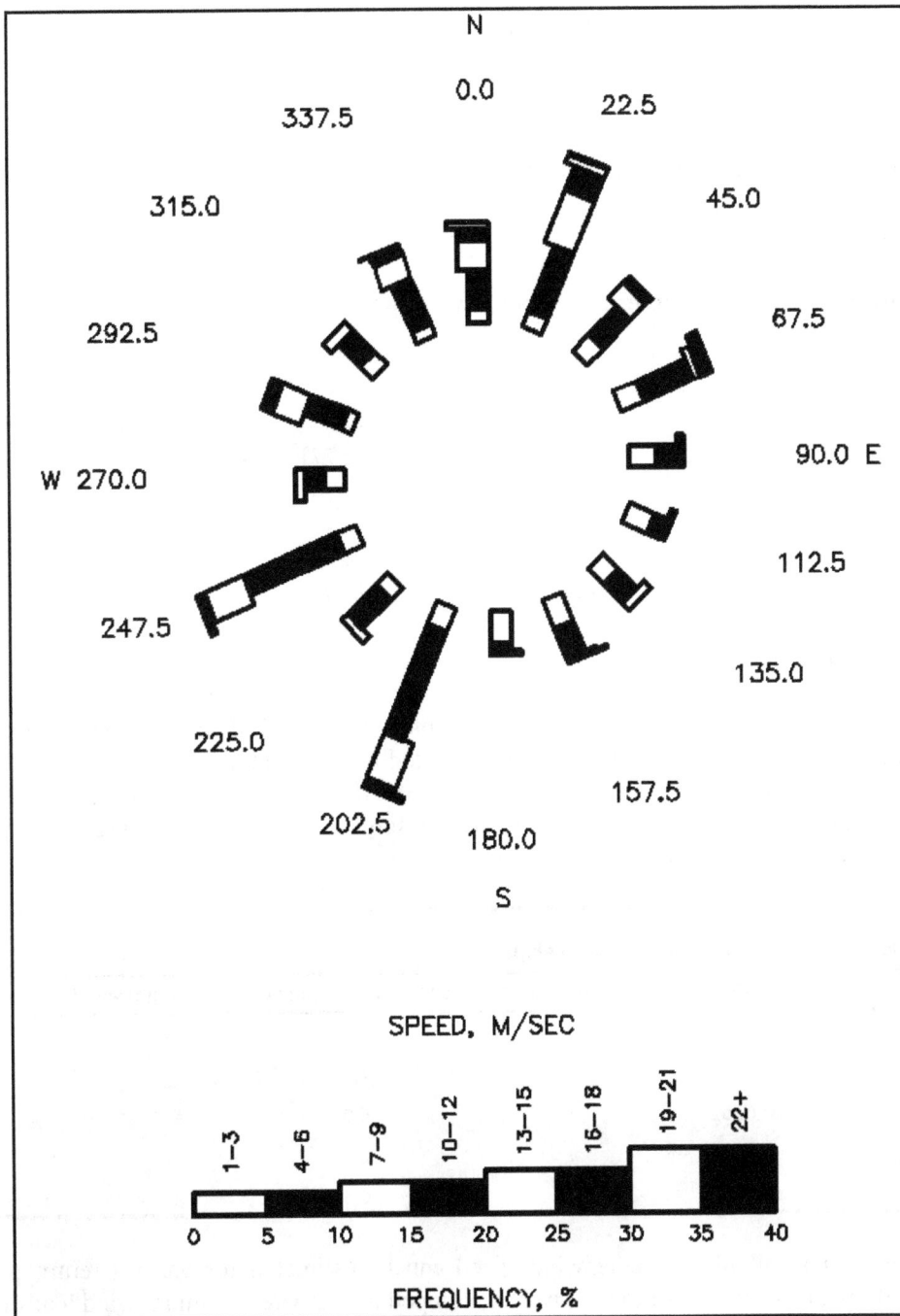

Figure II-8-6.   Example wind rose (Leffler et al. 1990)

(a) Extreme wind speeds can conveniently be analyzed with the ACES extremal significant wave height analysis application (using wind speeds in place of wave heights) or with more traditional graphical methods. However, wind records at project sites often cover a very limited time period, and extrapolation to rare events may be difficult. Extreme records from any nearby, long-term wind stations may be transferrable to the project site with due consideration of differences between locations. Also, a simple approach using *monthly* extreme wind speeds may be helpful in conjunction with limited data sets (Simiu and Scanlan 1986). By this approach, based on the assumption that extreme wind speeds follow the FT-I distribution function,

$$U_r = \bar{U}_m + 0.78 \; \sigma_m \; [\; \ln (12T_r) \; - \; 0.577 \;] \tag{II-8-3}$$

where

U_r = wind speed with r-year return period

$\bar{U}_m$ = mean value of maximum monthly wind speeds

σ_m = standard deviation of maximum monthly wind speeds

and

$$\sigma_{rm} = \sqrt{0.49 \; + \; 0.89 \; ln \; (12\bar{U}_m) \; + \; 0.67 \; [ln \; (12\bar{U}_m) \; - \; 0.577]^2} \; \; \frac{\sigma_m}{\sqrt{N_m}} \tag{II-8-4}$$

where

σ_{rm} = standard deviation of the sampling error in estimating U_r

N_m = number of months of data

(b) The parameter σ_{rm} can be related to confidence intervals using Table II-8-5. An integral number of years of data covering at least 3 years (36 months) is needed for this simple approach. Only populations of extreme events that are well-represented in the data can be effectively included in the long-term extreme estimates (e.g., if hurricane events do not appear or are sparsely represented in the data sample, they will not be effectively represented in the extreme estimates).

Table II-8-5
Standard Deviation and Confidence Interval Relationships

Confidence Level, %	Confidence Interval Bounds Around U_r	Probability of Exceeding Upper Bounds, %
80	$\pm1.28\sigma_{rm}$	10.0
85	$\pm1.44\sigma_{rm}$	7.5
90	$\pm1.65\sigma_{rm}$	5.0
95	$\pm1.96\sigma_{rm}$	2.5
99	$\pm2.58\sigma_{rm}$	0.5

(c) If data are unavailable, extreme wind speed can be estimated for various return periods with Figures II-8-7 through II-8-9. These figures were developed to estimate maximum wind loads for building design and are expected to be conservative for coastal engineering applications. Figure II-8-7 gives extreme fastest mile wind velocity data with a 50-year return period (annual probability of 0.02 that the wind speed is exceeded). Wind speed information was prepared from data collected at 129 weather stations (Simiu, Changery, and Filliben 1979), representing a 10-m elevation. Data were statistically reduced using extreme value analysis based on Fisher-Tippett Type-I distributions. Wind speed contours of Alaska are based primarily on data collected in open areas (Thom 1968-69). Wind speed contours in the hurricane-prone region, Atlantic and Gulf of Mexico coastlines, are based on Monte Carlo simulations of hurricane storms striking the coastal region (Batts, Russell, and Simiu 1980). Recurrence intervals for 25 and 100 years may be estimated by multiplying wind speed from Figure II-8-7 with the appropriate adjustment factor in Table II-8-6.

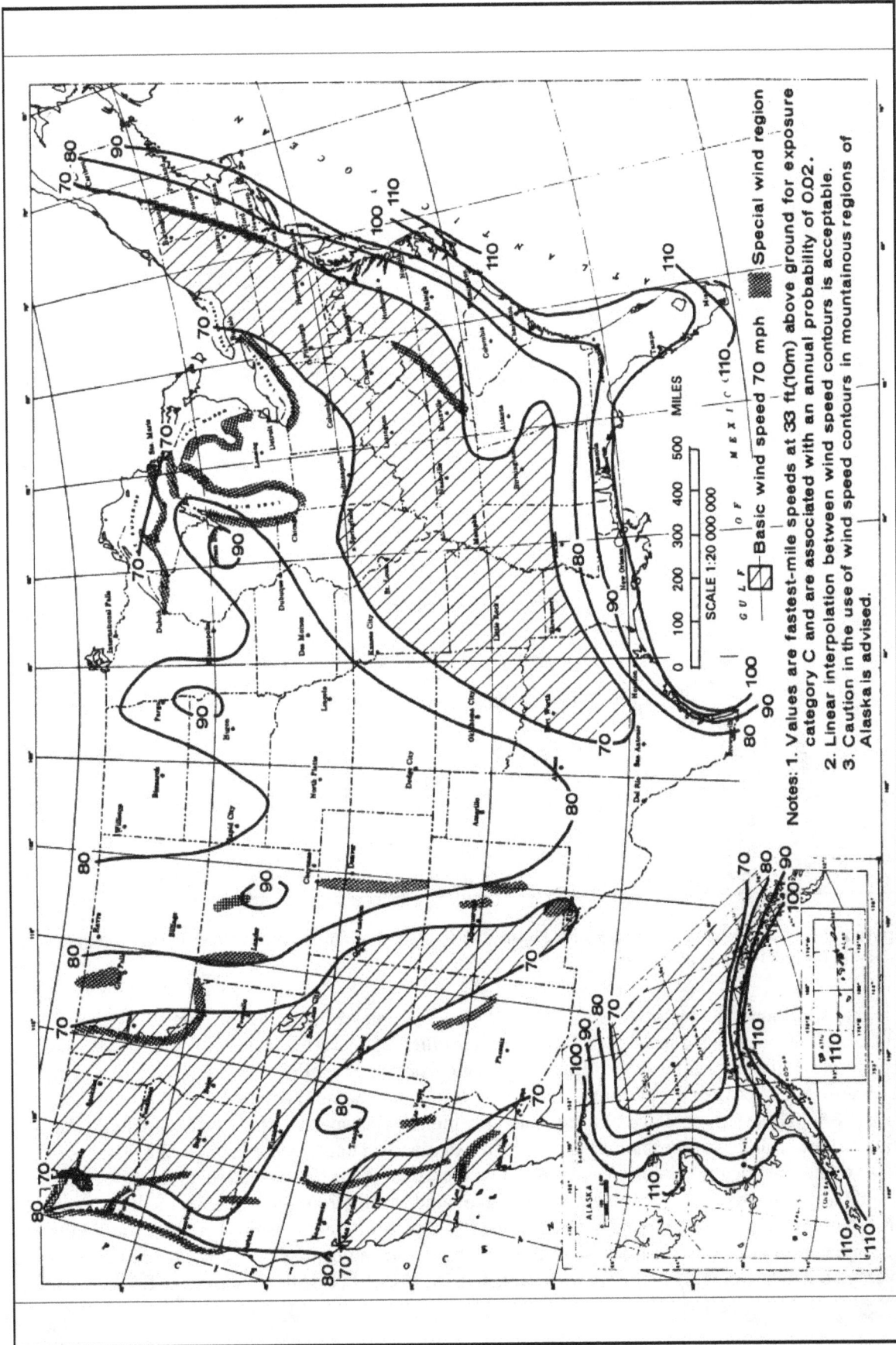

Figure II-8-7. Extreme fastest-mile wind speeds with 50-year return period across the United States at 10-m elevation (ASCE 1993)

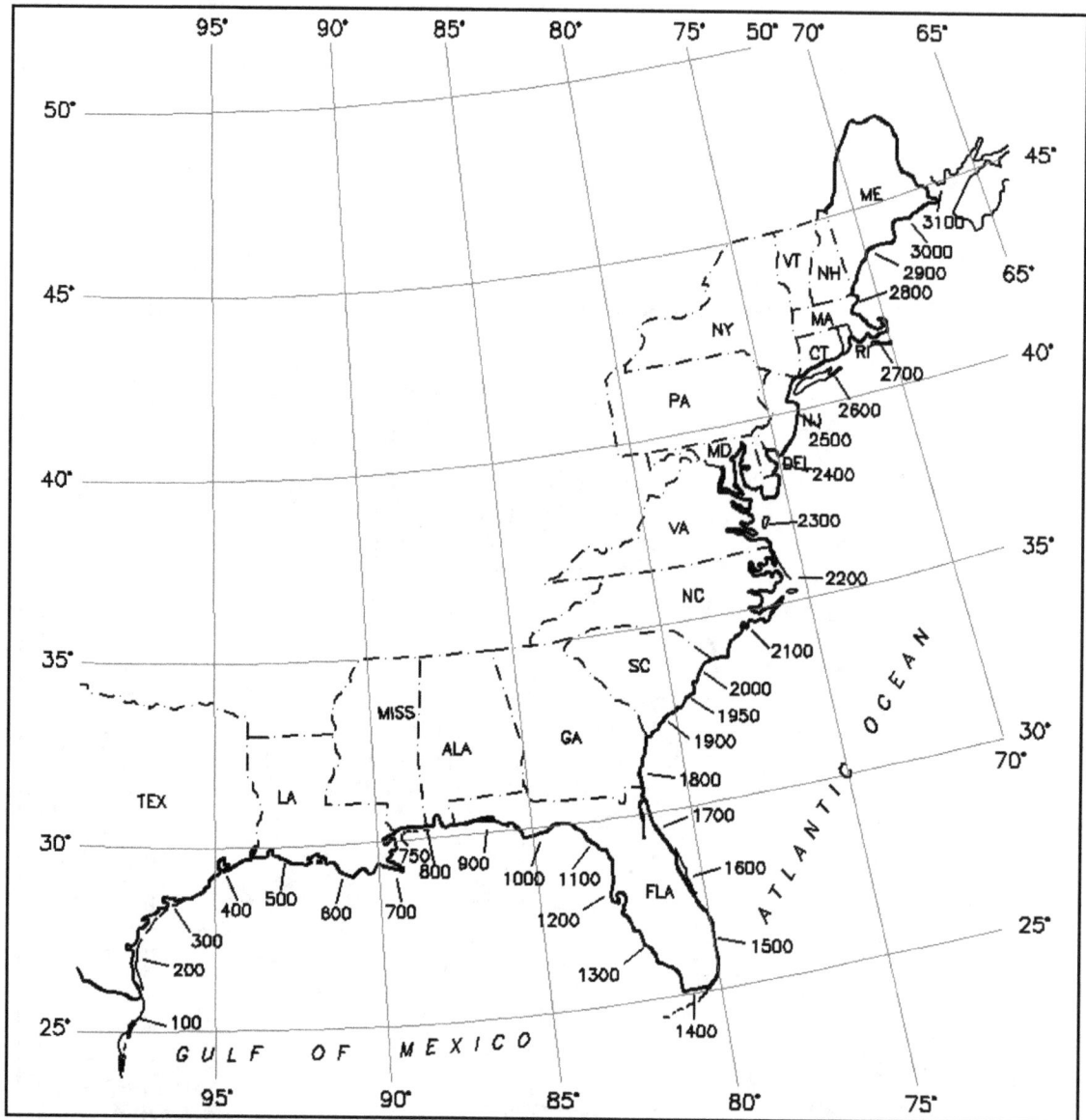

Figure II-8-8. Milepost map for use with Figure II-8-9; coastal distance intervals marked in nautical miles (1 nautical mile = 1.9 km)

(d) The adjustment factor at the hurricane-prone oceanline reflects the difference in probability distributions of hurricane wind speeds and wind speeds in other regions. Hurricane wind effects are assumed to be negligible at distances of more than 100 miles inland from the oceanline. The adjustment factor can be linearly interpolated between the oceanline and 100 miles inland. Some special wind regions are indicated in Figure II-8-7. These regions could have considerably higher wind speeds than indicated in the figure. All mountainous terrain and ocean promontories should be examined for unusual wind conditions. Figures II-8-8 and II-8-9 relate specifically to hurricane winds and are preferable for that application (Batts, Russell, and Simui 1980). It is important to recognize that extreme wind speeds from these figures must be converted from fastest mile to appropriate averaging time. Table II-8-7 provides relationships between fastest mile and 1-hr average wind speeds (also in Figure II-2-2). Averaging time adjustments, including conversion from 1-hr average to other averaging time, can be done using Table II-8-8, Figure II-2-1, or ACES.

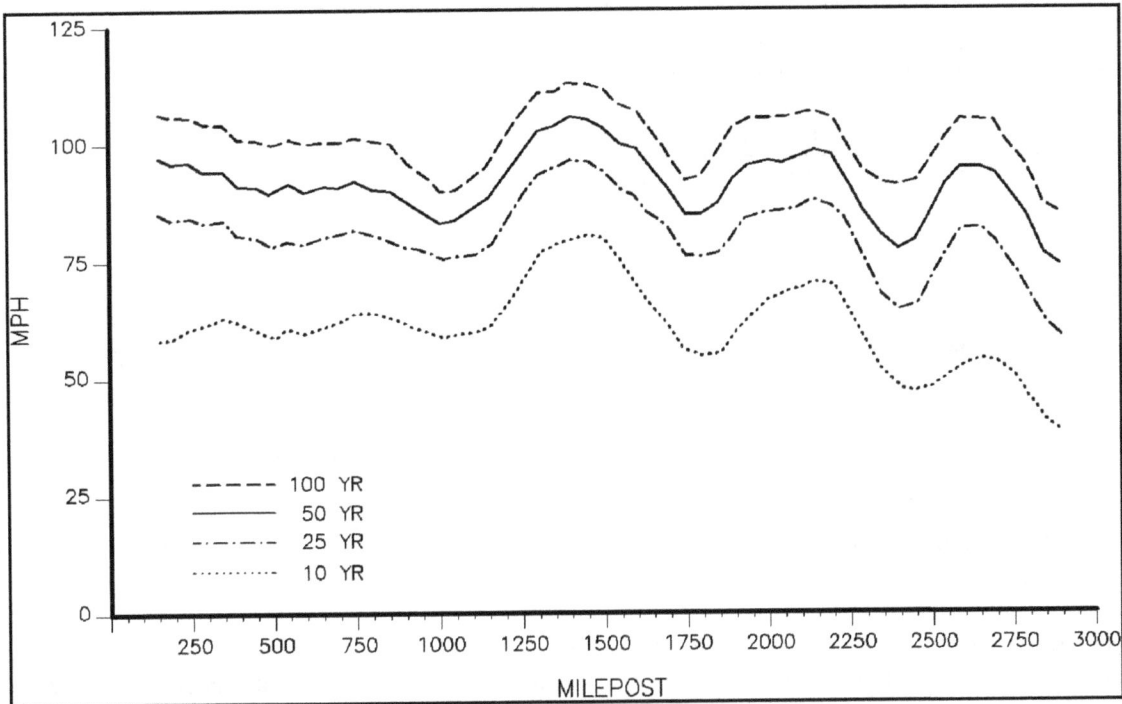

Figure II-8-9. Extreme fastest-mile hurricane wind speeds blowing from any direction at 10 m above ground in open terrain near the coastline for various return periods (after National Bureau of Standards (1980))

Table II-8-6
Return Period Adjustment Factor

Return Period, years	Adjustment Factor	
	Other Regions	Hurricane Region (Gulf & Atlantic)
25	0.95	1.00
50	1.00	1.05
100	1.07	1.11

Table II-8-7
Conversion of Extreme Wind Speeds from Fastest-Mile to 1-hr Average

Fastest Mile U, mph	U_{1hr}, mph	Conversion Factor[1]
50	41.0	0.82
80	62.4	0.78
100	77.0	0.77
120	91.2	0.76
150	111.0	0.74

[1] U_{1hr} = (factor) × (fastest mile U)

Table II-8-8
Averaging Time Adjustments for Extreme Wind Speeds[1]

t_{given}	t_{find}					
	1 min	5 min	10 min	1 hr	3 hr	6 hr
1 min	1.00	0.88	0.84	0.80	0.75	0.71
5 min	1.14	1.00	0.96	0.92	0.85	0.81
10 min	1.18	1.04	1.00	0.95	0.88	0.84
1 hr	1.24	1.09	1.05	1.00	0.93	0.88
3 hr	1.34	1.17	1.13	1.08	1.00	0.95
6 hr	1.41	1.23	1.19	1.13	1.05	1.00

[1] $U_{t,find}$ = (factor from table) × $U_{t,given}$

   c. Extreme waves. Part II-1 and II-4 discuss some aspects of extreme waves. Further information is provided in this chapter.

   (1) *Design importance.* Extreme wave conditions are almost always a major design concern in coastal engineering projects. Extreme significant wave heights are usually the most critical concern, but wave period, wave direction, and spectral shape (both frequency and direction) can be important as well. Possible secondary wave systems, an integral part of wave climate at exposed ocean sites, are generally not important in conjunction with an extreme event. Energetic extreme waves are usually a key factor causing coastal structure damage, vessel damage, beach erosion, channel sedimentation, and coastal flooding. Extreme waves may be generated by a winter storm, a tropical storm, or some combination of storms (see Part II-8-7).

   (2) *Deep water.*

   (a) Extreme values of H_s and associated T_p are usually obtained from measurements or hindcasts. Values of H_s, typically in the form of a partial duration series (see Part II-8-5-e-(1)), can be expected to give a reasonable fit to the FT-I, Weibull, FT-II, or log-normal distribution function (Figure II-8-2), from which design values of H_s can be estimated. FT-I or a form of the Weibull distribution function is generally preferred. Some of the following considerations may also apply:

   ● If extreme values are generated by more than one population of extreme events (e.g. winter storms and hurricanes) and if sufficient data are available from each population, it may be desirable to fit each population separately.

   ● If the largest value or several values deviate significantly from the general trend of the data, these outliers should be given special consideration as discussed in Part II-8-5-e-(2).

   ● Maximum H_s might be limited by sheltering or fetch constraints due to geography or consistent storm characteristics.

   ● Extreme wave measurements, observations, and hindcasts are often subject to larger errors than would normally be expected. For example, gauges can be affected by large, steep waves, breaking waves, severe winds, and ice accumulation. A floating accelerometer buoy gauge may tilt severely, stretch the mooring lines, or even break loose from the mooring. Wave height estimates may be modified in ways that are difficult to predict. Data loss from gauges, due to equipment damage or loss of power, is more likely during severe storms than during normal conditions.

EXAMPLE PROBLEM II-8-1

FIND:

Wind speed with 25- and 50-year return period (10-min average at 10-m elevation).

GIVEN:

Maximum wind speed by month (10-min average measurement at 10-m elevation) in Table II-8-9.

Table II-8-9
Maximum Wind Speed by Month, in m/s

Year	Month											
	Jan	Feb	Mar	Apr	May	Jun	Jul	Aug	Sep	Oct	Nov	Dec
1969	---	---	20	19	17	11	15	13	15	14	18	18
1970	22	16	18	18	12	14	12	13	11	21	19	22
1971	24	18	24	18	17	16	11	14	16	22	21	20
1972	20	19	18	20	16	15	14	14	20	15	17	23
1973	18	15	18	14	15	13	15	12	19	17	22	20
1974	24	18	23	15	12	17	20	10	11	12	16	22
1975	21	25	22	14	18	12	13	12	12	20	20	22
1976	18	19	22	13	12	13	13	10	11	13	11	13
1977	10	18	19	10	14	13	10	13	12	12	16	23
1978	21	13	11	15	14	10	10	10	10	9	13	15
1979	16	22	15	16	15	11	14	11	14	18	14	18
1980	20	15	23	15	11	11	15	10	11	13	20	21
1981	12	19	15	13	12	14	13	8	12	15	18	20
1982	14	13	7	10	11	10	12	9	11	17	14	21
1983	14	18	17	11	---	---	---	---	---	---	---	---

SOLUTION:

Three options are presented on the following pages.

(Sheet 1 of 4)

Example Problem II-8-1 (Continued)

Option 1: (use ACES; this option is generally preferred)

A series of annual maximum wind speeds is taken from the table of data. The data include 13 complete years (1970-1982) and two partial years (1969 and 1983). Combining January and February from 1982 with the March -December 1969 data gives another full year for analysis. The annual maximum series with 14 values is then:

20, 22, 24, 23, 22, 24, 25, 22, 23, 21, 22, 23, 20, 21   m/s

These values are entered as significant heights into the ACES extremal significant wave height analysis application. Appropriately, ACES provides a warning that a 14-year sample is too short to give a reliable estimate of the 50-year event. The FT-I distribution function is found to provide a good fit (Figure II-8-10). The corresponding 25- and 50-year wind speeds are

$$U_{25} = 25.5 \text{ m/s} \quad (10\text{-min average})$$

$$U_{50} = 26.3 \text{ m/s} \quad (10\text{-min average})$$

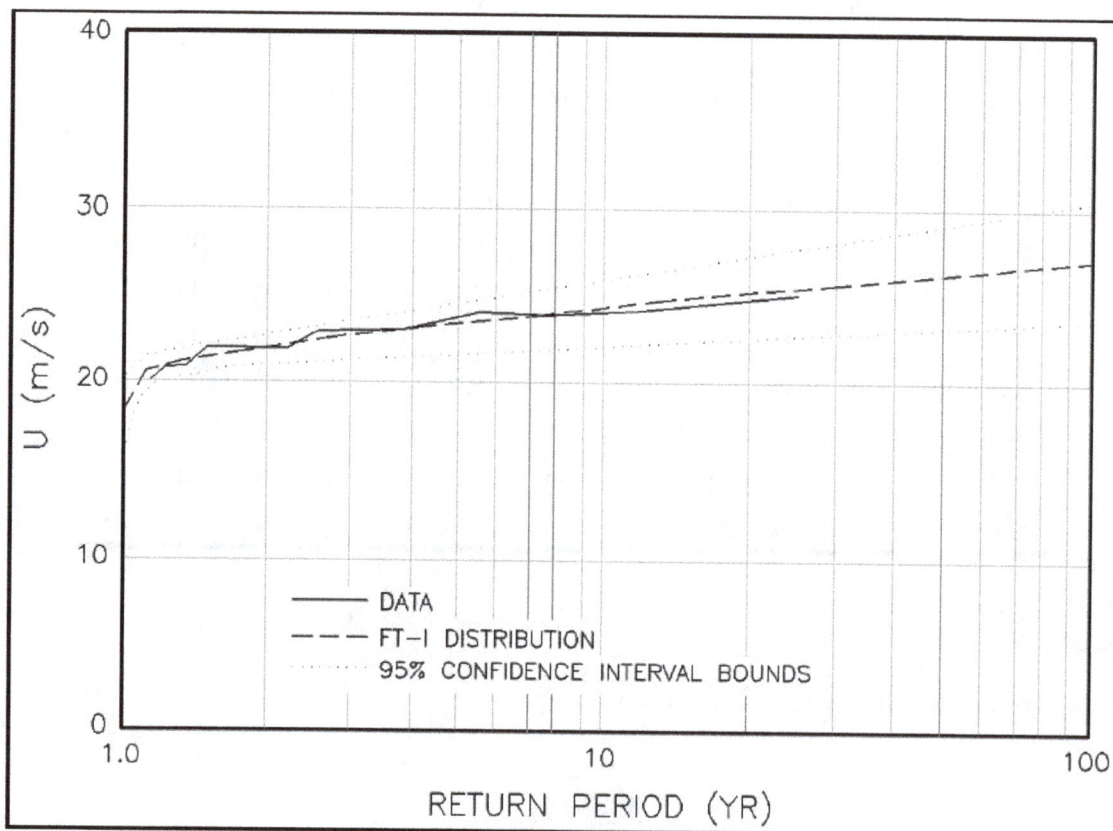

Figure II-8-10.   Example Problem II-8-1 probability distribution of wind speeds

(Sheet 2 of 4)

Hydrodynamic Analysis and Design Conditions

Example Problem II-8-1 (Continued)

Option 2: (use Equation II-8-3)

Equation II-8-3 can be applied to 3 or more complete years of data. Up to 14 years of data are available in this example. Normally, all full years would be used together. For illustration, the following solution treats four 3-year samples from the full record as well as the 14-year sample extending from March 1969 through February 1983. Means and standard deviations of maximum monthly wind speeds are given in Table II-8-10.

Table II-8-10
Wind Speed Statistics

Years	$\bar{U}_m$, m/s	σ_m , m/s
1971-73	17.50	3.32
1974-76	16.08	4.58
1977-79	14.03	3.62
1980-82	14.03	4.02
Mar 69 - Feb 83	15.53	4.06

U_{25} is calculated from Equation II-8-3 for the years 1971-73 as

$$U_{25} = \bar{U}_m + 0.78\,\sigma_m\,[\,ln(12\times25) - 0.577\,]$$
$$= \bar{U}_m + 4.00\,\sigma_m = 17.50 + 4.00\times3.32 = 30.8 \quad m/s$$

Similarly, U_{50} is calculated from Equation II-8-3 for the years 1971-73 as

$$U_{50} = \bar{U}_m + 0.78\,\sigma_m\,[\,ln(12\times50) - 0.577\,]$$
$$= \bar{U}_m + 4.54\,\sigma_m = 17.50 + 4.54\times3.32 = 32.6 \quad m/s$$

The standard deviation of the sampling error in estimating U_r for the years 1971-73 may be calculated from Equation II-8-4 as

$$\sigma_{rm} = \sqrt{0.49 + 0.89\,ln(12\bar{U}_m) + 0.67\,[ln(12\bar{U}_m) - 0.577]^2}\left(\frac{\sigma_m}{\sqrt{36}}\right)$$

$$= \sqrt{0.49 + 0.89\,ln(12\times17.50) + 0.67\,[ln(12\times17.50) - 0.577]^2}\left(\frac{3.32}{\sqrt{36}}\right)$$

$$= \sqrt{0.49 + 0.89\times5.347 + 0.67\times[5.347-0.577]^2}\left(\frac{3.32}{6}\right) = \sqrt{20.49}\times0.553 = 2.5 \ m/s$$

(Sheet 3 of 4)

Example Problem II-8-1 (Concluded)

These results and those calculated for the other samples are summarized in Table II-8-11.

Table II-8-11
Wind Speed at 25- and 50-year Return Period

Years	U_{25}, m/s	U_{50}, m/s	σ_{rm}, m/s
1971-73	30.8	32.6	2.5
1974-76	34.4	36.9	3.4
1977-79	28.5	30.5	2.6
1980-82	30.1	32.3	2.9
Mar 69 - Feb 83	31.8	34.0	1.4

Option 3: (use Figure II-8-7)

Figure II-8-7 is the least desirable option, since it does not make use of the available measurements at the site. However, it provides a reference to regional behavior against which the data analyses can be interpreted. For this example, the figure indicates that

$$U_{50} \approx 80 \text{ mph} = 35.8 \text{ m/s} \qquad \text{(fastest mile)}$$

From Table II-8-6,

$$U_{25} = 0.95 \times 35.8 = 34.0 \text{ m/s} \qquad \text{(fastest mile)}$$

These wind speeds are converted from fastest mile to 1-hr averages using Table II-8-7:

$$U_{25} = 0.78 \times 34.0 = 26.5 \text{ m/s} \qquad \text{(1-hr average)}$$

$$U_{50} = 0.78 \times 35.8 = 27.9 \text{ m/s} \qquad \text{(1-hr average)}$$

Finally, the 1-hr averages are converted to 10-min averages to be comparable to the results of Options 1 and 2. Using Table II-8-8,

$$U_{25} = 1.05 \times 26.5 = 27.8 \text{ m/s} \qquad \text{(10-min average)}$$

$$U_{50} = 1.05 \times 27.9 = 29.3 \text{ m/s} \qquad \text{(10-min average)}$$

Discussion: Option 1, which is generally preferred, gives the smallest values for U_{25} and U_{50}. Option 3 gives wind speeds about 10 percent higher and Option 2 over 20 percent higher than Option 1. Use of Option 1 with a partial duration series of wind speed maxima, which is recommended in practice rather than the annual maximum series, could be expected to give more realistic estimates.

(Sheet 4 of 4)

• The effect of errors on extreme wave height estimates can be significant. Errors can be expected to increase the width of confidence intervals and induce a systematic, artificial increase in H_s values at return periods of interest (Earle and Baer 1982). Errors can be as important as the finite number of years of record in limiting the reliability of extreme wave height estimates (Le Méhauté and Wang 1984).

(b) Design wave period is a period representative of extreme wave conditions. Along coasts exposed to the ocean, the design T_p is usually an intermediate period between the limits of mild local sea and long swell periods. At locations exposed to large swell, a design T_p representative of long-period swell conditions may be required. At sites sheltered from ocean swell or enclosed water bodies the size of the Great Lakes or smaller, the largest values of T_p can be associated with the largest values of H_s. At many locations, it may be reasonable to estimate design T_p with a scatter plot of peak storm H_s and associated T_p values. A regression line relating H_s and T_p is computed and then used to estimate T_p for any given H_s (e.g., Goda (1990)).

(c) In using a design T_p, it is assumed that this wave period is representative of the irregular or regular wave period needed in follow-on design calculations. Often this assumption is realistic, as high-energy wave events tend to be dominated by a single spectral peak. However, it may be preferable in some applications to consider more than one spectral peak or even a full design spectrum if follow-on calculations can make use of the information.

(d) Design wave direction is estimated based on measurements, hindcasts, and/or knowledge of extreme storm characteristics.

(3) Intermediate-depth water.

(a) When a coastal project is in intermediate water depth (that is, waves are affected by the bottom but depth-induced breaking has not begun), nearshore processes such as refraction and shoaling must be considered to transform from the measurement or hindcast site to the project site (Part II-3). Figure II-3-6 provides the simplest methodology. A more comprehensive approach would be to represent each wave condition as a TMA spectrum with appropriate energy, peak period, and direction, and compute transformation over straight, parallel bottom contours. More typically, a full numerical model representation of bathymetry and wave conditions is used, as discussed in Part II-3.

(b) Values of H_s, T_p, and wave direction in intermediate-depth water can be analyzed for design using the same procedures as for deepwater waves. The H_s and wave direction values are modified from the deep-water values because of nearshore bottom effects. Values of T_p are usually considered to be unchanged from the deepwater values by the transformation process. However, some spectral transformation techniques can predict changes in T_p. These changes are usually quite small.

(4) Shallow water (depth-limited).

(a) Extreme wave heights in coastal engineering applications are often limited by shallow-water depths. Thus, depending on the local water depth and wave climate, the distribution for significant wave heights can be expected to follow one of the appropriate functional forms in Figure II-8-2 up to a significant height of about 0.6 times the water depth (Equation II-4-10) and then increase more slowly beyond that point (Figure II-8-11). The probability at which the curve flattens depends on the local water depth and wave climate. The flattened curve can be expected to continue rising slowly, but in this region increases in significant height depend on other parameters such as wave steepness and water level rather than incident significant height. Water level can be expected to be the main controlling factor. The probability distribution for significant heights in this region may be essentially equivalent to the probability distribution of local water levels.

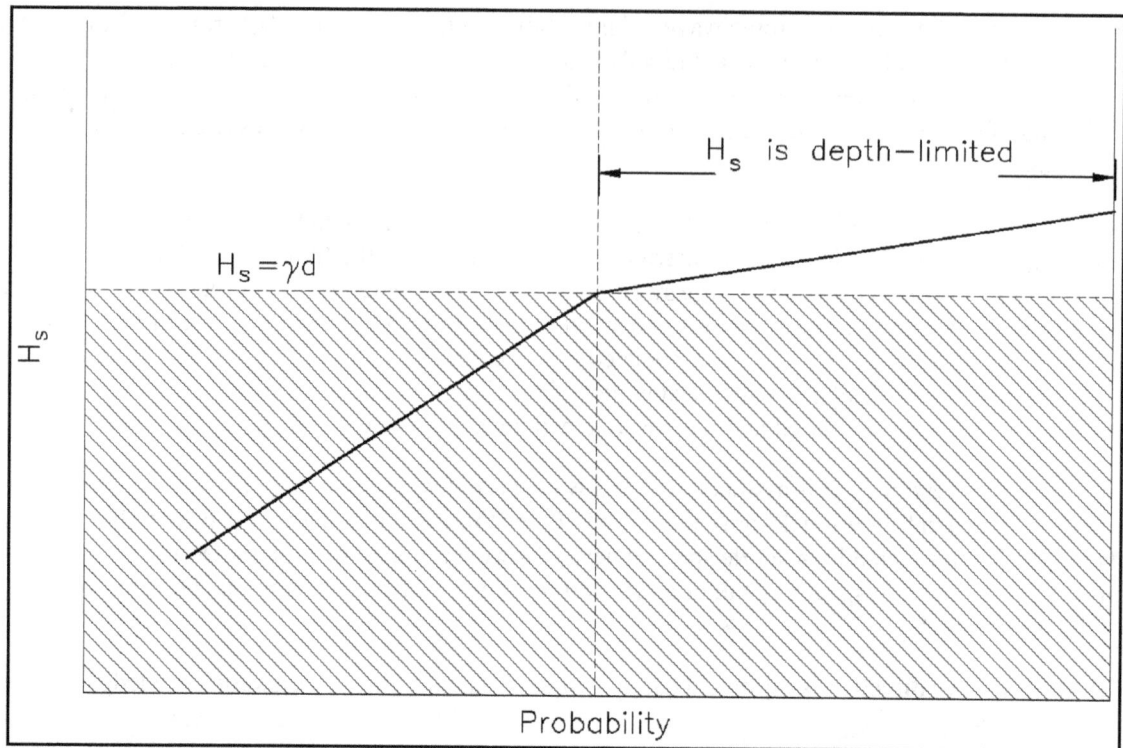

Figure II-8-11.   Probability distribution of H_s on shallow water

(b) Often extreme shallow-water waves are estimated with wave information from an offshore source. Extreme waves in shallow water are typically greatly transformed from incident (often deepwater) waves due to a variety of processes discussed in Part II-3 and 4. Consequently, the most extreme incident waves are not necessarily design conditions when transformed to the shallow-water site. A sufficient range of incident wave cases must be analyzed to ensure that the most extreme shallow-water cases are identified.

(c) In shallow water, the competing processes of shoaling (tending to increase wave height) and breaking (decreases wave height in the surf zone) often create a point at which significant height reaches a maximum value in the outer surf zone. That maximum value can be estimated from Figure II-8-12. It should be noted that these values are $H_{1/3}$ rather than the energy-based significant height H_{m0} (Part II-1). The water depth at which the maximum occurs is shown in Figure II-8-13.

(d) Design wave period can be estimated as a representative value for extreme wave conditions, as with waves in deeper water. However, allowances must be made for a range of wave periods accompanying a relatively fixed, depth-limited design H_s. In this case, it may be advisable to consider several design T_p values to adequately represent the range of possibilities impacting design.

(e) Design wave direction is estimated based on measurements, hindcasts, or knowledge of extreme storm characteristics. Due consideration must be given to shallow-water effects on wave direction (Part II-3).

(5) Extreme individual wave characteristics.

(a) Extreme individual waves can have heights on the order of $2H_s$. The Rayleigh distribution function (Part II-1) is usually sufficient for describing individual wave heights in coastal engineering applications, even in extreme (but nondepth-limited) conditions. With the Rayleigh distribution function for individual

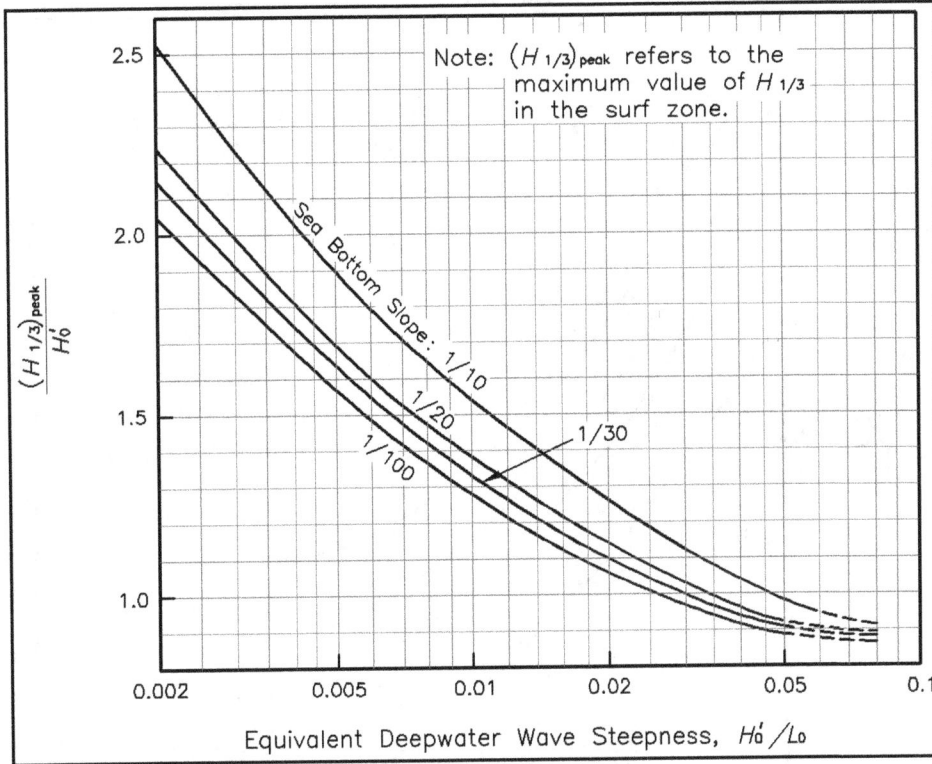

Figure II-8-12. Maximum value of H_s in the surf zone (Goda 1985)

wave heights in a given sea state (a given H_s) and a known joint distribution of H_s and T_p or T_z, an overall probability distribution of individual wave heights can be computed by a straightforward approach (Goda (1990), referring to Battjes (1972)).

(b) Extreme individual wave heights are strongly affected by depth-induced breaking. In the surf zone, the Rayleigh distribution can be expected to overestimate the higher individual wave heights. Practical approximations for extreme individual wave heights in this case are given by the Construction Industry Research and Information Association (CIRIA) (1991) as

$$H_1 = \frac{1.517 \, H_{1/3}}{\left(1 + \dfrac{H_{1/3}}{d} \right)^{\frac{1}{3}}}$$

(II-9-5)

$$H_{01} = \frac{1.859 \, H_{1/3}}{\left(1 + \dfrac{H_{1/3}}{d} \right)^{\frac{1}{2}}}$$

where $H_{1/3}$ and d are the local significant height and water depth.

(c) Extreme wave crest heights are sometimes a design consideration. Coastal field measurements indicate that maximum crest height above the local mean water level can be up to 80 percent of the maximum wave height (Goda 1985). Figure II-8-14, based on a combination of irregular wave tests in the laboratory

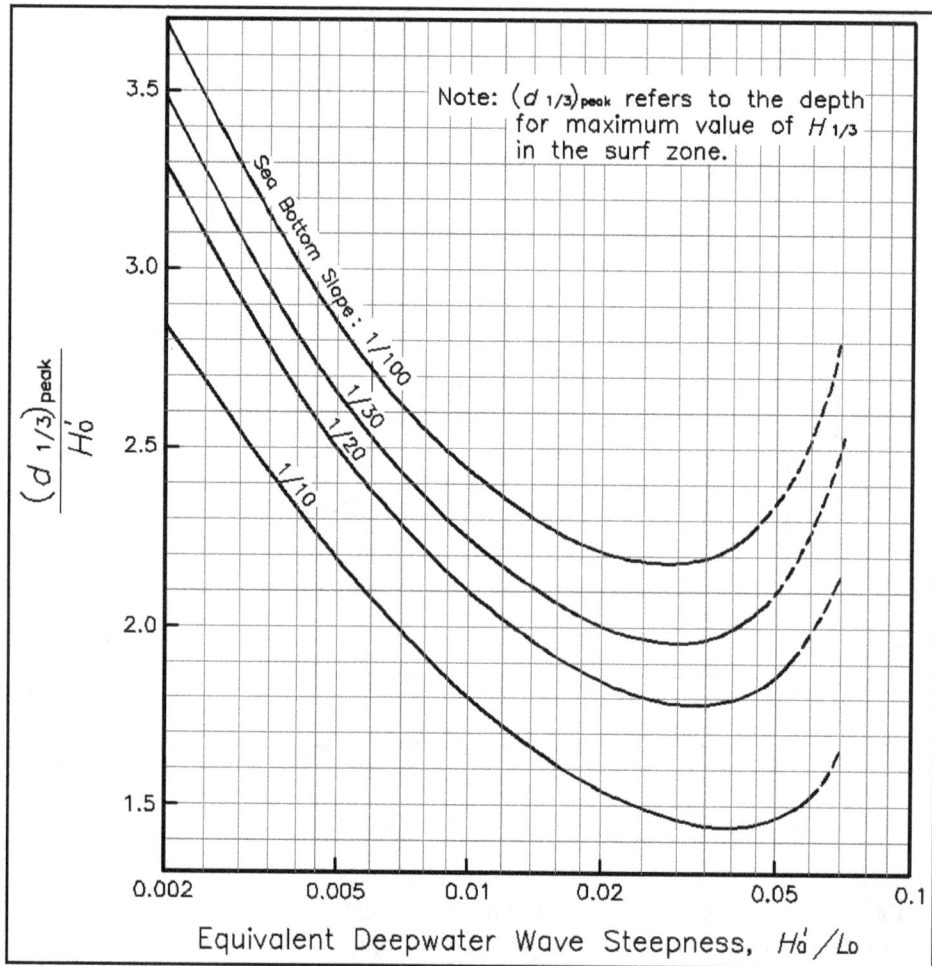

Figure II-8-13.   Water depth at which H_s is maximum in the surf zone (Goda 1985)

and stream function wave theory, can be used to estimate (with small conservatism) crest heights at the 2-percent probability level of exceedance.

(d) Grouping of individual high waves can influence functional design (e.g. runup and overtopping of a breakwater or revetment) and structural design. The recommended way to account for this effect is through physical model tests. The tests should ensure that design alternatives that may be prone to damage by wave groups include a realistic sampling of grouping in the incident waves.

d. Wave climate. Wave climate is discussed in Part II-2.

(1) Design importance. Wave climate affects functional performance of a project and operational activities. Its impact includes longshore and cross-shore sediment transport, harbor agitation, navigation, dredging, and surveying.

(2) General climate.

(a) General wave climate is the probabilistic mix of sea states occurring at a site. Key components of a sea state typically include H_s, T_p, and θ_p for each major wave system (such as a sea and one or more swell systems). Sea states along U.S. Atlantic, Pacific, and Gulf of Mexico coasts can be expected to include more

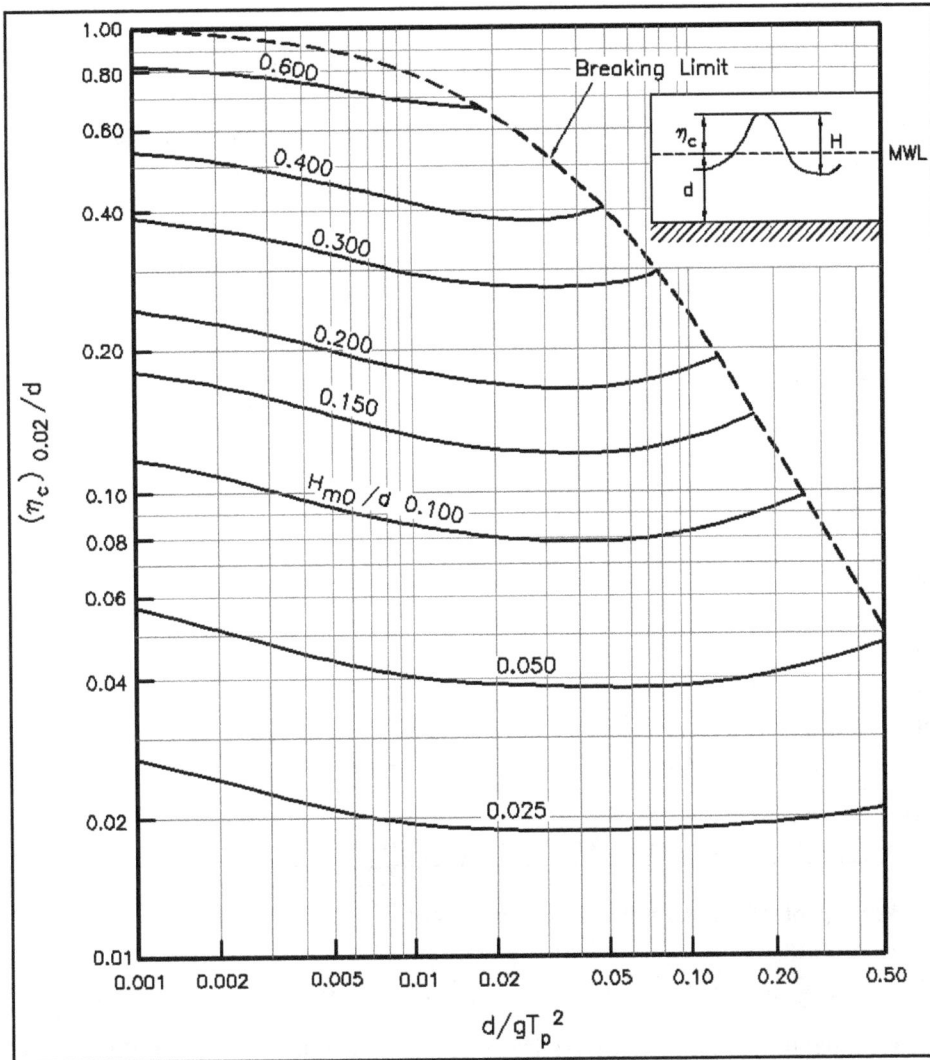

Figure II-8-14.   Crest elevation at the 2-percent probability level of exceedance (Seelig, Ahrens, and Grosskopf 1983)

than one major wave system about two-thirds of the time (Thompson 1980). Wave climate can vary greatly between seasons of the year, but variations between years are usually small. At least three full years of data are desirable for a stable estimate of wave climate. Figure II-8-15 is a typical table of wave climate information (from one directional sector) from the Wave Information Studies.

(b) The mean H_s and distribution of H_s are always of interest in describing wave climate. Both Weibull and log-normal distribution functions have been used to represent the distribution of H_s. However, the actual data usually provide a sufficiently stable estimate and the formal distribution functions are unnecessary. Means and distributions of T_p and θ_p are vital for some applications. Joint occurrence probabilities of H_s, T_p, and θ_p give a fairly complete representation of wave climate, which is needed in applications such as nonlinear refraction modeling. When these standard statistical summaries are formed, information about the simultaneous occurrence of more than one wave system is lost.

WIS ATLANTIC REVISION 1956 - 1975
LAT: 36.00 N, LONG: 75.25 W, DEPTH: 37M
OCCURRENCES OF WAVE HEIGHT AND PEAK PERIOD FOR 45-DEG DIRECTION BANDS

STATION: 55   (67.50 - 112.49) 90.0 DEG

Tp (sec)

Hmo (m)	3.0-4.9	5.0-6.9	7.0-8.9	9.0-10.9	11.0-12.9	13.0-14.9	15.0-16.9	17.0-18.9	19.0-20.9	21.0-LONGER	TOTAL
0.00 - 0.99	381	784	2655	1809	667	267	46	11	4	.	6624
1.00 - 1.99	58	1344	1327	1197	990	347	48	8	1	.	5320
2.00 - 2.99	.	44	429	180	161	67	.	.	.	.	881
3.00 - 3.99	.	.	55	156	54	30	5	.	.	.	300
4.00 - 4.99	.	.	3	64	33	9	5	.	.	.	114
5.00 - 5.99	.	.	.	6	21	4	2	.	.	.	33
6.00 - 6.99	.	.	.	.	6	5	.	1	.	.	12
7.00 - 7.99	.	.	.	.	.	3	2	.	.	.	5
8.00 - 8.99	.	.	.	.	.	7	3	.	.	.	10
9.00-GREATER	.	.	.	.	.	.	.	.	.	.	0
Total	439	2172	4469	3412	1932	739	111	20	5	0	13299

Figure II-8-15.   Wave climate summary (Hubertz et al. 1993)

(3) Storms.

(a) Storms are a natural part of the wave climate at a site. On the order of 20-50 storm events can be expected at a site during a year of record. The upper portion of the distribution of H_s is due to storms, either local or distant. Much of the storm wave climate is fairly consistent from year to year. Three years of data usually suffice for a reasonable representation of the storm portion of wave climate (excluding extreme events). Large storm events usually dominate any secondary wave systems present, and the sea state can be well-represented by one H_s, T_p, and θ_p parameter set. Tropical storms are generally a concern only in the extreme portion of wave climate, if at all, since even the more exposed sites are rarely affected by them.

(b) Some areas can experience changes in climate that systematically affect the incidence and intensity of severe storm waves over a time period of months or years. The reasons for climate change are not always easily understood. Short-term climate variation can be related to deviations from characteristic upper air flow patterns or large-scale ocean current patterns that persist over at least one storm season. One documented example is the influence of the El Niño-Southern Oscillation climatic anomaly on the occurrence of extreme waves along the California coast (Seymour et al. 1984). Both tropical and winter storms are affected. Long-term systematic climate changes can be generated by factors such as local subsidence and global temperature change. An example of long-term change is the increasing trend in annual mean significant wave height off the southwest tip of England as measured over a 25-year period at Seven Stones Light Vessel (Carter and Draper 1988).

(4) Persistence of high and low wave conditions.

(a) The duration of storm events is another important component of the wave climate. Storm duration in this context is usually defined as the length of time H_s persists above some fixed threshold value (Figure II-8-16). Storms with long duration are likely to be more damaging than storms with short duration, in part because they are likely to encompass one or more highs in astronomical tide. Storm duration decreases as threshold increases, but it appears to be fairly independent of storm intensity (Smith 1988). The choice of threshold value is based on the application. Smith (1987) suggested using an H_s value that is exceeded by 6 percent of the observations, which gave mean durations of about 25 hr at U.S. east and west coast hindcast locations. Thus these events can be expected to encompass two high tides along the U.S. east coast and one higher high water along the U.S. west coast.

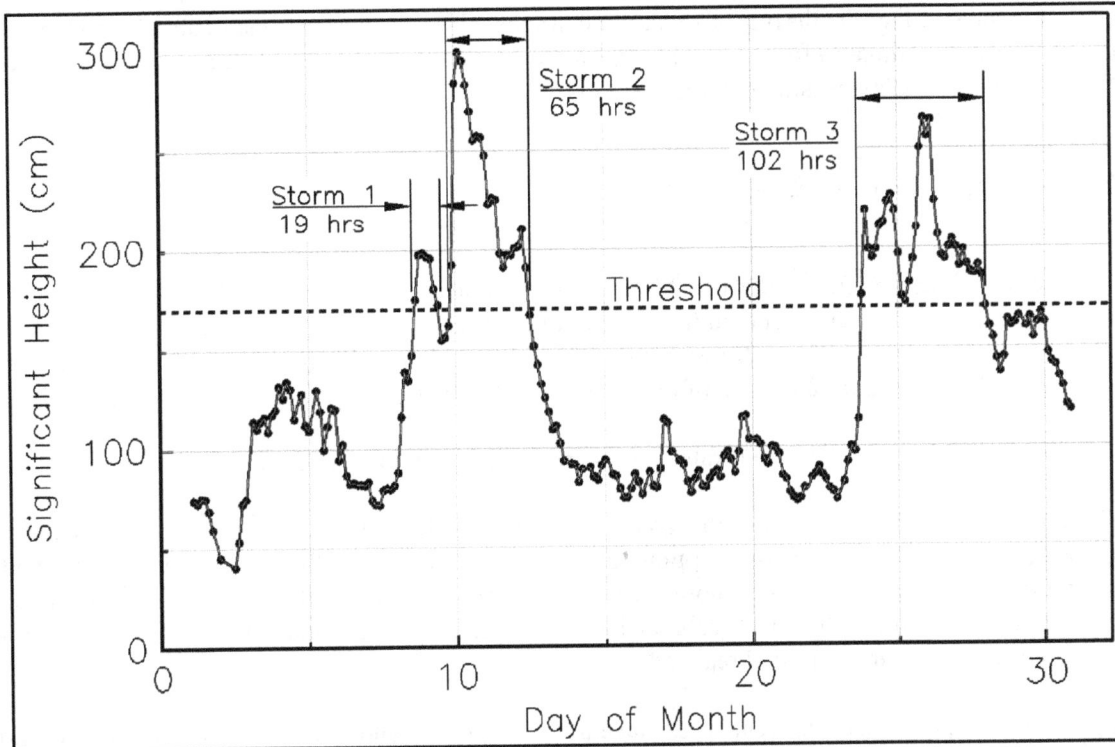

Figure II-8-16.   Persistence of storm waves

(b) Persistence of wave conditions *below* a threshold H_s can be an important operational concern, since it provides information on operational windows of low wave activity. It can also be a consideration in functional design of some coastal projects.

e. Long waves. Wave phenomena with periods between those of swell and tides are collectively termed *long waves*. In most coastal engineering applications, they have a limited role in design. Occasionally they can be a major design concern (e.g. harbor oscillations).

(1) Tsunamis. Tsunamis are briefly discussed in Part II-5. They are sufficiently rare and unpredictable that they become a concern for design return periods of about 100 years or longer. Since most coastal engineering works are designed for return periods of 50 years or less, tsunamis can generally be omitted from the design.

(2) Seiche. Seiche is discussed in Part II-7. It can be an important concern in harbor design or modification. Harbors in areas where energetic, long-period swell can occur are especially prone to seiching

problems. In such areas (e.g. the U.S. Pacific coast and Hawaii), seiche should be routinely considered in design.

(3) Infragravity waves. Infragravity waves are discussed in Part II-4-5. They can be an important component of surf zone processes, particularly during storms, and they are a forcing mechanism for harbor oscillations and other seiching phenomena. Methods for estimating infragravity waves and incorporating them into design are relatively immature at present. Infragravity waves can be considered in design by conducting physical model tests with irregular waves if long waves can be sufficiently controlled, a demanding task. They may also be estimated with some confidence from wind wave/swell conditions using theory, numerical modeling, and/or empiricism. For example, Bowers (1992) considered long waves at three coastal sites in intermediate depths typical of harbor entrances. He used theory to estimate a bound long wave H_s and empiricism to estimate a free long wave H_s (including both edge waves and leaky waves described in Part II-4-5). His general expression for free waves is

$$H_s \ (free \ long \ waves) \ = \ K \ \frac{H_s^\alpha \ T_p^\beta}{d^\gamma} \tag{II-9-6}$$

For his three sites, K ranged from 0.0041 to 0.0066 and overall best-fit values for $\alpha, \beta,$ and γ were 1.11, 1.25, and 0.25, respectively. Bowers observed that bound long waves increasingly dominate free long waves as wind wave/swell H_s increases. For a 10-year return period in the 12-m to 13-m depth, Bowers estimated total long wave H_s values of about 12 percent of the wind wave/swell H_s.

f. Extreme water level. Extreme water levels are discussed in Part II-5.

(1) Design importance. Extreme high water levels cause flooding. They also facilitate wave damage by raising the base level for runup and overtopping, by allowing increased depth-limited wave heights, and by shifting the zone of wave attack further shoreward such that waves can damage dunes and coastal structures. At some locations, extreme high-water levels can lead to pollution and health hazards when sewage treatment ponds or other containment areas are breached.

(2) Estimation procedures. Extreme water levels are caused by some combination of astronomical tides, storm surge (high wind stress, low atmospheric pressure, rainfall/runoff in enclosed or semi-enclosed areas), and wave setup. Probabilities must be estimated as a joint probability of the various combinations that can occur. Procedures for developing storm water level frequency-of-occurrence relationships are reviewed in Part II-5-5.b, including the historical method, synthetic method, and empirical simulation technique (EST). The EST is convenient for development of water level design criteria requiring the quantification of risk and uncertainty associated with the frequency predictions. Traditionally, the distribution of extreme water levels has been fitted to either a Pearson Type III (Engineer Manual (EM) 1110-2-1412) or log-Pearson Type III (U.S. Water Resources Council 1976) distribution function. The EST approach does not require a theoretical distribution function.

g. Water level climate. Water level climate is discussed in Part II-5.

(1) Design importance. The general water level climate at a site impacts navigation channel depths, harbor depths, currents, harbor flushing, and physical and biological processes in the intertidal zone, including marsh areas.

(2) Estimation procedures. The principal component of water level climate at most coastal sites is astronomical tide (Part II-5-3). In areas with little or no tide, particularly very shallow areas, wind,

atmospheric pressure, and rainfall can be the primary components of water level climate (Part II-5-5). In lakes, seasonal fluctuations in water level can be dominant, as in the Great Lakes (Part II-5-4.b.(2)).

(3) Long-term changes. Long-term changes in relative water level can be caused by climatological effects and secular fluctuations (such as melting of the polar ice caps, large-scale isostatic adjustments of the earth's crust, and local subsidence). These long-term changes, which operate on time scales ranging from semiannual to decades, may sufficiently shift the water level datum relative to project features to merit consideration in design.

h. Currents. Surf zone currents are discussed in Part II-4-6. Currents at inlets and harbors are treated in Part II-6 and 7, respectively.

(1) Design importance.

(a) Nearshore shelf. Currents over the continental shelf are important relative to rate and direction of transport of fluids and solids, such as river discharge into the ocean, sewer outfall discharge, movement of sediment from offshore dredge disposal sites, and movement of civic waste material from ocean dump sites. They may also be important for their effect on navigation into harbors, particularly for large vessels subjected to currents moving across the entrance channel. Currents are driven mainly by tides and winds, but temperature and salinity gradients, Coriolis effect, river discharges, and organized current systems (such as the Gulf Stream) can also be important. Currents can vary greatly between the surface and bottom.

(b) Surf zone. Surf zone currents are the driving force transporting sediments in both the longshore and cross-shore directions. As such, they are the key factor in beach erosion and accretion. They may also be important relative to scour and stability of breakwaters and revetments. Surf zone currents are driven by breaking waves and nearshore winds. Currents are very sensitive to wave direction. The magnitude of longshore transport can vary greatly over a time period of days, months, and even from year to year in response to natural variations in wind and wave climate (Figure II-8-17). At many sites, even the dominant direction during a single year (upcoast or downcoast) can deviate from the normal pattern. Thus an adequate sample of years is necessary for stable design estimates. Surf zone currents are discussed in detail in Part II-4-6. Nearshore sediment transport is covered in Part III.

(c) Inlets. Currents through inlets are the primary process affecting exchange of water and sediments between the bay and ocean. They impact water quality, bay ecology, and erosion and shoaling patterns. They can impede navigation by creating steepened, breaking waves when a strong ebb current opposes energetic ocean waves. They may cause scour along jetties and other inlet structures and affect structure stability. Tides, wind, and density differences are typical driving forces. Inlet currents are a necessary consideration in design of projects at inlets.

(d) Harbors. Currents through harbor entrances are generally important in terms of circulation and flushing of the harbor to maintain water quality and, in some cases, to reduce maintenance dredging. They can be driven by tides, winds, and river discharge within the harbor. Additional detail is given in Part II-7-6.

(2) Estimation procedures.

(a) Currents are estimated in terms of time-averaged mean speed and direction and, often, some measure of maximum current speed. For tidal currents, it is helpful to distinguish between ebb and flood tide maxima.

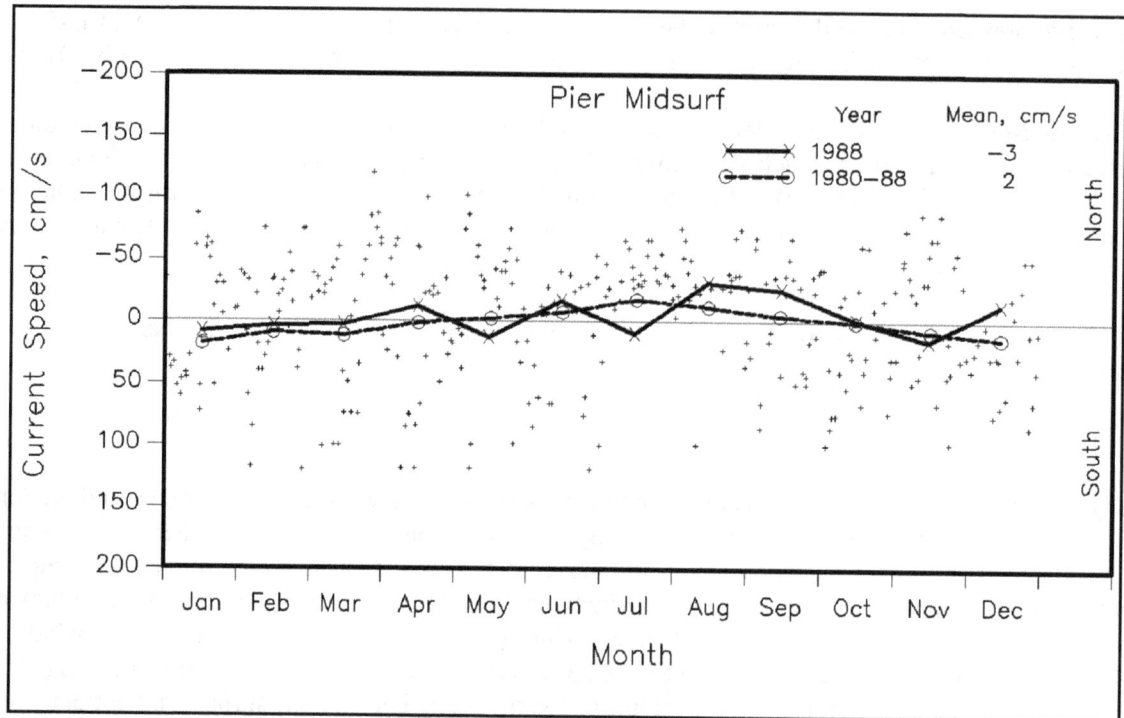

Figure II-8-17. Example of short- and long-term surf zone current data (Leffler et al. 1990)

Currents can vary significantly over short distances, especially around inlets, and some knowledge of the spatial current field can be useful for design.

(b) Currents are best estimated from measurements, numerical modeling, or physical modeling. A combination of both measurement and modeling typically yields the best estimates. Measurements provide boundary conditions and calibration/validation data for the model. The model provides full spatial current fields and a capability for estimating design conditions well beyond any measured events.

(c) More approximate information on currents can be obtained from several sources. Published tidal current tables for use by mariners are available for many U.S. areas important to navigation (Part II-8-7). As with winds, currents at one location can sometimes be transferred to another nearby location, with due consideration of differences between locations. A short measurement record from the desired location can be very helpful in estimating transfer relationships. ACES includes a simplified model for inlet hydraulics. Time-varying inlet currents can be calculated for given time-dependent sea level fluctuation.

  i. Design example.

(1) This section contains a detailed example of estimating hydrodynamic parameters for design. Because of the complexities and many variations of design environment possible in coastal engineering (for example, inner surf zone versus outer surf zone versus outside the surf zone), the example is more an illustration than a blueprint for coastal design. Data used in the example are more extensive than would typically be available. When measurements are lacking, the information on water levels and waves must be estimated from some combination of experience, nearby measurements, hindcasts/forecasts, and other physical and numerical modeling. Information must be properly transformed to the project site.

Hydrodynamic Analysis and Design Conditions

EXAMPLE PROBLEM II-8-2

FIND: Hydrodynamic parameters for design of the north jetty. The intended design life is 25 years.

GIVEN: Two jetties are being designed at a site exposed to energetic waves and currents (Figure II-8-18). The jetties extend from shore to a depth of -10 m MLLW. A representative bottom slope is 1/100. Water level and wave measurements near the site are available at 6-hr intervals over a period of 17 years.

Note: This period of measurement is unusually long for coastal engineering project sites. Typically information on water levels and waves must be generated by numerical modeling to supplement limited measurements. The level of confidence in any calculated results depends highly on the *quality* of available water level and wave information. The level of detail and refinement of calculations should be consistent with the quality of available information and project needs.

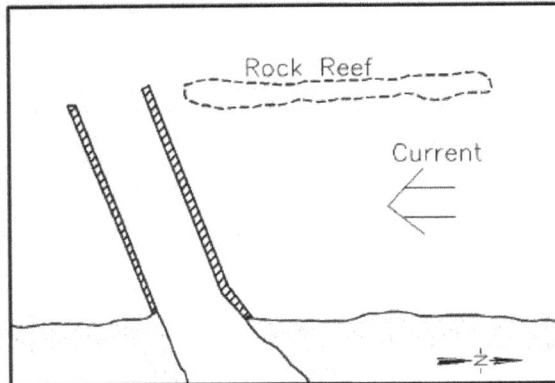

Figure II-8-18.   Jetty plan

SOLUTION:

WATER LEVEL:

Assume all storm populations that could affect design are well-represented (both number and intensity of events) in the data.

Note: At sites exposed to tropical storms, the assumption that storms are well-represented in the historical record at the site is often not justified. In such cases, a numerical model study is required to adequately represent the range of possibilities (Part II-5).

Datums. Datum information for the site (obtained from NOS tidal benchmark sheet; see Part II-8) is listed in Table II-8-12.

Table II-8-12 Tidal Datum Information		
Datum	**Abbreviation**	**Level (m)**
Mean higher high water	MHHW	2.55
Mean high water	MHW	2.34
Mean tide level	MTL	1.38
Mean sea level	MSL	1.36
National Geodetic Vertical Datum	NGVD	1.26
Mean low water	MLW	0.42
Mean lower low water	MLLW	0.00

Example Problem II-9-2 (Sheet 1 of 21)

Example Problem II-8-2 (Continued)

Astronomical tide. A probability distribution of astronomical tide levels (Figure II-8-19) can be obtained from: (1) Harris (1981) for selected tide stations; (2) harmonic reconstruction of a tidal series (Part II-5); or, (3) to a good approximation over all but the more extreme ends, statistical summarization of long-term water level measurements.

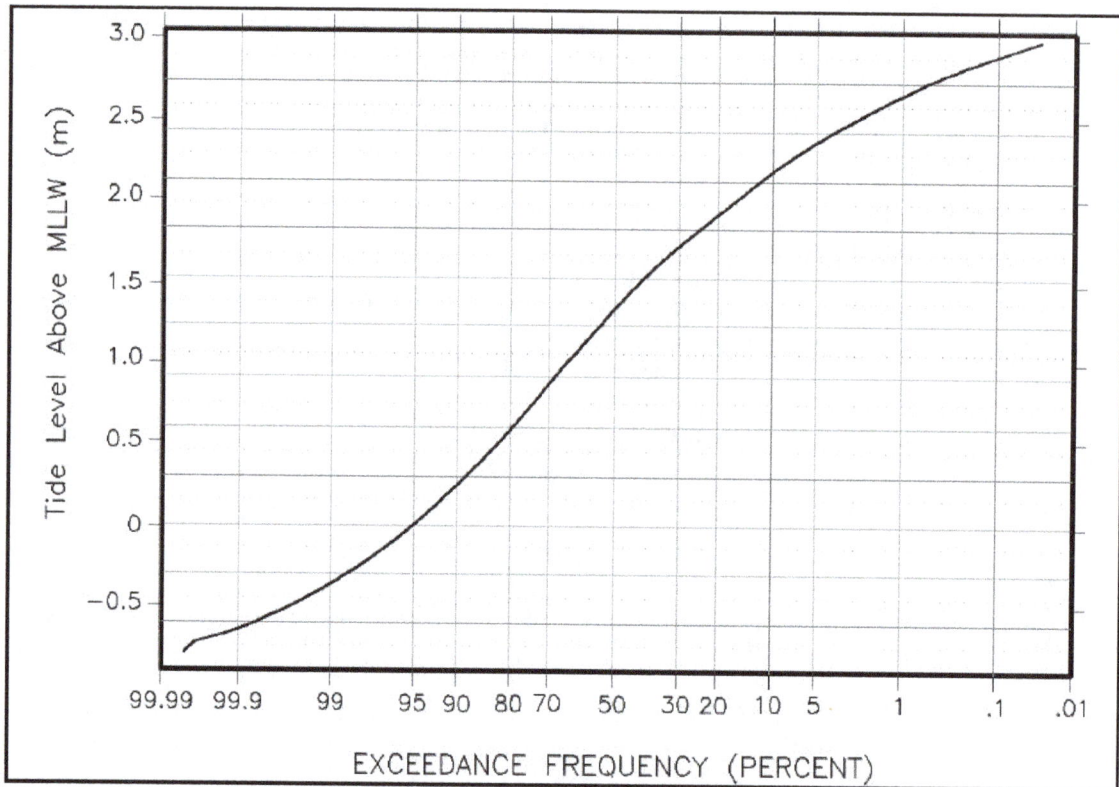

Figure II-8-19. **Probability distribution of astronomical tide levels**

Storm surge. Storm surge water levels (Figure II-8-20) can be estimated by the following steps:

(1) Develop criteria for identifying storm events, such as events with $H_s \geq 5$ m and separated from each other by 5 or more days.

(2) Identify all cases in the water level measurement record that meet the criteria in (1).

Example Problem II-9-2 (Sheet 2 of 21)

Example Problem II-8-2 (Continued)

(3) For each selected event, subtract the predicted astronomical tide and wave setup from measured water levels. The maximum difference during the event is considered the peak surge. Details of calculating wave setup are omitted here but discussed later in conjunction with wave analysis.

(4) Organize peak surges into a probability distribution. The ACES Extremal Significant Wave Height Analysis (using water levels in place of significant heights) may be a helpful tool for this and the following step.

(5) Fit a probability distribution function to the data to extrapolate to lower probabilities as needed.

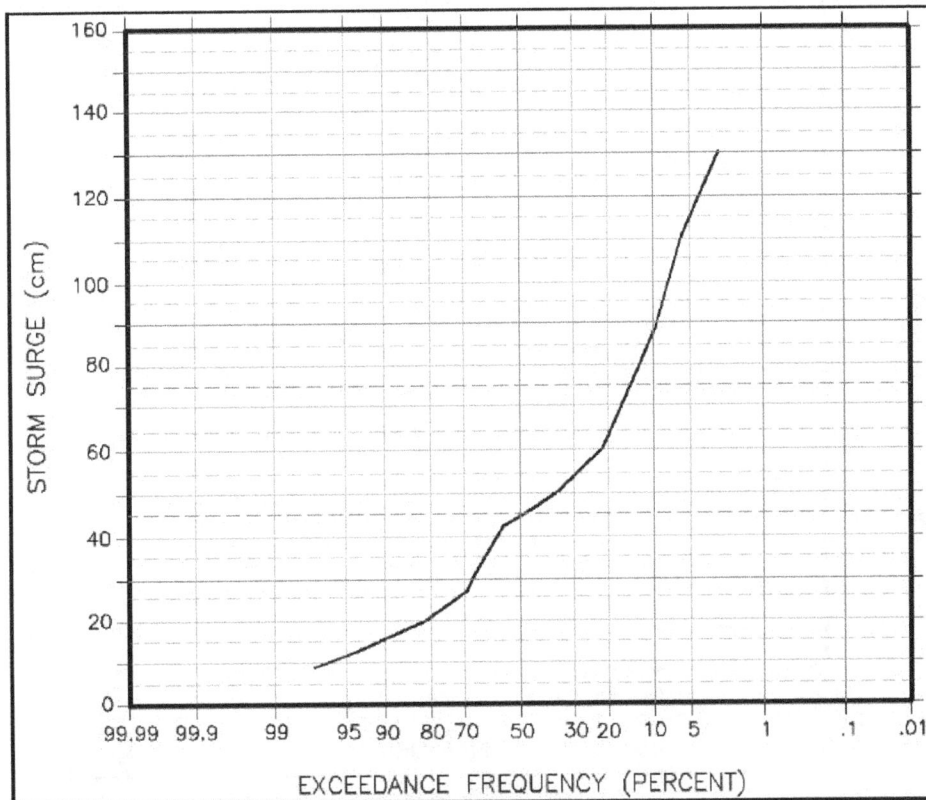

Figure II-8-20.   Probability distribution of storm surge

Combined tide and storm surge. A simplified analysis is used here to generate probabilities for combined tide and storm surge. With consideration of general knowledge about the duration of storms, typical storm surge hydrographs, and tidal variations at the site, it is reasonable (but somewhat conservative) to assume that peak water level events can be represented as the measured peak surges coinciding with a high tide.

Example Problem II-9-2 (Sheet 3 of 21)

Example Problem II-8-2 (Continued)

The distribution of high water tide elevations above mean sea level is given by Harris (1981) for a nearby location with similar tidal response (Figure II-8-21). For this example, the traditional *joint probability method* (as described by Harris (1981)) was used to combine high-water tide frequencies with storm surge probability expressed as frequency of occurrence *per year*. Combined tide and storm surge probabilities for design return periods are included in Figure II-8-21.

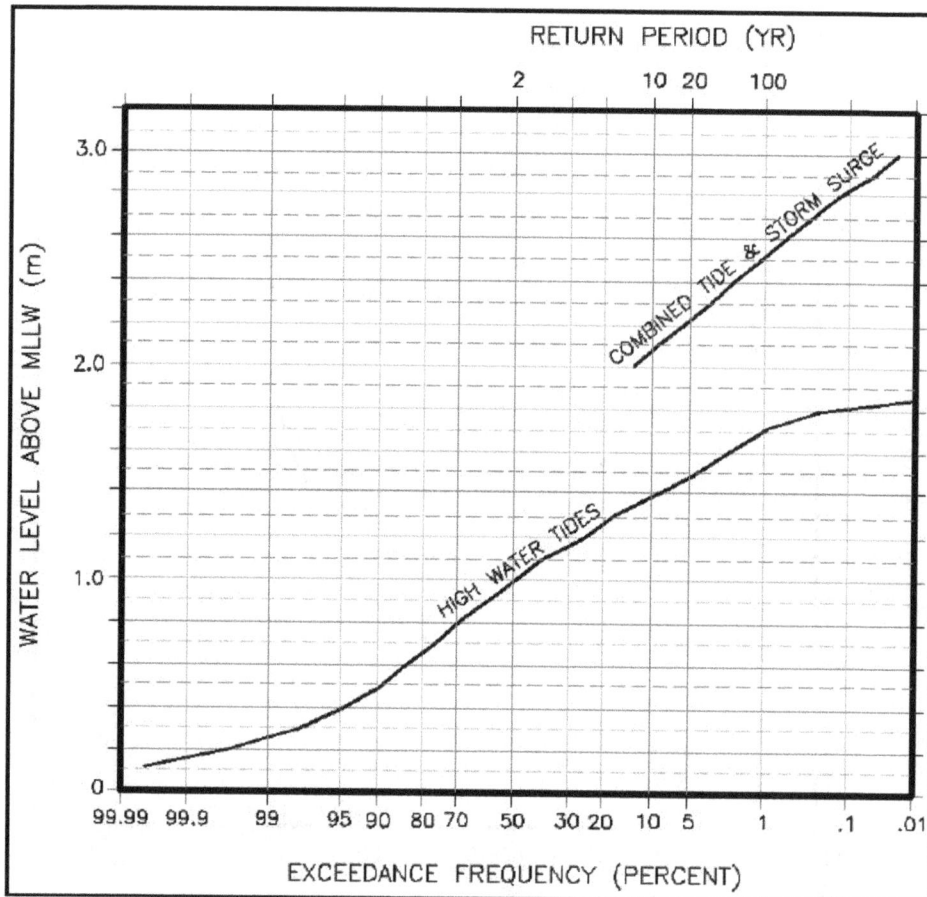

Figure II-8-21. Probability distribution of high tides and combined tide and storm surge

CURRENTS:

The north jetty will deflect a strong longshore current seaward. Therefore strong offshore-directed currents can be expected along the north side of the jetty. Because of the local bathymetry, principally a rocky reef parallel to shore, the current is expected to affect only the area near the jetty head. Local currents can be estimated based on experience at similar sites (if any exist), physical modeling, and, possibly, numerical modeling. For this example, moderate current speed is taken as 1.5 m/s and design current as 3.0 m/s. Tidal currents can also be significant, since the tide range is fairly large. Tidal currents will affect navigation in the entrance, but are not expected to influence jetty structural design.

Example Problem II-9-2 (Sheet 4 of 21)

Example Problem II-8-2 (Continued)

WAVES:

    Measurements. Measurements are available over a period of 17 years from a wave gauge located in -15.2-m depth MLLW. Events with H_s>6.1 m were selected for design analysis, a total of 33 cases. The maximum H_s during each event and the corresponding T_p, storm surge, and water depth at the gauge are given in Table II-8-13.

The data record represented in the table is quite long, and it is considered statistically representative of storm events to which the site is exposed. It is used as the basis for design. Measured values of T_p can be taken as representative of both the gauge and jetty locations. Measured values of H_s must be transformed between the gauge location and jetty. Also, the design water level at the jetty must be estimated because it strongly affects calculations of H_s.

Estimation of H_0'. The equivalent deepwater wave height H_0 is calculated as an intermediate step in estimating wave height at the jetty. Refraction between gauge and jetty locations is assumed to be negligible in this example, and values of H_0 estimated from gauge data are also applicable to the jetty location. Steps in estimating H_0 are listed below and calculation results are given in Table II-8-13. The ACES application "Irregular Wave Transformation" (Goda's Method) could be used to assist in these and subsequent calculations. It is advisable to spot-check any ACES calculations with some manual calculations.

    (1) Calculate L_0 from known values of T_p, $L_0 = (gT_p^2)/(2\pi) = 1.56gT_p^2$ (L_0 in m).

    (2) Calculate d_{gauge}/L_0.

    (3) Calculate H_s/L_0 as an initial estimate of H_0/L_0 (needed for using the curves in step (4)).

    (4) Get shoaling coefficient K_s from Figure II-8-22.

    (5) Calculate H_0 from significant height at the gauge H_{sgauge} (Table II-8-13) as

$$H_0' = \frac{H_{sgauge}}{K_s} \qquad \text{(II-9-7)}$$

    (6) Calculate H_0/L_0 to ensure that K_s from step (4) is valid. It may be necessary to repeat steps (4)-(6) to arrive at a final value of H_0.

Example Problem II-8-2 (Sheet 5 of 21)

Example Problem II-8-2 (Continued)

Table II-8-13
Significant Storm Events

Date	Measurements					Calculations			
	H_s(cm)	T_p(sec)	Surge(cm)	$d_{gauge}{}^1$(m)	L_0(m)	d_{gauge}/L_0	K_s	H_0'(cm)	H_0'/L_0
Feb 78	613	13.5	73	17.3	284	0.0609	1.00	613	0.0216
Jan 81	625	17.4	12	15.1	472	0.0320	1.18	530	0.0112
Nov 81	646	13.6	131	17.6	289	0.0609	1.00	646	0.0224
Dec 81	619	15.4	30	14.9	370	0.0403	1.11	558	0.0151
Dec 82	884	15.0	88	17.9	351	0.0510	1.08	819	0.0233
Jan 83	713	16.9	110	16.9	446	0.0379	1.15	620	0.0139
Feb 83	622	15.8	46	17.6	389	0.0452	1.07	581	0.0149
Apr 83	631	15.2	43	17.9	360	0.0497	1.04	607	0.0169
Feb 84	707	15.2	79	17.7	360	0.0492	1.05	673	0.0187
Nov 84	634	14.0	61	16.9	306	0.0552	1.01	628	0.0205
Dec 85	613	17.6	43	16.7	483	0.0346	1.13	542	0.0112
Feb 86	686	17.1	49	18.1	456	0.0397	1.10	624	0.0137
Mar 86	610	16.4	46	18.0	420	0.0429	1.08	565	0.0135
Apr 86	680	15.6	12	16.2	380	0.0426	1.10	618	0.0163
Apr 86	631	14.5	27	16.8	328	0.0512	1.03	613	0.0187
May 86	619	14.5	30	17.0	328	0.0518	1.03	601	0.0183
Sep 86	689	15.4	46	17.5	370	0.0473	1.06	650	0.0176
Nov 86	652	16.9	49	17.1	446	0.0383	1.12	582	0.0130
Jan 87	695	16.0	55	17.5	399	0.0439	1.09	638	0.0160
Feb 87	631	13.0	40	17.3	264	0.0655	0.98	644	0.0244
Apr 87	686	14.6	21	17.2	333	0.0517	1.04	660	0.0198
Sep 87	643	13.8	12	17.5	297	0.0589	1.00	643	0.0217
Nov 87	610	16.7	70	17.1	435	0.0393	1.10	555	0.0128
Dec 87	628	15.4	24	17.6	370	0.0476	1.05	598	0.0162
Mar 88	613	14.8	3	16.9	342	0.0494	1.05	584	0.0171
Apr 88	643	13.0	21	17.2	264	0.0652	0.98	656	0.0248
Nov 88	628	13.2	24	16.6	272	0.0610	1.00	628	0.0231
Jan 90	619	16.9	58	16.2	446	0.0363	1.14	543	0.0122
Nov 91	735	14.0	43	17.4	306	0.0569	1.00	735	0.0240
Dec 91	646	16.7	27	16.7	435	0.0384	1.12	577	0.0133
Jan 92	686	14.6	49	17.3	333	0.0520	1.03	666	0.0200
Sep 92	625	15.8	9	16.9	389	0.0434	1.09	573	0.0147
Jan 94	762	14.1	12	16.9	310	0.0545	1.03	740	0.0239

[1] Water depth at gauge including combined tide and storm surge.

Example Problem II-8-2 (Sheet 6 of 21)

Example Problem II-8-2 (Continued)

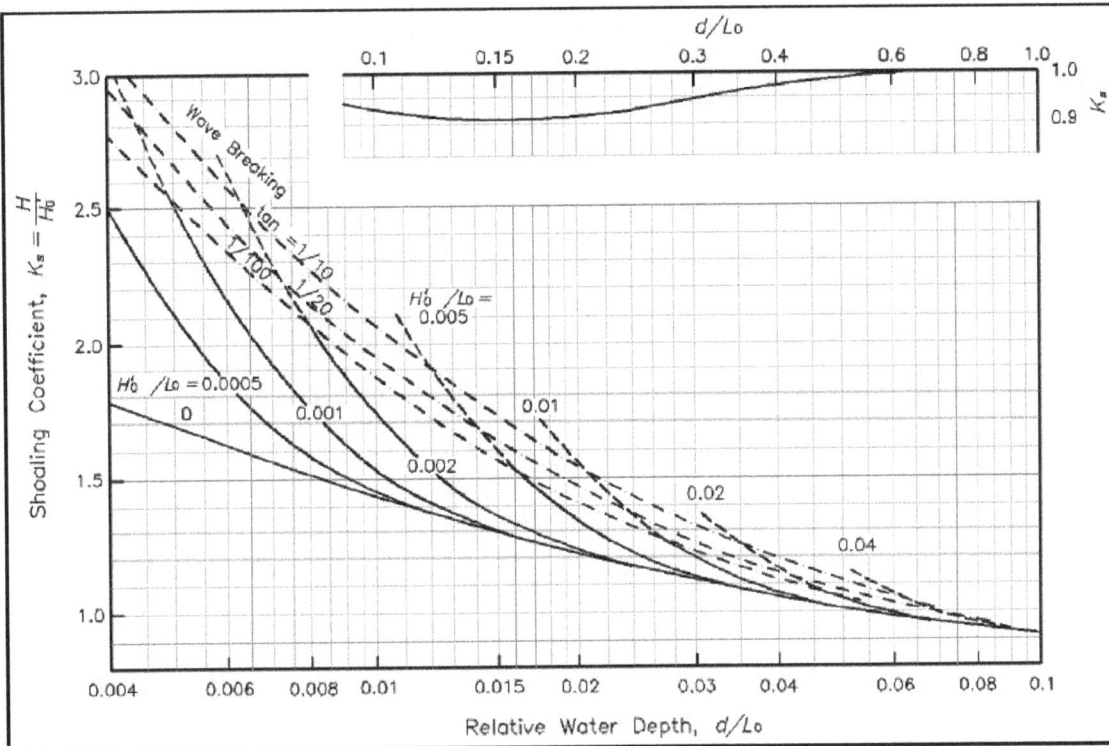

Figure II-8-22.   Estimation of shoaling coefficient (Goda 1985)

Combined tide and storm surge for design wave analysis at jetty. For design wave analysis, extreme combinations of tide, storm surge, and waves must be considered. Since extreme storm surge and waves are often highly correlated (both can be produced by intense storms), they should not be treated as independent processes. The relationship between storm surge and waves is embodied in the available measurements to an extreme occurrence of approximately once in 17 years.

For this example problem, combined tide, storm surge, and waves for design analysis are derived by the following approach. Each measured wave and surge event (represented by H_0', T_p, and storm surge height) is coupled with N high tide levels taken from the probability distribution of high tides (see Table B-37b in Harris (1981)). (Figure II-8-23 is a partial extraction of Table B-37b for use in this example). As before, it is assumed that the duration of extreme storm surge events is long enough that peak surge will coincide with a high tide, and high tide level is independent of storm surge level. The N high tide levels are determined by dividing the probability distribution of high tides into N equal probability increments and taking the tide level for the mid-probability of each increment. The number N should be large enough that the combined probability of the highest tide level and the most severe surge/wave event is lower than the design probability.

Example Problem II-8-2 (Sheet 7 of 21)

Example Problem II-8-2 (Continued)

Lower Limit of Class Interval Shown, All Heights are Normalized by Half the Diurnal Range 1.274 m [Extracted from Table B-37b (Harris 1981)]		
High Water		
Lower Limit	Frequency	Cumulative Frequency
1.4547	.0001	.0001
.	.	.
.	.	.
1.1018	.0110	.0989
1.0877	.0105	.1094
.	.	.
.	.	.
0.9183	0.0206	0.2891
0.9043	0.0229	0.3120
.	.	.
.	.	.
0.7771	0.0213	0.4919
0.7630	0.0187	0.5106
.	.	.
.	.	.
0.6218	0.0185	0.6923
0.6077	0.0156	0.7079
.	.	.
.	.	.
0.4101	0.0077	0.8896
0.3960	0.0108	0.9004

Figure II-8-23. Astronomical tide probability referenced to MSL

A value of $N = 5$, corresponding to probability increments of 0.2, was used for this problem. The mid-probabilities are 0.1, 0.3, 0.5, 0.7, and 0.9. High tide levels matching these probabilities, determined from Figure II-8-24, are in Table II-8-14. The probability of the most severe surge/wave event is the reciprocal of the number of possible measurements (at 6-hr intervals or 4 observations per day) during 17 years:

$$\frac{1}{(4/day) \times (365 \ days/yr) \times (17 \ yr)} = 0.000040$$

Similarly, the probability corresponding to the design life of 25 years is:

$$\frac{1}{(4/day) \times (365 \ days/yr) \times (25 \ yr)} = 0.000027$$

Example Problem II-8-2 (Sheet 8 of 21)

Hydrodynamic Analysis and Design Conditions

Example Problem II-8-2 (Continued)

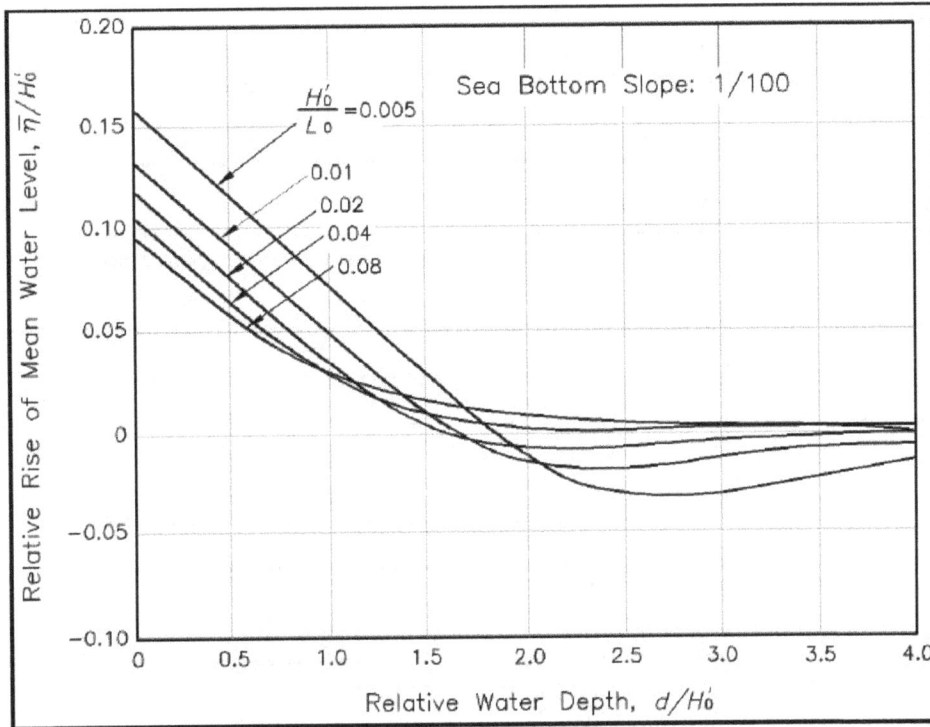

Figure II-8-24. Estimation of wave setup (Goda 1985)

A 1-in-17-year event coupled with a 0.1 probability high tide level corresponds to a combined probability of $0.000040 \times 0.1 = 0.000004$. Since this probability is well below the 25-year design probability, the number of tide increments is sufficient for illustration.

The jetty is divided into three design segments with bottom levels of -10 m MLLW (representing the jetty head), -6 m MLLW and -2 m MLLW (representing sections along the jetty trunk). Tidal water levels at the five probability levels (expressed as exceedance frequencies in percent) are summarized in Table II-8-14. Information in Table II-8-12 is used to convert from MSL to MLLW datum.

Table II-8-14
Tidal Water Levels for Jetty Design

Water Level	Exceedance Frequency (percent)				
	90	70	50	30	10
High tide level, in m MSL	+0.51	+0.78	+0.98	+1.16	+1.40
High tide level, in m MLLW	+1.87	+2.14	+2.34	+2.52	+2.76
Tidal water depth (m)					
Bottom at -10 m MLLW (jetty head)	11.9	12.1	12.3	12.5	12.8
Bottom at -6 m MLLW (jetty trunk)	7.9	8.1	8.3	8.5	8.8
Bottom at -2 m MLLW (jetty trunk)	3.9	4.1	4.3	4.5	4.8

Example Problem II-8-2 (Sheet 9 of 21)

Example Problem II-8-2 (Continued)

The combined effect of tide and storm surge for the first event (excluding wave setup) at the jetty head and nearshore (shallowest) trunk segment are given in Table II-8-15. These initial depth estimates are referenced as $d_{ijettyA}$ for the head and $d_{ijettyC}$ for the nearshore trunk segment. Similarly, combined tide and storm surge levels are computed for each of the 33 storm events at the three jetty design segments.

Table II-8-15
Calculation of d_{ijetty} for Storm Event 1

Water Level	Water Level Exceedance Frequency (percent)				
	90	70	50	30	10
Storm surge (m)	0.7	0.7	0.7	0.7	0.7
Jetty Head					
Tidal water depth (m)	11.9	12.1	12.3	12.5	12.8
$d_{ijettyA}$ (m)	12.6	12.8	13.0	13.2	13.5
Nearshore Jetty Trunk					
Tidal water depth (m)	3.9	4.1	4.3	4.5	4.8
$d_{ijettyC}$ (m)	4.6	4.8	5.0	5.2	5.5

Inclusion of wave setup. The effect of wave setup on water levels at the jetty can now be considered. Calculation steps are listed below and results for Event 1 at the jetty head and nearshore jetty trunk segment are given in Table II-8-16.

    (1) Retrieve H_0/L_0 from Table II-8-13.

    (2) Calculate d_{ijetty}/H_0'.

    (3) Estimate ratio of wave setup $\bar{\eta}$ to H_0' from Figure II-8-24.

    (4) Calculate $\bar{\eta}$ from ratio in step (3). If $\bar{\eta}$ is less than zero, a value of zero should be used for prudent design.

    (5) Calculate a final d_{jetty} as the sum of $\bar{\eta}$ and the previous d_{ijetty}, which included only tide and storm surge. If $\bar{\eta}$ is large, it may be necessary to return to step (1) using d_{jetty} in place of d_{ijetty} and repeat steps (1) through (5).

Example Problem II-8-2 (Sheet 10 of 21)

Example Problem II-8-2 (Continued)

Table II-8-16
Calculation of d_{jetty} for Storm Event 1

	Water Level Exceedance Frequency (percent)				
	90	70	50	30	10
H_0'/L_0	0.0216	0.0216	0.0216	0.0216	0.0216
Jetty head					
$d_{jjettyA}$ (m)	12.6	12.8	13.0	13.2	13.5
$d_{jjettyA}/H_0'$	2.06	2.09	2.12	2.15	2.20
$\bar{\eta}/H_0'$	<0	<0	<0	<0	<0
$\bar{\eta}$ (m)	0	0	0	0	0
d_{jettyA} (m)	12.6	12.8	13.0	13.2	13.5
Nearshore jetty trunk					
$d_{jjettyC}$ (m)	4.6	4.8	5.0	5.2	5.5
$d_{jjettyC}/H_0'$	0.75	0.78	0.82	0.85	0.90
$\bar{\eta}/H_0'$	0.053	0.050	0.046	0.043	0.040
$\bar{\eta}$ (m)	0.3	0.3	0.3	0.3	0.2
d_{jettyC} (m)	4.9	5.1	5.3	5.5	5.7

Estimation of $H_{1/3}$ at Jetty. Now significant wave heights at the jetty H_{sjetty} can be estimated. Calculation steps are listed below and results for Event 1 at the jetty head and nearshore trunk segment are given in Table II-8-17.

(1) Calculate d_{jetty}/H_0, where d_{jetty} is the combined tide, storm surge, and wave setup water level at the jetty (Table II-8-16).

(2) Estimate ratio of $H_{1/3}$ at the jetty H_{sjetty} to H_0 from Figure II-8-25.

(3) Calculate H_{sjetty} from ratio in step (2). Note in this example that this is a *breaking* wave. Also note that the higher H_0 values lead to higher $H_{1/3}$ values in the breaker zone, if d and H_0/L_0 are held constant (because of the gentle slope of the lines in the left portion of Figure II-8-25). Thus the extreme measured H_s values selected for design analysis can be expected to give extreme nearshore wave heights in the surf zone.

After these calculations are completed for all 33 storm events and 5 water levels, there are a total of $33 \times 5 = 165$ event values of H_{sjetty} at the jetty head and at each of the two trunk sections.

Example Problem II-8-2 (Sheet 11 of 21)

Example Problem II-8-2 (Continued)

Table II-8-17
Calculation of H_{sjetty} for Storm Event 1

	Water Level Exceedance Frequency (percent)				
	90	70	50	30	10
	Jetty Head				
d_{jettyA}/H_0'	2.06	2.09	2.12	2.15	2.20
$H_{sjettyA}/H_0'$	1.05	1.06	1.07	1.07	1.07
$H_{sjettyA}$ (cm)	644	650	656	656	656
	Nearshore Jetty Trunk				
d_{jettyC}/H_0'	0.80	0.83	0.86	0.90	0.93
$H_{sjettyC}/H_0'$	0.57	0.59	0.60	0.62	0.63
$H_{sjettyC}$ (cm)	349	362	368	380	386

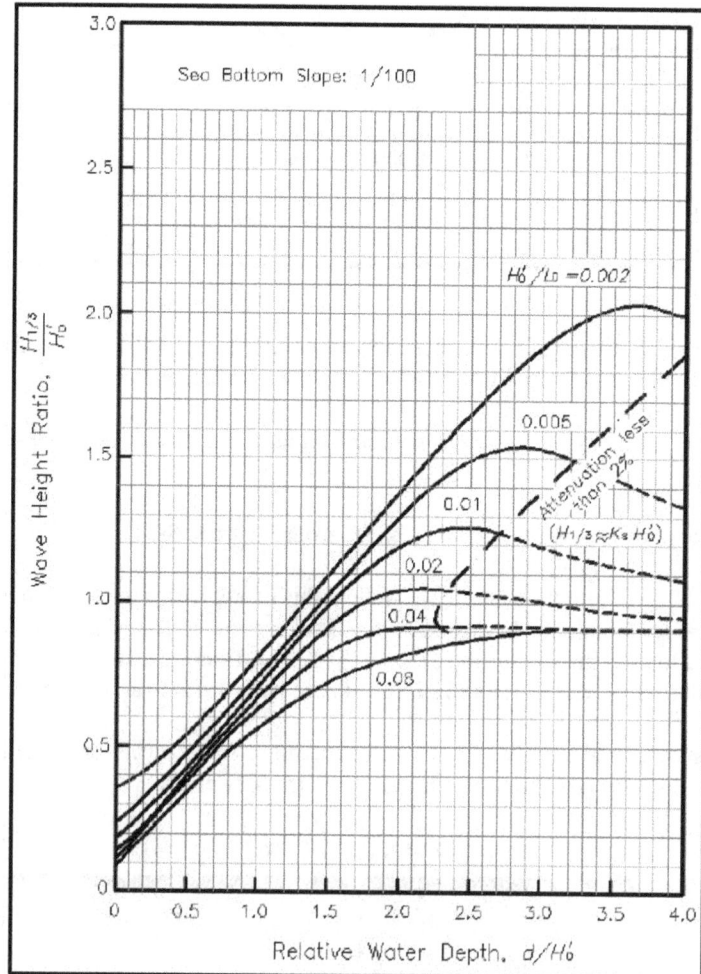

Figure II-8-25. Estimation of wave height at the jetty (Goda 1985)

Example Problem II-8-2 (Sheet 12 of 21)

Example Problem II-8-2 (Continued)

Importance of wave breaking. In estimating design wave conditions, it is important to know whether waves are breaking. Event values of H_0'/L_0 from Table II-8-13 plotted with the curve from Figure II-8-13 with bottom slope of 1/100 show that all values of d_{jetty} at the jetty head are very near or smaller than the depth at which H_{sjetty} reaches a maximum value during shoaling and breaking (Figure II-8-26). Thus, the jetty head can be considered as inside the surf zone and subject to breaking waves for all of the storm events. Since the jetty trunk is in shallower water than the head, it too will be subject to breaking waves for design.

Relationship to H_{m0}. The significant wave height analysis used in this example is based on statistics of crest-to-trough wave heights in a train of irregular waves. This wave height parameter can differ significantly from the energy-based significant wave height parameter H_{m0}, especially for low-steepness waves in shallow water. Some design applications may require H_{m0} instead of H_{sjetty}. If so, the event values of H_{sjetty} may be converted by the procedure described in Part II-1.

Extremal wave height analysis at the jetty. Extremal wave height analysis is presented here only for the jetty head. Similar analyses would be needed for jetty trunk sections, but they are omitted for brevity. Since the design waves are inside the surf zone, water level can be expected to be a key parameter in determining design wave heights. The equivalent deepwater wave steepness H_0'/L_0 and H_0' also influence H_{sjetty}. Design wave estimates should include due consideration of the influence of all three of these parameters: d_{jetty}, H_0'/L_0, and H_0'.

The ACES "Extremal Significant Wave Height Analysis" application was applied to the 165 event values of H_{sjetty} and the FT-I distribution function was selected as a good fit (Figure II-8-27). Confidence intervals of 95 percent were also computed, which is approximately equivalent to ±2 standard deviations.

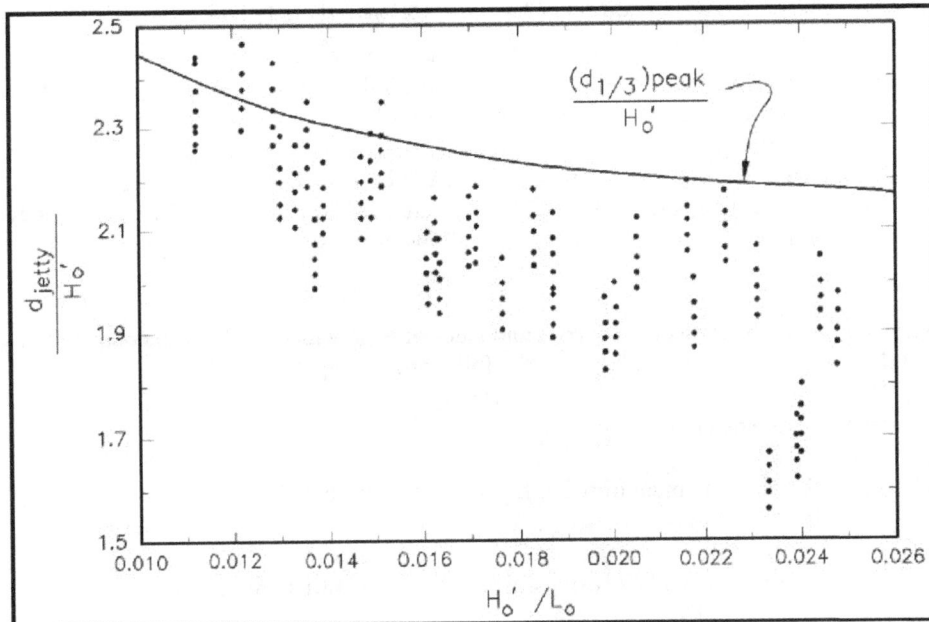

Figure II-8-26.   Identification of surf zone conditions

Example Problem II-8-2 (Sheet 13 of 21)

Example Problem II-8-2 (Continued)

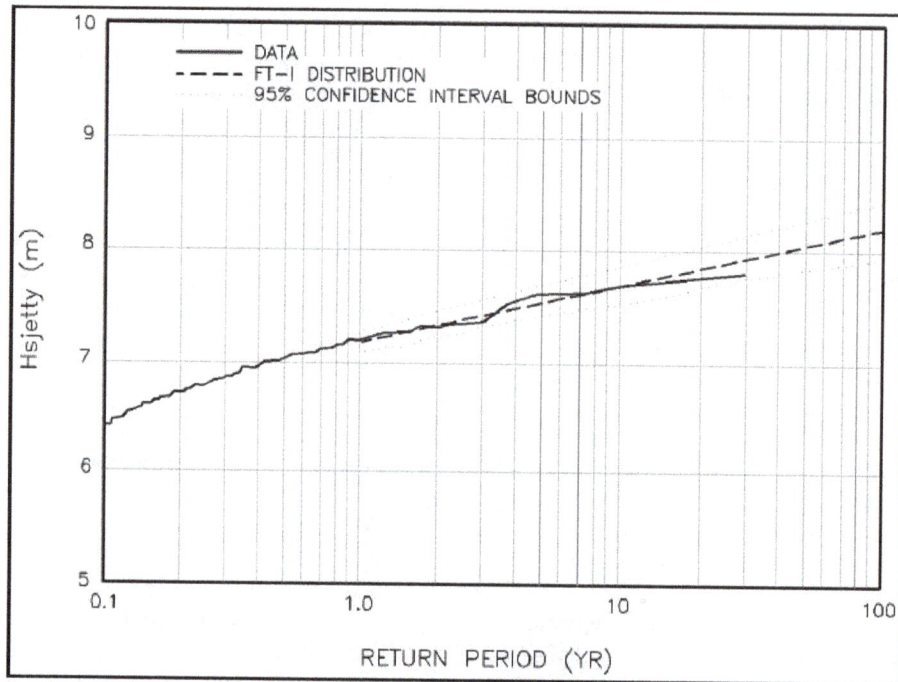

Figure II-8-27. Return period wave heights, 165 events, jetty head

Significant heights in Figure II-8-27 are representative of the extrapolated observed events, but they may have some important limitations. Extreme values of H_{sjetty} are strongly dependent on water depth, and the highest tide level considered in any of the 165 event cases was only at the 10-percent exceedance frequency. (This limit in the tide levels considered was a practical consequence of the manual method being used. If the analysis were done by computer, a much greater range of tide levels could have been included.) For longer return periods, the probability of an event coinciding with a tide level higher than the 10-percent exceedance frequency increases. This concern applies more to the jetty trunk, which is well inside the surf zone for design events, than the jetty head.

It is important in design to be aware of the maximum values of H_{sjetty} which might be encountered. Estimates of the upper bound value of H_{sjetty} were computed by the following two approaches:

(a) Parameters were determined as:

Tide level = maximum from Harris (1981) (Figure II-8-23)
= 1.4547×1.274 m
= +1.85 m MSL
= +1.85+1.36 m MLLW (using Table II-8-12) ≈ +3.2 m MLLW

Storm surge = value estimated from extremal analysis of event storm surges for each T_r (Figure II-8-20)

$H_0'/L_0 = 0.01$ (lower bound on event values in Table II-8-13)

Example Problem II-8-2 (Sheet 14 of 21)

Example Problem II-8-2 (Continued)

H_0' = value estimated from extremal analysis of H_0' event values (Table II-8-13) for each T_r (Figure II-8-28)

Parameters and calculations of H_{sjetty} are summarized in Table II-8-18.

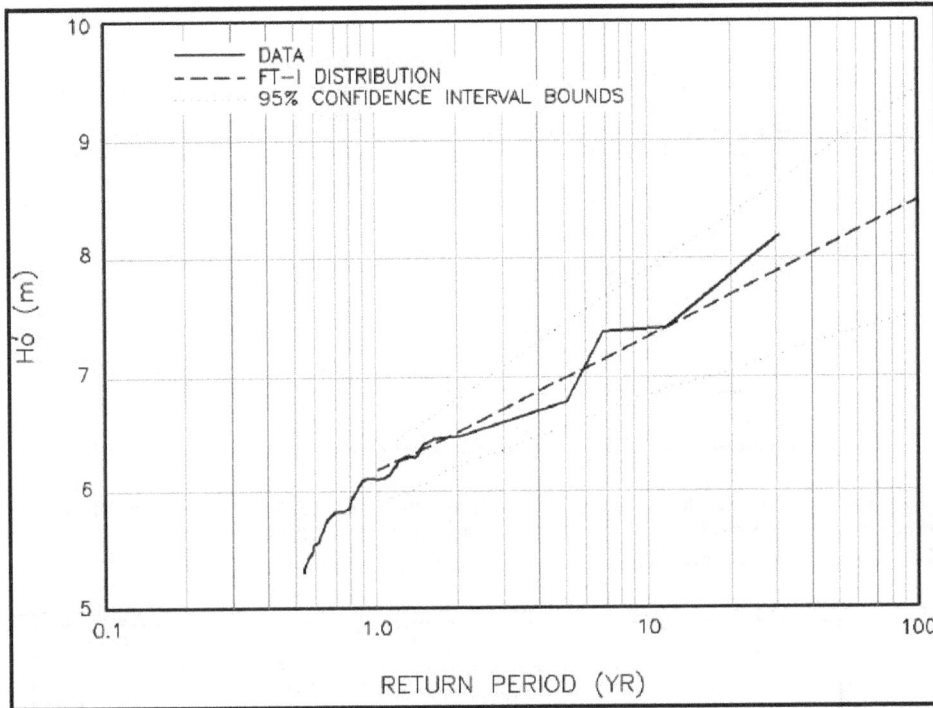

Figure II-8-28.   Return period wave heights, H_0'

Table II-8-18
Calculation of Upper Bound H_{sjetty} Based on Maximum Tide Level

T_r (yr)	H_0' (cm)	Storm Surge (m)	Total Water Level [1] (m MLLW)	d_{jetty} (m)	d_{jetty}/H_0'	H_{sjetty}/H_0'	H_{sjetty} (cm)
2	653	0.7	3.9	13.9	2.13	1.24	810
5	702	0.9	4.1	14.1	2.01	1.21	849
10	737	1.0	4.2	14.2	1.93	1.18	870
25	782	1.2	4.4	14.4	1.84	1.14	891
50	816	1.3	4.5	14.5	1.78	1.12	914
100	850	1.4	4.6	14.6	1.72	1.10	935

[1] Combined maximum tide level (+3.2 m) and storm surge

Example Problem II-8-2 (Sheet 15 of 21)

Example Problem II-8-2 (Continued)

(b) As a more conservative check on (a), the peak possible value of H_t during the shoaling/breaking process between deepwater and shore for given values of bottom slope H_0'/L_0 and H_0' was estimated from Figure II-8-12 using the same values of H_0'/L_0 and H_0' as in (a). Results are summarized in Table II-8-19.

Table II-8-19
Calculation of Upper Bound H_{sjetty} Based on Maximum During Shoaling

T_r (yr)	H_0' (cm)	H_{sjetty}/H_0' [1]	H_{sjetty} (cm)
2	653	1.27	829
5	702	1.27	892
10	737	1.27	936
25	782	1.27	993
50	816	1.27	1036
100	850	1.27	1080

[1] From Figure II-8-12 using bottom slope = 1/100 and H_0'/L_0 = 0.01.

Final design wave heights excluding current effects. Upper bound estimates of H_{sjetty} (Tables II-8-18 and II-8-19) are plotted relative to the 165 cases derived from the measured storm events (Figure II-8-29). The return period curve based on the 165 events is taken as the design curve for this example (Table II-8-20).

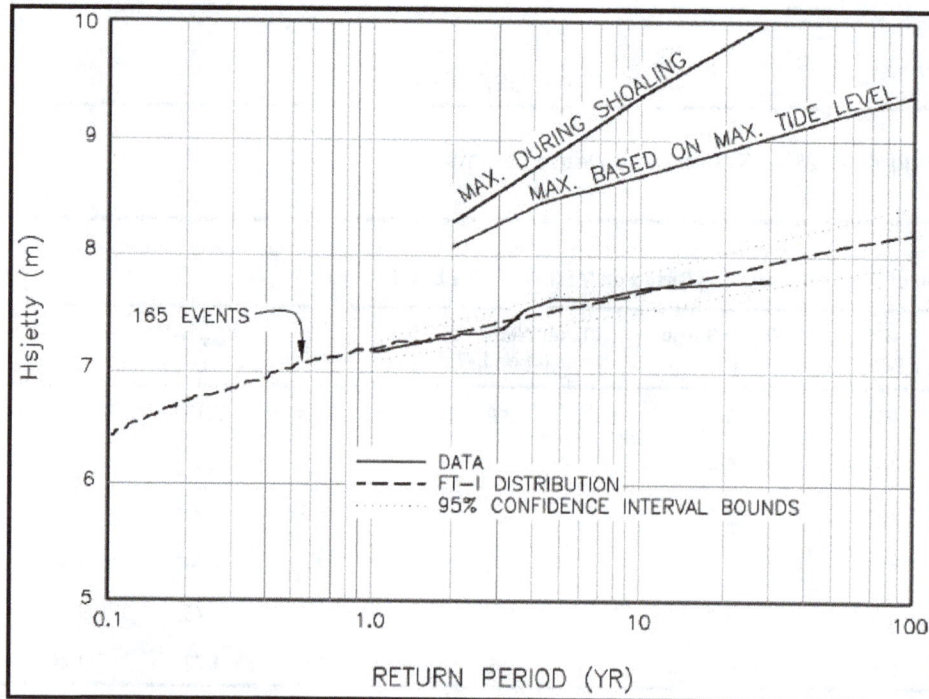

Figure II-8-29. Return period wave heights, 165 events including upper bounds, jetty head

Example Problem II-8-2 (Sheet 16 of 21)

Example Problem II-8-2 (Continued)

Table II-8-20
Design Significant Wave Heights at Jetty Head

T_r (yr)	Jetty Design H_{sjetty} (cm)	Toe Design H_{stoe} (cm)
2	732	612
5	753	622
10	768	628
25	789	637
50	804	644
100	819	650

If a design life longer than 25 years were needed, values of H_{sjetty} approaching the maximum based on maximum tide level might need to be considered. This maximum represents an increase of about 1 m in H_{sjetty} at any selected return period, which would significantly influence jetty design. There is some chance of the maximum occurring even in a 25-year design life. That possibility should be considered if risks are analyzed. Also it should be recognized that the great hydrodynamic energy associated with design events can significantly change nearshore bottom elevations. Storm-induced scour could subject the jetty to increased wave heights, possibly approaching the maximum during shoaling at the jetty head. This effect should also be considered in risk analysis.

Design wave heights - jetty toe. Past physical model studies of some jetty cross sections have shown that waves attacking at water levels of around MLLW are most likely to cause damage to the underwater portion of the structure, including the toe. Similar behavior is assumed in this example. Since this jetty is in a high wave energy environment with fairly large tide range, it is worthwhile to estimate extreme wave heights at MLLW. The lower wave heights (because of reduced depth) can be assessed relative to the lower stability of underwater armor units (due to less precise placement).

Values of H_0' and H_0'/L_0 for each of the observed events (Table II-8-13) are used with a water depth of 10 m (corresponding to MLLW at the jetty head), and bottom slope of 1/100 to estimate a significant wave height for toe design H_{stoe} for each event (illustrated in Table II-8-21 for one event). These 33 values of H_{stoe} were subjected to extremal analysis (Figure II-8-30). The shallow depth greatly limits wave heights, and H_{stoe} values are confined to a very narrow range between 5.8 m and 6.5 m. The Weibull distribution function with $k = 2.0$, which best simulates the shape of a capped distribution function (Figure II-8-3), provides the best fit. H_{stoe} values for design are summarized in Table II-8-20.

Example Problem II-8-2 (Sheet 17 of 21)

Example Problem II-8-2 (Continued)

Table II-8-21
Calculation of H_{stoe} for Storm Event 1[1] at Jetty Head

d_{jetty} (m)	10.0
d_{jetty}/H_0'	1.63
H_{sjetty}/H_0'	0.95
H_{sjetty} (cm)	582

[1] H_0'=613 cm; H_0'/L_0=0.0216

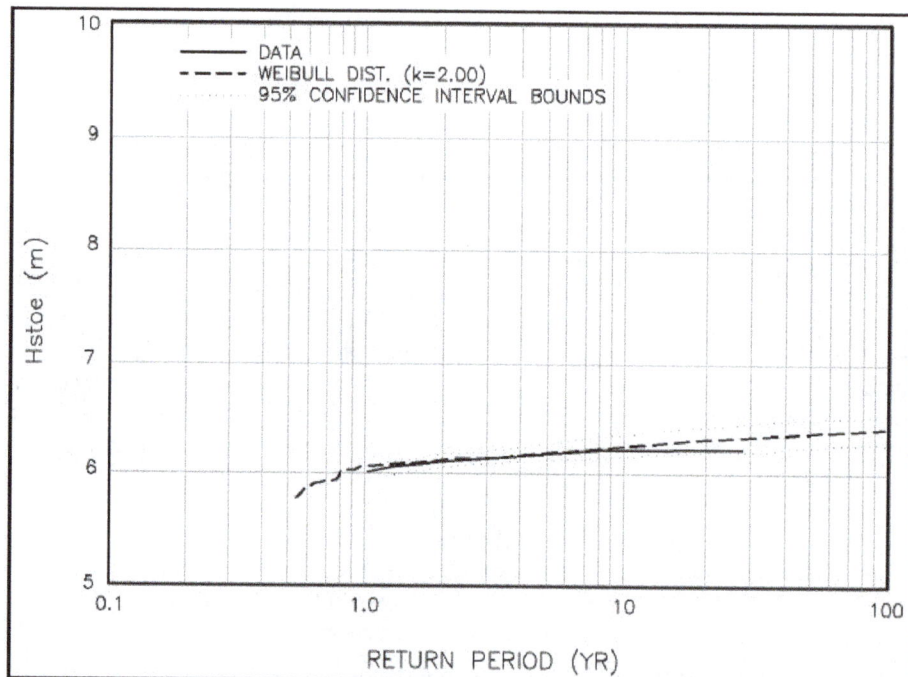

Figure II-8-30.   Return period wave heights, toe design, jetty head

Effect of currents on waves. Information on currents around the jetty head is not precise but indicates that strong currents of up to 3 m/s may be experienced during design storms. The currents must be considered in design for two reasons: 1) direct impact on bottom and jetty material stability; and 2) effect on waves. Part II-6-2.1 provides a simple method for estimating a factor for wave height modification by currents R_H. The following nondimensional parameters are required:

$$F = \frac{V \cos \theta}{\sqrt{gd}}$$
$$\Omega = \left(\frac{2\pi}{T}\right)\left(\frac{d}{g}\right)^{1/2} \tag{II-8-8}$$

where V is the current speed and θ is the angle between the current and a wave orthogonal. For this design problem, V= 3 m/s and θ is taken as -180 deg (current directly opposing the waves).

Example Problem II-8-2 (Sheet 18 of 21)

Example Problem II-8-2 (Continued)

Appropriate values for d and T are more subjective. An opposing current amplifies wave height. The amount of amplification increases as Ω increases (Figure II-8-31). A value for T can be determined by considering that any wave period consistent with the observed events is a likely possibility. Thus the smallest reasonable T_p (designated T_{pmin}), giving the largest likely value of Ω can be used. The largest observed equivalent deepwater wave steepness, 0.025 (Table II-8-13), can be used to estimate T_{pmin} as follows:

$$0.025 = \left(\frac{H_0'}{L_0}\right)_{max} = \frac{2\pi\, H_0'}{g\, T_{pmin}^2}$$

(II-8-9)

$$T_{pmin} = \sqrt{\frac{2\pi}{0.025\, g} H_0'} = 0.506\sqrt{H_0'} \qquad (H_0' \text{ in cm})$$

Return period values of H_0' are given in Tables II-8-18 and II-8-19. For each return period, a maximum depth (d_{jetty} from Table II-8-18) and a minimum depth (10 m, corresponding to MLLW) were used for calculation, as summarized in Table II-8-22. Maximum depth cases relate to jetty design and minimum depth cases to toe design.

Values of R_H ranging from 1.13 to 1.23 indicate that currents could increase wave heights at the jetty head by between 13 percent and 23 percent. A wave height increase of this magnitude has a major impact on design. Since available estimates of current speed are speculative and the methods used to assess their impact are highly simplified (uniform current field, waves coming in opposite direction from current, etc.), a site-specific physical model study would be required in practice to complete the hydrodynamic design.

Table II-8-22
Calculation of Wave Height Modification by Currents

T_r (yr)	H_0' (cm)	T_{pmin} (sec)	Jetty Design d_{jetty} (m)	F	Ω	R_H	Toe Design (d_{jetty}=10 m) F	Ω	R_H
2	653	12.9	13.9	-0.26	0.58	1.23	-0.30	0.49	1.18
5	702	13.4	14.1	-0.26	0.56	1.19	-0.30	0.47	1.17
10	737	13.7	14.2	-0.25	0.55	1.17	-0.30	0.46	1.16
25	782	14.1	14.4	-0.25	0.54	1.15	-0.30	0.45	1.15
50	816	14.5	14.5	-0.25	0.53	1.14	-0.30	0.44	1.14
100	850	14.8	14.6	-0.25	0.52	1.13	-0.30	0.43	1.13

Example Problem II-8-2 (Sheet 19 of 21)

Example Problem II-8-2 (Continued)

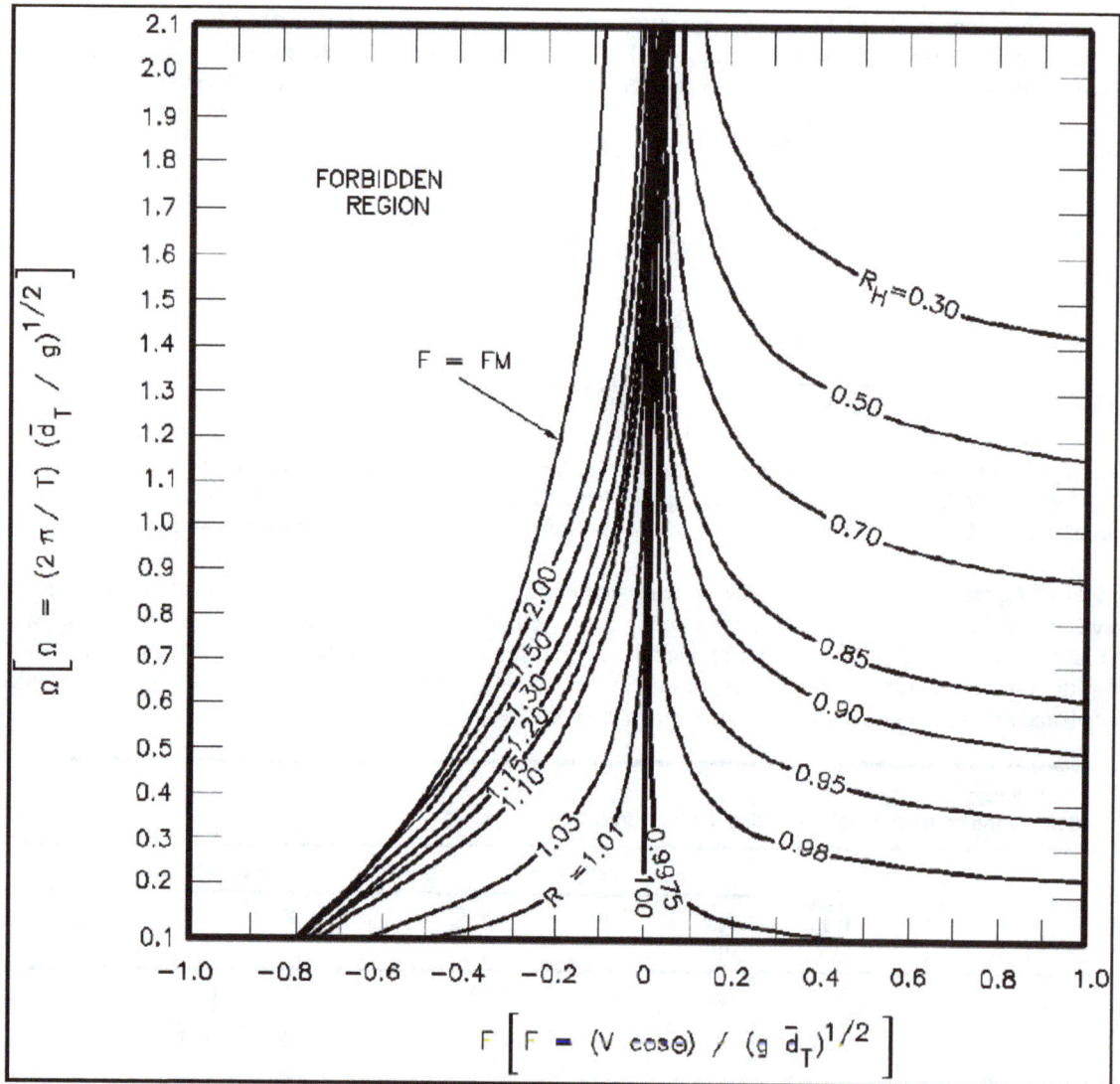

Figure 31. Wave height modification by currents

Example Problem II-8-2 (Sheet 20 of 21)

Example Problem II-8-2 (Concluded)

Design wave period. Design wave periods can be estimated from measured values of T_p (Table II-8-13). A scatter plot of T_p versus H_{sjetty} gives perspective on how wave periods are related to extreme wave heights (Figure II-8-32). Values of H_{sjetty} in the figure are for only the highest of the five tide levels used in conjunction with the 33 storm events. The figure indicates that the design wave period range of 14-18 sec is representative of the higher wave conditions of interest in design.

Confidence intervals. Confidence intervals or uncertainties in hydrodynamic design estimates should always be considered. Confidence intervals associated with statistical aspects of extremal analysis are included in this example. However, there are other sources of uncertainty. For example, the *quality* of available information is an important concern. Wave information may be from accurate, well-maintained gauges, lower quality measurements, hindcasts (which can vary in quality), or observations. Such concerns are not addressed in this example, but they are given some attention in Parts V and VI.

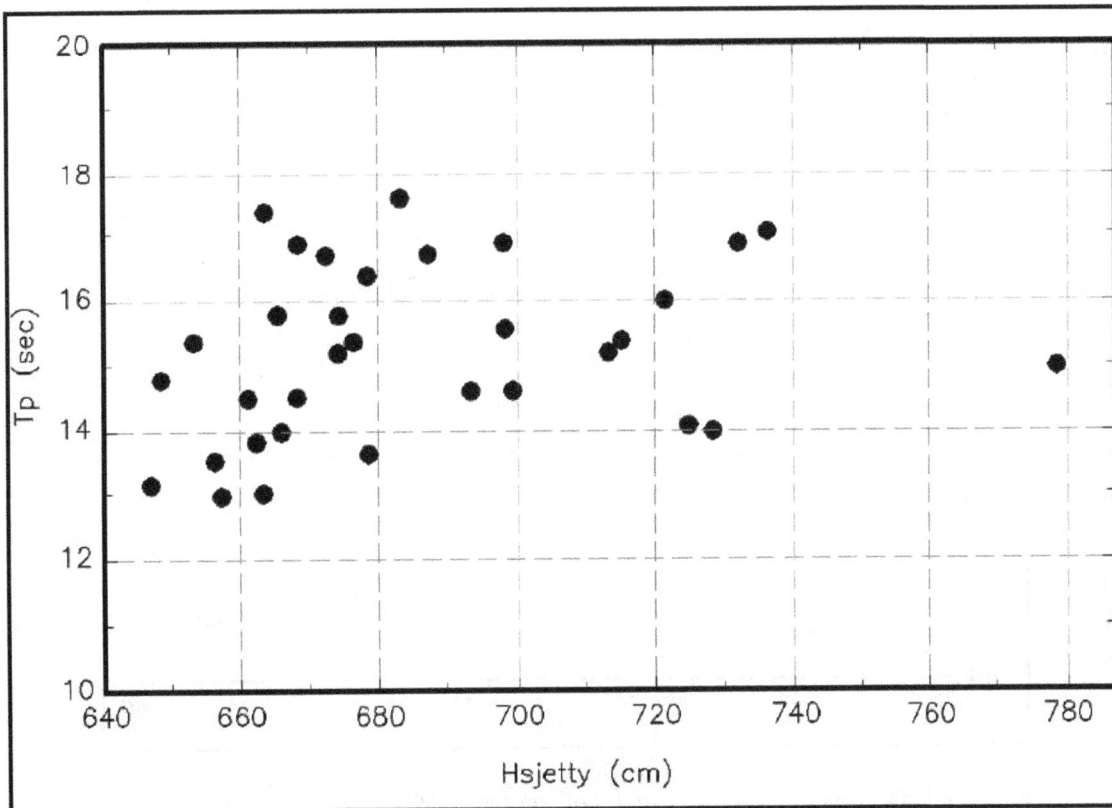

Figure II-8-32. Joint distribution of H_{sjetty} and T_p from 33 storm events, jetty head

Final design conditions. In final design of a jetty or any comparable major coastal engineering project, the complexity of variable bathymetry, breaking waves, currents, interaction between waves, currents, and structures, etc., cannot be adequately represented with the procedures used in this example. Even comprehensive numerical models are limited in their capabilities for reproducing the full design conditions. Standard practice is to construct a three-dimensional physical model and use it to determine the final project design.

Example Problem II-8-2 (Sheet 21 of 21)

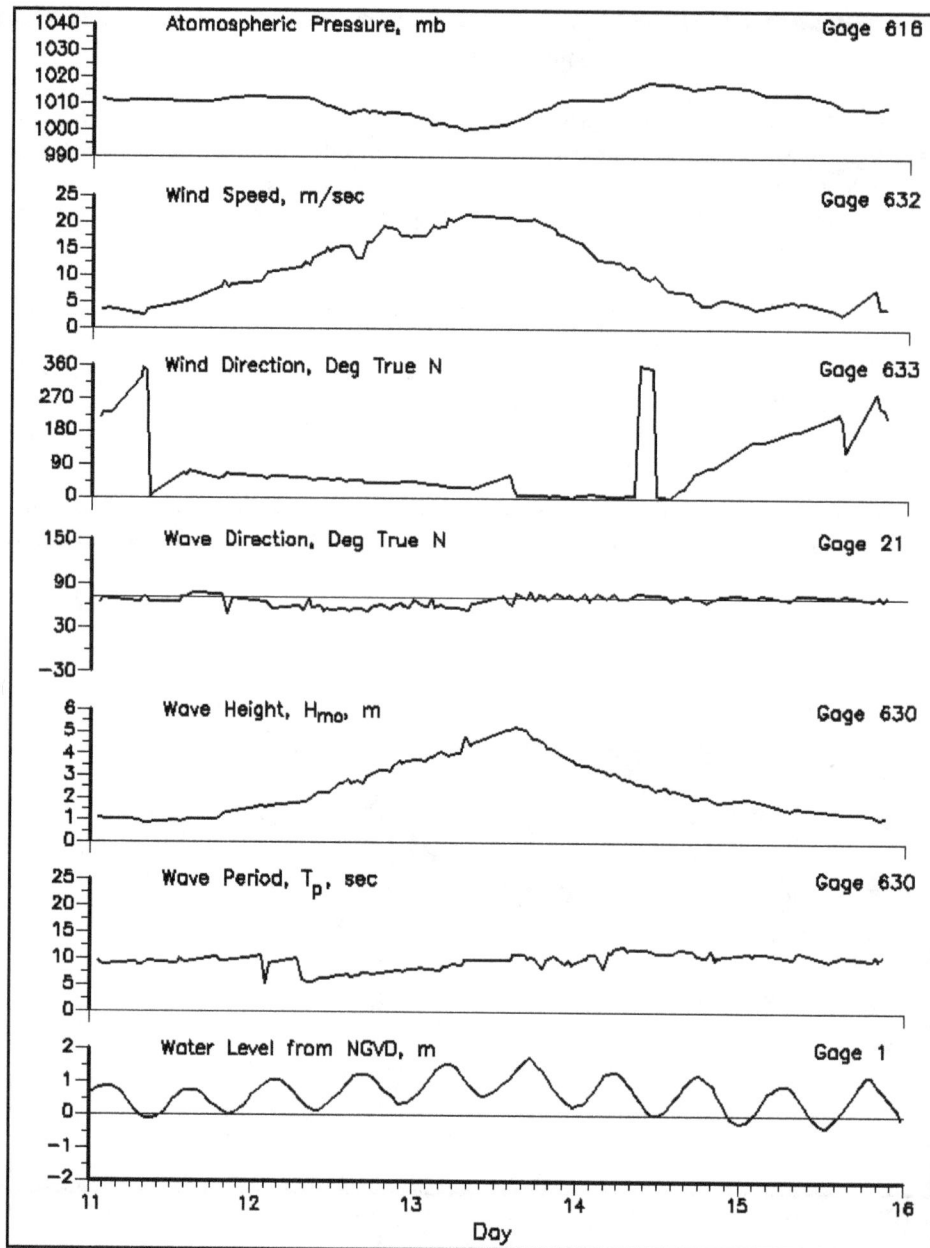

Figure II-8-33.   Example storm event (Leffler et al. 1990)

(2) In standard coastal engineering practice, problems such as the one presented here are usually solved with the help of computer programs. However, the example problem solution relies mainly on graphical methods. The graphical approach has two advantages here: 1) it helps convey an intuitive understanding of the solution (one can see in the graphs *how* parameters are interrelated); and 2) it provides a self-contained methodology independent of complex computer programs. Some graphical tools must be introduced in the example because earlier chapters in Part II relating to this problem presuppose the use of computers in practical work. Most graphical tools are taken from Goda (1985).

II-8-7. Interdependence of Processes During Severe Events

a. Design importance.

(1) The interdependence of processes during severe events is a critical consideration for design. For example, an extreme wave event is usually caused by a severe storm which also typically generates extreme occurrences of wind, water level, infragravity waves, and currents (Figure II-8-33). The true probability of occurrence of the combined extremes is higher than would be expected if each of these processes were analyzed separately and treated as independent of each other.

(2) Without knowledge of the interdependence of processes, designs would be based on a chosen nonexceedance probability for each process critical to the design. This design condition tends to be conservatively high, since extremes at the design level of occurrence are unlikely to all occur together.

b. Procedures for estimating realistic design events, probabilities, and return periods. The approach used in Example Problem II-8-2 illustrates one approach by which data may be used to consider both water levels and wave heights in design. Another simplified approach which considers wind, surge level, and H_s is presented in CIRIA/CUR (1991) (pp 223-225). The EST provides a convenient, and more comprehensive, method for taking advantage of historical information to estimate realistic design events. It is often important to consider storm duration in estimating design events.

II-8-8. References

American Society of Civil Engineers 1993
American Society of Civil Engineers. 1993. ASCE Standard, *Minimum Design Loads for Buildings and Other Structures*, ASCE 7-93, a revision of ANSI/ASCE 7-88, American Society of Civil Engineers, New York.

Battjes 1972
Battjes, J. A. 1972. "Long-Term Wave Height Distribution at Seven Stations around the British Isles," *Peutchen Hydr. Zeit,* Band 25, Hefty, pp 179-189.

Batts, Russell, and Simiu 1980
Batts, M. E., Russell, L. R., and Simiu, E. 1980. "Hurricane Wind Speeds in the United States," *J. Struct. Div., American Society of Civil Engineers,* Vol 100, No. STIO, pp 2001-2015.

Blom 1958
Blom, G. 1958. *Statistical Estimates and Transformed Beta-Variables*, Wiley, New York.

Bowers 1992
Bowers, E. C. 1992. "Low Frequency Waves in Intermediate Water Depths," *Proceedings of the 23rd International Conference on Coastal Engineering*, American Society of Civil Engineers, pp 832-845.

Carter and Draper 1988
Carter, D. J. T., and Draper, L. 1988. "Has the North-East Atlantic Become Rougher?" *Nature,* Vol 332, No. 7, p 494.

Construction Industry Research and Information Association (CIRIA)
CIRIA. 1991. "Manual on the Use of Rock in Coastal and Shoreline Engineering," Construction Industry Research and Information Association Special Publication 83, London, UK (also published as: Centre for Civil Engineering Research and Codes Report 154, Gouda, The Netherlands).

Earle and Baer 1982
Earle, M. D., and Baer, L. 1982. "Effects of Uncertainties on Extreme Wave Heights," *Journal of Waterway, Port, Coastal, and Ocean Engineering*, Vol 108, No. WW4, pp 456-478.

Farrar et al. 1994
Farrar, P. D., Borgman, L. E., Glover, L. B., Reinhard, R. D., Pope, J., Swain, A., and Ebersole, B. A. 1994. "Storm Impact Assessment for Beaches at Panama City, Florida," Technical Report CERC-94-11, U.S. Army Engineer Waterways Experiment Station, Vicksburg, MS.

Goda 1985
Goda, Y. 1985. *Random Seas and Design of Maritime Structures*, University of Tokyo Press, Tokyo, Japan.

Goda 1988
Goda, Y. 1988. "On the Methodology of Selecting Design Wave Height," *Proceedings of the 21st International Conference on Coastal Engineering*, American Society of Civil Engineers, pp 899-913.

Goda 1990
Goda, Y. 1990. "Distribution of Sea State Parameters and Data Fitting," *Handbook of Coastal and Ocean Engineering*, Vol 1, Gulf Publishing Co., pp 371-408.

Goda and Kobune 1990
Goda, Y., and Kobune, K. 1990. "Distribution Function Fitting for Storm Wave Data," *Proceedings of the 22nd International Conference on Coastal Engineering*, American Society of Civil Engineers, pp 18-31.

Gringorten 1963
Gringorten, I. I. 1963. "A Plotting Rule for Extreme Probability Paper," *Journal of Geophysical Research*, Vol 68, No. 3, pp 813-814.

Gumbel 1958
Gumbel, E. J. 1958. *Statistics of Extremes*, Columbia University Press, New York.

Harris 1981
Harris, D. L. 1981. "Tides and Tidal Datums in the United States," Special Report No. 7, U.S. Army Corps of Engineers, Fort Belvoir, VA., U.S. Government Printing Office (GPO Stock No. 008-022-00161-1).

Hubertz et al. 1993
Hubertz, J. M., Brooks, R. M., Brandon, W. A., and Tracy, B. A. 1993. "Hindcast Wave Information for the U.S. Atlantic Coast," WIS Report 30, U.S. Army Engineer Waterways Experiment Station, Vicksburg, MS.

Kite 1978
Kite, G. W. 1978. *Frequency and Risk Analysis in Hydrology*, Water Resources Publications, Fort Collins, CO.

Leffler et al. 1990

Leffler, M. W., Hathaway, K. K., Scarborough, B. L., Baron, C. F., and Miller, H. C. 1990. "Annual Data Summary for 1988, CERC Field Research Facility," Technical Report CERC-90-13, U.S. Army Engineer Waterways Experiment Station, Vicksburg, MS.

Le Méhauté and Wang 1984

Le Méhauté, B., and Wang, S. 1984. "Effects of Measurement Error on Long-Term Wave Statistics," *Proceedings of the 19th International Conference on Coastal Engineering*, American Society of Civil Engineers, pp 345-361.

Mathiesen et al. 1994

Mathiesen, M., Goda, Y., Hawkes, P. J., Mansard, E., Martin, M. J., Peltier, E., Thompson, E. F., and Van Vledder, G. 1994. "Recommended Practice for Extreme Eave Analysis," *Journal of Hydraulic Research*, Vol 32, No. 6, pp 803-814.

National Bureau of Standards 1980

National Bureau of Standards. 1980. "Hurricane Wind Speeds in the United States," *Building Science Series 124*, Gaithersburg, MD.

Scheffner and Borgman 1992

Scheffner, N. W., and Borgman, L. E. 1992. "Stochastic Time-Series Representation of Wave Data," *Journal of Waterway, Port, Coastal, and Ocean Engineering*, Vol 118, No. 4, pp 337-351.

Seelig, Ahrens, and Grosskopf 1983

Seelig, W. N., Ahrens, J. P., and Grosskopf, W. G. 1983. "The Elevation and Duration of Wave Crests," Miscellaneous Report 83-1, U.S. Army Engineer Waterways Experiment Station, Vicksburg, MS.

Seymour, Strange, Cayan, and Nathan 1984

Seymour, R. J., Strange, R. R., Cayan, D. R., and Nathan, R. A. 1984. "Influence of El Ninos on California's Wave Climate," *Proceedings of the 19th International Conference on Coastal Engineering*, American Society of Civil Engineers, pp 577-592.

Simiu and Scanlan 1986

Simiu, E., and Scanlan, R. H. 1986. *Wind Effects on Structures, An Introduction to Wind Engineering*, Wiley, New York.

Simiu, Changery, and Filliben 1979

Simiu, E., Changery, M. J., and Filliben, J. J. 1979. "Extreme Windspeeds at 129 Stations in the Contiguous United States," NBS BSS 118, National Bureau of Standards, U.S. Dept. of Commerce, Washington, DC.

Smith 1987

Smith, O. 1987. "Duration of Extreme Wave Conditions," Miscellaneous Paper CERC-87-12, U.S. Army Engineer Waterways Experiment Station, Vicksburg, MS.

Smith 1988

Smith, O. 1988. "Duration of Extreme Wave Conditions," *Journal of Waterway, Port, Coastal, and Ocean Engineering*, Vol 114, No. 1, pp 1-17.

Thom 1968-69
Thom, H.C.S. 1968-69. "New Distribution of Extreme Winds in the United States," *J. Sruct. Div.*, *American Society of Civil Engineers,* Vol 94, No. ST7, pp 1787-1801, July 1968; and Vol 95, No. ST8, pp 1769-1770, Aug. 1969.

Thompson 1980
Thompson, E. F. 1980. "Energy Spectra in Shallow U.S. Coastal Waters," Technical Paper 80-2, U.S. Army Engineer Waterways Experiment Station, Vicksburg, MS.

U.S. Department of Commerce 1992
U.S. Department of Commerce. 1992. "The Halloween Nor'easter of 1991," Natural Disaster Survey Report, National Oceanic and Atmospheric Administration, National Weather Service, Silver Spring, MD.

U.S. Water Resources Council 1976
U.S. Water Resources Council. 1976. "Guidelines for Determining Flood Flow Frequency," Bulletin No. 17 of the Hydrology Committee, Washington, DC.

II-8-9. Definitions of Symbols

σ	Standard deviation
σ_m	Standard deviation of maximum monthly wind speeds [length/time]
σ_{rm}	Standard deviation of the sampling error in estimating U_r (Equation II-8-4) [length/time]
d	Water depth [length]
F	Probability that the mth highest data value will not be exceeded (Figure II-8-5)
$f(x)$	Probability distribution function
G	Skew coefficient in the Pearson Type III distribution function (Figure II-8-2) [dimensionless]
$\hat{H}_s$	Design significant wave height [length]
$H_{1/3}$	Significant wave height [length]
k	Parameter in the Weibull distribution function (Figure II-8-5) [dimensionless]
k	Skew of ln x corrected for bias in the Log Pearson Type III distribution function (Figure II-8-2) [dimensionless]
L	Desired time period used in the encounter probability formula (Equation II-8-2) [years]
m	Rank of data value in descending order (= 1 for largest) in distribution function formulas (Figure II-8-5)
N	Number of events
N_m	Number of months of data used in computing the standard deviation of the sampling error in estimating U_r (Equation II-8-4)
$-_0$	The subscript 0 denotes deepwater conditions
$P(\hat{H}_s)$	Cumulative probability that $H_s \leq \hat{H}_s$
P_e	Encounter probability (Equation II-8-2)
t	Time interval associated with each data point [years]
T_p	Design wave period [time]
T_r	Return period (Equation II-8-1) [years]
$\overline{U}_m$	Mean value of maximum monthly wind speeds [length/time]
U_r	Wind speed with r-year return period (Equation II-8-3) [length/time]

II-8-10. Acknowledgments

Author of Chapter II-8, "Hydrodynamic Analysis and Design Conditions:"

Edward F. Thompson, Ph.D., Coastal and Hydraulics Laboratory, Engineer Research and Development Center, Vicksburg, Mississippi.

Reviewers:

Lee E. Harris, Ph.D., Department of Marine and Environmental Systems, Florida Institute of Technology, Melbourne, Florida.
Thomas Leung, U.S. Army Engineer District, Los Angeles, Los Angeles, California, (deceased).
Heidi P. Moritz, U.S. Army Engineer District, Portland, Portland, Oregon.
Arthur T. Shak, U.S. Army Engineer District, Los Angeles, Los Angeles, California.

www.ingramcontent.com/pod-product-compliance
Lightning Source LLC
Chambersburg PA
CBHW081346190326
41458CB00018B/6091